OPTICS

Related publications by the Association of Dispensing Opticians

The Principles of Ophthalmic Lenses	M. Jalie F.A.D.O.(Hons)
Introduction to Visual Optics	A.H. Tunnacliffe
Worked Problems in Optics	A.H. Tunnacliffe
Worked Problems in Ophthalmic Lenses	G.D. Janney and A.H. Tunnacliffe
Spectacle Frame Dispensing	G.H. Clayton B.Sc., F.A.D.O.(Hons)
Optics of Contact Lenses	A.G. Bennett
Bifocals Without Tears	L.E. Swift
Ocular Prostheses	H.S. Warren

OPTICS

A.H. TUNNACLIFFE

BA, Dip. Maths, DCLP, FBOA

Senior Lecturer in Optics, Bradford College

and

J.G. HIRST

BSc, M Phil, Grad. Cert. Ed.

Senior Lecturer in Physics, Bradford College

THE ASSOCIATION OF DISPENSING OPTICIANS
22 Nottingham Place
LONDON W1M 4AT

ISBN 0-900099-15-1

Printed in Great Britain by The Eastern Press Ltd, London and Reading

FOREWORD

One of the problems associated with writing any textbook is that the authors of such a book are necessarily more aware of the subject matter than the people who are most likely to use it - namely the students. It is not easy to write any book, much less a highly technical one, and most authors cannot avoid the situation of "talking above the heads" of their readers.

Alan Tunnacliffe and Gordon Hirst belong to the select band of authors who are able to write a textbook as though they were still students - and consequently the result is a highly readable textbook which even I can follow without difficulty. They are to be congratulated on their achievement and I have no hesitation in warmly recommending this work to all those interested in its subject matter.

T.A. Doyle
President. 1981.

PREFACE

This book has been written to meet the requirements of examinations in optics for the Optical Dispensing Diploma and for university undergraduate students of Optometry. It will also be found useful for Ophthalmologists, Ophthalmic Nurses, and Orthoptists, and will make a good introduction for those who are proceeding to further studies of Photographic Optics and Illumination Engineering.

Our intention, above all, has been to produce a readable teaching textbook containing clear diagrams together with many illustrative worked examples. In fact, the book contains in excess of 600 diagrams and photographs and more than 100 worked examples. In most instances a diagram will be found to be adjacent to the text which describes it, and it is hoped that this will lead to an easy appreciation of the subject matter. This approach has resulted from suggestions and discussions with College and University students of Optics over several years of teaching in the subject.

It is almost certain that, in the next few years, subjective examinations based on spatial frequencies will be in common use in the consulting room. With this in view, Chapter 16 includes a description of imagery in terms of spatial frequencies and an introduction to the laser and the mathematics of holography.

The Sign Convention used in this book is the one widely used in the optical profession and has advantages from the teaching point of view. Exercises are included at the end of each chapter and answers are given to the numerical questions.

We have included, in the Appendix, certain mathematical formulae and methods which are used throughout the book. In addition, the Appendix contains several derivations which, whilst meriting inclusion in the book, are not included in the main text so that re-reading the latter will be easier.

We are extremely grateful to Mr G.H. Clayton, B.Sc., F.A.D.O.(Hons) for reading the manuscript and for helpful suggestions during the preparation of this work. We should also like to acknowledge the help of Mr A.G. Tunnacliffe, B.A. with the photographs in this book. Our appreciation is also due to Mr R.A. Earlam, B.Sc., F.B.O.A., F.A.D.O., of the University of Wales Institute of Science and Technology for his excellent photograph of an Airy pattern (figure 14.2), and to Mr R. Adams of Bradford College Photographic Services Department for figures 8.27 and 8.34(a).

Bradford
April, 1981

A.H. Tunnacliffe
J.G. Hirst

CONTENTS

Greek letters

Α	α	alpha
Β	β	beta
Γ	γ	gamma
Δ	δ	delta
Ε	ε	epsilon
Ζ	ζ	zeta
Η	η	eta
Θ	θ	theta
Ι	ι	iota
Κ	κ	kappa
Λ	λ	lambda
Μ	μ	mu
Ν	ν	nu
Ξ	ξ	xi
Ο	ο	omicron
Π	π	pi
Ρ	ρ	rho
Σ	σ	sigma
Τ	τ	tau
Υ	υ	upsilon
Φ	φ	phi
Χ	χ	chi
Ψ	ψ	psi
Ω	ω	omega

Notation

Symbol	Meaning
$>$	greater than
$\gg$	much greater than
$<$	less than
$\ll$	much less than
$\lvert\ \rvert$	modulus or magnitude
$\simeq$	approximately equal to
$\equiv$	equivalent to
Δx	a small increment in a variable (here x)
$\sim$	approximately
∞	infinity
$\Rightarrow$	which leads to
$\pm$	plus or minus

1 PROPAGATION AND VELOCITY OF LIGHT

1.0 INTRODUCTION

Sources of light may be natural or artificial. The most important and beneficial natural light source is the sun. Many early experiments and measurements in optics, the science of light, were based on shadows formed by interruption of the sun's rays. Artificial sources of light may be produced by heating a solid body until it glows, or by forming an electrical discharge in a gas or vapour. For example, a domestic light bulb emits light from a tungsten filament which is heated by the passage of an electric current. Indoor tubular (fluorescent) lights and also most forms of outdoor street lighting are based on an electrical discharge through vapours obtained from the elements sodium or mercury. Details of light sources and their functions will be dealt with in Chapter 15.

Light sources which are very small are referred to as *point* sources, and those of appreciable size as *extended* sources.

1.1 THE RECTILINEAR PROPAGATION OF LIGHT

This heading means *"Light travels in straight line directions"*. A good illustration of the truth of this statement is to arrange a point source of light and three opaque screens containing pinholes, as shown in figure 1.1 .

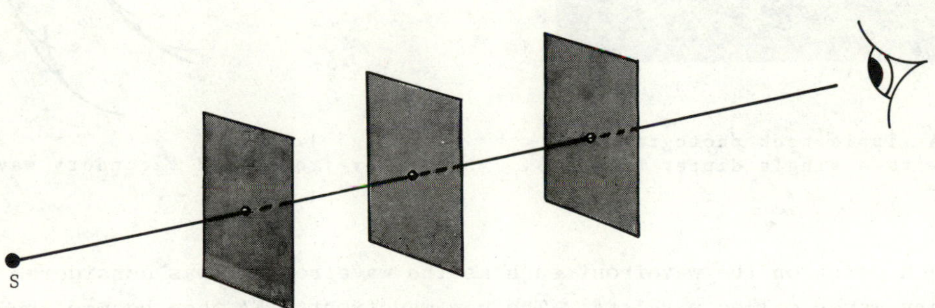

Fig. 1.1 Demonstration of rectilinear propagation of light.

Only when the pinholes and source, S, are precisely arranged in a straight line is it possible for the light from the source to reach the eye and render the source visible. Slight displacement of one screen cuts off the light.

It is also common experience that light cannot be seen round corners, whereas a source of sound may often be heard round a corner when it cannot be seen. This illustrates one of the fundamental differences between light and sound.

Let us consider the passage of light through a medium, such as air, which has the same optical properties in all directions. Such a type of medium is called *isotropic*, and we shall include in this category a vacuum which does not, as far as we know, contain any material substances. Also, many materials may be found to possess these same optical properties throughout the whole of their volume. As such, these materials are referred to as *homogeneous*.

In the 17th Century the Dutch physicist and mathematician Christian Huygens, and many other workers, had been investigating various optical phenomena. In order to explain certain effects Huygens suggested that light energy travels as a waveform, and published a treatise on light in 1690, five years before his death, which formed the basis of the *Wave Theory of Light*. Essentially, the wave theory describes light as spreading outwards from the source in a homogeneous isotropic medium in the form of *spherical waves*. Originally it was thought that the medium through which the light was travelling was actually formed into waves, rather like waves on water. However, later work by James Clerk Maxwell (1831 - 1879) confirmed that the waves are actually variations of an electromagnetic character and are not waves formed in any material substance. Figure 1.2 shows a two-dimensional illustration of spherical waves with a point source S in which the waves appear as circles called *wavefronts*.

Fig. 1.2(a) A ripple tank photograph with a single dipper (source).

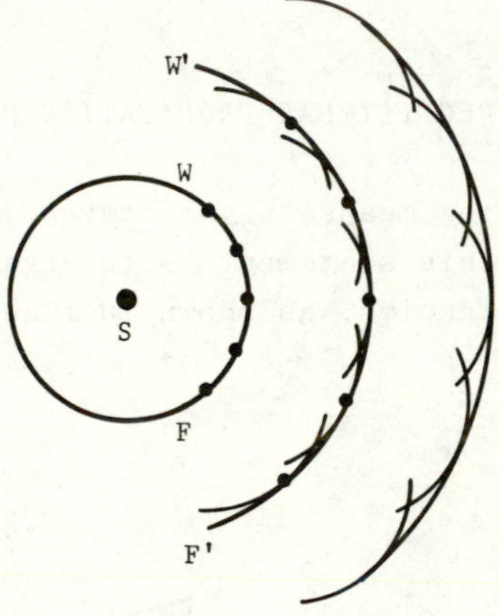

Fig. 1.2(b) Wavefronts and secondary wavelets.

At any instant each point on the wavefront such as the wavefront WF was considered to be a new source of secondary waves called wavelets. The new wavefront W'F' then became the line which envelopes the wavelets which created it. It will be seen that for all points on any particular wavefront the wave has travelled from S for the same time and over the same distance. Hence, all points on a particular wavefront are said to be in the *same phase*. This aspect of wave motion, together with phase differences resulting from two wave motions travelling different distances, will be more fully discussed in Chapters 10 and 11.

All wave motions possess the characteristics of *wavelength, velocity, frequency,* and *amplitude*. These features may be defined with reference to figure 1.3 which shows a wave profile and its successive wavefronts.

Wavelength (symbol λ) - this is the distance between successive points of similar phase. These points may be taken to be successive wave crests such as B and C on the wave profile, or successive troughs such as D and E. For light waves, the wavelengths are quite small and are most often expressed in nanometres ($1\,nm = 10^{-9}\,m$).

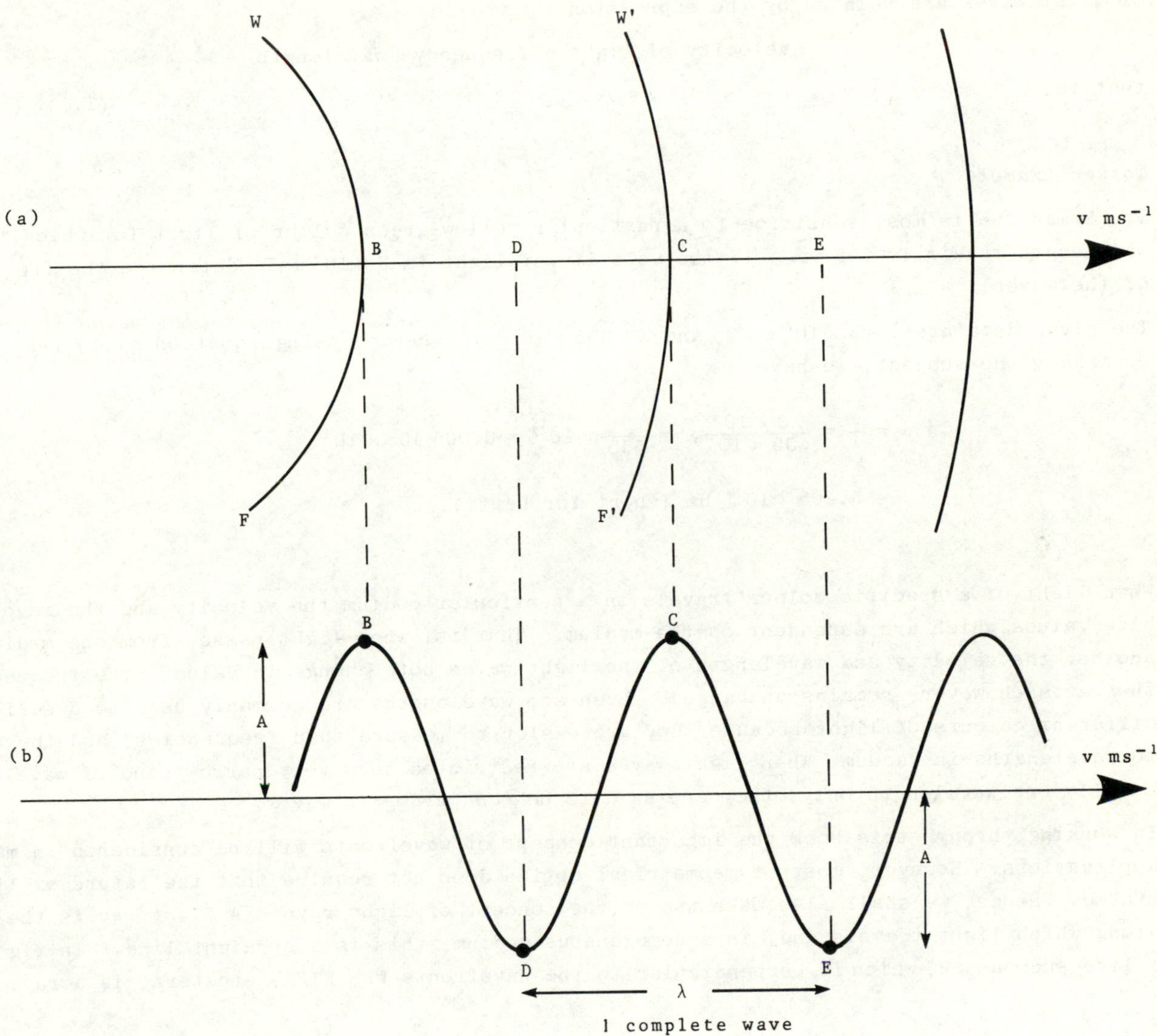

Fig. 1.3 (a) Wavefronts and (b) corresponding wave profile. Note the amplitude A.

Frequency (symbol ν*)* - this is the number of complete waves generated by the source in one second. Frequency is expressed in hertz, where 1 hertz means 1 complete wave per second. We shall see shortly that the freqencies of light waves are very large values and differ for different colours of light.

Amplitude (symbol A, or E when the wave is a disturbance in an electric field*)* - this is the peak value of the wave and is a measure of the height of the wavecrest or depth of the wave trough. It may be used as a measure of the intensity of a light wave, see section 10.5 .

Velocity (symbol v*)* - this is the rate (speed) at which the wave travels and is expressed in metres per second (ms^{-1}). For light waves travelling through vacuum the velocity is about

300 000 km s^{-1} (3×10^8 ms^{-1}) and this value is the same for all colours of light.

It can be shown, see section 10.2.2 , that the frequency(ν), wavelength(λ), and velocity(v) of light waves are related by the expression

velocity of light = frequency × wavelength

that is,

$$v = \nu\lambda \qquad (1.1).$$

Worked Example

The human eye is most sensitive to a particular yellow-green colour of light for which the wavelength in air is 555 nm. If the velocity of light is 3×10^8 ms^{-1} determine the frequency of the waves.

The given data are $v = 3 \times 10^8$ ms^{-1} and $\lambda = 555 \times 10^{-9}$ m. Hence, using equation (1.1) rearranged to make ν the subject, we have

$$\nu = \frac{v}{\lambda} = \frac{3 \times 10^8}{555 \times 10^{-9}} = \frac{3}{555} \times 10^{17} = 0.005\,405 \times 10^{17}$$

$$= 5.405 \times 10^{14} \text{ Hz (short for hertz).}$$

When light of a specific colour travels in a particular medium the velocity and the wavelength have values which are dependent on the medium. That is, when light passes from one medium to another the velocity and wavelength of the light waves both change in value. The frequency of the waves, however, remains unchanged. Even so, wavelengths are commonly used to describe different colours of light because they are easier to measure than frequencies, and these refer to wavelengths in vacuum. When light waves are restricted to a very narrow band of wavelengths, ideally one wavelength only, they are said to be *monochromatic waves.*

In working through this book the important concept of wavefronts will be considered in many applications. However, most of geometrical optics does not require that the nature of light be stated. Hence, we shall also make use of the concept of light rays. A *light ray* is the path along which light travels and, in a homogeneous medium, this is a straight line. In figure 1.4, a line such as SR, which is perpendicular to the wavefronts WF, W'F', etcetera, is a ray.

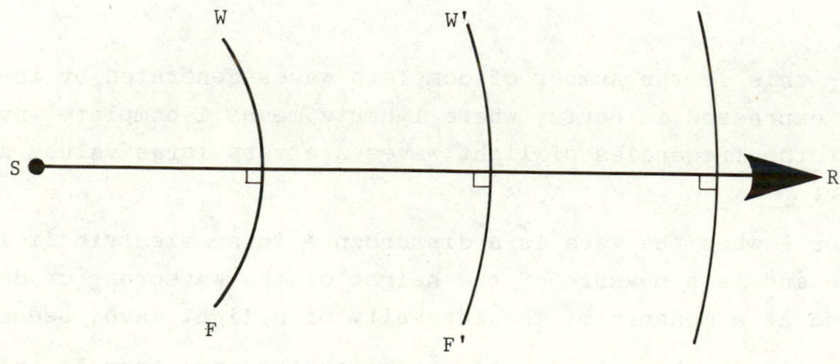

Fig. 1.4 A ray perpendicular to the wavefronts.

It does not exist in the material sense, but it will simplify our work considerably if we consider rays of light instead of waves. It must be clearly understood that when we speak of rays of light we mean specific paths along which waves are travelling.

A group of light rays is referred to as a *pencil of rays*. A pencil may be divergent, convergent, or parallel. Figure 1.5 illustrates pencils of rays together with their associated wavefronts. The central ray in each pencil is referred to as the *chief ray*.

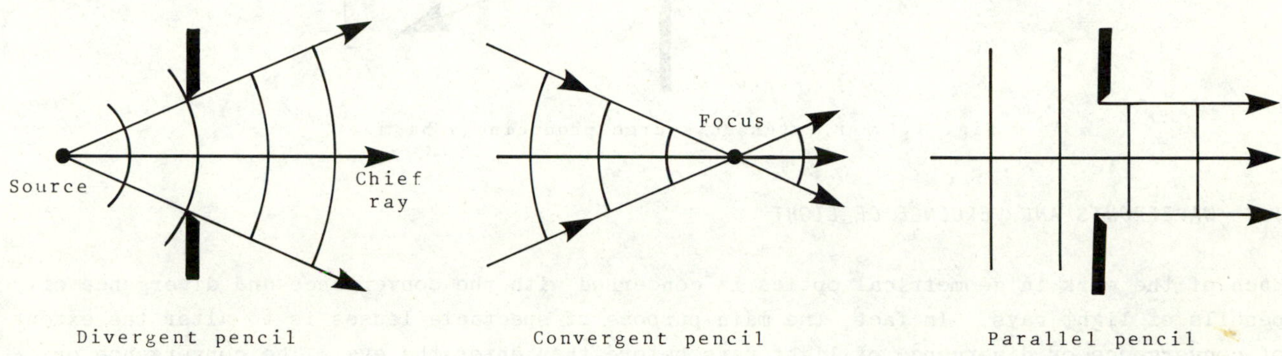

Fig. 1.5 Pencils of rays.

A pencil of rays may be rendered parallel by the use of a lens. This results in *plane wavefronts*. However, if a diverging pencil of rays is considered at a great distance from the source the rays will become essentially parallel and the wavefronts plane. Thus, light originating from a distant source or object may be represented as a parallel pencil of rays. Figure 1.6 indicates the flattening of the wavefronts with distance even though the object can hardly be described as distant in the diagram.

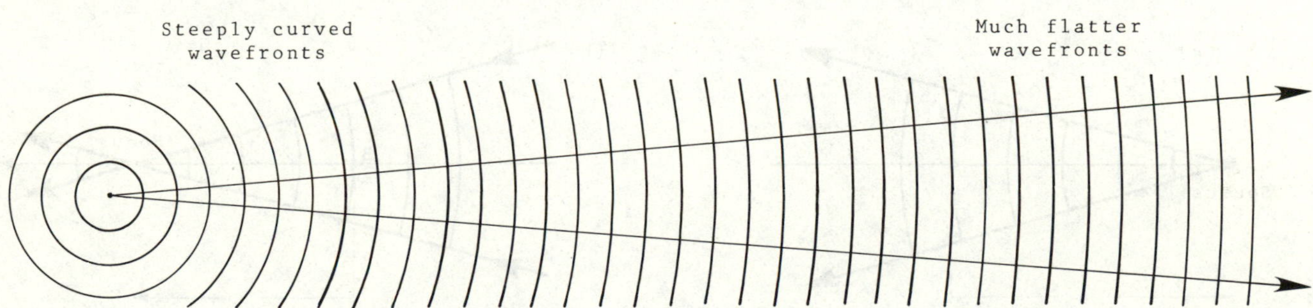

Fig. 1.6 The flattening of wavefronts with distance.

A collection of pencils forms a *beam of light*. This may typically originate from an extended source placed fairly close to an aperture in a screen, figure 1.7 .

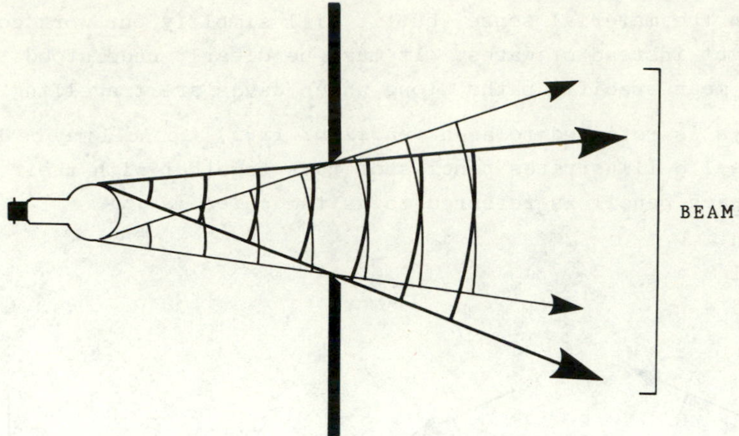

Fig. 1.7 An extended source producing a beam.

1.2 WAVEFRONTS AND VERGENCE OF LIGHT

Much of the work in geometrical optics is concerned with the convergence and divergence of pencils of light rays. In fact, the main purpose of spectacle lenses is to alter the extent of convergence or divergence of light rays before they enter the eye. The convergence or divergence of a pencil of light rays may be expressed by the general term *vergence* defined as follows.

1.2.1 VERGENCE

The vergence at a particular point in a pencil of rays travelling in air is the reciprocal of the distance from the point to the source or focus. Figures 1.8(a) and 1.8(b) show diverging and converging pencils of rays, together with their associated wavefronts.

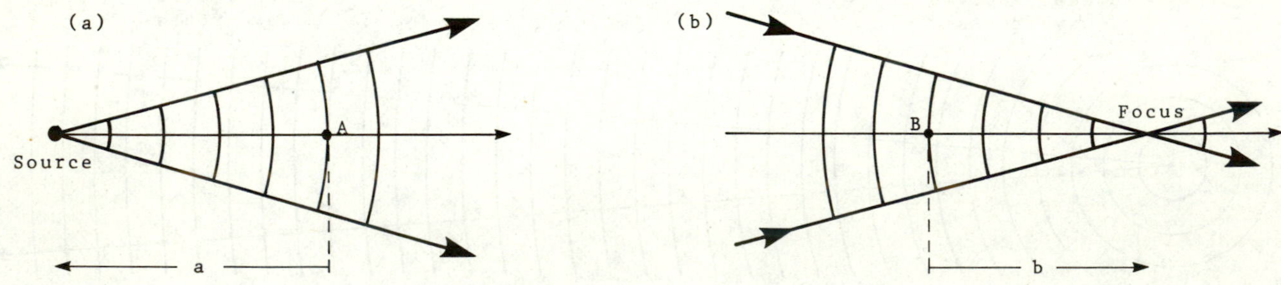

Fig. 1.8 Definition of vergence - see text.

In figure 1.8(a) the distance from the point A to the source is a. Hence, the vergence at A is $\frac{1}{a}$. In figure 1.8(b) the distance from the point B to the focus is b. Hence, the vergence at B is $\frac{1}{b}$.

The definition of vergence stated above does not distinguish between those points which are situated in converging pencils of rays and those points which are situated in diverging pencils. In order that we can distinguish between diverging and converging pencils, when studying the effects of optical components such as lenses, a *sign convention* must be employed. A sign convention is a set of very definite rules such that the value of any distance measured on an optical diagram may be given a positive or a negative sign. It should be noted that, at this stage, the sign convention described is only part of the full sign convention necessary for successful optical calculations. This will be considered in detail in section 2.2.3 . For the moment the abbreviated sign convention will be sufficient, and this must be studied very carefully.

1.2.2 SIGN CONVENTION

The sign convention for vergence has *two rules* as follows.

1 If the distance from the point in question to the source or focus is measured in the *same direction* as that in which the rays of light are directed, the numerical value of the distance is given a *positive sign* (+).

2 If the distance from the point in question to the source or focus is measured in the *opposite direction* to that in which the light is travelling, the numerical value of the distance is given a *negative sign* (-).

Let us again refer to figures 1.8(a) and (b). The distance from the point A to the source is measured in the opposite direction to the direction in which the light is travelling and we must apply a minus sign to the distance a. Hence, the vergence at A$\left(=\frac{1}{a}\right)$ will be a negative value. However, the distance from the point B to the focus is measured in the same direction as the direction in which the light is travelling and the value of the distance b takes a plus sign. Therefore, the vergence at B $\left(=\frac{1}{b}\right)$ will be a positive value.

These rules result in the important conclusion that at any point in a *converging* pencil of rays the value of the vergence of the light is *positive*, and at any point in a *diverging* pencil of rays the value of the vergence of the light is *negative*. It also follows that at any point in a parallel pencil of rays, for which the source or the focus may be imagined to be at infinity, the vergence will be $\frac{1}{\pm\infty} = 0$. That is, at any point in a *parallel* pencil of rays, the value of the vergence of the light is *zero*.

1.2.3 THE UNIT FOR VERGENCE

If the distance from a point in a pencil to the source or focus is expressed in metres, or a fraction of a metre, then the value of the vergence at that point is expressed in *dioptres*. We can define *one dioptre*, symbol 1D, as being the vergence in a pencil of rays in air at a point *one metre* to the source or focus.

1.2.4 SYMBOLS FOR VERGENCES

Let us now refer to figure 1.9(a) and (b) which shows a diverging pencil and a converging pencil of rays, respectively.

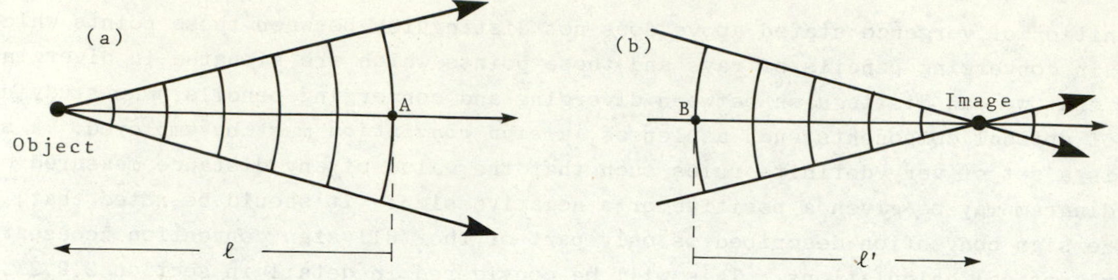

Fig. 1.9 (a) Object distance, ℓ. (b) Image distance, ℓ'.

In figure 1.9(a) the point from which the diverging pencil of rays actually originates will now be referred to as an *object*. In optics, the distance from a point such as A to an object is represented by the symbol ℓ. That is, for the point A, the distance to the object = ℓ where ℓ is measured in metres. The vergence at A is $\frac{1}{\ell}$, and this is given the symbol L (dioptres). That is,

$$L \text{ (dioptres)} = \frac{1}{\ell} \text{ (metres)} \qquad (1.2).$$

In figure 1.9(b) the point through which the converging rays pass will be referred to as an *image*. The distance from a point such as B to an image is represented by the symbol ℓ'. The vergence at B is $\frac{1}{\ell'}$, and this is given the symbol L' (dioptres). Thus,

$$L' \text{ (dioptres)} = \frac{1}{\ell'} \text{ (metres)} \qquad (1.3).$$

It is worth emphasising that the dioptre unit is really the reciprocal metre. That is $D \equiv \frac{1}{m}$, and the dioptre is merely a unit of convenience which is easier to say than 'reciprocal metre', but when considering units in some equations it will be necessary to think in terms of reciprocal metres. The name dioptre was chosen from dioptrics, the name given to the branch of optics dealing with refraction.

Worked Examples

1 Light is made to converge in air (by means of a lens) to a point 50 cm from the lens. Determine the vergence at the point in the pencil of rays (a) on leaving the lens, (b) at 15 cm from the lens, (c) at 40 cm from the lens, and (d) at 60 cm from the lens.

Let the four points in question be A,B,C, and D, respectively, as shown in the next figure. We must remember that the distance measurements are taken from each point to the source or to the focus.

(a) For A - the point where the rays are leaving the lens.

Vergence $L' = \frac{1}{\ell'}$, where $\ell' = +50$ cm $= +0.50$ m. $\therefore L' = \frac{1}{+0.50} = +2.00$ D.

(b) For B - the point 15 cm from the lens; i.e. 35 cm to the focus or image.

Vergence $L' = \frac{1}{\ell'}$, where $\ell' = +35$ cm $= +0.35$ m. $\therefore L' = \frac{1}{+0.35} = +2.86$ D.

(c) For C - the point 40 cm from the lens; i.e. 10 cm to the focus or image.

Vergence $L' = \frac{1}{\ell'}$, where $\ell' = +10$ cm $= +0.10$ m. $\therefore L' = \frac{1}{+0.10} = +10.00$ D.

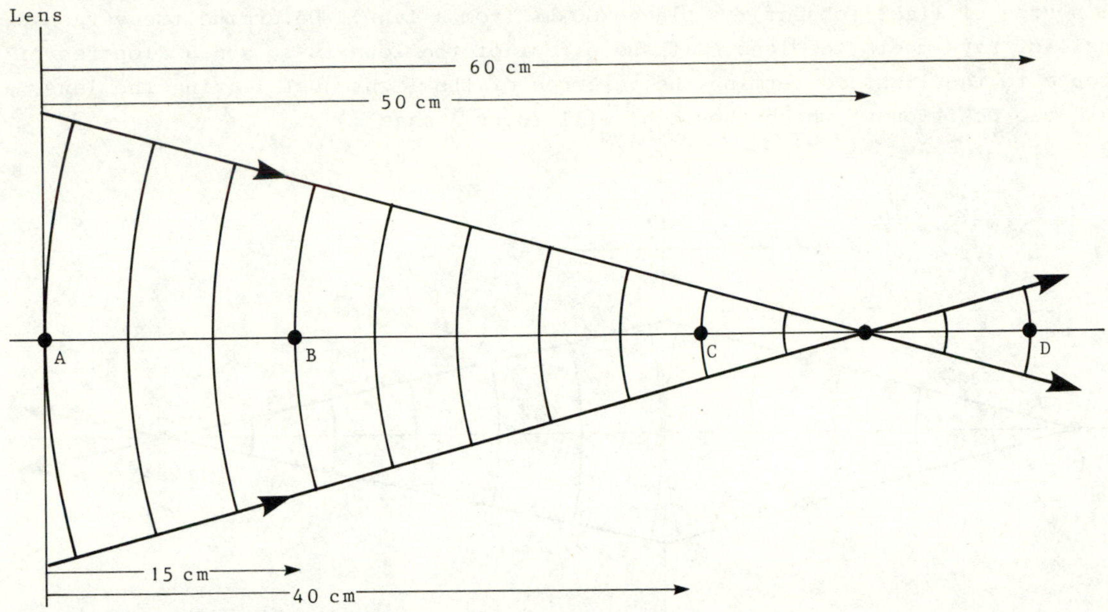

(d) For D - the point 60 cm from the lens.
In effect, the focus point (or image) in the pencil of rays now becomes the source (or object). Therefore, the distance from D to the source (or object) is 10 cm and the vergence is $L = \frac{1}{\ell}$, where $\ell = -10$ cm $= -0.10$ m. $\therefore L = \frac{1}{-0.10} = -10$ D.

These results may be represented pictorially as shown below.

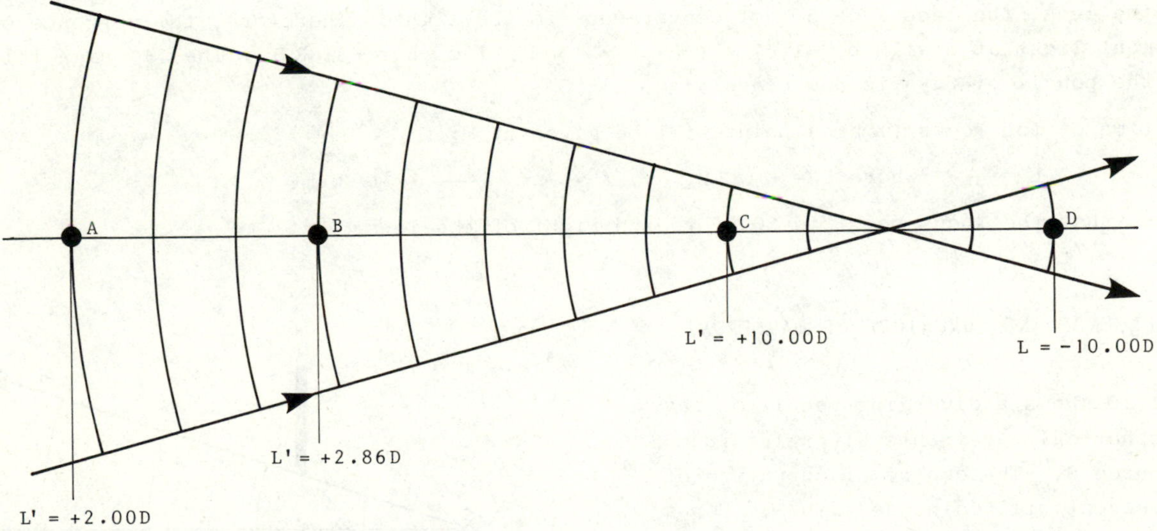

This example illustrates the important point we made in section 1.2.2; namely, that the vergence is positive at any point in a converging pencil and negative at points in a diverging pencil.

2 A small source of light (object) is placed 60 cm from a lens. Determine the vergence as the pencil of rays meets the lens. If the effect of the lens is to add 5 dioptres of convergence to the light, determine the vergence of the light just leaving the lens, and calculate the position at which the light will focus (image).

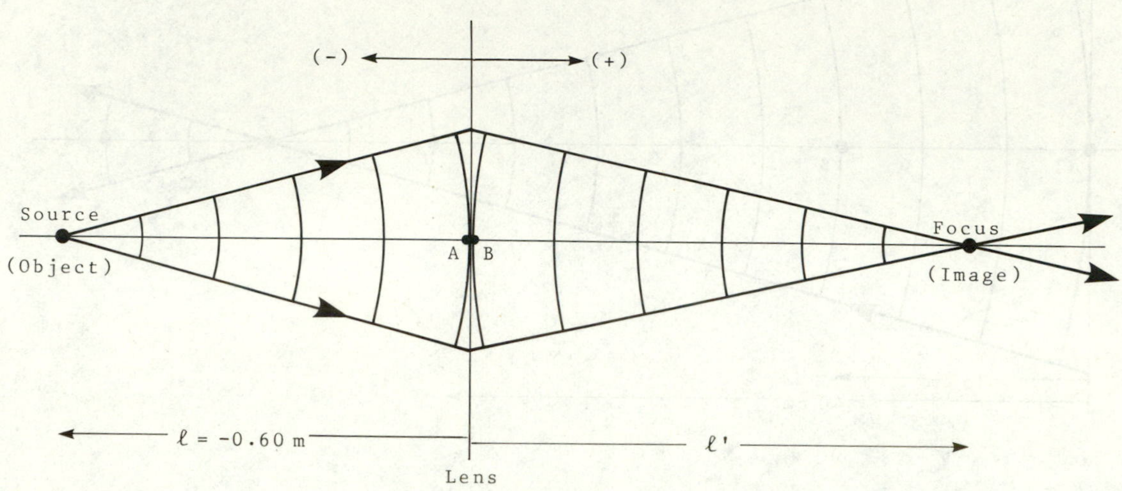

At A - the point where the pencil of rays meets the lens:

$$\text{vergence } L = \frac{1}{\ell}, \text{ where } \ell = -60 \text{ cm} = -0.60 \text{ m}. \quad \therefore L = \frac{1}{-0.60} = -1.67 \text{ D}.$$

Let us remind ourselves that the (-) sign represents a *diverging* pencil of rays.

Now, at the lens, the lens adds 5 D of convergence to the light. Therefore, the vergence of the emergent light at B will be -1.67 + (+5) = +3.33 D; i.e. L' = +3.33 D. The (+) sign tells us that the pencil of rays is now *converging*.

The position of the focus (image) point will be given by

$$L' = \frac{1}{\ell'}, \text{ where } L' = +3.33 \text{ D}. \quad \therefore \ell' = \frac{1}{L'} = \frac{1}{+3.33} = +0.30 \text{ m}.$$

This means that the focus point is 30 cm to the right of the lens.

1.2.5 VERGENCE AND CURVATURE OF WAVEFRONT

Figure 1.10 shows a diverging pencil of rays having spherical wavefronts diverging from a point source S. The centre of curvature of each wavefront will coincide with S. For a particular wavefront such as WF let the radius of curvature be r, as measured *from* the wavefront *to* the source.

Let us now define the *curvature of the wavefront*, symbol R, as the reciprocal of the radius of curvature; that is, $R = \frac{1}{r}$ (1.4).

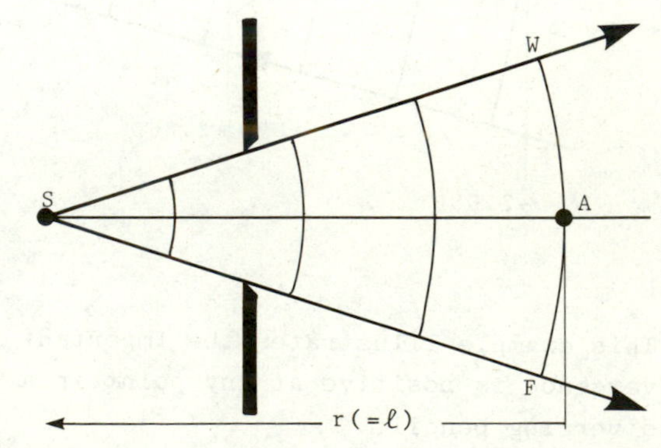

Fig. 1.10

This means that as the radius (r) of a wavefront increases the curvature (R) of the wavefront decreases.* But we have seen in figure 1.9(a) that the *vergence* at the point A on the wavefront is given by

$$L = \frac{1}{\ell}\,, \quad \text{using } \ell \text{ to represent the distance from A to the source.}$$

We therefore deduce from the foregoing the very important conclusion that

> *vergence at any point in a pencil of rays (in air) is equal to the curvature of the wavefront at that point.*

As the unit of measurement for vergence is the dioptre, then we shall also express the curvature of a wavefront in dioptres. This requires the radius of curvature to be expressed in metres, so that

$$R(\text{dioptres}) = \frac{1}{r(\text{metres})}\,.$$

It is useful to note that curvature is not confined only to wavefronts. The quantity may also be used in connection with the curved surfaces of lenses and mirrors. In all cases, whether these be wavefronts, or lenses, or curved mirrors, the curvature and the radius of curvature are related by the reciprocal expression in equation (1.4): $R(\text{dioptres}) = \frac{1}{r(\text{metres})}$.

As with most other optical quantities, a radius of curvature must be subjected to the rules of the Sign Convention. In order that we comply with our convention for vergences in section 1.2.2, the numerical value of the radius of curvature will be given a positive sign or a negative sign as follows.

> Measuring from the curve towards its centre of curvature, the radius value is given a *positive* sign if the measurement is in the *same direction* as that in which the light travels, but the radius value is given a *negative* sign if the measurement is in the *opposite direction* to that in which the light travels.

Figures 1.11(a), (b), and (c) illustrate converging, diverging, and parallel pencils of rays.

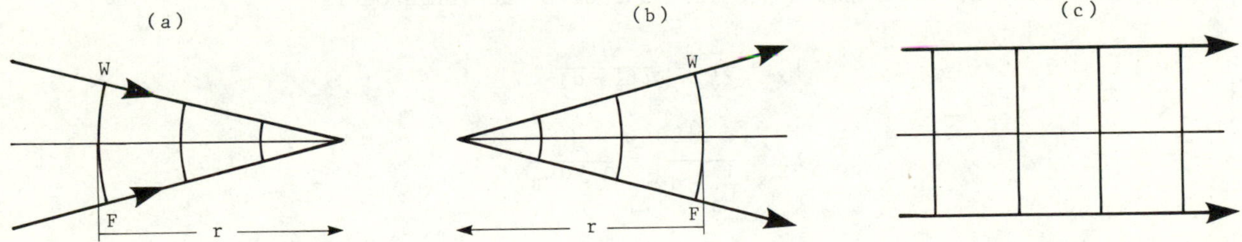

Fig. 1.11 Pencils of rays, associated wavefronts, radii of curvature and curvature.

In figure 1.11(a), the value of the radius of curvature r of the wavefront WF is given a (+) sign. Hence, the curvature $R\left(=\frac{1}{r}\right)$ will be positive. In figure 1.11(b), the value of the radius of curvature r of the wavefront WF is given a (−) sign and the curvature will be negative.

* This definition of curvature is taken from a more general definition. The derivation is shown in problem 5.1 of Worked Problems in Optics, A.H. Tunnacliffe.

In figure 1.11(c), the radius in infinite and the curvature is consequently zero. That is, *a plane wavefront has zero curvature*.

1.2.6 EFFECTIVITY

The principle of effectivity is very important in ophthalmic optics, and is concerned with the changes that occur in the vergence of a pencil of rays at different points along its path. Examples of particular importance concern the pencil of rays passing through the air space between a spectacle lens and the eye (section 3.1.5), and the light passing through the thickness of a thick lens (section 4.2.4.3).

Let us consider a converging pencil of rays travelling in air, as in figure 1.12 .

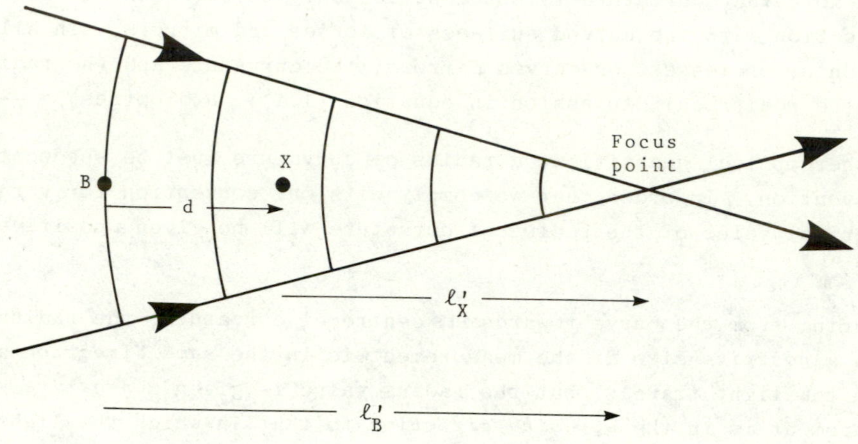

Fig. 1.12

At some point B the vergence of the pencil of rays is given by equation 1.3 : $L' = \frac{1}{\ell'}$. Now, at some other point X positioned a distance d from B the vergence is

$$L'_X = \frac{1}{\ell'_X} = \frac{1}{(\ell'_B - d)}$$

$$\therefore L'_X = \frac{1}{\frac{1}{L'_B} - d} = \frac{L'_B}{1 - dL'_B} \qquad (1.5).$$

Thus, L'_X is the vergence of the pencil of rays at X and is referred to as the *effectivity* of the pencil at X.

In using this expression all symbols, excepting d, are subject to the rules of the sign convention already stated in section 1.2.2 . d, then, will always be taken as a positive quantity. Some texts allow d to follow the sign convention where it is positive when measured to the right, as it would be when finding the vergence at X when already knowing the vergence at B, or where it is negative when measured to the left. The latter case assumes L'_X is known and L'_B is to be found. Allowing d to take a sign results in a single equation, but it also results in the questionable appearance of negative thicknesses for lenses, for example. Hence, this text will always choose to take d as a positive quantity. This means we can rearrange equation (1.5) for L'_B, which gives

$$L'_B = \frac{L'_X}{1 + dL'_X} \qquad (1.5a).$$

We can then refer to equation (1.5) as the *step-along* equation, that is, stepping along from left to right, and equation (1.5a) as the *step-back* equation, stepping back from right to left. Notice there is a certain symmetry about the two equations but with the minus sign in the denominator of the step-along equation and the plus sign in the step-back equation.

Worked Example

A pencil of rays has a vergence of +6.00 D at a certain point. Find the vergence of the pencil at points 20 mm on each side of the point.

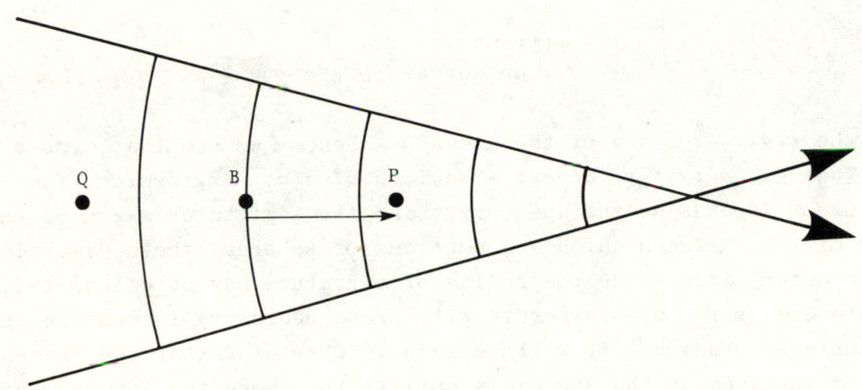

As the vergence is quoted as (+) the pencil of rays is converging. Let B represent the point where the vergence is +6.00 D; i.e. L'_B = +6.00 D.

At P, d = 20 mm = 0.02 m. The vergence at P is then given by the step-along equation

$$L'_P = \frac{L'_B}{1 - dL'_B} = \frac{6}{1 - (0.02 \times 6)} = \frac{6}{1 - 0.12} = \frac{6}{0.88} = +6.82\ \text{D}.$$

The vergence at P, the effectivity, is +6.82 D.

At Q, d = 20 mm = 0.02 m. The vergence at Q is then given by the step-back equation

$$L'_Q = \frac{L'_B}{1 + dL'_B} = \frac{6}{1 + (0.02 \times 6)} = \frac{6}{1 + 0.12} = \frac{6}{1.12} = +5.36\ \text{D}.$$

The vergence at Q, the effectivity, is +5.36 D.

1.2.7 APERTURE, RADIUS OF CURVATURE, AND SAGITTA

As stated in section 1.2.5, the curvature R of a spherical surface is given by $R = \frac{1}{r}$ where r is the radius of curvature. This is applicable to the spherical surface of a lens as well as to the spherical wavefronts in a pencil of rays. The circle shown in figure 1.13 is a section of a sphere taken through its centre. The shaded portion ABD represents a section through a wavefront, or a lens surface, forming part of a sphere.

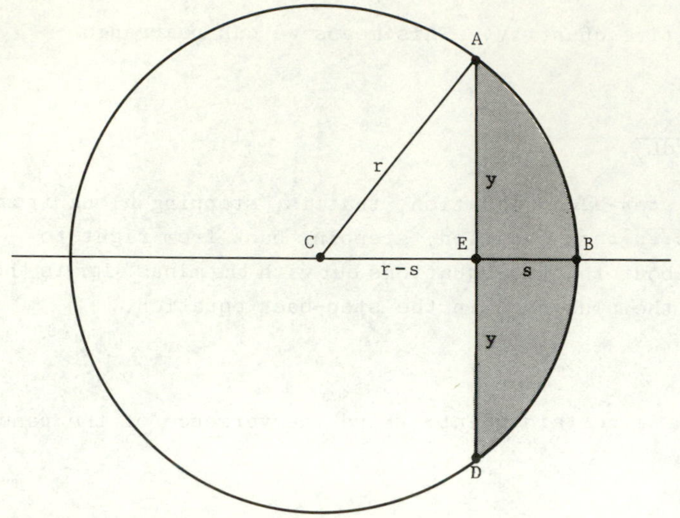

Fig. 1.13 Radius and sagitta (or sag).

The aperture of the wavefront, or lens surface, is 2y and the radius of curvature is r. Applying Pythagoras' theorem to the triangle CAE:

$$AC^2 = AE^2 + CE^2$$

$$\Rightarrow \quad r^2 = y^2 + (r-s)^2$$

$$\Rightarrow \quad r^2 = y^2 + r^2 - 2rs + s^2$$

giving $y^2 = 2rs - s^2$.

Now, if s is very small compared with r, then we may ignore the term s^2. For example, if s = 2 mm = 0.002 m, then s^2 = 0.0004 m which is a much smaller number! Hence, $y^2 = 2rs$, approximately. That is,

$$r = \frac{y^2}{2s} \qquad (1.6),$$

$$\text{or curvature } R = \frac{1}{r} = \frac{2s}{y^2} \qquad (1.7).$$

The small distance s is called the *sagitta or sag* of the arc ABD. Hence we see that, for a given small aperture, $R \propto s$. That is, *curvature of arc ∝ sagitta of arc*. Expression (1.6) shows that the radius of curvature r depends on the aperture y and the sagitta or sag s of the curve. In the case of lens and mirror surfaces which are portions of spheres, these dimensions are easily measured with a spherometer, from which the radius of curvature may be calculated. Since spherical lens surfaces are only parts of spheres it will prove necessary to measure the sag in order to calculate the surface's radius. It will be seen in Chapter 2 that the radius of a surface is a vital parameter involved in the surface's ability to change the vergence of light incident upon it. The approximate sag formula, equation (1.6), lies at the heart of the theory of the optician's lens measure which is used in practice for checking surface powers, see section 3.1.5.1 .

1.3 THE FORMATION OF SHADOWS

When the light from a source is partly restricted by an opaque body the formation of a shadow is explained as a consequence of the propagation of light in straight lines. The appearance of the shadow depends on the size of the light source. A point source of light gives rise to a well-defined shadow of an object. In figure 1.14(a), the body A is a sphere and its shadow U cast on a screen by a point source will be circular in shape and will have a sharp, clearly defined edge.

The region of total shadow U is referred to as the *umbra*, from the Latin word for shade. The diameter d of the umbra is obtained by reference to the pair of similar triangles SAB and SUV.* Since the triangles SAB and SUV are similar, then $\frac{d}{x} = \frac{h}{y}$. Hence, the umbra diameter is given by $d = \frac{xh}{y}$.

* The reader is directed to Appendix 4 in this book should he need to remind himself of the properties of similar triangles.

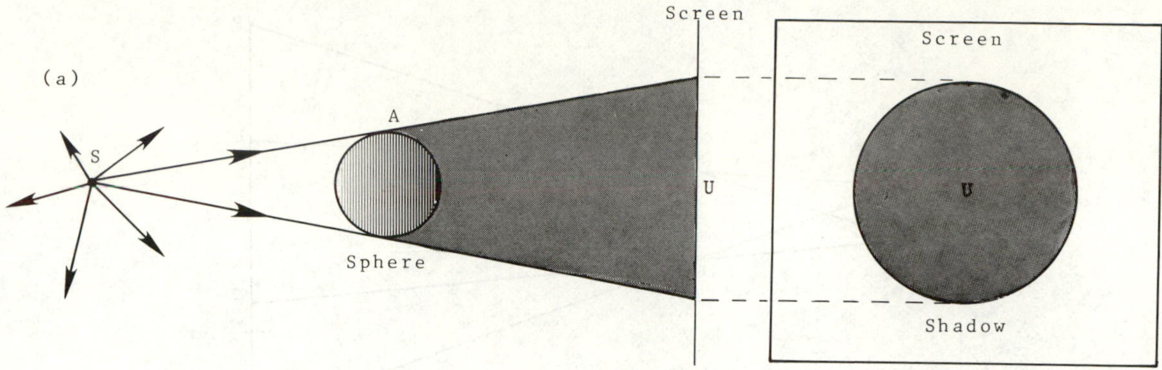

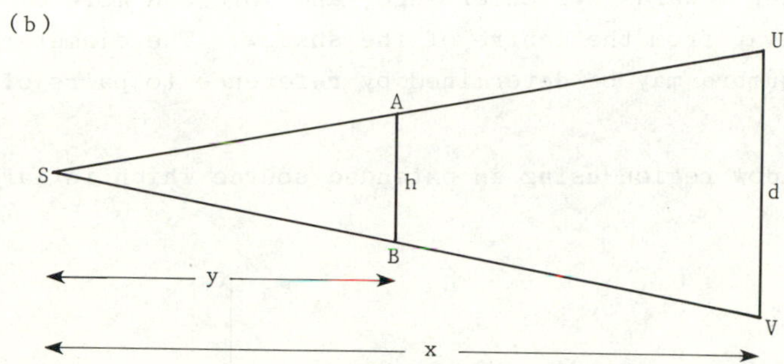

Fig. 1.14 (a) A point source, an obstacle, and its shadow. (b) The geometry.

If, however, the source is an extended one, such as a large pearl glass bulb, the shadow of the sphere will not be clearly defined. Figure 1.15 shows the form of the shadow produced when the extended source is smaller than the size of the sphere. In this case the area with the diameter UV, which is the darkest part of the shadow, is the *umbra*, whereas the area delimited by PQ , which is only partly shaded from the source, is referred to as the *penumbra* (from the Latin word paene meaning nearly).

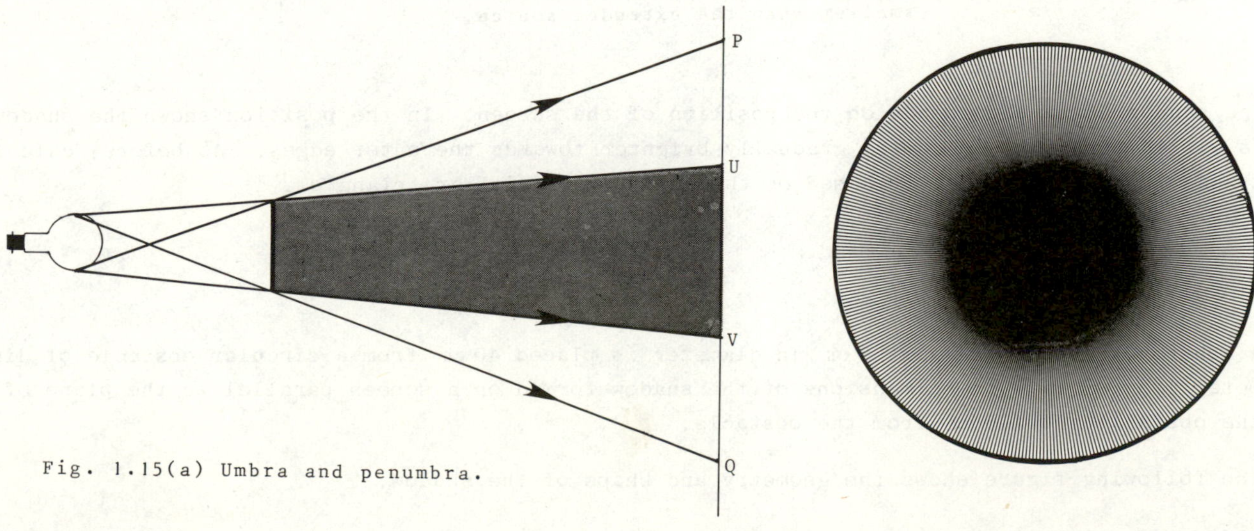

Fig. 1.15(a) Umbra and penumbra.

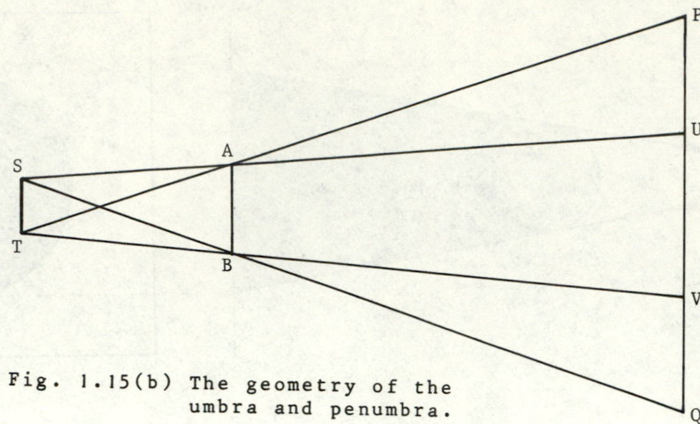

Fig. 1.15(b) The geometry of the umbra and penumbra.

The penumbra becomes gradually brighter towards its outer edge, and more and more of the source becomes visible as one moves away from the centre of the shadow. The diameter UV of the umbra and the width PU of the penumbra may be determined by reference to pairs of similar triangles in figure 1.15(b).

Figure 1.16 shows the form of the shadow region using an extended source which is larger than the obstacle.

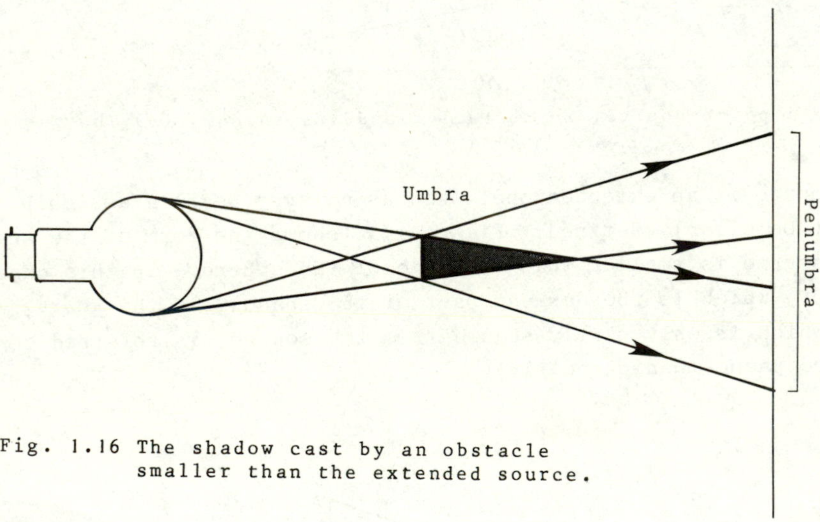

Fig. 1.16 The shadow cast by an obstacle smaller than the extended source.

The shadow obtained depends on the position of the screen. In the position shown the shadow is wholly penumbral, becoming gradually brighter towards the outer edges. As before, calculation of the shadow size is based on the geometry of similar triangles.

Worked Example

A spherical source of light 8 cm in diameter is placed 40 cm from a circular obstacle of diameter 12 cm. Find the dimensions of the shadow formed on a screen parallel to the plane of the obstacle and 100 cm from the obstacle.

The following figure shows the geometry and shape of the shadow.

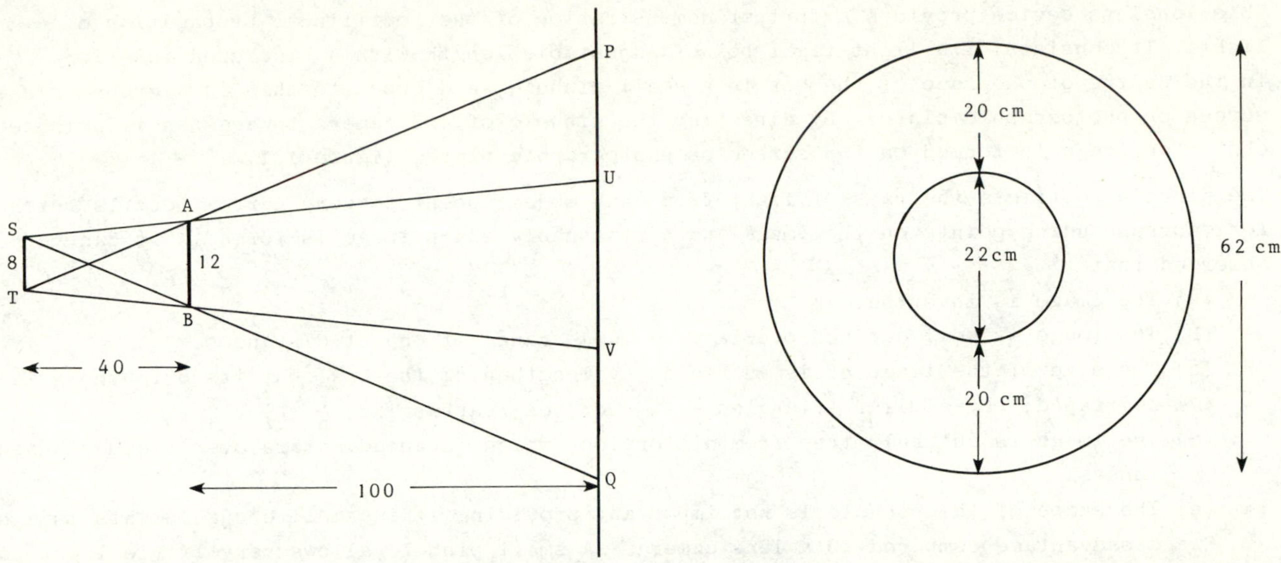

(i) To find the width PU of the penumbra ring we use the similar triangles AST and APU:

$$\frac{PU}{AU} = \frac{ST}{AS} \quad \Rightarrow \quad PU = \frac{AU}{AS} \times ST = \frac{100}{40} \times 8 = 20\,\text{cm}, \quad \text{since} \quad \frac{AU}{AS} = \frac{100}{40}\,.$$

(ii) To find the umbra diameter we use the similar triangles TPV and TAB:

$$\frac{PV}{TV} = \frac{AB}{TB} \quad \Rightarrow \quad PV = \frac{TV}{TB} \times AB = \frac{140}{40} \times 12 = 42\,\text{cm}, \quad \text{since} \quad \frac{TV}{TB} = \frac{140}{40}\,.$$

Therefore, the umbra diameter UV = PV - PU = 42 - 20 = 22 cm.

*1.4 THE PINHOLE CAMERA**

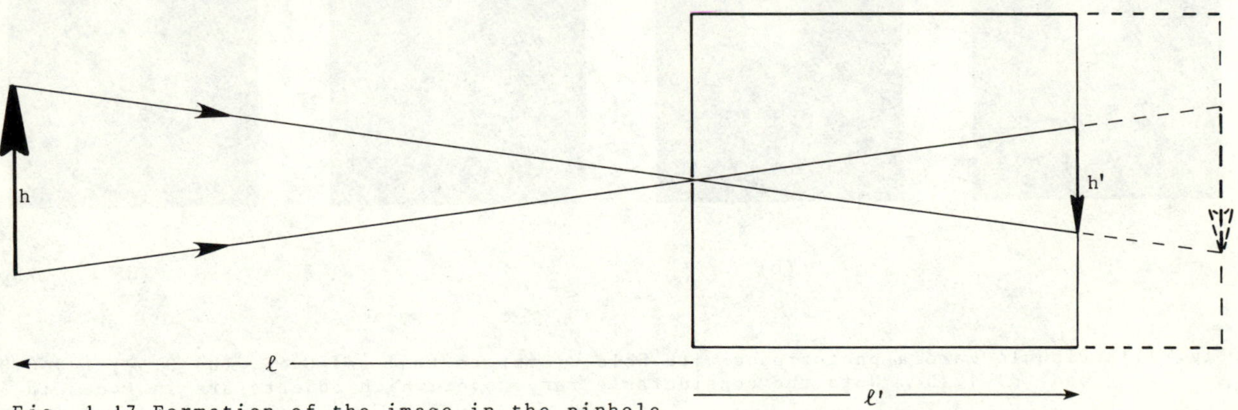

Fig. 1.17 Formation of the image in the pinhole camera.

* For any reader interested in the manufacture and theory of a pinhole camera, attention is drawn to a two-part article "Look No Lens" by George L. Wakefield, Amateur Photographer, August 31st, 1977, p.85, and September 7th, 1977, p.112.

This lensless device provides a further demonstration of the rectilinear propagation of light. It consists of a light-tight box of adjustable length with a blackened interior. In the centre of one face of the box is a small pinhole, and opposite this is a ground glass screen or photographic plate. By directing the pinhole of the camera towards an illuminated object an image is formed on the screen or photographic plate, figure 1.17 .

The pinhole restricts the rays of light from each object point to very narrow pencils which form corresponding points on the image and a reasonably sharp image is formed. It can be observed that:

(a) The image is inverted.

(b) The image is well defined over a very large range of object distances.

(c) The size of the image h' is increased by lengthening the box, but its brightness is decreased; by similar triangles $\frac{h'}{\ell'} = \frac{h}{\ell}$ or, on rearranging, $\frac{h'}{h} = \frac{\ell'}{\ell}$.

(d) The image is entirely free from distortion, which is an advantage over a camera using lenses.

(e) The shape of the pinhole is not important providing it is small enough. This is a big disadvantage compared to a lens camera. A small pinhole allows very little light to enter the camera. Consequently, the image will be faint and relatively long exposure times will be necessary when taking photographs.

(f) There is an optimum size of pinhole for maximum image sharpness. It has been found that this optimum hole diameter is proportional to the distance between the pinhole and the image screen.

(g) Increasing the size of the pinhole from the optimum value produces a brighter image, but one which is less distinct. This is due to the overlapping on the image of rays from different object points. As a consequence of this the image is blurred,

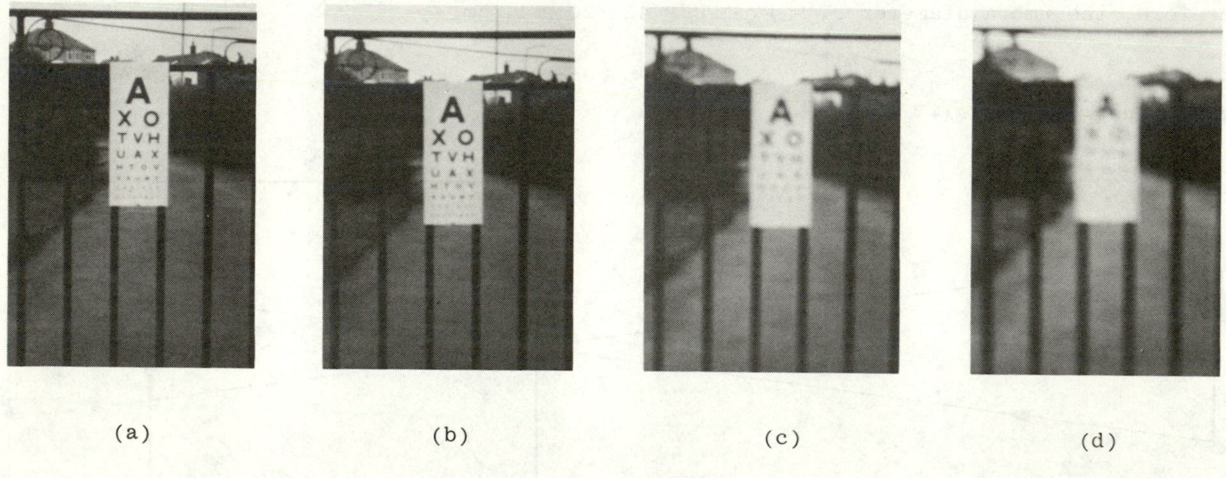

(a) (b) (c) (d)

Fig. 1.18 Pinhole camera photographs with hole diameters (mm) (a) 0.46, (b) 0.57, (c) 0.87, and (d) 1.15 . Note the considerable range over which objects are in focus in (a).

Figure 1.18(e) illustrates overlapping rays from different points on the object when the hole is too large. The effect is very obvious in the series of photographs in figure 1.18 (a) to (d) where increasing the hole beyond the optimum (a) creates a steadily increasing blur.

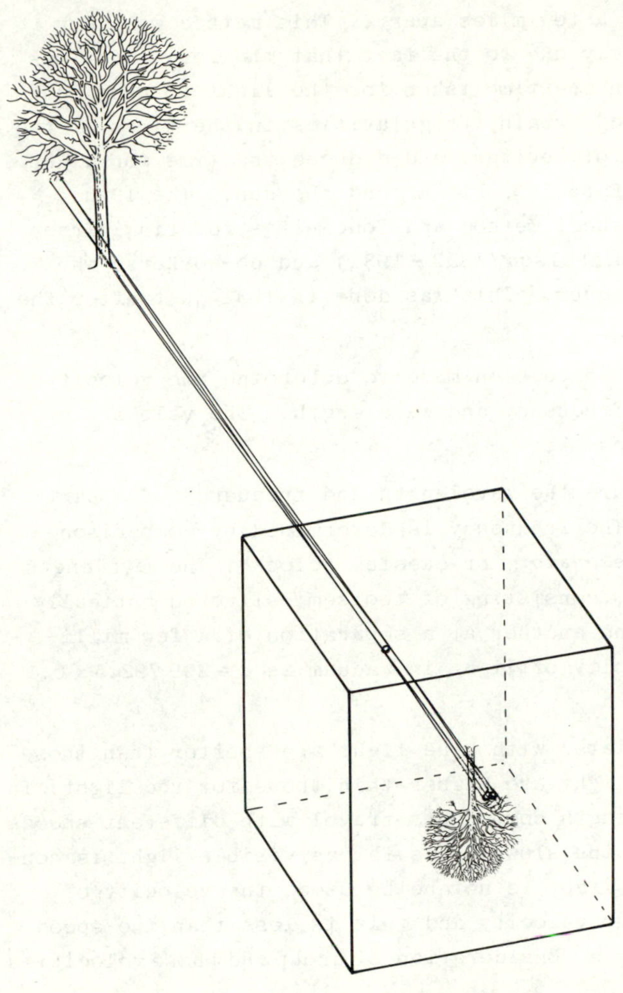

Decreasing the size of the pinhole from the optimum value again renders the image less clear. However, this is due to the effect of diffraction, which will be dealt with in Chapter 14.

In later chapters of this book (Chapters 10 to 14 and Chapter 16) we shall be considering the detailed wave nature of light. Then it will be seen that there is evidence of slight departures from rectilinear propagation and that light waves do spread very slightly into a shadow region cast by an obstacle. However, in geometrical optics we will base our considerations on true rectilinear propagation. Only when we wish to investigate the fine detail in the structure of an image does the method prove inadequate.

Fig. 1.18(e) Overlapping rays from different points on the object.

1.5 THE VELOCITY OF LIGHT

The velocity of light in vacuum, symbol c, is one of the most important and fundamental physical constants, and the precise determination of its value is of practical and theoretical importance. The electromagnetic theory of Maxwell predicted the existence of electromagnetic waves which travel in vacuum with a velocity which is independent of the frequency of the waves. This has been verified experimentally by many different methods for various frequencies of waves including gamma rays, light waves, and radio waves. Indeed, it was largely the theoretical predictions of Maxwell that led to the discovery and development of radio signals. More detail on the various types of electromagnetic waves will be found in section 10.6.2 .

In view of the universal nature of the velocity of light many of the methods for its determination are unrelated to optics. We shall consider very briefly the development of optical methods only, together with the details of some of these.

All the earlier methods for the determination of the velocity of light were based on the attempted measurement of the time taken for light to travel a known distance. The velocity is then calculated using the equation velocity = distance/time. The first known attempt was made by Galileo Galilei (1564 - 1642), the Italian mathematician, astronomer, and physicist, in about

1600, using two observers with lanterns on hills a few miles apart. This method was very inaccurate and inconsistent in its results, largely due to the fact that the reaction times of the two obervers were found to be greater than the time taken for the light to pass between them. Later, in 1676, Römer investigated certain irregularities in the times of the eclipses of Ganymede, one of the brightest moons of Jupiter, and deduced the time for light to travel a distance equal to the radius of the Earth's orbit around the sun. The 19th Century saw the development of Fizeau's toothed wheel method and Foucault's rotating mirror method. The latter method was improved by A.A. Michelson (1852 - 1931) and co-workers, who eventually determined the velocity of light in vacuum. This was done in 1931 just after the death of Michelson.

More recently, since about 1950, further attempts have been made to determine the velocity of electromagnetic radiation by measurements of frequency and wavelength. The velocity c is then calculated using equation (1.1), namely $c = \nu\lambda$.

One interesting method, in principle, is to measure the wavelength and frequency of a particular monochromatic light emitted by a laser. The frequency is determined by comparison with the frequency standard called the *caesium resonator*, or caesium "clock". The wavelength is measured with the aid of a Fabry-Perot Etalon, consisting of two semi-silvered optically flat glass plates fixed accurately parallel to one another at a separation of a few millimetres. The present accepted value for the velocity of light in vacuum is $c = 299\,792.5 \pm 0.1$ km/s.

Observations indicate that the wavelengths associated with blue light are shorter than those for red light, and that the frequencies of blue light are higher than those for red light. In any medium other than a vacuum the various wavelength components travel with different speeds, with red light the fastest and blue/violet light the slowest, as far as visible light is concerned. It must be noted that, if monochromatic light is not being used, the velocity of light in such a medium is referred to as the group velocity and this is less than the speed of the slowest wavelength or phase velocity. For a consideration of group and phase velocities see sections 10.3, 10.7, and 11.2.1 .

1.5.1 FIZEAU'S TOOTHED WHEEL EXPERIMENT

In 1849 the French physicist, Armand Fizeau, who incidentally was a wealthy amateur experimentalist, suggested a method which involved the measurement of a very small interval of time. His apparatus is illustrated in figure 1.19 .

Light from the bright source S is reflected by the plane glass plate P through the space between two teeth on the toothed wheel W. The light is then rendered parallel by the lens L_1 and transmitted over a known distance (about 5 miles) to lens L_2. By means of a suitably positioned plane mirror M, the light returns along its own path. An image of the source S can then be seen through the plate P using a lens L_3. The wheel W is now set in motion about the axis A. For slow speeds of rotation a flickering effect is observed as the teeth on the wheel successively interrupt the light, and produce a succession of images on the retina.

Eventually, as the rotation speed of W is increased, the stage is reached where the light passing through a gap between two teeth travels the distance to M and back in time to meet the adjacent tooth. In this situation the field of view becomes dark. Hence, it follows that the time taken for the light to travel from W to M and back is equal to the time taken for a

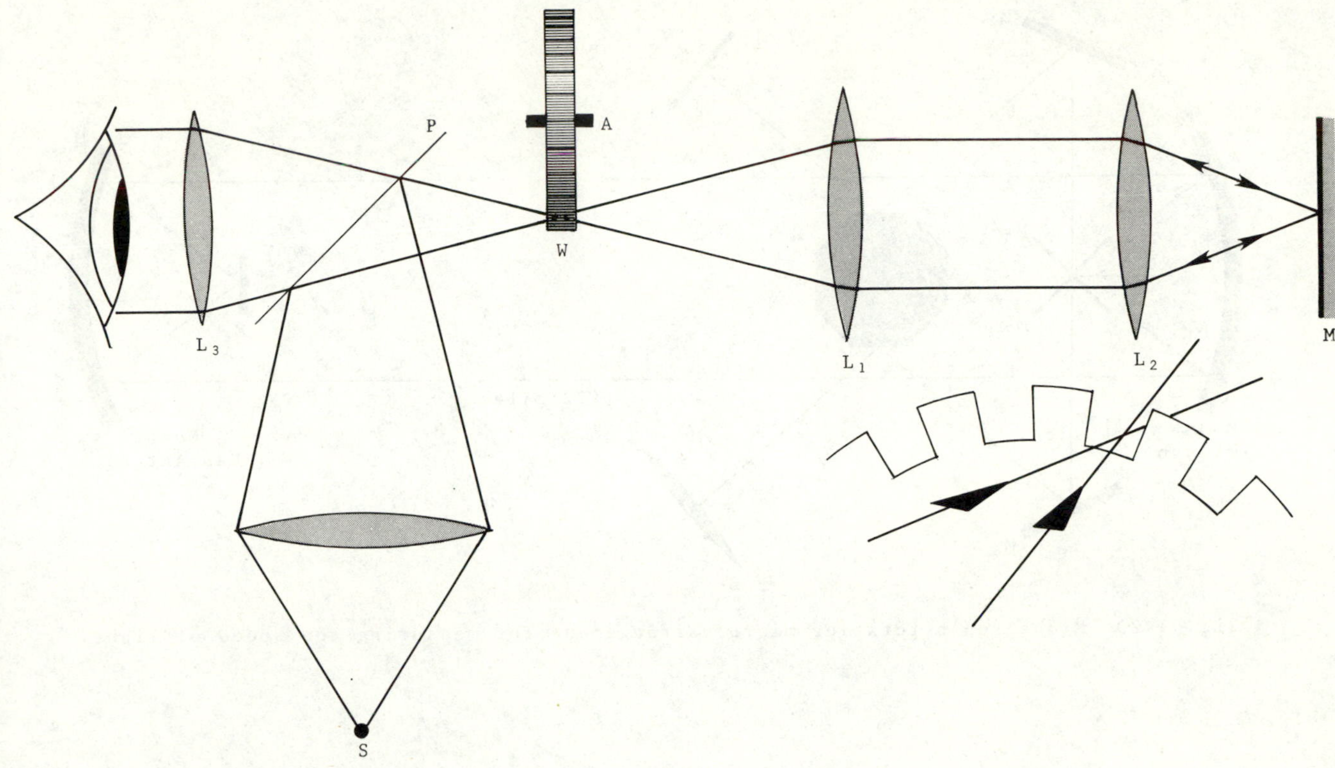

Fig. 1.19 Fizeau's experimental arrangement for measuring the speed of light.

tooth to move and occupy the position of the adjacent gap.

With a wheel of 720 teeth revolving at 12.6 revolutions per second to achieve a dark field of view, and with the distance WM being 8650 m, Fizeau obtained a value for the velocity of light as $c = 3.14 \times 10^8\ ms^{-1}$.

1.5.2 MICHELSON'S ROTATING MIRROR EXPERIMENT

The American astronomer and physicist Michelson carried out many determinations of the velocity of light and, in 1926, used a rotating octagonal mirror as a means of measuring the very small time interval for light to travel a distance of about 44 miles. The apparatus is shown in figure 1.20 .

Light from a powerful arc lamp S placed behind a slit is reflected from one face of an octagonal mirror M_1. Using mirrors M_2 and M_3 the light is directed on to a large concave reflector M_4. By careful positioning , a parallel pencil of light was directed from Mount Wilson in California to a large concave reflector M_5 22 miles away on Mount San Antonio. A plane mirror M_6 reflected the light back to M_5 and hence to M_4. It was then reflected at mirrors M_7 and M_8 on to the back face of the octagonal mirror M_1 and finally to an eyepiece E where an image of the slit was formed.

If the mirror M_1 is rotating, the image seen in the eyepiece will be in a different position from that which it would occupy if the mirror were stationary. At one particular speed of rotation, however, the time taken for the light to travel from M_4 to M_5 and back will be the

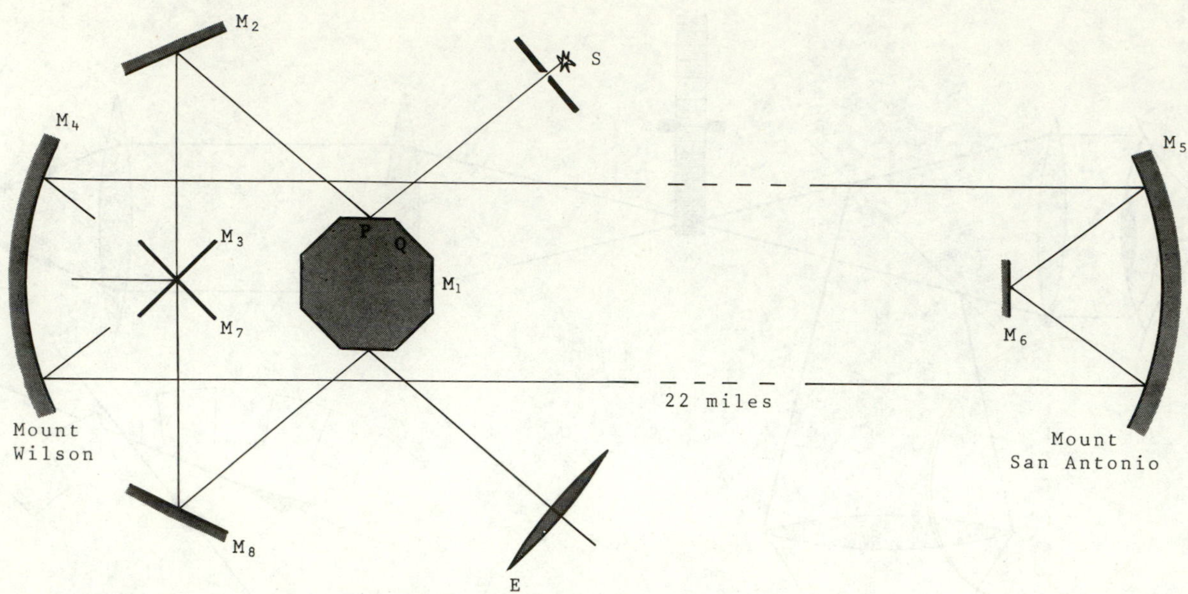

Fig. 1.20 Michelson's rotating mirror arrangement for measuring the speed of light.

same as the time taken for one face of the octagonal mirror, say face P, to be replaced by the adjacent face Q. When this is so, the image at E will be seen in the same position as when M_1 was stationary. Michelson found that a rotation speed of about 528 rev/sec was necessary. With the distance between M_4 and M_5 being accurately surveyed, the velocity of light quoted by Michelson based on many measurements was

$$c = 299\,796 \pm 4 \text{ km/s}.$$

In determining this result allowance had to be made for the variable optical properties of the air over the 22 miles range. The exact conditions of pressure, temperature, and humidity could not be measured or controlled. Nevertheless, the value obtained was in fairly good agreement with the current accepted value.

1.5.3 BERGSTRAND'S METHOD USING THE KERR CELL

Before considering this modern method for the determination of the velocity of light the reader is advised to acquaint himself with section 9.11.3 on the photoelectric cell and with Chapter 12 on polarisation.

The principle of the Kerr cell is that it acts as an electrically operated light shutter and, in effect, replaces the rotating toothed wheel of Fizeau's method. The cell contains a substance which becomes doubly refracting when a voltage is applied across electrodes which are immersed in the substance. When the cell is placed between crossed polaroids, which normally allow no light to pass through, the application of a modulating voltage allows light to pass through the polaroid analyser. Figure 1.21 shows the schematic arrangement of the Kerr cell, and figure 1.22 illustrates the basic arrangement of Bergstrand's apparatus.

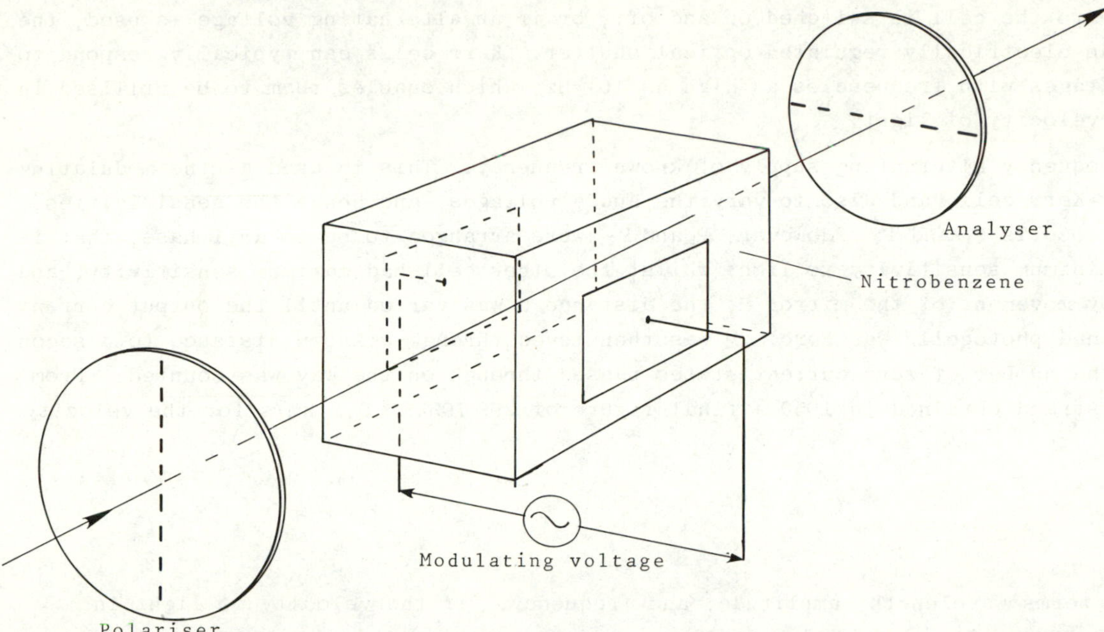

Fig. 1.21 The Kerr Cell.

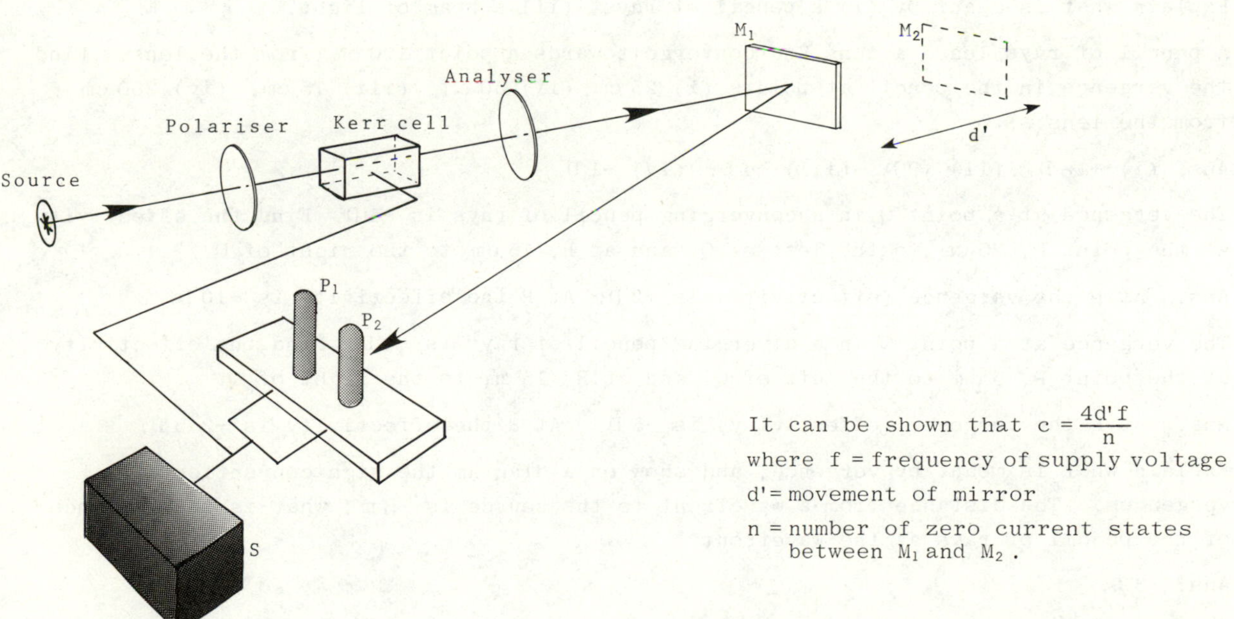

Fig. 1.22 Bergstrand's apparatus (schematic).

If the voltage to the cell is switched on and off, or if an alternating voltage is used, the cell becomes an electrically regulated optical shutter. Kerr cells can typically respond to modulating voltages with frequencies as high as 10^{10} Hz, which enables them to be utilised in measuring the velocity of light.

S is a high frequency alternating supply of known frequency. This is used as the modulating voltage to the Kerr cell, and also to vary the anode voltages, and hence the sensitivities, of the two photocells P_1 and P_2. However, P_1 and P_2 were arranged to be in antiphase, that is one cell had minimum sensitivity to light whilst the other cell had maximum sensitivity, and vice versa. By movement of the mirror M, the distance d was varied until the output current from the combined photocells was zero. M was then moved through a known distance to a second position and the number of zero current states passed through on the way was counted. From this data Bergstrand obtained in 1950 a final figure of 299 793.1 ± 0.2 km/s for the velocity of light.

EXERCISES

1 Define the terms wavelength, amplitude, and frequency. If the velocity of light in vacuum is 3×10^8 ms^{-1}, what will be the wavelengths corresponding to the following frequencies? (i) 4.5×10^{14} Hz and (ii) 6.96×10^{14} Hz.

Ans. (i) 667 nm and (ii) 431 nm. See Worked Problems in Optics, by A.H. Tunnacliffe, page 1.

2 Define the terms wavefront, crest, and trough. 100 water wave crests, each $\frac{1}{5}$ m from the next, pass a fixed post in 300 seconds. What is the speed of propagation?

Ans. $\frac{1}{15}$ ms^{-1}.

3 Explain what is meant by (i) a pencil of rays, (ii) a beam of light.

4 A pencil of rays leave a lens and converge towards a point 100 cm from the lens. Find the vergence in the pencil at points (i) 25 cm, (ii) 50 cm, (iii) 75 cm, (iv) 200 cm from the lens.

Ans. (i) +1⅓ D. (ii) +2 D. (iii) +4 D. (iv) -1 D.

5. The vergence at a point Q in a converging pencil of rays is +4 D. Find the effectivity at the point P, 25 cm to the left of Q, and at R, 15 cm to the right of Q.

Ans. At P the vergence (effectivity) is +2 D. At R the effectivity is +10 D.

6 The vergence at a point Q in a diverging pencil of rays is -4 D. Find the effectivity at the point P, 5 cm to the left of Q, and at R, 15 cm to the right of Q.

Ans. At P the vergence (effectivity) is -5 D. At R the effectivity is -2.5 D.

7 Explain what is meant by vergence, and show on a diagram the sign convention for vergences. The distance from a wavefront to the source is -⅓ m ; what is the vergence of the pencil of rays at the wavefront?

Ans. -3 D.

8. Light from a very distant object point is converged by a lens. If the lens adds +3 D of convergence, how far from the lens is the image focus?

Ans. +⅓ m.

9 What is meant by the term sagitta (sag)? Archaeologists partially uncover what is

thought to be a large spherical chamber. If the uncovered portion rises 1 m above the excavated ground level, and is 10 m across, what is the approximate diameter of the chamber?

Ans. 25 m.

10 What is the exact diameter of the chamber in question 9?

Ans. 26 m.

11 A spherical source of light 10 cm in diameter is placed 50 cm from a circular aperture of 16 cm diameter in an opaque screen. Find the size and nature of the patch of light on a white screen 150 cm from and parallel to the plane of the aperture.

Ans. The diameter of the fully illuminated area is 34 cm, and the diameter of the whole patch is 94 cm. See W.P.O., page 2.

12 A man 1.83 m tall stands 12 m from a street lamp which is 6 m above ground level. Find the length of the man's shadow on the ground.

Ans. 5.27 m.

13 In a pinhole camera the distance between the object and the image is 5 m. Find (i) the length of the camera if the image is one fiftieth the size of the object, and (ii) the distance of the object from the pinhole.

Ans. See W.P.O., page 3. Length of camera = 10 cm. The object is 4.90 m from the pinhole.

14 A pinhole camera produces an image 2.25 cm tall, and when the screen is withdrawn 3 cm further from the pinhole the length of the image increases to 2.75 cm. What was the original distance between the pinhole and the screen?

Ans. See W.P.O., page 3. 13.5 cm.

15 In a Fizeau's method experiment for measuring the speed of light, the toothed wheel rotates 5590 times per minute. The wheel has 100 teeth and 100 spaces and the mirror is 8.05 m distant. Find the speed of light.

Ans. See W.P.O., page 97. 2.999967×10^5 km.s^{-1}.

16 Suppose the experiment in question 13 were conducted in a tube which is now filled with water of refractive index 4/3. What is the number of revolutions per minute through which the wheel must now rotate to first cut off the light?

Ans. See W.P.O., page 98. 4192.5 rev/min.

17 In Michelson's rotating octagonal mirror experiment the speed of rotation was 528 rev/sec when the image was first reflected to its initial position. If the distance between Mount San Antonio and Mount Wilson is 35.5 km, find the speed of light.

Ans. 2.99904×10^5 km.s^{-1}.

18 A spherical light source, 20 cm in diameter, is 2 m from a circular opaque disc which is 4 m from and parallel to a white screen. If the opaque disc is 40 cm in diameter find the size of the shadow on the screen.

Ans. Diameter of umbra = 80 cm. Diameter of total shadow = 160 cm.

19 Red light, $\lambda = 600$ nm in vacuum, enters a glass block whereupon the wavelength reduces to 400 nm. Find the speed of light in the glass if the speed of light in vacuum is 3×10^8 metres per second. Find the ratio of the speed of light in vacuum to the speed in glass.

Ans. $2\times10^8 ms^{-1}$; 1.5 .

20 Describe the pinhole camera and the factors which determine the formation of a clear image. Since this camera produces an image free from distortion, why is it not used in place of expensive lens systems?

21 Explain the gradual change in the penumbra in figure 1.15(a) from dark to light from the centre outwards.

2 REFRACTION AT PLANE AND CURVED SURFACES

2.0 INTRODUCTION

When a ray of light in air is incident on the surface of a transparent medium such as glass, some of it is reflected according to the laws of regular (specular) reflection, see section 5.1.1, whilst the remainder is transmitted. No medium is perfectly transparent and some absorption of light always occurs, the energy being converted to heat. This latter effect will be ignored. The direction of the ray inside the medium is different to that of the incident ray unless the incident ray is normal to the surface. This *change of direction* of the ray at the surface is called *refraction*.

Common refracting materials are glass, in particular high grade optical quality glass, quartz, and many plastics. These have been chosen not only for their effect on the direction of light rays passing through them, but also for their transparency, homogeneity, and their resistance to atmospheric corrosion.

Optical glass is commonly used in the manufacture of prisms and lenses. Two main types of optical glass are available: these are crown and flint. The former is a compound of silica (sand) and salts of sodium and potassium. In addition, small quantities of other materials such as barium and zinc oxides may be present. Flint glasses, in addition to the constituents above, contain oxides of lead and are heavier than crown glasses.

Certain plastics materials are now increasingly being used for ophthalmic lenses. In this category is the thermosetting material allyl diglycol carbonate, commonly known as CR-39 (the CR standing for Columbia Resin).

2.1 REFRACTION AT PLANE SURFACES

Figure 2.1 shows an incident ray of light being refracted at the plane boundary between air and glass. Note the weak reflected ray; this is more pronounced as i increases.

It is seen that, as the ray of light passes from air into the glass, it is deviated towards the normal. The glass is said to be *optically denser* than air. i is the angle of incidence and i' is the angle of refraction. Note that both angles are measured between the normal and the appropriate ray.

The bending of light as it passes from air to glass is due to the wave nature of light and the fact that the light waves in the optically denser medium are travelling more slowly than in air. Figure 2.2(a) illustrates a parallel pencil of monochromatic light with plane

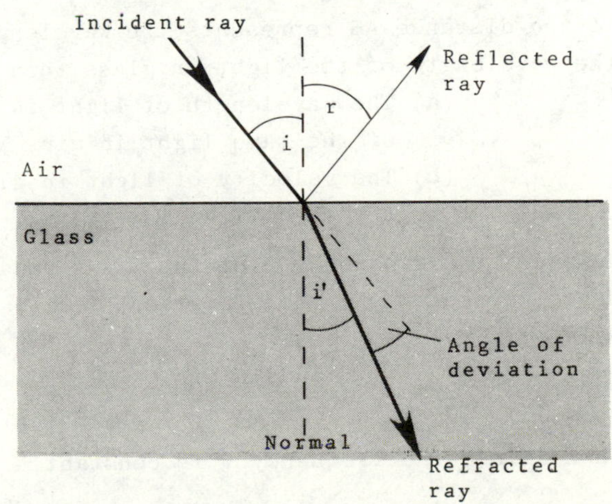

Fig. 2.1 Refraction at a plane air-glass boundary.

wavefronts, undergoing refraction at a glass surface. When the pencil of light meets the glass surface its waves are slowed down. Thus, as the wave in air travels the distance AB, the same wave in glass travels the shorter distance CD.

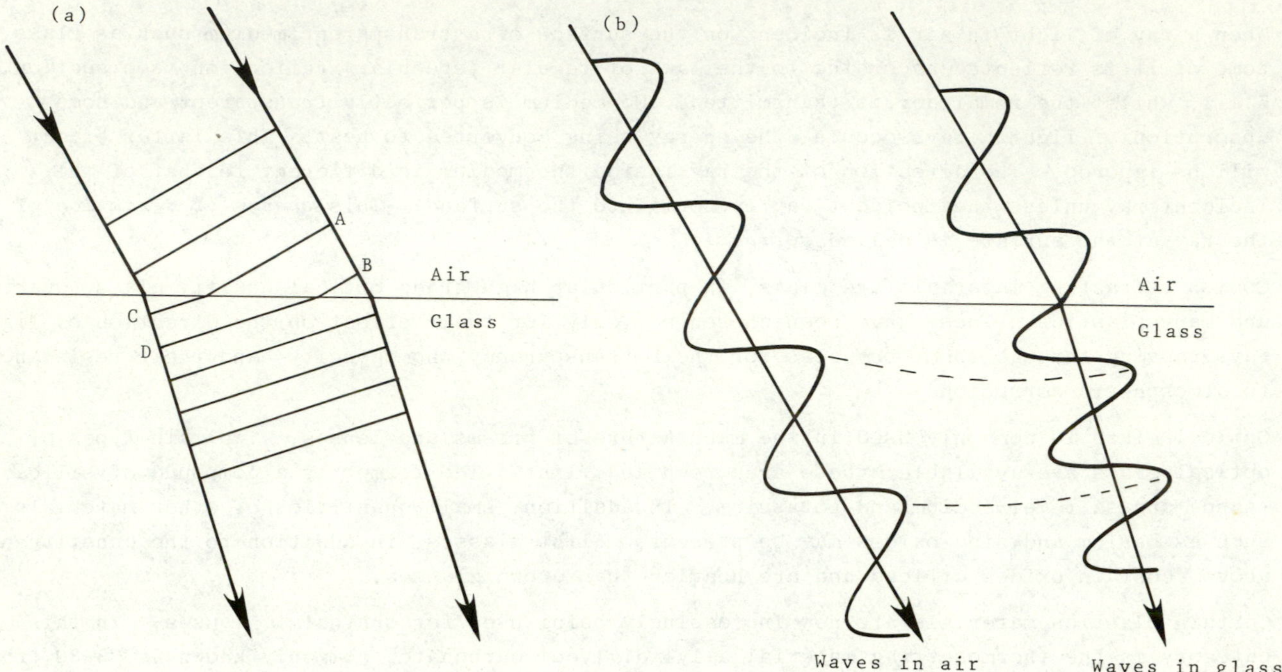

Fig. 2.2 (a) Wavefronts and rays bending at the air-glass boundary.
(b) Monochromatic waves bunched together on entering the glass.

This has two effects on the waves:

(i) the pencil of light rays changes direction , bending towards the normal

and (ii) the waves are bunched closer together in the glass (figure 2.2(b).

If the distance AB represents the wavelength of the light in air and the distance CD represents the wavelength of the light in glass then it is seen that:

(a) The wavelength of light in glass, λ_g, is smaller than the wavelength of the same light in air, λ_a.

(b) The velocity of light in glass, v_g, is smaller than the velocity of the same light in air, v_a.

However, the frequency ν of the waves remains constant as the light passes from air into glass. Hence, since $v = \nu\lambda$, then in glass

$$v_g = \nu\lambda_g$$

and in air

$$v_a = \nu\lambda_a.$$

Then, since the frequency ν is constant,

$$\frac{v_a}{v_g} = \frac{\lambda_a}{\lambda_g} \qquad (2.1).$$

That is, $\dfrac{\text{velocity of light in air}}{\text{velocity of light in glass}} = \dfrac{\text{wavelength of light in air}}{\text{wavelength of light in glass}}$.

2.1.1. THE LAWS OF REFRACTION

Experimental measurements of refraction at a water surface were made as long ago as the second century A.D. by Ptolemy and others, but no relationship between the angles of incidence and refraction was discovered. Law 1 below was formulated in the 11th Century and it was not until 1621 that Willebrord Snell, professor of mathematics at the University of Leiden, finally deduced the relationship between the angles of refraction and incidence for various pairs of media. This is given below as Law 2. Although this is usually referred to as *Snell's Law* some credit for its formulation should go to the French philosopher René Descartes.

Law 1. The incident ray and the refracted ray lie in one plane which is normal to the refracting surface at the point of refraction.

Law 2. (Snell's law). The ratio of the sine of the angle of incidence to the sine of the angle of refraction is a constant for any pair of media; that is,

$$\frac{\sin i}{\sin i'} = \text{constant}.$$

This expression applies rigidly only to isotropic media, that is, those where the optical properties do not vary in different directions through the medium, and for light of a single colour (monochromatic light).

2.1.2 REFRACTION OF LIGHT WAVES BY HUYGENS' PRINCIPLE

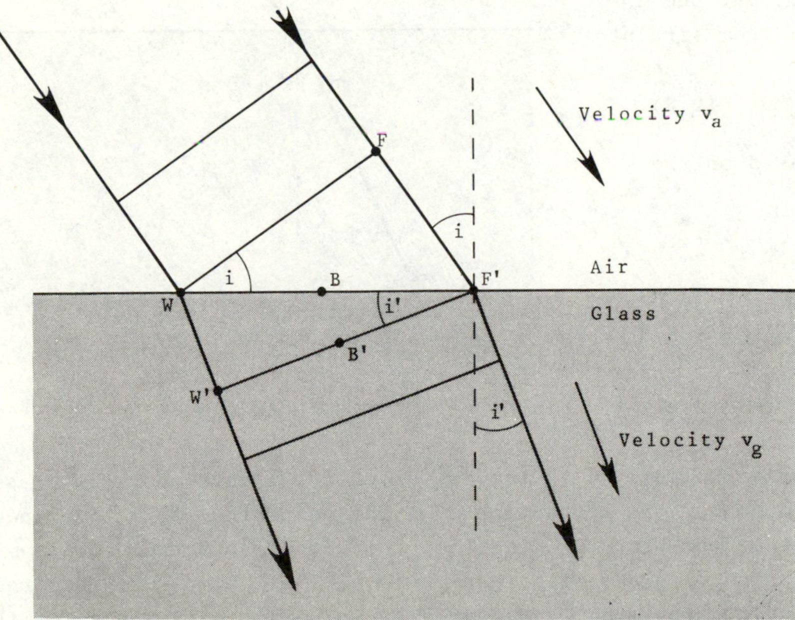

Fig. 2.3 Snell's law derived from the wave theory.

Huygens showed that the wave theory of light could account for Snell's law. From simple geometry it follows that

$$\widehat{FWF'} = i$$

and $\widehat{WF'W'} = i'$.

In figure 2.3, let us suppose that a plane wavefront WF is incident on a plane boundary separating air from glass. The secondary wave in the glass arising from W will have reached some point W' when the point

F arrives at F'. Similarly, the secondary wavelet originating from the point B will have reached a point B'. The new wavefront is the straight line W'B'F' drawn tangential to these wavelets.

Now, as light travels in glass with a different velocity from that in air, the distances WW' (in glass) and FF' (in air) are unequal. Hence, the refracted wavefront and the incident wavefront are not parallel. Let v_a = the velocity of light in air, and v_g = the velocity of light in glass. Since distance travelled = velocity × time, we have

in air $FF' = v_a \times t$ and, in glass, $WW' = v_g \times t$, where t is the time for W to travel to W' and for F to travel to F'.

Thus, in triangle WFF' we have $\sin i = \frac{FF'}{WF'}$, and in triangle WW'F' we have $\sin i' = \frac{WW'}{WF'}$.

Hence, $\frac{\sin i}{\sin i'} = \frac{FF'/WF'}{WW'/WF'} = \frac{FF'}{WF'} \times \frac{WF'}{WW'} = \frac{FF'}{WW'} = \frac{v_a \times t}{v_g \times t}$

$$\text{giving} \quad \frac{\sin i}{\sin i'} = \frac{v_a}{v_g} = \text{constant.}$$

This is Snell's law of refraction, derived from the wave theory.

2.1.3 REFRACTIVE INDEX

Snell's law, $\frac{\sin i}{\sin i'}$ = a constant, may be written for a given pair of media and for light of any one colour. This constant is referred to as the *refractive index*, symbol n, of one medium relative to the other. In figure 2.4 it is the *relative refractive index* of glass with respect to air. We denote this by the symbol $_an_g$, the subscript letters a and g indicate that the value of n is for light travelling from air into glass:

$$\text{that is,} \quad \frac{\sin i}{\sin i'} = {_an_g} \qquad (2.2).$$

We have shown in section 2.1.2, based on the wave theory of light, that

$$\frac{\sin i}{\sin i'} = \frac{v_a}{v_g},$$

where v_a and v_g are the velocities in air and in glass, respectively.

Therefore,

$$_an_g = \frac{v_a}{v_g} \qquad (2.3)$$

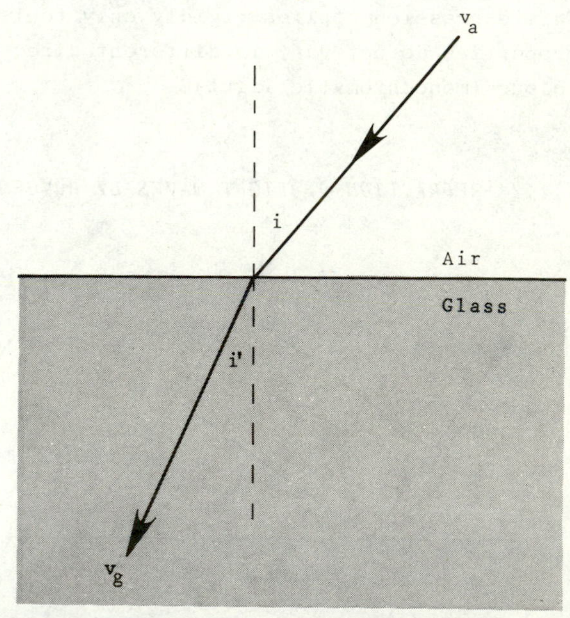

Fig. 2.4 Snell's law and refractive index.

The value of a refractive index varies slightly with the colour of the light. For the present we shall ignore the variation and consider the refraction of light using the refractive index value for yellow light which is the colour near the middle of the visible spectrum. This mean value for the relative refractive index of glass with respect to air will simply be referred to as the *refractive index of glass*, and the symbol used will be $_an_g$. Typical values for the refractive index of common optical and other materials are given in table 2.1 . Those materials of higher value are said to be *optically denser* than those of lower value.

TABLE 2.1

Material	Refractive index
Water	1.333
Transpex plastics	1.490
CR-39 plastics	1.499
Spectacle crown glass	1.523
Dense flint glass	1.613
Diamond	2.417 Highest value known

The refractive index of a substance with respect to a vacuum is referred to as the *absolute refractive index* of the substance, given by the expression

$$\text{absolute refractive index of substance} = \frac{\text{velocity of light in a vacuum}}{\text{velocity of light in substance}}.$$

The absolute refractive index of vacuum is therefore 1, and this is the lowest value possible for all substances. The absolute refractive index of air is about 1.000 27 which is so close to the value for vacuum that the difference can usually be ignored. Similarly, it is not usually necessary to distinguish between the refractive index of a material relative to air and its absolute refractive index. Note that the more optically dense a material, the smaller is the velocity of light in that medium and the greater is its absolute refractive index. This means that a ray travelling from a less dense to a more dense medium is bent towards the normal after refraction, and vice versa.

2.1.4 GRAPHICAL RAY TRACING FOR REFRACTION

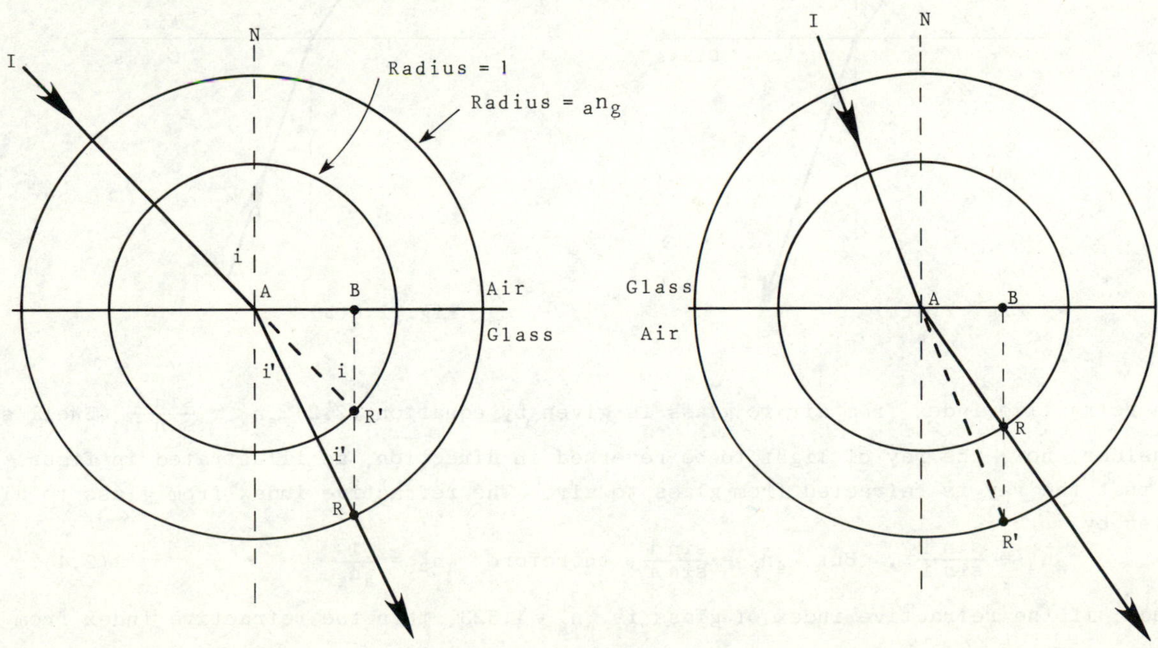

This scale drawing method is used to construct the refracted ray corresponding to an incident ray IA at a plane boundary between air and glass. The method may also be used for refractions at curved surfaces and has been used by optical designers as a check when forming new designs.

In figure 2.5(a), IA represents a ray incident in air on a plane glass surface. Two circles are drawn, centred at A, having radii in the ratio 1 to ${}_an_g$. For example, if ${}_an_g = 1.520$ the radii of the circles would be 1 cm and 1.52 cm, or 5 cm and 7.6 cm, etcetera. Ray IA is produced until it intersects the smaller circle at R'. The line RR' is drawn parallel to the normal AN to intersect the larger circle at R. AR is the required direction of the refracted ray in the glass. Figure 2.5(b) shows the construction for a ray passing in the reverse direction from glass into air: the instructions are identical!

The construction works simply because it satisfies Snell's law. Consider figure 2.5(a), for example:

$$\frac{\sin i}{\sin i'} = \frac{AB/AR'}{AB/AR} = \frac{AR}{AR'} = \frac{{}_an_g}{1} = {}_an_g .$$

2.1.5 REFRACTIVE INDICES AND THE REVERSIBILITY OF LIGHT

The *Reversibility Principle* in optics states that if a reflected or refracted ray be reversed in direction it will retrace its original path. Figure 2.6(a) shows a ray of light refracted from air into glass with the ray being deviated towards the normal.

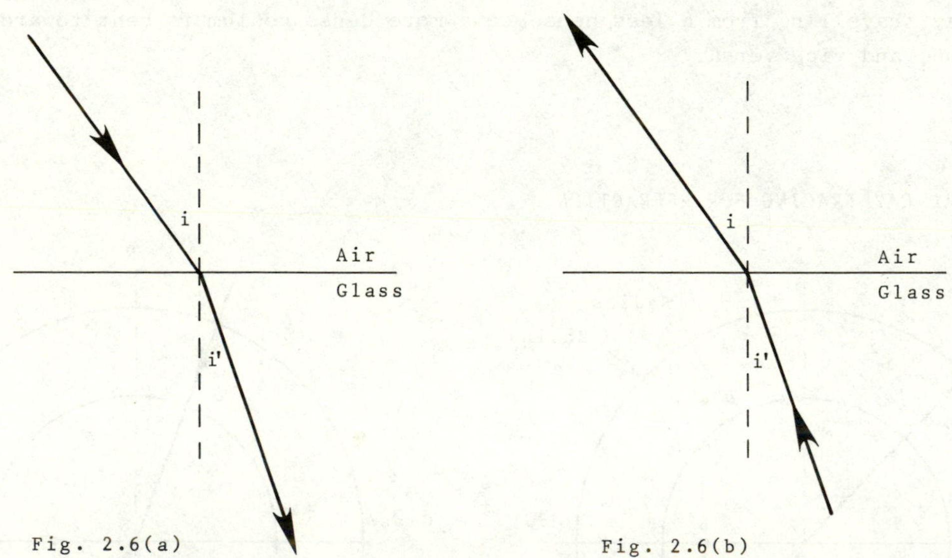

Fig. 2.6(a) Fig. 2.6(b)

The refractive index from air to glass is given by equation (2.2) ${}_an_g = \frac{\sin i}{\sin i'}$, Snell's law. Consider, now, the ray of light to be reversed in direction, as illustrated in figure 2.6(b) so that the ray is refracted from glass to air. The refractive index from glass to air is given by

$${}_gn_a = \frac{\sin i'}{\sin i} . \quad \text{But} \quad {}_an_g = \frac{\sin i}{\sin i'}, \text{ therefore } {}_gn_a = \frac{1}{{}_an_g} \qquad (2.4).$$

Hence, if the refractive index of glass is ${}_an_g = 1.523$, then the refractive index from glass to air is ${}_g n_a = \frac{1}{{}_a n_g} = \frac{1}{1.523} = 0.657$. Similarly, if the refractive index of water is ${}_a n_w = 1.333$, then the refractive index from water to air is ${}_w n_a = \frac{1}{{}_a n_w} = \frac{1}{1.333} = 0.750$.

2.1.6 REFRACTION BY A RECTANGULAR GLASS BLOCK

Figure 2.7 shows the transmission of light through a parallel sided glass plate. This is an important example of refraction.

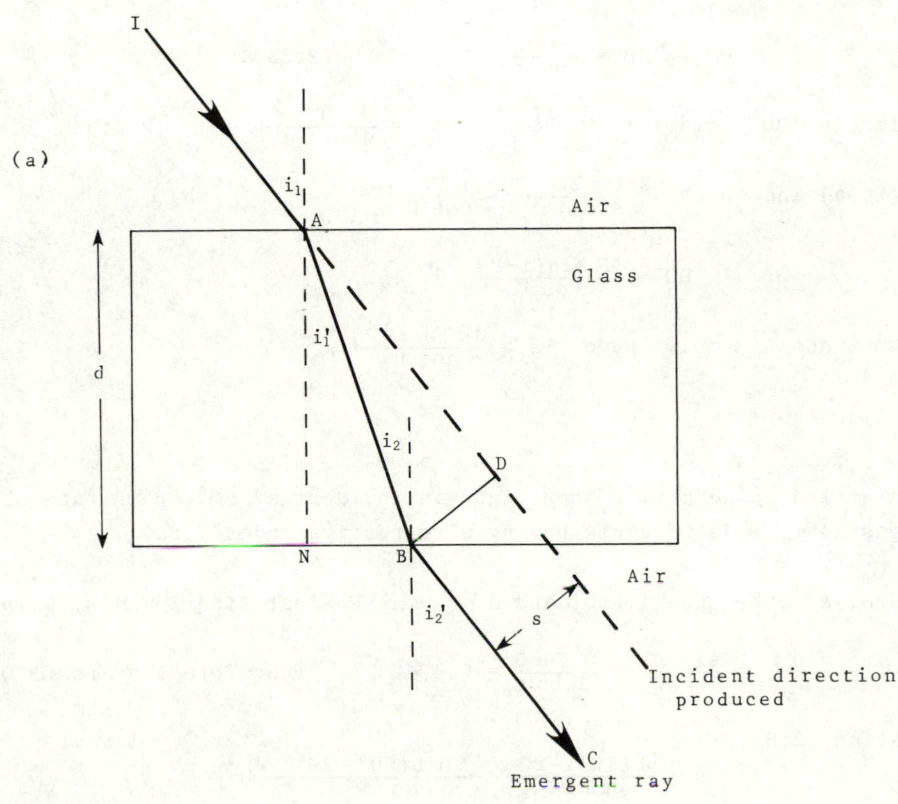

Fig. 2.7 Refraction by a rectangular block.

At A, the ray of light is passing from air into the optically denser medium glass and is deviated towards the normal. By Snell's law, the refraction at A is given by ${}_a n_g = \frac{\sin i_1}{\sin i_1'}$. At B, the ray is passing from glass into the less dense air and is deviated away from the normal. The refraction at B is given by ${}_g n_a = \frac{\sin i_2}{\sin i_2'}$. As the glass block has parallel sides, the normals at A and B will be parallel to each other and $i_1' = i_2$, by alternate angles. Since, from equation (2.4), ${}_g n_a = \frac{1}{{}_a n_g}$, it follows that $i_2' = i_1$. That is, the angle of emergence and the angle of incidence are equal. Hence, *the ray emerges in a direction BC parallel to its incident direction IA*. This result can easily be confirmed experimentally using aligned pins or a ray box.

However, although the direction of the ray is not altered it is displaced sideways by a distance s. This sideways displacement of the light is referred to as the *lateral displacement* of the ray. The lateral displacement s must not be confused with the distance NB. The value of s may be determined from a scale drawing of the glass block, with the ray path calculated and drawn as accurately as possible, or by using the formula

$$\text{lateral displacement} \quad s = \frac{d\sin(i-i')}{\cos i'} \qquad (2.5)$$

where d is the thickness or width of the glass block and i and i' are the angles of incidence and refraction at the first face of the block.

To derive this expression for lateral displacement consider figure 2.7 .

By inspection $\widehat{BAD} = (i-i')$, and in triangle ANB $\cos i' = \frac{AN}{AB}$, without subscripts,

$$\text{so} \quad AB = \frac{AN}{\cos i'} \qquad (2.5a).$$

In triangle ABD, $\sin(i-i') = \frac{BD}{AB}$, so $AB = \frac{BD}{\sin(i-i')}$ (2.5b).

From (2.5a) and (2.5b) $\frac{BD}{\sin(i-i')} = \frac{AN}{\cos i'}$

$$\text{or} \quad BD = \frac{AN\sin(i-i')}{\cos i'},$$

whence, since BD = s and AN = d, we have $s = \frac{d\sin(i-i')}{\cos i'}$.

Example

Calculate the lateral displacement s when light is incident at 50^0 on one face of a rectangular block of thickness 3 cm, made of glass having a refractive index 1.550 .

The given data are: $i = 50^0$, ${}_an_g = 1.550$, and d = 3 cm. We must find i' and s. Using ${}_an_g = \frac{\sin i}{\sin i'}$, we have

$$\sin i' = \frac{\sin i}{{}_an_g} = \frac{\sin 50^0}{1.550} = \frac{0.7660}{1.550} = 0.4942 . \qquad \text{Therefore } i' = \arcsin 0.4942 = 29^0 37'.$$

Then, using equation (2.5),

$$s = \frac{d\sin(i-i')}{\cos i'} = \frac{3\sin(50^0 - 29^0 37')}{\cos 29^0 37'}$$

$$= \frac{3\sin 20^0 23'}{\cos 29^0 37'}$$

$$= \frac{3 \times 0.3483}{0.8694}$$

$$\therefore s = 1.202 \text{ cm.}$$

2.1.7 REFRACTION BY PARALLEL LAYERS OF MATERIALS

In the last section we saw two refractions at air-glass and glass-air boundaries. This occurs in practice when light traverses a lens, although the surfaces are curved of course. A more complicated situation arises when there is a succession of surfaces such as in the eye. Refraction at a succession of spherical surfaces will be dealt with in Chapter 4 and for the moment we will content ourselves with a look at refractions at successive plane boundaries.

Let us consider a ray of light traversing a layer of water followed by a thickness of glass. Figure 2.8 illustrates these materials with plane parallel boundaries. Experiment shows that if the incident and emergent rays are in the same medium then $i_3' = i_1$, and the emergent ray is parallel to the incident ray. Also, by inspection, $i_1' = i_2$ and $i_2' = i_3$.

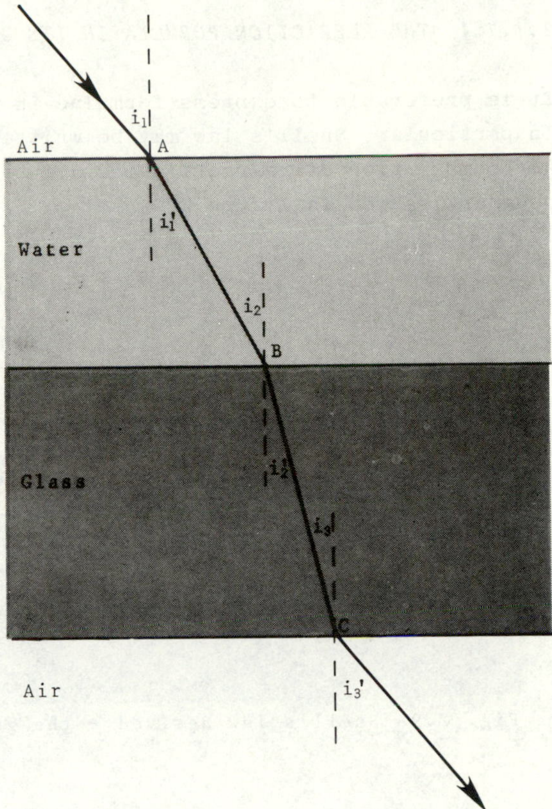

Fig. 2.8 Refraction by parallel layers of materials.

Now,

at A $\quad {}_a n_w = \dfrac{\sin i_1}{\sin i_1'}$

and at B $\quad {}_w n_g = \dfrac{\sin i_2}{\sin i_2'}$

and at C $\quad {}_g n_a = \dfrac{\sin i_3}{\sin i_3'}$.

Multiplying together:

$${}_a n_w \times {}_w n_g \times {}_g n_a = \frac{\sin i_1}{\sin i_1'} \times \frac{\sin i_2}{\sin i_2'} \times \frac{\sin i_3}{\sin i_3'} = 1,$$

since $\sin i_1' = \sin i_2$, etcetera. Therefore,

$${}_w n_g = \frac{1}{{}_a n_w \times {}_g n_a}, \text{ or since } {}_a n_g = \frac{1}{{}_g n_a},$$

$${}_w n_g = \frac{{}_a n_g}{{}_a n_w} \qquad (2.6).$$

For example, to determine the refractive index of glass with respect to water (the light travelling from water to glass), the refractive indices of water and glass with respect to air being 1.333 and 1.523, respectively, we use ${}_w n_g = \dfrac{1.523}{1.333} = 1.143$.

In the preceding treatment appeal was made to the experimental result that $i_3' = i_1$ when the first and last media are the same. We can show this analytically by the use of the absolute refractive indices of the various media. Recall that absolute refractive index is defined as

$$\text{absolute refractive index of substance} = \frac{\text{velocity of light in a vacuum}}{\text{velocity of light in substance}}.$$

In symbols, we can write $n_s = \dfrac{c}{v_s}$, where c is the velocity of light in vacuum. We can show that it is possible to express the refractive index of water with respect to air in terms of the absolute refractive indices of air and water, as follows.

$${}_a n_w = \frac{v_a}{v_w} = \frac{v_a}{c} \times \frac{c}{v_w} = \frac{1}{n_a} \times n_w = \frac{n_w}{n_a}.$$

Then, in the example of figure 2.8,

$$\frac{\sin i_1}{\sin i_1'} \times \frac{\sin i_2}{\sin i_2'} \times \frac{\sin i_3}{\sin i_3'} = {}_a n_w \times {}_w n_g \times {}_g n_a = \frac{n_w}{n_a} \times \frac{n_g}{n_w} \times \frac{n_a}{n_g} = 1$$

since all the absolute refractive index terms cancel in pairs.

Now, by inspection of figure 2.8, $i_1' = i_2$ and $i_2' = i_3$ (alternate angles), so $i_3' = i_1$ if the above expression is to equal unity. Hence, the emergent ray is parallel to the incident ray when the first and last media are the same.

2.1.7.1 THE REFRACTION FORMULA IN ITS GENERAL FORM

It is preferable to express formulae in ways which are easily used for calculation purposes. In particular, Snell's law may be written in a general form making it convenient to apply to any combination of refracting media. Let us again consider refraction at a plane water/glass boundary, shown in figure 2.9 .

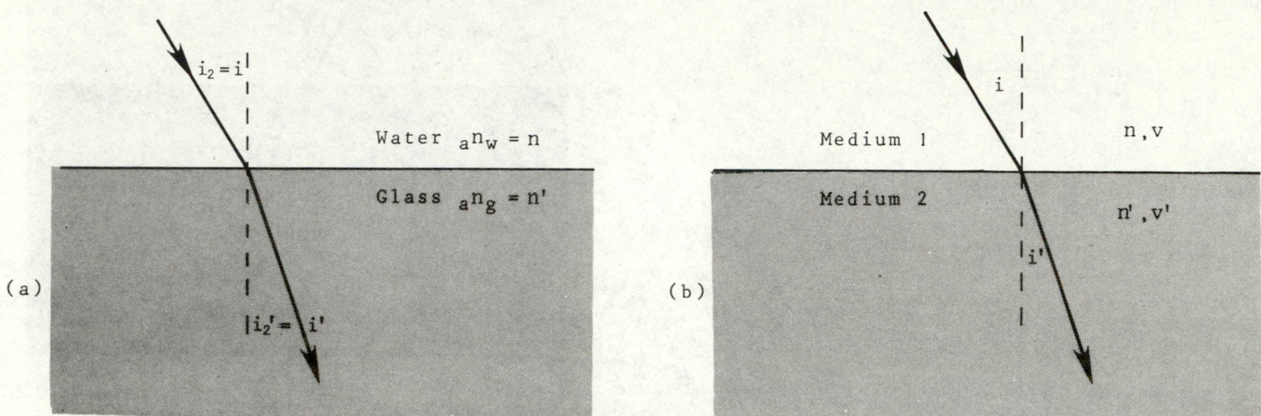

Fig. 2.9 Snell's law derived - the general form.

Let i_2 and i_2' in figure 2.8 be replaced here by i and i', respectively. We have shown that

$$ {}_wn_g \left(= \frac{\sin i}{\sin i'}\right) = \frac{{}_an_g}{{}_an_w} \qquad (2.6). $$

Now let ${}_an_w$ and ${}_an_g$ be represented by n and n', respectively. Therefore, we can rewrite the expression above as

$$ {}_wn_g = \frac{\sin i}{\sin i'} = \frac{n'}{n} $$

$$ \text{or} \quad n \sin i = n' \sin i' \qquad (2.7). $$

This is the most useful form of the refraction formula, and is referred to as the *Snell's Law Formula*. It can be applied to any number of plane parallel transparent layers of materials in the form:

$$ n \sin i = n' \sin i' = n'' \sin i'' = n''' \sin i''' \quad \text{etcetera} $$

First medium	Second medium	Third medium	Fourth medium

Note that we have used n and n' to represent the refractive indices of the media relative to air, equations (2.6) and (2.7) above. However, if we write equation (2.6) again, as below,

$$ {}_wn_g = \frac{{}_an_g}{{}_an_w} = \frac{n_g/n_a}{n_w/n_a} = \frac{n_g}{n_w} $$

we see that we can just as easily write $n_w = n$ and $n_g = n'$ with the result that the refractive indices in equation (2.7) can be the absolute refractive indices of the materials.

It is also useful to note the relationship between the refractive indices of a pair of media and the velocities of light in the media. Referring to figure 2.9(b) where i, n, and v represent for the first medium the angle made by the ray to the normal, the refractive index of the medium, and the velocity of light in the medium, respectively. With similar meanings for the dashed symbols in the second medium, we have

$$n \sin i = n' \sin i' \quad \text{or} \quad \frac{\sin i}{\sin i'} = \frac{n'}{n} .$$

But, from the wave theory, section 2.1.2, we have $\dfrac{\sin i}{\sin i'} = \dfrac{v}{v'}$,

$$\text{therefore} \quad \frac{n'}{n} = \frac{v}{v'} \qquad (2.8),$$

or
$$\frac{\text{refractive index of second medium}}{\text{refractive index of first medium}} = \frac{\text{velocity of light in first medium}}{\text{velocity of light in second medium}} .$$

Worked Examples

1 A ray of light in glass of refractive index 1.523 meets a glass/air surface at an angle of 35^0. Calculate the angle of emergence (refraction) in the air and the deviation.

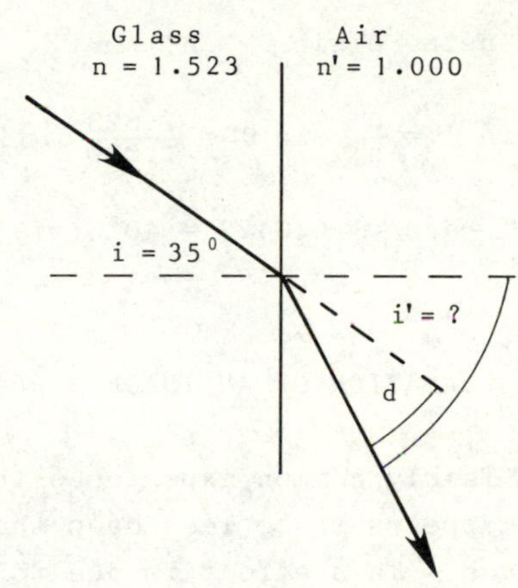

The data are: $i = 35^0$, $n = 1.523$ (glass)
$i' = ?$, $n' = 1.000$ (air).

Using $n \sin i = n' \sin i'$,

$$1.523 \sin 35^0 = 1.000 \sin i'$$

$$\therefore \sin i' = \frac{1.523}{1.000} \sin 35^0 = \frac{1.523}{1.000} \times 0.5736$$

$$= 0.8736$$

$\therefore \quad i' = \arcsin 0.8736 = 60^0 52'$.

Deviation $d = i' - i = 60^0 52' - 35^0 = 25^0 52'$.

2 If the velocity of light in air is 3×10^8 ms^{-1}, find the velocity in spectacle crown glass of refractive index 1.523 .

The data are: $v = 3 \times 10^8$ ms^{-1}, $n = 1.000$ (air), $n' = 1.523$ (glass), $v' = ?$

Using $\dfrac{n'}{n} = \dfrac{v}{v'}$, $v' = \dfrac{nv}{n'} = \dfrac{1.000 \times 3 \times 10^8}{1.523} = 1.97 \times 10^8$ ms^{-1}.

3 Monochromatic light travels at velocities of 2.25×10^8 and 2×10^8 ms^{-1} in water and glass, respectively. Find the relative refractive index ${}_wn_g$.

The absolute refractive indices of water and glass can be found from the definition in section 2.1.7 . Thus,

$$n_g = \frac{c}{v_g} \quad \text{and} \quad n_w = \frac{c}{v_w}, \quad \text{whence} \quad {}_wn_g = \frac{n_g}{n_w} = \frac{c/v_g}{c/v_w} = \frac{v_w}{v_g} = \frac{2.25 \times 10^8}{2.00 \times 10^8} = 1.125 .$$

4 A layer of olive oil of refractive index 1.460 rests on a layer of glass of refractive index 1.523 . A ray of light in air meets the surface of the oil at an angle of incidence of 40^0, and after refraction at each boundary emerges into air again. Calculate the path taken by the ray.

At A, using $n \sin i = n' \sin i'$ rearranged,

$$\sin i' = \frac{n}{n'} \sin i = \frac{1.000}{1.460} \sin 40^0 = 0.4403$$

so $i' = \arcsin 0.4403 = 26^0 7'$

At B, using $n' \sin i' = n'' \sin i''$

$$\sin i'' = \frac{n'}{n''} \sin i' = \frac{1.460}{1.523} \sin 26^0 7' = 0.4220$$

so $i'' = \arcsin 0.4220 = 24^0 59'$.

At C, using $n'' \sin i'' = n''' \sin i'''$

$$\sin i''' = \frac{n''}{n'''} \sin i'' = \frac{1.523}{1.000} \sin 24^0 59' = 0.6427$$

so $i''' = \arcsin 0.6427 = 40^0$ (=i)!

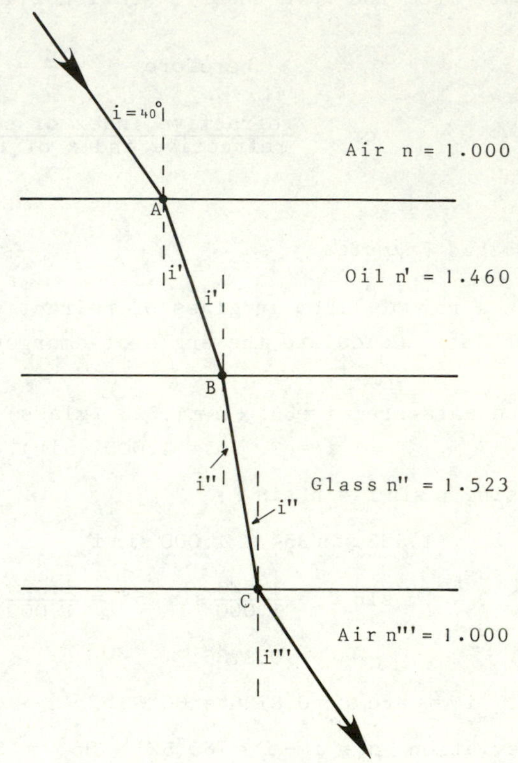

2.1.8 FORMATION OF AN IMAGE - REAL AND APPARENT DEPTH

It is fairly common experience that a depth of water appears to be less deep than is actually the case. This effect is due to the refraction of light rays at the water surface. For the same reason an object, such as a stick, partly immersed in water appears to bend at the water surface. In figure 2.10 two rays of light are shown originating from the immersed end of a stick. After refraction these rays appear to diverge from a point nearer the water surface. This point will be the apparent position of the end of the stick and is a *virtual image*. We should remember that a virtual image is one from which the rays of light only appear to arise.

Similarly, a piece of glass appears to be less thick than it really is and an object viewed through a thickness of glass appears to be nearer the observer due to refraction at the glass/air boundary. When the glass thickness is relatively small, as with windows for example, the effect goes unnoticed in ordinary circumstances but it can be important when accurate measurements are required.

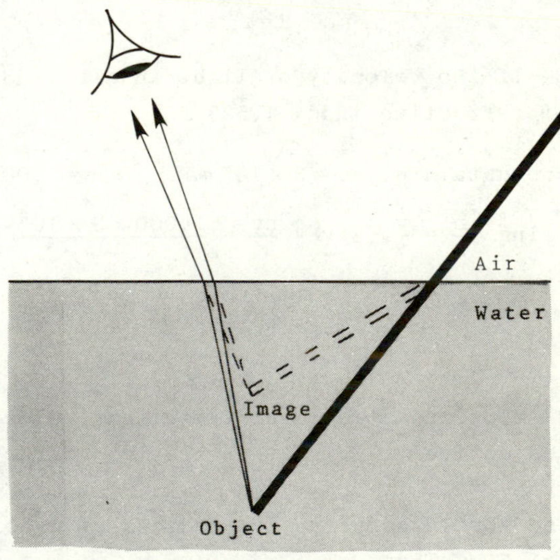

Fig. 2.10 Apparent bending of a stick partially immersed in water.

In figure 2.11, two rays are shown from a point B in the refracting medium. The ray BMN normal to the surface leaves the medium undeviated, but the ray BPR is deviated away from the normal. An eye in the second medium will receive both these rays if M and P are close together, and they will appear to have come from the point B' on NB and on RP produced. B' is nearer to the surface of the medium than is B, and the glass appears less thick than it actually is.

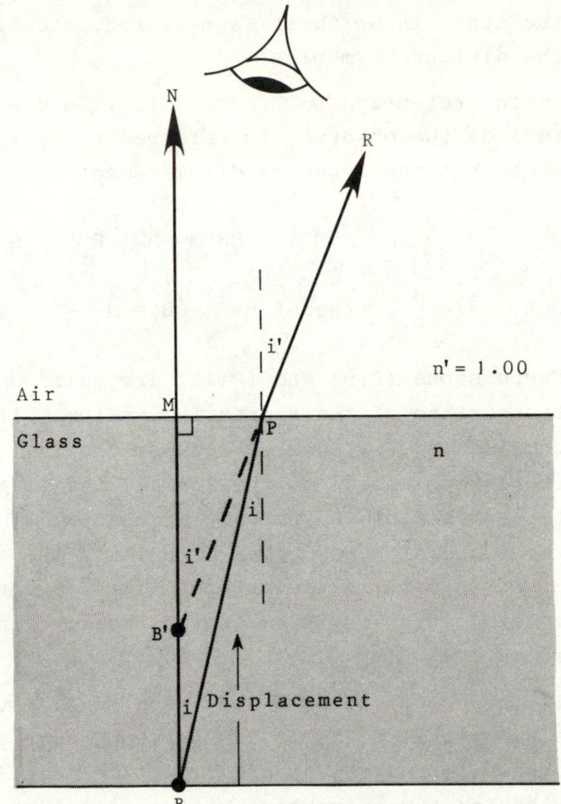

Fig. 2.11 See text for details.

The distance BM is the *real thickness* d of the refracting material, and the distance B'M is referred to as the *apparent thickness*, or the *reduced thickness* ($\bar{d}$) of the material.

Now, as the pupil diameter of the eye is only a few millimetres, the distance MP cannot be greater than this, and will probably be much less. If MP is very small, it can be shown that

$$\text{Refractive index of medium} = \frac{BM}{B'M} = \frac{\text{Real thickness}}{\text{Reduced thickness}},$$

that is, $n_g = \frac{d}{\bar{d}}$ (2.9).

Equation (2.9) is derived by considering the refraction at P in figure 2.11 .

By inspection $M\widehat{B}P = i$ (alternate angles) and $M\widehat{B'}P = i'$ (corresponding angles).

Using the Snell's law formula, equation (2.7), we have

$$n \sin i = n' \sin i', \quad \text{therefore} \quad \frac{n}{n'} = \frac{\sin i'}{\sin i}.$$

Now, if P is very close to M, the angles i and i' will be small. For any small angle $\sin i \simeq \tan i$ and $\sin i' \simeq \tan i'$. Therefore, $\frac{n}{n'} = \frac{\tan i'}{\tan i}$.

In triangle MB'P, $\tan i' = \frac{MP}{MB'}$, and in triangle MBP, $\tan i = \frac{MP}{MB}$. $\therefore \frac{n}{n'} = \frac{MP/MB'}{MP/MB} = \frac{MB}{MB'}$.

Now, MB is the real thickness of the medium = d, and MB' is the apparent or reduced thickness = $\bar{d}$. If the upper medium is air then n' = 1.000, and writing $n = n_g$, we have $n_g = \frac{d}{\bar{d}}$.

In many instances in this book the concept of the reduced thickness of a refracting medium will be met. Let us note that the reduced thickness $\bar{d}$ (read as "d reduced") of a refracting substance is, therefore, given by

$$\bar{d} = \frac{d}{n} \qquad (2.10)$$

where d is the real thickness of a substance and n is its refractive index. The reduced thickness represents the optically equivalent air thickness of the substance as far as the vergence is concerned. This is because the rays emerging into the air appear to have come from the point B', as if this point were also in air. Any real distance divided by the refractive index of the space in which it is measured is called a reduced distance and denoted by a "bar" over the distance symbol.

Again, referring to figure 2.11, the distance BB' by which the object is apparently moved towards the observer is referred to as the *apparent displacement* of the object. It can be seen that the apparent displacement BB' is given by

$$BB' = BM - B'M = d - \bar{d} \, .$$

But, $\bar{d} = \frac{d}{n}$, therefore $BB' = d - \frac{d}{n}$ or $BB' = d\left(1 - \frac{1}{n}\right)$ (2.11).

Expressions (2.9) and (2.11) are valid only if the object is viewed in a direction normal to the surface of the refracting medium. If viewed from any other direction the problem is more complicated. It can be seen by drawing other rays of light that, as the object is viewed at increasing angles to the normal, the image is displaced nearer to the observer and nearer to the surface of the refracting medium. This is shown for one such pair of rays in figure 2.12, the image of B now being at B_1'. If the image positions, such as B_1', are plotted as the eye is moved to various viewing positions, a curved plot called a *caustic curve* is obtained. The caustic curve is in two parts, meeting in a *cusp* at the position of the image B' when viewed normally.

If the object point B is placed some distance behind the thickness of the refracting medium, as in figure 2.13, it can be shown that the object again appears to be displaced towards the observer, and the apparent displacement BB' of the object is independent of the position of B behind the refracting medium, see Appendix 3. As before, the apparent displacement is given by expression (2.11): $BB' = d\left(1 - \frac{1}{n}\right)$, where n is the refractive index of the refracting medium and d is its thickness.

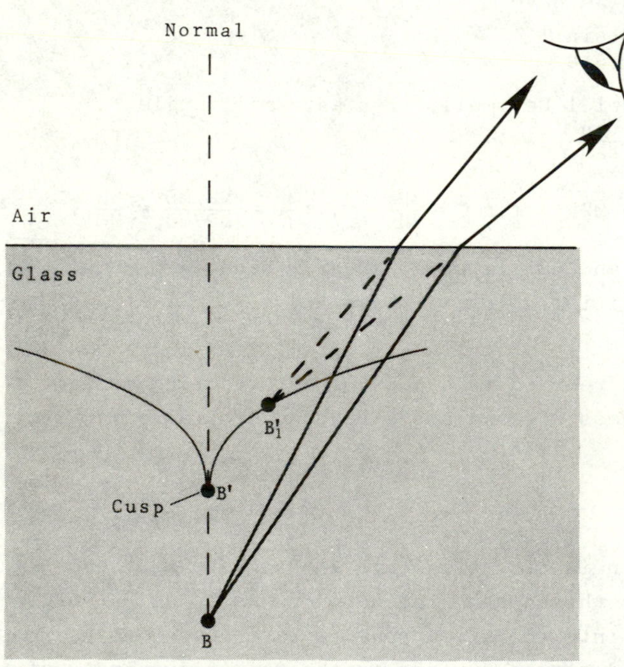

Fig. 2.12 A caustic curve.

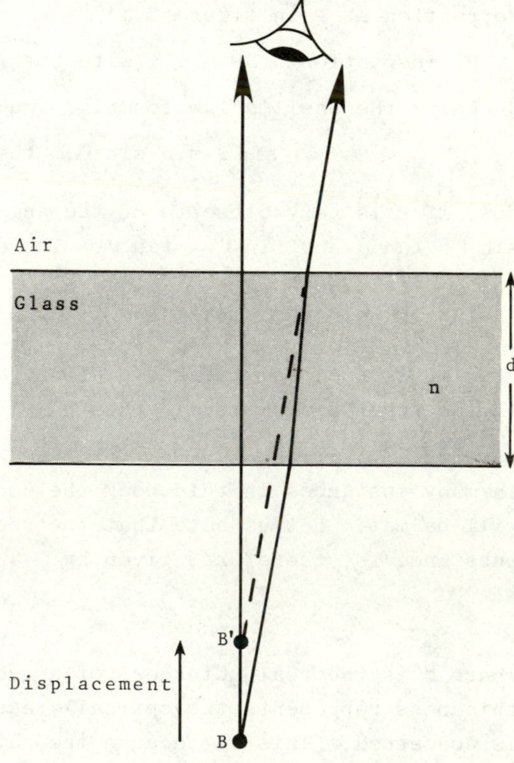

Fig. 2.13 Displacement seen through a glass block.

In cases where an object is viewed through two or more layers of refracting materials the total displacement of the object is found by considering each layer separately.

Worked Example

The base of a tank, made of glass of refractive index 1.55, is 5 cm thick. Water, of refractive index 1.33, fills the tank to a depth of 15 cm. Find the apparent position of a mark on the lower surface of the glass when viewed normally from above through the water.

For the glass only, B is the position of the mark and B_1' is the apparent position. The apparent displacement BB_1' is given by expression (2.11):

$$BB_1' = d_1\left(1 - \frac{1}{n_1}\right)$$

$$= 5\left(1 - \frac{1}{1.55}\right)$$

$$= 5(1 - 0.6452)$$

$$= 1.774 \text{ cm.}$$

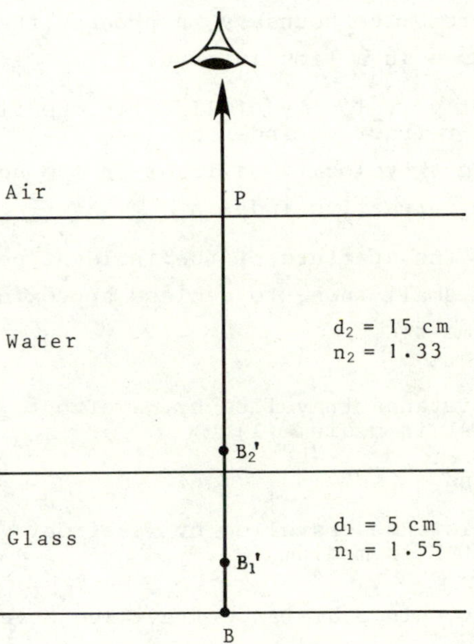

Now, for the water only, B_1' acts as the object, and B_2' is the apparent position. The apparent displacement $B_1'B_2'$ is given by

$$B_1'B_2' = d_2\left(1 - \frac{1}{n_2}\right)$$

$$= 15\left(1 - \frac{1}{1.33}\right)$$

$$= 15(1 - 0.7519)$$

$$= 3.722 \text{ cm.}$$

Hence, the total apparent displacement = 1.774 + 3.722 = 5.496 cm. Therefore, the mark appears to be 5.496 cm above the lower glass surface.

It is interesting to find an expression for the distance PB_2' which is the apparent thickness of the two layers.

By inspection, $PB_2' = PB - BB_2'$

$$= (d_1 + d_2) - \left(\left(d_1 - \frac{d_1}{n_1}\right) + \left(d_2 - \frac{d_2}{n_2}\right)\right)$$

$$= \frac{d_1}{n_1} + \frac{d_2}{n_2}.$$

In words, this result states that the apparent thickness, or reduced thickness, is the sum of the reduced thicknesses of the individual layers. This statement is true of any number of layers. In the example above, the apparent position of the mark is $\frac{d_1}{n_1} + \frac{d_2}{n_2} = \frac{5}{1.55} + \frac{15}{1.33}$ which is 14.504 cm from the top, or 5.496 cm from the bottom!

2.1.8.1 *CHANGE IN WAVEFRONT CURVATURE AT A PLANE REFRACTING SURFACE*

Figure 2.14 shows the refraction of a *narrow* pencil of rays from the object point B by a plane boundary separating two media of refractive indices n and n', with n' < n.

WAF and W'A'F' represent the same wavefront at the start and at the end of its passage across the boundary, a process that takes place in a time interval t.

Let v = velocity of light in the first medium of refractive index n,
and v' = velocity of light in the second medium of refractive index n'.

If the aperture of the incident pencil of rays is small then, to a close approximation, in time t:

distance travelled by wavefront WAF in medium (1) = sag = s

and

distance travelled by wavefront W'A'F' in medium (2) = sag = s'

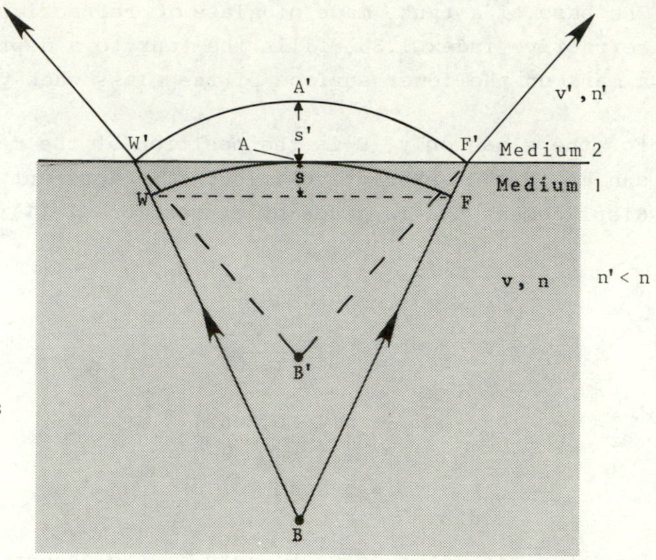

Fig. 2.14 Change in wavefront curvature by refraction at a plane surface.

But, since distance travelled = velocity × time, we have

in medium (1) $s = v \times t$, and in medium (2) $s' = v' \times t$.

Therefore, $\frac{s}{s'} = \frac{v}{v'}$. But the curvature of a wavefront is proportional to the sag, as shown in section 1.2.6 ; that is, $\frac{1}{r} \propto s$, where r is the radius of curvature of the wavefront.

Therefore, $\frac{1}{AB} \propto s$ and $\frac{1}{A'B'} \propto s'$, so that $\frac{s}{s'} = \frac{A'B'}{AB} = \frac{v}{v'}$. Now, since $\frac{v}{v'} = \frac{n'}{n}$ (equation (2.8)), we can write $\frac{A'B'}{AB} = \frac{n'}{n}$. But, AB is the radius of curvature of the incident wavefront (=r) and A'B' is the radius of curvature of the emergent wavefront (=r') , and therefore $\frac{r'}{r} = \frac{n'}{n}$. This important result may be written

$$\frac{r'}{n'} = \frac{r}{n} \qquad (2.12).$$

That is, the reduced radius of curvature of the emergent wavefront is equal to the reduced radius of curvature of the incident wavefront. Hence, we conclude that, when light is refracted at a *plane* boundary between two media, *the reduced radius of curvature of the wavefront remains constant*.

We shall see shortly that this result is of great practical importance in calculations involving wavefronts. It will turn out that we are more interested in curvatures of wavefronts than in their radii of curvatures, and the reciprocal of the reduced radius of curvature will figure in many calculations. Clearly, since the reduced radius of curvature is constant on refraction at a plane boundary, its reciprocal will also remain constant.

2.1.9 TOTAL REFLECTION AND THE CRITICAL ANGLE

We have seen that when light enters an optically denser medium the rays deviate towards the normal. Since light is reversible, it follows that as it passes from a denser to a less dense, or rarer, medium it will deviate away from the normal. This means that the angle of refraction is now larger than the angle of incidence. Figure 2.15 shows a luminous point S situated in a transparent medium such as glass. Five rays are shown travelling towards the boundary between the glass and the air.

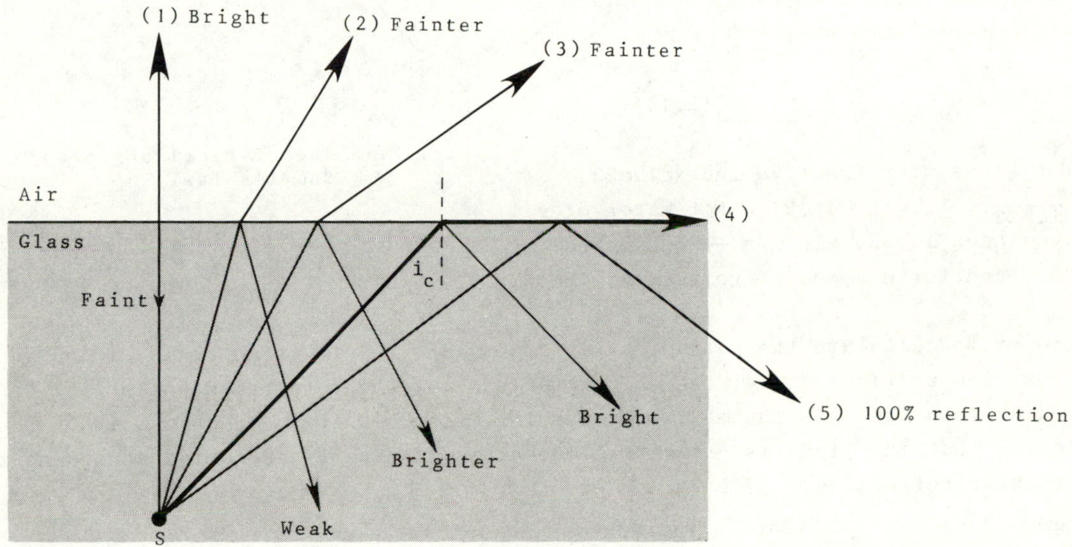

Fig. 2.15 Refraction and reflection at various angles of incidence.

Ray (1), incident normally at the surface, is transmitted into air without deviation. A very faint reflected ray is produced inside the glass.

Rays (2) and (3) are incident at the surface with increasing angles of incidence. It is seen that the refracted ray approaches the surface and becomes fainter whilst the reflected ray inside the glass becomes brighter.

Ray (4) shows a particular value for the angle of incidence, such that the refracted ray just escapes and passes along the surface of the glass and is very faint, whilst the reflected ray inside the glass is bright.

If, as with ray (5), the angle of incidence is further increased very slightly, the refracted ray disappears and all the energy is reflected back inside the glass. This process is known as *total internal reflection*. It occurs only when rays of light are incident on a surface separating an optically dense from an optically less dense medium at angles of incidence which are sufficiently large. The minimum angle of incidence, measured in the denser medium, at which total internal reflection occurs is called the *critical angle*. This is the angle i_c in figure 2.15 .

There is a simple relationship between the critical angle i_c and the refractive indices of the media, because when the angle of incidence in the optically denser medium is i_c, the angle of refraction in the rarer medium is 90^0.

Figure 2.16 shows the *critical ray*, this being ray (4) in figure 2.15 . Using Snell's law for the refraction,

$n \sin i = n' \sin i'$ where $i' = 90^0$.

$n \sin i_c = n' \sin 90^0$. But $\sin 90^0 = 1$, so

$n \sin i_c = n' \times 1$

$$\text{OR} \quad \sin i_c = \frac{n'}{n} \qquad (2.13).$$

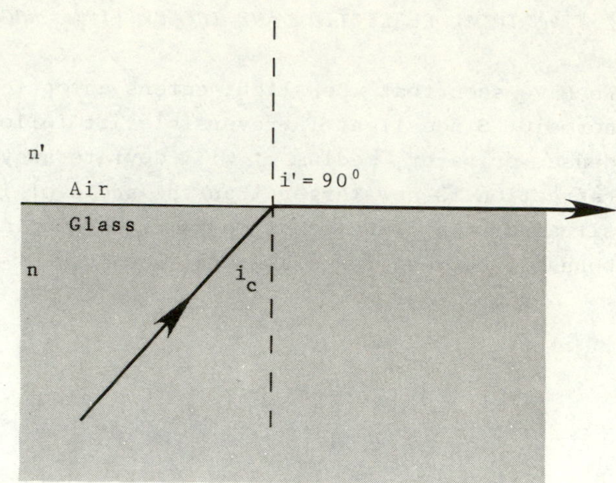

Fig. 2.16 The critical angle and Snell's law.

Now, n is the refractive index of the optically denser medium, and when $n' = 1.000$ (air), we have

$$\sin i_c = \frac{1}{n} \qquad (2.14).$$

Hence, for glass of refractive index 1.523, $\sin i_c = \frac{1}{1.523}$. $\therefore i_c = 41^0 2'$. For water of refractive index 1.333, $\sin i_c = \frac{1}{1.333}$ so that $i_c = 48^0 36'$, and for diamond of refractive index 2.417, $\sin i_c = \frac{1}{2.417}$ whence $i_c = 24^0 26'$.

If we attempt to calculate the path of a ray from glass to air at an angle of incidence greater than the critical angle, the calculated sine of the refraction angle will be found to be greater than one. As the maximum value for a sine is unity (1) then this is not possible, and indicates that the light is not refracted but is internally reflected according to the laws of regular reflection.

For example, if a ray of light is incident on a glass/air boundary at an angle of incidence of 43^0 as in the figure opposite, then using Snell's law

$$n \sin i = n' \sin i'$$

$$\sin i' = \frac{n}{n'} \sin i = \frac{1.523}{1.000} \sin 43^0$$

$$= 1.0387$$

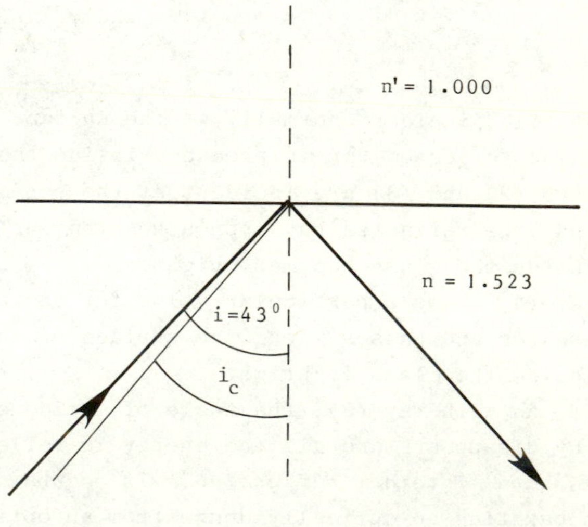

which is impossible. Note that $i = 43^0$ is greater than the critical angle ($41^0 2'$) for this pair of media.

The phenomenon of critical angle is of great importance in practical methods of measuring refractive index. It should be carefully noted that total internal reflection can only occur when the light is originally travelling in the medium of higher refractive index and is meeting the boundary with a rarer medium. The effect cannot occur when the light travels from the rarer medium into the optically denser medium, as a refracted ray is then always obtained. The phenomenon is not just of academic interest but is put to practical use in a number of interesting applications, some of which we will consider in this text.

2.1.9.1 EXAMPLES OF TOTAL INTERNAL REFLECTION

(a) Reflecting prisms

Reflecting prisms are often used instead of plane mirrors to change the direction of light, and have certain advantages over them. These prisms act as reflectors by total internal reflection, and are discussed in section 2.1.10.6 .

(b) The Mirage

A mirage is an image formed by the reflection of light rays at layers of air which are at different temperatures. The effect is frequently seen near the surface of a road when the weather is hot. The thin layer of air next to the road surface is warmed by the road and this decreases its refractive index. This provides the necessary condition for total internal reflection to occur. Figure 2.17 illustrates the effect in which light rays from the sky undergo total reflection producing an image which has the appearance of a 'pool' of water apparently on the road ahead of the observer.

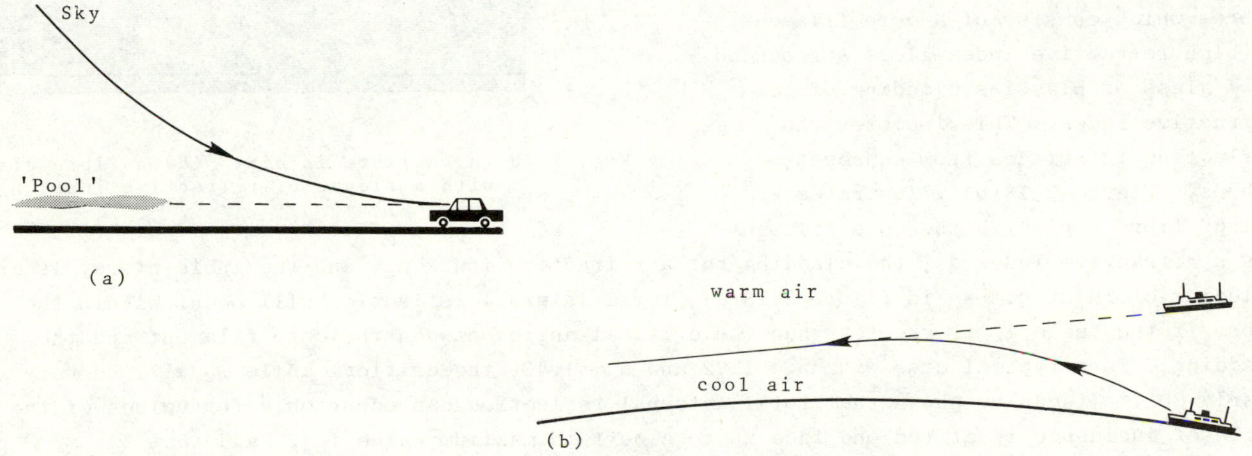

Fig. 2.17 (a) A mirage. (b) A looming.

The diagram is somewhat simplified since the boundary between the cooler air and the warmer air is not so clearly defined. If the object of the mirage is a building or a tree, say, then the image is inverted. However, some light enters the eye directly from the object so that a double image is observed - the erect direct image and the inverted mirage.

Over a body of water a cool, dense layer of air can lie below a heated layer of air. This is the reverse situation which produced the mirage, and in this case light rays originating from an object such as a ship can be reflected downwards to an observer enabling the latter to see a ship which might be below the horizon and otherwise invisible. The phenomenon is known as *looming*.

(c) Fibre Optics

In recent years, techniques have improved for the efficient transfer of light from one place to another using transparent glass fibres called *fibre optics*. In the case of fibres with diameters much larger than the wavelength of light, the passage of light is achieved by total internal reflection within the fibres, and may be described in terms of geometrical optics. Figure 2.18(a) shows the path of a ray along a glass fibre of refractive index n_f situated in air. The angle of incidence on the end face of the fibre is i_1, the angle of refraction is i_1', and then the ray meets the wall of the fibre at an angle i_2. Total internal reflection will occur if i_2 is greater than the critical angle between glass and air. If, as is often the case, glass fibres are grouped together, then some leakage between the adjacent touching fibres is likely, reducing the efficiency of light transfer along the fibres. In a typical fibre optic there may be up to 100 000 fibres in a bundle. To overcome these losses it has become common practice to use fibres which consist of a core filament of high refractive index glass surrounded by a glass or plastics cladding of low refractive index. This isolates the reflecting interfaces from adjacent fibres. Figure 2.18(b) illustrates a "clad" fibre for which the core filament has a refractive index n_f, the cladding has a refractive index n_c, and the incident ray is in a medium of index n. As in figure 2.18(a), total internal reflection will occur within the fibre if the angle i_2 is greater than the critical angle between the core filament and the cladding. In a typical case with $n_f = 1.72$ and $n_c = 1.49$, the critical angle i_c will be very nearly 60^0. It can be shown that total internal reflection can occur only for values of the angle of incidence i_1 at the end face up to a certain maximum value i_{max}, and that

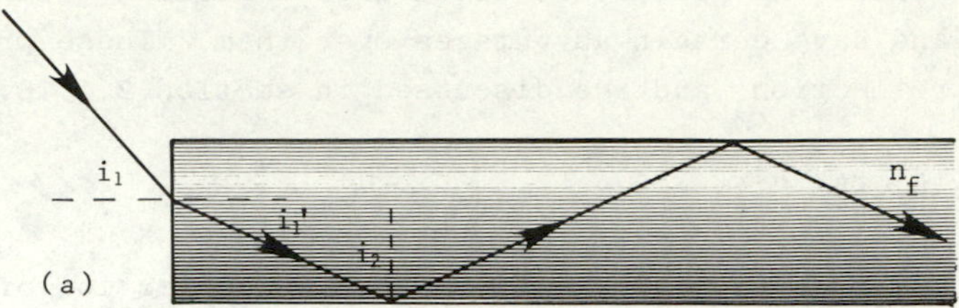

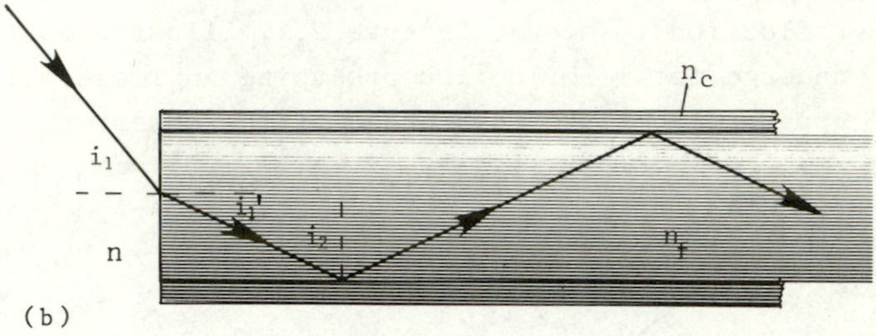

Fig. 2.18 (a) A fibre in air. (b) A fibre clad with a sleeve of refractive index n_c.

$$\sin i_{max} = \frac{1}{n}\left(n_f^2 - n_c^2\right)^{\frac{1}{2}} \qquad (2.15).$$

The above theory is based on a straight fibre which, in practice, is not always the case. With curved fibres the angles of incidence are changed along the length of the fibre and the conditions for total internal reflection may be lost. It has been found that a curve in a fibre can be tolerated if the radius of curvature is not less than about 30 times the fibre radius.

More recently, tapered or conical fibres have been employed to increase the illuminance of an image, particularly with the group of medical instruments known as fibrescopes. Advances in fibre optics research using laser light now enables the conveyance of luminous energy along fibres several kilometres in length without appreciable loss of intensity.

2.1.10 REFRACTION BY PRISMS

2.1.10.0 Introduction

A solid piece of glass or plastics bounded by plane polished faces is termed a *prism*. Often a prism is of triangular section. Refracting prisms have many important uses in optics, such as for producing dispersion in spectroscopy and acting as beam splitters in photometry and interferometry. Ophthalmic prisms are employed in the correction of muscular defects in binocular vision, such as heterophoria. In the present discussion on refracting prisms we shall ignore the effects of dispersion. This is the splitting of white light into its component colours, and is a result of the different wavelengths which make up white light having different velocities in a medium such as glass. Although dispersion is an extremely important effect, we shall illustrate the refracting behaviour of prisms using monochromatic light; that is, light of a single colour. Dispersion and colour effects are considered in Chapter 6.

Figure 2.19 shows a typical refracting prism. In many cases one face of the prism is left unpolished. This face may be considered as the base of the prism. The section ABC is called a *principal section* of the prism and this representation of a prism will be used in constructing ray diagrams. The two faces of the prism which contain AB and AC are the *refracting faces*. A principal section is perpendicular to the refracting faces.

The angle of the principal section at A is called the *apical angle* of the prism, and this is opposite the base. The angle is represented by the symbol a.

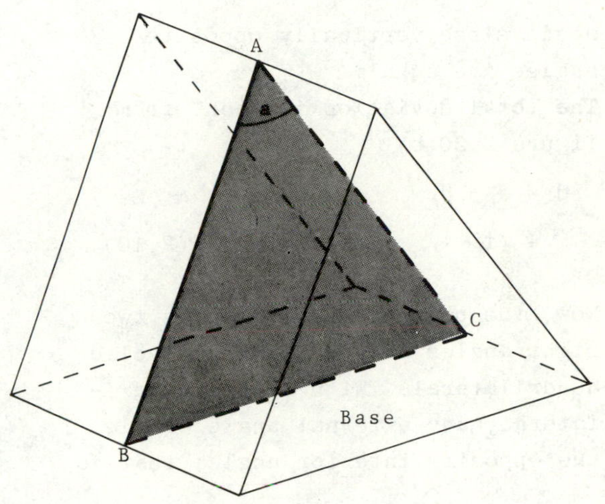

Fig. 2.19 The principal section of a triangular prism.

The importance of the triangular prism lies mainly in the fact that the deviation of light produced by the first refracting surface is further increased by the second surface and is not annulled, as it would be using a parallel sided glass block. To begin with, we shall consider the theory and behaviour of prisms having large apical angles which means angles larger than about 10^0. Then the theory will be applied to ophthalmic prisms which are prisms having small apical angles.

2.1.10.1 THE DEVIATION OF LIGHT BY A PRISM OF LARGE APICAL ANGLE

Figure 2.20 illustrates a prism ABC of refractive index n_p and apical angle a, surrounded by air of refractive index $n_a = 1.000$. Figure 2.20(a) shows the angles of incidence and refraction at each surface whilst figure 2.20(b) shows the angles of deviation.

The incident ray of monochromatic light at the first surface is OE. After refraction the ray travels in the direction EF to undergo further refraction at the second surface, emerging in the direction FP.

i_1 and i_1' are the angles of incidence and refraction at the first surface whilst i_2 and i_2' are the angles of incidence and refraction at the second surface.

d_1 and d_2 are the deviations which take place at the first and second surfaces respectively, and d is the total deviation.

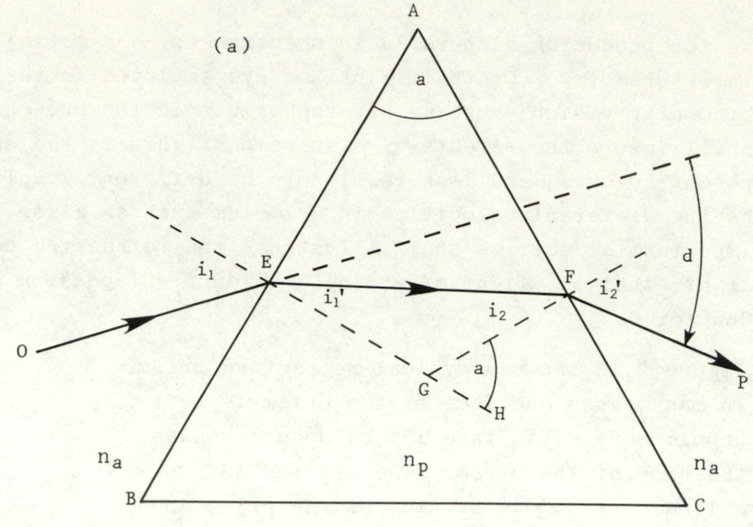

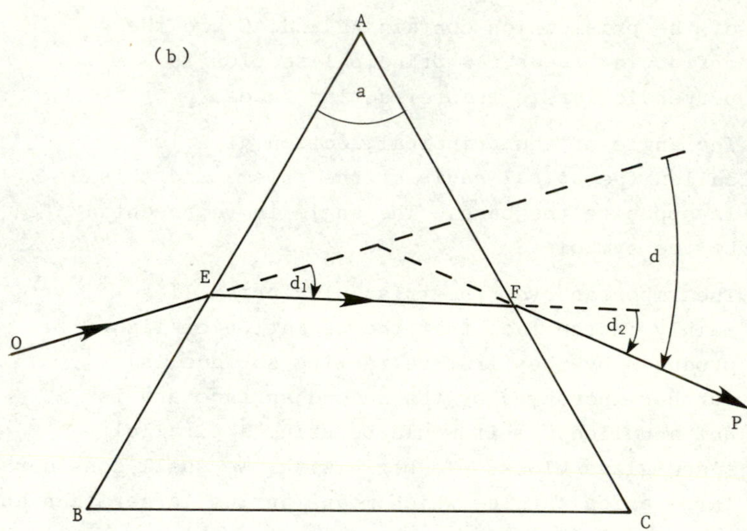

Fig. 2.20 Deviation of light by a prism of large apical angle.

At the first refraction at E the deviation of the ray is

$$d_1 = (i_1 - i_1')$$

since vertically opposite angles are equal.

At the second refraction at F the deviation is

$$d_2 = (i_2' - i_2)$$

again since vertically opposite angles are equal.

The total deviation is then, from figure 2.20(b)

$$d = d_1 + d_2$$

$$= (i_1 - i_1') + (i_2' - i_2) \qquad (2.16).$$

Now, the polygon AEGF contains two right angles and is termed a cyclic quadrilateral. With such a quadrilateral, any external angle equals the opposite interior angle; that is

$$\widehat{FGH} = a \;.$$

But, $\widehat{FGH}$ is an external angle of triangle EFG, and as such equals the sum of the two opposite interior angles.

Hence, $\widehat{FGH} = \widehat{GEF} + \widehat{EFG}$,

or $\quad a = i_1' + i_2 \qquad (2.17).$

Rearrangement of expression (2.16) above gives

$d = i_1 + i_2' - (i_1' + i_2)$. But $i_1' + i_2 = a$, so that

$$d = i_1 + i_2' - a \qquad (2.18).$$

So, the deviation produced depends on the apical angle a and on the angles of incidence and emergence.

Worked Example

A prism having a 40^0 apical angle is made of glass of refractive index $n_g = 1.580$. The light is incident at 50^0 to the normal to the first face as shown in the accompanying figure. Calculate the angles of emergence and deviation. (Note here that the subscripts 1 and 2 refer to the first and second surfaces, and undashed and dashed terms relate to quantities in the media to the left and right of the surface, respectively).

To find the angle of emergence i_2'

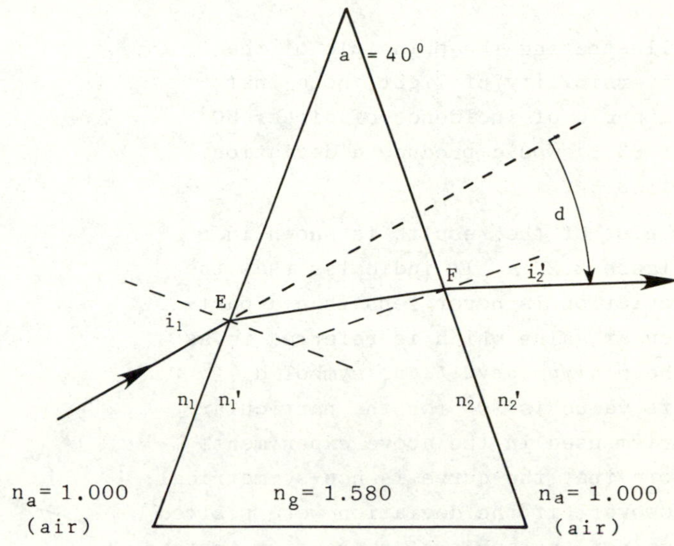

At E, $i_1 = 50^0$ (given). $n_1 = n_2' = n_a = 1.000$ and $n_1' = n_2 = n_g = 1.580$.

Using $n_1 \sin i_1 = n_1' \sin i_1'$ (Snell's law)

$$\sin i_1' = \frac{n_1}{n_1'} \sin i_1 = \frac{1.000}{1.580} \sin 50^0$$

$$= \frac{0.7660}{1.580} = 0.4848,$$

so $i_1' = \arc\sin 0.4848 = 29^0 00'$.

Now, using equation (2.17) $a = i_1' + i_2$,

then $i_2 = a - i_1' = 40^0 - 29^0 = 11^0$.

At F, $i_2 = 11^0$, $n_2 = n_g = 1.580$ and $n_2' = n_a = 1.000$.

Using $n_2 \sin i_2 = n_2' \sin i_2'$

$$\sin i_2' = \frac{n_2}{n_2'} \sin i_2 = \frac{1.580}{1.000} \sin 11^0 = 1.580 \times 0.1908 = 0.3015, \text{ whence } i_2' = \arcsin 0.3015 = 17^0 33'.$$

That is, the angle of emergence is $17^0 33'$.

To find the angle of deviation d

Now, using $d = i_1 + i_2' - a$ (expression (2.18)),

$$d = 50^0 + 17^0 33' - 40^0 = 67^0 33' - 40^0 = 27^0 33'.$$

That is, the angle of deviation is $27^0 33'$.

2.1.10.2 THE VARIATION OF DEVIATION - MINIMUM DEVIATION

From the equations given in section 2.1.10.1, or by experiment, it is possible to determine the angle of deviation d for various values of the angle of incidence i_1 at the first face of the prism. In practice, both i_1 and i_2' are measured for a particular deviation, since, using the general principle of the reversibility of light, both i_1 and i_2' represent angles of incidence which correspond to the same value for d, as shown in figure 2.21 . That is, i_1 and i_2' are interchangeable as far as the deviation is concerned. In the laboratory, an instrument known as the spectrometer is used to measure the apical angle of the prism, the angle of incidence, and the resulting deviation. With a prism with a 60^0 apical angle and made from glass of refractive index 1.520, the following results were obtained. Notice that a deviation of about 40.5^0 is produced with either a 40^0 or a 60^0 angle of incidence (approximately).

Angle of incidence i_1	30^0	40^0	50^0	60^0	70^0	80^0
Angle of deviation d	53.3^0	40.6^0	38.9^0	40.5^0	44.4^0	50.7^0
Angle of emergence i_2'	83.3^0	60.6^0	48.9^0	40.5^0	34.4^0	30.7^0

Table 2.2

Illustrating the principle of the reversibility of light, note that an angle of incidence of either 30^0 or 83.3^0 would produce a deviation of 53.3^0.

A plot of the results is shown in figure 2.22 . It indicates that the deviation is never less than a particular value which is referred to as the *minimum deviation*, symbol d_{min}. Its value is 39^0 for the particular prism used in the above experiment. Note that the curve is non-symmetrical. However, if the deviation were plotted against the angle of refraction instead of the angle of incidence, the graph would be symmetrical.
As the deviation approaches the minimum value, the two angles of incidence i_1 and i_2' which give the same deviation

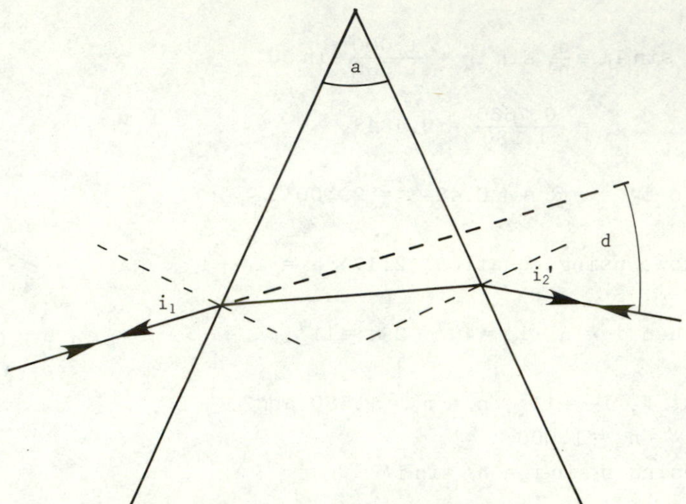

Fig. 2.21 See text.

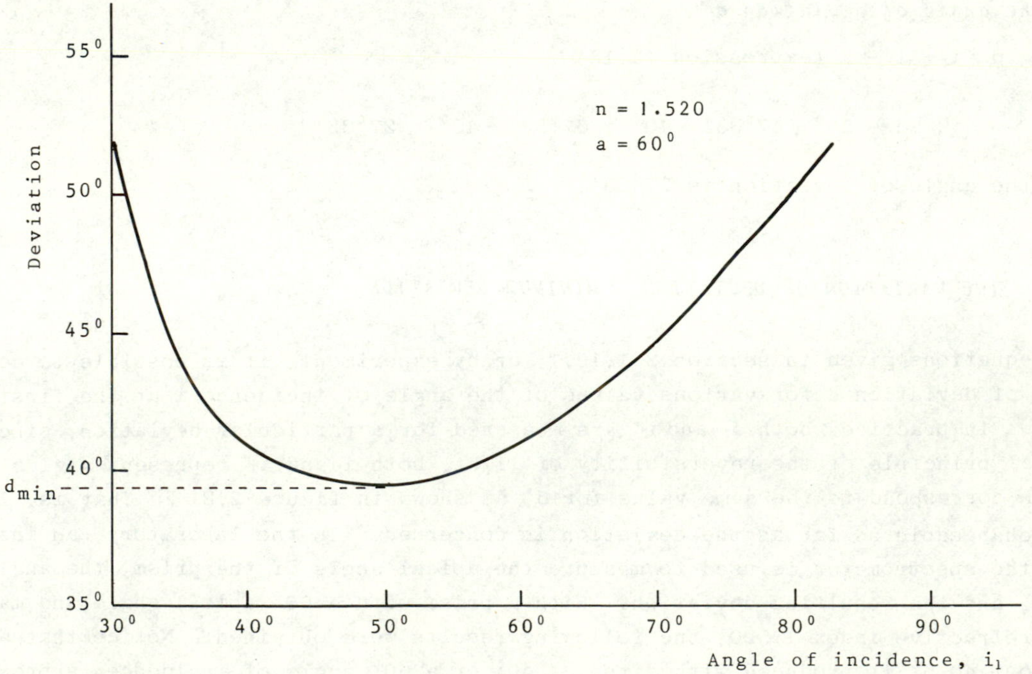

Fig. 2.22

approach one another. At the actual minimum deviation position they are equal. This means that, when the deviation is a minimum, the angles of incidence (i_1) and emergence (i_2') are equal, and hence, the ray of light passes *symmetrically* through the prism.

This is shown in figure 2.23 where the prism of refractive index n_p is immersed in the surrounding medium of refractive index n_s. We can obtain a special equation for the minimum deviation position which expresses the refractive index of the prism in terms of the apical angle and the minimum deviation d_{min}. It can be shown, as follows, that

$$\frac{n_p}{n_s} = \frac{\sin\frac{1}{2}(a + d_{min})}{\sin\frac{1}{2}a} \qquad (2.19).$$

This equation is the basis for one of the most accurate methods of determining the refractive index of a transparent substance in the shape of a triangular prism. Using a hollow prism where the sides are constructed of plane parallel-sided glass sheets, the deviating effect of liquids or gases may be investigated. The glass plates will not result in any deviation of their own. The measurements are made using the spectrometer (see section 8.3.5).

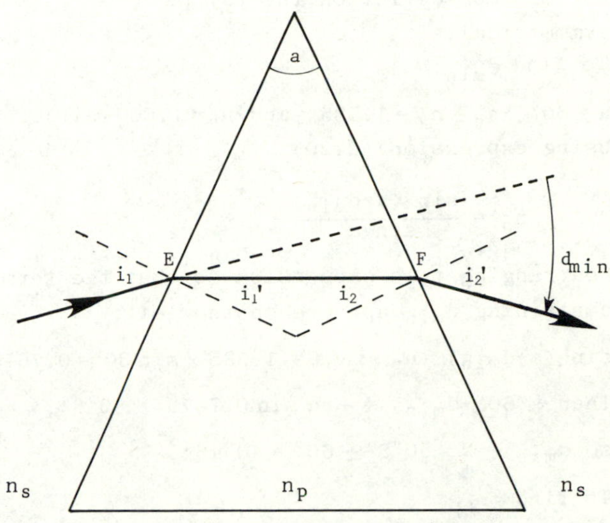

Fig. 2.23 Minimum deviation - EF is parallel to the base.

Let us now consider the symmetrical ray diagram for minimum deviation, shown in figure 2.23 . At minimum deviation $i_1 = i_2'$ and $i_1' = i_2$. Since, in general, $d = i_1 + i_2' - a$ from equation (2.18), then, putting $i_2' = i_1$ when the deviation $d = d_{min}$ gives

$$d_{min} = i_1 + i_1 - a = 2i_1 - a$$

$$\text{OR} \qquad i_1 = \tfrac{1}{2}(a + d_{min}).$$

Also, in general, using equation (2.17) $a = i_1' + i_2$. Then, putting $i_2 = i_1'$ when the deviation is a minimum gives $a = i_1' + i_1'$,

$$\text{OR} \qquad i_1' = \tfrac{1}{2}a.$$

Now, i_1 and i_1' are related by the Snell's law expression (2.7) for the refraction at E,

$$\text{namely,} \quad n_1 \sin i_1 = n_1' \sin i_1'.$$

$$\text{Then} \quad \frac{n_p}{n_s} = \frac{n_1'}{n_1} = \frac{\sin i_1}{\sin i_1'} = \frac{\sin\frac{1}{2}(a+d_{min})}{\sin\frac{1}{2}a} \qquad (2.19).$$

If the prism is in air, $n_s = 1.000$ and we get the more common form of equation (2.19),

$$n_p = \frac{\sin\frac{1}{2}(a+d_{min})}{\sin\frac{1}{2}a} \qquad (2.20).$$

Worked Example

What will be the minimum deviation produced by a prism of apical angle 60^0 and made from glass of refractive index $n_g = 1.535$? What will be the angle of incidence when the ray is at minimum deviation?

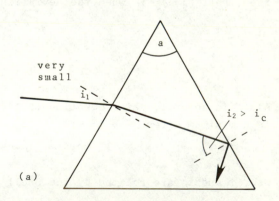

For minimum deviation the ray path is symmetrical.

To find d_{min}:

$a = 60^0$, $n_p = n_g = 1.535$, and $n_s = 1.000$ (air).

Using expression (2.20)

$$n_p = \frac{\sin\frac{1}{2}(a+d_{min})}{\sin\frac{1}{2}a}$$

Rearranging this expression so that the term containing d_{min} appears on the left:

$\sin\frac{1}{2}(a+d_{min}) = n_p \sin\frac{1}{2}a = 1.535 \times \sin 30^0 = 0.7675$.

Then $\frac{1}{2}(60^0+d_{min})$ = arc sin 0.7675 = $50^0 8'$,

so $d_{min} = 2 \times 50^0 8' - 60^0 = 40^0 16'$.

To find i_1:

In general, $d = i_1 + i_2' - a$ from equation (2.18) . For minimum deviation $i_2' = i_1$ so we can write

$$d_{min} = i_1 + i_1 - a \quad \text{or} \quad i_1 = \frac{1}{2}(a+d_{min}), \text{ whence } i_1 = \frac{1}{2}(60^0 + 40^0 16') = 50^0 8'.$$

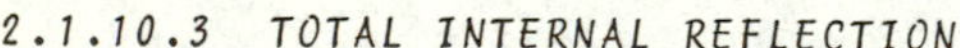

2.1.10.3 TOTAL INTERNAL REFLECTION

If the angle of incidence i_1 at the first face of a prism is small, then it follows that the angle of incidence i_2 inside the glass at the second surface will be large, assuming that the apical angle of the prism is sufficiently large. This situation provides the possibility for total internal reflection to occur at the second face of the prism, for if i_2 exceeds the value of the critical angle i_c, then no light will be refracted out of the second face of the prism. Figure 2.24(a) shows the case in which a very small angle of incidence causes total internal reflection to occur at the second surface.

Fig. 2.24(a) Total internal reflection.

In figure 2.24(b) we see the case where the angle of incidence i_1 is such that the light is refracted out of the second face of the prism at grazing emergence. That is, $i_2' = 90^0$ and $i_2 = i_c$.

It is possible to derive a formula to calculate the smallest value of i_1 such that the light is refracted from the second face of a prism. However, it is preferable that ray traces are calculated from first principles, using, for example, the Snell's law formula. The following example will illustrate this. Before continuing, look at figure 2.24(c) which shows a diverging pencil of rays incident on a prism. Note initially the reflections at the first face!

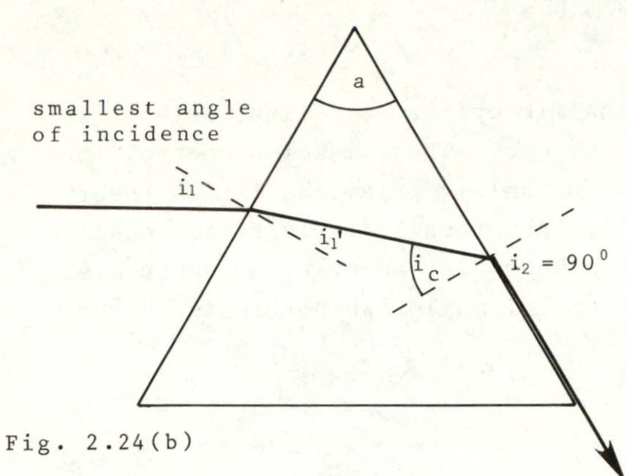

Fig. 2.24(b)

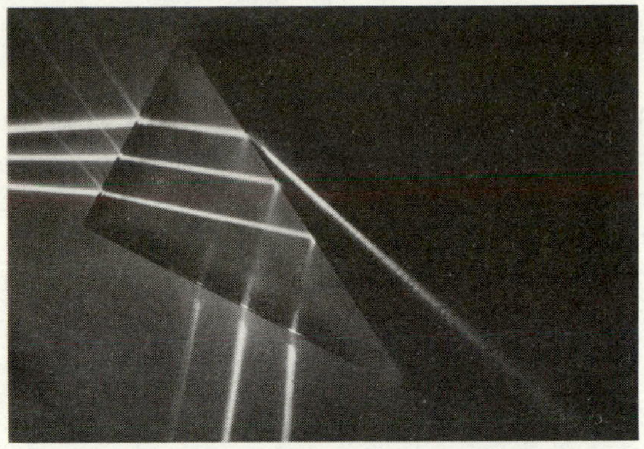

Fig. 2.24(c) See text for details.

Since the pencil of rays is divergent, the rays meet the second face at different values of i_2 and one is seen to undergo refraction and emerge into the air. The other two are totally internally reflected. Note that some light from the ray which is refracted is also reflected at the second face: this should justify figure 2.15 .

Worked Example

Determine the smallest angle of incidence at the first face of an equilateral triangular prism of refractive index $n_p = 1.65$ for light to pass through the second face when $n_s = 1.000$ (air).

In this case the apical angle is 60^0. The smallest angle of incidence i_1 at the first face corresponds to an angle of emergence $i_2' = 90^0$ at the second face of the prism. Applying Snell's law to the refraction at F,

$$n_2 \sin i_2 = n_2' \sin i_2'$$

where $i_2' = 90^0$, $i_2 = i_c$, $n_2' = 1.000$, and $n_2 = 1.65$.

Hence,

$$1.65 \sin i_c = 1 \times \sin 90^0$$

$$\Rightarrow \sin i_c = \frac{1}{1.65} = 0.6061$$

whence $i_c = i_2 = \arcsin 0.6061 = 37^0 18'$.

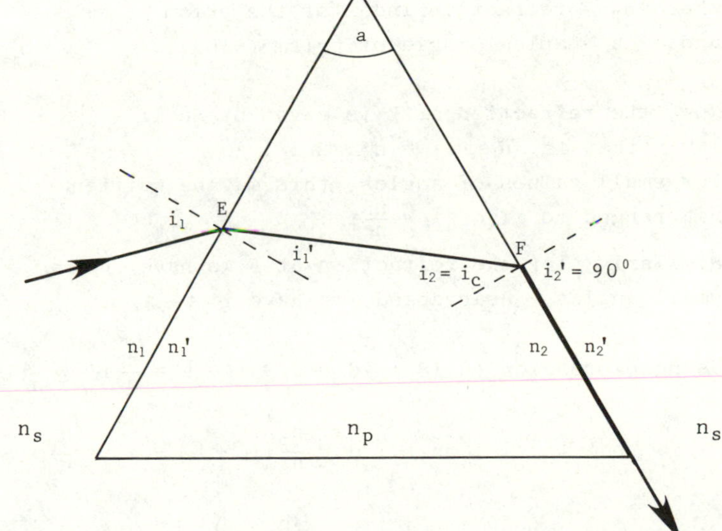

Now, using expression (2.17) $i_1' + i_2 = a$, $i_1' = a - i_2 = 60^0 - 37^0 18' = 22^0 42'$.
Applying Snell's law to the refraction at E,

$$n_1 \sin i_1 = n_1' \sin i_1', \text{ where } n_1 = 1.000,\ n_1' = 1.65, \text{ and } i_1' = 22^0 42'.$$

Hence, $1 \times \sin i_1 = 1.65 \times \sin 22^0 42'$, or $\sin i_1 = 1.65 \times 0.3859 = 0.6367$. Then $i_1 = 39^0 33'$. This is the smallest angle of incidence for light to pass through the prism.

2.1.10.4 OPHTHALMIC PRISMS

A prism, or prismatic effect, is commonly used in ophthalmic optics to relieve defects in the external, or extrinsic, muscle system attached to the eye. These muscles control the rotation of the eye. The most important feature of an ophthalmic prism, as far as theory is concerned, is that the apical angle is small, usually less than 10^0. This fact enables the general theory of prisms to be simplified somewhat. There is, however, no sharp distinction between ophthalmic prisms and prisms of larger apical angle, when this is in the region of 10^0.

Figure 2.25 shows a ray of light refracted through an ophthalmic prism. If all the angles i_1, i_1', i_2, and i_2' are small values, then we will see that the apical angle of the prism will also be small, since $a = i_1' + i_2$, from equation (2.17).

Now, the general expression for the deviation by a prism is, from equation (2.18),

$$d = i_1 + i_2' - a\,.$$

It can be shown as follows that, only for ophthalmic prisms situated in air, this equation becomes modified to

$$d = (n_p - 1)a \qquad (2.22),$$

where n_p = refractive index of the prism
and a = apical angle of prism ($<10^0$).

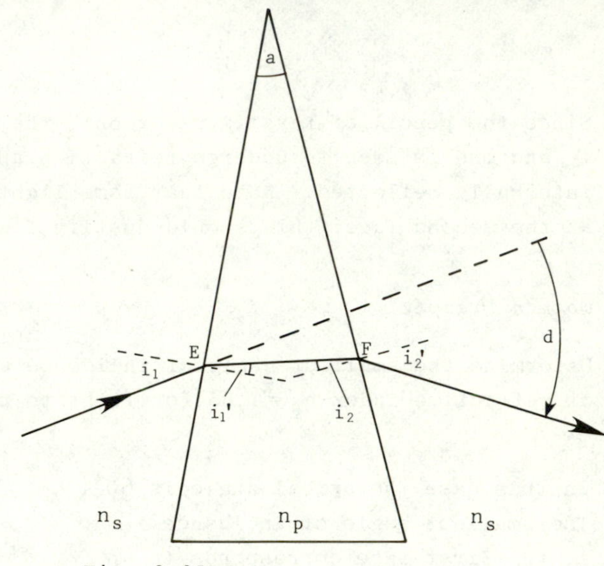

Fig. 2.25

Now, the refraction at E is given by Snell's law, that is $n_1 \sin i_1 = n_1' \sin i_1'$.

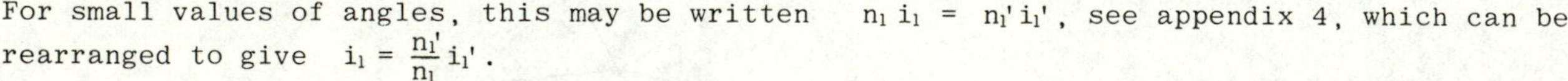

For small values of angles, this may be written $n_1 i_1 = n_1' i_1'$, see appendix 4, which can be rearranged to give $i_1 = \frac{n_1'}{n_1} i_1'$.

Similarly, for the refraction at F we have $n_2 \sin i_2 = n_2' \sin i_2'$, which gives $n_2 i_2 = n_2' i_2'$ for small angles. Rearranged, we have $i_2' = \frac{n_2}{n_2'} i_2$.

Using expression (2.18): $d = i_1 + i_2' - a = \frac{n_1'}{n_1} i_1' + \frac{n_2}{n_2'} i_2 - a = \frac{n_p}{n_s} i_1' + \frac{n_p}{n_s} i_2 - a$

$$\text{or} \qquad d = \frac{n_p}{n_s}\left(i_1' + i_2\right) - a = \frac{n_p}{n_s} a - a\,, \quad \text{since } i_1' + i_2 = a,$$

$$\text{whence} \qquad d = \left(\frac{n_p}{n_s} - 1\right)a \qquad (2.21).$$

If the prism is in air, $n_s = 1.000$ and

$$d = (n_p - 1)a \qquad (2.22).$$

We often drop the subscript p (for prism) and write

$$d = (n - 1)a \qquad 2.22a)$$

where n is the refractive index of the prism surrounded by air. This expression for an ophthalmic prism shows that the deviation depends only on the shape and the material of the prism and is independent of the angle of incidence, if this is small.

For example, the deviation produced by an ophthalmic prism of apical angle 6^0, made of glass of refractive index 1.523, will be given by equation (2.22a)

$$d = (n-1)a = (1.523-1)6 = 0.523 \times 6 = 3.138^0.$$

This deviation will be effectively constant for all angles of incidence between 0^0 and 10^0 but will be approximately correct for slightly larger values.

Ophthalmic prisms are usually marked with their deviation value in degrees or, more commonly, in *prism dioptres*. A prism is said to have a *deviating power* (P) of one prism dioptre, symbol 1^Δ, if it produces a displacement of 1 cm on a screen positioned 100 cm (= 1 m) from the prism, figure 2.26 .

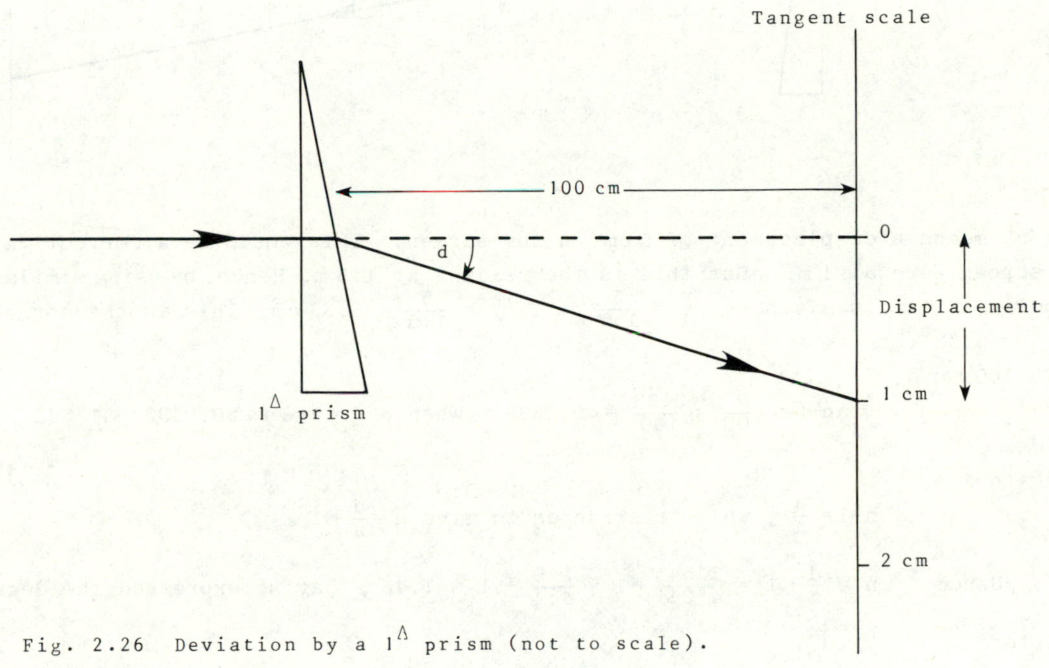

Fig. 2.26 Deviation by a 1^Δ prism (not to scale).

The power P^Δ and the deviation produced d^0 are related by

$$P^\Delta = 100 \tan d^0 \qquad (2.23).$$

This is the basis of the *tangent scale* from which the deviating power of a prism may be read directly. For example, an ophthalmic prism may be labelled 5^Δ. The deviation which this prism produces is given by expression (2.23):

$$P = 100 \tan d, \text{ so that } \tan d = \frac{P}{100} = \frac{5}{100} = 0.05, \text{ whence } d = \arctan 0.05 = 2^0 52'.$$

For further detailed information on the design and use of ophthalmic prisms and the tangent scale, the reader is directed to "*The Principles of Ophthalmic Lenses*", by M. Jalie.

Worked Example

The deviation produced by an ophthalmic prism is read off as 5^{Δ} when measured on a tangent scale designed to be used at 1 m but used in error at 1.5 m. If the apical angle of the prism is $4^{\circ}24'$ determine its refractive index.

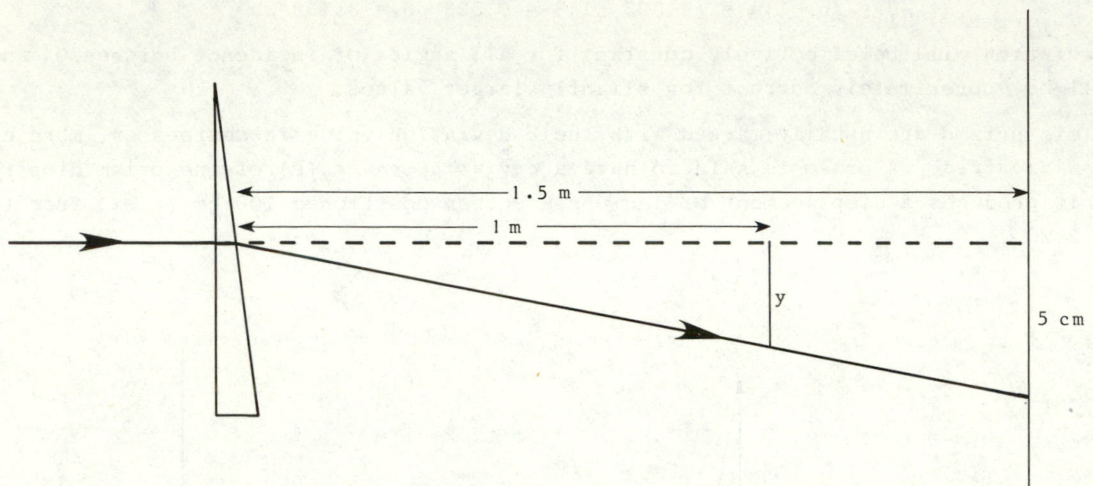

Reading 5^{Δ} means a displacement of 5 cm on the screen. This would be a correct value only if the screen were at 1 m. But this is the reading at 1.5 m. Hence, by using similar triangles the reading at 1 m is given by $\frac{y}{1} = \frac{5}{1.5}$. $\therefore$ $y = \frac{5}{1.5} = 3.33$ cm. This is the correct value; that is $P = 3.33^{\Delta}$.

Using $P = 100 \tan d$,

$$\tan d = \frac{P}{100} = \frac{3.33}{100} = 0.0333, \quad \text{whence } d = \arctan 0.0333 = 1^{\circ}54'.$$

Using $d = (n-1)a$,

$$n-1 = \frac{d}{a}, \text{ which rearranges to give } n = \frac{d}{a} + 1 .$$

Hence, $n = \frac{d}{a} + 1 = \frac{1^{\circ}54'}{4^{\circ}24'} + 1 = \frac{1.9}{4.4} + 1 = 1.43$, having expressed the degrees in decimal form.

It is occasionally useful to realise that the prism dioptre unit is only a unit of convenience. In S.I. units* the prism dioptre is really the centimetre per metre! That is, if θ° is the deviation produced by a thin (ophthalmic) prism, then θ^{Δ} is given by

$$\theta^{\Delta} = 100 \tan \theta^{\circ} = 100 \frac{y}{x}$$

where $\tan \theta^{\circ} = \frac{y}{x} \frac{\text{(metres)}}{\text{(metres)}}$, y being the displacement at a distance of x metres. This can be rewritten by writing 100y(metres) $\equiv$ y (centimetres), whence $\theta^{\Delta} = \frac{y}{x} \frac{\text{(cm)}}{\text{(m)}}$, and we see that the prism dioptre $\equiv$ cm/m.

This interpretation is sometimes useful clinically when a tangent scale is used for measuring the deviation of one eye relative to the other in determining heterophorias.

* Système International d'Unités.

2.1.10.5 THE IMAGE FORMED BY A PRISM

A prism will only form a clear image if the rays of light from the object pass through the prism in the minimum deviation path. The curve shown in figure 2.22 indicates that there is a range of angles of incidence near this position for which the deviation is approximately constant. In figure 2.27 only the central emergent ray is exactly in the minimum deviation direction (remember that this is the symmetrical path). However, the other emergent rays are so close to this path that they undergo approximately the same deviation. Hence, the emergent rays, when produced backwards, meet at a point I. This is the image position.

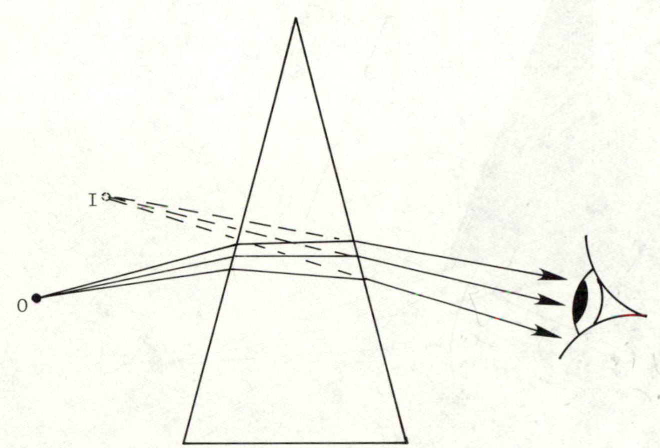

Fig. 2.27 A narrow pencil of rays close to the minimum deviation path producing a point image.

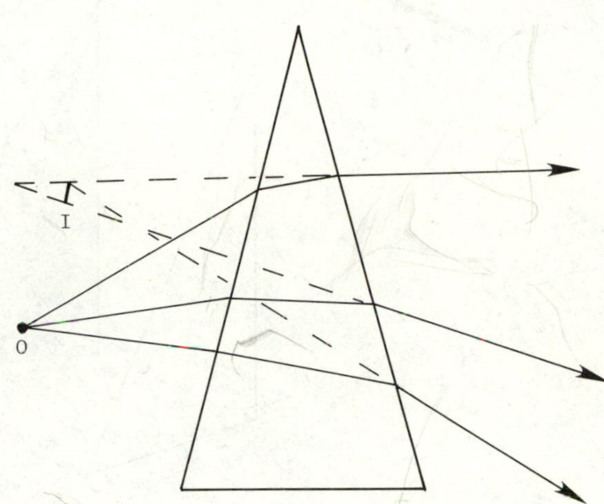

Fig. 2.28 A wide pencil of diverging rays and an astigmatic (non-point) image.

If, as in figure 2.28, a pencil of rays is drawn from O, with the central ray not coinciding with the minimum deviation path, or with the rays having widely differing angles of incidence at the first face, the emergent rays would not suffer equal deviations and, when produced backwards, would not meet at a point. In fact, the image of a point object will be a short line. This is an example of astigmatism and is inherent in the geometry of the prism. It can be overcome by arranging for the incident pencil of rays to be parallel. This necessitates the use of a collimator lens, which is an essential part of the spectrometer (see section 8.3.5 .

Worked Problem

Suppose a subject views a long horizontal line through a prism held with its base down in front of one eye. Describe the appearance of the line if the subject is allowed to look at the line by scanning from side to side and thereby looks through other than principal sections of the prism.

In moving the eye the subject will effectively be looking through a prism of varying apical angle, the angle increasing the more oblique the direction is to the direction which involves a principal section. Looking through a principal section will present a minimum apical angle.

However, looking through the prism in some other section will effectively present a larger value of apical angle. The figure below illustrates the situation.

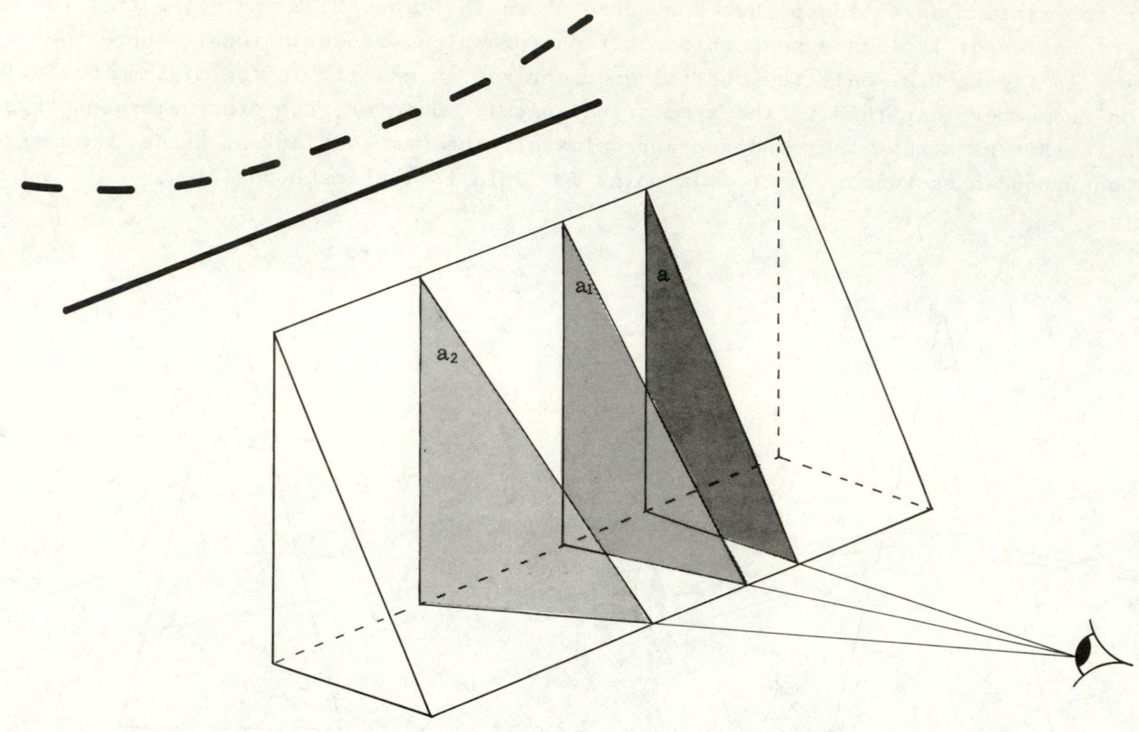

Clearly, $a_2 > a_1 > a$, where a is the apical angle for the principal section. Each section will deviate the light by an amount which is proportional to its 'apical angle'. In ophthalmic practice the prisms deviate light by an amount given by $d = (n-1)a$ and the line will therefore be seen curving upwards away from the centre.

2.1.10.6 REFLECTING PRISMS

Reflecting prisms are commonly used instead of plane mirrors to change the direction of propagation of rays of light, or to alter the orientation of an image. Although they are more expensive, reflecting prisms have certain advantages over mirrors. For a prism to act as a reflector of light the rays are introduced into the prism in such a way that at least one internal reflection takes place, with the angle of incidence at the reflecting surface inside the glass exceeding the critical angle. Hence, the reflecting surface need not be silvered. In contrast to a back silvered plane mirror, reflecting prisms do not cause the formation of multiple images (see figure 5.15), and unlike front silvered mirrors, there is no unprotected reflecting layer which would need frequent renewal.

The overall reflecting ability of a reflecting prism is about 90%. The losses are due to the weak reflections as the light enters and leaves the prism and some absorption in the glass itself. This is slightly inferior to the reflectance of a freshly prepared front silvered mirror, but the prism has the great advantage of maintaining its reflecting ability indefinitely assuming it is kept clean.

It is desirable that a reflecting prism does not cause dispersion into a spectrum. It is shown in Appendix 3 that the deviation produced by a triangular reflecting prism is independent of the wavelength of the incident light, and is given by

$$d = 2i_1 + a \qquad (2,24)$$

where a is the apical angle and i_1 is the angle of incidence at the first face. Since the reflection is without colour preference, the prism is said to be achromatic.

Figures 2.29 to 2.32 illustrate a few of the large number of widely used reflecting prisms. Let us assume in each case that the refractive index of the glass is 1.55, so that the critical angle (see section 2.1.9) is $i_c = 40^0 11'$.

The Porro Prism

The prism section has angles $45^0/45^0/90^0$. For normal incidence on a selected face the light inside the prism meets the reflecting face(s) at 45^0. Since this exceeds the critical angle of $40^0 11'$ this ensures that total internal reflection will occur.

The Porro prism orientated as in figure 2.29(b) is used in pairs in the Prism Binocular (see section 8.3.4.3). Notice that the beam is deviated by 180^0 on emerging from the prism. Used as in figure 2.29(a) it acts as an inverter with the reflecting surface behaving as a plane mirror.

The Dove Prism

Illustrated in figure 2.30, this is a truncated version of the Porro prism which saves material and weight. Looking back along the direction of propagation it is not difficult to see the prism's inverting property.

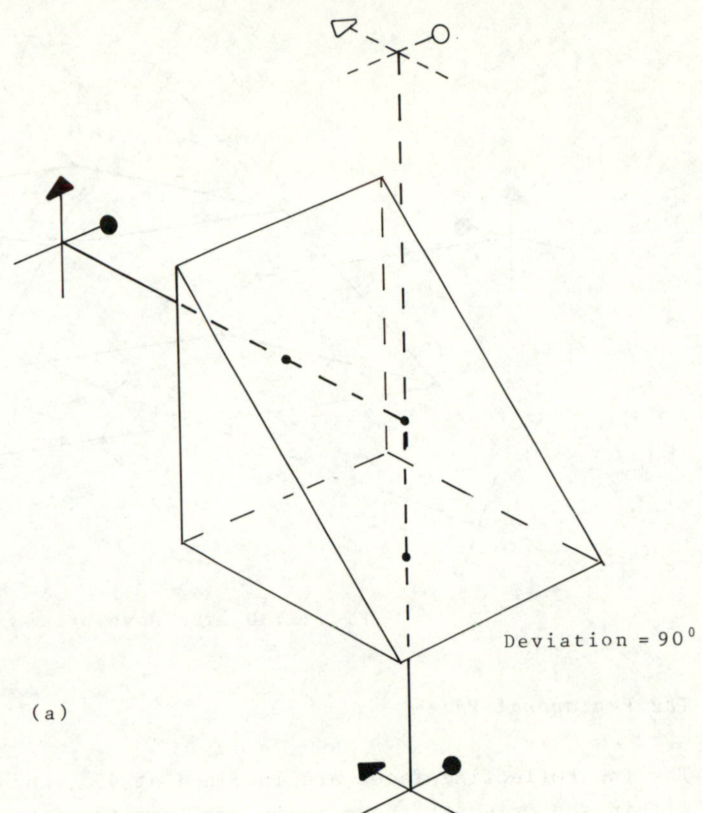

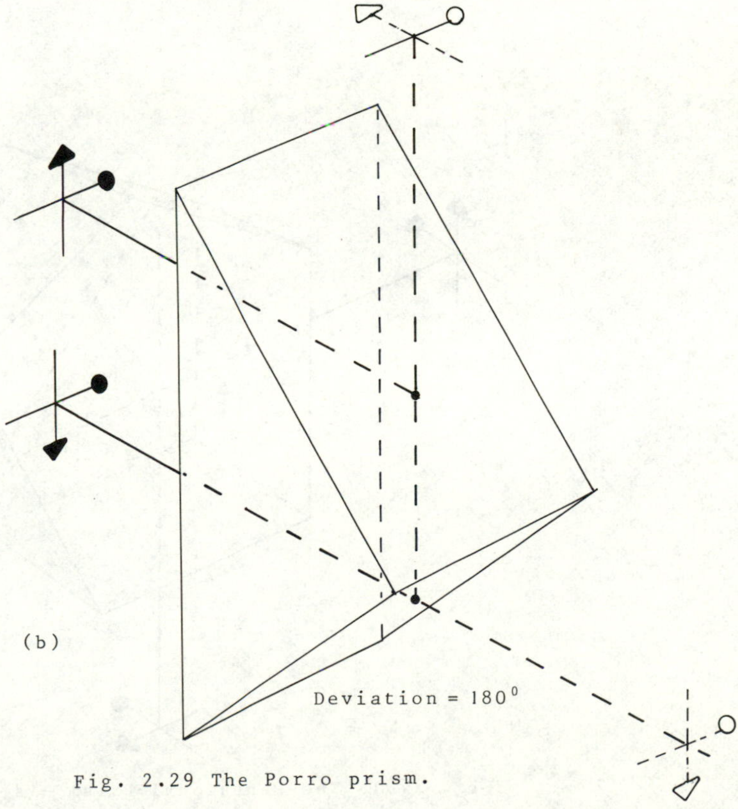

Fig. 2.29 The Porro prism.

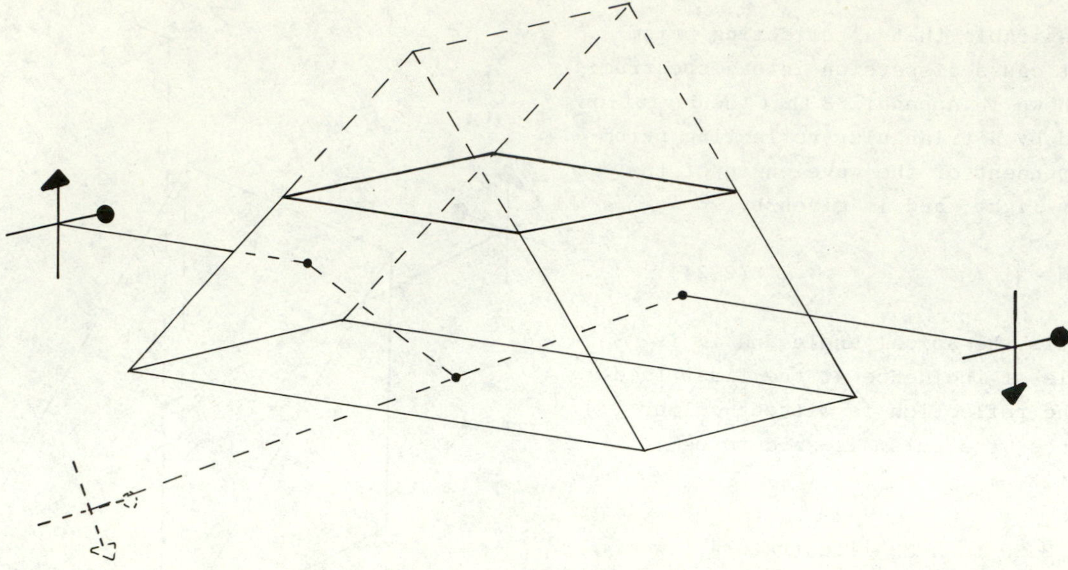

Fig. 2.30 The Dove prism

The Pentagonal Prism

The two reflecting faces are inclined at 45°, and are silvered, as the angles of incidence within the prism at these faces are less than the critical angle. However, the silvered faces can be coated with protective lacquer. The Pentagonal prism is equivalent to a pair

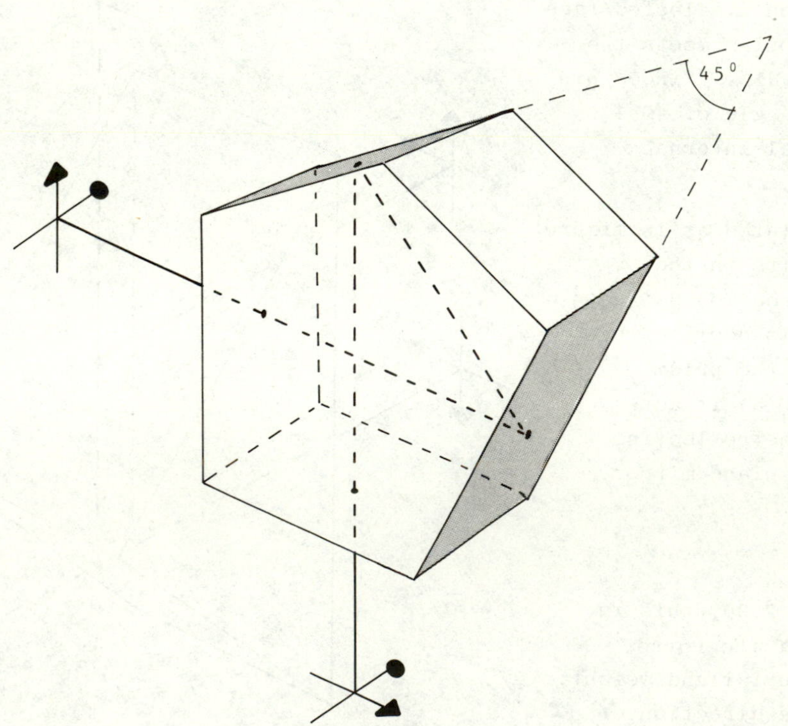

Fig. 2.31 The Pentagonal Prism - silvered faces shown shaded.

of plane mirrors inclined at 45^0 (see figure 5.12) and, as such, produces a constant deviation of 90^0. The prism may be used, together with a moveable plane mirror, as the view finder of a single lens reflex camera. It utilises the rays of light which actually pass through the camera lens system in producing a focused image on the film.

The Fresnel Prism-Lens

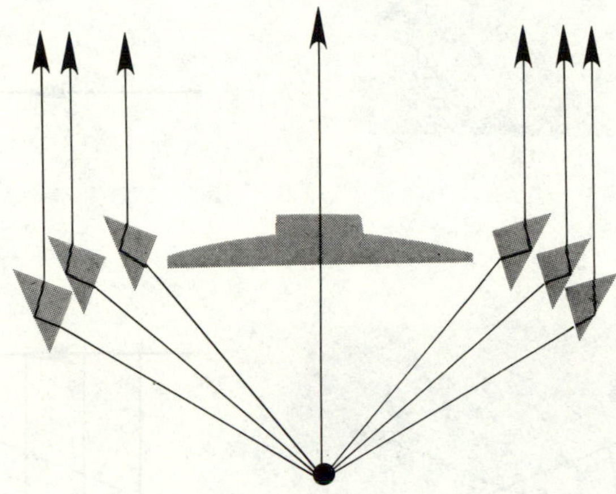

Fig. 2.32 The Fresnel prism-lens.

The structure is built up from a series of prismoidal rings with each ring having a 45/45/90 prism shape as its cross-section. Each prism section is carefully aligned so that rays from the source undergo total internal reflection producing essentially a parallel emergent pencil of rays. The system is equivalent to a solid glass lens, but is much lighter in weight. This is clearly an advantage, particularly if a large aperture lens is required. The prism-lens was originally developed for use as a lighthouse lens, and is still used for that purpose.

2.2 REFRACTION AT CURVED SURFACES

2.2.0 *Introduction*

In view of its comparative ease of manufacture, the *spherical surface* is by far the most common form of curved refracting surface. Many optical devices use lenses in which the surfaces are parts of spheres. Non-spherical (aspherical) surfaces, in particular cylindrical and toric surfaces, have special application, together with spherical surfaces, in spectacle lens design.

A spherical boundary which separates two transparent media has similar properties to a lens. It can form real and virtual images. Although it is not often that single interfaces are used for image formation, their study, however, is still of fundamental importance.

2.2.1 CONVERGING AND DIVERGING REFRACTING SURFACES

Figures 2.33 (a) and (b) show a spherical interface between two media of refractive indices n

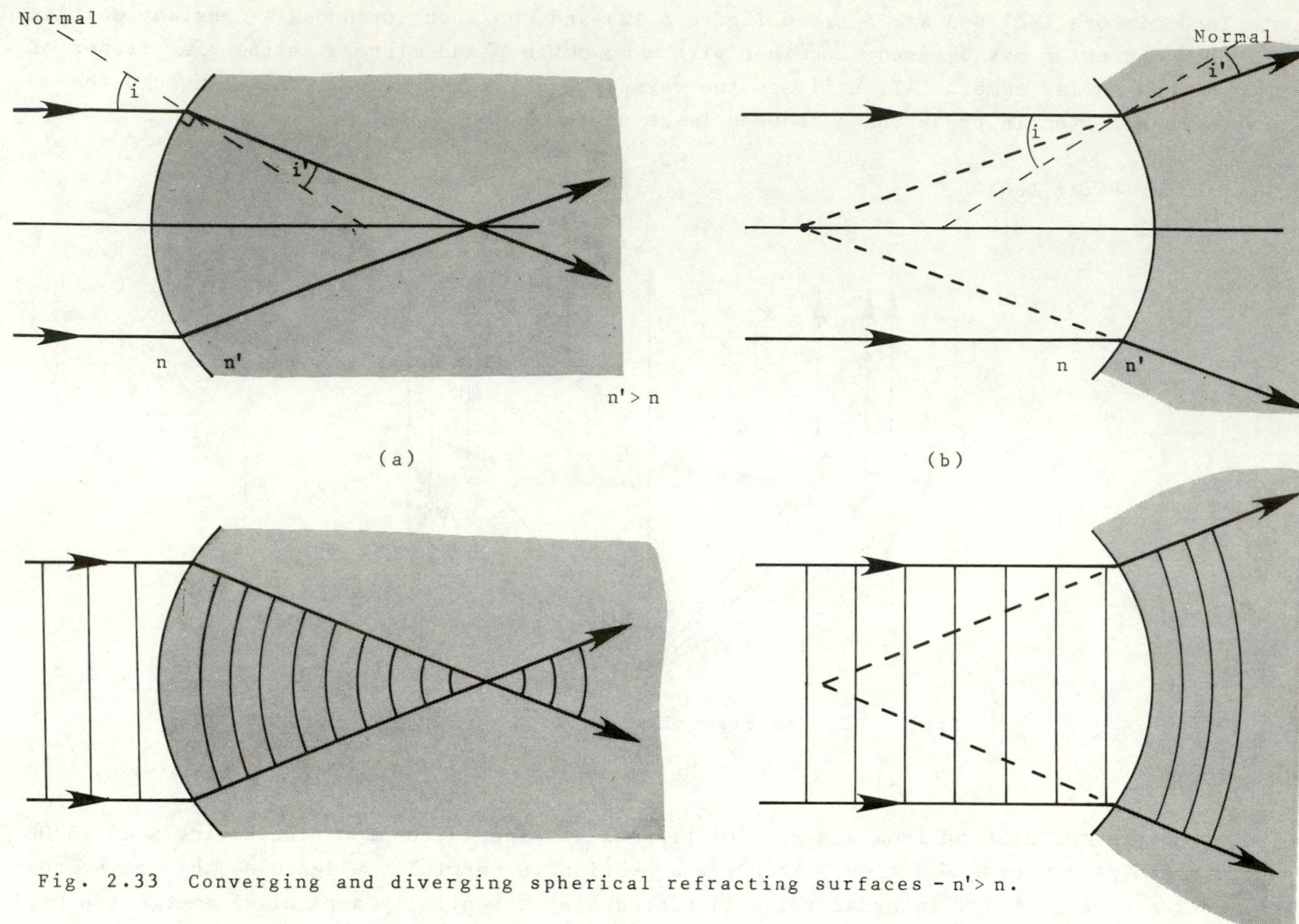

Fig. 2.33 Converging and diverging spherical refracting surfaces - n' > n.

and n'. Let us assume that n' > n; that is, the rays are travelling from an optically less dense to an optically more dense medium. It follows from Snell's law that the rays deviate towards the normal in the higher index medium.

Also illustrated is the reshaping of the wavefronts at each refracting surface. For convenience the incident rays are drawn as parallel pencils having plane wavefronts. Note that as the rays deviate towards the normal the wavefronts become closer together in the optically denser medium. This is due to the fact that the light waves are travelling more slowly in the medium with the higher refractive index. Figure 2.33(a) represents a *converging refracting surface*. Such a surface can be recognised by the fact that it is *convex** to the less optically dense medium. That is, it *bulges towards the rarer medium*. Figure 2.33(b) represents a *diverging refracting surface*. This can be recognised by the fact that it is *concave*** to the rarer medium. That is, the surface *bulges away from the rarer medium*.

2.2.2 REFRACTION BY A SINGLE SPHERICAL SURFACE

Figure 2.34 depicts a wide parallel pencil of rays incident on a single convex spherical refracting surface which separates media of indices n and n', with n' > n.

* From the Latin convexus = arched. **From concavus = hollow.

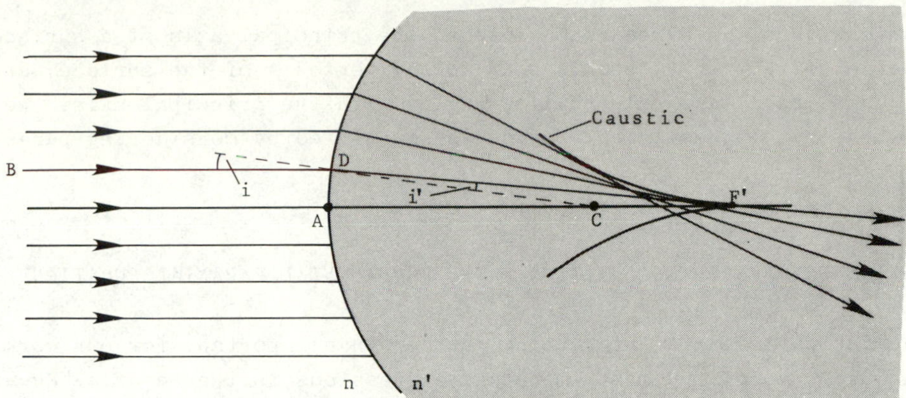

Fig. 2.34 Refraction by a spherical surface - marginal and paraxial rays and the caustic curve.

A is the midpoint of the aperture (diameter) of the surface and is called the *vertex* of the surface. C is the *centre of curvature* of the surface. The line joining A and C, produced if necessary in each direction, is the *principal axis* or *optical axis* of the surface.

Now, each ray of light, on crossing the surface, obeys the laws of refraction (section 2.1.1) and deviates towards its respective normal. This is with the exception of the ray incident along the axis which, meeting the surface normally, is undeviated. For example, the ray BD is incident at the angle i and is refracted at the angle i', where $n \sin i = n' \sin i'$. This refracted ray takes the path from D through F'. However, it is seen that the various refracted rays do not intersect in a single point. Instead, the rays near to the axis of the surface intersect each other further from the vertex than rays refracted near the edge of the surface. This inability of a wide aperture convex spherical refracting surface to converge the light to a single focus point is known as *spherical aberration*. It is an unavoidable defect inherent in spherical surfaces, and is also exhibited by a wide aperture lens having spherical surfaces. Rays reflected from a wide aperture spherical reflector also suffer the same defect (section 5.2.2.1).

As we saw in section 2.1.8 in connection with plane surface refraction, the rays refracted at the spherical surface are enveloped by a curved line, called a *caustic curve*. This is shown in figure 2.34 . The pointed part of the curve, called a cusp, coincides with the focus point formed by rays which are close to the axis. A similar effect occurs with a wide aperture concave spherical refracting surface. In this case the caustic curve will envelope, or outline, the virtual ray directions. This is shown in figure 2.35 .

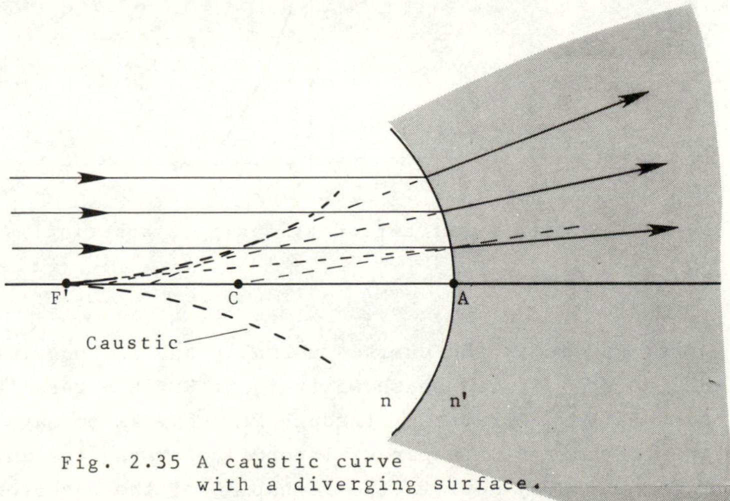

Fig. 2.35 A caustic curve with a diverging surface.

Rays of light which travel in directions close to the principal axis of a surface or a lens and, hence, undergo refraction at points near to the vertex A of the surface, are termed *paraxial rays*. These rays make only shallow angles with the principal axis. We shall see that much of the theory of geometrical optics is simplified by considering paraxial rays only.

2.2.3 THE EQUATION FOR THE IMAGE POSITION - THE FUNDAMENTAL PARAXIAL EQUATION

Certain formulae derivations are interesting rather than important for our work in geometrical optics. However, one of the most important expressions is the *Paraxial Equation*, and we shall now discuss its derivation for one particular case of refraction at a spherical surface. The general derivation, which deals with all cases of refraction at a spherical surface, is included in Appendix 3.

Let us consider a diverging pencil of rays, arising from a point B at the bottom of an object, incident on a convex refracting surface having a radius of curvature r. The surface separates two transparent media having refractive indices n and n', with n' > n. C is the centre of curvature of the surface and A is the vertex of the surface.

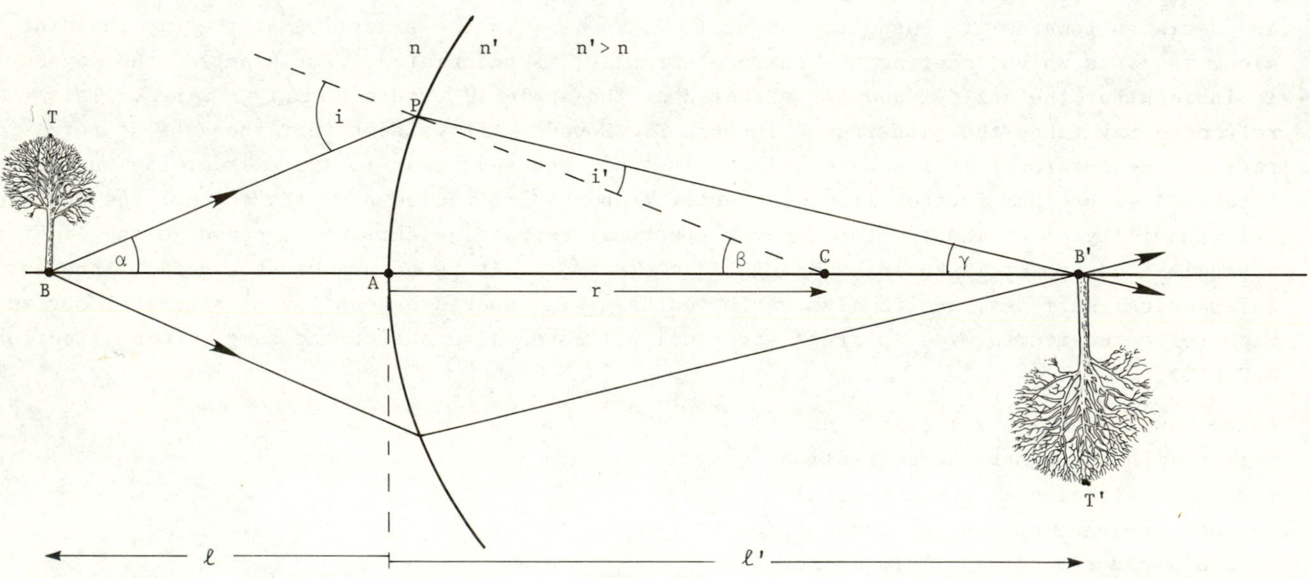

Fig. 2.36(a) Refraction at a single spherical surface - the paraxial equation.

A ray BAC meets the surface normally and is undeviated. Another ray BP, after refraction, cuts BAC at B'. It can be shown that, if BP is a paraxial ray then all paraxial rays from B, will pass, after refraction, through B'. That is to say, B' is the image of B. Figure 2.36(a) shows the refraction of a paraxial pencil of rays (the angles are exaggerated for clarity), and figure 2.36(b) depicts the reshaping of the wavefronts caused by the refraction. Note that the central part of the wavefront shown traversing the surface is slowed down whilst the marginal part is travelling faster, and this causes the wavefront to be reshaped.

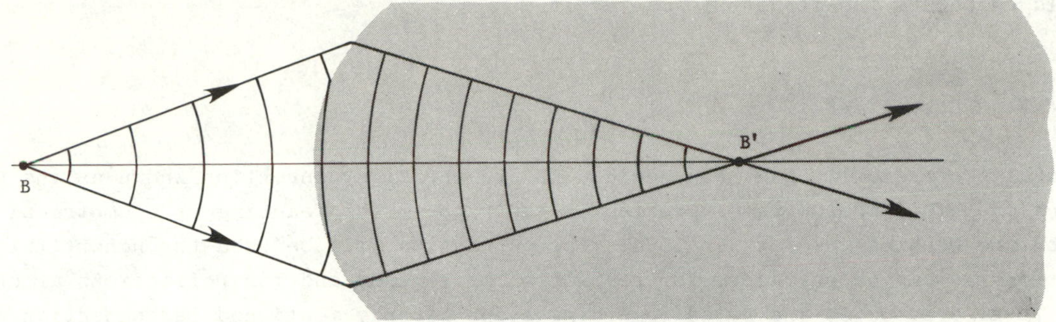

Fig. 2.36(b) The reshaping of wavefronts on refraction at a spherical surface.

For the refraction at P in figure 2.36(a), $n \sin i = n' \sin i'$. Now, if P is close to A, then i and i' will be *small angles*. In Appendix 4, it is shown that, for small angles, $\sin i \simeq i$ when i is measured in radians. Hence, the refraction at P may be written

$$n\,i = n'i' \qquad \text{(i)}$$

Now, in triangle BPC,

$$i = \alpha + \beta$$

and in triangle B'PC,

$$\beta = i' + \gamma$$

OR

$$i' = \beta - \gamma$$

since in any triangle an external angle equals the sum of the two interior opposite angles.

Substituting in (i) for i and i' gives

$$n(\alpha+\beta) = n'(\beta-\gamma)$$

OR

$$n\alpha + n'\gamma = (n'-n)\beta \qquad \text{(ii).}$$

Now, since the angles α, β, and γ are also small, $\alpha \simeq \tan\alpha$, etcetera (see Appendix 4). Therefore, equation (ii) may be written as

$$n\tan\alpha + n'\tan\gamma = (n'-n)\tan\beta,$$

or, if P is very close to A then PA is very nearly a straight line perpendicular to the axis and we can write

$$n.\frac{PA}{AB} + n'.\frac{PA}{AB'} = (n'-n).\frac{PA}{AC}$$

OR

$$\frac{n}{AB} + \frac{n'}{AB'} = \frac{n'-n}{AC} \qquad \text{(dividing through by PA).}$$

Now, in order that this expression may be used for all refractions at convex surfaces and concave surfaces, it is necessary to apply a *sign convention* for the distances AB, AB', and AC. The use of a sign convention is vitally important and must be studied carefully. It also enables the direction (to the left or to the right of the vertex A) in which a calculated distance lies to be easily determined. The worked examples which follow shortly will help to make this clear. Without a sign convention it would be impossible, or at least very difficult, to deal with most optical problems. In fact, we would need several equations instead of the one paraxial equation we are in the process of deriving. The sign convention also gives us information about the nature of the image such as whether it is erect or inverted, all of which should encourage the student to master it.

The Optical Sign Convention

In the optical profession it has been decided to use the sign convention known as the *Cartesian Sign Convention*. This choice was suggested in the report "The Teaching of Geometrical Optics", published by the Physical Society in 1934. The convention will be used throughout this book for refractions at curved surfaces, for refraction by lenses, and for reflections at curved surfaces. It also embodies the rules regarding signs already mentioned in connection with vergence in section 1.2.2 .

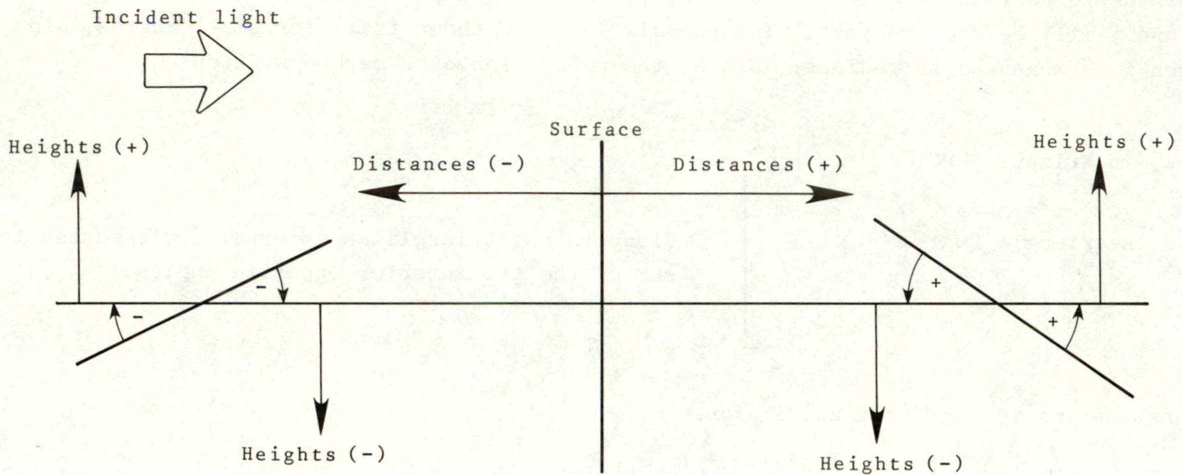

Fig. 2.37 The Cartesian Sign Convention.

With reference to figure 2.37, the details of the sign convention are as follows:

1 *For distances measured along the principal axis*
   (a) the incident light is drawn for convenience travelling from left to right,
   (b) all distances are measured FROM the vertex A of the surface,
   (c) the numerical value of a distance which is measured in the SAME DIRECTION as that in which the light is travelling is given a POSITIVE SIGN (+),
   (d) the numerical value of a distance which is measured in the OPPOSITE DIRECTION to that in which the light is travelling is given a NEGATIVE SIGN (-).

2 *For distances and heights measured perpendicular to the axis*
   (a) all heights are measured FROM the axis,
   (b) the numerical value of a height which is measured ABOVE the axis is given a POSITIVE SIGN (+),
   (c) the numerical value of a height which is measured BELOW the axis is given a NEGATIVE SIGN (-).

3 *For angles*
   (a) measurements are taken as the acute angle FROM the ray to the axis

(b) CLOCKWISE angles are given a NEGATIVE SIGN (-),

(c) ANTICLOCKWISE angles are given a POSITIVE SIGN (+).

Returning now to our derivation, we have shown that $\frac{n}{AB} + \frac{n'}{AB'} = \frac{n'-n}{AC}$. Let a, b, c be the *numerical values* for AB, AB', and AC, respectively.

$$\therefore \quad \frac{n}{a} + \frac{n'}{b} = \frac{n'-n}{c}.$$

Now, AB is the *object distance*, represented by the symbol ℓ,

AB' is the *image distance*, represented by the symbol ℓ',

and AC is the *radius of curvature* of the surface, represented by r.

By the rules of the sign convention:- $r = +c$, $\ell' = +b$, but we must write $\ell = -a$, because the numerical value of the object distance is measured in the *opposite direction* to the direction of the light rays. So, we now have

$$\frac{n}{-\ell} + \frac{n}{\ell'} = \frac{n'-n}{r}$$

or, by a slight rearrangement,

$$\frac{n'}{\ell'} - \frac{n}{\ell} = \frac{n'-n}{r} \qquad (2.25).$$

This is the *FUNDAMENTAL PARAXIAL EQUATION* for refraction at spherical surfaces. It relates object distance, image distance, and radius of curvature with the refractive indices of the two media. The equation does not involve the angle α (figure 2.36(a)), and hence, the image distance ℓ' is the same value for all small values of the angle α. This means that *ALL THE PARAXIAL RAYS* from the object point B must pass, after refraction, through the image point B'.

2.2.3.1 THE POWER OF A REFRACTING SURFACE

The quantity $\frac{n'-n}{r}$ in equation (2.25) is defined as the *power* of the surface. It is a measure of the ability of the surface to alter the curvature of the wavefronts of the incident light; that is, to alter the vergence of the incident light. The power of a surface is denoted by F where

$$F = \frac{n'-n}{r} \qquad (2.26).$$

If the value of the radius of curvature r is given in metres, or a fraction of a metre, the value of the power F will be expressed in dioptres. Hence, we see that the power of a surface is expressed in the same units as the values of vergences (section 1.2.3). This would seem to logical if we remember that the power of a surface is a measure of the change in vergence which that surface imposes on a pencil of rays.

A refracting surface forming the interface between two given media has only one value for its power, even though light may be incident from either side of a surface (see the first worked example in the next section). Similarly, we shall see that a lens, in a given environment, has only one value for its power, whichever way round the light passes through the lens. This statement will need to be qualified when we come to consider lenses which do not have a negligible thickness, for then we shall see that the lens power can be defined in three ways all of which are useful and two of these powers are dependent on the manner in which the light is incident.

However, it is true to say that for thin lenses (see Chapter 3) that the power is independent of the manner in which the light passes through the lens.

2.2.3.2 *REDUCED VERGENCES*

We have seen in section 1.2.4 that the quantities $\frac{1}{\ell}$ and $\frac{1}{\ell'}$ represent vergences of ray pencils measured in air, where ℓ is an object distance and ℓ' is an image distance.
That is,

$L = \frac{1}{\ell}$ is the vergence to an object, in air,

and $L' = \frac{1}{\ell'}$ is the vergence to an image , in air.

We have also seen in section 2.1.8 that, when light passes through a thickness t of a transparent medium of refractive index n, the quantity $\frac{t}{n} = \bar{t}$ is called the apparent or reduced thickness of the medium. Hence, we can write the fundamental paraxial equation in a simplified form using the notation of *reduced vergences*. If $\frac{n'}{\ell'}$ is written as $\frac{1}{\ell'/n'}$, this represents the reciprocal of the reduced image distance. This is called the *reduced vergence to the image* and is denoted by $\bar{L}'$. Similarly, if $\frac{n}{\ell}$ is written as $\frac{1}{\ell/n}$, this quantity represents the reciprocal of the reduced object distance. This is called the *reduced vergence to the object*, and is denoted by $\bar{L}$. The values of the reduced vergences $\bar{L}'$ and $\bar{L}$ are expressed in dioptres if the values of ℓ' and ℓ are stated in metres.

Hence, since $\frac{n'}{\ell'} = \frac{1}{\ell'/n'} = \bar{L}'$, and $\frac{n}{\ell} = \frac{1}{\ell/n} = \bar{L}$, and Power $= \frac{n' - n}{r} = F$, the paraxial equation can be written in its simplest form, namely

$$\bar{L}' - \bar{L} = F \qquad (2.27)$$

and it is seen that the surface power represents the change in the reduced vergence of the pencil of rays which the refracting surface produces.

Note here that only the actual object and image distances ℓ and ℓ' can be depicted on a diagram. The reduced distances $\bar{\ell}\left(=\frac{\ell}{n}\right)$ and $\bar{\ell}'\left(=\frac{\ell'}{n'}\right)$ cannot be displayed.

2.2.3.3 *LINEAR MAGNIFICATION OF THE IMAGE*

Figure 2.38 shows the image B' T' of an object BT formed by a convex refracting surface. A ray from the tip T of the object, incident at the vertex A of the surface , will be refracted through the tip T' of the image (see section 2.2.5).

Let the angle of incidence at A be i and the angle of refraction i'. For paraxial rays, i and i' are small angles. Using n sin i = n' sin i', we can simplify this to ni = n'i', whence

$$\frac{i}{i'} = \frac{n'}{n} .$$

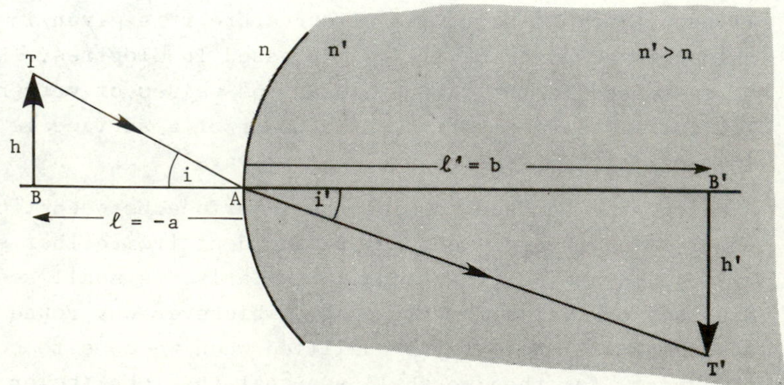

Fig. 2.38 Diagram for the derivation of the linear magnification equation.

Now, $\frac{i}{i'} \approx \frac{\tan i}{\tan i'} = \frac{BT/a}{B'T'/b} = \frac{BT}{B'T'} \times \frac{b}{a}$. $\therefore \frac{n'}{n} = \frac{BT}{B'T'} \times \frac{b}{a}$.

We shall now introduce the optical symbols ℓ, ℓ', h, and h', and to do this we must make very careful reference to our sign convention (section 2.2.3). So, we have object height h = BT, but we must write image height h' = -B'T', where B'T' is regarded as the magnitude of the image which must take a negative sign since T' is below the axis. Also, the distance to the image from the vertex is $\ell' = b$, but we must write $\ell = -a$, since the numerical value a of the object distance from the vertex is measured against the direction in which the incident light is travelling. Hence,

$$\frac{n'}{n} = \frac{h}{-h'} \times \frac{\ell'}{-\ell} \text{, which on rearranging gives } \frac{h'}{h} = \frac{\ell'}{n'} \times \frac{n}{\ell}$$

$$\text{OR} \quad \frac{h'}{h} = \frac{\ell'/n'}{\ell/n} \quad \left(= \frac{\bar{\ell}'}{\bar{\ell}} \right).$$

Now, the *linear magnification* (m) of an image is defined by the ratio

$$\text{linear magnification} \quad m = \frac{\text{height of image}}{\text{height of object}} = \frac{h'}{h}$$

$$\therefore \quad m = \frac{h'}{h} = \frac{\bar{\ell}'}{\bar{\ell}} \qquad (2.28).$$

Using the notation of reduced vergences, the magnification formula can be written in its simplest form, viz

$$m = \frac{\bar{L}}{\bar{L}'} \qquad (2.29),$$

since $\bar{L} = \frac{1}{\bar{\ell}}$ and $\bar{L}' = \frac{1}{\bar{\ell}'}$. In keeping with our sign convention, if the sign of the image height, or the magnification, is *NEGATIVE*, the image is *INVERTED* with respect to the object. Conversely, if these quantities are *POSITIVE*, the image is *UPRIGHT*.

Note: in the equations used in this book, the plain symbols and the "dashed" symbols refer to the object space and the image space respectively.

Worked Examples

1 Calculate the powers of the following refracting surfaces:

(a) a converging surface, radius 200 mm, separating air of refractive index 1.000 from glass of refractive index 1.523 ,

(b) a diverging surface, radius 200 mm, separating air and glass with refractive indices as in (a),

(c) a diverging surface, radius 500 mm, separating media of refractive indices 1.333 and 1.555,

(d) as in (a) but with the curve and the media reversed.

From section 2.2.1 we recall that a converging surface is a convex surface, that is it bulges towards the less dense medium. A diverging surface is a concave surface, bulging away from the less dense medium.

(a) The given data are n = 1.000, n' = 1.523, r = +200 mm = +0.200 m. The first medium is air, and the second medium is glass, as shown in the diagram on the right. From section 2.2.2, the power of the surface is given by

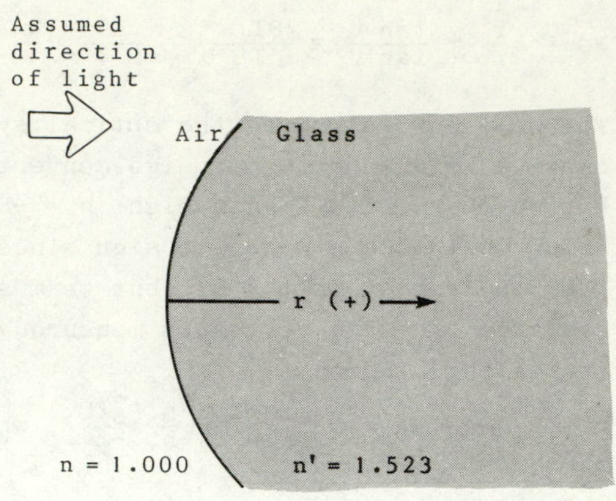

$$F = \frac{n' - n}{r}$$

$$= \frac{1.523 - 1.000}{+0.200}$$

$$= \frac{0.523}{0.2}$$

$$= +2.615 \text{ D} .$$

The surface has a power of +2.615 D . The positive sign confirms the converging property of the surface.

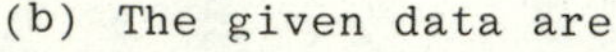

(b) The given data are

$n = 1.000$
$n' = 1.523$
$r = -200 \text{ mm} = -0.200 \text{ m}$

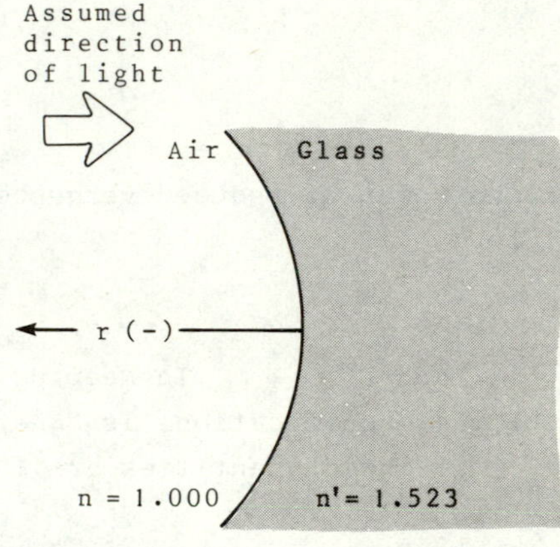

whence, the surface power is

$$F = \frac{n' - n}{r}$$

$$= \frac{1.523 - 1.000}{-0.200}$$

$$= \frac{0.523}{-0.2}$$

$$= -2.615 \text{ D}.$$

The surface has a power of -2.615 D and the negative sign confirms the surface's diverging power.

(c) The given data are

$n = 1.333$
$n' = 1.555$
$r = -500 \text{ mm} = -0.500 \text{ m}.$

whence the surface power is

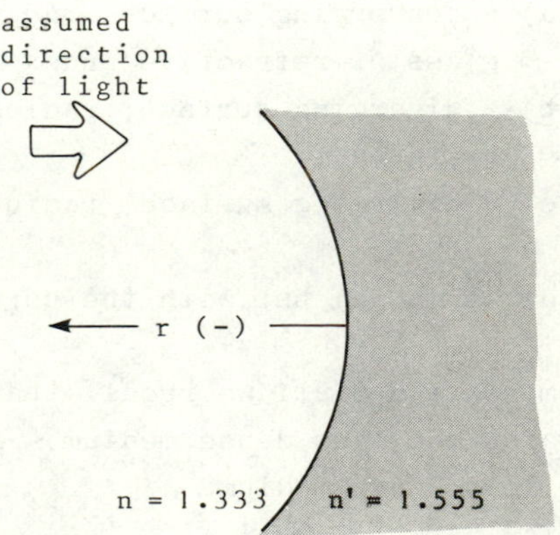

$$F = \frac{n' - n}{r}$$

$$= \frac{1.555 - 1.333}{-0.500}$$

$$= \frac{0.222}{-0.5}$$

$$= -0.444 \text{ D}.$$

The surface power is -0.444 D.

(d) This case is the surface in (a) turned around so that the glass is to the left. The data are then: $n = 1.523$, $n' = 1.000$, $r = -0.200$ m.

$$\text{So,} \quad F = \frac{n' - n}{r} = \frac{1.000 - 1.523}{-0.200} = \frac{-0.523}{-0.200} = +2.615 \text{ D, the same as for (a).}$$

2 A surface separates two media of refractive indices 1.389 and 1.640 and has a power of -1.25 D. Calculate the radius of curvature.

Now, since the the surface power is negative, we must draw the surface concave, or bulging away from the optically less dense medium. Then, $n = 1.389$, $n' = 1.640$, and $F = -1.25$ D.

So,

$$r = \frac{n' - n}{F} = \frac{1.640 - 1.389}{-1.25}$$

$$= \frac{0.251}{-1.25}$$

$$= -0.201 \text{ m}$$

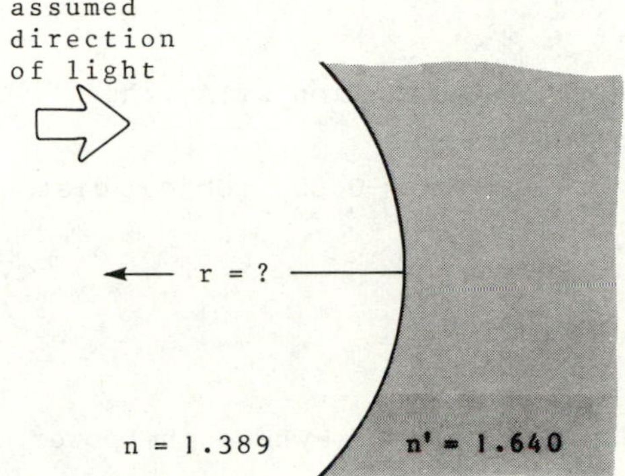

3 An eye can be considered as a single refracting surface, a cornea, of radius 5.50 mm with a medium in the eye of refractive index 1.33 . If the distance from the cornea to the retina is 24 mm, at what distance from the eye will objects be clearly seen?

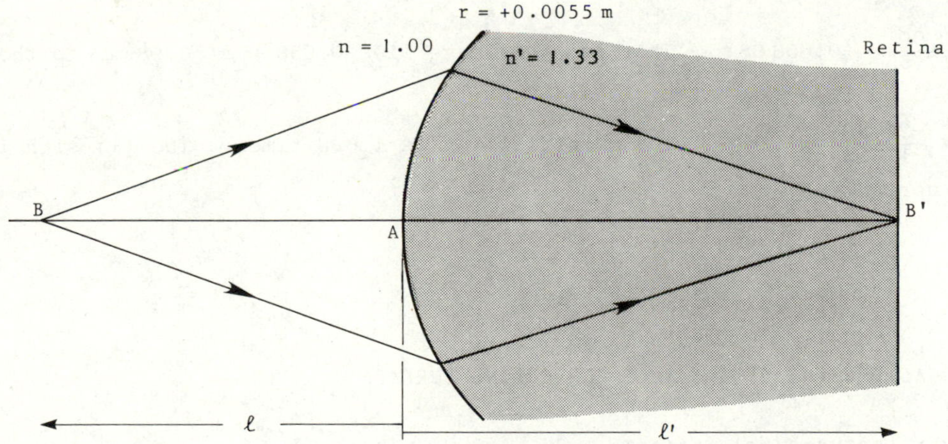

For an object to be seen clearly, the image B' must be formed exactly on the retina. Let B be the corresponding object position. As is usual, we will position the object to the left of the cornea so that the light is travelling left to right. Hence, as drawn, the radius of the cornea lies in the positive direction. For the refraction at the surface the given data are: $\ell = ?$ (unknown object position), $\ell' = +24$ mm $= +0.024$ m, $n = 1.00$ (air), $n' = 1.33$ (interior of eye), and $r = +0.0055$ m.

Using $\bar{L}' - \bar{L} = F$, where $F = \frac{n'-n}{r} = \frac{1.33 - 1.00}{+0.0055} = +60.00$ D, and $\bar{L}' = \frac{1}{\ell'/n'} = \frac{n'}{\ell'} = \frac{1.33}{+0.024} = +55.42$ D,

hence, the vergence to the object is $\bar{L} = \bar{L}' - F = 55.42 - 60.00 = -4.58$ D. But $\bar{L} = \frac{1}{\ell/n} = \frac{n}{\ell}$, whence $\ell = \frac{n}{\bar{L}} = \frac{1.00}{-4.58} = -0.2183$ m. That is, the object position is 0.2183 m to the left of the surface: the minus sign indicates the direction according to the sign convention.

4 A magnifying device consists of a plano-convex block of plastics material of refractive index 1.49 and maximum thickness of 5 cm. The curved surface has a radius of curvature of 4 cm. If an object is placed in contact with the plane surface, find the position and magnification of the image.

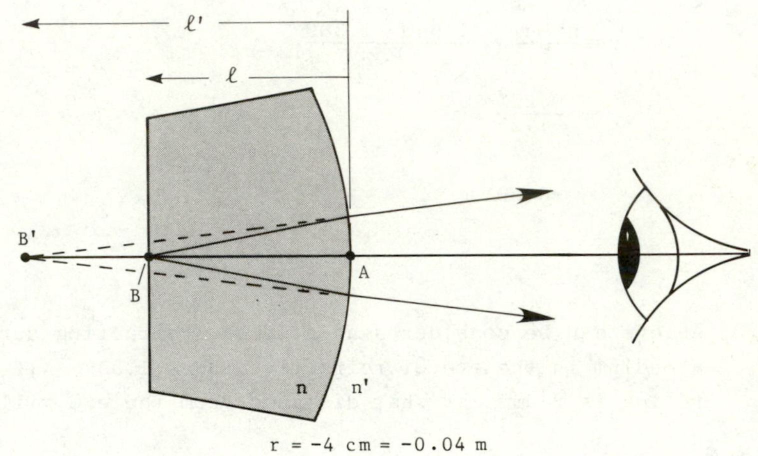

For the refraction at A, the given data are:

$\ell = -5$ cm $= -0.05$ m (object distance)

$\ell' = ?$

$n = 1.49$

$n' = 1.00$

$r = -0.04$ m.

Using $\bar{L}' - \bar{L} = F$, where the power of the surface is

$$F = \frac{n'-n}{r} = \frac{1.00 - 1.49}{-0.04} = \frac{-0.49}{-0.04} = +12.25\text{D}.$$

Then $\bar{L} = \frac{1}{\ell/n} = \frac{n}{\ell} = \frac{1.49}{-0.05} = -29.80$ D.

So, $\bar{L}' = \bar{L} + F = -29.80 + 12.25 = -17.55$ D.

Using $\bar{L}' = \frac{1}{\ell'/n'} = \frac{n'}{\ell'}$, we have

$\ell' = \frac{n'}{\bar{L}'} = \frac{1.00}{-17.55} = -0.05698$ m. That is, the image lies 0.056 98 m (56.98 mm) to the left of the vertex A.

Using $m = \frac{\bar{L}}{\bar{L}'} = \frac{-29.80}{-17.55} = +1.698$, the magnification is 1.698 times. The (+) sign indicates that the image is upright.

2.2.4 THE FOCAL LENGTHS OF A SINGLE REFRACTING SURFACE

2.2.4.1 *Convex (converging) Surface*

Let us consider a pencil of rays originating from an object point at an infinite distance from a convex spherical refracting surface. In such a case the pencil of rays will effectively be parallel when reaching the refracting surface. Figures 2.39(a) and (b) show the refraction together with the reshaping of plane wavefronts into spherical wavefronts.

For an object at an infinite distance, $\ell = -\infty$. Using the equation (2.25) $\frac{n'}{\ell'} - \frac{n}{\ell} = \frac{n'-n}{r}$ gives $\frac{n'}{\ell'} - \frac{n}{-\infty} = \frac{n'-n}{r}$, or $\frac{n'}{\ell'} + 0 = \frac{n'-n}{r}$. Rearranging, we have $\ell' = \frac{n'r}{n'-n}$.

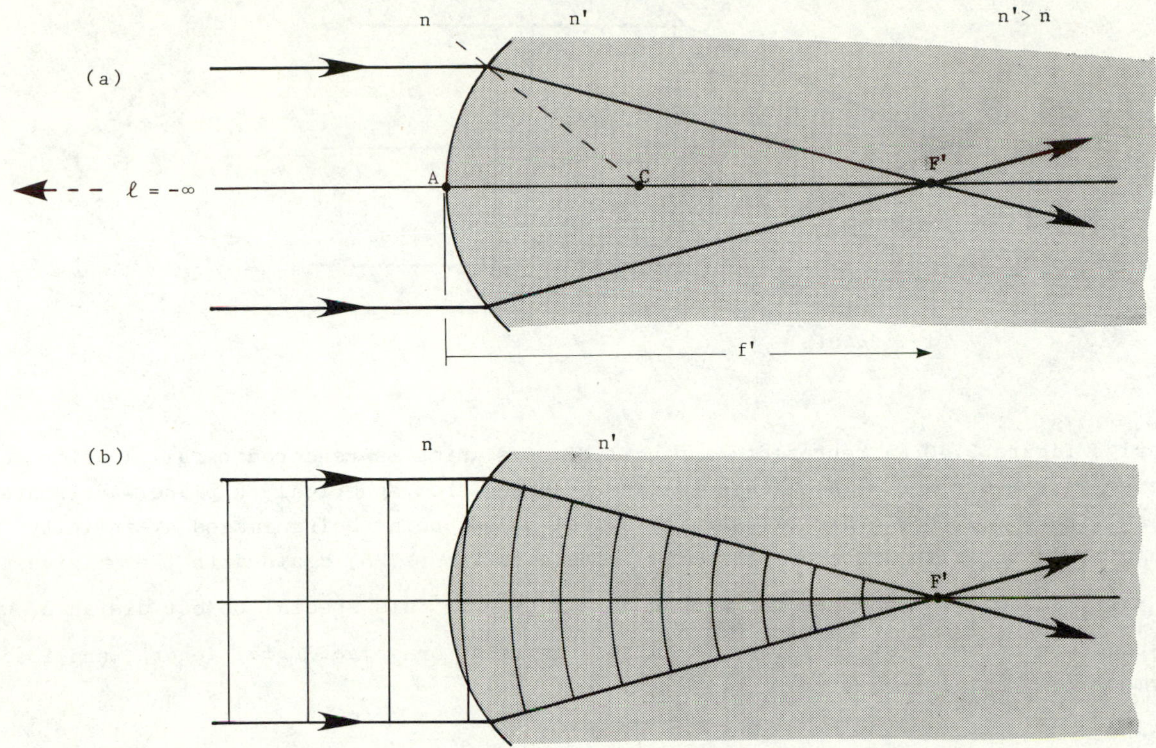

Fig. 2.39 The second focal length f' of a converging surface.

This special image distance, corresponding to an object at infinity, is called the *second* or *image focal length* and is represented by the symbol f'. That is, $f' = \frac{n'r}{n' - n}$. The axial point F' is known as the *second principal focus* of the surface. Now, since the power of the surface is $F = \frac{n' - n}{r}$, then it follows that $f' = \frac{n'}{F}$, or

$$F = \frac{n'}{f'} \qquad (2.30).$$

Since, this may be written $F = \frac{1}{f'/n'}$, and f'/n' represents the reduced second focal length, we can see that the power of a refracting surface is equal to the reciprocal of the reduced second focal length of the surface.

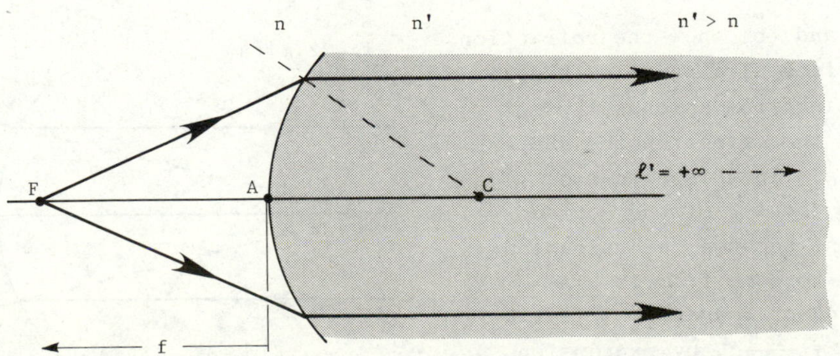

Fig. 2.40(a) The first focal length f of a converging surface.

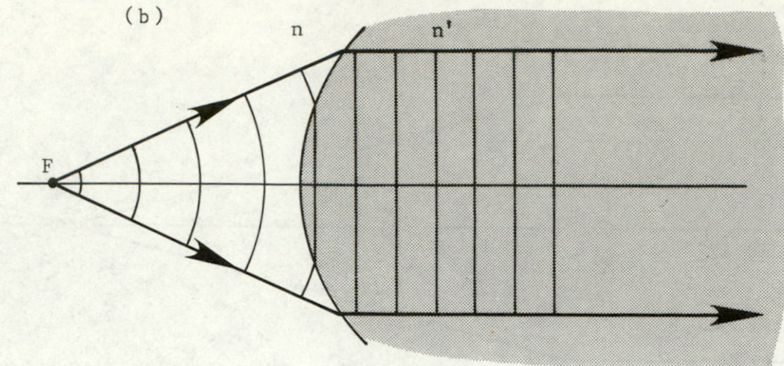

Fig. 2.40(b)

Similarly, figure 2.40(a) represents a pencil of rays which is rendered parallel after being refracted at the surface. The wavefronts are reshaped from spherical to plane wavefronts as shown in figure 2.40(b). This illustrates a real object point being imaged at infinity. For an image at an infinite distance, $\ell' = +\infty$. Using equation (2.25) again with $\ell' = +\infty$ gives $\frac{n'}{\infty} - \frac{n}{\ell} = \frac{n'-n}{r}$, or $0 - \frac{n}{\ell} = \frac{n'-n}{r}$, whence $\ell = -\frac{nr}{n'-n}$. This special object distance, which corresponds to an image at infinity, is called the *first* or *object focal length*, and is given the symbol f. That is,

$$f = -\frac{nr}{n'-n}\ .$$

The axial point F is referred to as the *first principal focus* of the surface.

Since the surface power is $F = \frac{n'-n}{r}$, then it follows that $f = -\frac{n}{F}$,

$$\text{or} \quad F = -\frac{n}{f} \qquad (2.31).$$

Once again, note that $\frac{n}{f}$ can be written as $\frac{1}{f/n}$ and that the power is given by minus the reciprocal of the reduced first focal length.

So, we have the surface power $F = \frac{n'}{f'}$ and also $F = -\frac{n}{f}$. Thus, it will be noted that the two focal lengths are not the same, but that $\frac{n'}{f'} = -\frac{n}{f}$

$$\text{or} \quad \frac{f}{f'} = -\frac{n}{n'} \qquad (2.32).$$

2.2.4.2 *Concave (diverging) Surface*

Figures 2.41(a) and (b) show the refraction by a concave surface of a parallel pencil of rays originating from an object point at infinity, and the effect on the shape of the wavefronts. In this case the *second principal focus* F' of the surface is a virtual point because the refracted rays only appear to come from it. As before, the *second* or *image focal length* f' is given by $f' = \frac{n'r}{n'-n}$, by expression

Fig. 2.41(a)

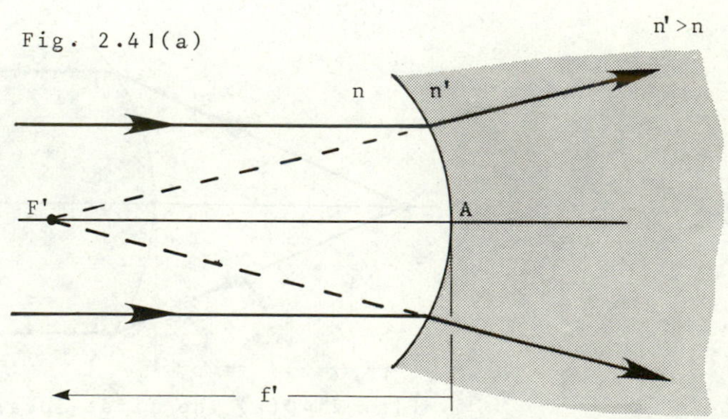

(2.30), and as before, the surface power is

$$F = \frac{n'}{f'} .$$

That is, the power is equal to the reciprocal of the reduced second focal length. In this case, we see (figure 2.41(a)) that the second focal length f' is measured in the medium of refractive index n. However, the rays of light which are refracted by the surface, and appear to originate from F', actually travel in the medium of index n'. Thus the reduced second focal length is, as before, f'/n', and $F = \frac{1}{f'/n'} = \frac{n'}{f'}$, as stated.

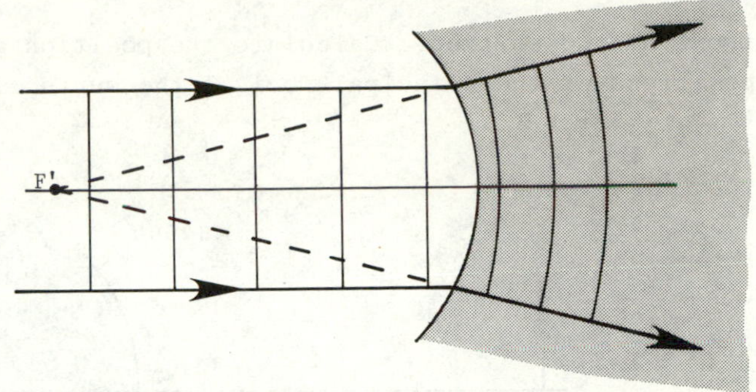

Fig. 2.41(b)

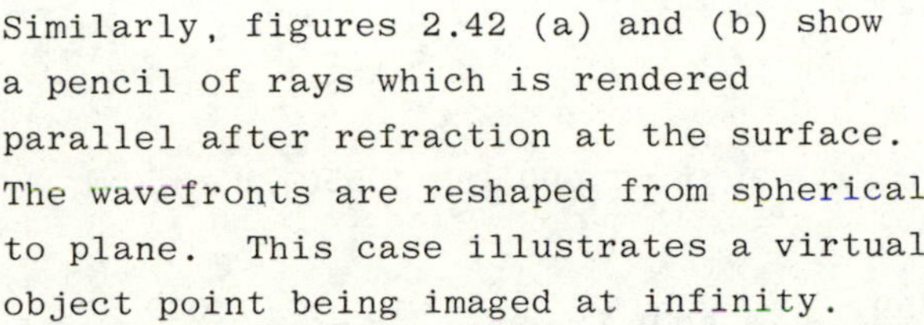

Similarly, figures 2.42 (a) and (b) show a pencil of rays which is rendered parallel after refraction at the surface. The wavefronts are reshaped from spherical to plane. This case illustrates a virtual object point being imaged at infinity.

Here, the *first principal focus* F of the surface is a virtual point because, although the rays are converging towards it, they do not actually pass through it.

As before, the *first* or *object focal length* f is given by

$$f = -\frac{nr}{n' - n}$$

by expression (2.31), and once again the surface power is

$$F = -\frac{n}{f} .$$

We again see (figure 2.42) that the first focal length f is measured in the medium of refractive index n'. But, as the rays of light which converge towards F are actually in the medium of index n, then the reduced first focal length is still f/n, and the power is

$$F = -\frac{1}{f/n} = -\frac{n}{f}$$

as previously stated.

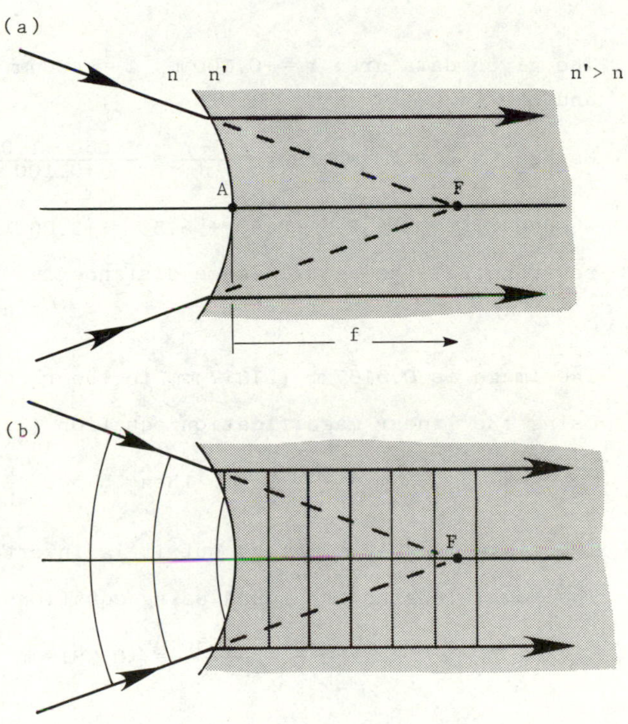

Fig. 2.42 The first principal focal length of a diverging surface.

Expression (2.32) also applies to a concave surface, namely $\frac{f}{f'} = -\frac{n}{n'}$.

Worked Example

A convex spherical surface has a radius of curvature of 100 mm and separates air from glass of refractive index 1.550 . An object (symbolised by an arrow) 5 mm high is positioned 400 mm in front of the surface. Calculate the position and size of the image. Determine, also, the positions of the principal foci of the surface.

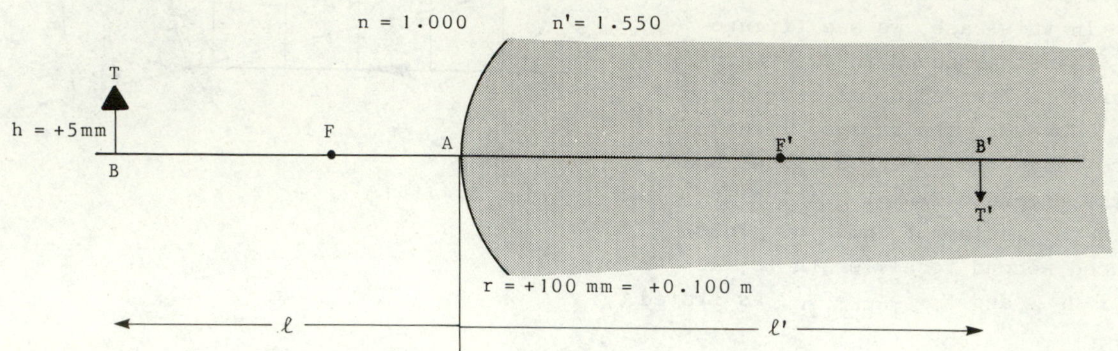

The given data are: $r = +0.100$ m, $\ell = -400$ mm $= -0.400$ m, $\ell' = ?$, $n = 1.000$, $n' = 1.550$, $h = +5$ mm, and $h' = ?$

Using $\bar{L}' = \bar{L} + F$, where $F = \dfrac{n'-n}{r} = \dfrac{1.550 - 1.000}{+0.100} = \dfrac{0.550}{+0.100} = +5.50$ D, and $\bar{L} = \dfrac{1}{\ell/n} = \dfrac{n}{\ell} = \dfrac{1.000}{-0.40} = -2.5$ D

whence $\bar{L}' = \bar{L} + F = -2.50 + 5.50 = +3.00$ D. Hence, $\bar{\ell}' = \dfrac{1}{\bar{L}'} = \dfrac{1}{+3.00}$ m. Now $\bar{\ell}' = \dfrac{\ell'}{n'}$, so on rearranging, the actual image distance is

$$\ell' = n'\,\bar{\ell}' = 1.550 \times \frac{1}{+3.00} = +0.5167 \text{ m}.$$

The image is 0.5167 m (516.7 mm) to the right of the vertex.

Using the linear magnification equation (2.29),

$$m = \frac{h'}{h} = \frac{\bar{L}}{\bar{L}'}, \quad \text{then} \quad h' = h \times \frac{\bar{L}}{\bar{L}'} = 5 \times \frac{(-2.5)}{+3.0} = -4.17 \text{ mm}.$$

The image is 4.17 mm high, but it is inverted.

The focal lengths are found using equations (2.30) and (2.31):

$$f' = \frac{n'}{F} = \frac{1.550}{5.5} = +0.2818 \text{ m} = +281.8 \text{ mm}$$

$$\text{and} \quad f = -\frac{n}{F} = -\frac{1.000}{5.5} = -0.1818 \text{ m} = -181.8 \text{ mm}.$$

2.2.4.3 NEWTON'S EQUATION

It is normal practice when considering refraction at a single spherical surface to measure the distances of the object and image from the vertex A of the surface. Similarly, it is often convenient, when studying the refraction of light by a thick lens, to express object and image distances from the front vertex and the back vertex respectively (see Chapter 4).

However, it is sometimes more useful and more practical to express object and image distances as measured from the first and second principal foci, respectively, instead of from the surface(s). Figure 2.43(a) shows how the two principal focal points F and F' of a refracting surface provide convenient reference points for object and image distances. It should be noted that the two principal focal points are not conjugate points.

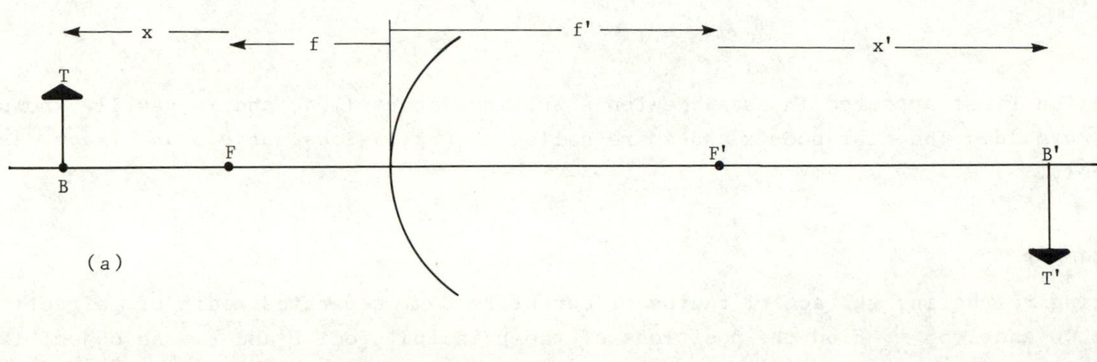

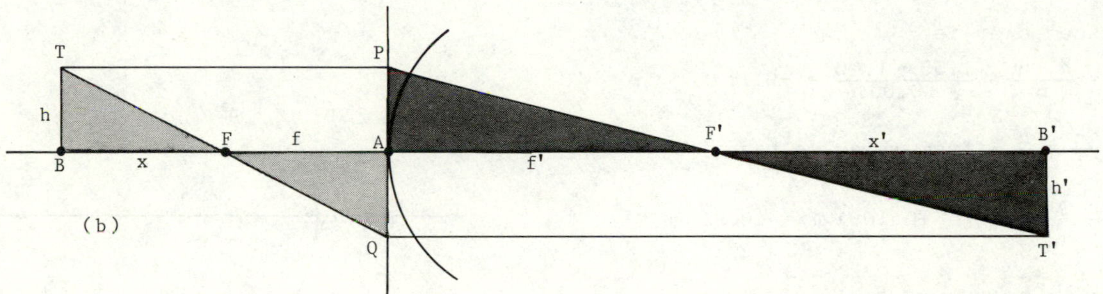

Fig. 2.43 Extra-focal object distance, x, and extra-focal image distance, x'.

Let x represent the distance from the first principal focus F to the the bottom B of the object, and x' the distance from the second principal focus F' to the bottom B' of the image. BT is the object height h, and B' T' is the height h' of the image. The first and second focal lengths of the surface are f and f', respectively.

Now, AP = h, which implies the measurement of h from A to P which is upwards from the principal axis and therefore positive. In similar vein, AQ = h' which will be negative since it is measured downwards from A to Q. Also, we have AF' = f', AF = f, F' B' = x', and FB = x. Note that AF = f and FB = x will be negative since they are measured against the direction of the incident light which is assumed to be coming from the left.

In the similar triangles FBT and FAQ (light shading in figure 2.43(b),

$$\frac{BT}{BF} = \frac{QA}{FA}, \quad \text{that is,} \quad \frac{h}{-x} = \frac{-h'}{-f}$$

$$\text{or} \quad \frac{h'}{h} = -\frac{f}{x} \qquad (2.33).$$

In the similar triangles APF' and F' B' T',

$$\frac{AP}{AF'} = \frac{T'B'}{F'B'}, \quad \text{that is,} \quad \frac{h}{f'} = \frac{-h'}{x'}$$

$$\text{or} \quad \frac{h'}{h} = -\frac{x'}{f'} \qquad (2.34).$$

Hence, since the left hand sides of equations (2.33) and (2.34) are equal, we can equate the right hand sides thus $-\frac{f}{x} = -\frac{x'}{f'}$;

$$\text{or} \quad xx' = ff' \qquad (2.35).$$

This equation first appeared in Isaac Newton's writings about 1704, and is usually known as *Newton's Formula*. The distances x and x' are called the *extra-focal object and image distances*, respectively.

Worked Example

A converging refracting surface of radius of curvature 6 cm separates media of refractive indices 1.00 and 1.55 . Find the positions of the principal foci F and F'. An object 12 mm high is placed 10 cm outside F in the first medium. Find the position and size of the image.

The surface power is

$$F = \frac{n'-n}{r} = \frac{1.55 - 1.00}{+0.06} = +9.17 \text{ D}.$$

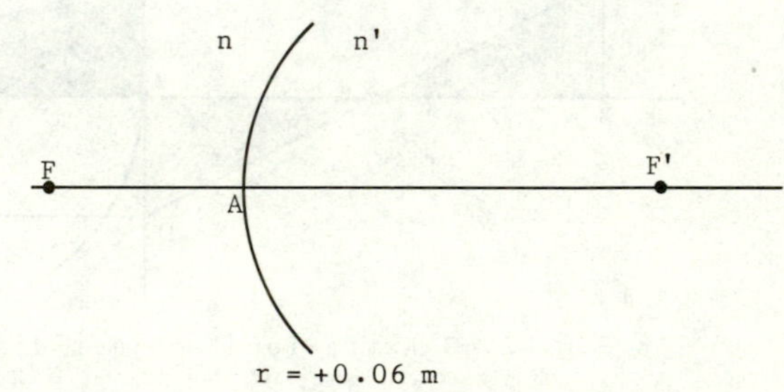

Now, $F = -\frac{n}{f} = \frac{n'}{f'}$, from which we have

$$f = -\frac{n}{F} = -\frac{1.00}{9.17} = -0.1091 \text{ m} = -10.91 \text{ cm},$$

and

$$f' = \frac{n'}{F} = \frac{1.55}{9.17} = +0.1690 \text{ m} = +16.90 \text{ cm}.$$

So, F is 10.91 cm to the left of A, and F' is 16.90 cm to the right of A.

Now, with the object 10 cm outside F, x = -10 cm. Using Newton's formula, xx' = ff', we have

$$x' = \frac{ff'}{x} = \frac{(-10.91)(+16.90)}{-10} = +18.44 \text{ cm (to the right of F')}.$$

The image is therefore f' + x' = (+16.90) + (+18.44) = +35.34 cm from the vertex A.

The image size is given by either equation (2.33) or (2.34):

By equation (2.33)

$$h' = -h \cdot \frac{f}{x} = -(+1.2)\left(\frac{-10.91}{-10}\right) = -1.309 \text{ cm},$$

and by equation (2.34)

$$h' = -h \cdot \frac{x'}{f'} = -(+1.2)\left(\frac{+18.44}{+16.90}\right) = -1.309 \text{ cm}.$$

The image is 1.309 cm high and the minus sign indicates that it is inverted.

The value of equations (2.33), (2.34), and (2.35) cannot be overemphasised. There are some occasions when they offer by far the simplest method of tackling an optical problem.

2.2.5 GRAPHICAL CONSTRUCTION FOR IMAGES

A graphical construction is a scale drawing by which the position and size of an image may be determined. In addition to its application now to be described for refraction at a single surface, the method may also be used in connection with lenses (section 3.1.4.1) and with curved mirrors (section 5.2.5.1).

Let us refer back to figure 2.36(a). If the line BACB' is rotated through a small angle about the point C, the object point B and the image point B' will both trace out small arcs. Although the position of A on the surface changes, the distances ℓ and ℓ' will be unaltered.

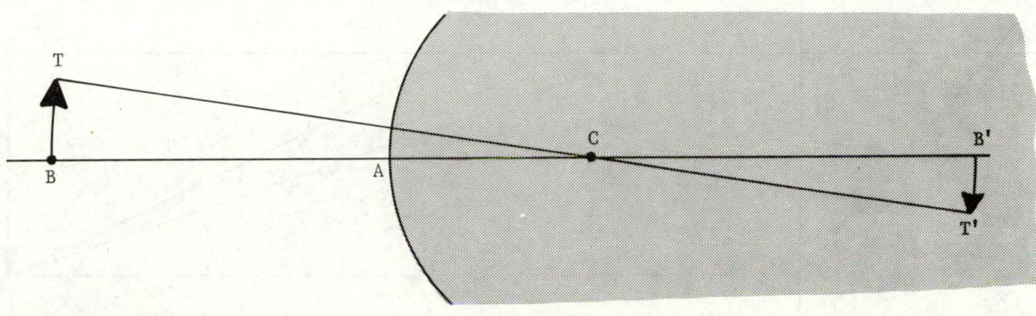

Fig. 2.44

In figure 2.44, if the angle of rotation is small the arc BT and the arc B'T' will be almost perpendicular to the axis BACB'. Thus, it can be said that, if the object is fairly small and perpendicular to the axis, refraction will give rise to an image also perpendicular to the axis.

The object and image are said to lie in *conjugate planes*. Conjugate means interchangeable. It follows from the principle of reversibility of light (section 2.1.5) that an object and its image may be interchanged with each other.

Procedure for Graphical Construction

(a) Using a convenient horizontal scale, draw the principal axis of the refracting surface and mark in the positions of the surface, the centre of curvature C, and the two principal focal points F and F'. Since only the portion of the surface near the axis is being considered its position is represented by a straight line perpendicular to the axis, figure 2.45 .

(b) Mark in the object BT (assumed standing on the axis for convenience) using as large a vertical scale as possible, bearing in mind the probable size of the image.

(c) Draw any *two* of the following construction rays from the tip of the object. Construction rays are certain rays which take definite predictable paths.

 (1) A ray from T parallel to the principal axis, is refracted through F'.

 (2) A ray from T through C is undeviated since it meets the surface normally.

(3) A ray from T, passing through F, is refracted parallel to the principal axis.

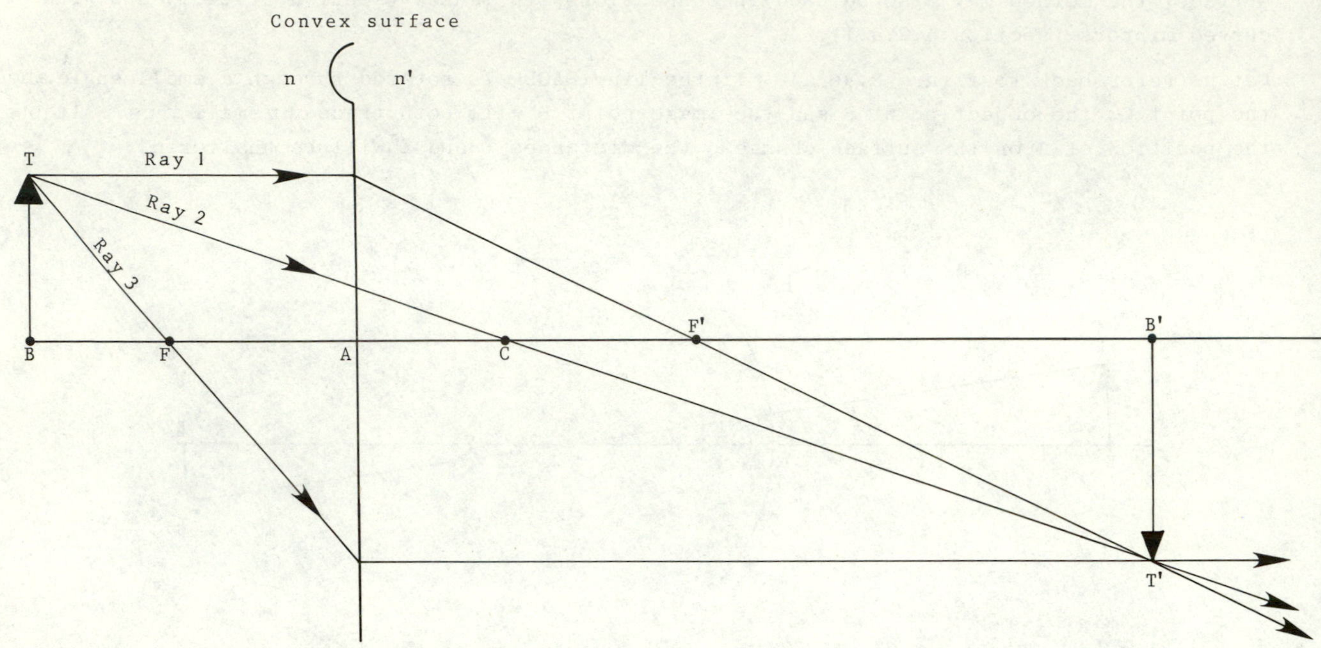

Fig. 2.45 Graphical construction.

Similar constructions are employed for all other cases involving convex or concave surfaces.

EXERCISES

1 Calculate the angle of refraction for a ray incident at 30^0 on a block of crown glass of refractive index 1.523 .

Ans. 19.47^0 .

2 The refractive index of a glass block relative to air ($n_a = 1.000\,29$) is $_an_g = 1.5$. Find the absolute refractive index of the glass.

Ans. $n_g = 1.500\,435$.

3 A ray of yellow light is incident at 40^0 on the surface of (i) glass of refractive index 1.5 , (ii) diamond of refractive index 2.417 . Find the deviation of the ray in each case.

Ans. (i) 14.63^0, (ii) 24.58^0.

4 Find the lateral displacement of a monochromatic ray incident at 45^0 on a glass cube of edge 10 cm and refractive index 1.5 . What happens to the displacement as the edge length of the cube approaches zero?

Ans. 3.29 cm. The lateral displacement approaches zero: this is applied to thin lenses in Chapter 3 .

5 A parallel beam of monochromatic light strikes a parallel sided glass block at a non-zero angle of incidence. Part of the beam reflects off the top surface (angle of incidence = angle of reflection) and part off the bottom. Show that the two beams going back into the air at the upper surface are parallel.

Ans. See W.P.O., page 12, question 3.

6 A coin on the bottom of a 4 m deep swimming pool appears to be 3 m below the surface of the water when viewed from above. What is the absolute refractive index of the water if the absolute refractive index of air is 1.000 29?

Ans. 1.333 72 .

7 Explain what is meant by the term reduced distance. A small stain on the table beneath a whisky tumbler is viewed from above. If the base of the glass ($n_g = 1.5$) is 2 cm thick, and the tumbler is filled to a depth of 3 cm with whisky ($n_w = 1.351$), find the apparent position of the stain below the surface of the whisky.

Ans. 3.55 cm.

8 Define the term critical angle with respect to an air-glass boundary. A point source of light lies at the bottom of a pool of water 2 m deep. A thin cork mat floats on the water with its centre immediately above the source. Find the shape and minimum size of the mat such that an observer above the surface cannot see the light; assume $n_w = 1.33$.

Ans. Circular shape of diameter 4.54 m. See W.P.O., page 12, question 4 .

9 ABCD are corners on one face of a glass cube. One face of the cube, with edge AB, is covered with white absorbent paper and is brightly illuminated with monochromatic yellow light; a dark light absorbing screen covers the face DA. E is a point on the face BC. When the white paper is moistened with a liquid, an observer looking into the face BC along a direction FE sees a boundary between light and dark areas. Explain this. When the liquid on face AB is water, $n_w = 1.33$, the angle FEC is $37^0 23'$; when the liquid is glycerol the angle FEC is $45^0 15'$. Calculate the refractive indices of the glass and the glycerol.

Ans. Glass: $n = 1.56$. Glycerol: $n = 1.39$. See W.P.O., page 13, question 5.

10 Derive equation (2.15).

Ans. See W.P.O., page 15, question 8.

11 A ray of monochromatic light passes through a triangular glass prism. In what circumstances is the deviation a minimum? An equilateral prism of principal section ABC transmits a ray of sodium yellow light, the light entering the face AB and leaving through the face AC. When the deviation is a minimum the angle of incidence is 58.63^0. Calculate the refractive index of the glass and the angle of minimum deviation.

Ans. $n_g = 1.707$ and $d_{min} = 57.26^0$. See W.P.O., page 15, question 7.

12 Show that if a prism of refractive index n is to give a minimum deviation d_{min}, its refracting angle a is given by

$$\tan\frac{a}{2} = \frac{\sin(d_{min}/2)}{n - \cos(d_{min}/2)} .$$

Ans. See W.P.O., page 17, question 9.

13 Monochromatic light is incident at 30^0 on one face of an equilateral prism on the side of normal nearer the base. If the refractive index of the prism is 1.52, trace its path

through the prism and calculate the total deviation of the emergent ray.

Ans. 120^{0}.

14 A monochromatic ray of yellow light from a helium gaseous discharge lamp just emerges from the second face of a prism with apical angle 50^{0} and refractive index 1.535 for this wavelength. What is the angle of incidence at the first face of the prism?

Ans. 14.438^{0}.

15 A thin prism displaces a ray of monochromatic light by 18 cm on a tangent scale at 3 m from the prism. Express the power of the prism in prism dioptres and calculate the apical angle if $n_p = 1.523$.

Ans. $P = 6^{\Delta}$. $a = 6.57^{0}$.

16 Show that $1^{\Delta} \simeq 1.75^{0}$.

17 What is the minimum refractive index for a Porro prism?

Ans. $n > \sqrt{2}$.

18 A concave spherical surface separates water, $n_w = 1.333$, and glass, $n_g = 1.533$. Find the image of a point source in the water on the principal axis 1 m from the surface if r = -5 cm.

Ans. $\ell' = -28.75$ cm.

19 The left end of a long glass rod of refractive index 1.7 is ground with a convex spherical surface of radius 3.5 cm. An object 1 cm high is situated in air on the principal axis 20 cm from the vertex of the surface. Calculate (i) the focal lengths of the surface, (ii) the power of the surface, (iii) the image position, and (iv) the image size.

Ans. (i) f = -5 cm and f' = +8.5 cm, (ii) F = +20 D, (iii) $\ell' = +6.67$ cm, (iv) h' = -0.25 cm.

20 A goldfish's eye is 4 cm from the surface of a spherical goldfish bowl of radius 10 cm. Neglecting the thickness of the glass, find the apparent position and linear magnification of the eye to an observer if the refractive index of the water is 1.333 .

Ans. $\ell' = -3.334$ cm and m = 1.111 .

21 A spherical surface of radius +5 cm is polished on the end of a glass rod of refractive index 1.550 . Find its power when placed in (i) air, (ii) water of refractive index 1.333. (iii) oil of index 1.550 , and (iv) liquid of index 1.650 .

Ans. (i) +11 D, (ii) +4.34 D, (iii) 0 D, (iv) -2 D.

22 A parallel pencil of monochromatic light enters the rod in question 21 from the side with the index stated in parts (1) to (iv). Where will it focus in each case?

Ans. (i) f' = +14.09 cm, (ii) f' = +35.71 cm, (iii) +∞ , (iv) f' = -77.5 cm.

23 Show that the curvature of a circle is 1/r, where r is the radius of the circle.

Ans. See W.P.O., page 19, question 1.

24 A spherical refracting surface, separating air from glass, forms a real image twice the size of the real object. If the image is 6 times as far from F' as the object is from F, find the refractive index of the glass.

Ans. 1.5 .

3 THIN LENSES

3.0 INTRODUCTION

Undoubtedly lenses are the most widely used optical devices. We can describe a lens as being an optical system of two refracting interfaces where one or both of these is curved. This description is often referred to as a simple lens. Compound lenses are made up from several component parts and contain more than two interfaces.

A lens may be classified according to the nature of its surfaces. The most common group of lenses have spherical surfaces, that is, their surfaces are portions of the surfaces of spheres. Other important forms of lenses include plane, cylindrical, and toroidal surfaces.

Alternatively, a lens may be described as converging or diverging based on its effect on light rays, and may be classed as thin or thick depending on whether its thickness is neglible or not.

In the present chapter we shall be considering the theory of thin lenses only. Thick lenses will be discussed in Chapter 4.

3.1 SPHERICAL SURFACED LENSES

As already mentioned in section 2.2.0 , spherical surfaces are comparatively easy to manufacture accurately, and lenses having surfaces of this form are by far the most common. Most of the theory in this chapter is based on spherical surfaced thin lenses. Lenses for which one, or both, surfaces are not spherical will be discussed towards the end of the chapter.

3.1.1 LENS FORMS

A thin lens is one where the aperture is small compared with the radii of curvature of its faces, and where the thickness can be ignored. As a rough guide, in the case of spectacle lenses having a diameter of say 50 mm, we shall be restricted to thicknesses less than about 5 mm. As will be discussed in Chapter 4, certain meniscus forms of thin lens, including contact lenses, are included as thick lenses for a satisfactory treatment of their optical properties.

Figure 3.1 shows different forms of a thin lens, all of which are referred to as *converging* lenses.

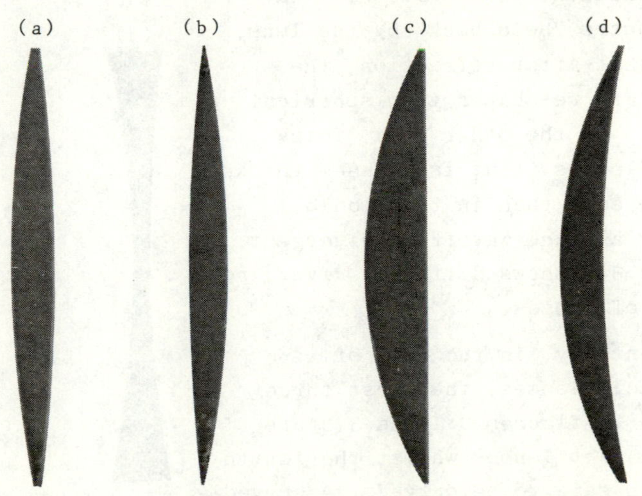

Fig. 3.1 Converging lenses.
(a) Equi-convex. (b) Double or bi-convex
(c) Plano-convex. (d) Meniscus.

It will be noticed that all of these forms are thicker in the middle than at the edge. If these lenses are made of an optically denser material than their surroundings, as is usual, with the velocity of light in the lens being less than in air, then it is evident that all thin-edged lenses will retard light passing through the centre more than light passing through the edge. Figure 3.2 shows this effect with a parallel pencil of rays incident on a biconvex lens. As described in Chapter 1, a parallel pencil of rays consists of waves having plane wavefronts.

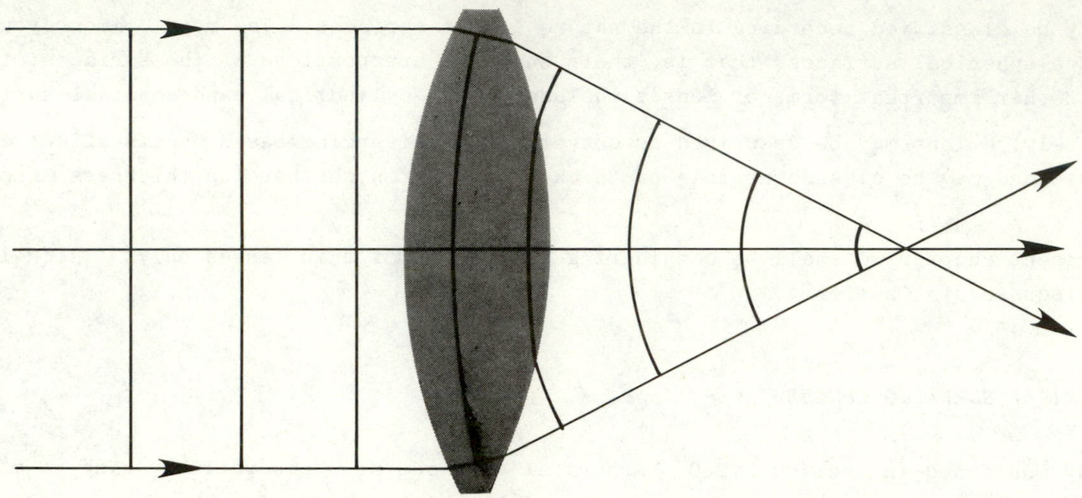

Fig. 3.2 Refraction of a parallel pencil of rays by a converging lens. Note the curvature of the wavefronts is changed at each surface.

Each incident plane wave will have its centre "held back" by the lens, such that after refraction, the waves are re-shaped into spherical waves. On the other hand, thick edged lenses, that is, lenses thicker at the edge than in the middle, render a plane wavefront divergent. Figure 3.3 shows different *diverging* forms of lenses.

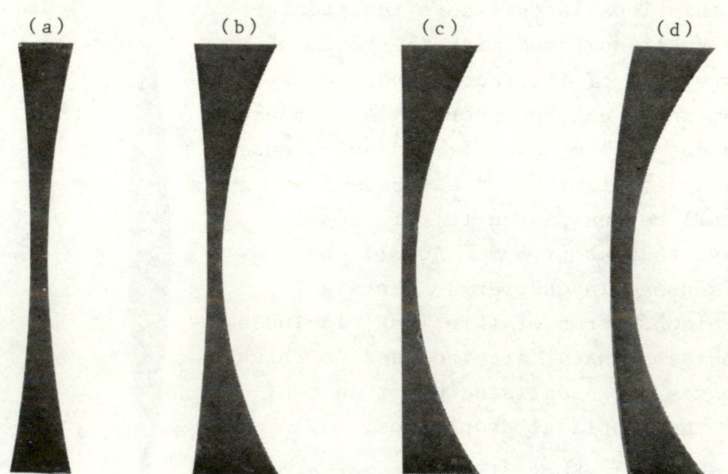

Fig. 3.3 Diverging lenses.
(a) Equi-concave. (b) Double or bi-concave.
(c) Plano-concave. (d) Meniscus.

Incidentally, in the case of spectacle lenses, the first three lenses in figures 3.1 and 3.3 are called *flat* lenses whilst the fourth lens is said to be *curved*. A curved lens is one which is wholly convex on one surface and wholly concave on the other. When the two surfaces of a curved spectacle lens are spherical then the lens form is called *meniscus* which means crescent shaped.

Figure 3.4 depicts the reshaping of wavefronts on passing through a diverging lens.

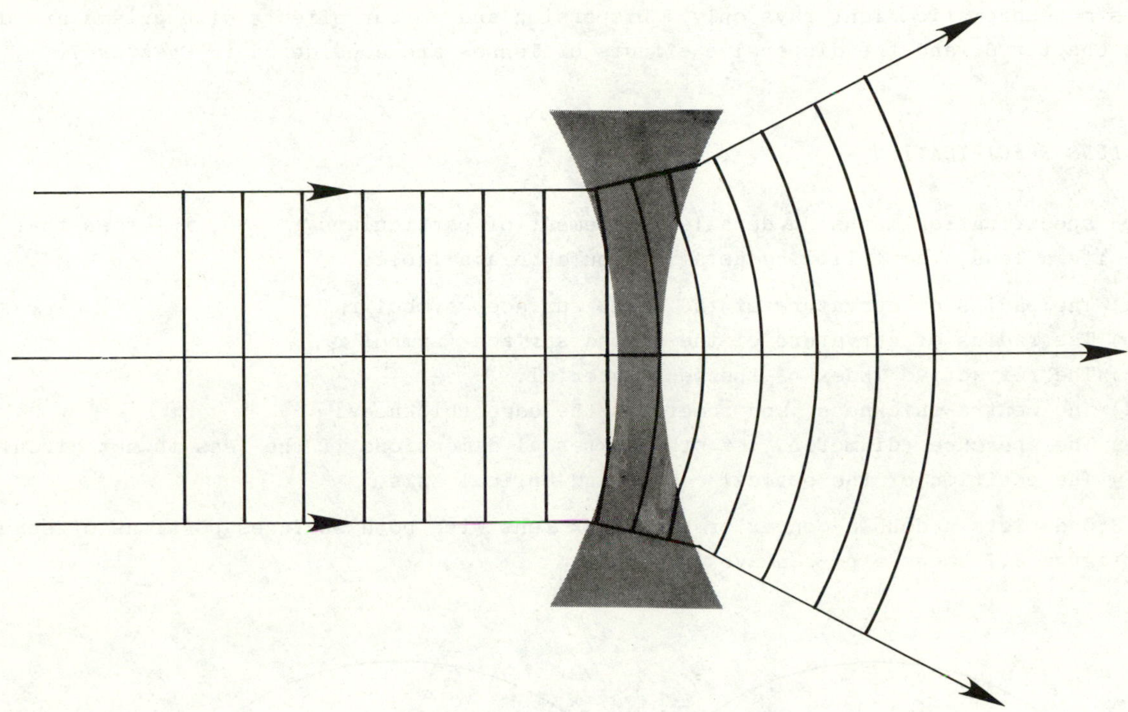

Fig. 3.4 Refraction of a parallel pencil of rays by a diverging lens. Note the divergence is increased at each surface in this lens.

The converging and diverging effects of thin lenses may also be explained in a rather simplified way by representing a lens as a pair of prisms. Figure 3.5(a) shows the way in which a converging lens may be represented as a pair of prisms fitted base to base. For a small region around the centre of the lens the surfaces are parallel and a ray through it is undeviated.

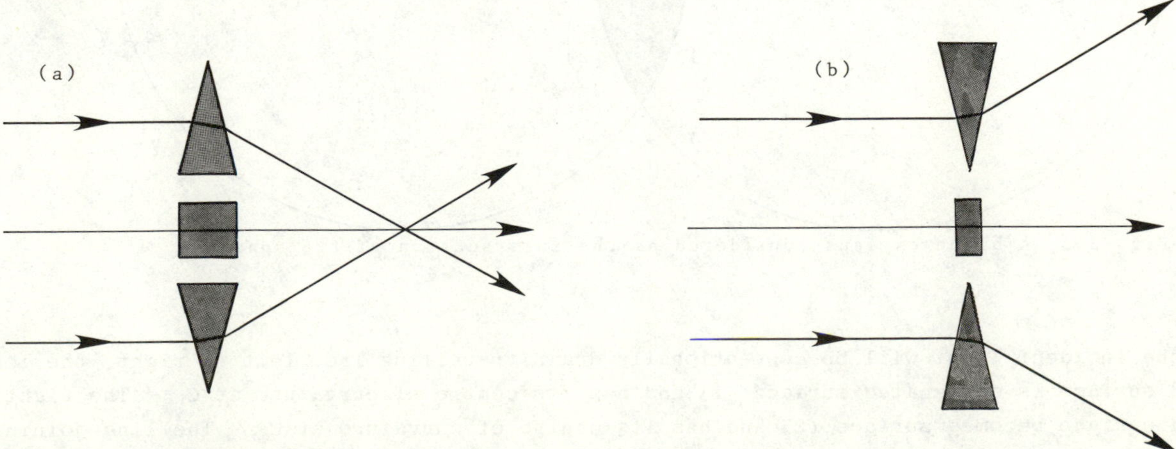

Fig. 3.5 The prismatic representation of converging and diverging lenses.

Figure 3.5(b) illustrates a diverging lens represented as two prisms fitted apex to apex. The small area near the lens centre again acts as a parallel sided piece of material.

It was mentioned in section 2.1.10.0 that the dispersing effect of a prism on white light was being ignored. Dispersion also occurs with lenses, but will also, for the time being, be ignored. Hence, unless otherwise stated, we must note that the present theory of lenses involves monochromatic light rays only. Dispersion and colour effects with prisms are dealt with in Chapter 6, and the dispersive effects of lenses are considered in Chapter 7.

3.1.2 LENS SPECIFICATION

The word specification means " a detailed statement of particulars ". So, in order that we may specify a lens, the following details should be available.

(a) The radius of curvature of the first surface, symbol r_1.
(b) The radius of curvature of the second surface, symbol r_2.
(c) The refractive index of the lens material.
(d) The centre thickness, and sometimes the edge thickness.
(e) The aperture (diameter), or other external dimensions if the lens is not circular.
(f) The position of the optical centre and optical axis.

Figure 3.6 depicts a double-convex or bi-convex lens with both surfaces portions of spheres, whilst figure 3.7 shows a bi-concave lens.

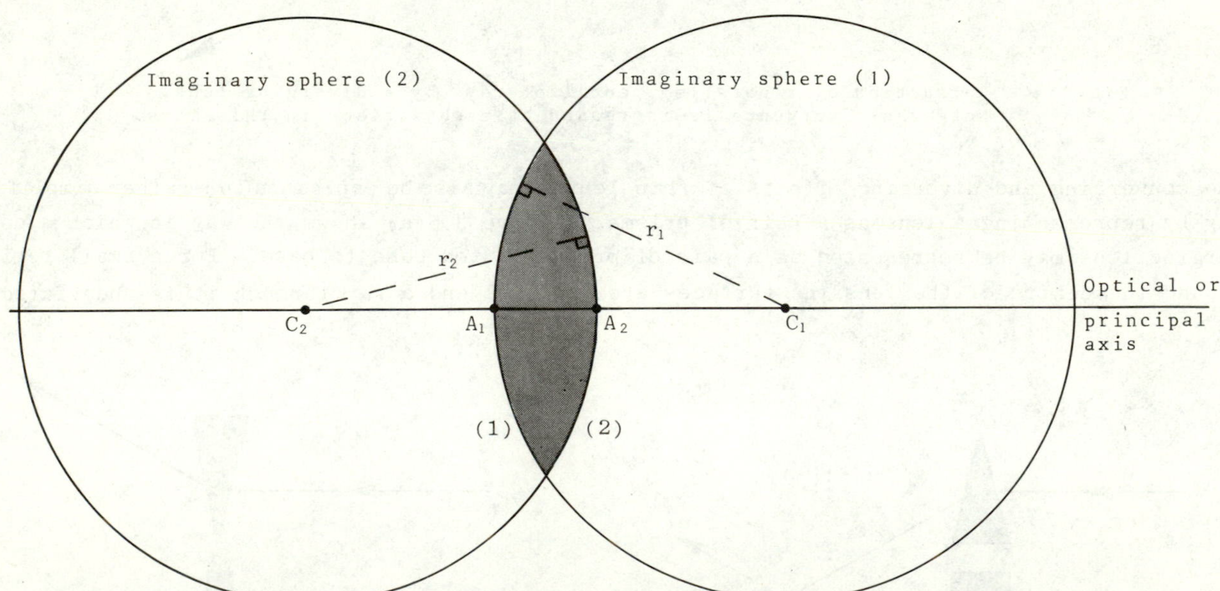

Fig. 3.6 A biconvex lens considered as the intersection of two spheres.

As the incident light will be conventionally drawn travelling from left to right, the left hand surface is designated surface (1) and has its centre of curvature at C_1. The right hand surface becomes surface (2) and has its centre of curvature at C_2. The line joining C_1 and C_2, produced in either direction if required, is termed the *optical axis* or *principal axis* of the lens. This line crosses the front surface at the point A_1 called the *front vertex* of the lens, and cuts the back surface at A_2 called the *back vertex* of the lens. The separation t of the vertices A_1 and A_2 is the *centre thickness* of the lens. Figure 3.7 shows these features for a diverging lens.

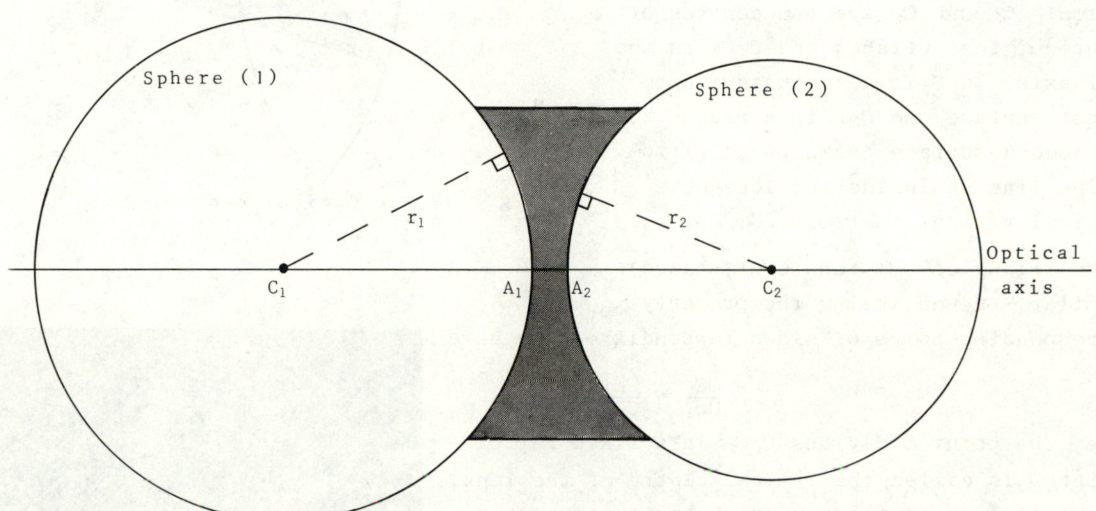

Fig. 3.7 The surfaces of a bi-concave lens imagined as parts of two spheres.

With modern ophthalmic lenses having such a wide variety of shapes it is more than likely that the optical axis will not coincide with the geometrical axis which passes through the centre of the lens aperture. This is shown in figure 3.8 for a converging meniscus form.

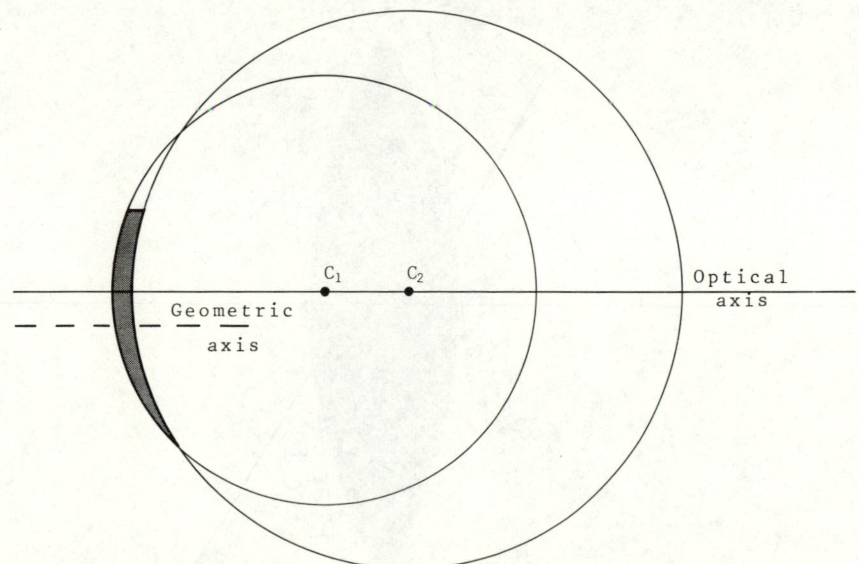

Fig. 3.8 A lens with its optical axis non-coincident with the axis through the centre of the aperture (the geometric axis).

Referring now to figure 3.9 it is seen that a ray of light travelling along the optical axis will strike both lens surfaces normally and will therefore be undeviated. However, there are other directions along which a ray may travel and yet suffer no deviation. C_1 and C_2 are the centres of curvature of the surfaces and C_1C_2 is the optical axis. C_1D is a radius drawn to the first surface and C_2E is a radius to the second surface drawn parallel to C_1D. The line DE inside the lens cuts the optical axis at the point O.

Fig. 3.9 The optical centre, O.

Now, the triangles C_1DO and C_2EO (shaded) are similar. Hence, using the property of corresponding pairs of sides (appendix 4), we have

$$\frac{OC_1}{OC_2} = \frac{C_1D}{C_2E} = \frac{r_1}{r_2}.$$

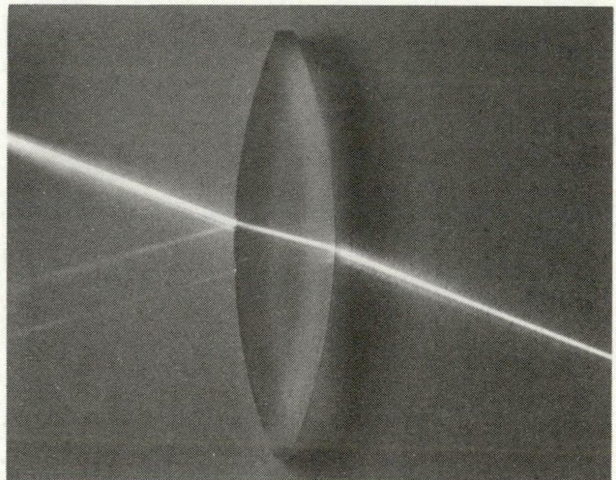

That is, the point O divides C_1C_2 in a fixed ratio. The point O is called the *optical centre* of the lens. In the case of an equiconvex or an equiconcave lens the two radii of curvature, r_1 and r_2, are identical. Consequently, $OC_1 = OC_2$ and the optical centre is at the geometrical centre of the lens. In general, however, a lens will not be equicurved and the radii will be different values. Figure 3.10 shows a ray RDER' which passes through O. The lines DP and EQ

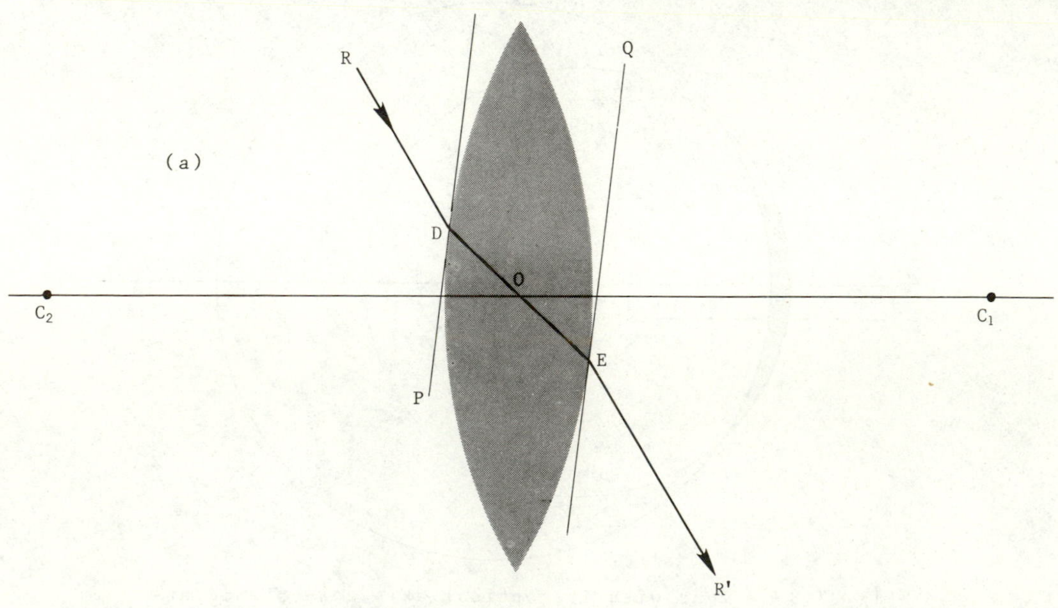

Fig. 3.10 A ray through the optical centre is undeviated.

are tangents to the two surfaces at D and E, and they will be parallel to each other. This follows from the fact that C_1D is parallel to C_2E and a tangent is perpendicular to the radius. Now we have seen (section 2.1.6) that any ray entering another medium will emerge in a direction parallel to that on entering provided that the surfaces on entry and emergence are also parallel. Thus, it follows that any ray passing through the optical centre is laterally displaced, but not deviated. As we are considering the case of thin lenses only, the displacement is so small that it can be ignored.

So, we can define the optical centre O of a lens as that point on the lens axis, such that for an undeviated ray, the part of the ray within the lens passes through the point or, when projected, appears to pass through the point.

As already stated, the optical centre of an equicurved lens lies inside the lens midway between the vertices A_1 and A_2. For thin lenses, A_1, A_2, and O may be considered to be a single point at O. However, with other lens forms the optical centre will be elsewhere on the optical axis, and may even be outside the lens altogether. Figure 3.11 shows the positions of the optical centre O for various forms of lenses where the thickness t is not neglible.

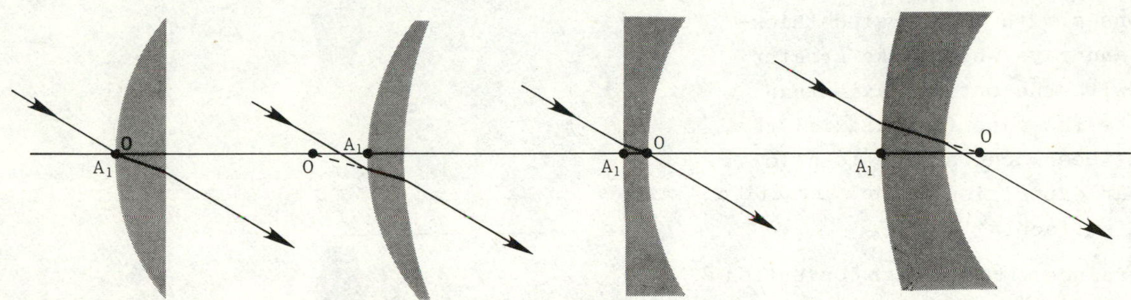

Fig. 3.11 The position of the optical centre O for various lens shapes.

It can be shown that the distance of the optical centre from the front vertex is given by $A_1O = tF_2/(F_1 + F_2)$, where t is the centre thickness and F_1 and F_2 are the surface powers, see equation (4.29).

3.1.3 *PRINCIPAL FOCI AND FOCAL LENGTHS OF A LENS*

We recall (section 2.2.2) that a wide aperture spherical refracting surface possesses the defect known as spherical aberration, and this shows as the inability of the surface to converge all rays to a single point focus. We also recall that in order that we may ignore this troublesome defect, and also simplify the theory, only paraxial rays are considered. The same defect is present with spherical surfaced lenses of appreciable aperture. For the present theory we shall confine ourselves to paraxial rays; that is, rays which lie close to the optical axis and make small angles with it. Hence, we are making the approximation that all paraxial rays from a point object are focused to form a point image.

For any lens there are two points on the optical axis which need special attention. Figure 3.12(a) shows a pencil of paraxial rays from an object at an infinite distance from a converging lens. As such the pencil of rays will be effectively parallel on reaching the lens. These rays are refracted through the point F' which is called the *second principal focus* of the lens. The special distance OF' is called the *second* or *image focal length* of the lens and is given the symbol f'.

Similarly, in figure 3.12(b), there is the point F from which paraxial rays, after refraction by the lens, emerge parallel to the axis. F is referred to as the *first principal focus* of the lens, and the distance OF is referred to as the *first* or *object focal length* of the lens and is given the symbol f.

If the lens is thin, and the medium on each side of the lens is the same, then the numerical values of f and f' will be identical (see also section 3.1.4.3). Note that the diagrams show lenses with exaggerated thicknesses and rays which make greater angles with the optical axis than should be the case for paraxial rays. This has been done simply to allow the clear exposition of the refraction at each surface.

In accordance with the Sign Convention in section 2.2.3, the numerical value of the second focal length f', measured from O to F', is given a *positive* sign whereas the value of the first focal length f, measured from O to F, is given a *negative* sign.

That is $f = -f'$.

In the case of a diverging lens, the principal foci are both virtual points as the refracted rays do not actually pass through them. This is exactly analogous with the definitions of the first and second focal lengths of single surfaces in section 2.2.4.

Figure 3.13(a) shows the point F, the first principal focus of a diverging lens, together with the first or object focal length f. As before, if the lens is thin and the media on each side of the lens are identical in refractive index, then $f = -f'$.

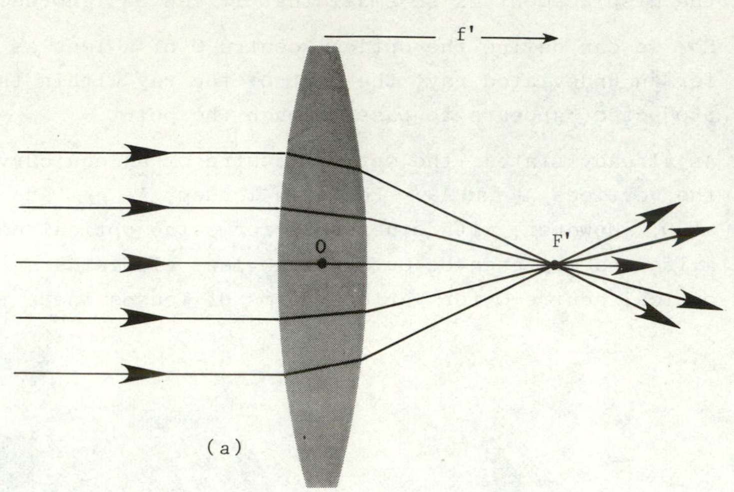

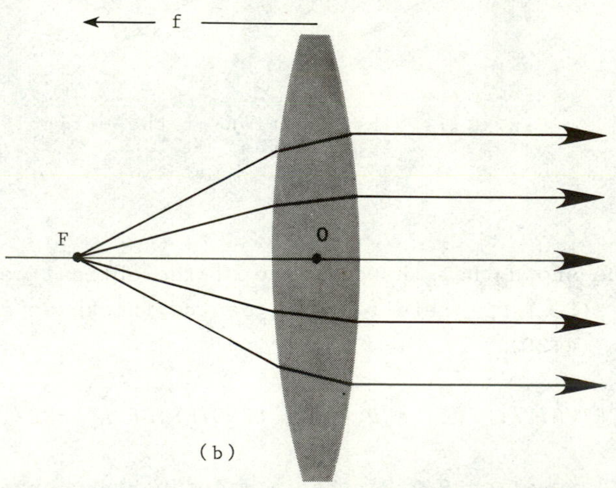

Fig. 3.12 A converging lens.
(a) The second principal focus, F', and the second focal length, f'.
(b) The first principal focus, F, and the first focal length, f.

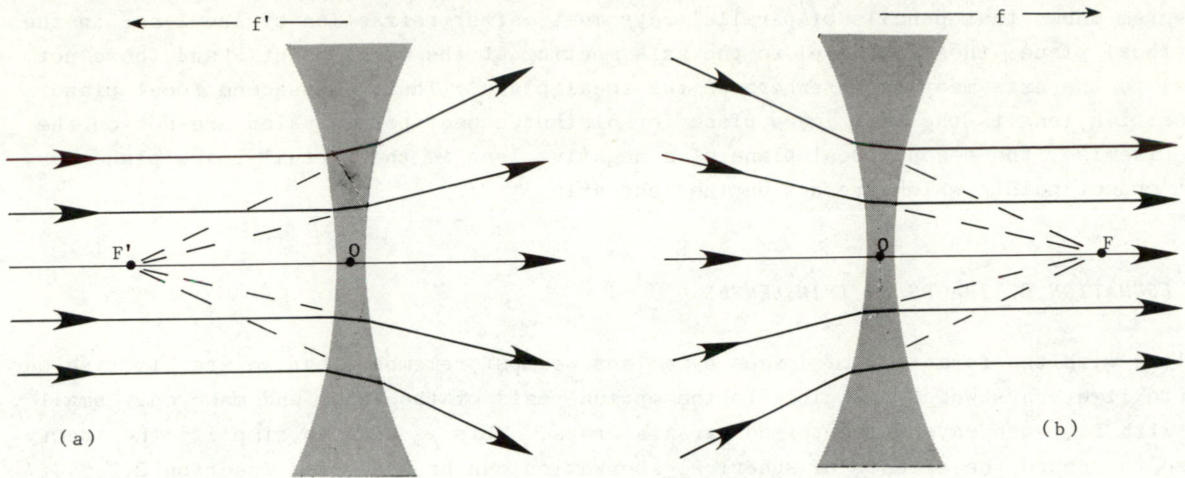

Fig. 3.13 A diverging lens.
(a) The second principal focus, F', and the second focal length, f'.
(b) The first principal focus, F, and the first focal length, f.

When *"the focal length"* of the lens is stated without specifying if it be the first or second focal length, it must always be inferred that this means the *second focal length* f' of the lens. Thus, using ordinary everyday language, a *converging lens* may be described as having a *positive focal length* since the value of f' is positive. For this reason such a lens is referred to as a *positive lens*. Similarly, a *diverging lens* is described as having a *negative focal length* since the value of f' is negative. This type of lens is, therefore, called a *negative* lens.

The planes through the principal foci, drawn perpendicular to the optical axis of a lens, are called the *focal planes* of the lens. Use can be made of them in problems involving the tracing of rays through a lens, or lens system. The focal plane through F' of a positive lens is shown in figure 3.14 .

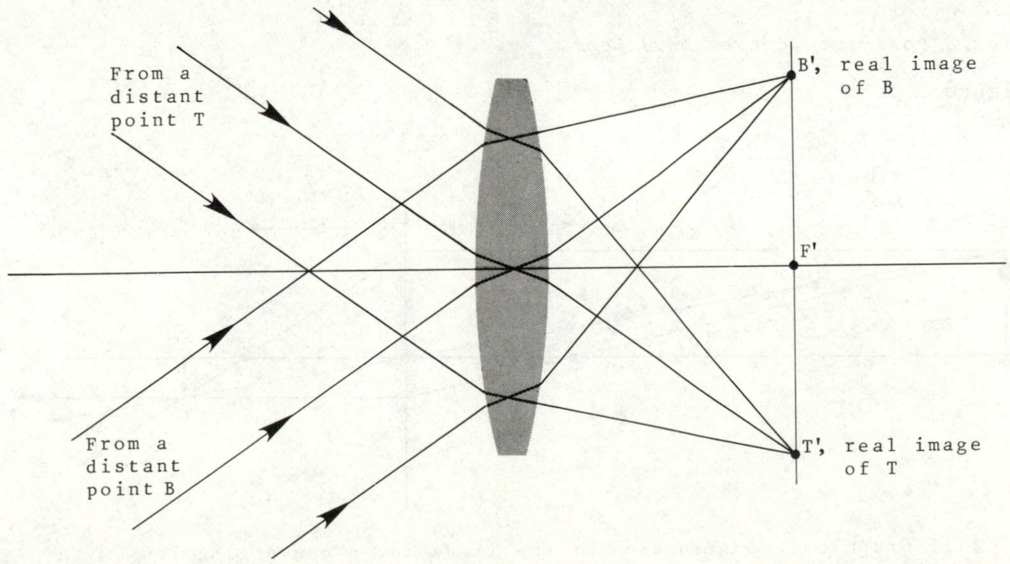

Fig. 3.14 Images of distant off-axis object points.

The diagram shows that pencils of parallel rays meet, after refraction by the lens, in the second focal plane, those parallel to the axis meeting at the focal point F' and those not parallel to the axis meeting elsewhere on the focal plane. Thus, the second focal plane of a positive lens is the real image plane for distant object points which are not on the axis. Likewise, the second focal plane of a negative lens is the virtual image plane for distant object points which are not on the lens axis.

3.1.4 FORMATION OF IMAGES BY THIN LENSES

In dealing with the formation of images by a lens we must remember that we are limiting our theory to light rays which are close to the optical axis of the lens, and make only small angles with it, such rays being termed paraxial rays. This is done to simplify the theory and also to ensure the effects of spherical aberration can be neglected (section 2.2.2).

Essentially, a lens is a device used to alter the convergence or divergence of a pencil of rays. More precisely a lens affects the vergence, or curvature of the wavefronts, of the light incident upon it.

When a lens forms an image, the position and nature of the image, and its size relative to the object size will depend on the lens itself, and on the object position. Two methods are available for the evaluation of the image. These are

(a) by scale drawing - called a graphical construction,
(b) by calculation - called the analytical method, using formulae.

3.1.4.1 *Graphical Construction for Images with a Thin Lens*

This scale drawing method involves the use of *construction rays*. These are selected rays of light which take definite predictable paths following refraction. For simplicity the rays chosen will be confined to one plane only, the plane containing the object and the optical axis of the lens. Rays in other planes are termed *skew rays*, and these are beyond the scope of this book.

Procedure for a Positive (converging) Lens

Refer to figure 3.15 .

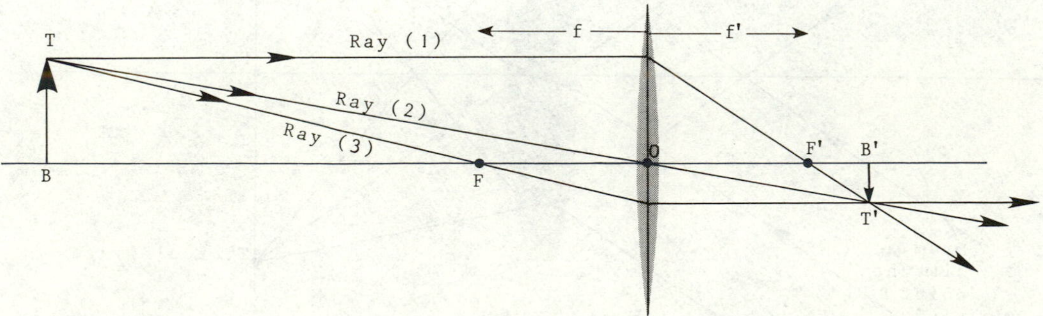

Fig. 3.15 Graphical construction of the image for a converging lens with the object at a finite distance from the lens and outside F.

(a) Using a convenient horizontal scale, draw the optical axis of the lens and mark in the positions of the lens, and the principal foci F and F'. The lens has a straight line through the optical centre, perpendicular to the axis. This line represents the *principal plane* of the lens and rays are shown to deviate there. This assumes a negligible thickness and also simplifies the construction.

(b) Mark in the upper half of the object, using as large a vertical scale as possible. bearing in mind the probable size of the image.

(c) Draw any *two* of the following construction rays from the tip of the object. After refractions the intersection of these two rays indicates the tip of the image.

Construction Rays

(1) A ray from T, parallel to the optical axis, is refracted through F'. This follows from figure 3.12(a).

(2) A ray from T, passing through F, is refracted parallel to the optical axis. In effect, ray (3) is the reverse of ray (1) and follows from figure 3.12(b).

(3) A ray from T, passing through the optical centre O, is undeviated.

The following constructions show the variety of images formed by a positive (converging) lens. The details of the images are summarised in Table 3.1 .

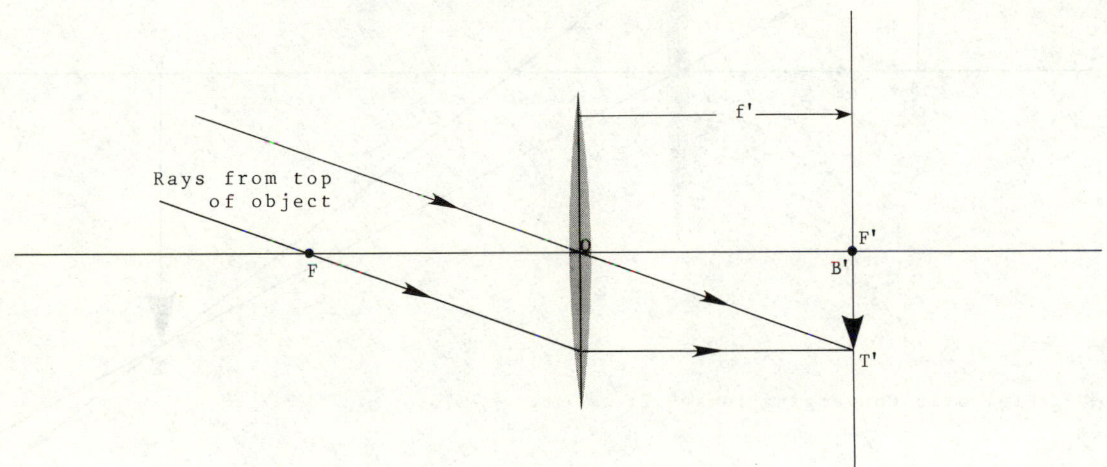

Fig. 3.16 A converging lens - object standing on optical axis at $-\infty$.

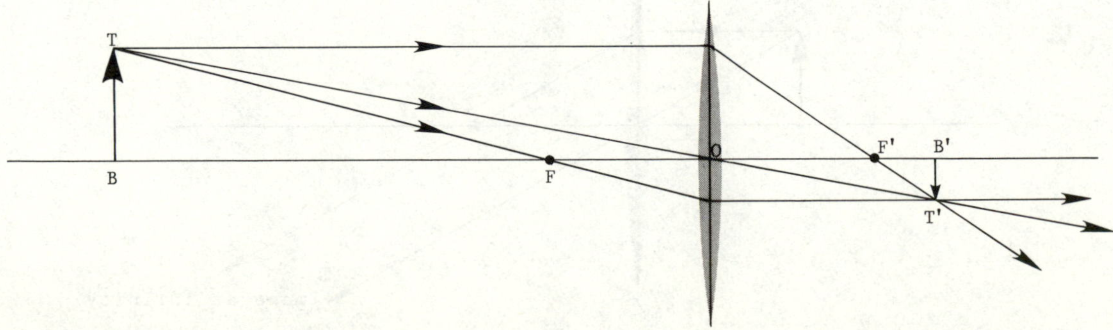

Fig. 3.17 A converging lens - object further from the lens than 2f but closer than infinity.

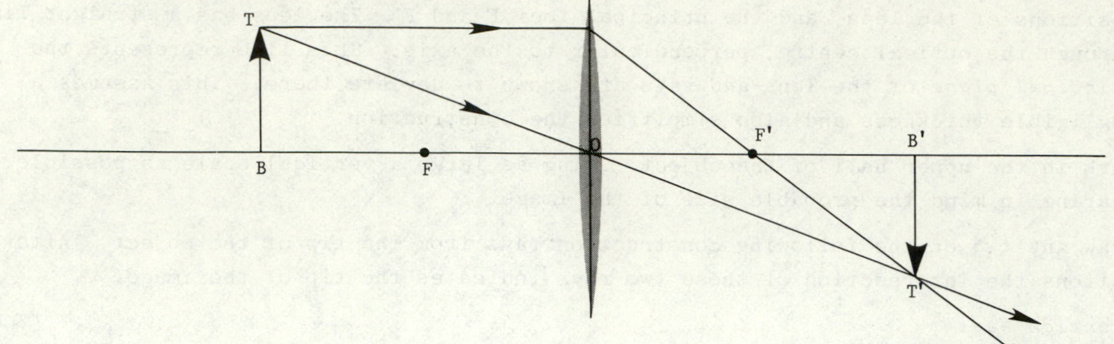

Fig. 3.18 Object at a distance of 2f from the converging lens.

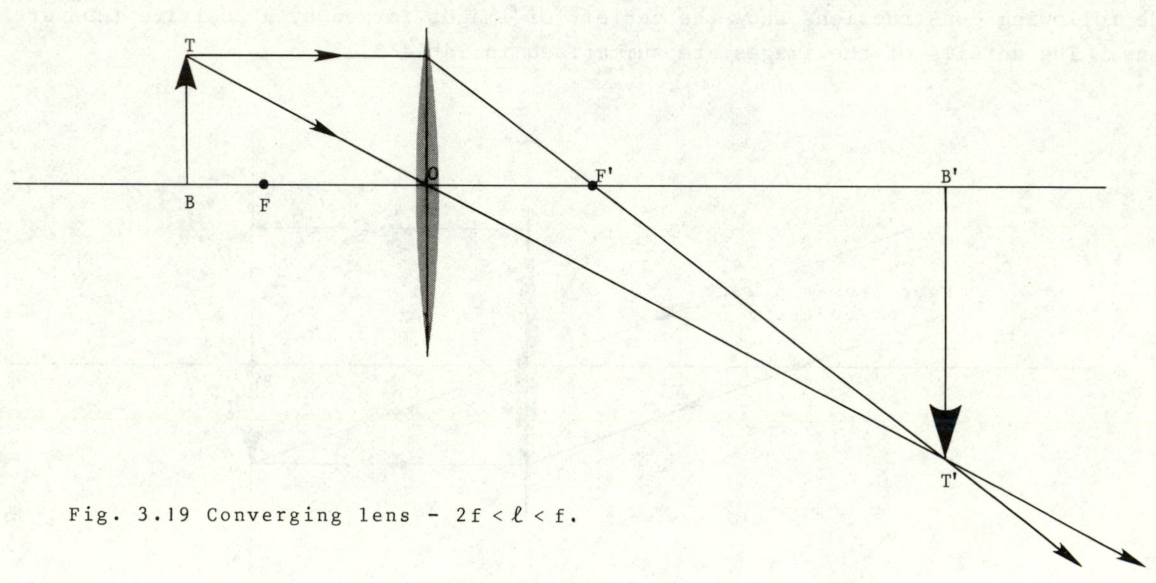

Fig. 3.19 Converging lens - $2f < \ell < f$.

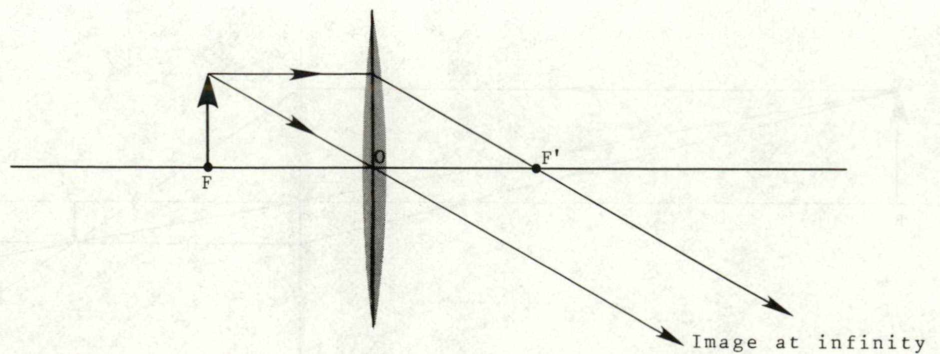

Fig. 3.20 Converging lens - object at F.

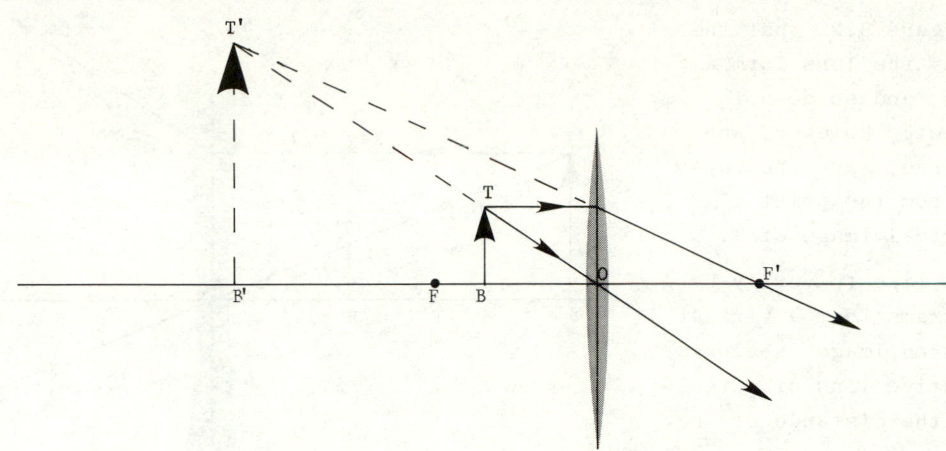

Fig. 3.21 Converging lens - object closer than F. Note that the emergent rays do not intersect; they only appear to come from what is a virtual image.

Table 3.1 Summary of image details.

Real object position	Image position	Nature	Size relative to object
At infinity	At F'	Real, inverted	-
Between infinity and 2f	Between f' and 2f'	Real, inverted	Diminished
At 2f	At 2f'	Real, inverted	Same size
Between 2f and f	Between 2f' and infinity	Real, inverted	Enlarged
At F	At infinity	-	-
Between F and O	Same side of lens as object	Virtual, upright	Enlarged

Procedure for a Negative (diverging) Lens

(a) As described for a positive (converging) lens, draw the optical axis of the negative lens and mark in the positions of the lens, and its principal foci F and F', using a convenient horizontal scale.

(b) Mark in the upper half of the object, using as large a vertical scale as possible, bearing in mind the probable size of the image.

(c) Draw any *two* of the following construction rays from the tip of the object. Following refractions, the intersection of these ray directions indicates the tip of the image.

Construction Rays

(1) A ray from T, parallel to the optical axis, is refracted as if from F'. This follows from figure 3.13(a).

(2) A ray from T, passing through the optical centre O, is undeviated.

(3) A ray from T, directed towards F, is refracted parallel to the optical axis. This follows from figure 3.13(b).

It is seen in figure 3.22 that the rays refracted by the lens form a diverging pencil, and so do not actually intersect. However, when received by the eye, say, the rays appear to come from the point T', which is the virtual image of T.

The reader might like to verify by a series of diagrams that a virtual, upright, diminished image is always formed by a negative lens of a real object whatever the distance of the object from the lens.

Table 3.2 summarises the image characteristics.

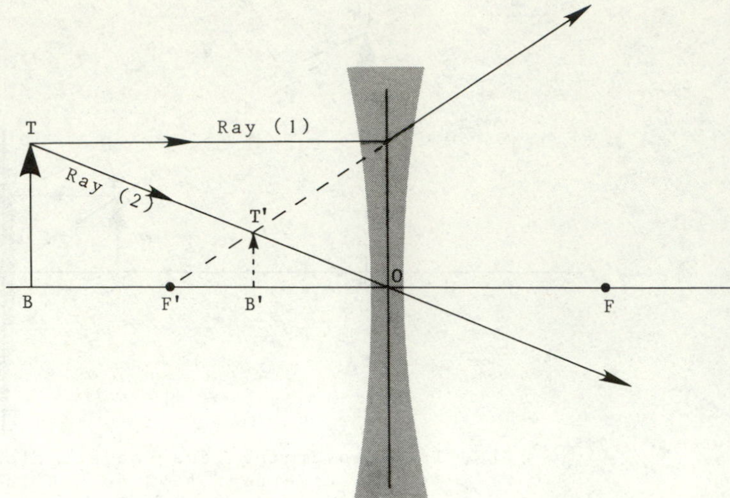

Fig. 3.22 A diverging lens - graphical construction.

Table 3.2 The characteristics of an image formed by a negative lens

Real object position	Image position	Nature	Size relative to object
At infinity	At F'	Virtual	
Between infinity and O	Between F' and O	Virtual, and upright	Diminished

3.1.4.2 *Calculations for Images - The Conjugate Foci Formula*

The position, nature, and magnification of the images formed by thin lenses may be determined by calculation using lens formulae. Compared to the graphical method the analytical method gives a greater degree of accuracy and the answers are generally obtained more quickly. However, a ray diagram drawn to scale provides a "picture" of the problem and this is always a help in obtaining a full appreciation of the principles involved. Perhaps the best solution is that, wherever possible, a calculated answer should be accompanied by a graphical solution, even if this is only done roughly.

Since a lens is two surfaces separated by some medium other than air, we might expect the formulae associated with single refracting surfaces to be involved in the analysis of the lens. This is indeed the case. However, since there are two surfaces, it will be necessary to use subscript numbers 1 and 2 to identify terms related to the first and second surface, respectively. Thus, ℓ_1, ℓ_2, ℓ_1', and ℓ_2' will refer to the object and image distances of the first and second surfaces. Clearly, similar terminology will refer to the object and image vergences to the two surfaces. So, $\bar{L}_1$ and $\bar{L}_2$ will be the symbols for the object (or incident) vergences to the first and second surface respectively. Note the tendency to drop the more exact terminology of "reduced vergence". This is a common event in practice. However, the use of the reduced bar notation will always distinguish between vergence and reduced vergence. Finally, image or emergent vergences and surface powers will also bear a labelling subscript.

In deriving the general formula we shall consider a thin lens as being one refracting surface followed by a second refracting surface with a negligible separation between them. Figure 3.23 shows a biconvex thin lens drawn as two separate refracting surfaces.

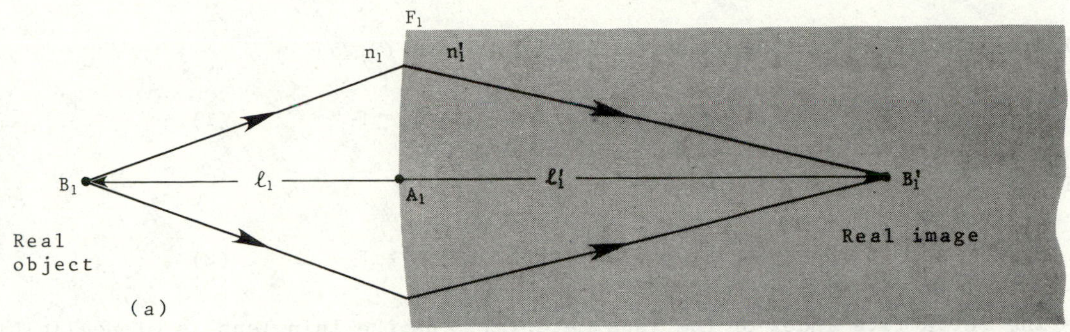

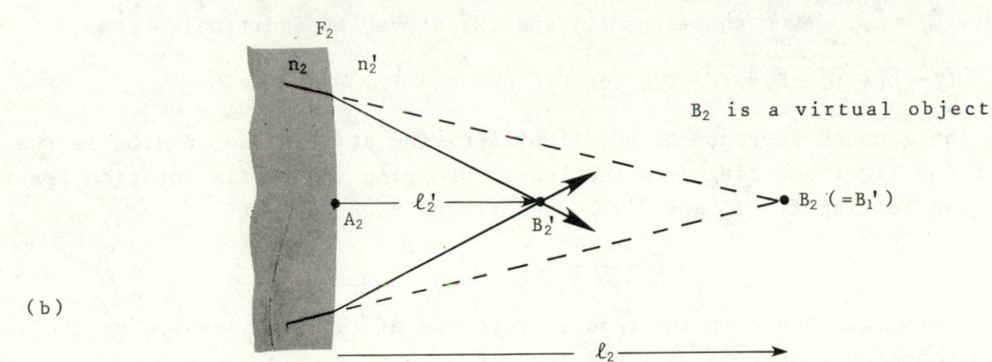

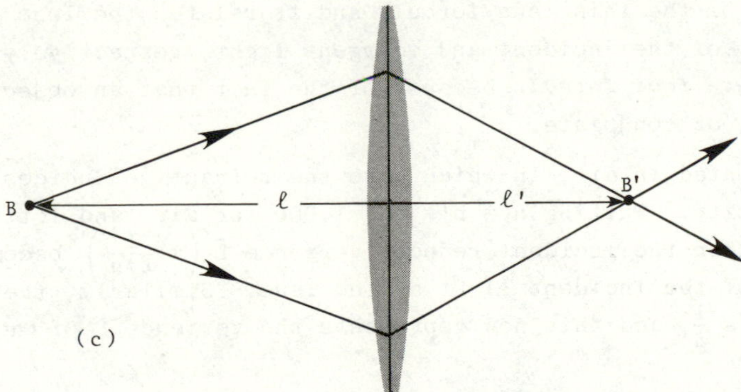

Fig. 3.23 A thin converging lens considered as two successive refracting surfaces separated by a neglible thickness of glass. Note that in real objects and images the rays intersect. A real image can be formed on a screen.

In figure 3.23(a), B_1' is the "image" formed by refraction at the first surface. This acts as the "object" B_2 for the second surface refraction, figure 3.23(b), forming the final image B_2'. It should be noted in figure 3.23(c) that the two separate surface refractions are shown as a single refraction taking place in one plane. As was stated in section 3.1.4.1, this plane is referred to as the principal plane of the thin lens. This procedure is nothing more than a convenient simplification. Here the object and image for the lens are denoted by B and B' respectively. Clearly, B is the same as B_1, the object for the first surface, and B' is the same as B_2', the image formed by refraction at the second surface.

For refraction at the first surface we have, using equation (2.25)

$$\frac{n_1'}{\ell_1'} - \frac{n_1}{\ell_1} = \frac{n_1' - n_1}{r_1}, \quad \text{or in reduced vergence notation} \quad \bar{L}_1' - \bar{L}_1 = F_1 \qquad (1).$$

For refraction at the second surface we have

$$\frac{n_2'}{\ell_2'} - \frac{n_2}{\ell_2} = \frac{n_2' - n_2}{r_2}, \quad \text{or in reduced vergence notation} \quad \bar{L}_2' - \bar{L}_2 = F_2 \qquad (2).$$

Now, $n_1' = n_2 =$ the refractive index of the lens material. As the thin lens is of negligible thickness, the reduced vergence of the light emerging from surface (1) inside the lens will be the same value as the reduced vergence of the light arriving at surface (2) inside the lens; i.e. $\bar{L}_1' = \bar{L}_2$. From equations (1) and (2) above, by addition we get

$$\bar{L}_1' - \bar{L}_1 + \bar{L}_2' - \bar{L}_2 = F_1 + F_2, \quad \text{or } \bar{L}_2' - \bar{L}_1 = F_1 + F_2.$$

Now, $\bar{L}_1$ is the reduced vergence of the light arriving at the lens, and $\bar{L}_2'$ is the reduced vergence of the light emerging from the lens. Dropping the suffix notation for the reduced vergences, and letting $\ell_2' = \ell'$ and $\ell_1 = \ell$, see figure 3.23, we get

$$\bar{L}' - \bar{L} = F_1 + F_2.$$

$\bar{L}' - \bar{L}$ represents the change in the reduced vergence of the light caused by the lens, and hence, this represents the *power of the lens*, F. That is,

$$\bar{L}' - \bar{L} = F \qquad (3.1) \qquad \text{where} \qquad F = F_1 + F_2 \qquad (3.2).$$

Equation (3.1) is the general form of the thin lens formula and it relates the lens power F with the reduced vergences $\bar{L}$ and $\bar{L}'$ of the incident and emergent light, respectively, at the lens. It is called the *conjugate foci formula* because of the fact that an object and its real image are interchangeable, or conjugate.

In many cases the lens will be situated in air, in which case the refractive indices of the object and image spaces are both unity. Putting $n_1 = n_2' = n_a = 1.000$ for air, and letting $n_1' = n_2 = n_g$ for the lens material, then the incident reduced vergence $\bar{L}\left(= \frac{1}{\ell/n_1}\right)$ becomes $\frac{1}{\ell}$ and now represents the vergence L of the incident light at the lens. Similarly, the emergent reduced vergence $\bar{L}'\left(= \frac{1}{\ell'/n_2'}\right)$ becomes $\frac{1}{\ell'}$, and this now represents the vergence L' of the emergent light from the lens. Hence, in air

$$\bar{L} = L = \frac{1}{\ell} \qquad \text{and} \qquad \bar{L}' = L' = \frac{1}{\ell'}.$$

When the lens is in air, see figure 3.24, the conjugate foci formula may be written in its simpler form

$$L' - L = F \qquad (3.3).$$

Also, from equation (3.2) the lens power $F = F_1 + F_2$, where F_1 is the front surface power and F_2 is the back surface power, and from equation (2.26) we have the following relationships

$$F_1 = \frac{n_1' - n_1}{r_1} = \frac{n_g - 1}{r_1} \qquad (3.4a) \qquad \text{and} \qquad F_2 = \frac{n_2' - n_2}{r_2} = \frac{1 - n_g}{r_2} \qquad (3.4b).$$

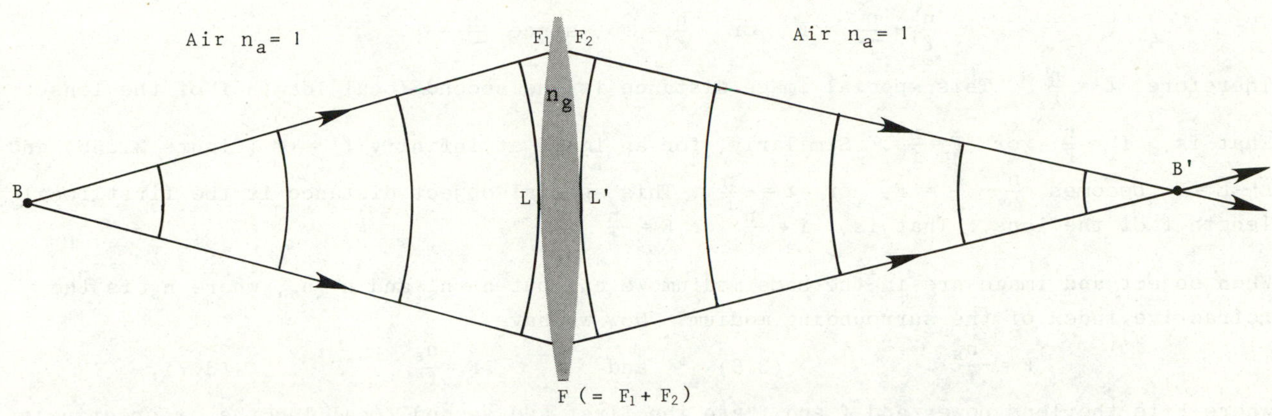

Fig. 3.24 The power of a thin lens $F = F_1 + F_2 = L' - L$.

It follows that we can write the equation for the power of the lens in air as

$$F = \frac{n_g - 1}{r_1} + \frac{1 - n_g}{r_2} = \frac{n_g - 1}{r_1} - \frac{n_g - 1}{r_2} = (n_g - 1)\left(\frac{1}{r_1} - \frac{1}{r_2}\right) \qquad (3.5a).$$

Writing $\frac{1}{r_1} = R_1$ and $\frac{1}{r_2} = R_2$, equation (3.5a) can be written in terms of the surface curvatures thus

$$F = (n_g - 1)(R_1 - R_2) \qquad (3.5b).$$

3.1.4.3 Focal Lengths and Power of a Thin Lens

We have seen (figure 3.12a) that when a parallel pencil of rays, originating from an object at infinity, is incident on a converging lens, the rays are focused at the second principal focus F' of the lens. The distance from the lens to the point F' is termed the second focal length f' of the lens, figure 3.25a.

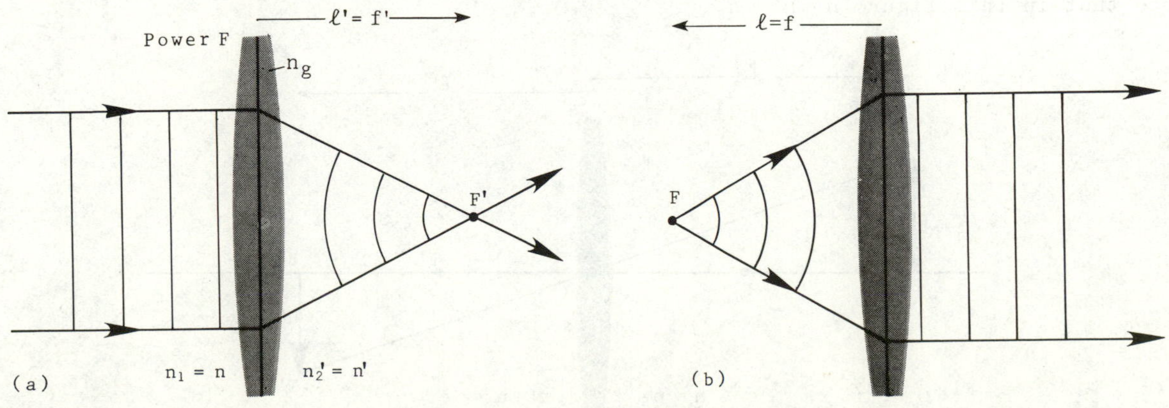

Fig. 3.25 The focal lengths of a thin converging lens.

We have, for an object at infinity, $\ell = -\infty$. Now, using the conjugate foci formula (3.1):

$$\bar{L}' - \bar{L} = F \quad \text{or} \quad \frac{1}{\ell'/n_2'} - \frac{1}{\ell/n_1} = F .$$

If we now write $n_2' = n'$ and $n_1 = n$ for the media to the left and right of the lens, when the object is at $-\infty$ this now becomes

$$\frac{n'}{\ell'} - \frac{n}{-\infty} = F \quad \text{or} \quad \frac{n'}{\ell'} = F \text{ , since } \frac{n}{-\infty} = 0 .$$

Therefore, $\ell' = \frac{n'}{F}$. This special image distance is the second focal length f' of the lens.

That is, $f' = \frac{n'}{F}$, or $F = \frac{n'}{f'}$. Similarly, for an image at infinity $\ell' = +\infty$ (figure 3.25b) and $\bar{L}' - \bar{L} = F$ becomes $\frac{n'}{+\infty} - \frac{n}{\ell} = F$, or $\ell = -\frac{n}{F}$. This special object distance is the first focal length f of the lens. That is, $f = -\frac{n}{F}$ or $F = -\frac{n}{f}$.

When object and image are in the same medium we can put $n = n_s$ and $n' = n_s$, where n_s is the refractive index of the surrounding medium. Now we have

$$F = -\frac{n_s}{f} \quad (3.6) \qquad \text{and} \qquad F = \frac{n_s}{f'} \quad (3.7),$$

where F is the lens power and f and f' are the first and second focal lengths, respectively.

So we can see that when a thin lens is surrounded by the same medium, and therefore when it is in air, the first and second focal lengths of the lens are numerically equal, figure 3.26. When the lens is in air $n_s = 1.000$ and equations (3.6) and (3.7) become the particular equations

$$F = -\frac{1}{f} \quad (3.8) \qquad \text{and} \qquad F = \frac{1}{f'} \quad (3.9).$$

These relationships between the power of a lens and its focal lengths hold for both converging and diverging lenses.

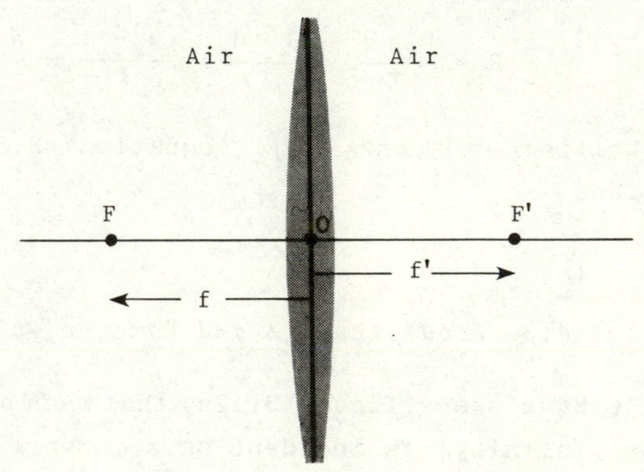

Fig. 3.26 A thin converging lens in air: $f = -f'$ or $|f| = |f'|$.

3.1.4.4 *Image Magnification*

Figure 3.27 shows a positive (converging) lens forming a real image B'T' of a real object BT. This is just one of the possible image formations by a positive lens (see figures 3.16 to 3.21). Note that in this figure $n = n' = n_s$.

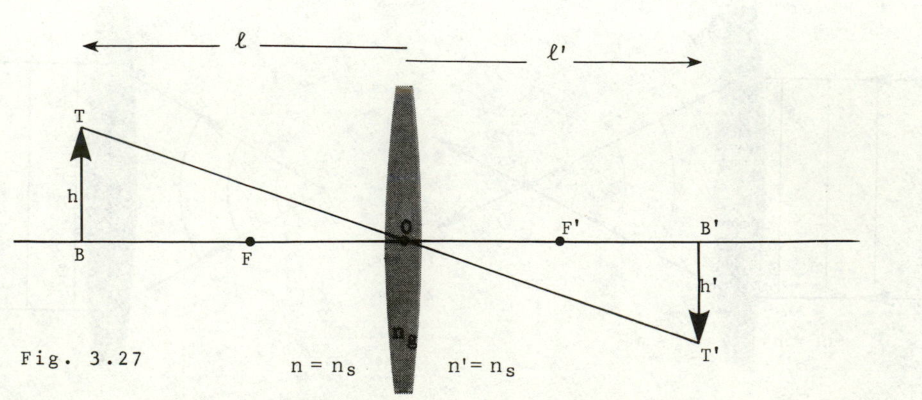

Fig. 3.27

The incident reduced vergence of the light at the lens is $\bar{L} = \frac{1}{\ell/n} = \frac{n}{\ell}$ and the emergent reduced vergence is $\bar{L}' = \frac{1}{\ell'/n'} = \frac{n'}{\ell'}$, from which $\ell = \frac{n}{\bar{L}}$ and $\ell' = \frac{n'}{\bar{L}'}$.
Therefore we can now write

$$\frac{\ell'}{\ell} = \frac{n'/\bar{L}'}{n/\bar{L}} = \frac{\bar{L}}{\bar{L}'} \quad \text{when } n = n' .$$

Now, the triangles BTO and B'T'O are similar. Therefore, we can write

$$\frac{T'B'}{OB'} = \frac{BT}{BO} \quad \text{or} \quad \frac{-h'}{\ell'} = \frac{h}{-\ell}, \text{ giving } \frac{h'}{h} = \frac{\ell'}{\ell} .$$

Hence we have $\frac{h'}{h} = \frac{\bar{L}}{\bar{L}'}$. But $\frac{h'}{h}$ represents the ratio $\frac{\text{image size}}{\text{object size}}$ and this is the linear magnification m of the image. Therefore, the linear magnification of the image is given by

$$m = \frac{\text{incident reduced vergence}}{\text{emergent reduced vergence}} = \frac{\bar{L}}{\bar{L}'} \qquad (3.10).$$

In working problems we quite often quote the composite equation $m = \frac{h'}{h} = \frac{\bar{L}}{\bar{L}'}$ (3.10a).
It should be noted that this equation also works when the lens is not surrounded by the same medium. In effect, the thin lens then acts as a single surface and a ray directed towards the optical centre (defined with the lens in a surrounding medium) is not transmitted undeviated.

3.1.5 *MEASUREMENT OF CURVATURE - THE SPHEROMETER*

The spherometer is an accurate measuring device which requires some practice for effective use. It consists of a three-legged frame, with the legs arranged at the corners of an equilateral triangle. In the centre of the frame a moveable leg works on a screw thread and this may be screwed up or down. The instrument is fitted with two scales, the fixed vertical scale usually being marked in millimetres, whilst the horizontal scale is usually calibrated in 1/100ths of a millimetre. In use the spherometer is first placed on a perfectly flat surface and a check made that, when all four feet are in contact with the surface, both scale readings are zero. Figure 3.28 shows a spherometer placed on a lens with the distance s through which the central leg is raised (or lowered) so that the

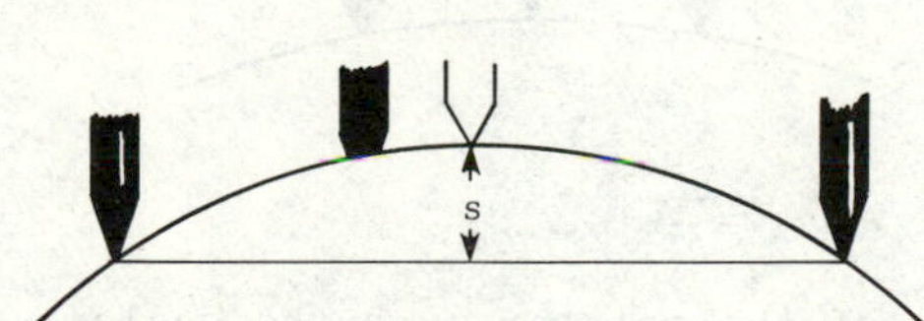

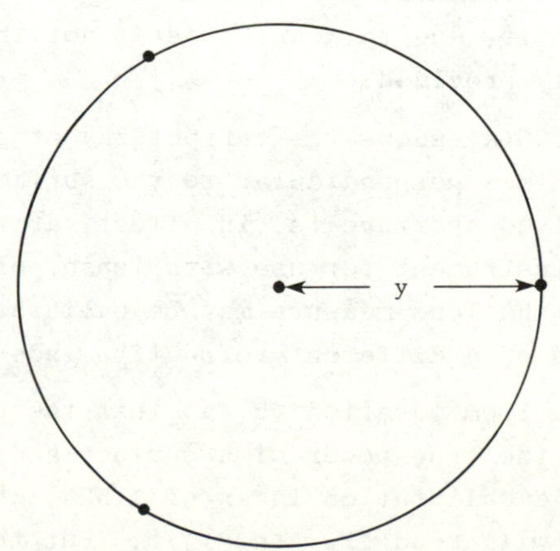

Fig. 3.28 The spherometer - see text.

four feet are again in contact with the surface being measured. s is the sag of the arc of the surface. The distance y may be determined from an imprint of the four feet on a sheet of paper, and is taken as the average distance between the central leg and the three fixed legs.

Using equation (1.6), the radius of curvature of a spherical surface may be calculated.

3.1.5.1 Measurement of Power - the Lens Measure

The lens measure is an instrument, based on the spherometer, used to give a direct reading of the approximate power of a surface. It has two fixed legs A and B (figure 3.29(b)) with a third spring-loaded central leg C which is free to move in or out of the instrument. The movement of the centre leg is the means of measuring the sag of a surface over an arc of fixed aperture.

(a)

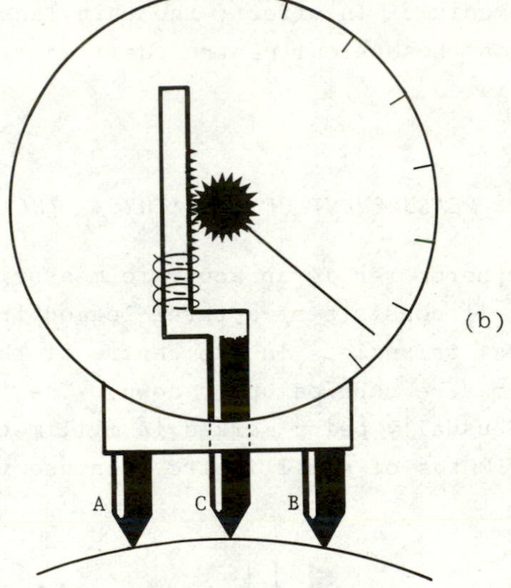

(b)

Fig. 3.29 The Lens Measure

The movement of the centre leg is recorded as a rotation of the pointer over the scale. When all three legs are in contact with a perfectly flat surface the scale reading should be zero. If, due to wear, this is not the case, some adjustment of one of the outer legs is usually provided.

Figure 3.29(a) shows the correct way of placing the lens measure against a lens surface, with the legs perpendicular to the surface. Measurement of the sag of a surface over an arc of fixed aperture is, in effect, a measurement of surface curvature. However, by restricting the instrument for use with lenses of one particular refractive index, usually 1.523, the scale of the lens measure may be calibrated in units of surface power (dioptres). When used on a lens of a different refractive index the instrument will give an incorrect reading.

We recall from equation (3.4a) that the power of a lens surface in air is given by $F = \frac{n_g - 1}{r}$. If F_T is the true power of a surface and F_R is the power recorded by the lens measure, then assuming a calibration index of 1.523, when the instrument is placed on a surface of this index it will read $F_R = (n_g - 1)/r$. But the true power is F_T associated with some other index say n for which $F_T = (n-1)/r$. Therefore, eliminating r which is common to both these expressions, we have

$$\frac{F_T}{F_R} = \frac{n-1}{0.523} \quad \text{or} \quad F_T = \frac{n-1}{0.523} \cdot F_R \qquad (3.11).$$

That is, true power $= \frac{n-1}{0.523} \times$ actual indicated reading.

3.1.6 *THE EFFECTIVITY OF A LENS*

The concept of effectivity was introduced in section 1.2.6 . Let us recall that effectivity concerns the changes that occur in the vergence at different points along a pencil of rays. We will now consider its application to the "effect" produced by a lens. In particular, let us investigate how two different lenses may produce the same effect as one another. Figure 3.30(a) shows a thin lens in air of power F forming an image of a distant object.

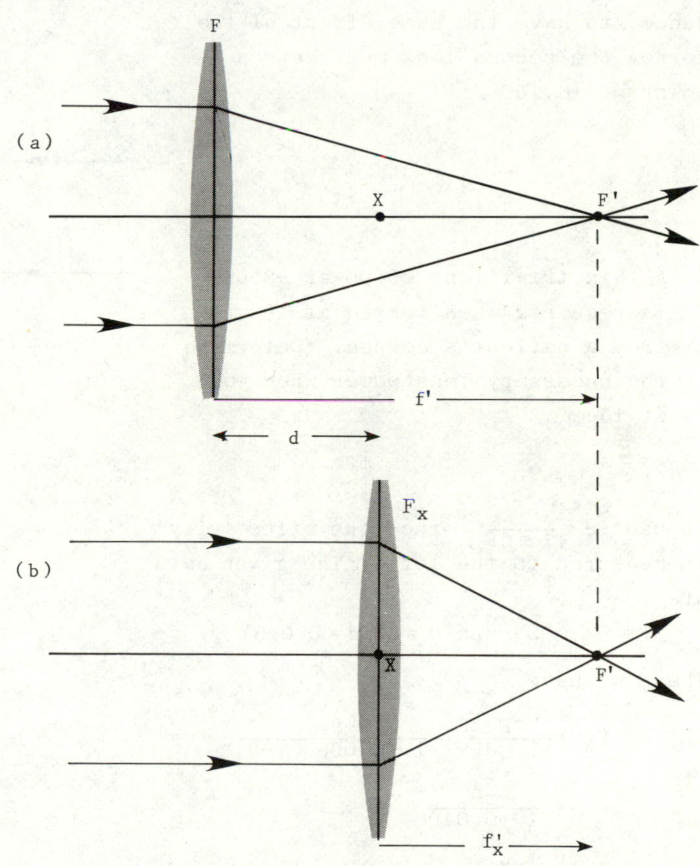

Fig. 3.30 The effectivity of a lens.

The image is formed at the second principal focus F' of the lens. Let us now assume that a second lens is positioned at X, nearer to F', but at the same time is required to maintain the position of the focus F' at the same place. That is, figure 3.30(b), the second lens is required to have the *same effect* as the original lens, namely to focus the light to the position F'. It is clear that the focal length f'_x of the new lens will be shorter than that of the first lens, since the extent of convergence of the light rays needs to be increased. If f' is the focal length of the first lens and f'_x is the focal length of the second lens, then

$$f'_x = f' - d .$$

The power F_x of the second lens is given by equation (3.9)

$$F_x = \frac{1}{f'_x} = \frac{1}{f' - d} = \frac{1}{\frac{1}{F} - d} = \frac{F}{1 - dF} ,$$

That is, $$F_x = \frac{F}{1 - dF} \qquad (3.12).$$

F_x is the *effectivity* or *effective power* of the *first* lens at the point X. We may define the effectivity of the lens as the power of a lens which when placed a distance d from the position of the original lens (power F) has the same effect on the vergence of the light as the original lens. It should be noted that if we wish to consider the effectivity of the lens at some point to the left of the lens of power F, then the equation will be $F_x = F/(1+dF)$ (3.12a).

Worked Examples

1 A thin trial case lens of power +6.00D corrects a person's vision when worn at 20 mm from the cornea. What power of lens will be required if it is worn at 16 mm from the cornea?

This may be illustrated as shown, where the second lens worn at 16 mm must have the same effect at the cornea as the trial lens at 20 mm. Using $F_x = F/(1-dF)$ with the given data F = +6.00D, d = 4 mm = 0.004 m, then

$$F_x = \frac{1}{1-dF} = \frac{6}{1-0.004\times 6}$$

$$= \frac{6}{1-0.024} = \frac{6}{0.976} = +6.15\text{D}.$$

Hence, to have the same effect at the cornea the second lens must have a power of +6.15D .

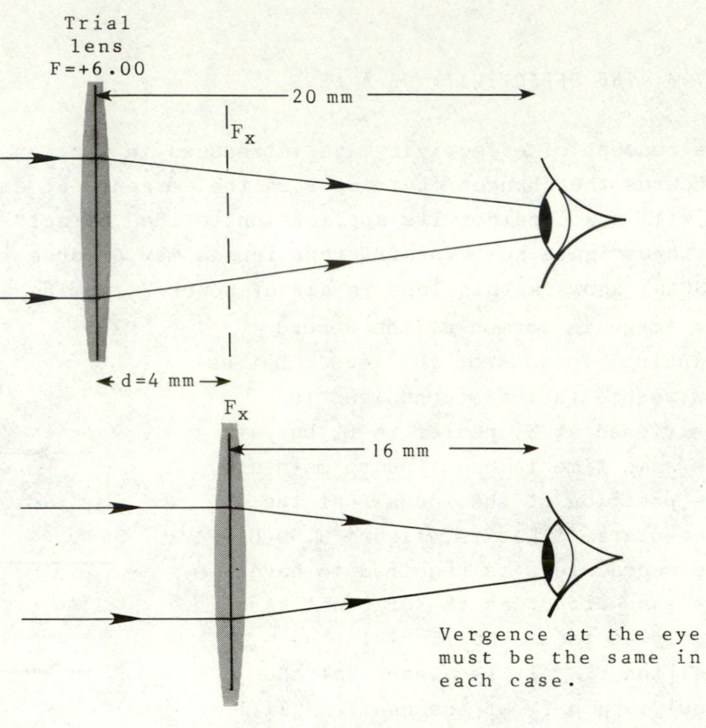

2 A thin trial lens of power -8.00D is required when tested at 10 mm from a patient's cornea. Determine the necessary lens power when worn at 15 mm.

We use $F_x = \frac{F}{1+dF}$ since the effectivity is measured to the left. The given data are

F = -8.00D and d = 5 mm = 0.005 m.

Then, we have

$$F_x = \frac{F}{1+dF} = \frac{-8}{1+0.005\times(-8)}$$

$$= \frac{-8}{1-0.040}$$

$$= \frac{-8}{0.96}$$

$$= -8.33\text{D}$$

Hence, the second lens needs to have a power of -8.33D. That is, the effectivity of the first (trial) lens at 15 mm from the eye is -8.33D.

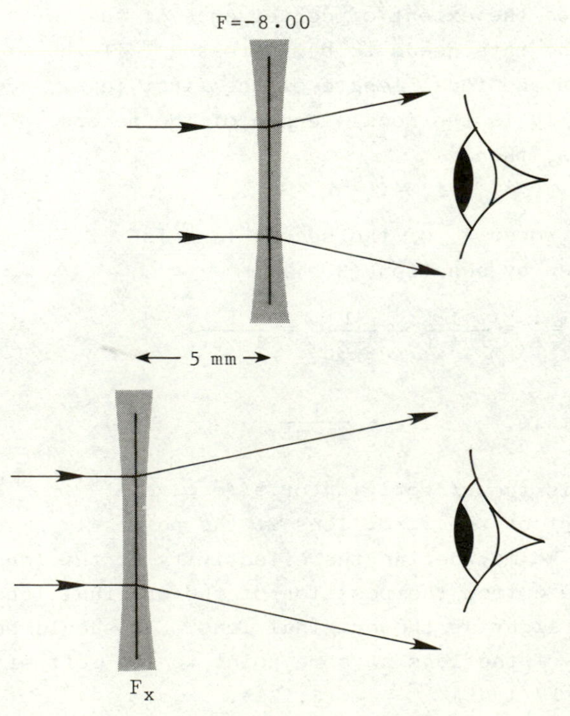

3.1.7 THE SMALLEST SEPARATION OF OBJECT AND REAL IMAGE

When forming a real image with a positive lens the object and image screen must be separated by a distance which is at least four times the focal length of the lens. It is then evident that for any separation between object and real image greater than 4f', where f' is the focal length of the lens, there will be two positions of the lens which will give conjugate foci. This is because the object and real image positions are interchangeable. Figure 3.31 shows that if, with the lens placed at a distance a from the object, an image is formed a distance b from the lens, then if the lens is placed a distance b from the object, the image will be at a distance a from the lens.

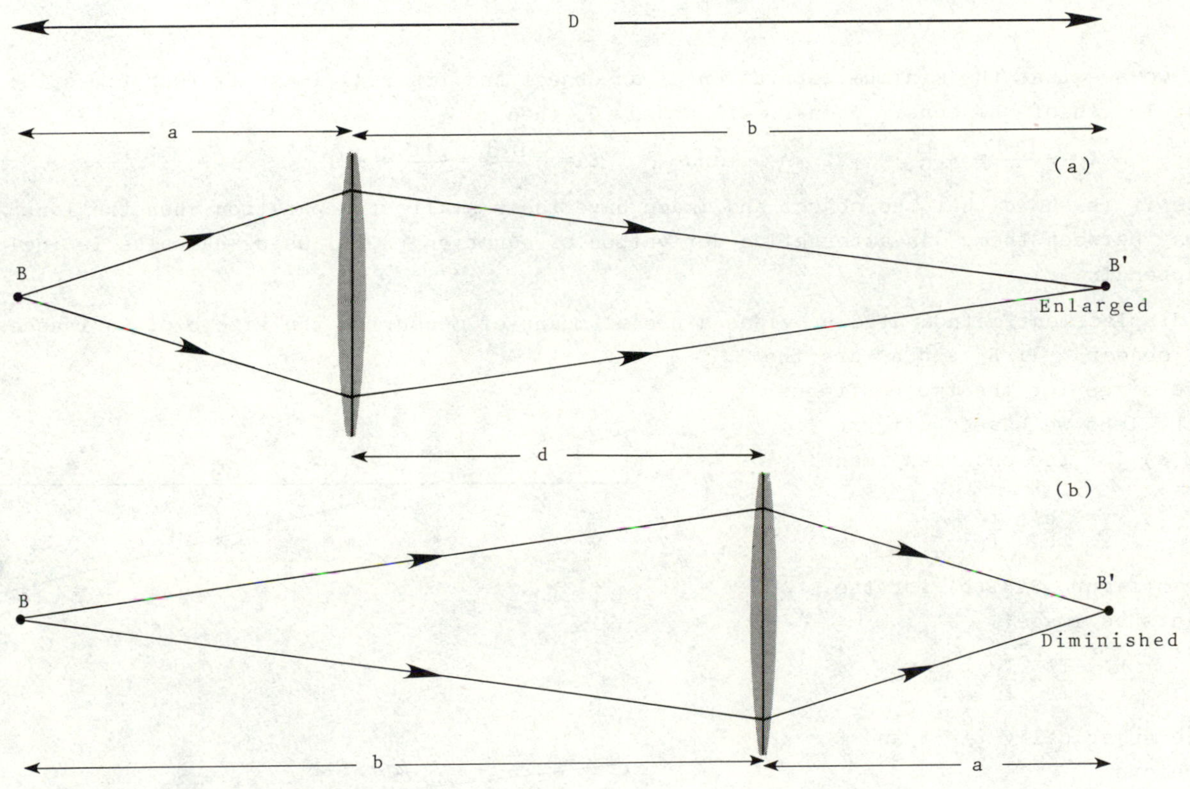

Fig. 3.31 See text for details.

For a thin lens in air, using equations (3.3) and (3.8):

$L' - L = F = \frac{1}{f'}$, that is, $\frac{1}{\ell'} - \frac{1}{\ell} = \frac{1}{f'}$ where ℓ is the object distance from the centre of the lens and ℓ' is the image distance from the centre of the lens. From figure 3.31 we see that $D = a+b$ and $d = D-2a = b-a$.

Adding these relationships, $D+d = 2b \Rightarrow b = \frac{D+d}{2}$

and subtracting them gives, $D-d = 2a \Rightarrow a = \frac{D-d}{2}$.

Now, in figure 3.31(a), object distance $= \ell = -a = -\frac{D-d}{2}$ and image distance $= \ell' = +b = +\frac{D+d}{2}$.

Then using $\frac{1}{\ell'} - \frac{1}{\ell} = \frac{1}{f'}$, $\frac{1}{\frac{D+d}{2}} - \frac{1}{-\frac{D-d}{2}} = \frac{1}{f'}$ which rearranges to give

$$f' = \frac{D^2 - d^2}{4D} \qquad (3.13)$$

where D is the separation of the object and image, and d is called the displacement of the lens. Equation (3.13) shows that the focal length f' can be determined without making measurements to the centre of the lens (see also section 3.1.8). This is particularly useful if the lens is inaccessible, such as the lens of a microscope eyepiece mounted inside a tube, or if the lens is thick when the "centre" of the lens has no precise significance. Equally, the focal length of a lens system may be determined where the measurement of the displacement d, through which the system is moved, can be made using any convenient reference mark on the lens system.

Now, the smallest separation of the object and real image, that is the minimum value for D, may be obtained from equation (3.13) by putting $d = 0$. Therefore, $f' = \frac{D^2}{4D}$, or

$$D = 4f' \qquad (3.14).$$

Thus we see that the minimum separation of an object and its real image is four times the focal length of the lens. When $D = 4f'$ and $d = 0$, then

$$\ell = -\frac{D-d}{2} = -\frac{4f'}{2} = -2f' \qquad \text{and} \qquad \ell' = +\frac{D+d}{2} = +\frac{4f'}{2} = +2f' .$$

Hence it is shown that the object and image have their smallest separation when the lens is midway between them. An alternative derivation of equation (3.14) using calculus is included in Appendix 3.

The displacement method also provides a useful means of measuring the size h of an inaccessible object. If h_1' and h_2' are the image sizes for the two positions of the lens we have in figure 3.32(a) for the enlarged image:

$$\frac{h_1'}{h} = \frac{\ell_1'}{\ell_1}$$

and in figure 3.32(b) for the diminished image:

$$\frac{h_2'}{h} = \frac{\ell_2'}{\ell_2} = \frac{\ell_1}{\ell_1'}$$

since numerically $\ell_2' = \ell_1$ and $\ell_2 = \ell_1'$. Therefore

$$\frac{h_2'}{h} = \frac{h}{h_1'}$$

or

$$h = \sqrt{h_1' h_2'} \qquad (3.15).$$

(a)

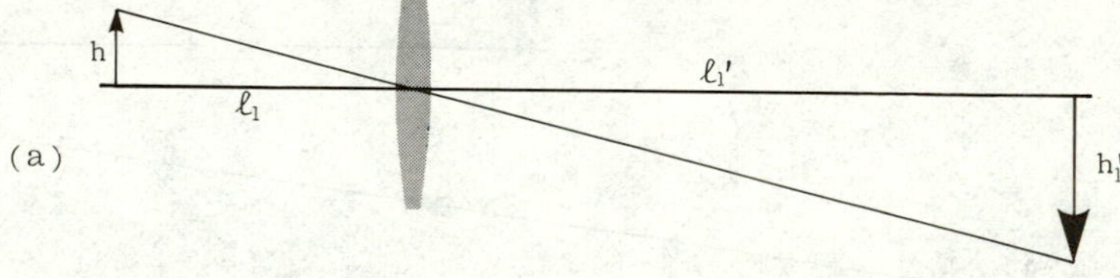

(b)

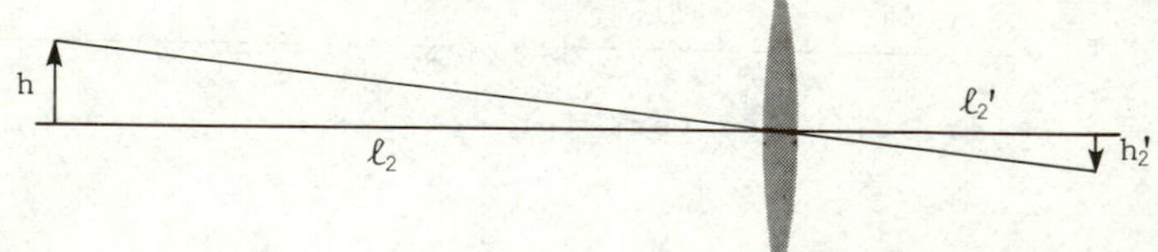

Fig. 3.32 Image sizes in the 'displacement method'.

3.1.8 NEWTON'S EQUATIONS FOR A THIN LENS

In section 2.2.4.3 we saw that when light is refracted by a single refracting surface it may be convenient to express the object and image distances from the first and second principal foci respectively. Under these conditions Newton's formula, equation (2.35), applies, that is $xx' = ff'$ where f and f' are the first and second principal focal lengths respectively of the surface. Now, for a thin lens in air, figure 3.26, the first and second principal focal lengths are equal, such that $f = -f'$ and Newton's formula may be written in the form which is normally used for calculations, viz $xx' = -(f')^2$ or

$$(f')^2 = -xx' \qquad (3.16)$$

Figure 3.33 shows the real image formed by a positive thin lens.

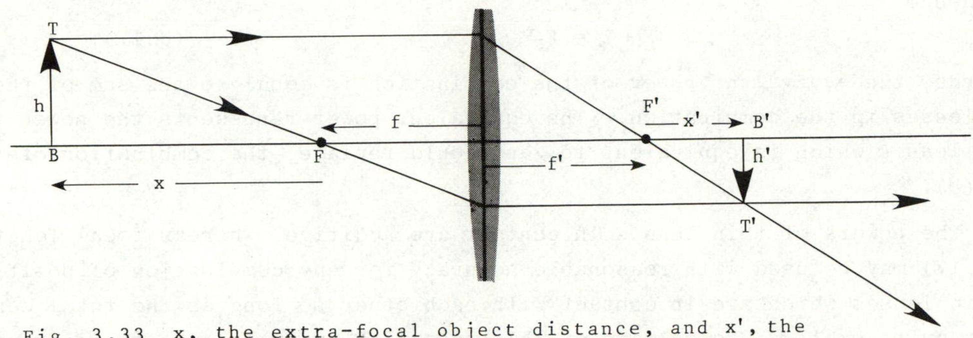

Fig. 3.33 x, the extra-focal object distance, and x', the extra-focal image distance in Newton's equations.

x represents the distance from the first principal focus F to the object B. That is, x = FB. x' is the distance from the second principal focus F' to the image B'; that is, x' = F'B'. The size of the image may be determined using equation (2.33) or equation (2.34) which are

$$\frac{h'}{h} = -\frac{f}{x} = -\frac{x'}{f'}.$$

Newton's formula, equation (3.16) provides the means for determining the focal length f' of an inaccessible lens where it may not be possible to make measurements to the lens centre.

3.1.9 TWO THIN LENSES IN CONTACT

If two thin lenses are placed in contact we can obtain an expression for the *equivalent power* of the combination by considering the refractions in turn. In figure 3.34(a) B_1' is the real image of the object point B formed by the lens L_1 alone.

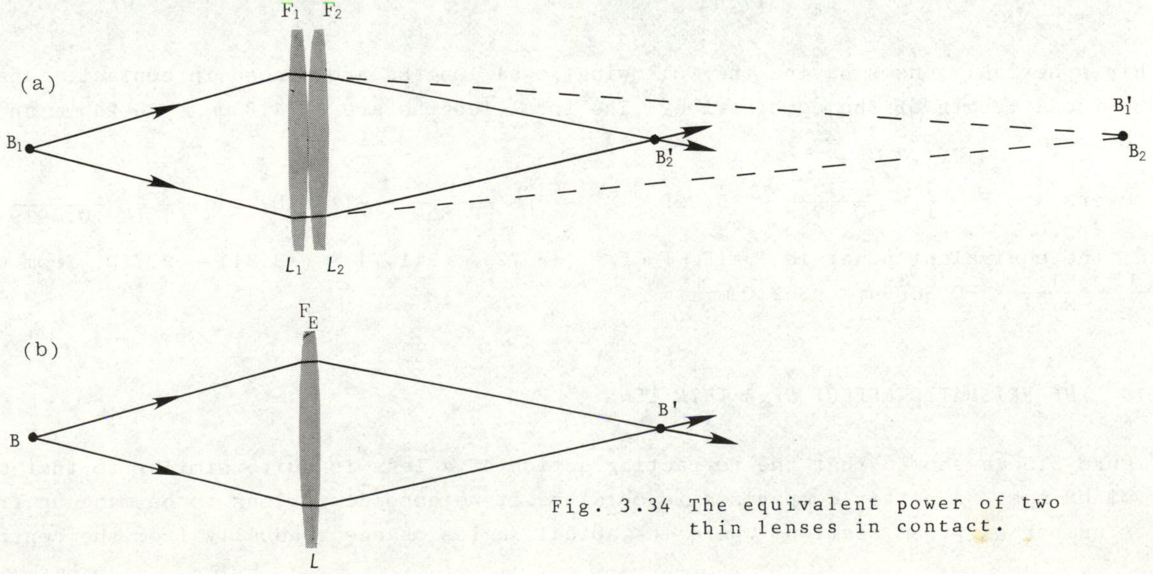

Fig. 3.34 The equivalent power of two thin lenses in contact.

On placing the second lens L_2 in contact with L_1, B_1' now becomes a virtual object B_2 for the lens L_2, and the final image is now formed at B_2'. Now, for the first lens we have $L_1'-L_1=F_1$ and for the second lens $L_2'-L_2=F_2$. Adding these equations gives $L_1'-L_1+L_2'-L_2=F_1+F_2$. But since $\ell_1'=\ell_2$, then $L_1'=L_2$ and we get $L_2'-L_1=F_1+F_2$. But L_2' and L_1 are the emergent and incident vergences of the combination, so if this combination has an equivalent power F_E, then $L_2'-L_1=F_E$ whence

$$F_1+F_2=F_E \qquad (3.17).$$

In other words, the *equivalent power* of the combination is equal to the sum of the powers of the two lenses in the combination. The equivalent power represents the power of the single thin lens L which is equivalent to, and could replace, the combination of thin lenses, figure 3.34(b).

We see that the powers of thin lenses in contact are additive, whereas focal lengths are not. Equation (3.17) may be used with reasonable accuracy for any combination of positive and negative thin lenses which are in contact with each other so long as the total centre thickness is kept quite small. Use is made of this fact with an optician's trial frame, where a number of thin lenses may be arranged very nearly in contact with each other in front of the patient's eyes. If three trial case lenses together are found to be suitable then the equivalent power is given approximately by an extended version of equation (3.17):

$$F_E=F_1+F_2+F_3\,.$$

Again, an optician can make a rough check on the power of a lens by selecting a lens from the trial case which, when held in contact with the test lens, just cancels its refraction. In this case, the power of the trial case lens will be numerically equal to but opposite in sign to the test lens. The process is known as *neutralisation*.

Worked Examples

1 The following thin spherical lenses in air are placed in contact: +1.50D, +5.75D, -0.25D. Calculate the equivalent power. What is the equivalent focal length in millimetres?

The equivalent power is given by $F_E=F_1+F_2+F_3=(+1.50)+(+5.75)+(-0.25)=+7.00\text{D}$. The equivalent focal length is $f_E'=\frac{1}{F_E}=\frac{1}{+7}=+0.1429\text{ m}=+142.9\text{ mm}$.

2 Thin spherical lenses having the following focal lengths are placed in contact. What is the focal length of the combination? The focal lengths are +174.8 mm, -85.24 mm, and +301.9 mm.

The powers are $F_1=\frac{1}{f_1'}=\frac{1}{+0.1748}=+5.72\text{D}$, $F_2=\frac{1}{f_2'}=\frac{1}{-0.852}=-11.74\text{D}$, and $F_3=\frac{1}{f_3'}=\frac{1}{+0.3019}=+3.31\text{D}$.

Hence, the equivalent power is $F_E=F_1+F_2+F_3=(+5.72)+(-11.74)+(+3.31)=-2.71\text{D}$, from which $f_E'=\frac{1}{F_E}=\frac{1}{-2.71}=-0.3690\text{ m}=-369.0\text{ mm}$.

3.1.10 THE PRISMATIC EFFECT OF A THIN LENS

In figure 3.5 we showed that the refracting action of a lens is quite similar to that of a pair of prisms. A better comparison is obtained if we consider a lens to be made up from a larger number of prism sections where the apical angles change gradually from the centre of the lens towards the edge. Figure 3.35 shows how a converging lens may be "built up" from a number of annular prisms where the apical angles increase towards the edge of the lens.

The prism section P_1 further from the principal axis has a larger refracting angle and produces a larger deviation on a ray of light than the section P_2 nearer the axis of the lens. The result is that all rays of light incident parallel to the principal axis converge to the second principal focus F'. This is assuming that the lens is free of all defects, in particular spherical aberration.

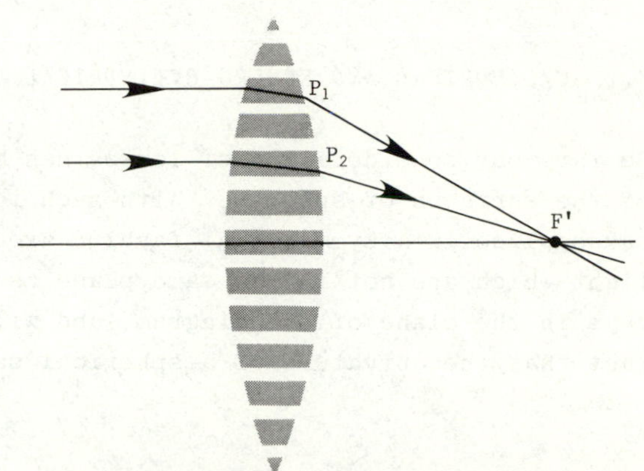

Fig. 3.35 A lens considered as a stack of prisms with apical angles increasing towards the margin.

The deviation ability of a lens at a particular point off the principal axis is referred to as the *prismatic effect* at that point and this may be defined as the power of the prism which would replace the lens at that point.

Following the worked example in section 2.1.10.4 we discussed the deviation produced by a prism, and showed that the prism deviation θ^Δ expressed in prism dioptres is given by $\theta^\Delta = \frac{y}{x} \ \frac{(cm)}{(m)}$, where y cm is the displacement of a ray on a tangent screen at a distance of x metres from the prism. This is illustrated in figure 3.36. It shows that the prism dioptre is equivalent to the cm/metre.

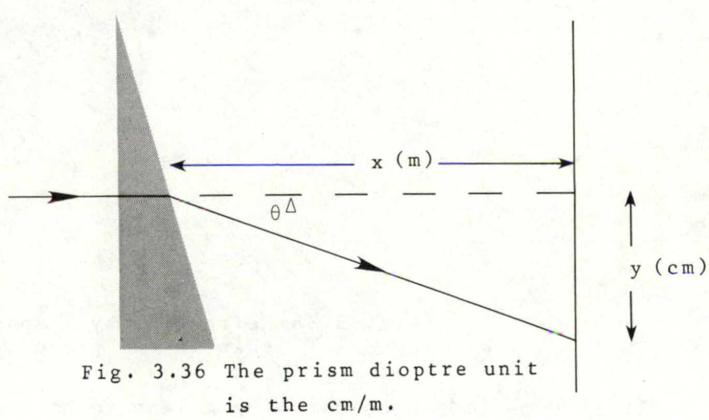

Fig. 3.36 The prism dioptre unit is the cm/m.

Figure 3.37 shows a ray of light, parallel to the principal axis of a converging lens, incident at a distance c from the axis. At the distance c from the axis, the prismatic deviation θ^Δ is given by

$$\theta^\Delta = \frac{c}{f'} \ \frac{(cm)}{(m)} \ .$$

The prismatic deviation θ^Δ is often given the symbol P, and since $\frac{1}{f'} = F$ where F is the lens power, we can write

$$P = cF \qquad (3.18).$$

Equation (3.18) shows that the prismatic deviation produced at a point in a lens depends on the distance c of the point from the lens axis, and on the lens power F. A spectacle lens may purposely be decentred in front of the eye in order that a prismatic effect is introduced to correct a muscular defect in addition to the power correction of

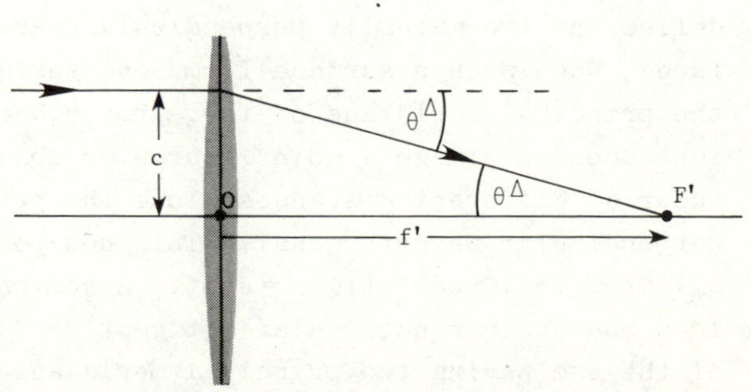

Fig. 3.37 The derivation of Prentice's rule.

the lens for refractive error.

Note that the prismatic effect at the optical centre of a lens is zero as c = 0. This is because the central section of the lens is essentially a parallel sided piece of material.

3.2 *CYLINDRICAL AND SPHERO-CYLINDRICAL LENSES*

So far, our consideration of lenses has been with those having surfaces which are portions of the surfaces of spheres. With such lenses refraction of light from an axial object point takes place in a symmetrical fashion around the optical axis. Hence, we know that rays of light which are not in the same plane as the diagram behave in a similar fashion to those rays in the plane of the diagram, and will all intersect in the image. This is due to the fact that the curvature of a spherical surface is the same in all meridians, figure 3.38.

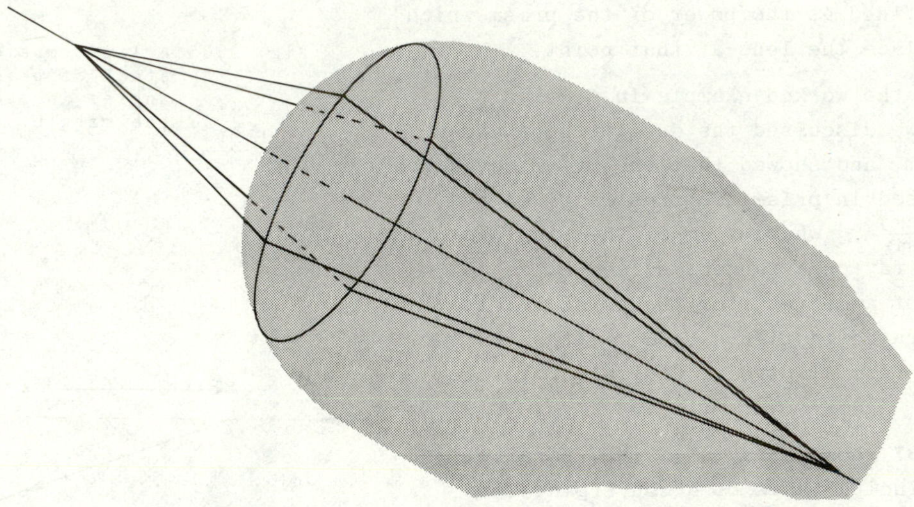

Fig. 3.38 Refraction by a spherical surface - rays in planes other than in the plane of the paper.

It is sometimes necessary for a lens to have different curvatures in different meridians. However, in such cases there are two mutually perpendicular meridians which will bring an incident pencil of rays to a focus such that the power may be defined as $\bar{L}' - \bar{L}$ for those meridians. Considering a single surface, those meridians with maximum and minimum curvature define the two mutually perpendicular meridians known as the *principal meridians* of the surface. When such a surface forms one surface of a lens, the other being a spherical surface, the principal meridians of the surface become the principal meridians of the lens. Such a lens does not image a point source or object in a point image but forms two *focal lines* which occur at different distances along the principal axis of the lens and are themselves mutually perpendicular or orthogonal. This non-pointlike image pencil is known as an *astigmatic* pencil from the Greek stigma = spot. A common example of this requirement is in the case of a lens used to correct ocular astigmatism. This eye defect is mainly as a result of the cornea of the eye having two principal meridians of curvature with the consequent formation of an astigmatic pencil when an incident pencil from a point object is refracted by the eye. This results in an indistinct retinal image which can only be made stigmatic by the application of a correcting lens which, combined with the eye's refraction, produces a stigmatic pencil eventually focusing as a point on the retina when the object is a point.

Lenses intended for the correction of astigmatism are often referred to as *astigmatic* lenses and always incorporate at least one surface which has two principal meridians.

3.2.1 THE CYLINDRICAL LENS

The cylindrical surface is the simplest form of astigmatic surface, and the cylindrical lens is the simplest form of astigmatic lens. Figure 3.39 shows how a cylindrical lens is derived from a portion of a cylinder.

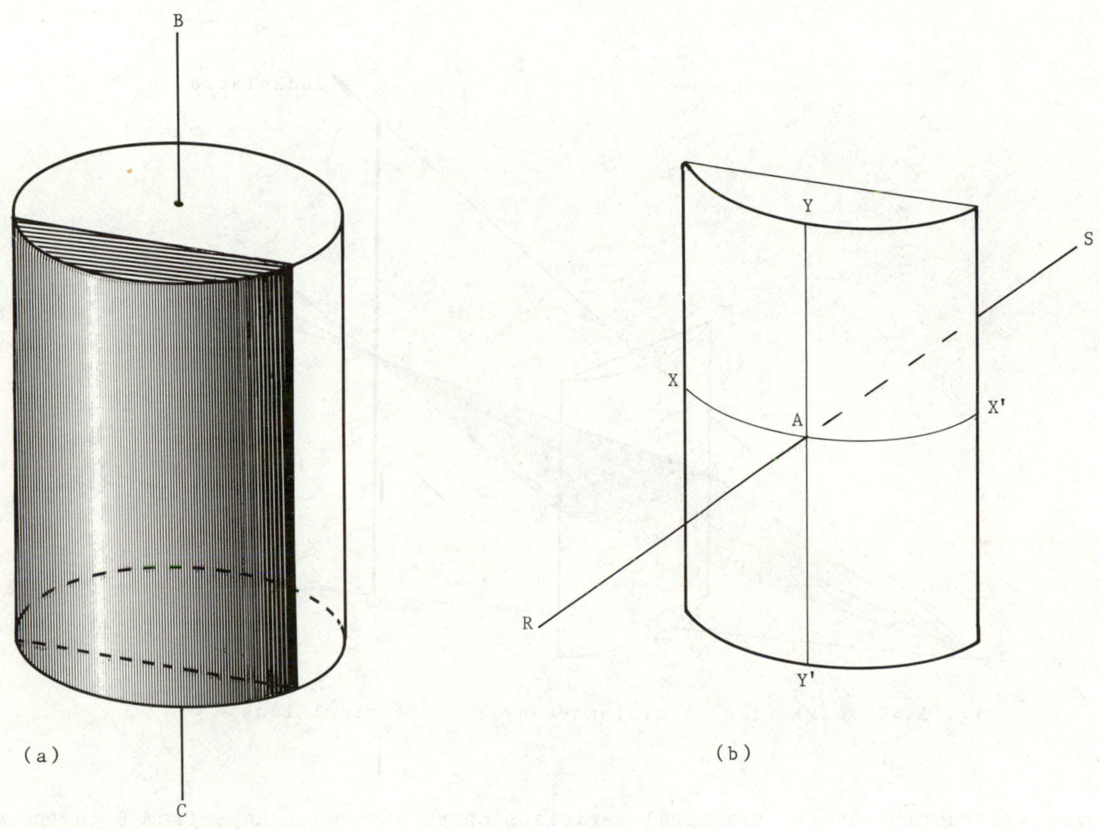

Fig. 3.39 (a) A glass cylinder. (b) A plano-convex cylindrical lens.

The actual lens shown is a positive powered plano-cylindrical variety. The line BC is the axis of the cylinder. Any line on the cylindrical surface, such as YY', which is parallel to the cylinder axis is termed an *axis meridian*. Along an axis meridian it is seen that the surface has zero curvature. Any meridian, such as XX', which intersects an axis meridian perpendicularly is termed a *power meridian*. Since only pencils of rays in principal meridians can be brought to a focus it is only possible to define power for those two meridians. In the case of a plano-cylindrical lens the minimum curvature is along an axis meridian and this is plane. This implies the radius of curvature is infinite. Using the surface power equation $F = (n'-n)/r$ the power is clearly zero since $r = \infty$. However, in the power meridian the radius is finite which results in a finite power. In figure 3.39(b), if the radius along meridian

XX' is r then the power is F = (n' - n)/r. For glass of refractive index n_g with air to the left of the surface the power is F = (n_g - 1)/r.

The optic axis of the lens is a line such as RS which is perpendicular to both surfaces of the lens. In fact, any line parallel to RS and intersecting YY' is an optic axis direction. Because there is no unique optic axis there can be no unique optical centre on a plano-cylindrical lens. However, for a convenient reference point, the intersection of the two principal meridians at the geometric centre of the lens may be regarded as the optical centre. This is the point A in figure 3.39 .

It can now be seen that the focusing properties of a cylindrical surface are not the same as for a spherical surface. Let us consider figure 3.40 which shows the formation of the image of a point object by a plano-cylindrical lens.

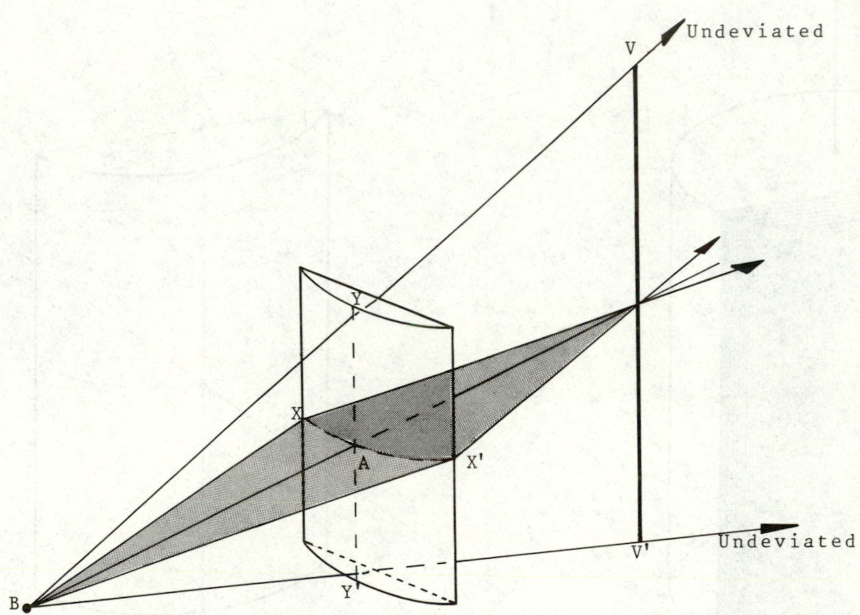

Fig. 3.40 Refraction by a plano-convex cylindrical lens.

For clarity, refraction at the principal meridians only is shown. Rays from B in the vertical plane BYY' are incident on the axis meridian, along which the surface has no curvature, and hence no power. Therefore, rays in the vertical plane are undeviated by the lens (note that a little displacement is shown due to the thickness of the lens). Rays of light in the horizontal plane BXX' are incident on the power meridian of the lens and are refracted according to the power of the cylindrical surface in this meridian. The combined effect of the refractions in the two principal meridians, together with the refractions in other meridians (not shown) is to image the object as a *line image* VV'.

Rays from B in meridians oblique to YY' and XX' and incorporating the point A do not focus in a point. They undergo skew refraction. For example, imagine a meridian on the lens which occupies the quadrants XAY and X'A Y'. Then a ray from B incident at the lens on this meridian in the first quadrant will pass through the line focus VV' in the upper half of this line. The diametrically opposite ray in the quadrant X' AY' will pass through VV' in the lower half. Figure 3.41 is an alternative version of figure 3.40 and perhaps this shows the focal properties

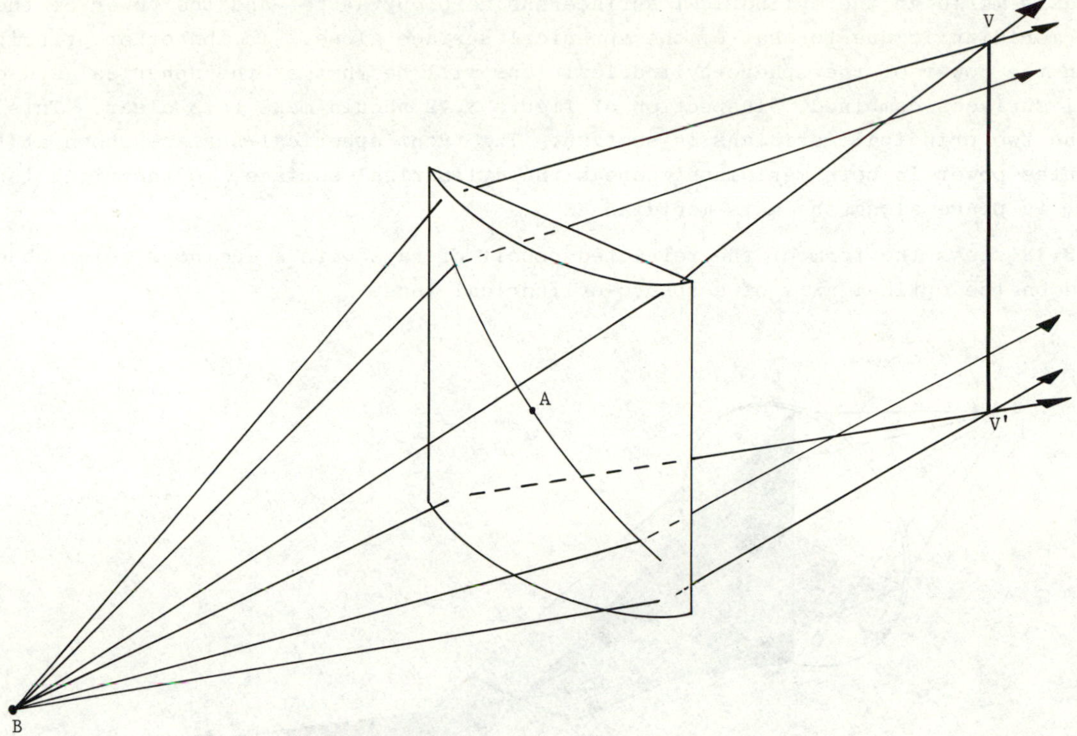

Fig. 3.41 Refraction at a cylindrical surface showing skew refraction in an oblique meridian.

of the cylindrical surface better.

Cylindrical lenses can have a variety of forms and may be described as plano-convex, plano-concave, convexo-concave, etcetera. In all cases it is useful to remember that the power of a cylinder lies in the plane which is at right angles to the axis of the cylinder.

3.2.2 THE SPHERO-CYLINDRICAL LENS

As the name suggests, a sphero-cylindrical lens is one which possesses both spherical and cylindrical surfaces, and typically may be used to provide spherical power correction for an eye together with cylindrical power correction for astigmatism.

The principal meridians of a sphero-cylindrical lens are defined in a similar fashion to the case of a plano-cylindrical lens. Hence, the *axis meridian* of a sphero-cylindrical lens is the direction parallel to the axis of the cylinder, whilst the *power meridian* is perpendicular to this. However, a sphero-cylindrical lens possesses power in both these meridians, either of which may be positive or negative.

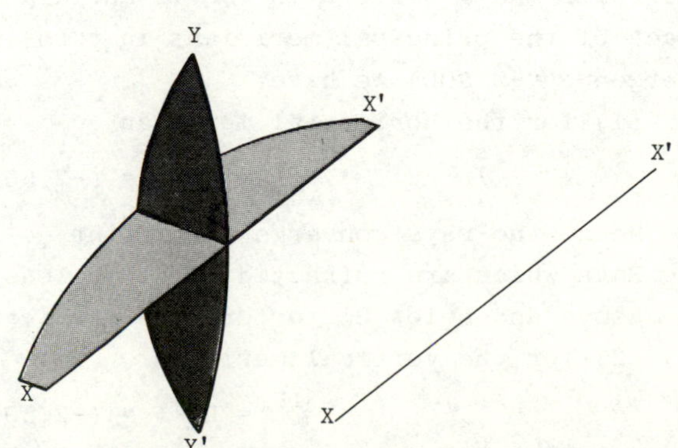

Fig. 3.42 Principal sections through a sphero-cylindrical lens.

In the axis meridian the cylindrical surface has zero curvature, and the power of the lens in this meridian is due to that of the spherical surface alone. In the other principal meridian the power of the sphero-cylindrical lens will be that of the spherical and cylindrical surfaces combined. Inspection of figure 3.42 should make this clear. This figure shows the two principal meridians in section. The front spherical surface shown at the left contributes power in both meridians whereas the cylindrical surface, to the right in the diagram, is plane along the axis meridian XX'.

Figure 3.43 shows the form of the refracted pencil of rays with a luminous point object B situated on the optical axis of a sphero-cylindrical lens.

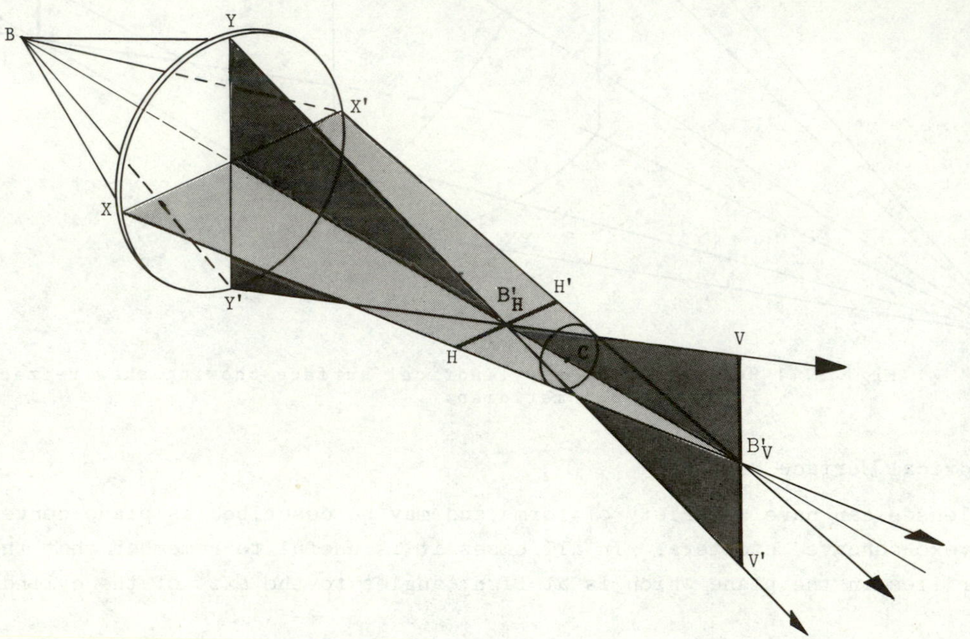

Fig. 3.43 The astigmatic pencil produced by a sphero-cylindrical lens.

Let us assume that the lens is made up from a +4.00D spherical surface and a +1.00D cylindrical surface with the cylinder axis horizontal. Hence, in this case F = +4.00D in the XX' meridian and F = +5.00D in the YY' meridian. Applying the conjugate foci formula $L' - L = F$ to each of the principal meridians in turn, and assuming an object distance of 40 cm (incident vergence = -2.50D) we have:

(1) for the horizontal meridian

$$L' = L + F = (-2.50) + (+4) = +1.50\text{D}.$$

Hence the rays converge to a point $\frac{1}{+1.50} = 0.6667\text{ m} = 66.67\text{ cm}$ to the right of the lens. Rays which are refracted by other lens sections parallel to XX' will be converged to points above and below B'_V to form the *vertical focal line* VV'.

(2) for the vertical meridian

$$L' = L + F = (-2.50) + (+5) = +2.50\text{D}.$$

The rays converge to a point $\frac{1}{+2.50}\text{ m} = +40\text{ cm}$ from the lens (to the right). Rays which are refracted by other lens sections parallel to YY' will be converged to points on either side of B'_H to form the *horizontal line focus* HH'.

When the refracted pencil of rays has the form described, in which a point object is imaged as two perpendicular focal lines, it is termed an *astigmatic pencil* which, as already stated, comes from the Greek meaning "not a point". Assuming that the lens has a circular aperture then the cross-sectional shape of the astigmatic pencil will, in general, be an ellipse, becoming focal lines at right angles to each other at the two positions described.

At some point between the two focal lines the cross-section of the astigmatic pencil will be circular and this may be considered to be the nearest approach to a point image. This position, where the pencil has the smallest cross-section is termed the *circle* or *disc of least confusion*. The centre of the circle of least confusion is denoted by the point C in figure 3.43 .

3.2.2.1 *The Positions of the Line Foci and Circle of Least Confusion*

Figure 3.44 shows side and top views of the astigmatic pencil depicted in figure 3.43 . In each case only one of the two focal lines is clearly seen, as these are orthogonal or mutually perpendicular. The aperture of the lens is assumed to be circular and of diameter D.

The positions of the two focal lines may be determined by using the formula $L' = L + F$ separately on each of the principal meridians of the lens. In addition, the size and position of the circle of least confusion, and the lengths of the two focal lines can be obtained from the geometry of figure 3.44 .

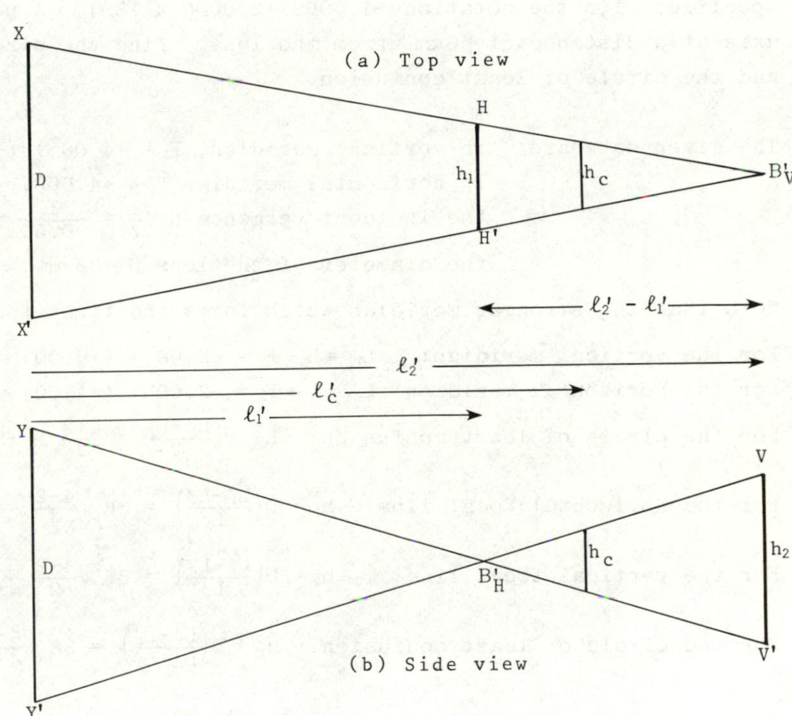

Fig. 3.44 Horizontal and vertical sections through the astigmatic pencil of figure 3.43 .

In figure 3.44(a), the triangles $XX'B'_V$ and $HH'B'_V$ are similar. Hence, we have for the horizontal focal line:

$$\frac{h_1}{\ell_2' - \ell_1'} = \frac{D}{\ell_2'} \quad \text{or} \quad h_1 = D\left(\frac{\ell_2' - \ell_1'}{\ell_2'}\right) \qquad (3.19).$$

Alternatively, using vergence notation, where $\ell_2' = \frac{1}{L_2'}$, and so on, we get

$$h_1 = D\left[\frac{\frac{1}{L_2'} - \frac{1}{L_1'}}{\frac{1}{L_2'}}\right] = DL_2'\left[\frac{L_1' - L_2'}{L_1' L_2'}\right] = D\left[\frac{L_1' - L_2'}{L_1'}\right] \qquad (3.20).$$

This is the length of the first (nearer to the lens) focal line. It is left to the reader to show that the following relationships apply.

Length of the second focal line $h_2 = D\left(\frac{\ell_2' - \ell_1'}{\ell_1'}\right)$ (3.21) or $h_2 = D\left[\frac{L_1' - L_2'}{L_2'}\right]$ (3.22)

The position of the circle of least confusion and its size (diameter) are denoted by ℓ_c' and

h_c, respectively. The linear and vergence equations for these quantities are

$$\ell_c' = \frac{2\ell_1'\ell_2'}{\ell_1'+\ell_2'} \quad (3.23) \quad \text{and} \quad \ell_c' = \frac{2}{L_1'+L_2'}$$

$$\text{or} \quad L_c' = \frac{L_1'+L_2'}{2} \quad (3.24).$$

The diameter of the circle of least confusion is

$$h_c = D\left(\frac{\ell_2'-\ell_1'}{\ell_1'+\ell_2'}\right) \quad (3.25) \quad \text{or} \quad h_c = D\left(\frac{L_1'-L_2'}{L_1'+L_2'}\right) \quad (3.26).$$

Worked Example

A lens is 38 mm diameter and comprises a spherical surface of power +4.00D and a cylindrical surface of power +2.00D with the axis meridian horizontal. (This form of the lens is usually specified with the notation +4.00DS/+2.00DC x 180). A point object is placed on the optical axis at a distance of 50 cm from the lens. Find the sizes and positions of the focal lines and the circle of least confusion.

The given data are: in vertical meridian, $F_1 = +4.00 + (+2.00) = +6.00\text{D}$
in horizontal meridian $F_2 = +4.00\text{D}$.
the incident vergence $L = \frac{1}{\ell} = \frac{1}{-0.50} = -2.00\text{D}$
the diameter of the lens $D = 38$ mm .

Note that the stronger meridian which forms the first focal line has been given the subscript 1.

For the vertical meridian: $L_1' = L + F_1 = -2.00 + (+6.00) = +4.00\text{D}$. $\therefore\ \ell_1' = +25$ cm.

For the horizontal meridian: $L_2' = L + F_2 = -2.00 + (+4.00) = +2.00\text{D}$. $\therefore\ \ell_2' = +50$ cm.

For the circle of least confusion: $L_c' = \frac{L_1'+L_2'}{2} = \frac{4+2}{2} = +3.00\text{D}$. $\therefore\ \ell_c' = +33\tfrac{1}{3}$ cm.

For the horizontal focal line: $h_1 = D\left(\frac{L_1'-L_2'}{L_1'}\right) = 38 \times \frac{4-2}{4} = 19$ mm.

For the vertical focal line: $h_2 = D\left(\frac{L_1'-L_2'}{L_2'}\right) = 38 \times \frac{4-2}{2} = 38$ mm.

For the circle of least confusion: $h_c = D\left(\frac{L_1'-L_2'}{L_1'+L_2'}\right) = 38 \times \frac{4-2}{4+2} = 12\tfrac{2}{3}$ mm.

3.2.3 THE TORIC LENS

Toric lenses are produced to provide a choice of form for astigmatic lenses in an analogous way to which meniscus forms are often a more suitable form of spherical powered lens than a flat form for correcting visual defects. Most often then, a toric spectacle lens is a curved rather than a flat lens although it should be borne in mind that a toric may be of flat form.

A toroidal surface is defined as the surface generated by the revolution of an arc of a circle about an axis in the same plane as the arc, figure 3.45 . Three forms of just such a toroidal surface are possible according as to whether the centre of curvature of the arc C lies to the right of the arc and the axis of revolution as in the barrel form, or between the arc and the axis of revolution as in the tyre form, or to the left of the arc and the axis of revolution as in the capstan form. These forms are shown as (a), (b), and (c) respectively in figure 3.45 .

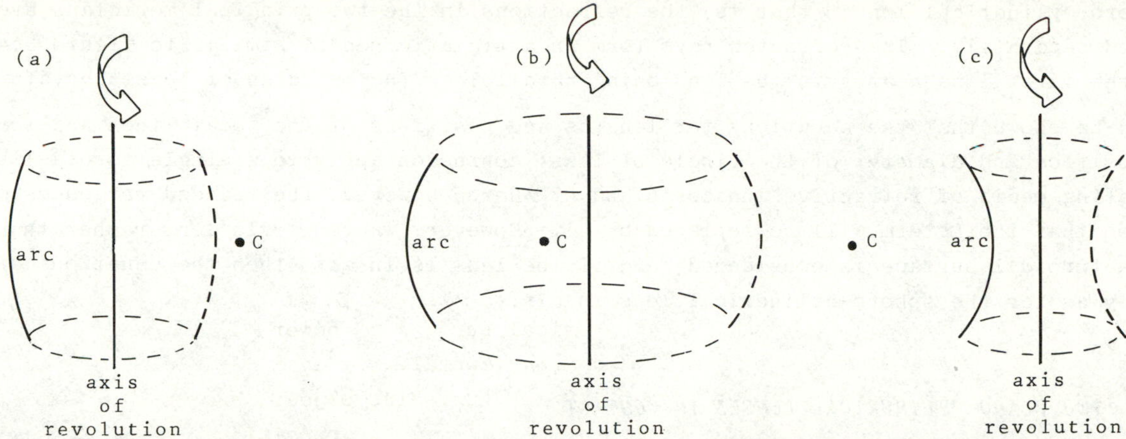

Fig. 3.45 The three forms of toroidal surface. (a) Barrel form, (b) tyre form, and (c) capstan form.

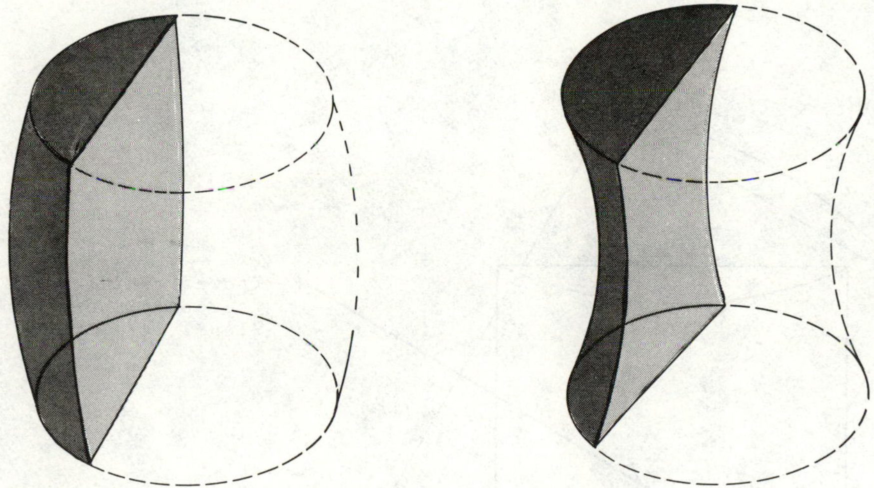

Fig. 3.46 Two forms of toroidal surface - barrel and capstan.

Figure 3.46 shows how a toric lens surface may be regarded as a portion of the whole toroidal surface. The three forms of complete surfaces bear the general name torus from which the words toric and toroidal derive.

We recall that the cylindrical surface has two principal meridians, one of which has zero power along the axis meridian. Similarly, a toroidal surface has two principal meridians of differing power, but the smaller power is not zero. Like a sphero-cylindrical lens, the purpose of a toric lens is to provide spherical power correction together with cylindrical power correction for astigmatism. This is achieved in the lens by working one surface toroidal and the other as a sphere. Less frequently both surfaces will be toroidal when the lens is described as a bitoric.

The refracting properties of a toroidal surfaced lens may be considered in the same way as for a sphero-cylindrical lens; that is, the refractions in the two principal meridians are considered separately. The refracted rays form an astigmatic pencil similar to figure 3.43, with the first line nearer to the lens being parallel to the meridian of least curvature.

It can be shown that the equations for lengths and positions of the focal lines and for the position and diameter of the circle of least confusion apply to a single toroidal surface separating media of refractive indices n and n'. Here, however, the reduced vergences must be used so that the L' term will be replaced by $\bar{L}'$. However, where a toric lens rather than just a toroidal surface is considered, and if the lens is in air, then the equations are exactly as for the sphero-cylindrical lens in air.

3.2.4 TWO PLANO-CYLINDRICAL LENSES IN CONTACT

We know that the placing of two thin spherical lenses in contact results in an equivalent spherical lens given by $F_E = F_1 + F_2$, equation (3.17). This suggests we should investigate the result of placing two thin plano-cylindrical lenses in contact, as illustrated in figure 3.47.

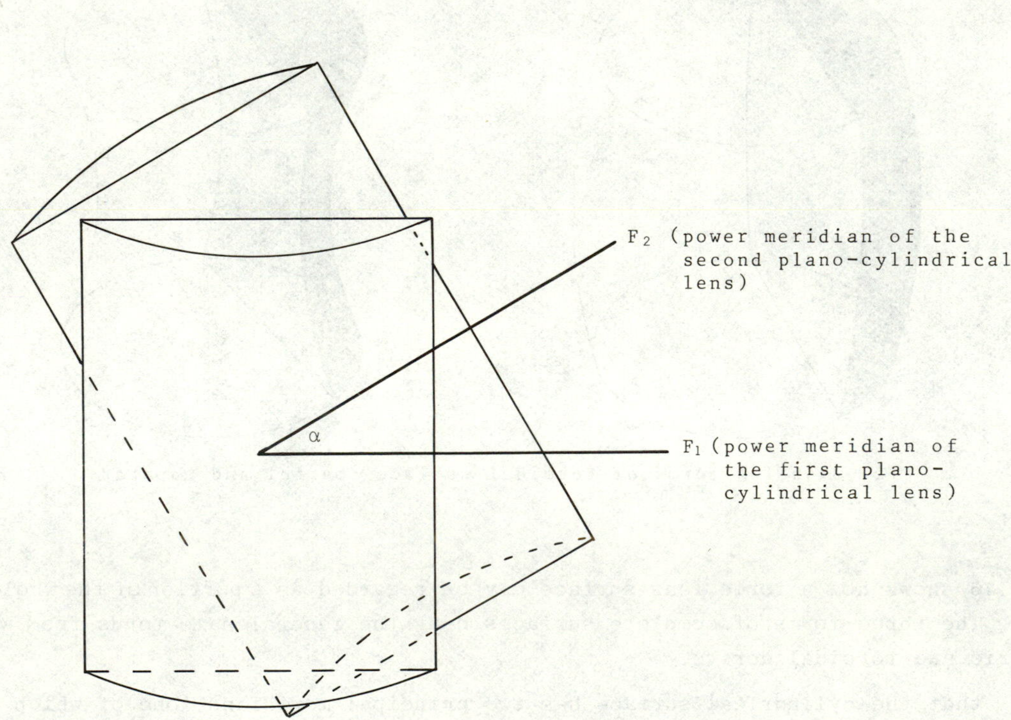

Fig. 3.47 Two convex plano-cylindrical lenses in contact with their power meridians inclined at an angle α.

In order to analyse the effect produced by placing two thin plano-cylindrical lenses in contact we need to borrow from mathematics a result about curvature at a point on a surface. Consider the point O on the surface indicated in figure 3.48, where the z-axis is normal to the surface at O.

Suppose the surface has maximum and minimum curvatures at the point O along the meridians designated by the x- and y-axes. That is, either the curve OB is a maximum curvature and the curve OE is a minimum curvature, or vice versa. Denoting these curvatures by R_x and R_y, then the curvature of the curve OD, occupying a vertical plane ODC, is given at the point O by

$$R_\theta = R_x \cos^2\theta + R_y \sin^2\theta .$$

This is the result borrowed from mathematics.

Now, imagine the surface in figure 3.48 is a convex cylindrical surface with R_x the power meridian and R_y the axis meridian. Then $R_y = 0$ and the curvature at O in the θ-meridian is

$$R_\theta = R_x \cos^2\theta = R \cos^2\theta,$$

having dropped the subscript x. R is the curvature in the power meridian of the cylindrical surface. The power of the cylindrical surface is $F = (n_g-1)R$, so we may write

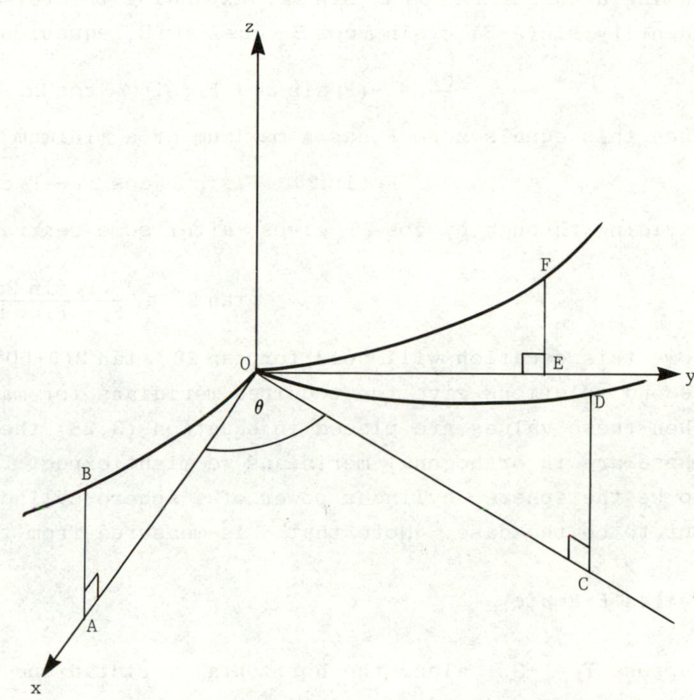

Fig. 3.48 A surface with principal meridians OB and OF. Oz is normal to the surface at O, and OD is a curve on the surface such that the angle AOC = θ.

$F_\theta = (n_g-1)R_\theta = (n_g-1)R\cos^2\theta = F\cos^2\theta$. That is,

$$F_\theta = F\cos^2\theta \qquad (3.27).$$

Note that F_θ is not a power in the sense that a pencil of rays will be changed in vergence by F_θ. In figure 3.41 we showed that a pencil of rays, in a meridian other that the axis or the power meridian of a cylindrical lens, undergoes skew refraction. Therefore, since the refracted rays do not converge towards or diverge from a point, we cannot define a power $F = L' - L$ since L' does not exist. However, we can make use of equation (3.27) to analyse the effect of two plano-cylindrical lenses in contact. Suppose we have two such lenses of powers F_1 and F_2 with their power meridians inclined at an angle α, figure 3.49 . Then along the θ meridian we have

$$F_\theta = F_{1\theta} + F_{2\theta} = F_1\cos^2\theta + F_2\cos^2(\theta-\alpha),$$

or, dropping the θ subscript,

$$F = F_1\cos^2\theta + F_2\cos^2(\theta-\alpha) \qquad (3.28).$$

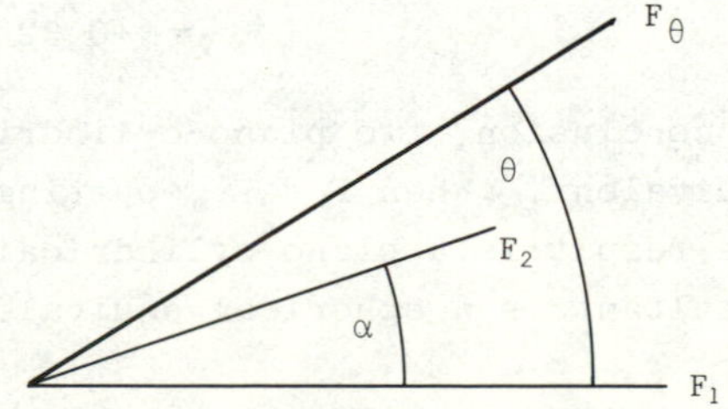

Fig. 3.49

We can find θ for the maximum and minimum values of F by differentiating equation (3.28) with respect to θ, since F is a function of θ for all θ.

$$\text{Thus,}\quad \frac{dF}{d\theta} = 2F_1(-\sin\theta)\cos\theta + 2F_2(-\sin(\theta-\alpha))\cos(\theta-\alpha)$$

$$= -F_1\sin 2\theta - F_2\sin 2(\theta-\alpha) \qquad (3.29),$$

having used $2\sin A\cos A = \sin 2A$. Expanding the term $\sin 2(\theta-\alpha)$, using the trigonometrical identity $\sin(A-B) = \sin A\cos B - \cos A\sin B$, equation (3.29) becomes

$$\frac{dF}{d\theta} = -(F_1\sin 2\theta + F_2(\sin 2\theta\cos 2\alpha - \cos 2\theta\sin 2\alpha)).$$

When this equals zero F has a maximum or a minimum value. That is, when

$$F_1\sin 2\theta + F_2\sin 2\theta\cos 2\alpha - F_2\cos 2\theta\sin 2\alpha = 0 .$$

Dividing through by $\cos 2\theta$ gives, after some rearrangement,

$$\tan 2\theta = \frac{F_2\sin 2\alpha}{F_1 + F_2\cos 2\alpha} \qquad (3.30).$$

Now, this equation will hold for $\tan 2\theta$, $\tan 2(\theta+90^0)$, $\tan 2(\theta+180^0)$, etc. The first and second solutions give the required meridians for maximum and minimum F; i.e. θ and $\theta+90^0$. When these values are placed in equation (3.28) the values of F are F_{max} and F_{min}. Since these are in orthogonal meridians we might expect F_{min} to be the sphere power, and F_{max} to be the sphere + cylinder power of a sphero-cylindrical equivalent lens, and this turns out to be the case. Note that θ is measured from the F_1 meridian.

Worked Example

Suppose $F_1 = +2$ D along the horizontal meridian and $F_2 = +3$ D along the 30^0 meridian. Then using equation (3.30)

$$\tan 2\theta = \frac{F_2\sin 2\alpha}{F_1 + F_2\cos 2\alpha} = \frac{3\sin 60^0}{2 + 3\sin 60^0} = 0.7423 ,$$

whence $\theta = 18.29^0$, measured from the F_1 meridian.

The maximum power lies along 18.29^0, so inserting this value in equation (3.28)

$$F_{max} = F_1\cos^2 18.29^0 + F_2\cos^2(18.29^0 - 30^0)$$

$$= +4.68\text{ D along } 18.29^0.$$

Then,

$$F_{min} = F_1\cos^2(18.29^0+90^0) + F_2\cos^2((18.29^0+90^0)-30^0)$$

$$= 2(-0.3138)^2 + 3(0.2030)^2$$

$$= +0.32\text{ D along } \theta+90^0 = 118.29^0.$$

In conclusion, two plano-cylindrical lenses in contact result generally in a sphero-cyl equivalent. When $F_1 = F_2$, putting $\alpha = 0^0$ and $\alpha = 90^0$ gives two particular results. When $\alpha=0^0$ the result is a plano-cylindrical equivalent lens equal to $2F_1$ (or $2F_2$). When $\alpha = 90^0$ the resultant is a spherical equivalent of power F_1 (or F_2).

EXERCISES

1 Explain the change in curvature of the wavefronts shown in figure 3.2 .

2 Define the term principal or optical axis when referring to a thin lens.

3 Find the positions of the optical centres in the following lenses:
(a) $F_1 = +10$ D, $F_2 = +10$ D, $t = 6$ mm, (b) $F_1 = +8$ D, $F_2 = 0$ D, $t = 5$ mm,
(c) $F_1 = +10$ D, $F_2 = -4$ D, $t = 3$ mm, (d) $F_1 = 0$ D, $F_2 = -9$ D, $t = 2$ mm,
(e) $F_1 = +4$ D, $F_2 = -14$ D, $t = 2$ mm.
Ans. (a) $A_1O = +3$ mm. (b) $A_1O = 0$. (c) $A_1O = -2$ mm. (d) $A_1O = +2$ mm. (e) $A_1O = +2.8$ mm.

4 Why are thin converging and diverging lenses also called positive (or plus) and negative (or minus) lenses, respectively?

5 A thin converging lens of focal length 10 cm forms an image of a 2 cm high object on the principal axis 50 cm from the lens. Find the position and size of the image by graphical construction.

Ans. The image is on the right of the lens, 12.5 cm away, 0.5 cm long, and inverted.

6 A 5 mm high object is placed on the principal axis 20 cm from a thin positive lens of focal length ⅓ m. Use a graphical construction to find the position, size, and nature of the image.

Ans. The image is erect and virtual, 50 cm from the lens and on the same side of the lens as the object, and 12.5 mm tall.

7 A thin diverging lens of focal length 20 cm forms an image of a real, 1 cm high object placed on the principal axis 1 m from the lens. Find graphically the position, size, and nature of the object.

Ans. The image is 16.67 cm to the left of the lens, erect, and 1.67 cm high.

8 For a thin lens in air show that $\frac{h'}{h} = 1 - \frac{\ell F}{1+\ell F}$. Use this relationship to show that a thin diverging lens always produces an erect but diminished image of a real, finite object.

9 Calculate the linear magnification produced when an object is placed 26⅔ cm from a biconvex lens with front and back surface radii of 10 cm and 20 cm, and $n_g = 1.5$.

Ans. -1 .

10 An object 16⅔ cm in front of a lens has its image formed 33⅓ cm on the other side of the lens. Find the focal length and the power of the lens.

Ans. +11.11 cm and +9 D.

11 If the lens in question 9 is immersed in water of refractive index 1.33 , calculate the linear magnification produced and the focal length of the lens.

Ans. $m = +1.47$ and $f' = +52.22$ cm.

12 A plano-concave lens is made of glass of refractive index 1.65 . Calculate the radius of curvature of the concave surface necessary to make the lens power -10 D.

Ans. +6.5 cm.

13 A parallel pencil of rays is incident on a +5 D thin lens. Where must a -10 D thin lens be placed with respect to the first lens to ensure parallel light emerges from the diverging lens. Illustrate with a diagram.
Ans. 10 cm to the right of the first lens.

14 A photographic transparency 35 mm × 25 mm is to be projected on to a screen 2.1 m from the the slide and to produce an image 700 mm × 500 mm. Calculate the power of the lens required and its position relative to the transparency.

Ans. $F = +10.5$ D. $\ell = -10$ cm.

15 A virtual object is 20 cm to the right of a thin lens of focal length +50 cm. Find the position of the image.

Ans. $\ell' = +14.29$ cm. See W.P.O., page 26, question 1(a).

16 A real object is 50 cm to the left of a thin diverging lens of focal length 33⅓ cm. If the object is 2 cm high, find the position, size, and nature of the image by calculation.

Ans. $\ell' = -20$ cm; $h' = +0.8$ cm. The image is virtual. See W.P.O., page 26, question 1(b).

17 Rework the last two questions using Newton's equations $xx' = -(f')^2$ and $m = -\frac{f}{x}$.

Ans. See W.P.O., page 27, question 1(c).

18 An eye requires a vergence of −4.76 D incident upon the cornea in order to produce a clear image on the retina. What thin lens worn at 10 mm from the cornea will provide this vergence when viewing a distant object through the lens?

Ans. $F = -5$ D. See W.P.O., page 28, question 2.

19 A lamp and a screen are 1 m apart and a +4.5 D thin lens is mounted between them. Where must the lens be placed in order to produce a sharp image on the screen, and what will be the magnification?

Ans. The lens may be placed either at ⅓ m or ⅔ m from the lamp. The magnification is then −2 or −½, respectively.

20 Calculate the focal length of the lens required to produce an image of a given object on a screen with linear magnification 3.5, the object being situated 20 cm from the lens.

Ans. $f' = +15.56$ cm. See W.P.O., page 31, question 5 .

21 The difference in the positions of the image when the object is first at a great distance (at infinity) and then at 5 m from the first focal point of a thin converging lens is 3 mm. Find the focal length of the lens.

Ans. $f' = +12.25$ cm. See W.P.O., page 32, question 6.

22 A real, inverted image of an object is formed on a screen by a thin lens, and the image and object are equal in size. When a second thin lens is placed in contact with the first the screen must be moved 2 cm nearer the lenses in order to obtain a clear image, and the size of this image is ¾ that of the first image. Find the focal lengths of the two lenses.

Ans. +4 cm and +24 cm. See W.P.O., page 32, question 7.

23 An image formed on a screen by a thin positive lens is 10 cm long. Without moving either the object or the screen, which are 300 cm apart, a second image can be produced which is 40 cm long. Show how this is possible, and find the power of the lens and the size of the object.

Ans. $h = 20$ cm and $F = +1.50$ D. See W.P.O., page 8, question 8.

24 A thin positive lens produces a real image of a real object. Show that the shortest possible distance between the object and the image is 4f', and that this occurs when $\ell' = 2f'$ and $\ell = -2f'$. See W.P.O., page 36, question 9.

25 A thin converging lens of focal length f' is to cast a real image N times greater than the object's size. Show that $\ell' = (N+1)f'$.

Ans. See W.P.O., page 39, question 12.

26 An object 2 cm high is situated 50 cm in front of a lens the front surface of which is convex with a radius of curvature of 50 cm. If a real image 4 cm high is produced by the lens, what is the radius of curvature of the second surface if $n_g = 1.5$?

Ans. $r_2 = -25$ cm. See W.P.O., page 13, question 13.

27 A thin lens of refractive index n_g has a focal length f_a' in air. Show that its focal length f_w' when immersed in water of refractive index n_w is given by

$$f_w' = n_w\left(\frac{n_g - 1}{n_g - n_w}\right)f_a'.$$

28 A meniscus lens of power +6 D is made of glass of refractive index 1.7 but is worked with tools calibrated for refractive index 1.523. If the tool used for the back surface is marked -5 D, what tool must be used for the front surface?

Ans. The tool must be +8.74 D.

29 An object 5 cm from the first focal point of a thin converging lens produces a real image 20 cm from the second focal point. If the radius of the first surface is +0.2 m and the refractive index of the glass is 1.5, find the radius of the second surface and the linear magnification.

Ans. $r_2 = -6.667$ cm and $m = -2$.

30 A ray is deviated through 5^0 by a thin converging lens when incident at a point 1 cm above the principal axis. What is the power of the lens?

Ans. +8.749 D.

31 A ray of light parallel to and 5 mm above the principal axis is incident on the plane surface of a thin plano-concave lens. If the angle of incidence at the second surface is 5.7248^0, find the power of the lens.

Ans. F = -10.06 D.

32 A thin lens is made with a convex spherical surface of radius +20 cm and a convex cylindrical surface of radius -10 cm in the horizontal meridian. What is the power of the lens in the horizontal and vertical meridians if $n_g = 1.5$?

Ans. Horizontal meridian: +7.5 D. Vertical meridian: +2.5 D.

33 A lens measure placed on a cylindrical surface reads +1.00 D. When rotated through 30^0 it reads +3.00 D. If neither position of the lens measure is along the power meridian, find the power of the cylindrical surface.

Ans. +4.00 D.

34 Find the equivalent lens when two +2 D plano-cylindrical lenses are placed in contact with their power meridians making an angle α equal to (i) 60^0, (ii) 90^0, (iii) 0^0. Assume the power meridian of one cylinder is along the zero meridian.

Ans. (i) +1 D sphere with a +2 D cylinder with its power meridian along 30^0.
(ii) +2 D sphere.
(iii) Plano/+4 D cylinder with its power meridian along 0^0.

35 Two plano-cylindrical lenses, F_1 and F_2, placed in contact are equivalent to a sphero-cylindrical lens with principal powers F_{min} and F_{max}.
Show that (i) $F_{min} + F_{max} = F_1 + F_2$, and (ii) if S and C represent the sphere and cylinder powers of the equivalent lens, then $2S + C = F_1 + F_2$.

36 A thin converging toric lens with a circular shape of diameter 40 mm is made with a plane front surface and a back surface with horizontal and vertical radii of curvature of 6.5 cm and 13 cm, respectively. Find the positions and dimensions of the two line foci and the circle (disc) of least confusion when the object is a luminous point source on the principal axis ⅓ m from the lens. Assume $n_g = 1.5$. Which line focus is vertical?

Ans. $h_1 = 28.57$ mm. $h_2 = 100$ mm. $h_c = 22.22$ mm. The line focus nearer the lens is vertical.

4 LENS SYSTEMS

4.0 INTRODUCTION

In Chapter 3 we discussed the optical behaviour of single thin lenses where the thickness of the lens material was small enough to be ignored. From this it can be said that the vergence of the light leaving the first surface of the lens (inside the lens material) is unchanged when the light meets the second surface of the lens. This led us to neglect the distance between the surfaces. In this chapter, however, we shall be considering those cases where the distances between thin lenses or between surfaces cannot be ignored.

In the first part of the present chapter we shall be considering *simple lens systems* made up of two separated thin lenses in air. In all cases the lenses forming the system will be assumed to be *co-axial* or *centred*; that is, the principal axes of the component lenses will be coincident. The applications of lens systems to optical instruments will be dealt with in Chapter 8.

The second part of this chapter is devoted to the optical theory of *thick lenses* where the thickness of the lens material may not be ignored. Much of the theory of simple lens systems is applicable to thick lenses.

Then, the last two sections of the chapter will consider the general case of an optical system in which the refractive indices of the image and object spaces are different. The schematic eye is an important example of this type of system.

Once again, we will ignore lens aberrations and only consider the paraxial region of the lenses and surfaces.

4.1 SIMPLE LENS SYSTEMS - Two thin lenses in air

A simple lens system is understood to be made up from two centred thin lenses in air, and separated by a specified distance. We may recall from section 3.1.4.3 that a single thin lens in air has a single value for its refracting power, but that it has two numerically identical focal lengths. These focal lengths, called the first and second focal lengths, are both measured from the centre of the lens.

If two thin lenses are combined in contact, theoretically in the same plane, the equivalent power F_E of the combination is given by equation (3.17), namely $F_E = F_1 + F_2$, where F_1 and F_2 are the individual lens powers; see figure 4.1.

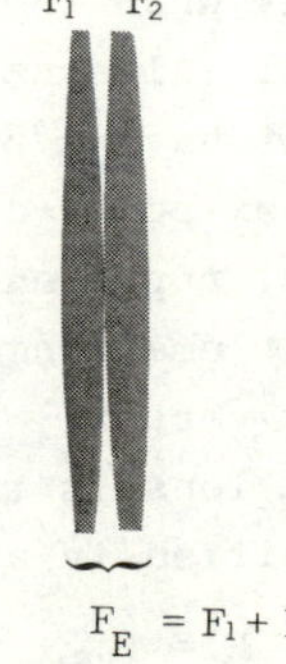

Fig. 4.1

When the two thin lenses are separated by a certain distance the equivalent power of the system changes, the value now depending on the powers F_1 and F_2 and also on the separation d of the two lenses. The equivalent power of such systems will be considered in sections 4.1.2.2 and 4.1.6.1.

It is also useful, and particularly convenient in the case of spectacle lenses, to express the refracting ability of a system (or a spectacle lens) in terms of its *Back Vertex Power (BVP)* and also its *Front Vertex Power (FVP)*.

4.1.1 VERTEX POWERS AND VERTEX FOCAL LENGTHS

Figure 4.2 illustrates a simple lens system made up of two centred positive lenses in air, separated by a distance d. A_1 is called the *front vertex* of the system and A_2 is the *back vertex* of the system.

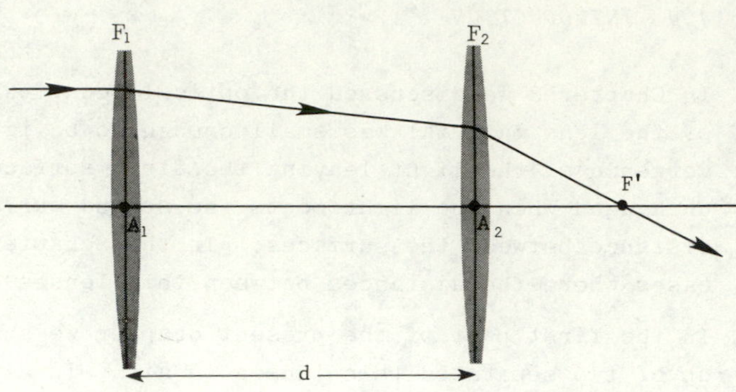

Fig. 4.2 Two thin co-axial lenses separated by a distance d.

A ray of light parallel to the principal axis is incident from the left and is refracted by each lens, finally crossing the axis at the point F'. As the incident light is parallel to the axis, F' will be the second principal focal point of the system. Now let us imagine that the lens F_1 is moved through the distance d until it is in contact with the lens F_2. However, in order that the principal focal point F' of the system remains fixed in the same position, the power of F_1 must now be changed to a new value. We can consider this procedure with the aid of the concept of effectivity (see section 3.1.6).

Hence, using equation (3.12) we can write the effectivity of lens F_1 in the plane of F_2 as

$$F_{1x} = \frac{F_1}{1 - dF_1} .$$

Thus, the lens system can now be illustrated as shown in figure 4.3, with two thin lenses of powers F_{1x} and F_2 effectively in contact with each other at the back vertex A_2. The total power of the system *in the plane of* A_2 is now $F_{1x} + F_2$.

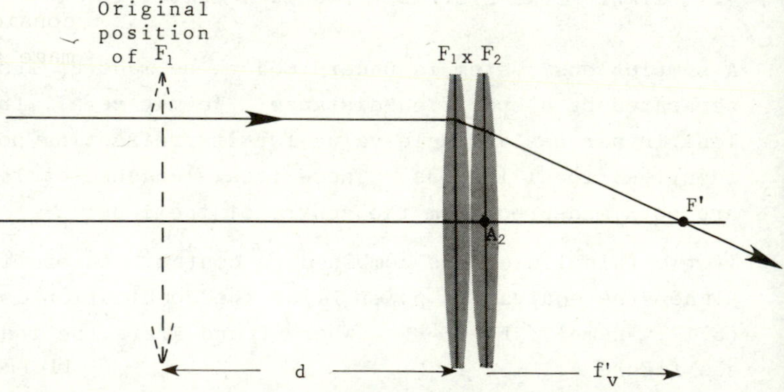

Fig. 4.3 The power of two lenses separated by a distance d, measured in the plane through the back vertex A_2.

The total power of a lens system, referred to the plane through the back vertex, is called the *Back Vertex Power* of the system. The back vertex power is designated F'_v. It represents the change in the vergence of light passing through the system from left to right when the incident light is parallel. Alternatively, with the incident light having zero vergence, the back vertex power of a system, or thick lens, represents the vergence of the light emerging from the plane of the back vertex. The back vertex power of a lens is the power which is specified in a spectacle prescription.

So, $$F'_v = F_{1x} + F_2 = \frac{F_1}{1-dF_1} + F_2 = \frac{F_1}{1-dF_1} + \frac{F_2.(1-dF_1)}{1-dF_1} = \frac{F_1 + F_2(1-dF_2)}{1 - dF_1} = \frac{F_1 + F_2 - dF_1 F_2}{1 - dF_1} .$$

That is, the back vertex power is

$$F'_v = \frac{F_1 + F_2 - dF_1 F_2}{1 - dF_1} \qquad (4.1),$$

where d is expressed in metres.

The *Back Vertex Focal Length*, symbol f'_v, is the distance from the back vertex of the system to the second principal focal point. This is the distance A_2F' and, *in air*, is given by

$$A_2F' = f'_v = \frac{1}{F'_v} \qquad (4.2).$$

The *Front Vertex Power* of a system can now be considered in a similar manner. To do this, let us firstly reverse the lenses so that F_1 becomes the second lens and F_2 the first lens, and then repeat the procedure as before.

Note that the point F is the first principal focal point of the system, but that it is now on the right hand side of the system as the lenses have been temporarily reversed.

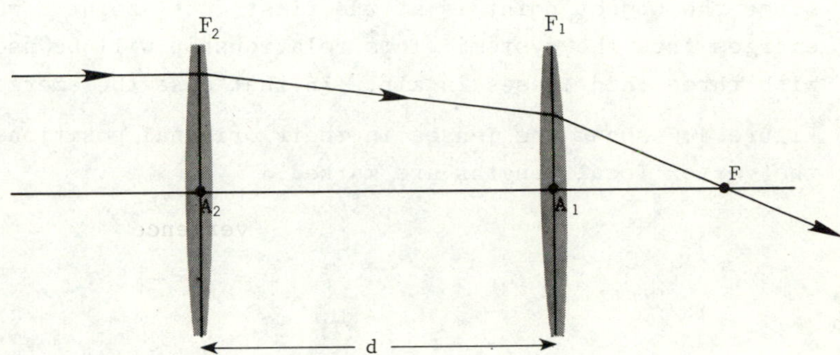

Fig. 4.4 The lens system in figure 4.2 turned around so that the lens F_2 is on the left.

Using equation (3.12), we can write the effectivity of the lens F_2 in the plane of F_1 as

$$F_{2x} = \frac{F_2}{1 - dF_2}.$$

The lens system may now be drawn as illustrated in figure 4.5 with two thin lenses, of powers F_{2x} and F_1, effectively in contact with each other at the front vertex A_1.

The total power of the system in the plane of F_1 is now $F_{2x} + F_1$ as the lenses are in contact.

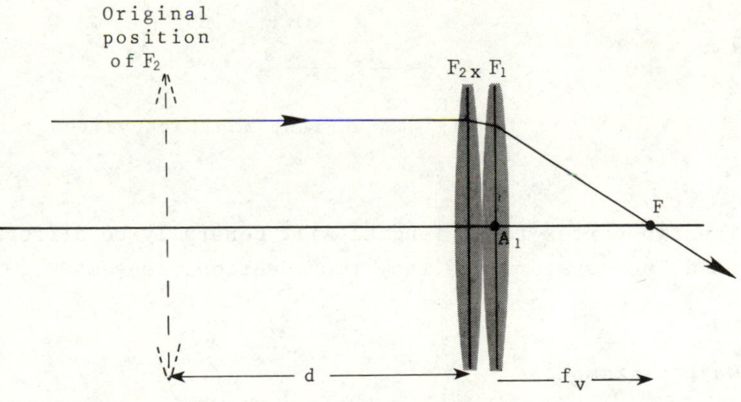

Fig. 4.5 Diagram used for the derivation of front vertex power.

The total power of the lens system, referred to the plane through the front vertex, is called the *Front Vertex Power* of the system. The front vertex power is represented by the symbol F_v, the subscript v is for vertex, of course. Note that the front vertex power is measured at the left of the system, in object space, and is an undashed symbol, whereas the back vertex power is measured in image space and carries a dash on its symbol.

The expression for the front vertex power of the system is derived from the effectivity thus:

$$F_v = F_{2x} + F_1 = \frac{F_2}{1-dF_2} + F_1 = \frac{F_2}{1-dF_2} + \frac{F_1(1-dF_2)}{1-dF_2} = \frac{F_2 + F_1(1-dF_2)}{1-dF_2} = \frac{F_1 + F_2 - dF_1F_2}{1-dF_2}.$$

$$\text{That is,} \quad F_v = \frac{F_1 + F_2 - dF_1F_2}{1-dF_2} \qquad (4.3).$$

The *Front Vertex Focal Length*, symbol f_v, is the distance from the front vertex of the system to the first principal focal point. This is the distance A_1F ; that is, $A_1F = f_v$. If we now reverse the lenses back to their original positions we must say that $F_v = -\frac{1}{f_v}$,

$$\text{or} \qquad f_v = -\frac{1}{F_v} \qquad (4.4),$$

as the distance f_v is now measured against the direction of the incident light. This is exactly analagous to the thin lens in air where the focal lengths are related to the thin lens power *in air* thus: $f = -\frac{1}{F}$ and $f' = \frac{1}{F}$.

When the lenses are in their correct orientation, i.e. F_1 is to the left of F_2, and the emergent vergence $L_2' = 0$, the incident vergence is

$$L_1 = \frac{n_1}{\ell_1} = \frac{1}{\ell_1} = \frac{1}{f_v} = -F_v$$

since the object point is at the first focal point F when parallel light (zero vergence) emerges from the system. This relationship will be used in section 4.4 to solve a problem with three thin lenses in air. In that case the emergent vergence will be $L_3' = 0$.

Figure 4.6 shows the lenses in their original positions and illustrates the way in which the vertex focal lengths are marked.

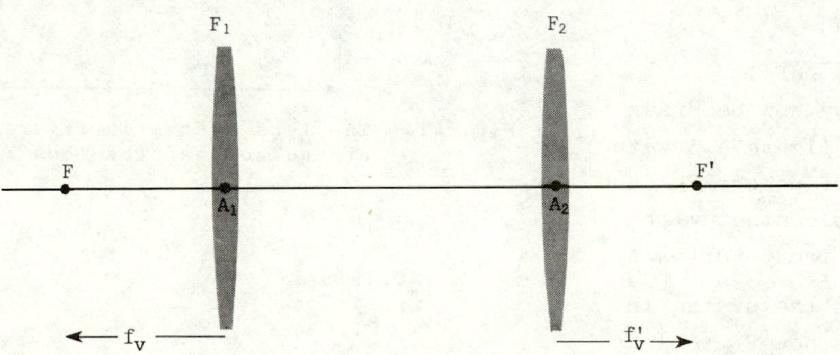

Fig. 4.6 Front and back vertex focal lengths.

The two vertex focal lengths will generally be different values, except in the special case of a lens system made from two identical lenses.

Worked Examples

1 A simple lens system consists of two thin lenses in air of powers +4.00 D and +6.00 D separated by 75 mm. By determination of the front and back vertex focal lengths, locate the positions of the two principal foci.

The front vertex power (FVP) is $F_v = \dfrac{F_1 + F_2 - dF_1F_2}{1 - dF_2}$

$$= \frac{4 + 6 - 0.075 \times 4 \times 6}{1 - 0.075 \times 6}$$

$$= \frac{10 - 1.8}{0.55} = +14.91\ \text{D}.$$

The front vertex focal length is then

$$f_v = -\frac{1}{F_v} = -\frac{1}{(+14.91)} = -0.0671\text{ m} = -67.1\text{ mm}.$$

This locates F at 67.1 mm to the left of A_1.

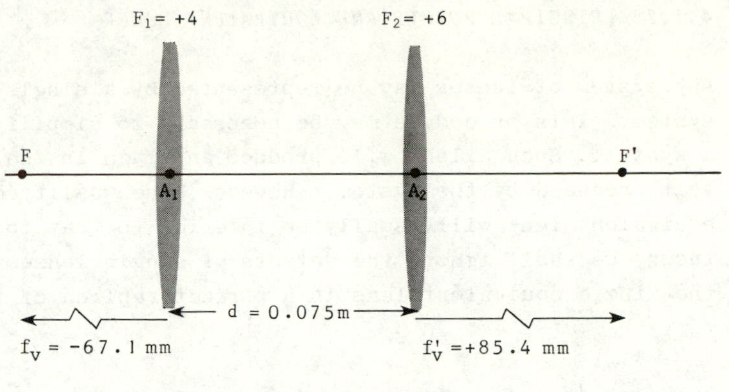

Next, the back vertex power (BVP) is

$$F'_v = \frac{F_1 + F_2 - dF_1 F_2}{1 - dF_1}$$

$$= \frac{8.2}{1 - 0.075 \times 4}$$

$$= \frac{8.2}{0.7} = +11.71\text{ D}$$

The back vertex focal length is then $f'_v = \frac{1}{F'_v} = \frac{1}{(+11.71)} = +0.0854\text{ m} = +85.4\text{ mm}$. This puts the second principal focus at

$$A_2 F' = +85.4\text{ mm}$$

or 85.4 mm to the right of the back vertex A_2.

Note here that d was expressed in metres. It is important to consider why this should be. Consider the term $dF_1 F_2$, and recall that the dioptre unit is really the reciprocal metre (m^{-1}). In the equations for FVP and BVP all the terms must have the same units of dioptres or reciprocal metres. For this to be the case for $dF_1 F_2$ we must have a units relationship

$$dF_1 F_2 \Rightarrow \text{units } m.m^{-1}.m^{-1} = m^{-1} = D.$$

With d in cm, say, we would have

$$dF_1 F_2 \Rightarrow \text{units } (cm).m^{-1}.m^{-1} = (100m).m^{-1}.m^{-1} = 100m^{-1}$$

which is not the dioptre (m^{-1}).

2 Find the positions of F and F' for a Huygens eyepiece consisting of two lenses of powers +10 D and +20 D separated in air by 75 mm.

The FVP $= F_v = \frac{F_1 + F_2 - dF_1 F_2}{1 - dF_2}$

$$= \frac{10 + 20 - 0.075 \times 10 \times 20}{1 - 0.075 \times 20}$$

$$= \frac{30 - 15}{-0.5}$$

$$= -30\text{ D}.$$

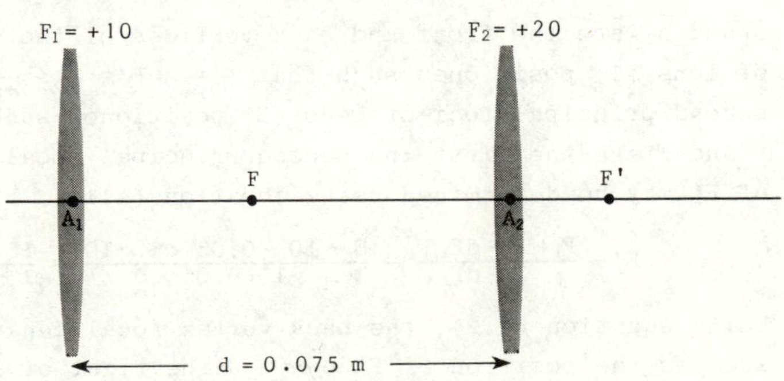

Hence, $f_v = -\frac{1}{F_v} = -\frac{1}{(-30)} = +0.0333\text{ m}$

and $A_1F = +33.3$ mm.

This locates F at a distance 33.3 mm to the right of A_1.

Next, the BVP $= F'_v = \frac{F_1 + F_2 - dF_1 F_2}{1 - dF_1} = \frac{15}{1 - 0.075 \times 10} = \frac{15}{0.25} = +60\text{ D}.$

The back vertex focal length is therefore $f'_v = \frac{1}{F'_v} = \frac{1}{+60} = +0.01667\text{ m} = +16.67\text{ mm}$, which places F' 16.67 mm to the right of A_2.

4.1.2 PRINCIPAL POINTS AND EQUIVALENT POWER

Any system of lenses may be represented by a single thin lens which is equivalent to the system. This procedure may be necessary to simplify a consideration of the behaviour of a system. Such a lens will produce an image in the same position and of the same size as that produced by the system. However, the quality of the image produced by the single equivalent lens will usually be inferior to that formed by the system. For the present theory we shall ignore the defects of single lenses and assume that the image formed by the single equivalent lens is a perfect replica of that produced by the lens system.

4.1.2.1 *Positions of the Equivalent Thin Lens*

Figure 4.7 shows a centred lens system consisting of two thin lenses in air having powers +5 D and +10 D, separated by 5 cm.

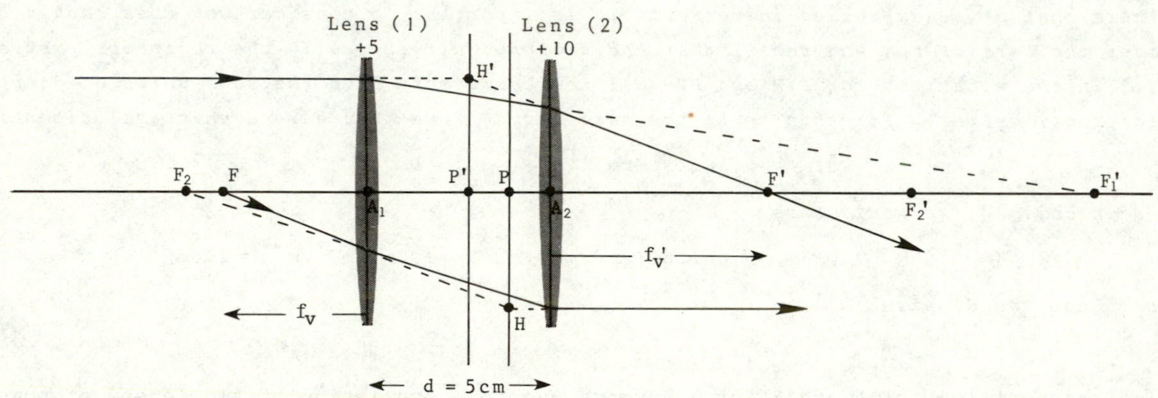

Fig. 4.7 The principal points and the equivalent thin lens positions.

A_1 and A_2 are the front and back vertices of the system. F_1' is the second principal focus of lens (1) positioned such that $f_1' = A_1F_1' = \frac{1}{F_1} = \frac{1}{+5}$ m = +20 cm. F_2 and F_2' are the first and second principal foci of lens (2) positioned such that $f_2' = A_2F_2' = +10$ cm and $f_2 = A_2F_2 = -10$ cm. F and F' are the first and second principal focal points of the lens system. The position of F' may be determined using equation (4.1). From equation (4.1) we have for the BVP

$$F_v' = \frac{F_1 + F_2 - dF_1F_2}{1 - dF_1} = \frac{5 + 10 - 0.05 \times 5 \times 10}{1 - 0.05 \times 5} = \frac{15 - 2.5}{1 - 0.25} = \frac{12.5}{0.75} = +16.67 \text{ D}.$$

Using equation (4.2), the back vertex focal length is $f_v' = A_2F' = \frac{1}{F_v'} = \frac{1}{+16.67} = +0.06$ m. This locates the position of F' 6 cm to the right of A_2.

Similarly, the position of F may be determined using equations (4.3) and (4.4). From equation (4.3)

$$F_v = \frac{F_1 + F_2 - dF_1F_2}{1 - dF_2} = \frac{5 + 10 - 0.05 \times 5 \times 10}{1 - 0.05 \times 10} = \frac{12.5}{0.5} = +25 \text{ D},$$ whereupon using equation (4.4) gives

$$A_1F = f_v = -\frac{1}{F_v} = -\frac{1}{(+25)} = -0.04 \text{ m} = -4 \text{ cm},$$ so that F is 4 cm to the left of A_1.

We can now locate the positions of the single thin lens which is equivalent to the system by means of a scale drawing. Figure 4.7 should be drawn to scale by the reader and constructed as follows.

(1) *Let a ray of light, parallel to the axis, be incident on lens (1). This ray undergoes refraction towards F_1', but on its way it strikes lens (2) and suffers further refraction through F', the second principal focus of the lens system.*

(2) *By producing the incident ray forwards and the emergent ray backwards until they intersect, we can see that the system behaves as if it were a single thin lens placed in the plane P'H'. The plane P'H' is called the SECOND PRINCIPAL PLANE of the lens system and this is the position of the single equivalent lens as far as the emergent light is concerned. The point P', where the plane P'H' cuts the principal axis of the system, is referred to as the SECOND PRINCIPAL POINT of the system.*

(3) *Similarly, let a ray, parallel to the axis, be incident from the right on lens (2). This ray is refracted towards F_2, then undergoes a further refraction to pass through F, the first principal focus of the system. The direction of this ray may now be considered in reverse so that it travels left to right.*

(4) *As before, by producing the incident ray forwards and the emergent ray backwards until they intersect each other, we now locate the plane PH. It is seen that the system is now behaving as if a single thin lens were placed in this plane. The plane PH is called the FIRST PRINCIPAL PLANE of the system and this is the position of the single equivalent lens as far as incident light is concerned. The point P where the plane PH cuts the principal axis of the system is referred to as the FIRST PRINCIPAL POINT of the system.*

It was Gauss (1840) and Listing (1851) who introduced the idea of principal points to greatly simplify calculations on image formation by thick lenses and lens systems. The positions of the principal points P and P', and hence the positions of the principal planes PH and P'H', may be determined by a scale drawing (as just described) or by means of calculation, see section 4.1.5.3 .

4.1.2.2 *Equivalent Power and Equivalent Focal Lengths*

We have seen that a lens system can be replaced by a single equivalent lens and that this lens may be imagined to be positioned at P as far as the incident light is concerned, but at P' as far as emergent light is concerned. The power of the single equivalent lens represents the power of the lens system and is called the *equivalent power* of the system. It is given the symbol F_E. Figure 4.8 is a simplified version of figure 4.7 and shows the positions of the principal points P and P', and the principal focal points F and F' of the lens system.

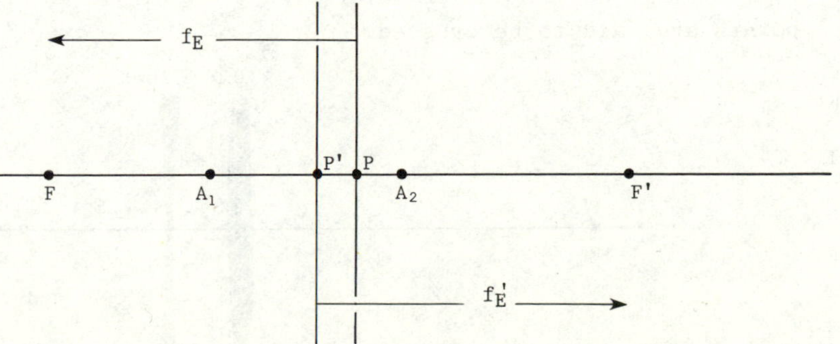

Fig. 4.8 The focal lengths of the equivalent thin lens.

Since P' is the position of the single equivalent lens with respect to light emerging from the system, then the distance P'F' represents the *second equivalent focal length* of the system, denoted by the symbol f_E'; that is, $f_E' = P'F'$.

Similarly, as P is the position of the single equivalent lens with respect to light entering the system, the distance PF represents the *first equivalent focal length* of the system and is denoted by the symbol f_E; that is, $f_E = PF$. It will be found (check this from your scale drawing of figure 4.7) that for a lens system in air the distance PF is the same as the distance P'F'; that is

the magnitude or modulus of P'F' = the magnitude or modulus of PF

or, in symbols, $|P'F'| = |PF|$.

However, when taking into account the sign convention for optical distances, we note that $f_E' = P'F'$ is a positive distance, but that $f_E = PF$ is a negative distance since it is measured against the incident light. Hence, we can say that $P'F' = -PF$, or

$$f_E = -f_E' \qquad (4.5)$$

and this applies to any simple lens system in air, as it does for a thin lens.

The equivalent power F_E of the lens system in air is thus related to the first and second equivalent focal lengths f_E and f_E' by the thin lens expressions

$$F_E' = \frac{1}{f_E'} \qquad (4.6) \qquad \text{and} \qquad F_E = -\frac{1}{f_E} \qquad (4.7).$$

4.1.2.3 *Examples of Principal Planes of Lens Systems*

Figure 4.9 shows the positions of the principal points and also the principal foci of some simple lens systems. As the reader gains more experience in the treatment of lens systems it will soon become apparent that the principal points of lens systems can lie in a variety of positions depending on the powers of the component lenses, and on their spacing. In those cases where P' lies to the left of P (see figures 4.8, 4.9(a) and (b)) the principal points are said to be crossed.

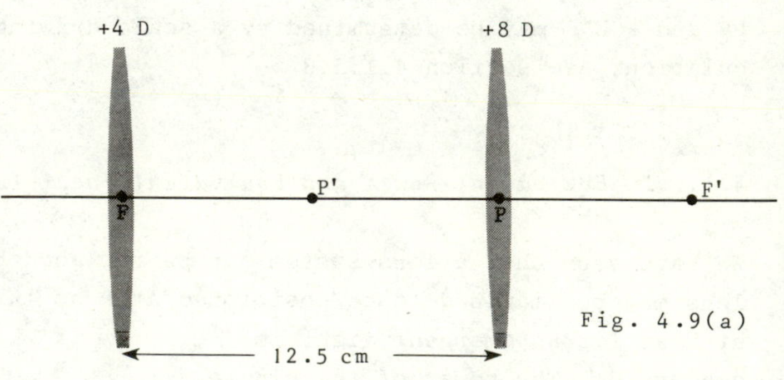

Fig. 4.9(a)

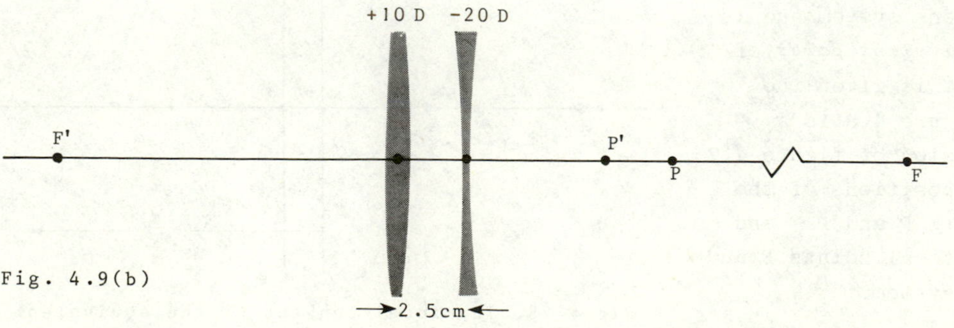

Fig. 4.9(b)

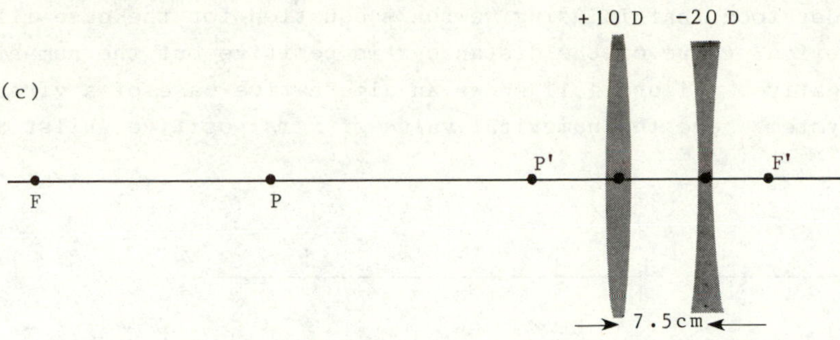

Fig. 4.9 Positions of principal points and principal foci of some simple lens systems. (The scales are drawn differently in each case).

4.1.3 OBJECT AND IMAGE POSITIONS, AND MAGNIFICATION

When a lens is used in combination with another lens to form an image, we may determine the position and size of the image by various methods. The choice of method is often one of convenience, and may be selected from the following:

4.1.3.1 Newton's Formulae

In our consideration of a single thin lens we saw (section 3.1.8) that it is possible to specify the positions of conjugate planes, such as an object BT and its image B'T', by the distances x and x' from the appropriate focal points. As such, Newton's formula applies and is given again here for reference: $(f')^2 = -xx'$ or $ff' = xx'$.

Newton's formula may also be applied to a lens system in the form

$$(f_E')^2 = -xx' \quad \text{or} \quad f_E f_E' = xx' \qquad (4.8)$$

where f_E and f_E' are the first and second equivalent focal lengths of the system, x is the extra-focal object distance FB (meaning measurement from F to B), and x' is the extra-focal image distance F'B' (meaning measurement from F' to B'). This is shown in figure 4.10 where f_E' is the distance P'F'.

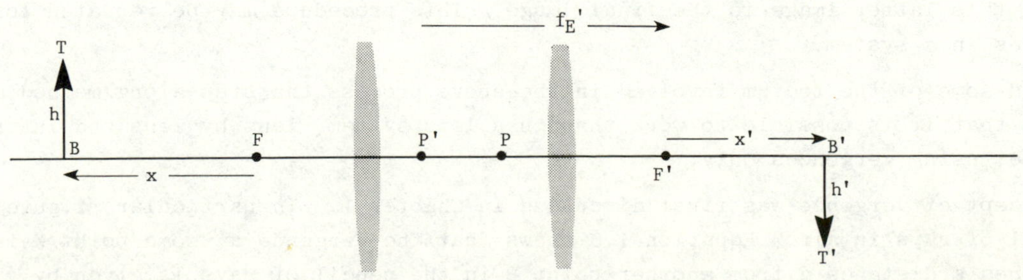

Fig. 4.10 Newton's formula with a real object and real image.

It should be clearly understood that in using Newton's equation for the case illustrated in figure 4.10, the numerical value of the distance x' is positive but the numerical value of the distance x is negative. Figure 4.11 shows an alternative case of a virtual image B'T' formed by a lens system where the numerical value of x is positive whilst x' has a negative value.

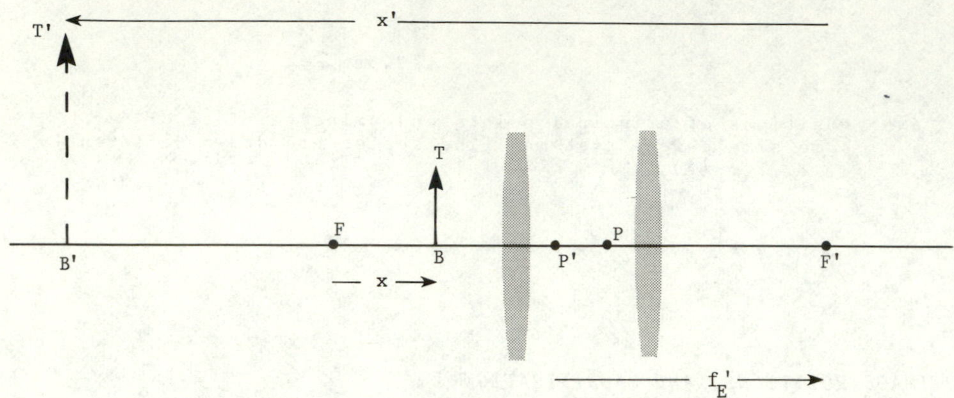

Fig. 4.11 Newton's formula with a real object and a virtual image.

The magnification of the image may be determined from Newton's magnification relationship since from section 3.1.8 we have

$$\text{linear magnification } m = \frac{h'}{h} = -\frac{f_E}{x} = -\frac{x'}{f_E'} = -x'F_E$$

where h is the height of the object and h' is the height of the image.

As in the case of a single thin lens, an upright image results in a positive value for the magnification, whilst an inverted image corresponds to a negative value for the magnification.

4.1.3.2 Step-along Vergences

The "step-along" procedure is a systematic and useful method whereby the effect of each lens in a lens combination is considered in turn. In terms of distances, the basic idea is that the position of the image produced by the first lens alone is computed, disregarding the second lens. This image is then considered as the object for the second lens for which the image produced by the second lens alone may then be located. In the case of a two lens system, this latter image is the final image. This procedure may be repeated for any number of lenses in a system.

To avoid some of the tedium involved in the above process the step-along method may be developed so that it is possible to work through a lens system, lens by lens and intervening space by space, using vergences only.

The concept of vergence was first discussed in Chapter 1. In particular, figure 1.12 showed a pencil of rays in air. Equation 1.5 shows that the vergence at some point X in figure 1.12 positioned a distance d from another point B in the pencil of rays is given by

$L'_X = L'_B/(1-dL'_B)$. Now, if we let B represent the point where light emerges from the *first* lens in a system, L'_B may be written as L'_1. Similarly, if the point X represents the position of the *second* lens in the system, L'_X will be the vergence arriving a this *second* lens, and can be denoted by L_2.

Hence, we may re-write equation (1.5) in the vergence form suitable for use with lenses in a system, namely

$$L_2 = \frac{L_1'}{1 - dL_1'} \qquad (4.9)$$

where d is the spacing of the lenses (measured in metres), L_2 is the vergence arriving at the second lens, and L_1' is the vergence leaving the first lens.

Figure 4.12 shows the symbols used to denote the lens powers, together with the vergences arriving at and leaving each lens in turn in a two lens system.

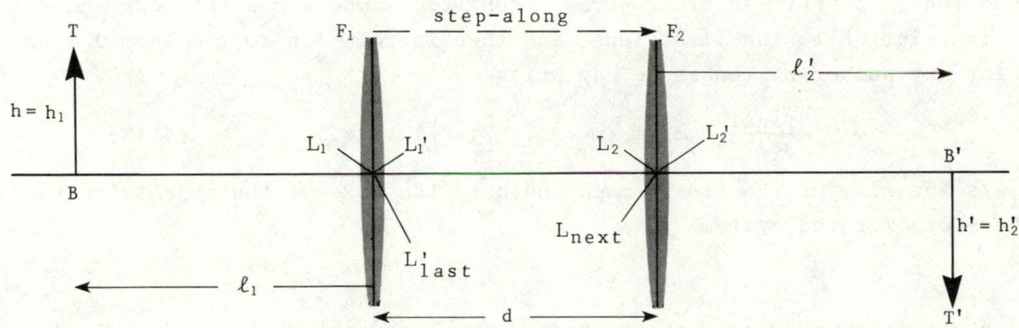

Fig. 4.12 The symbols used in the Step-along Vergences method. Note that h is the object for the system but also the object when considering the first lens alone. It is therefore denoted by h_1. Similarly, h' is the image formed by the system and by refraction at the second lens and is also labelled h_2'.

So, to summarise the step-along method for finding the image position:

(1) Given the object distance ℓ_1 *from the first lens in air, write* $L_1 = \frac{1}{\ell_1}$.

(2) Find L_1'*, the vergence emerging from the first lens, by using* $L_1' = L_1 + F_1$.

(3) Step-along to the second lens to find the vergence arriving there, given by

$L_2 = \frac{L_1'}{1 - dL_1'}$.

(4) Find L_2'*, the vergence leaving the second lens, using* $L_2' = L_2 + F_2$.

(5) If this is the final lens, obtain the image distance from the second lens using $\ell_2' = \frac{1}{L_2'}$.

In general terms, a useful way of expressing the step-along equation (equation (4.9)), in particular where more than two lenses, or surfaces, are present is

$$L_{next} = \frac{L'_{last}}{1 - dL'_{last}} \qquad (4.10)$$

where L_{next} means the vergence of the light arriving at the next lens, and L'_{last} means the vergence of the light leaving the last (previous) lens. These terms are illustrated in the lower part of figure 4.12 .

The magnification of the image formed by a single lens is given by equation (3.10), namely

$$m = \frac{\bar{L}}{\bar{L}'} \text{, which reduces to } m = \frac{L}{L'} \text{, in air.}$$

This is only applicable for objects and images at finite distances from a lens, such that $L \neq 0$ and $L' \neq 0$. For each lens in a two lens system in air the individual magnifications are

$$m_1 = \frac{h_1'}{h_1} = \frac{L_1}{L_1'} \quad \text{and} \quad m_2 = \frac{h_2'}{h_2} = \frac{L_2}{L_2'} .$$

The overall magnification for the system is then given by

$$m = \frac{\text{final image size}}{\text{object size}} = \frac{h_2'}{h_1} = \frac{h_2'}{h_2} \times \frac{h_1'}{h_1} = m_2 \times m_1 = m_1 \times m_2 = \frac{L_1}{L_1'} \times \frac{L_2}{L_2'} \qquad (4.11)$$

where it should be noted that, since the image formed by the first lens is then the object for the second lens, $h_1' = h_2$. In other words, these two symbols describe one and the same thing, first in relation to the first lens, and then in relation to the second lens.
In general, for any number of lenses we may write

$$m = \frac{h'_{final}}{h_1} = m_1 \times m_2 \times m_3 \times \ldots = \frac{L_1}{L_1'} \times \frac{L_2}{L_2'} \times \frac{L_3}{L_3'} \times \ldots \qquad (4.12)$$

where h'_{final} is the size of the final image and h_1 is the size of the object for the first lens, and therefore for the system.

When the object is at a very large distance from the lens system (the object is described as being located at infinity) equation (4.11) cannot be used since L_1 is zero. Instead, let us consider figure 4.13 .

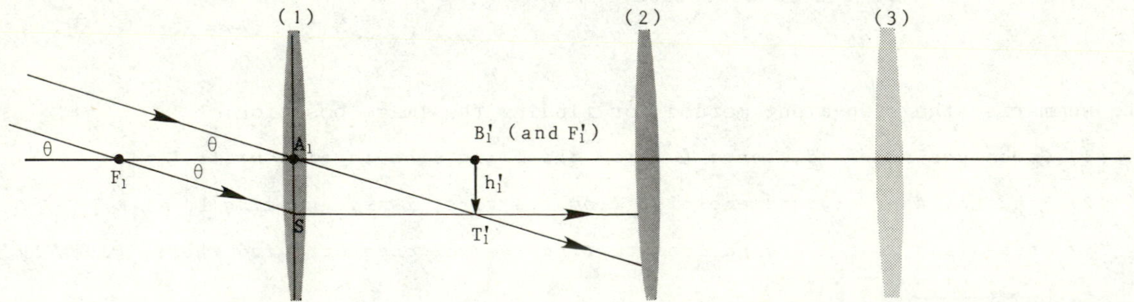

Fig. 4.13 Object at infinity - figure used to derive the magnification equation.

Two rays are shown originating from the tip of an object at an infinite distance from the lens system. These rays are essentially parallel to each other when meeting the first lens, and each makes an angle θ with the principal axis of the system. Since the ray F_1S passes through F_1, which is the first principal focus of lens (1), the refracted ray ST_1' will be parallel to the axis of the system. Hence, $A_1S = h_1'$, where h_1' is the height of the real image $B_1'T_1'$ formed by the first lens. (N.B. On many occasions this image will be virtual if it falls beyond the second lens. The theory is the same for both cases).

Now in triangle $F_1 A_1 S$,

$$\frac{A_1 S}{A_1 F_1} = \tan\theta .$$

But $A_1 S = h_1'$ and $A_1 F_1 = f_1$, the first principal focal length of the first lens, so we can write

$$\frac{h_1'}{f_1} = \tan\theta \qquad \text{or} \qquad h_1' = f_1 \tan\theta \qquad (4.13).$$

Note that equation (4.13) correctly makes h_1' an inverted image, since f_1 is negative and $\tan\theta$ is positive. This results in a negative value for h_1' in this case.

Finally, if we magnify the size of the first image h_1' by the magnification factor L/L' at each succeeding lens, we obtain the final image size $h'_{final} = h_1' \times \frac{L_2}{L_2'} \times \frac{L_3}{L_3'} \times \ldots$; that is,

$$h'_{final} = f_1 \tan\theta \times \frac{L_2}{L_2'} \times \frac{L_3}{L_3'} \times \ldots \qquad (4.14)$$

4.1.3.3 The Conjugate Foci Relationship

As an alternative to the use of Newton's formula, in which the conjugate object and image distances are measured from the points F and F', respectively, it is sometimes more convenient to measure the conjugate distances from the principal points at P and P'. Knowledge of the positions of the principal points of a lens system greatly simplifies a consideration of the optical properties of the system.

Hence, as the first principal plane PH of a lens system represents the position of the single equivalent lens, as far as the incident light is concerned, we may measure the distance ℓ to the object point B from this plane. That is, $\ell = PB$ (meaning the measurement is from P to B) in figure 4.14 .

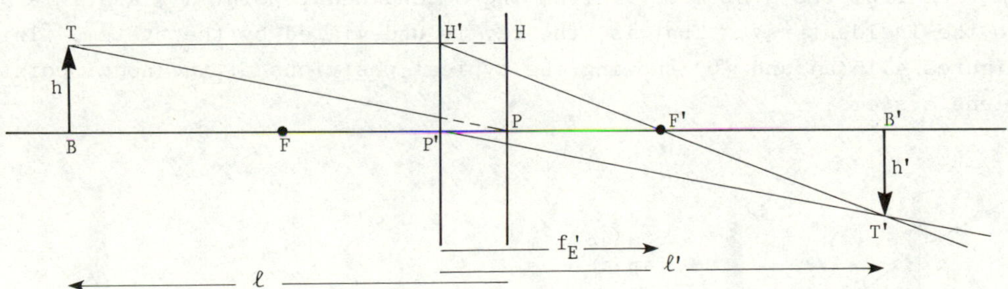

Fig. 4.14 The principal planes and object and image distances for a two lens system. Note the lenses do not need to be shown since the distances are measured from the principal planes.

Similarly, the second principal plane P'H' of a lens system represents the position of the single equivalent lens with respect to the emergent light. Consequently, we can express the distance ℓ' to the image point B' from this plane. That is, $\ell' = P'B'$ (meaning the measurement is from P' to B').

Under these conditions the simple conjugate foci formula (equation (3.3)) and the magnification equation (3.10) apply directly to a lens system as well as to a single thin lens.

Figure 4.14 shows the positions of the principal planes and the focal points of a simple lens system. The conjugate planes at BT and B'T' represent an object and its image formed by the system (which is not shown). Certain construction rays are shown in the figure. Imagine for light arriving at the system that the system is represented by its equivalent thin lens in the first principal plane PH. Then the ray TP is a ray to the optical centre of the equivalent lens. For rays emerging from the system the equivalent thin lens lies in the second principal plane P'H', and the ray P'T' emerges from the optical centre at P' and is parallel to TP. Similarly, a ray TH parallel to the principal axis emerges from the second principal plane at H' (PH = P'H') and passes through the second principal focus as the ray H'T'.

For the system as a whole we have :

$\ell = PB$, the object distance, giving $L = \frac{1}{\ell}$, and $\ell' = P'B'$, the image distance, giving $L' = \frac{1}{\ell'}$.

Since $P'F' = f'_E$ is the second equivalent focal length of the lens system, the conjugate foci expression (equation (3.3)) may be written

$$L' - L = F_E \qquad (4.15)$$

where $F_E = \frac{1}{f'_E}$, the equivalent power of the system.

The linear magnification m of the image produced by the system is given by equation (3.10), namely

$$m = \frac{L}{L'}, \text{ or in air, } \quad m = \frac{h'}{h} = \frac{L}{L'} = \frac{\ell'}{\ell} = \frac{P'B'}{PB}$$

4.1.4 THE NODAL POINTS AND OPTICAL CENTRE OF A SIMPLE LENS SYSTEM

All optical systems possess a pair of special points, known as the *nodal points* which, if their positions are known, enable scale drawings of the image formation by the lens system to be easily constructed. The nodal points are defined as those two points on the principal axis of a lens system, or thick lens, such that a ray of light directed towards the first nodal point N, emerges from the system as if from the second nodal point N', and in a direction parallel to the incident ray. That is, the ray is undeviated by the system. This is illustrated in figures 4.15(a) and (b) showing the typical positions of the nodal points for two different lens systems.

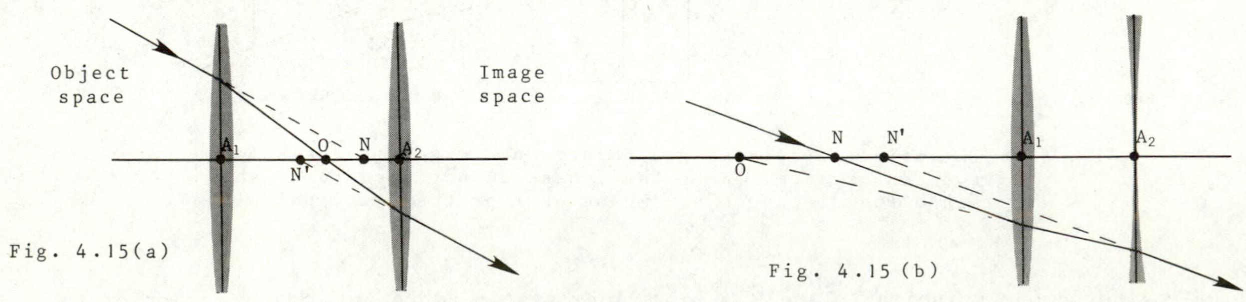

Fig. 4.15(a)

Fig. 4.15 (b)

The point where the nodal ray, as it passes between the lenses, crosses or appears to cross the principal axis of the system is the *optical centre* O of the system. If the two lenses are identical the point O lies midway between the vertices A_1 and A_2. In other cases O may lie outside the lenses.

It can be shown that the position of the optical centre is given by

$$A_1O = \frac{dF_2}{F_1 + F_2} \qquad (4.16).$$

This will be seen to follow from a more general equation for a thick lens in air, which will be dealt with in section 4.2.5 .

The effect may be compared with the transmission of a ray through the optical centre of a single thin lens, where as before, the ray is undeviated: see figure 4.15(c).

Thus, it may be considered that with a single thin lens the nodal points (and the principal points) coincide with the centre of the lens, and the principal planes coincide in the plane through the middle of the lens.

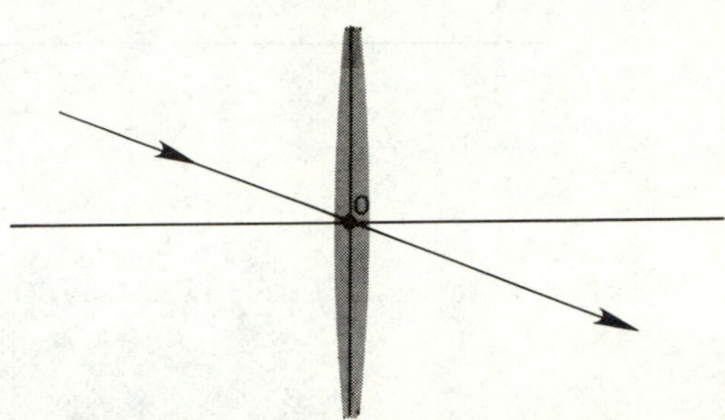

Fig. 4.15(c)

It can be shown that when the object space and the image space have the same refractive index, the nodal points N and N' coincide with the principal points P and P', respectively. If the object and image spaces have different refractive indices the nodal points do not coincide with the principal points, but are displaced towards the medium having the higher refractive index. This will be dealt with in section 4.2.6.2 .

The two nodal points N and N', the two principal points P and P', and the two principal foci F and F' of a system are collectively referred to as the *cardinal points* of the system.

4.1.4.1 *Ray Tracing using Nodal Points*

If the positions of the cardinal points N and N', P and P', and F and F' are known for a lens system, it is possible to construct a ray diagram showing the image formation by the system. The procedure is similar to that adopted for a single thin lens with two suitable construction rays drawn from the tip of the object. In the case of a lens system the construction rays are selected from the following, and can be followed from figures 4.16(a) and (b). With the lenses in air the nodal points coincide with the principal points, as indicated.

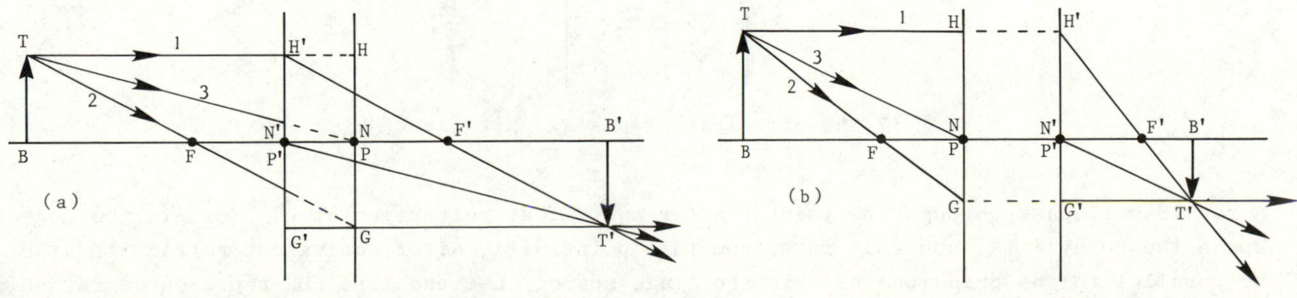

Fig. 4.16 Ray tracing where the nodal points coincide with the principal points.

Figure 4.16(c) illustrates the general case where the nodal points do not coincide with the principal points, and is included here for comparison. This example will be considered in greater detail in section 4.2.6.2 .

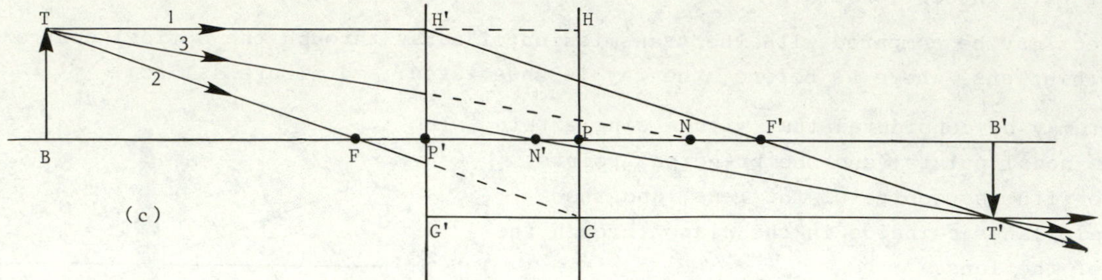

Fig. 4.16 Ray tracing using the six cardinal points.

(a) Ray (1), parallel to the principal axis, traverses the system to the first principal plane at H. It leaves the system as though refracted at H' in the second principal plane, and passes through the second principal focal point F'.

(b) Ray (2), which passes through the first principal focus F to meet the first principal plane at G, leaves the system from G' in the second principal plane and travels parallel to the axis.

(c) Ray (3), directed towards the first nodal point N, emerges as if from the second nodal point N', being parallel to its incident direction.

4.1.5 UNIT PLANES

Figure 4.17 shows the positions of the principal planes PH and P'H' of a typical lens system in air.

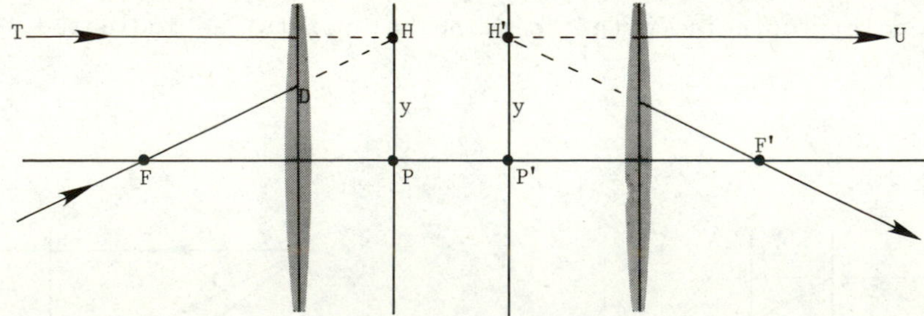

Fig. 4.17 The principal planes as unit planes.

A ray TH will pass through the point F' after equivalent refraction at H'. Now, in the object space the point F is such that rays from this point will, after equivalent refraction at H, be parallel to the principal axis in the image space. Let the line FDH represent a ray which after refraction emerges from the system along the line H'U.

H is the point of intersection of two rays in object space, and H' is the corresponding point of intersection of the two rays in image space. The points H and H' may be described as conjugate points with the planes PH and P'H' regarded as object and image, respectively. It is clear that, since the direction THH'U is a straight line, the planes PH and P'H' are of the same size, and the magnification is unity (1). For this reason, the planes PH and P'H' are often referred to as *unit planes*.

One extremely important property of these principal (unit) planes is that any ray in the object space passing through a point H on the first principal plane at a specific distance y from the axis, must pass through the conjugate point H' on the second principal plane. As the magnification in these planes is unity, H' will be situated at the same distance y from the axis. *It is very important to realise that this paragraph refers to construction rays using the cardinal points and the principal planes. When the cardinal points have been determined in relation to the system's lenses, the latter may be ignored for the purpose of constructing the image.*

4.1.6 FORMULAE FOR SIMPLE (TWO LENS) SYSTEMS

When accuracy is required in the determination of the optical characteristics and behaviour of a lens system, calculations are to be recommended in preference to scale drawings, although the latter method may well be sufficiently accurate in many cases. Calculation procedures are based on lens formulae used in conjunction with the sign convention. The formulae included in this section are for the determination of the equivalent power, the vertex powers, and the positions of the principal points.

4.1.6.1 *The Equivalent Power*

As mentioned in section 4.1.2.2 , the equivalent power of a lens system, symbol F_E, is the power of the single thin lens which is equivalent to the system. The expression for the equivalent power of a lens system can be derived from the geometry of figure 4.18 which shows the deviation of a ray by each lens in a simple system. As the incident ray on the system is parallel to the principal axis, the emergent ray passes through F', the second principal focal point of the system. P'H' is the position of the single equivalent lens for this ray.

In figure 4.18, let $A_1A_2 = d$, the spacing of the two lenses in air.

From equations (4.1) and (4.2), we have

$$F_V' = \frac{1}{f_V'} = \frac{F_1 + F_2 - dF_1F_2}{1 - dF_1} \quad \text{(i)}.$$

In the triangles H'P'F' and EA_2F', which are similar,

$$\frac{f_E'}{f_V'} = \frac{P'H'}{A_2E} = \frac{y_1}{y_2}$$

$$\text{so that} \quad f_E' = f_V' \cdot \frac{y_1}{y_2} \quad \text{(ii)}.$$

Now, since $f_E' = \frac{1}{F_E}$ and $f_V' = \frac{1}{F_V'}$, expression (ii) above may be rewritten $\frac{1}{F_E} = \frac{1}{F_V'} \cdot \frac{y_1}{y_2}$ which

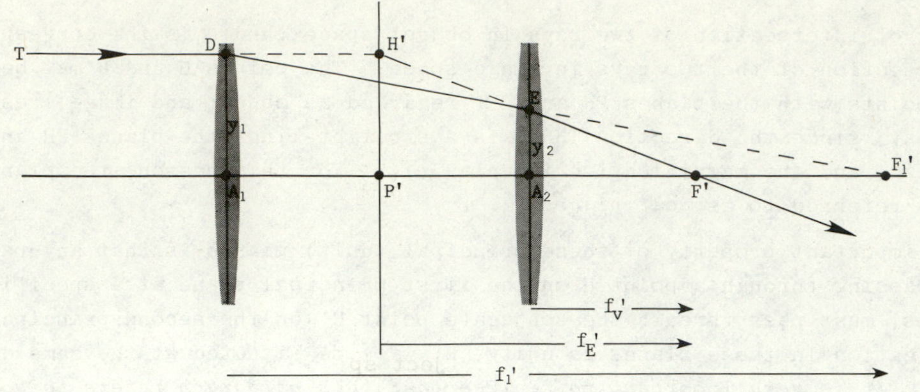

Fig. 4.18 Diagram used for the derivation of the equivalent power formula. d is the distance between the lenses.

upon taking reciprocals gives $$F_E = F_V' . \frac{y_2}{y_1} \qquad \text{(iii)}.$$

Now, from the triangles EA_2F_1' and DA_1F_1', which are similar,

$$\frac{y_2}{y_1} = \frac{f_1' - d}{f_1'} = 1 - \frac{d}{f_1'}$$

and since $\frac{1}{f_1'} = F_1$, $$\frac{y_2}{y_1} = 1 - dF_1 \qquad \text{(iv)}.$$

Substituting for $\frac{y_2}{y_1}$ from equation (iv) into equation (iii), we have

$$F_E = F_V'(1 - dF_1) = \frac{F_1 + F_2 - dF_1F_2}{1 - dF_1} \times (1 - dF_1)$$

giving $$F_E = F_1 + F_2 - dF_1F_2 \qquad (4.17).$$

Thus, the equivalent power of a simple lens system, i.e. a system of two thin lenses in air, depends on the individual lens powers and on the spacing of the lenses.

4.1.6.2 The Vertex Powers

We have already seen the derivation of the formulae for the back and front vertex powers. These equations are given here again for convenience along with the corresponding vertex focal length relationships:

$$F_V' = \frac{F_1 + F_2 - dF_1F_2}{1 - dF_1} \quad (4.1) \qquad \text{and} \qquad f_V' = \frac{1}{F_V'} \quad (4.2)$$

$$F_V = \frac{F_1 + F_2 - dF_1F_2}{1 - dF_2} \quad (4.3) \qquad \text{and} \qquad f_V = -\frac{1}{F_V} \quad (4.4)$$

The numerator for expressions (4.1) and (4.3) will be recognised as equation (4.17) for the equivalent power of the lens system. Hence, they may be rewritten in the form

$$F_V' = F_E/(1-dF_1) \quad (4.18) \qquad \text{and} \qquad F_V = F_E/(1-dF_2) \quad (4.19).$$

4.1.6.3 The Positions of the Principal Points

Figure 4.19 shows the positions of the principal points P and P', and the principal foci F and F', for a typical simple two lens system like figure 4.8 .

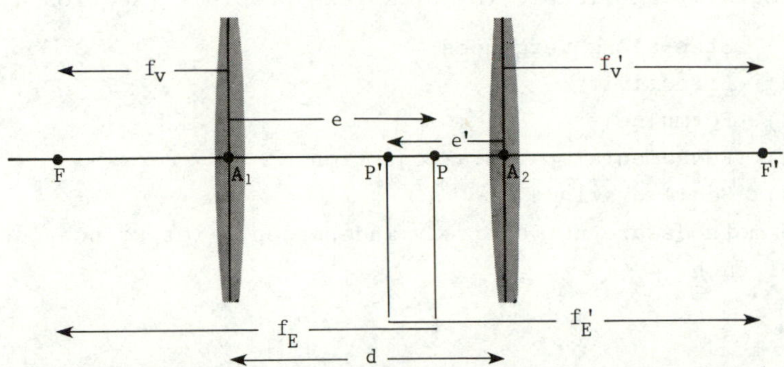

Fig. 4.19 The positions of the principal points.

Let the positions of the principal points P and P' be specified by the distances e and e' of those points from the front and back vertices, respectively. Hence,

$$e = A_1P \quad \text{(that is, measurement from } A_1 \text{ to P)}$$

and $$e' = A_2P' \quad \text{(that is, measurement from } A_2 \text{ to P').}$$

By taking into account the sign convention, for which the illustrated distances e', f_E, and f_v have negative numerical values, we have

$$e = (-f_E) - (-f_v), \quad \text{giving} \quad e = A_1P = f_v - f_E \qquad (4.20)$$

and $$-e' = f_E' - f_v' \quad \text{giving} \quad e' = A_2P' = f_v' - f_E' \qquad (4.21).$$

Alternatively, we may express equations (4.20) and (4.21) in terms of powers, rather than focal lengths, as follows.

From equation (4.20), we have $e = f_v - f_E = \left(-\frac{1}{F_v}\right) - \left(-\frac{1}{F_E}\right)$. But from equation (4.19),

$F_v = \frac{F_E}{1-dF_2}$, so $e = -\frac{(1-dF_2)}{F_E} + \frac{1}{F_E} = \frac{dF_2 - 1}{F_E} + \frac{1}{F_E} = \frac{dF_2 - 1 + 1}{F_E} = \frac{dF_2}{F_E}$.

That is, $$e = d\frac{F_2}{F_E} \qquad (4.22).$$

Also, from equation (4.21) we have $e' = f_v' - f_E' = \frac{1}{F_v'} - \frac{1}{F_E} = \frac{1}{F_E/(1-dF_1)} - \frac{1}{F_E}$ which rearranges in the manner above to give

$$e' = -d\frac{F_1}{F_E} \qquad (4.23).$$

The reader should complete the rearrangement before continuing. This completes the theory of simple lens systems but section 4.1.7 is a review of the methods of solving simple lens system problems, together with an example involving all the methods covered.

4.1.7 METHODS OF SOLUTION OF PROBLEMS

When faced with a numerical problem involving a lens system the reader is recommended to take time in consideration of the most suitable method for the problem. In general, five methods are available, but some of these have limitations as to their use, and these are detailed below. Experience will assist in making the choice of method. The comments made about each method are equally applicable to thick lens problems (section 4.2).

The methods are:
(1) Step-along vergences
(2) Effectivity
(3) Formulae
(4) Fundamental paraxial equation
(5) Scale drawing.

Although some of these methods are not entirely independent, let us now consider the advantages and limitations of each.

(1) Step-along Vergences

This is a systematic and logical method in which each lens is considered in turn. It may be used for the determination of the front and back vertex powers, and for the determination of the position and size of the image. It is not possible, by this method, to locate the positions of the principal points P and P'.

(2) Effectivity

This is really the step-along method with $L_1 = 0$. Then $L_1' = L_1 + F_1 = 0 + F_1 = F_1$. The vergence arriving at the next lens is $L_2 = \frac{L_1'}{1-dL_1'} = \frac{F_1}{1-dF_1}$, and this will be recognised as the effectivity of F_1 in the plane of F_2. If we continue, then $L_2' = L_2 + F_2 = \frac{F_1}{1-dF_1} + F_2 = \frac{F_1 + F_2 - dF_1 F_2}{1 - dF_1}$ which will be recognised as the expression for the back vertex power.

The method, as demonstrated in section 4.1.1 is suitable for determining the vertex powers, and therefore for determining the positions of the principal foci F and F'. It has particular application when considering lenses at different distances from the eye. As with the step-along method, the effectivity principle may not be used to determine the positions of the principal points.

(3) Formulae

The various formulae have been dealt with in earlier sections and are listed again here for reference.

Equivalent power $F_E = F_1 + F_2 - dF_1 F_2$, where $f_E' = \frac{1}{F_E}$ and $f_E = -\frac{1}{F_E}$.

Back vertex power $F_V' = \frac{F_E}{1 - dF_1}$, where $f_V' = \frac{1}{F_V'}$.

Front vertex power $F_V = \frac{F_E}{1 - dF_2}$, where $f_V = -\frac{1}{F_V}$.

Principal points $e = f_V - f_E$ and $e' = f_V' - f_E'$

or $e = d\frac{F_2}{F_E}$ and $e' = -d\frac{F_1}{F_E}$.

Newton's formulae $(f_E')^2 = -xx'$ (or $f_E f_E' = xx'$) and $m = \frac{h'}{h} = -\frac{f_E}{x} = -\frac{x'}{f_E'}$.

Conjugate foci formulae $L' - L = F$ and $m = \frac{h'}{h} = \frac{L}{L'}$, where object and image distances are measured to the first and second principal planes respectively.

The above formulae allow the positions of all the cardinal points of a simple lens system in air to be located, and are widely used.

(4) *The Fundamental Paraxial Equation*

This equation, $\frac{n'}{\ell'} - \frac{n}{\ell} = F$, is cumbersome and is applied to each refracting element in turn. It has perhaps more application in certain thick lens systems, such as the schematic eye. We shall not use it since the step-along vergences method is its vergence analogue. It was used more extensively before the advent of inexpensive electronic calculators.

(5) *Scale Drawing*

Any scale drawing is only as good as the skill of the drawer and quality of the drawing equipment will allow. In general, a scale drawing is not as accurate as the use of formulae. The main advantage of a diagram is that it presents a "picture" of the optical behaviour of a lens system, usually providing a greater appreciation of the purpose of the system. The method may be used to locate the principal points P and P', and also to determine the position and size of the image formed by the system. It is therefore a useful supplement to the step-along and effectivity methods.

Worked Example

The following example illustrates the use of formulae, including Newton's method, step-along vergences, and a graphical solution.

A lens system is made up from a +5.00 D lens and a +4.00 D lens separated by 8 cm. An object 2 cm tall is situated on the principal axis 40 cm in front of the first lens. Determine the position of the cardinal points and the position and size of the image formed by the system.

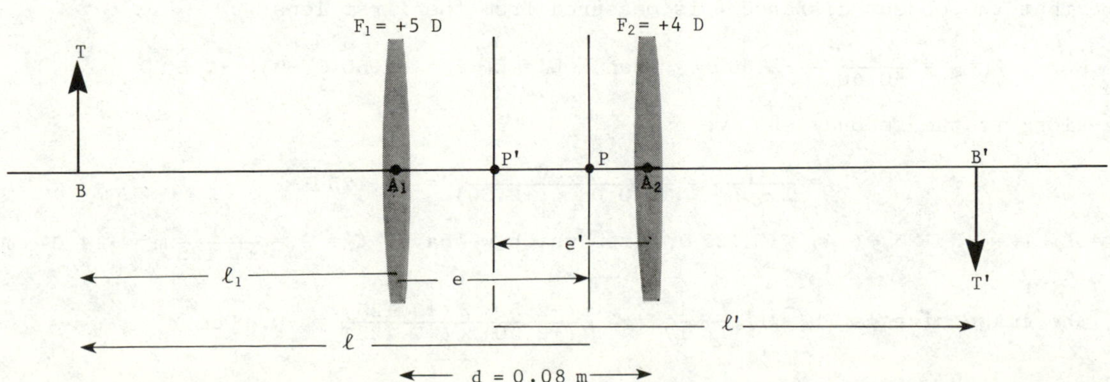

To find the equivalent power, $F_E = F_1 + F_2 - dF_1 F_2 = 5 + 4 - 0.08 \times 5 \times 4 = 9 - 1.6 = +7.40$ D.

The position of P is given by $A_1P = e = d\frac{F_2}{F_E} = 8 \times \frac{4}{7.4} = +4.32$ cm.

The position of P' is given by $A_2P' = e' = -d\frac{F_1}{F_E} = -8 \times \frac{5}{7.4} = -5.41$ cm.

Note that in the calculation of e and e' we may put d in any convenient unit we wish, in this case in centimetres. This is simply because the units of the terms F_2/F_E and F_1/F_E cancel.

Image Position and Size by the Conjugate Foci Method

Remember, the object and image distances are measured as in the previous figure! So, measuring the distances ℓ and ℓ' from P and P', respectively:

$$\ell = PB = \ell_1 - e = -40 - 4.32 = -44.32 \text{ cm}, \quad \text{whence } L = \frac{1}{\ell} = \frac{1}{-0.4432} = -2.256 \text{ D}.$$

Note that $\ell_1 = -40$ cm is the object distance measured from the first lens, and that ℓ was converted to metres in order to calculate the vergence L.

Now, the power of the single equivalent lens is $F = F_E = +7.40$ D, from above.

Hence, $L' = L + F = -2.256 + (+7.40) = +5.144$ D, whereafter we have $\ell' = \frac{1}{L'} = \frac{1}{+5.144}$ m which gives $\ell' = P'B' = +19.44$ cm. But $e' = -5.41$ cm, so measuring the image distance from the second lens, we have

$$\ell_2' = \ell' + e' = +19.44 + (-5.41) = +14.03 \text{ cm}.$$

That is, the image is 14.03 cm to the right of the second lens. We can now calculate the size of the image:

$$h' = h \times \frac{L}{L'} = 2 \times \frac{(-2.256)}{(+5.144)} = -0.88 \text{ cm}.$$

Notice, once again, the units of h in this expression may take any convenient unit we wish. This is because the units of L/L' have cancelled. Further, notice that because L' is positive the image is real. Since its height carries a minus sign, we also infer that it is inverted.

Image Position and Size by the Step-along Vergences Method

Note here that the object distance ℓ_1 is measured from the first lens.

$$\text{Now, } L_1 = \frac{1}{\ell_1} = \frac{1}{-0.40} = -2.50 \text{ D}. \quad \text{Then, } L_1' = L_1 + F_1 = -2.50 + (+5) = +2.50 \text{ D}.$$

Stepping-along to the second lens, we have

$$L_2 = \frac{L_1'}{1 - dL_1'} = \frac{+2.50}{1 - 0.08 \times (+2.50)} = +3.125 \text{ D}.$$

Next, $L_2' = L_2 + F_2 = +3.125 + (+4) = +7.125$ D, from which we have $\ell_2' = \frac{1}{L_2'} = \frac{1}{+7.125}$ m $= +14.04$ cm.

Finally, the image size is $h_2' = h_1 \times \frac{L_1}{L_1'} \times \frac{L_2}{L_2'} = 2 \times \frac{(-2.50) \times (+3.125)}{(+2.50) \times (+7.125)} = -0.88$ cm.

We can now compare the results of these two methods. Firstly, at least to two decimal places, the image sizes agree. The distance of the image from the second lens agrees to one part in ten thousand. This small disagreement is due to rounding errors.

Image Position and Size by Newton's Formulae

The first focal point is a distance $PF = f_E = -\frac{1}{F_E} = -\frac{1}{(+7.40)}$ m = -13.51 cm from P.

The second focal point is a distance $P'F' = f_E' = \frac{1}{F_E} = \frac{1}{(+7.40)}$ m = +13.51 cm from P'.

Since the object is a distance $\ell = -44.32$ cm from P, the extra-focal object distance x is $x = \ell - f_E = -44.32 - (-13.51) = -30.81$ cm from F. We can now use Newton's expression for x':

$$\text{thus,} \quad x' = \frac{f_E f_E'}{x} = \frac{(-13.51)(+13.51)}{-30.81} = +5.92 \text{ cm.}$$

That is, the image is 5.92 cm to the right of the second principal focus, F'. This places it at a distance

$$\ell' = f_E' + x' = (+13.51) + (+5.92) = +19.43 \text{ cm from P'.}$$

The image size is given by either $\frac{h'}{h} = -\frac{f_E}{x}$ or $\frac{h'}{h} = -\frac{x'}{f_E'}$. Using the latter, we have

$$h' = h \times \left(-\frac{x'}{f_E'}\right) = 2 \times \left(-\frac{(+5.92)}{(+13.51)}\right) = -0.88 \text{ cm.}$$

Image Position and Size by Graphical Construction

Finally, let us solve the problem by a graphical construction. The figure below shows the construction rays considered in section 4.1.4.1 .

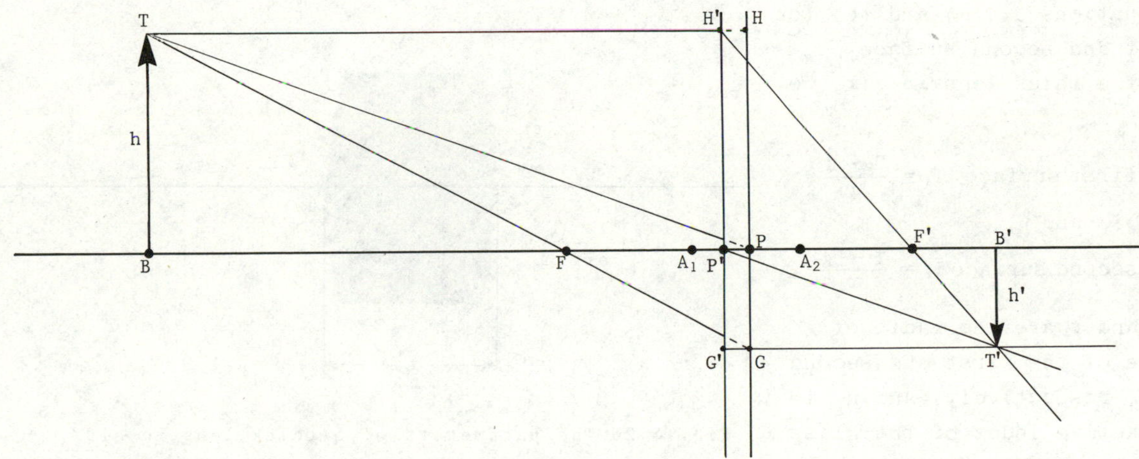

The positions of P and P' are computed as before. Similarly, the positions of F and F' are calculated and these four points, plus the vertices A_1 and A_2 are entered on the principal axis. In this case, the horizontal scale is 1:4 . The object is drawn in with a scale of 2:1 . The ray TH is drawn parallel to the principal axis to meet the first principal plane in H. The ray leaves the system in the direction H'F', where H' is the same height as H above the axis. Next, the ray TFG through the first principal focus meets the first principal plane in G. It leaves the system parallel to the principal axis in the direction G'T', where PG = P'G'. Finally, P'B' and B'T' can be measured and converted into centimetres, say. *(Because of problems involving size reduction in the printing, we are not able to quote the exact scale with this construction).*

4.2 *THICK LENSES*

The preceding part of this chapter dealt with paraxial theory as applied to lens systems made up from two thin spherical lenses. When the thickness of a lens cannot be regarded as negligible some of the single thin lens formulae become no longer applicable, and the lens is then classified as thick.

It is also useful to note that some forms of spectacle lenses, including contact lenses, are classed as thick lenses, even though their actual thickness may be quite small. In the case of typical positive powered scleral contact lens, the centre thickness may be as little as 0.6 mm, and in common with many lenses met in ophthalmic practice, the lens will often be of meniscus form. In these cases the front vertex power will differ from the back vertex power, and from the equivalent power. Consequently, the lenses must be treated with thick lens theory.

4.2.1 *SURFACE POWERS AND REDUCED THICKNESS*

Much of the theory of simple thin lens systems may be easily modified so as to apply to thick lenses. In effect, a thick lens may be considered to be like a simple lens system in which the space between the lenses is filled with the lens material, say glass. In this way it will be necessary to consider the powers of the two surfaces individually, and to replace the thickness of the lens by the optically equivalent air distance (the reduced distance).

Using equations 3.4(a) and (b) for the first and second surface powers of a thick lens *in air*, we have:

Power of first surface $F_1 = \dfrac{n_g - 1}{r_1}$,

and

power of second surface $F_2 = \dfrac{1 - n_g}{r_2}$,

where r_1 and r_2 are the radii of curvature of the first and second surfaces, respectively, and n_g is the refractive index of the glass. In these equations, the 1 is the refractive index of the surrounding air, and r is measured in metres.

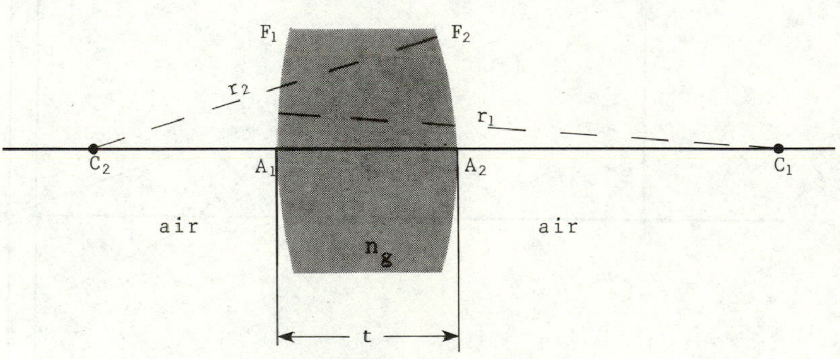

Fig. 4.20 The parameters of a thick lens in air.

A lens of appreciable thickness is illustrated in air in figure 4.20 , in which the two surfaces of powers F_1 and F_2 are separated by a thickness t between the vertices, known as the axial thickness. The centres of curvature of the two surfaces are at C_1 and C_2, respectively.

A thick lens possesses an equivalent power dependent on its surface powers, and on the refractive index and thickness of its material. The lens also possesses front and back vertex powers which generally are of different values except where the lens surfaces have equal powers.

In order that the formulae used for simple lens systems may be applied to a thick lens it is necessary to replace the thickness t of the lens by a thickness of air which is optically equivalent to it. In section 2.1.8 it was shown that the refractive index of a piece of glass is given by

$$n_g = \frac{d}{\bar{d}}$$

where d is the actual thickness of the material and $\bar{d}$ (read as "d reduced") is the apparent or reduced thickness of the material, from which we have

$$\frac{d}{n_g} = \bar{d} .$$

The reduced thickness $\bar{d}$ of a material represents the equivalent air distance of the material as far as vergence is concerned. Thus, it follows that for a thick lens of axial thickness t, the optically equivalent air thickness, or reduced thickness, will be

$$\bar{t} = \frac{t}{n_g}$$

where n_g is the refractive index of the lens material.

The value of the reduced thickness $\bar{t}$ of a thick lens may now be substituted for the spacing d between the lenses in the formulae derived for a two-lens systems in air.

4.2.2 THICK LENS FORMULAE

The formulae for simple lens systems were listed in section 4.1.7 . Replacing the lens spacing d by the reduced thickness $\bar{t}$, the corresponding expressions for thick lenses in air are as follows:

Equivalent power $F_E = F_1 + F_2 - \bar{t}F_1 F_2$ (4.24)

and $f_E = -\frac{1}{F_E}$ and $f_E' = \frac{1}{F_E}$.

Back vertex power $F_V' = \frac{F_E}{1 - \bar{t}F_1}$ (4.25)

and $f_V' = \frac{1}{F_V'}$.

Front vertex power $F_V = \frac{F_E}{1 - \bar{t}F_2}$ (4.26)

and $f_V = -\frac{1}{F_V}$.

The positions of the principal points are given by

$$e = A_1P = f_V - f_E$$

or $e = \bar{t}\cdot\frac{F_2}{F_E}$ (4.27)

and $e' = A_2P' = f_V' - f_E'$

or $e' = -\bar{t}\cdot\frac{F_1}{F_E}$ (4.28).

It is useful to note that if the lens is thin, t is negligible (=0) and equation (4.24) reduces to $F_E = F_1 + F_2$, a result expressed in equation 3.2 for a thin lens. Also, in a similar manner, equations (4.25) and (4.26) reduce to $F_V' = F_E$ and $F_V = F_E$, whereupon we see that $F_V = F_V' = F_E = F_1 + F_2$ for a thin lens.

In stating these thick lens equations ((4.24) to (4.28)) we have merely replaced d by $\bar{t}$. If the reader finds this none-the-less correct procedure somewhat less than satisfying, the proofs of the equations in section 4.1 may be reworked. Proofs for F_V' and F_E will be found in the elementary text "Introduction to Visual Optics", by A.H. Tunnacliffe. F_V' is found by using the step-along vergences method with $\bar{L}_1 = 0$. The step-along relationship from the first surface, where the vergence leaving it is $\bar{L}_1'$, can easily be proved to be

$$\bar{L}_2 = \frac{\bar{L}_1'}{1 - \bar{t}\bar{L}_1'} \ .$$

Start with the definition of the vergence arriving at the second surface: $\bar{L}_2 = \frac{n_2}{\ell_2}$. Then, writing $\ell_2 = \ell_1' - t$ and noting that $n_1' = n_2 = n_g$, the transformation to the step-along equation follows after some tedious manipulation.

In attempting a reworking of the proof for the equivalent power the only thing to pay particular attention to is the replacement of f_1' with the power of the first surface. This is simply $f_1' = \frac{n_1'}{F_1} = \frac{n_g}{F_1}$.

4.2.3 FOCAL POINTS AND PRINCIPAL POINTS OF A THICK LENS IN AIR

Figures 4.21(a) and (b) illustrate the characteristics of the two principal foci F and F' of a thick lens. Rays diverging from the first principal focus F are refracted by each surface of the lens, emerging parallel to the principal axis, figure 4.21(a), whilst incident rays which are parallel to the axis, after refraction by each surface, pass through the second principal focus F', figure 4.21(b).

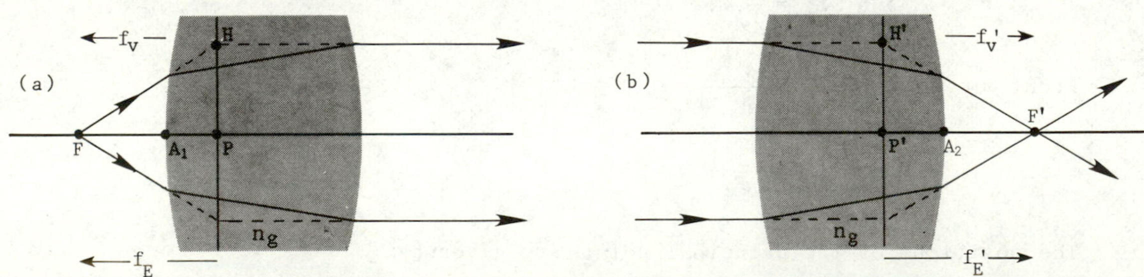

Fig. 4.21 The principal planes of a thick lens.

In a similar way to our treatment of two thin lenses in air (section 4.1.2.1), the incident ray is produced forwards and the emergent ray backwards until they intersect, thus locating the first principal plane PH and the second principal plane P'H'. We may recall that these planes are the positions of the single equivalent thin lens with respect to the incident and emergent light, respectively, and that these planes are planes of unit magnification (see also section 4.1.5).

The first equivalent focal length of the lens is represented by $f_E = PF$ (measurement from P to F), whilst the second equivalent focal length is represented by $f_E' = P'F'$ (measurement from P' to F').

The front and back vertex focal lengths are represented by $f_V = A_1F$ and $f_V' = A_2F'$, respectively. If the media are the same on both sides of the lens, then $f_E = -f_E'$, a statement which will be substantiated in section 4.2.6.1 by inspection of equation (4.32).
In general, the vertex focal lengths f_V and f_V' will have different values unless the powers of the lens surfaces are identical.

4.2.3.1 *Examples of Principal Planes of Thick Lenses*

A thick lens of specific fixed power may be manufactured in various forms. The variation in the form of a lens, whilst maintaining a fixed back vertex power, is a process known as *bending*. This may be done to reduce aberrations or to change the magnification. It is found that the positions of the principal points vary as a thick lens has its form changed. Figures 4.22(a) and (b) illustrate the bending of a lens and the subsequent variation in the positions of P and P', starting with a lens of equi-curved form in each case.

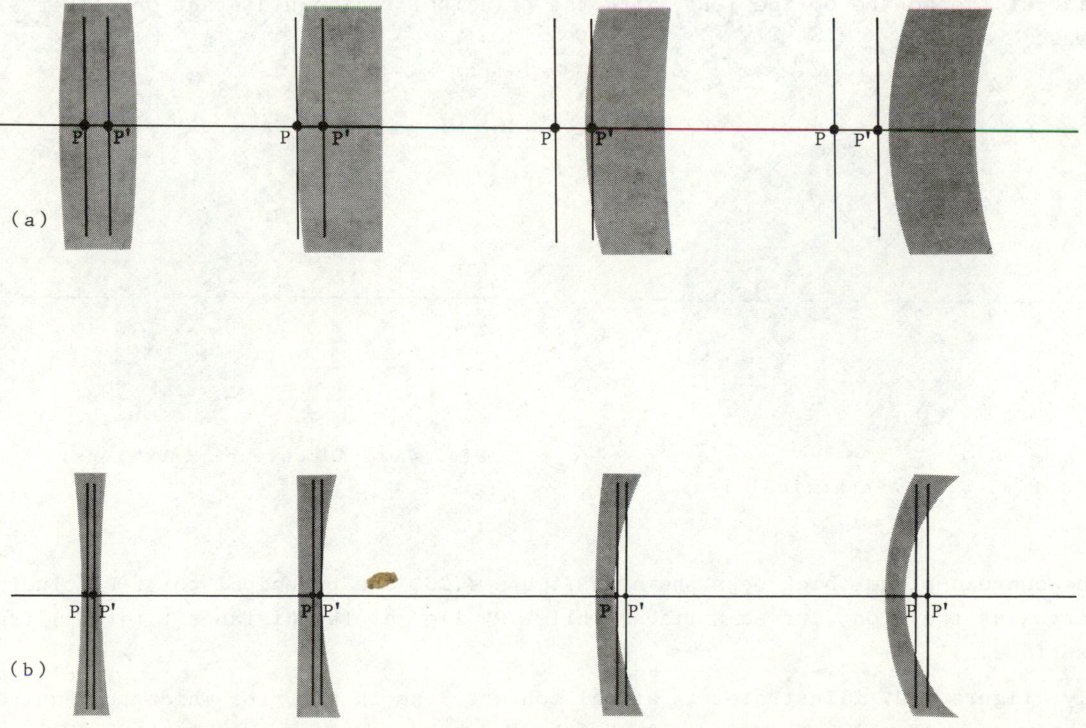

Fig. 4.22 (a) Converging and (b) diverging thick lenses. The movement of the principal planes with increased 'bending' of the lenses.

Certain special thick lenses are of some interest. Figure 4.23 shows a thick lens in meniscus form with surfaces having equal radii, and hence equal powers in air. When surrounded by a medium of lower refractive index the lens has a small but positive power. The principal planes are in front of the lens with their spacing equal to the lens thickness t.

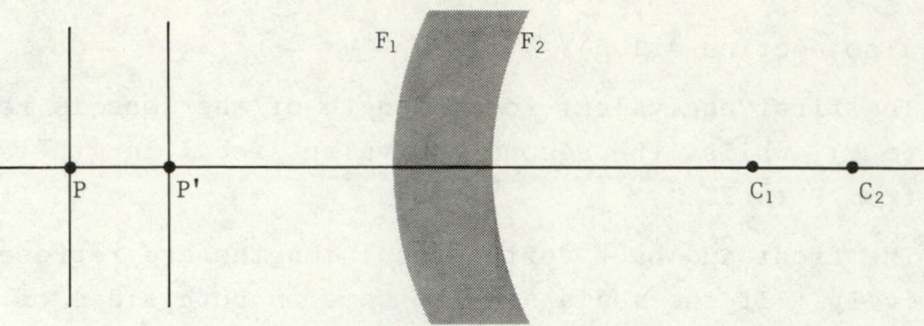

Fig. 4.23 A thick lens with equal surface radii.

Figure 4.24 illustrates a concentric lens in which both surfaces have the same centre of curvature. The radii of curvature differ by the thickness of the lens, and the lens surfaces will be parallel to each other. If this lens is surrounded by a medium of lower index, it has a small but negative power. The two principal planes are behind the lens, and both coincide with the common centre of curvature.

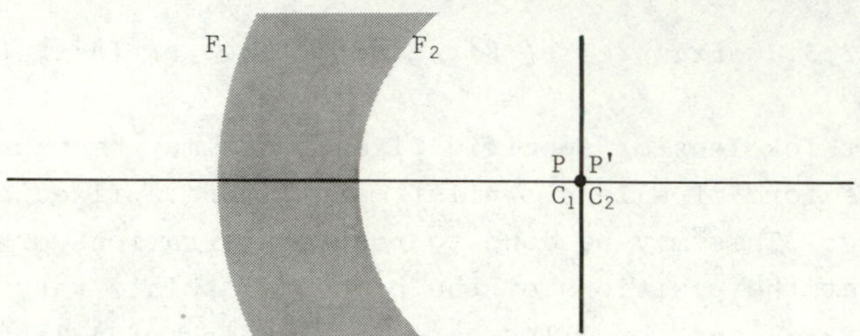

Fig. 4.24 A concentric lens.

The lens depicted in figure 4.25 is a glass sphere. In this case the radii r_1 and r_2 are numerically the same, and the lens thickness is $2r_1$. As such, the principal points P and P' coincide at the centre of the lens, with the principal foci equidistant on either side of the lens.

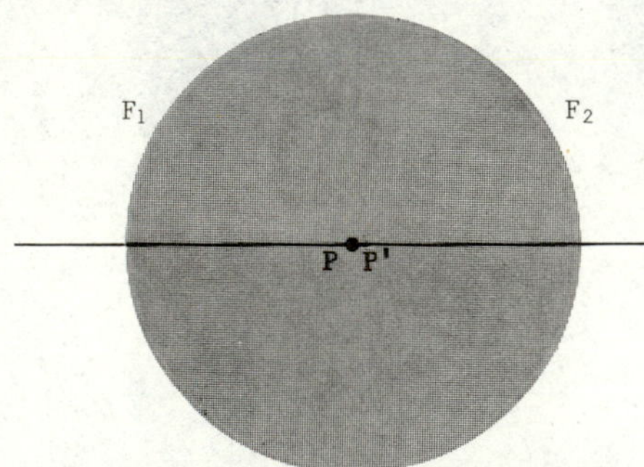

Fig. 4.25 A spherical lens.

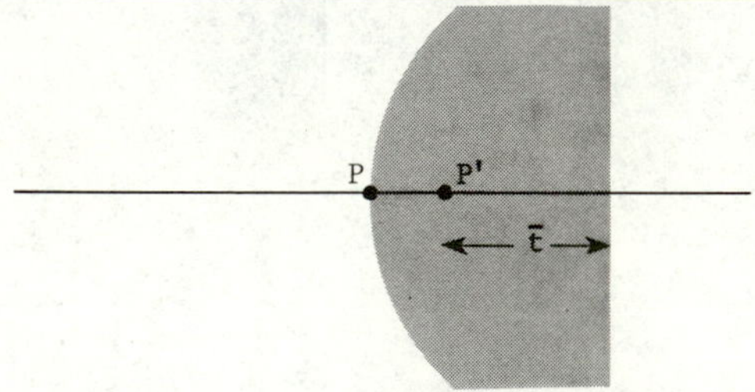

Fig. 4.26 A convexo-plane lens.

For the convexo-plane thick lens shown in figure 4.26, the principal point P coincides with the vertex at the front curved surface, whilst P' lies at the distance $\bar{t}$ ($=t/n_g$) from the back surface.

Finally, figure 4.27 illustrates an afocal contact lens in air, for which the equivalent power is zero. Afocal means literally "without focus", and corresponds to both incident rays and emergent rays forming parallel pencils. In such a case the principal points and the focal points are at infinity.

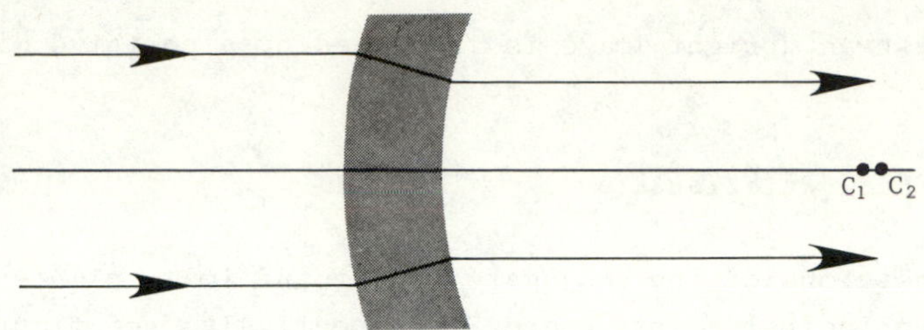

Fig. 4.27 An afocal lens.

4.2.4 OBJECT AND IMAGE POSITION AND MAGNIFICATION

In general, the methods for image location which are applicable to simple lens systems may also be used for thick lenses. If the positions of the cardinal points are known, the determination of the image position and size is made relatively easy. Ray tracings are included in section 4.2.6.2 . For the present discussion we shall again consider the methods previously mentioned for simple lens systems in section 4.1.3 .

4.2.4.1 *Newton's Formulae*

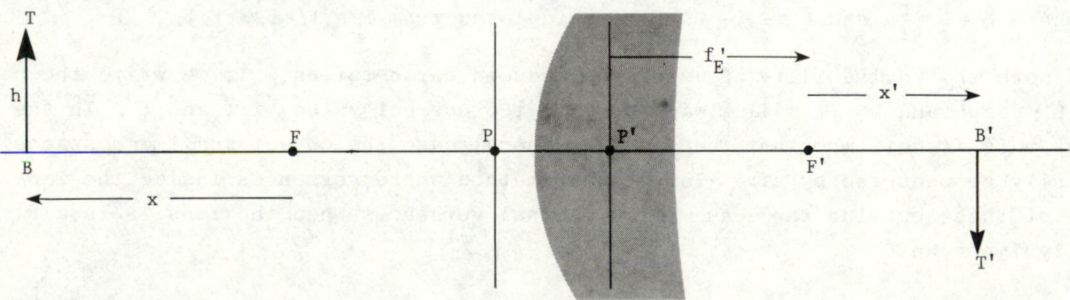

Fig. 4.28 Extra-focal object and image distances applied to a thick lens.

The object BT and the real image B'T' form a pair of conjugate planes. As such their positions may be specified from the appropriate principal focal points F and F', as shown in figure 4.28 . Using equation 4.8, Newton's formulae may be written in the form

$$(f_E')^2 = -xx'$$

where f_E' is the second equivalent focal length of the thick lens, and x and x' are the extra-focal object and image distances, respectively.

The magnification of the image may be determined using equations 2.33 and 2.34:

$$m = \frac{h'}{h} = -\frac{f_E}{x} = -\frac{x'}{f_E'}$$

Where h and h' are the object and image sizes respectively, and f_E is the first equivalent focal length of the thick lens. An inverted image results in a negative value for the magnification, whilst an upright image is indicated by a positive magnification.

4.2.4.2 Conjugate Foci Relationship

In this case the positions of the conjugate object and image planes are specified by their distances from the principal points P and P', respectively; see figure 4.29 .

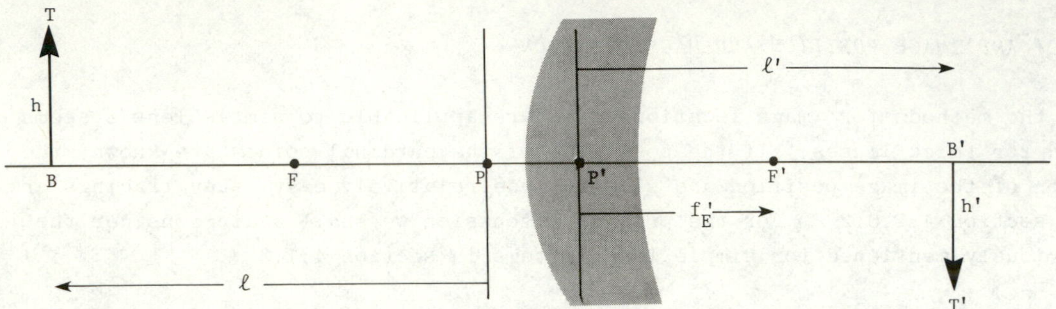

Fig. 4.29 The conjugate foci relationship for a thick lens.

Hence, with $\ell = PB$ and $\ell' = P'B'$ we may apply equation 3.1 to the case of a thick lens, namely $\bar{L}' - \bar{L} = F_E$, where $\bar{L} = \frac{1}{\bar{\ell}} = \frac{n}{\ell}$ and $\bar{L}' = \frac{1}{\bar{\ell}'} = \frac{n'}{\ell'}$. F_E is calculated from $F_E = F_1 + F_2 - \bar{t}F_1 F_2$.

Let us note here the advisability of using the reduced bar notation. If we write the bar over L and L', as $\bar{L}$ and $\bar{L}'$, it will always produce the correct value of ℓ and ℓ'. In the present case this is not important, but in the general case the object and image spaces need not necessarily be occupied by air. In the next method where vergences inside the lens are reduced whilst those outside the lens will be actual vergences when the lens is in air, it is especially important.

The image size in the conjugate foci method is calculated using equation 3.10: $m = \frac{h'}{h} = \frac{\bar{L}}{\bar{L}'}$

4.2.4.3 Step-along Vergences

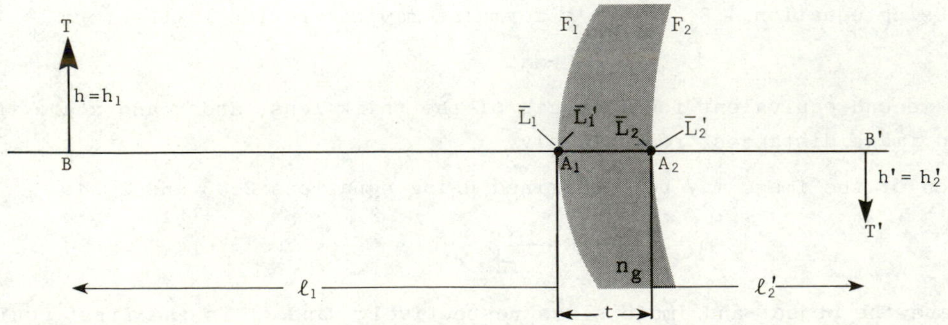

Fig. 4.30 Step-along vergences method.

Using vergences, the step-along method may be used for a thick lens in the same way as for a lens system made up of thin lenses (see section 4.1.3.2). For convenience, the various steps in the procedure for a thick lens are summarised below.

(1) Given the object distance ℓ_1 *from the first surface, write* $\bar{L}_1 = \frac{n_1}{\ell_1}$.

(2) Find $\bar{L}_1'$*, the emergent vergence from the first surface, using* $\bar{L}_1' = \bar{L}_1 + F_1$.

(3) Step-along to the second surface to find the vergence of the light arriving there, given by

$$\bar{L}_2 = \frac{\bar{L}_1'}{1 - \bar{t}\bar{L}_1'} .$$

(4) Find $\bar{L}_2'$*, the emergent vergence from the second surface, using* $\bar{L}_2' = \bar{L}_2 + F_2$.

(5) Obtain the final image distance from the second surface, using $\ell_2' = \frac{n_2'}{\bar{L}_2'}$.

For an object at a finite distance from the lens, the magnification is given by:

$$m = \frac{h'_{final}}{h_1} = \frac{h_2'}{h_1} = \frac{\bar{L}_1}{\bar{L}_1'} \times \frac{\bar{L}_2}{\bar{L}_2'}$$

where h_2' is the final image size in this two surface case, and h_1 is the object size for the first surface and therefore for the lens.

When the object is at an infinite distance from the lens we must use equation (4.14) to determine the size of the final image, that is

$$h_2' = f_1 \tan\theta \times \frac{\bar{L}_2}{\bar{L}_2'}$$

where $f_1 \left(= -\frac{n_1}{F_1}\right)$ is the first focal length of the first surface of the lens. Let us recall that the term $f_1 \tan\theta$ in the above expression represents the size of the image formed by the first surface, and that this is further magnified at the second surface by the factor $\bar{L}_2/\bar{L}_2'$.

Once again, let us reiterate that where we are hopping from one medium to another it is advisable to remain in the reduced bar notation. This is a constant reminder of the influence of the various refractive indices.

4.2.5 THE NODAL POINTS AND OPTICAL CENTRE OF A THICK LENS

The nodal points of a thick lens are those two points on the principal axis of the lens such that an incident ray directed towards the first nodal point N emerges from the lens as if from the second nodal point N', and in a direction parallel to the incident ray. This was discussed in section 4.1.4 for a system of two thin lenses in air.

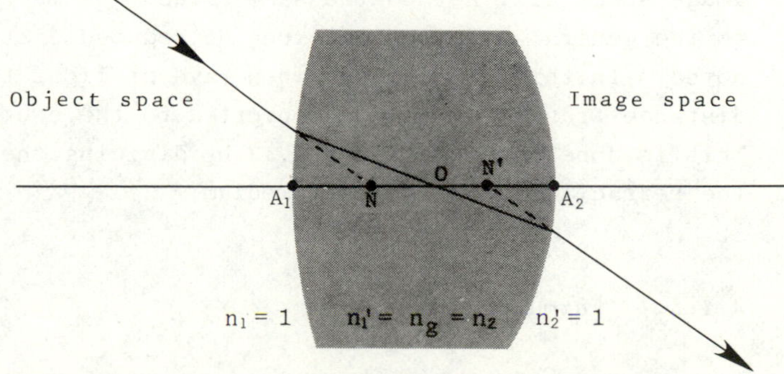

Fig. 4.31 A thick lens in air. When $n_1 = n_2'$ then the first nodal point N coincides with P, and the second nodal point N' with P'.

Figure 4.31 shows the typical positions of the nodal points N and N' for a double convex lens in air. When the media are the same in the object and image spaces, the nodal points coincide with the principal points P and P', respectively.

The point O where the nodal ray crosses the axis of the lens is the optical centre of the lens. With certain lens forms the nodal points and the optical centre will be outside the lens. The position of the optical centre of a thick lens is given by

$$A_1O = t \cdot \frac{F_2}{F_1 + F_2} \qquad (4.29).$$

The derivation of this expression is as follows.

At the first surface, N is a virtual object and O is a real image point.

$$\therefore \quad \ell_1' = A_1O \,, \quad \text{with } \bar{L}_1' = \frac{n_1'}{\bar{\ell}_1'} = \frac{n_g}{A_1O}$$

$$\text{and} \quad \ell_1 = A_1P = A_1N = \bar{t} \cdot \frac{F_2}{F_E} \,, \text{ using equation (4.27), with } \bar{L}_1 = \frac{n_1}{\ell_1} = \frac{1}{\ell_1} = \frac{1}{\bar{t}F_2/F_E} = \frac{F_E}{\bar{t}F_2}\,.$$

Now, $\bar{L}_1' = \bar{L}_1 + F_1$, so substituting for the incident and emergent vergences, using the two expressions above, we have

$$\frac{n_g}{A_1O} = \frac{F_E}{\bar{t}F_2} + F_1 = \frac{\bar{t}F_1F_2 + F_E}{\bar{t}F_2} = \frac{\bar{t}F_1F_2 + F_1 + F_2 - \bar{t}F_1F_2}{\bar{t}F_2}$$

$$\text{Therefore,} \quad \frac{A_1O}{n_g} = \frac{\bar{t}F_2}{F_1 + F_2} \,, \text{ on simplifying and taking reciprocals.}$$

Since $\bar{t} = \frac{t}{n_g}$, if we now multiply both sides by n_g, equation (4.29) results.

The general case of a thick lens bounded by media of different refractive indices is dealt with in section 4.2.6.2 , where it will be seen that the nodal points do not coincide with the principal points.

4.2.6 THE GENERAL CASE OF A THICK LENS

In the most general case of a thick lens the refractive indices of the object space and the image space will not be the same value. We may apply equations (4.24) to (4.28) inclusive to the general case, but the use of reduced distances in these equations must be carefully noted. In this connection, when rays of light travel through a medium other than air, the distance traversed must be converted to the equivalent air distance, or reduced distance. This is done (see section 4.2.1) by dividing the real distance traversed in the medium by the refractive index of that medium.

4.2.6.1 *Formulae*

Figure 4.32 shows a pair of centred spherical surfaces forming a thick lens, separating media of refractive indices n_1, n_g, and n_2'. The positions of the principal focal points F and F', and the principal points P and P' may be located by rewriting equations (4.24) to

(4.28) inclusive in the general forms.

So we have:

Power of first surface $F_1 = \frac{n_1' - n_1}{r_1} = \frac{n_g - n_1}{r_1}$

Power of second surface $F_2 = \frac{n_2' - n_2}{r_2} = \frac{n_2' - n_g}{r_2}$

Equivalent power $F_E = F_1 + F_2 - \bar{t}F_1 F_2$

and $f_E = -\frac{n_1}{F_E}$ (4.30)

and $f_E' = \frac{n_2'}{F_E}$ (4.31).

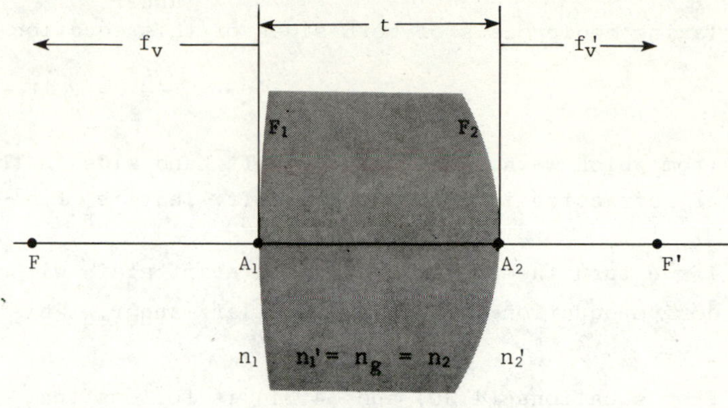

Fig. 4.32 The general case of a thick lens: $n_1 \neq n_2'$.

The occurrence of the object and image space refractive indices n_1 and n_2' in equations (4.30) and (4.31) may be determined by deriving the equivalent power equation from first principles, as we did for equation (4.17). Alternatively, consider the following:

Let f_E be the first focal length of the equivalent refracting element that will provide the same size of image as the thick lens in figure 4.32 . Notice that we have not stated whether this refracting element is a thin lens or a single surface! Then the image size produced by this single element, for an object at infinity subtending an angle θ, will be $f_E \tan\theta$. The thick lens will produce an image of size $f_1 \tan\theta \times \frac{\bar{L}_2}{\bar{L}_2'}$. Since these image sizes must be the same we equate them:

$$f_E \tan\theta = f_1 \tan\theta \times \frac{\bar{L}_2}{\bar{L}_2'} \qquad \text{(i).}$$

Now, since the object is at infinity, $\bar{L}_1 = 0$, from which it follows that $\bar{L}_1' = \bar{L}_1 + F_1 = 0 + F_1 = F_1$. Using the step-along equation for $\bar{L}_2$, we have

$$\bar{L}_2 = \frac{L_1'}{1 - \bar{t}\bar{L}_1'} = \frac{F_1}{1 - \bar{t}F_1} \qquad \text{(ii),}$$

then for $\bar{L}_2'$ we have

$$\bar{L}_2' = \bar{L}_2 + F_2 = \frac{F_1}{1 - \bar{t}F_1} + F_2 = \frac{F_1 + F_2 - \bar{t}F_1 F_2}{1 - \bar{t}F_1} \qquad \text{(iii).}$$

Next, substituting for $\bar{L}_2$ and $\bar{L}_2'$ from equations (ii) and (iii) into equation (i), dividing through by tan θ, and after some simplification, we have

$$f_E = f_1 \times \frac{F_1}{F_1 + F_2 - \bar{t}F_1 F_2} .$$

If we now multiply both sides of this equation by $-\frac{1}{n_1}$, and noting that $F_1 = -\frac{n_1}{f_1}$, we have

$$-\frac{f_E}{n_1} = -\frac{f_1}{n_1} \times \frac{F_1}{F_1 + F_2 - \bar{t}F_1 F_2} = \frac{1}{F_1 + F_2 - \bar{t}F_1 F_2} .$$

Taking reciprocals of both sides of this equation we obtain

$$-\frac{n_1}{f_E} = F_1 + F_2 - \bar{t}F_1 F_2$$

from which we notice that the left hand side is the power of a single surface with a medium of refractive index n_1 to the left. That is, $F_E = -\frac{n_1}{f_E}$, which is equation (4.30).

If we turn the system around and start again with $-f_E' \tan\theta = f_2 \tan\theta \times \frac{L_2}{\bar{L}_2'}$, it is possible to derive equation (4.31) in a similar manner. This is left as an exercise for the reader.

From equations (4.30) and (4.31) it follows that

$$\frac{f_E'}{f_E} = -\frac{n_2'}{n_1} \qquad (4.32).$$

Thus we see that the first and second equivalent focal lengths of the thick lens are not the same value when $n_1 \neq n_2'$. In this case the equivalent power represents a single surface with a medium of refractive index n_1 to its left, and a medium of refractive index n_2' to its right. When $n_1 = n_2'$ the equivalent power represents a thin lens in a surrounding medium $n_s = n_1 = n_2'$.

Continuing with the formulae, we also have

$$\text{Back vertex power} \qquad F_v' = \frac{F_E}{1 - \bar{t}F_1}$$

$$\text{and} \qquad f_v' = \frac{n_2'}{F_v'} \qquad (4.33).$$

This locates the point F' since $f_v' = A_2F'$. (See the worked problems in section 4.4 for the derivation with a more complicated system).

$$\text{Front vertex power} \qquad F_v = \frac{F_E}{1 - \bar{t}F_2}$$

$$\text{and} \qquad f_v = -\frac{n_1}{F_v} \qquad (4.34).$$

This locates the position of the point F since $f_v = A_1F$.

For the positions of the principal points we have

$$e = f_v - f_E \qquad \text{but} \qquad \frac{e}{n_1} = \bar{t}\cdot\frac{F_2}{F_E} \qquad (4.35)$$

$$\text{and} \qquad e' = f_v' - f_E' \qquad \text{but} \qquad \frac{e'}{n_2'} = -\bar{t}\cdot\frac{F_1}{F_E} \qquad (4.36).$$

Equation (4.35) is derived by putting $f_v = -\frac{n_1}{F_v} = -n_1 / \left(\frac{F_E}{1 - \bar{t}F_2}\right)$ and $f_E = -\frac{n_1}{F_E}$ in the expression $e = f_v - f_E$.

It is left to the reader to perform the transformation. A similar state of affairs applies to the derivation of equation (4.36) - simply replace the terms f_v' and f_E' with their power relationships.

4.2.6.2 *Nodal Points and Paraxial Ray Tracing*

The definition of the nodal points of a thick lens was given in section 4.2.5 . When the object and image spaces have different refractive indices the locations of the nodal points relative to the principal foci and principal points may be found by the construction indicated in figure 4.33 .

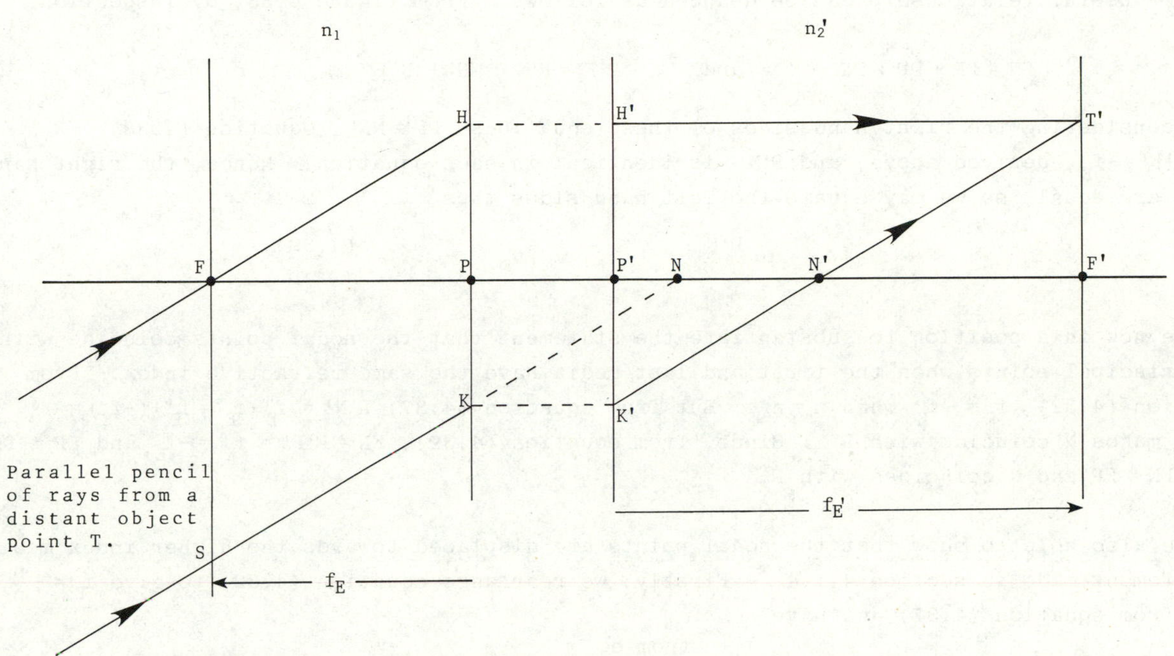

Fig. 4.33 The cardinal points for a thick lens where $n_1 \neq n_2'$.

The ray FH is drawn from the first principal focus F to intersect the first principal plane at H. The ray emerges as if from the second principal plane at H' and runs parallel to the principal axis, to intersect the second focal plane at T', with PH = P'H'. Now T' is an image point corresponding to an incident parallel pencil of rays inclined to the axis and the direction of the pencil must be the direction FH, since this is a ray which passes through T' after refraction. Consequently, if an emergent ray is drawn through T', parallel to FH, this ray will be parallel to its original direction before refraction. Thus, T'N' drawn parallel to FH cuts the axis at N', the second nodal point. Now produce T'N' to meet the second principal plane at K'. Then K, at the same distance as K' from the axis but in the first principal plane, is the point where the incident ray which emerges as K'N'T' meets the first principal plane. Draw SKN through K parallel to K'N'T' cutting the axis at N and meeting the first focal plane at S. The incident ray SK becomes the emergent ray K'T' , and clearly N is the first nodal point.

Figure 4.33 may also be used to deduce several useful relationships. We shall deduce three relationships, equations (4.37), (4.38), and (4.39), from which we shall be able to make further deductions about statements which have been made during the course of this chapter. The statements concern the positions of the nodal points, both when the object and image space indices are identical, and when they differ.

The triangles FHP and N'T'F' are congruent. Therefore, $N'F' = FP = -f_E$. Now, $P'N' = P'F' - N'F' = f_E' - (-f_E)$ or

$$P'N' = f_E' + f_E \qquad (4.37).$$

From the parallelogram NN'K'K, NN' = KK'. But KK' = PP' from figure 4.33, so

$$NN' = PP' \qquad (4.38).$$

Another useful relationship can be deduced as follows. From figure 4.33, by inspection

$$FN = FP + PP' + P'N \qquad \text{and} \qquad P'F' = P'N + NN' + N'F'$$

Now, considering the right hand sides of these equations, PP' = NN', equation (4.38), $FP = N'F' = -f_E$, derived above, and P'N' is identical in each equation. Hence, the right hand sides are equal, so we may equate the left hand sides thus

$$FN = P'F' \qquad (4.39).$$

We are now in a position to substantiate the statement that the nodal points coincide with the principal points when the first and last media have the same refractive index. From equation (4.32) $f_E = -f_E'$ when $n_1 = n_2'$. But from equation (4.37) $P'N' = f_E' + f_E = f_E' + (-f_E') = 0$, which makes N' coincide with P'. Since, from equation (4.39), $FN = P'F' = f_E' = -f_E$ and $FP = -f_E$, then FN = FP and N coincides with P.

We are also able to show that the nodal points are displaced towards the higher index medium, a statement made in section 4.1.4. Firstly, we rearrange equation (4.32) to give $f_E = -\frac{n_1}{n_2'} f_E'$. Then from equation (4.37), we have

$$P'N' = f_E' + f_E = f_E' - \frac{n_1}{n_2'} f_E' = f_E' \left(1 - \frac{n_1}{n_2'}\right)$$

Inspection of the last term in brackets shows that P'N' is positive when $n_2' > n_1$, and negative when $n_1 > n_2'$. So, N' will be to the right of P' when $n_2' > n_1$, and to the left of P' when $n_1 > n_2'$!

The image forming properties of the thick lens depicted in figure 4.33 may be shown by a scale drawing, making use of the known positions of the cardinal points F, F', P, P', N, and N'. The construction is illustrated in figure 4.34 with an object BT situated on the principal axis.

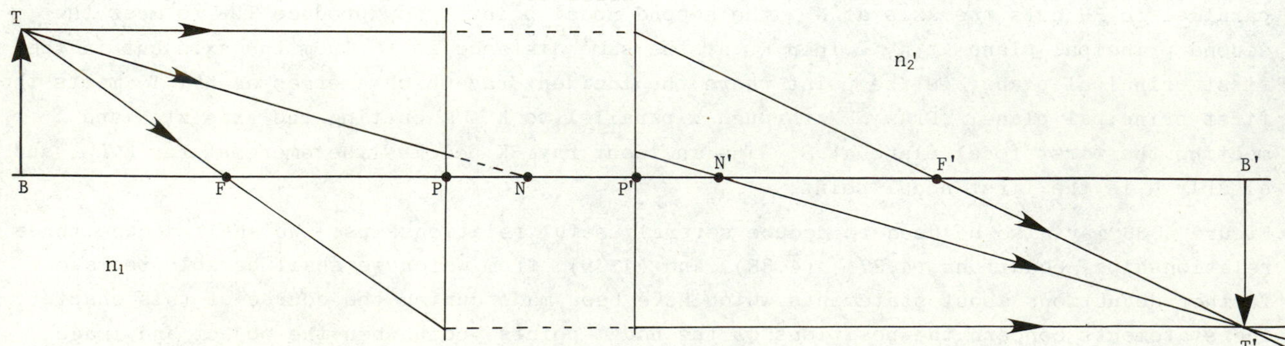

Fig. 4.34 Graphical construction for a thick lens where $n_2' > n_1$.

Worked Examples

1 One version of the Schematic Eye has the following constants relating to the cornea and the aqueous humour:

Corneal thickness	0.5 mm
Front surface radius	7.8 mm
Back surface radius	6.8 mm
Corneal refractive index	1.375
Humour refractive index	1.335

Treating the cornea as a thick lens, find its equivalent power and the first and second equivalent focal lengths.

Power of front surface:

$$F_1 = \frac{n_1' - n_1}{r_1} = \frac{1.375 - 1.000}{+0.0078} = +48.08\text{ D.}$$

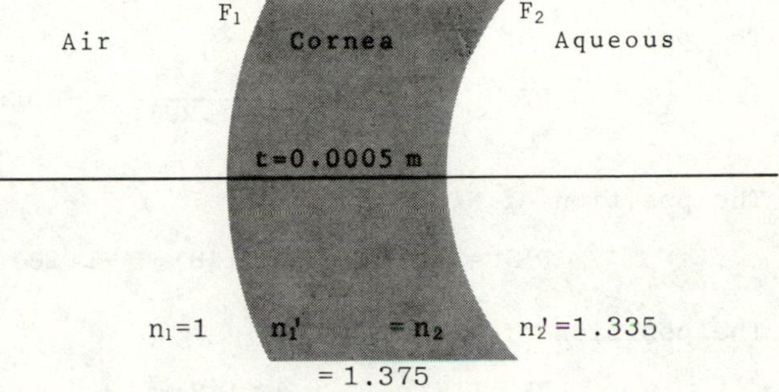

Power of back surface:

$$F_2 = \frac{n_2' - n_2}{r_2} = \frac{1.335 - 1.375}{+0.0068} = -5.88\text{ D.}$$

Equivalent power:

$$F_E = F_1 + F_2 - \bar{t}F_1 F_2$$

$$= (+48.08) + (-5.88) - \frac{0.0005}{1.375} \times (+48.08) \times (-5.88)$$

$$= (+42.20) + (+0.103) = +42.30\text{ D.}$$

The first equivalent focal length is given by $f_E = -\frac{n_1}{F_E} = -\frac{1.000}{(+42.30)} = -0.0236\text{ m} = -23.6\text{ mm}.$

The second equivalent focal length is given by $f_E' = \frac{n_2'}{F_E} = \frac{1.335}{(+42.30)} = +0.0316\text{ m} = +31.6\text{ mm}.$

2 A double convex lens of glass of refractive index 1.532 has surfaces of radii 20 cm and 15 cm, respectively, and axial thickness 2 cm. The first surface is in contact with water of refractive index 1.330, and the second surface is in contact with a liquid of refractive index 1.488 . Find the positions of the cardinal points.

Power of first surface:

$$F_1 = \frac{n_1' - n_1}{r_1} = \frac{1.532 - 1.330}{+0.20} = +1.01\text{ D.}$$

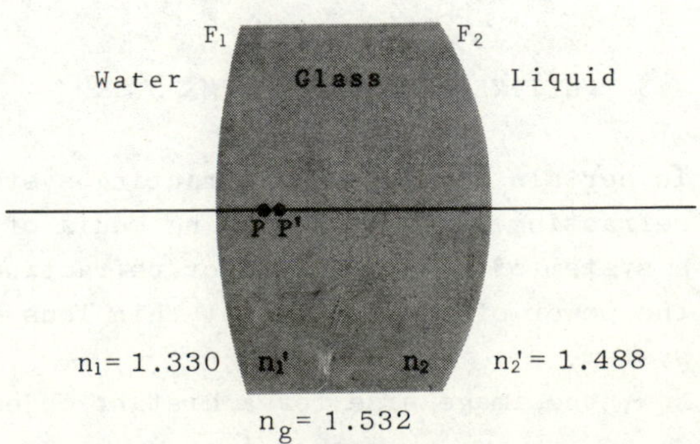

Power of second surface:

$$F_2 = \frac{n_2' - n_2}{r_2} = \frac{1.488 - 1.532}{-0.15} = +0.29\text{ D.}$$

Note the radii must be given a sign!

Equivalent power $F_E = F_1 + F_2 - \bar{t}F_1 F_2 = (+1.01) + (+0.29) - \frac{0.02}{1.532} \times (+1.01) \times (+0.29) = +1.296$ D.

The position of P from equation (4.35):

$$A_1P = e = n_1 \bar{t} \cdot \frac{F_2}{F_E} = 1.330 \times \frac{20}{1.532} \times \frac{(+0.29)}{(+1.296)} = +3.89 \text{ mm.}$$ Note t entered in mm.

The position of P' from equation (4.36):

$$A_2P' = e' = -n_2' \bar{t} \cdot \frac{F_1}{F_E} = 1.488 \times \frac{20}{1.532} \times \frac{(+1.01)}{(+1.296)} = -15.1 \text{ mm.}$$

The position of F':

$$P'F' = f_E' = \frac{n_2'}{F_E} = \frac{1.488}{+1.296} = +1.148 \text{ m} = +1\,148 \text{ mm.}$$

The position of F:

$$PF = f_E = -\frac{n_1}{F_E} = -\frac{1.330}{(+1.296)} = -1.026 \text{ m} = -1\,026 \text{ mm.}$$

The position of N':

$$P'N' = f_E' + f_E = (+1\,148) + (-1\,026) = +122 \text{ mm.}$$

The position of N:

$$FN = P'F' = f_E' = +1\,148 \text{ mm.}$$

Remember that FN means measurement from F to N.

3 Find the positions of the principal and nodal points of a single spherical surface of radius r, with media of refractive indices n and n' to the left and right of the surface.

Since the system has zero thickness - there is only one refracting surface - P and P' lie at the vertex of the surface. This is seen by putting $t = 0$ in equations (4.35) and (4.36).

Now,

$$P'N' = f_E' + f_E = f' + f = \frac{n'}{F} + \left(-\frac{n}{F}\right) = \frac{n' - n}{F} = r\,.$$

Therefore N' lies at the centre of curvature of the surface since P' is at the vertex.

Since $PP' = 0 = NN'$, N also lies at the centre of curvature. This forms the basis of the single surface reduced eye model where the two coincident principal points are labelled with the single letter P, and the two coincident nodal points are labelled with the single letter N at the surface's centre of curvature.

4.3 MULTIREFRACTING SYSTEMS

In certain instances a refracting system will consist of three or more separate lenses, or refracting surfaces, bounding media of differing refractive indices. The equivalent power of a system of x thin lenses or refracting surfaces, with air as the first and last medium, is the power of the equivalent thin lens in air which will produce the same size of image as the system.

Now, the image size for a distant object subtending an angle θ at the first component of the system is given by equation (4.14):

$$h'_{final} = f_1 \tan\theta \times \frac{\bar{L}_2}{\bar{L}'_2} \times \frac{\bar{L}_3}{\bar{L}'_3} \times \ldots\ldots\ldots\ldots \times \frac{\bar{L}_x}{\bar{L}'_x} \qquad \text{(i)}.$$

The term $f_1 \tan\theta$, equation (4.13), is the image size produced by the first component when the object is at infinity and subtends an angle θ, and this is then magnified by successive lenses or surfaces in the system. The terms $\bar{L}/\bar{L}'$ are the magnifications at the lenses or surfaces denoted by the subscripts they carry.

This final image size is equated to the image produced by the equivalent thin lens

$$f_E \tan\theta \qquad \text{(ii)}.$$

Equating (i) and (ii), and cancelling the $\tan\theta$ term, gives

$$f_E = f_1 \times \frac{\bar{L}_2}{\bar{L}'_2} \times \frac{\bar{L}_3}{\bar{L}'_3} \times \ldots\ldots\ldots\ldots \times \frac{\bar{L}_x}{\bar{L}'_x}.$$

But, in air, $f_E = -\frac{1}{F_E}$ and $f'_E = \frac{1}{F_E}$,

$$\text{so} \qquad -\frac{1}{F_E} = -\frac{1}{F_1} \times \frac{\bar{L}_2}{\bar{L}'_2} \times \frac{\bar{L}_3}{\bar{L}'_3} \times \ldots\ldots\ldots\ldots \times \frac{\bar{L}_x}{\bar{L}'_x}.$$

Inverting both sides (applying the reciprocal function!), and cancelling the two minus signs:

$$F_E = F_1 \times \frac{\bar{L}'_2}{\bar{L}_2} \times \frac{\bar{L}'_3}{\bar{L}_3} \times \ldots\ldots\ldots\ldots \times \frac{\bar{L}'_x}{\bar{L}_x} \qquad (4.40)$$

In general, when the indices of the first and last media are not the same, the equivalent power F_E represents the power of the equivalent single refracting surface. Examples of this type include the schematic eye, and the system made up from a contact lens and the contact fluid. F_E represents an equivalent thin lens only when the first and last media have the refractive index.

We should note that $F_E = -\frac{n_1}{f_E} = \frac{n'_x}{f'_E}$, equations (4.30) and (4.31).

The determination of the positions of the cardinal points follows essentially the same method as for the general case of a thick lens, and will be demonstrated in section 4.4.

4.3.1 THE HUMAN EYE

Measurements taken on large numbers of eyes have enabled certain average values to be determined for the optical constants of the eye system. An eye where the specifications are based on these average or mean values is termed a *schematic eye*. In modern parlance such an eye would be called a model, which is perhaps a better term since it immediately suggests its purpose. Many suggestions for a schematic eye have been put forward with that due to the Swedish ophthalmologist Allvar Gullstrand (1862 - 1930) being the most often quoted.

Because of its complex layer structure, the eye's crystalline lens is considerably simplified and is assumed to consist of a central biconvex lens surrounded by a larger biconvex lens of slightly lower refractive index. Figure 4.35 depicts the features of the refracting components of a schematic eye, slightly modified from Gullstrand's specification to make the eye emmetropic. The positions of the cardinal points are also shown.

Also shown is the equivalent single surface eye in its image space position at P'. For object space measurements the equivalent surface should be at P, and for image space measurements it will be at P'. However, in practice, the distance PP' is very small, being about 0.25 to 0.30 mm, and we usually place both points at P'. Similarly, N and N' are placed at the centre of curvature of the single surface. It is then usual to label them P and N.

Given this simplification, this *reduced eye* becomes a model in its own right, its devolution being somewhat disguised. Also, the first principal focal point F is moved a distance PP' nearer the equivalent surface.

In figure 4.35(a) the data of the schematic eye are:

Radii of curvature of surfaces (mm):

$r_1 = +7.7$, $r_2 = +6.8$, $r_3 = +10.0$,
$r_4 = +7.91$, $r_5 = -5.76$, $r_6 = -6.0$.

Refractive indices:

$n_1 = 1.000$, $n_2 = 1.376$, $n_3 = n_6' = 1.336$
$n_4 = n_6 = 1.386$, $n_5 = 1.406$.

Distances from front vertex of cornea (mm):

$A_1A_2 = 0.5$, $A_1A_3 = 3.6$, $A_1A_4 = 4.15$,
$A_1A_5 = 6.57$, $A_1A_6 = 7.20$.

Calculation gives the corneal power as +43.01 D and the total (equivalent) power of the eye as +58.64 D.

Figure 4.35(b) shows the positions of the cardinal points, for which

$A_1P = +1.35$ mm, $A_1P' = +1.60$ mm,
$A_1F = -15.70$ mm, $A_1F' = +24.38$ mm,
$A_1N = +7.08$ mm, $A_1N' = +7.33$ mm.

The equivalent focal lengths are $f_E = -17.05$ mm and $f_E' = +22.78$ mm.

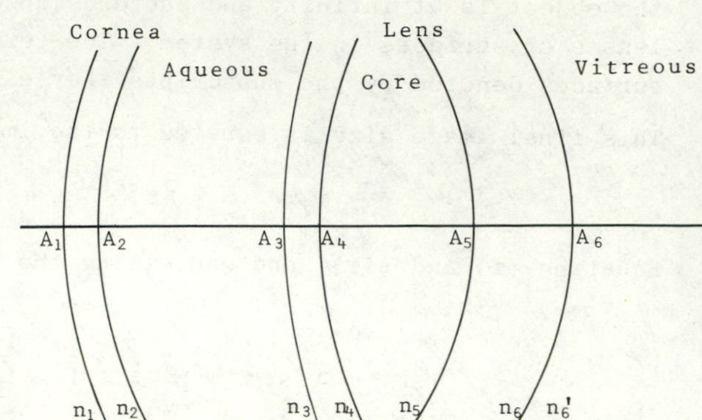

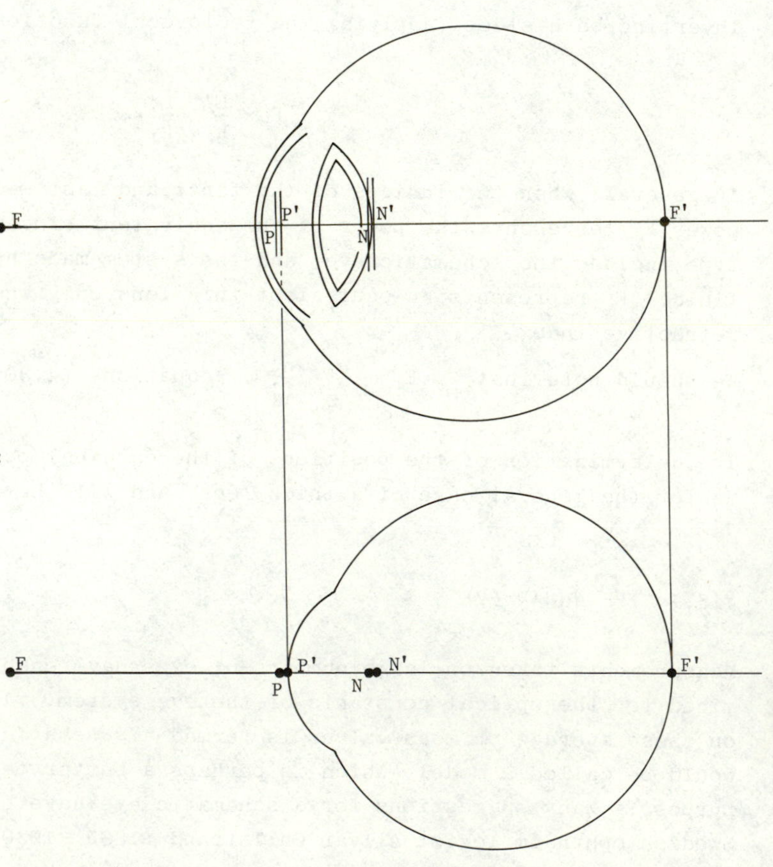

Fig. 4.35 (a) The surfaces and media of a modeleye.
(b) The cardinal points of the model eye.
(c) The single surface equivalent eye.

The reduced eye model used in visual optics calculations has the following constants:

$F_e = +60$ D $n = 1.000$ $n' = 1.333$

where e stands for eye. Note the simplification of the value of n'. Calculations provide the following data:

$r = +5.55$ mm $f_e = -16.67$ mm $f_e' = +22.22$ mm.

As shown in the third worked example of section 4.2.6.2, the nodal points coincide at the centre of curvature.

4.4 WORKED PROBLEMS

The three worked problems in this section are not simply worked examples. They illustrate the application of the theory already met, but at the same time they introduce a few new ideas.

4.4.1 A lens system consists of three separated lenses in air:

$F_1 = +8.00$ D ; $d_1 = 2$ cm; $F_2 = -12.50$ D; $d_2 = 4$ cm; $F_3 = +6.00$ D.

Find the cardinal points of the system.

Firstly, let us remember that the nodal points will coincide with the principal points since the first and last media are the same (air); that is, in this case $n_1 = n_3' = 1$.

The procedure is:

(i) Find the BVP and the FVP, F_V' and F_V. This allows us to find the vertex focal lengths $f_V = -\frac{n_1}{F_V}$ and $f_V' = \frac{n_3'}{F_V}$, whence we locate the first and second principal focal points of the system.

(ii) Calculate the equivalent power F_E. Then find the equivalent focal lengths

$$f_E = -\frac{n_1}{F_E} \quad \text{and} \quad f_E' = \frac{n_3'}{F_E}.$$

Knowing these focal lengths we can calculate the positions of the principal points P and P' using

$$A_1P = e = f_V - f_E \qquad \text{and} \qquad A_3P' = e' = f_V' - f_E'$$

where A_1 and A_3 are the front and back vertices of the system.

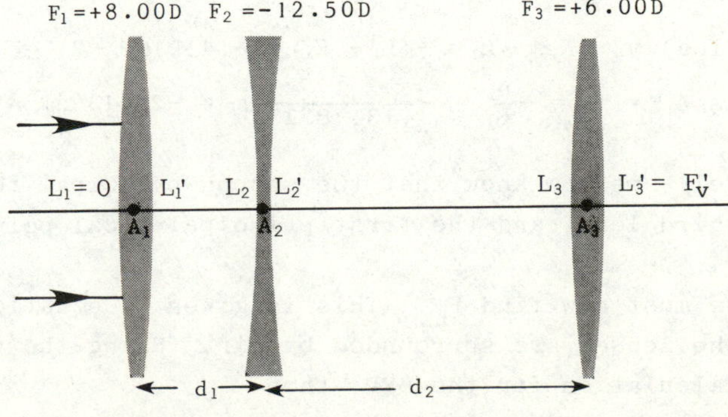

Fig. 4.36 When $L_1 = 0$ then $L_3' = F_V'$.

Figure 4.36 shows the data.

By definition, when parallel incident light from an axial object point (at $-\infty$!) enters the system, the vergence leaving the system equals the back vertex power; i.e. in this case BVP = $F_V' = L_3'$. Note that since the system is entirely in air all the vergences are actual vergences, and we do not need the reduced bar notation.

We must put $L_1 = 0$, then $L_1' = L_1 + F_1 = 0 + 8 = +8$ D.

Stepping along to the second lens: $L_1 = \frac{L_1'}{1 - d_1 L_1'} = \frac{8}{1 - (0.02 \times 8)} = \frac{8}{1 - 0.16} = +9.5238$ D.

At the second lens: $L_2' = L_2 + F_2 = 9.5238 + (-12.5) = -2.9762$ D.

Stepping along to the third lens: $L_3 = \frac{L_2'}{1 - d_2 L_2'} = \frac{-2.9762}{1 - (0.04 \times (-2.9762))} = \frac{-2.9762}{1 - (-0.11905)}$

$$= \frac{-2.9762}{1 + 0.11905} = -2.6596 \text{ D}.$$

At the third lens: $L_3' = L_3 + F_3 = -2.6596 + (+6) = +3.3404$ D.

So, the back vertex power is $F_V' = L_3' = +3.3404$ D, and the back vertex focal length is $f_V' = \frac{n_3'}{F_V'} = \frac{1}{(+3.3404)}$ m $= +29.94$ cm. Therefore, the second principal focal point F' is 29.94 centimetres to the right of the third lens.

To find the front vertex power we can either turn the lens system around and allow parallel light to be incident upon the third lens (now at the left), and then equate the vergence leaving the system with the FVP, or we can leave the system alone, allow parallel light to leave the system and thereafter calculate the vergence incident at the first lens. In this case the FVP $= F_V = -L_1$. The system is redrawn for this calculation in figure 4.37.

Now, $L_3 = L_3' - F_3 = 0 - (+6) = -6$ D.

Stepping back from the third lens to the second:

$L_2' = \frac{L_3}{1 + d_2 L_3} = \frac{-6}{1 + (0.04 \times (-6))} = \frac{-6}{1 - 0.24}$

$= -7.8947$ D.

Then, $L_2 = L_2' - F_2 = -7.8947 - (-12.5)$

$= +4.6053$ D.

Stepping back to the first lens:

$L_1' = \frac{L_2}{1 + d_1 L_2} = \frac{4.6053}{1 + (0.02 \times 4.6053)}$

$= \frac{4.6053}{1 + 0.09211} = +4.2169$ D.

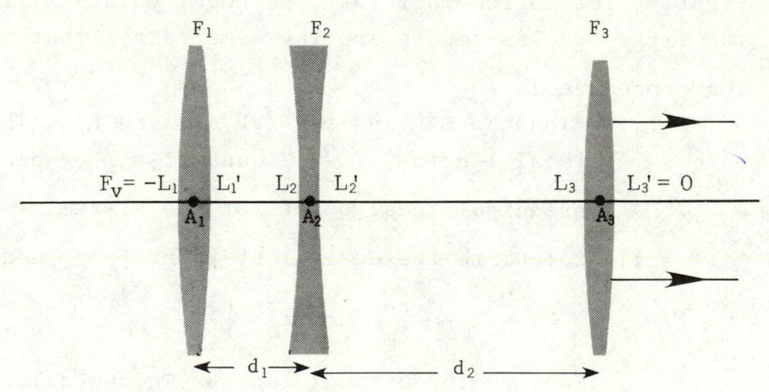

Fig. 4.37 When $L_3' = 0$ then $F_V = -L_1$.

Finally, $F_V = -L_1 = -(L_1' - F_1) = -(4.2169 - 8) = +3.7831$ D. Hence, the front vertex focal length is $A_1F = f_V = -\frac{n_1}{F_V} = -\frac{1}{(+3.7831)}$ m $= -26.43$ cm.

Well, we now know that the second principal focal point F' is 29.94 cm to the right of the third lens, and the first principal focal point is 26.43 cm to the left of the first lens.

We must now find F_E. This is given by equation (4.40), without the reduced bars since all the lenses are surrounded by air. Using the values of the vergences when $L_1 = 0$, from the calculation for the BVP, then

$F_E = F_1 \times \frac{L_2'}{L_2} \times \frac{L_3'}{L_3} = (+8) \times \frac{(-2.9762)}{(+9.5238)} \times \frac{(+3.3404)}{(-2.6596)} = +3.1400$ D. Then $f_E' = \frac{n_3'}{F_E} = \frac{1}{+3.1400}$ m $= +31.85$ cm

and $f_E = -\frac{n_1}{n_3'} \cdot f_E' = -\frac{1}{1} \cdot (+3185) = -31.85$ cm. Now, P is a distance A_1P from the first lens, given by

$$A_1P = e = f_v - f_E = (-26.43) - (-31.85) = +5.42 \text{ cm} .$$

Similarly, the position of P' is given by

$A_3P' = e' = f_v' - f_E' = (+29.94) - (+31.85) = -1.91$ cm. That is, P' is 1.91 cm to the left of the third lens.

The points are shown in figure 4.38 .

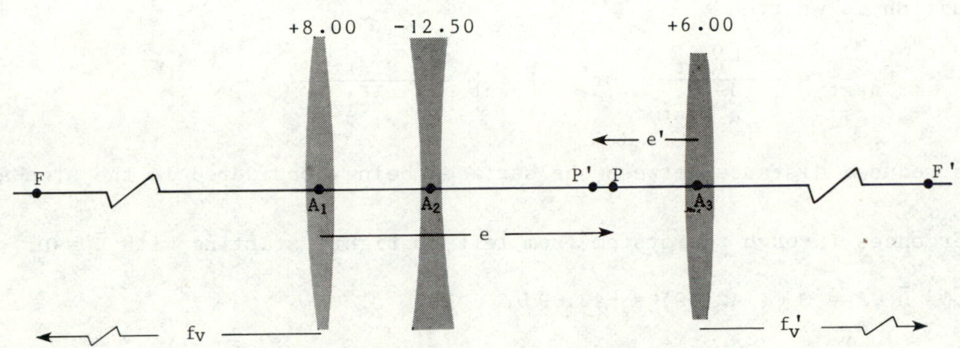

Fig. 4.38 The positions of the cardinal points: $f_v = -26.43$ cm, $f_v' = +29.94$ cm, $e = +5.42$ cm, and $e' = -1.91$ cm.

4.4.2 The data of a simplified, three-surface model eye are as follows:
Corneal radius 7.8 mm, separating air from aqueous humour of refractive index 1.333 . Depth of aqueous to lens 3.6 mm. Anterior lens surface radius 10.0 mm, posterior lens surface radius 6.0 mm, lens thickness 3.5 mm, and refractive index of the lens 1.415 . Refractive index of the vitreous humour 1.333 .
Find the positions of the cardinal points.

The data are tabulated here for convenient reference:

$n_1 = 1.000$ $\qquad r_1 = +0.0078$ m

$n_1' = n_2 = 1.333$ $\qquad r_2 = +0.010$ m

$n_2' = n_3 = 1.415$ $\qquad r_3 = -0.006$ m

$n_3' = 1.333$

$d_1 = 3.6$ mm and $\bar{d}_1 = \frac{d_1}{n_1'} = \frac{0.0036}{1.333} = 0.002701$ m

$d_2 = 3.5$ mm and $\bar{d}_2 = \frac{d_2}{n_2'} = \frac{0.0035}{1.415} = 0.002473$ m.

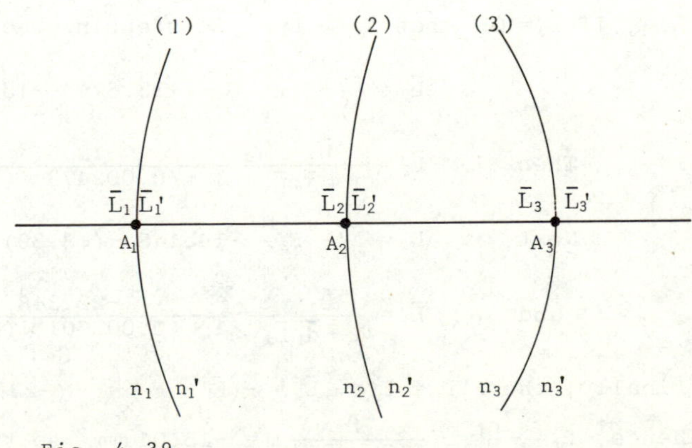

Fig. 4.39

The surface powers will be needed as we use the step-along vergences method, so we calculate these first.

Surface (1) (cornea):

$$F_1 = \frac{n_1' - n_1}{r_1} = \frac{1.333 - 1.000}{+0.0078} = +42.69\text{ D.}$$

Surface (2) (lens):

$$F_2 = \frac{n_2' - n_2}{r_2} = \frac{1.415 - 1.333}{+0.010} = +8.20\text{ D.}$$

Surface (3) (lens): $$F_3 = \frac{n_3' - n_3}{r_3} = \frac{1.333 - 1.415}{-0.006} = +13.67\text{ D.}$$

Before proceeding to the step-along calculations we should note that this problem differs from the last one in that $n_3' \neq n_1$, and all but one of the media have refractive indices greater than one. Obviously, the nodal points will not coincide with the principal points. Also, we use the reduced bar notation since most of the vergences are not in air. Further, the step-along equation is written

$$\bar{L}_{next} = \frac{\bar{L}'_{last}}{1 - \frac{d}{n}\bar{L}'_{last}} \quad \text{or} \quad \bar{L}_{next} = \frac{\bar{L}'_{last}}{1 - \bar{d}\bar{L}'_{last}}$$

where $\bar{d} = \frac{d}{n}$ is the reduced distance between the surfaces being considered in the step-along.

Now, working in vergences through the system from left to right, starting with $\bar{L}_1 = 0$:

At surface (1): $\bar{L}_1' = \bar{L}_1 + F_1 = 0 + (+42.69) = +42.69$ D.

Stepping along to surface (2): $$\bar{L}_2 = \frac{\bar{L}_1'}{1 - \bar{d}_1 \bar{L}_1'} = \frac{42.69}{1 - (0.002701 \times 42.69)} = +48.25\text{ D.}$$

At surface (2): $\bar{L}_2' = \bar{L}_2 + F_2 = +48.25 + (+8.20) = +56.45$ D.

Stepping along to surface (3): $$\bar{L}_3 = \frac{\bar{L}_2'}{1 - \bar{d}_2 \bar{L}_2'} = \frac{56.45}{1 - (0.002473 \times 56.45)} = +65.61\text{ D.}$$

At surface (3): $\bar{L}_3' = \bar{L}_3 + F_3 = +65.61 + (+13.67) = +79.28$ D. This is the vergence emerging from the third surface and equates with the BVP. The position of the second principal focus is therefore

$$A_3F' = \ell_3' = f_v' = \frac{n_3'}{\bar{L}_3'} = \frac{1.333}{+79.28}\text{ m} = +16.81\text{ mm}.$$

That is, the point F' is 16.81 mm to the right of the third surface.

Now, if $\bar{L}_3' = 0$, then $F_v = -\bar{L}_1$. So stepping back with $\bar{L}_3' = 0$:

$$\bar{L}_3 = \bar{L}_3' - F_3 = 0 - (+13.67) = -13.67\text{ D.}$$

Then $$\bar{L}_2' = \frac{\bar{L}_3}{1 + \bar{d}_2 \bar{L}_3} = \frac{-13.67}{1 + (0.002473 \times (-13.67))} = \frac{-13.67}{1 - 0.03381} = -14.148\text{ D.}$$

Next, $$\bar{L}_2 = \bar{L}_2' - F_2 = -14.148 - (+8.20) = -22.348\text{ D.}$$

and $$\bar{L}_1' = \frac{\bar{L}_2}{1 + \bar{d}_1 \bar{L}_2} = \frac{-22.348}{1 + (0.002701 \times (-22.348))} = \frac{-22.348}{1 - 0.06036} = -23.784\text{ D.}$$

Finally, the FVP = $F_v = -\bar{L}_1 = -(\bar{L}_1' - F_1) = -((-23.784 - (+42.69)) = +66.474$ D.

Hence, $f_v = -\frac{n_1}{F_v} = -\frac{1}{(+66.474)}\text{ m} = -15.04\text{ mm}$, so that F is 15.04 mm to the left of the first surface.

Now, using the vergences when $\bar{L}_1 = 0$, the equivalent power is

$$F_E = F_1 \times \frac{\bar{L}_2'}{\bar{L}_2} \times \frac{\bar{L}_3'}{\bar{L}_3} = (+42.69) \times \frac{(+56.45)}{(+48.25)} \times \frac{(+79.28)}{(+65.61)} = +60.35\ \text{D}.$$

Hence, the equivalent focal lengths are:

$$f_E' = \frac{n_3'}{F_E} = \frac{1.333}{+60.35}\ \text{m} = +22.09\ \text{mm} \quad \text{and} \quad f_E = -\frac{n_1}{F_E} = -\frac{1}{(+60.35)} = -16.57\ \text{mm}.$$

This places P' at $F'P' = -f_E' = -22.09$ mm from F', and P at $FP = -f_E = -(-16.57) = +16.57$ mm from F. These are shown on figure 4.40 .

To find the position of the second nodal point we use equation (4.37), which is quite general. Thus, $P'N' = f_E' + f_E = (+22.09) + (-16.57) = +5.52$ mm, so that N' is 5.52 mm to the right of P'. Then, by inspection of figure 4.33, and taking account of the sign convention,

$$\begin{aligned} PP' &= \text{thickness of system} - A_1P - P'A_2 \\ &= A_1A_2 - e - (-A_2P') \\ &= A_1A_2 - e + e' . \end{aligned}$$

Now, in the general case, where the thickness of the system with x surfaces is A_1A_x,

$$PP' = A_1A_x - e + e'$$

which gives for $x = 3$, in this case,

$$\begin{aligned} PP' &= A_1A_3 - (f_v - f_E) + (f_v' - f_E') \\ &= (3.6 + 3.5) - (-15.04 - (-16.57)) + (+16.81 - (+22.09)) \\ &= 7.1 - (1.53) + (-5.28) \\ &= +0.29\ \text{mm}. \end{aligned}$$

But, $NN' = PP' = +0.29$ mm, which places N 0.29 mm to the left of N' which has already been located. The six cardinal points are shown in figure 4.40 .

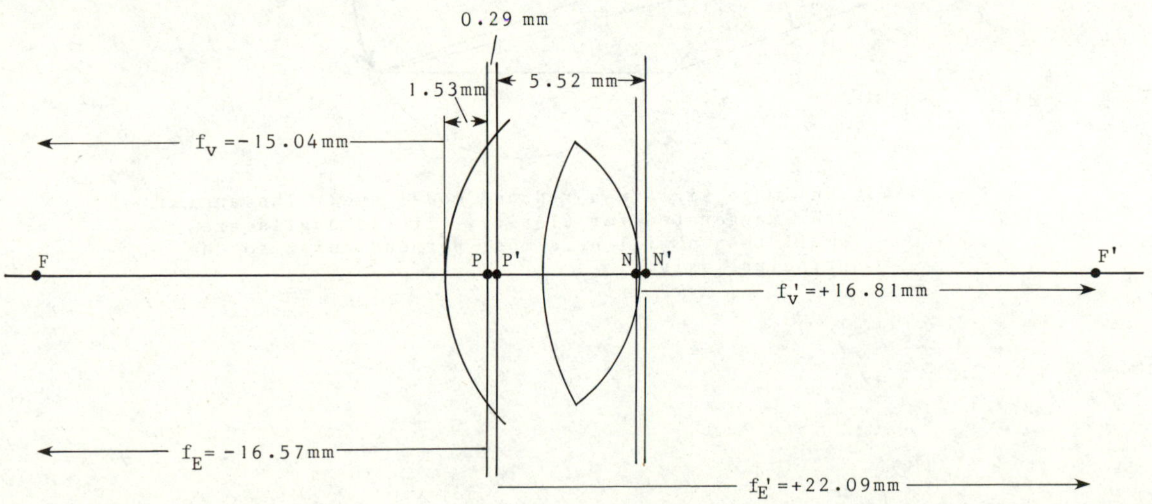

Fig. 4.40 The positions of the cardinal points of a simplified model eye.

If P and P', which are only 0.29 mm apart, are made coincident at the vertex of the surface, the equivalent (single surface) system is a reduced eye of power $F_e = F_E = +60.35$ D. If the eye is emmetropic the retina (the screen) will be placed at the second principal focus F', and the axial length of the reduced eye will be $f_e' = f_E' = +22.09$ mm.

Note that the reduced surface is $A_1P' = A_1P + PP' = 1.53 + 0.29 = +1.82$ mm from the front surface of its three-surface 'parent'. Also, its radius of curvature is $r = \frac{n'-n}{F_e} = \frac{1.333 - 1}{+60.35}$ m = +5.52 millimetres, which places its centre of curvature at N' in the three-surface eye.

4.4.3 Use the three-surface model eye from the last problem and its reduced eye derivative to find the size of the image of a distant object subtending 3^0.

(i) Using the three-surface eye and the vergences following $\bar{L}_1 = 0$ from the last problem,

the final image size $= h_3' = f_1 \tan\theta \times \frac{\bar{L}_2}{\bar{L}_2'} \times \frac{\bar{L}_3}{\bar{L}_3'} = -\frac{n_1}{F_1} \times \tan\theta \times \frac{\bar{L}_2}{\bar{L}_2'} \times \frac{\bar{L}_3}{\bar{L}_3'}$

$$= -\frac{1}{(+42.69)} \times 0.0524 \times \frac{(+48.25)}{(+56.45)} \times \frac{(+65.61)}{(+79.28)}$$

$$= -0.000868 \text{ m} = -0.868 \text{ mm}.$$

(ii) In the reduced eye we have

$$h' = f_e \tan\theta = -16.57 \times 0.0524 = -0.868 \text{ mm}.$$

where $\tan\theta = \tan 3^0 = 0.0524$, and $f_e = f_E$, the first equivalent focal length of the three-surface model.

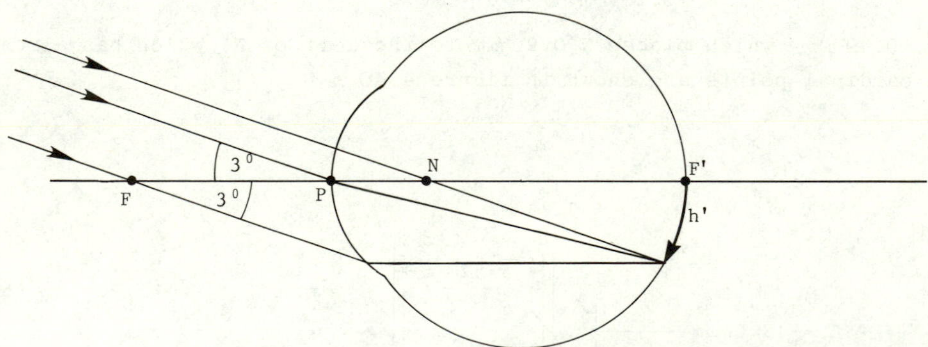

Fig. 4.41 The image size in a reduced model eye. The angles are exaggerated for clarity. If the angles are small then h' will be almost perpendicular to the principal axis.

EXERCISES

1 Define the terms back vertex power and front vertex power. Two thin lenses of powers $F_1 = +5$ D and $F_2 = +2$ D are separated by a distance of 10 cm. Find the back and front vertex powers.

Ans. B.V.P. = +12 D and F.V.P. = +7.5 D.

2 Calculate the vertex focal lengths in question 1.

Ans. $f_v' = +8.33$ cm and $f_v = -13.33$ cm.

3 Calculate the equivalent power and equivalent focal length of the system of two lenses $F_1 = +2$ D and $F_2 = +4$ D separated by 4 cm.

Ans. $F_E = +5.68$ D and $f_E' = +17.61$ cm.

4 Find the position, size, and nature of the image of a 2 cm tall object on the principal axis of a centred two lens system consisting of two thin lenses of focal lengths +50 cm and +20 cm separated by 30 cm, if the object is 1 m from the first lens.

Ans. $\ell_2' = +15.56$ cm. $h_2' = -0.44$ cm. The image is inverted, diminished, and real (since $L_2' > 0$.

5 Two thin lenses $F_1 = +4$ D and $F_2 = -10$ D are separated by 20 cm. Find the back and front vertex powers and focal lengths. Calculate the position and size of the image of a 3 cm tall object 2 m in front of the first lens.

Ans. $F_v' = +10$ D, $F_v = +\frac{2}{3}$ D, $f_v' = +10$ cm, $f_v = -1.5$ m. $\ell_2' = +60$ cm, $h_2' = -2.99$ cm.

6 Find the positions of the principal points of a centred system of two thin lenses of focal lengths $+33\frac{1}{3}$ cm and +50 cm, separated by $16\frac{2}{3}$ cm.

Ans. $A_1P = +8.33$ cm and $A_2P' = -12.5$ cm.

7 An object 1.5 cm high is situated $33\frac{1}{3}$ cm from P in the lens system of question 6. Find the position and size of the image.

Ans. The image is 1 m to the right of P', and it is inverted and 4.5 cm long.

8 Two thin lenses of power +2 D are separated by 25 cm. Find the position of the image of an object 50 cm to the left of the first focal point of the system.

Ans. $x' = +\frac{2}{9}$ cm.

9 Find the magnification of the image in the last question.

Ans. $m = -\frac{3}{2}$.

10 A real object 20 cm from the first focal point of a centred system of two thin lenses produces a real image 45 cm from the second focal point. If the focal length of the first lens is +12.5 cm, and the lens separation is 4 cm, find the power of the second lens.

Ans. $F_2 = -6.82$ D.

11 A telephoto lens in a camera consists of a +5 D lens separated by 15 cm from a -10 D lens, the latter being nearer the film. Show that this is equivalent to a thin lens of focal length +40 cm. What is the advantage of such a system compared to a camera with a single lens of focal length +10 cm, say?

Ans. See W.P.O., page 66, question 9.

12 Two thin converging lenses of focal lengths f_1' and f_2', where $f_1' = 3f_2'$, are separated by a distance $(f_1'+f_2')/2 = 2f_2'$. Find the positions of F, F', P, and P'.

Ans. $A_1P = 3f'$. $A_2P' = -f'$. $A_2F' = f'/2$. $A_1F = \frac{3}{2}f'$. See W.P.O., page 71, question 15.

13 A thin converging lens of power $3\frac{1}{3}$ D is 25 cm in front of a thin diverging lens of power 20 D. What is the equivalent power of the system?

Ans. $F_E = 0$ D. See W.P.O., page 73, question 16.

14 A thin convex lens of 20 cm focal length lies 10 cm in front of a thin convex lens of 40 cm focal length. Assuming the system lies in air, where must an object be placed in relation to the first lens in order to produce an image of the same size and (a) erect (b) inverted?

Ans. (a) $\ell_1 = +4$ cm (virtual); (b) $\ell_1 = -28$ cm.

15 Two identical thin convex lenses are fixed in a short tube 0.4 times the focal length of either lens apart. At some distance from an object they produce a real image magnified two times. When the tube is moved 25 cm further from the object the magnification is 0.4. Find (a) the focal length of the combination, and (b) the focal length of each lens.

Ans. $f_E' = +15.625$ cm and $f_1' = f_2' = +25$ cm.

16 A thin converging lens of power F_1 is placed a distance f_1' in front of a second lens of power F_2. Find the equivalent power of the combination.

Ans. $F_E = F_1$.

17 Find the position of the principal points in the system in question 16.

Ans. $A_1P = (f_1')^2 F_2$. $A_2P' = -f_1'$.

18 In the system of question 16, show that $f_v = f_1'(f_1'F_2 - 1)$.

19 A thick lens, axial thickness 6 cm and $n_g = 1.5$, has surface powers

(i) $F_1 = +5$ D and $F_2 = 0$ D

(ii) $F_1 = +10$ D and $F_2 = -5$ D.

Calculate the back vertex power in each case.

Ans. (i) +6.25 D and (ii) +11.66 D.

20 A thick lens with parameters $F_1 = +4$ D, $F_2 = +1$ D, $t = 3$ cm, $n_g = 1.5$, images (i) a distant object subtending 2^0, (ii) an object 1.164 0 cm tall at $33\frac{1}{3}$ cm from the front surface of the lens. Find the image position relative to the back surface and the image size in each case.

Ans. (i) $\ell_2' = +18.699$ cm, $h_2' = -0.709\,8$ cm. (ii) $\ell_2' = +49.495$ cm, $h_2' = -1.763\,6$ cm.

21 A 3 cm thick plano-convex lens is made of glass of refractive index 1.5, with the front surface +6.00 D in air. If it is placed in water, $n_w = 1.33$, find the front and back vertex powers, the equivalent power, and the positions of the focal, principal, and nodal points.

Ans. $F_v' = +2.127$ D, $f_v' = A_2F' = +62.53$ cm.

$F_v = +2.04$ D, $f_v = A_1F = -65.2$ cm.

$F_E = +2.04$ D, $A_1P = e = 0$, $A_2P' = e' = -2.66$ cm.

See W.P.O., page 57, question 2.

22 An object 2 cm high is placed in water 2.66 m from the lens in question 21. Find the position and size of the image using (i) step-along (ii) Newton's formulae.

Ans. (i) $\ell_2' = +83.7$ cm; $h_2' = -0.649$ cm. (ii) $x' = +21.17$ cm.
See W.P.O., page 59, question 3.

23 The angular size of the sun is 0.5^0. Find the diameter of its image formed on a screen by the lens which has $F_1 = +3$ D and $F_2 = +2$ D. The centre thickness is 1 cm and the refractive index is 1.5 .

Ans. $h_2' = -1.759$ mm. See W.P.O., page 61, question 4.

24 A gypsy's glass ball, diameter 10 cm, is used to focus the sun's rays. They focus 2.5 cm behind the ball. Find the refractive index of the ball.

Ans. 1.5 . See W.P.O., page 63, question 6.

25 The axial thickness of a lens is 5 cm and the vertex focal lengths are $f_v = -2$ cm and $f_v' = +3$ cm. When an object is 4 cm from the front vertex the real image formed is twice the size of the object. Find the positions of the principal points if the lens is in air.

Ans. $A_1P = +2$ cm and $A_2P' = -1$ cm. See W.P.O., page 65, question 7.

26 In question 25, find the surface powers if the refractive index of the lens is 1.5 .

Ans. $F_1 = +7.5$ D and $F_2 = +15$ D. See W.P.O., page 65, question 8.

27 A biconcave lens has front and back surface radii of 20 cm and 10 cm, respectively, and axial thickness 5 cm. Describe the image of a 1 cm tall object placed 12.5 cm from the first vertex. $n_g = 1.5$.

Ans. $\ell_2' = -7.82$ cm and $h_2' = +0.464$ cm.

28 What kind of glass lens in air will have an equivalent power independent of its thickness?

Ans. When one surface is plane. See W.P.O., page 69, question 12.

29 Suppose an object is located in the first principal plane of a thick meniscus lens. Determine the position of the image and its magnification.
Hint: use Newton's relationships.

Ans. Image in second principal plane and $m = +1$. Recall the unit planes!

30 A biconvex lens has its cardinal points located. An axial object point B lies to the left of the first focal point. Find its image B' by graphical construction.

Ans. See W.P.O., page 70, question 14.

31 An equiconvex lens made of glass of refractive index $n_g = 1.5$, with axial thickness 6 cm, has surface curvatures of 10 D. It is bounded by air on one side and water, $n_w = 1.33$, on the other. Find the positions of the cardinal points.

Ans. $A_1F = -14.66$ cm. $A_2F' = +16.73$ cm. $A_1P = +1.07$ cm. $A_2P' = -4.18$ cm.
$FN = +20.91$ cm. $F'N' = -15.72$ cm.

32 An afocal contact lens is made of plastics material of refractive index 1.49, with $r_1 = +8.00$ mm and $t = 0.5$ mm. Find the powers of the sufaces.

Ans. $F_1 = +61.25$ D and $F_2 = -62.54$ D.

33 A lens in air has equal radii $r_1 = r_2 = +4.0$ cm and $n_g = 1.5$. The centre thickness is 4.5 cm. Find its equivalent power, and the front and back vertex powers.

Ans. $F_E = +4.6875$ D, $F_V' = +7.5$ D, and $F_V = +3.409$ D.

34 A concentric lens has radii $r_1 = +8$ cm and $r_2 = +5$ cm. If $n_g = 1.5$ find F_E, F_V', and F_V.

Ans. $F_E = -2.5$ D, $F_V' = -2.86$ D, and $F_V = -2.08$ D.

35 A lens system consists of three separated thin lenses in air; $F_1 = +10$ D, $F_2 = -16.5$ D, and $F_3 = +8.33$ D. The separations are 2 cm between the first and second lenses and 5 cm between the second and third lenses. Find the positions of the points F, F', P, and P'.

Ans. $A_3F' = +20$ cm, $P'F' = +20.83$ cm, $A_1F = -12.69$ cm, and $PF = -20.83$ cm. So, P is 8.14 cm to the right of A_1 and P' is 0.83 cm to the left of A_3.

36 A three-surface model eye has radii $r_1 = +8.2$ mm, $r_2 = +10.5$ mm, $r_3 = -5.5$ mm. If the refractive indices of the aqueous, lens, and vitreous are respectively 1.336, 1.415, and 1.336, calculate the size of the clear retinal image of a distant object subtending 3^0. Assume the second principal focal point is on the retina.

Ans. $h_3' = -0.895$ mm.

37 Find the positions of the cardinal points in the eye in question 36.

Ans. $A_3F' = +17.37$ mm. $A_1F = -15.35$ mm. $P'F' = +22.08$ mm. $PF = -17.07$ mm. $FN = P'F' = +22.08$ mm. $P'N' = f_E' + f_E = 22.08 + (-17.07) = +5.01$ mm.

5 PLANE AND CURVED MIRRORS

5.0 INTRODUCTION

A ray of light may be turned through an angle by being reflected at a suitable surface. Highly polished glass will reflect only approximately 4% of the incident light near to normal incidence, whereas highly polished silver reflects about 95% of the incident light. Many other reflecting materials may be used, but in all cases, some loss of energy results. One exception to this exists concerning total internal reflection, where the actual reflection efficiency is 100%. This effect was considered in section 2.1.9 .

The production of optical reflecting surfaces gained impetus in 1840 when von Liebig achieved success with a chemical deposit of silver on glass. This process was used to produce the first silver-on-glass mirrors for reflecting telescopes. In more recent times optical reflectors have been produced by depositing a layer of aluminium or tin amalgam on the back of a sheet of glass by vacuum evaporation. This layer is then coated or painted to protect it from atmospheric corrosion. However, one of the main drawbacks to these back-reflecting mirrors is the formation of multiple images. This is shown in figure 5.15. For this reason, back-reflecting mirrors are never used in high grade optical equipment and instruments.

The glass used for front-reflecting mirrors need not be of the best quality because the light does not actually pass through it. The glass acts as a carrier for the reflecting surface. However, the deposit cannot be protected against corrosion and may have to be renewed at frequent intervals.

Mention may be made also of very high reflectance mirrors consisting of multilayer dielectric materials which are used in certain specialised applications, particularly where laser light is concerned. In the infrared region of the spectrum, quartz mirrors with coatings of calcium fluoride, germanium, or silicon are used.

5.1 REFLECTION OF LIGHT AT PLANE SURFACES

Reflection at a surface may be classified as *regular (or specular)* reflection or as *diffuse* reflection. When a parallel pencil of rays of light is incident on a highly polished metal surface it is reflected in a single definite direction. If the pencil of rays falls on a sheet of non-glossy white paper it is reflected, or scattered, in all directions. Reflection in one definite direction is termed *regular* or *specular* reflection, see figure 5.1(a).

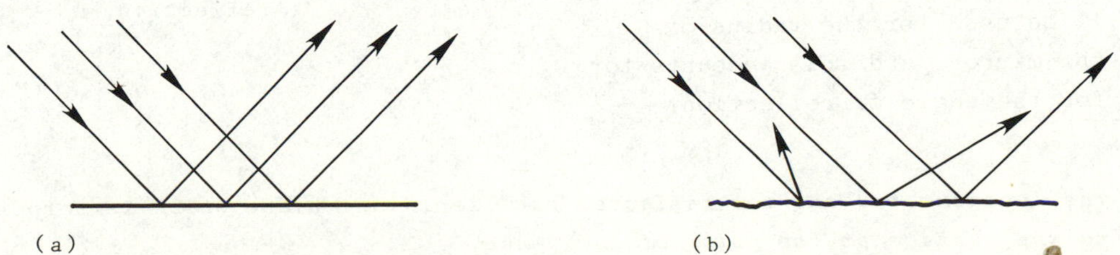

Fig. 5.1 (a) Regular or specular reflection. (b) Diffuse reflection.

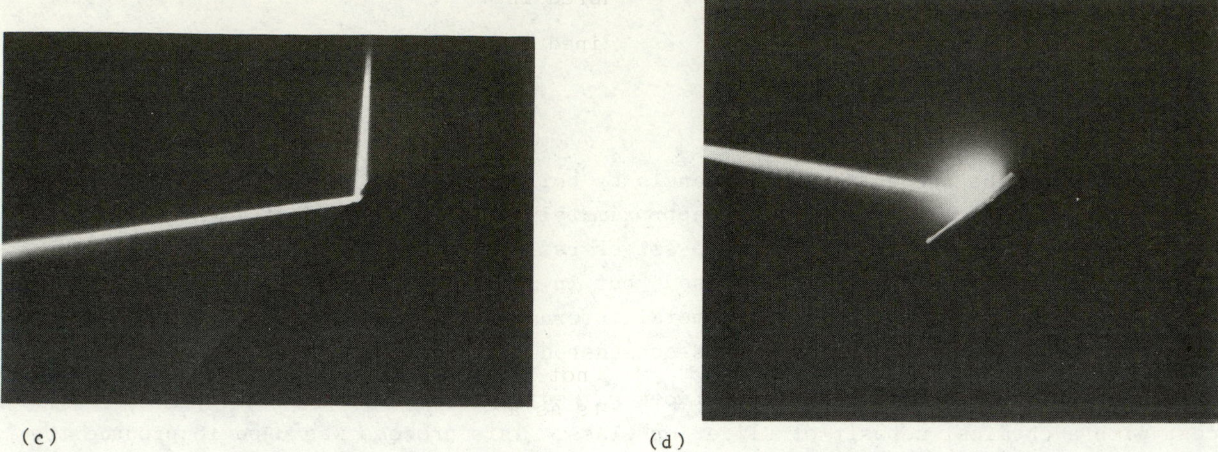

(c) (d)

Fig. 5.1 (c) and (d) Photographs of regular and diffuse reflection, respectively.

Reflection in many directions is referred to as *diffuse* reflection or *scattering*, figure 5.1(b). Figures 5.1 (c) and (d) are photographs taken in the laboratory using a ray box. Regular reflection obeys definite laws of reflection which are stated in the next section. Diffuse reflection occurs at rough surfaces and may be considered to be a large number of regular reflectors arranged in a random fashion.

5.1.1 THE LAWS OF REGULAR REFLECTION

Figure 5.2(a) shows a ray AB incident on a mirror surface and its being regularly reflected along the path BC. These rays are called the incident and the reflected ray, respectively. The angles between the incident ray and the normal and between the reflected ray and the normal are called the angle of incidence and the angle of reflection, respectively. They are labelled i and i' in the diagram. It should be noted that the use of the symbol r for the angle of reflection was met in figure 2.1. There, of course, the symbol i' was used for the angle of refraction. Later, when we come to consider spherical mirrors, r will be used for the radius of curvature of the mirror, and this accounts for the use of i' for the angle of reflection.

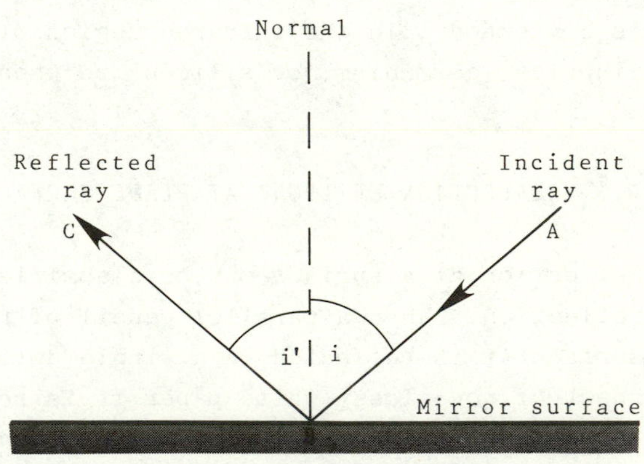

Fig. 5.2(a) Angles of incidence, i, and reflection, i'.

Law 1 *The incident ray and the reflected ray lie in one plane which is perpendicular to the surface at the point of incidence.*

Law 2 *The angle of reflection is equal to the angle of incidence: i' = i .*

These laws apply equally to plane and curved reflectors.

5.1.2 REFLECTION OF LIGHT WAVES BY HUYGENS' PRINCIPLE

Huygens showed how the wave theory could account for the reflection of light. Consider a plane wavefront WF incident on a plane reflecting surface, figure 5.2(b). The secondary wave arising from F will have reached some point F' when the point W has arrived at the mirror surface. Since the waves are in the same medium the distance FF' which the secondary wave has travelled is the same as the distance WW' which the original wave has travelled in the same time. The reflected wavefront is the line W'F', drawn tangentially to the wavelet at F'.

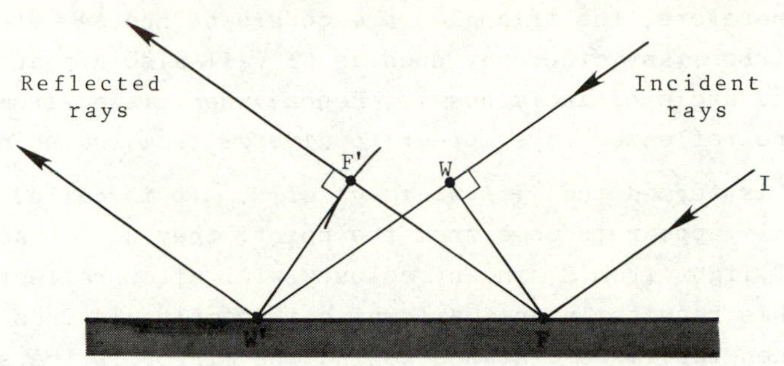

Fig. 5.2(b) Construction for Huygens' principle.

Now, the triangles WFW' and W'F'F are congruent since WW' = FF', W'F is common, and the angles $W'\widehat{F'}F = W'\widehat{W}F = 90^0$. Therefore, $W\widehat{F}W' = F'\widehat{W'}F$. These are the angles made by the incident and reflected wavefronts, respectively, with the mirror surface. Since the incident and reflected rays, such as IF and FF', are normal to the wavefronts, these rays also make equal angles with the mirror surface. Hence, it follows that the angles of incidence and reflection are equal. This is the law of reflection derived from the wave theory.

5.1.3 REVERSIBILITY OF A LIGHT RAY

In figure 5.2(a) a ray of light AB is shown reflected along the direction BC. Similarly, a ray directed along CB will be reflected in the direction BA. This is the *Reversibility Principle* and is applicable to the reflection and refraction of any ray, pencil, or beam, in all cases involving mirrors, prisms, and lenses.

5.1.4 THE FORMATION OF AN IMAGE BY A PLANE MIRROR

Figure 5.3 shows a point source of light S placed in front of a plane mirror. Three of the rays leaving S are shown incident on the mirror at A, B, and C. The ray SA, incident on the mirror normally, will be reflected back along the direction AS. The rays reflected at B and C will, according to the laws of reflection, travel along the reflected directions BD and CE, respectively.

Let SA and DB be produced backwards until they meet at the point S'. At B, let $S\widehat{B}N = i$ and $N\widehat{B}D = i'$ $(= i)$.

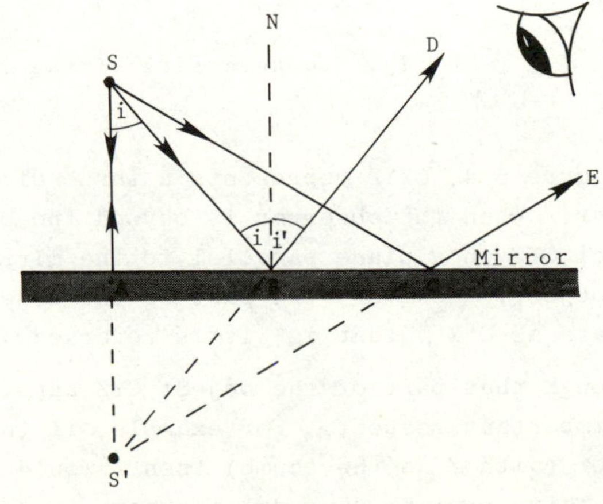

Fig. 5.3 An image S' in a plane mirror.

Since SAS' is parallel to BN, $\widehat{ASB} = i$ (alternate angles), and $\widehat{AS'B} = i$ (corresponding angles). In triangles SAB and S'AB,

$$\widehat{SAB} = \widehat{S'AB} \quad (=90^0), \quad \widehat{ASB} = \widehat{AS'B} \quad (=i), \text{ and AB is a common side.}$$

Therefore, the triangles are congruent and SA = S'A. Now, since B is any point on the mirror, any other ray such as CE will also appear to diverge from S', irrespective of the angle of incidence i. Hence, when viewed from in front of the mirror as shown, all the reflected rays appear to diverge from the point S'.

S' is termed the *virtual image* of S. It is called a virtual image because the rays of light only appear to come from the point; they do not actually intersect in the image. Any ray of light from S, in any colour, will after reflection appear to originate from the point S'. This results in an image which is completely free from aberrations (defects). Since SA=S'A, then the image distance behind the mirror is the same as the object distance in front of the mirror.

5.1.5 THE REVERSION OF AN EXTENDED IMAGE IN A PLANE MIRROR

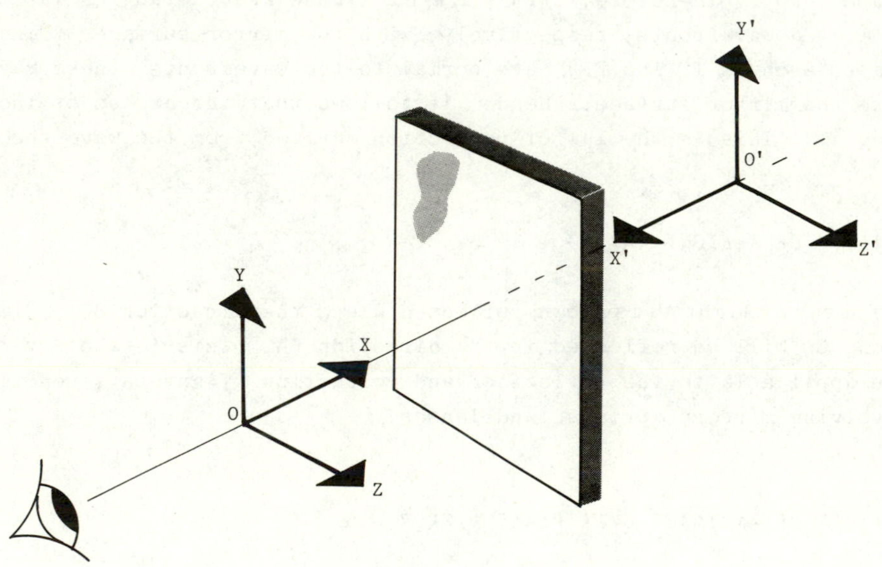

Fig. 5.4 Reversion of the image in a plane mirror.

In figure 5.4, OXYZ represents a three-dimensional object in front of a vertical plane mirror. When the observer is beyond the object, as shown, the image of the part of the object OYZ in a plane parallel to the mirror is O' Y' Z', and its appearance is similar to the object. However, the part of the object OX, perpendicular to the plane of the mirror appears as O'X'; that is, it is reversed. The effect is termed *reversion*.

Although that part of the object OYZ appears very similar in the image it does differ in one important respect. For example, if the object were a left hand held palmwards to the mirror (with Z as the thumb) then Z' would represent the thumb of a right hand facing it. This left-to-right reversal is known as *lateral inversion*. This everyday phenomenon gives rise to some interesting party questions such as "If you are left-to-right reversed why are you not upside-down?", or "Scratch your left ear with your right hand and turn with your back to the mirror. Which ear are you scratching in the mirror?". Try them on your non-optical friends!

When an observer is positioned between the object and the mirror lateral inversion becomes most evident since it is possible to view the object and its image successively. For example, if the object is a printed notice the printing appears backwards in the image, as in figure 5.5 .

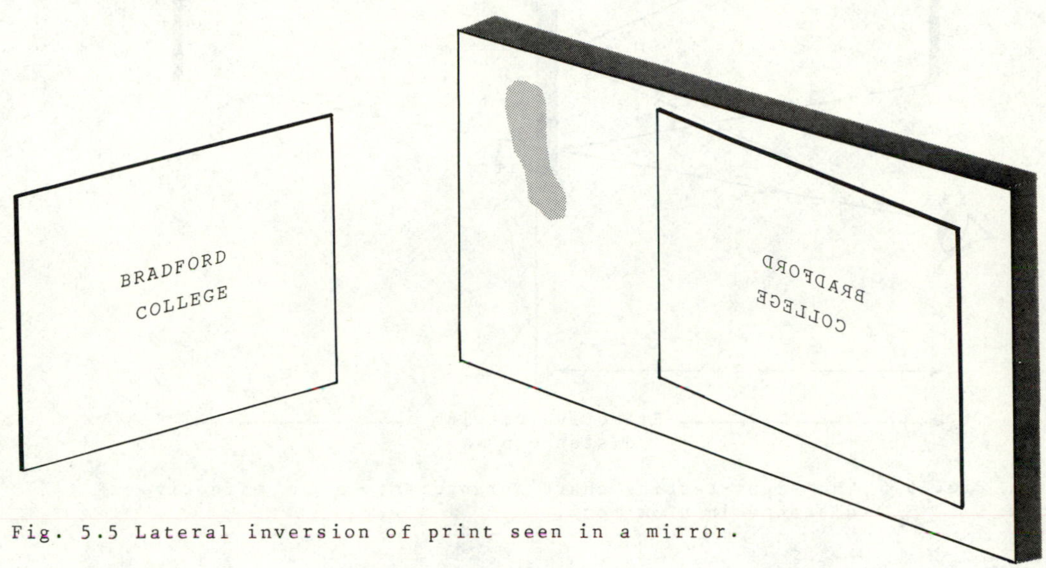

Fig. 5.5 Lateral inversion of print seen in a mirror.

5.1.6 *Summary of the Characteristics of the Image formed by a Plane Mirror*

When describing an image the characteristics of the image are:

(i) its position, (ii) its size, (iii) its nature (real or virtual), and
(iv) its appearance relative to the object (inverted, reversed, or upright).

With regard to a plane mirror, the image characteristics are:

(i) position: (a) the image is as far behind the mirror as the object is in front,
(b) the line joining the object and the image is perpendicular to the mirror surface.

(ii) size: the image is the same size as the object.

(iii) nature: the image is virtual.

(iv) appearance: the image is reversed, laterally inverted, and upright.

5.1.7 *THE SIGHT-TESTING CHART*

The sight-testing chart used by an optometrist or an ophthalmic optician may sometimes involve the use of a plane mirror. Under standard conditions the letter chart used in the examination should be positioned 6 m away from the patient. In this case the letters are printed the normal way round. However, where space is limited, the patient may be positioned

3 m from a plane mirror on the wall, and the chart is positioned just behind or above him facing the mirror (figure 5.6). Hence, the patient sees a virtual image of the chart at a distance of 6 m. In this case, the letters on the chart will have to be printed in reverse so as to appear the correct way round on the image.

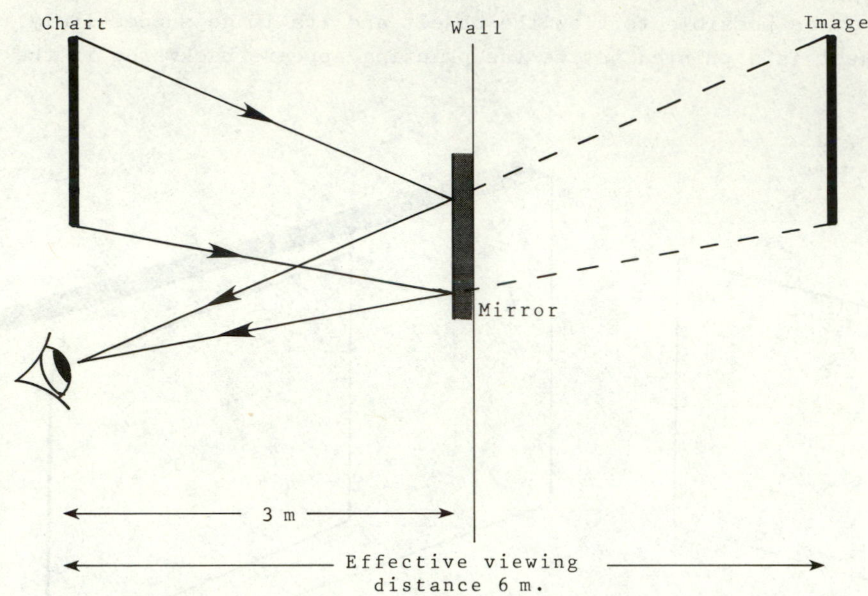

Fig. 5.6 The sight-testing chart arrangement - a 6 m effective distance in a 3m room.

Worked Example

A sight-testing chart measures 100 cm by 50 cm. Its image is viewed in a plane mirror, which is 3.5 m from the chart, by a patient who is 2.5 m from the mirror. Calculate the dimensions of the smallest mirror which can be used.

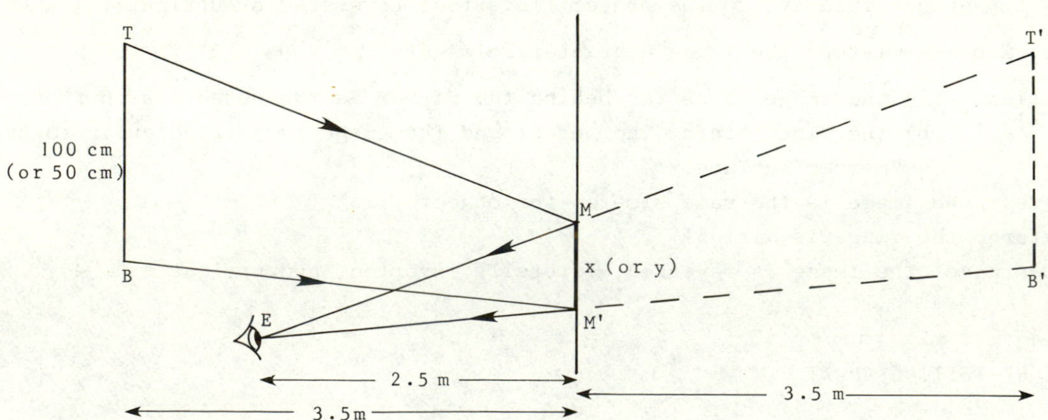

Let E represent the patient's eye and x the minimum height of the mirror, with the effective viewing distance to the image = (3.5 + 2.5) = 6 m. In the similar triangles MM'E and T'B'E

$\frac{T'B'}{EB'} = \frac{MM'}{EM'}$. $\therefore$ $MM' = T'B' \times \frac{EM'}{EB'}$. But $\frac{EM'}{EB'} = \frac{2.5}{6}$ and $T'B' = TB = 100$ cm.

$$\text{Therefore,} \quad MM' = x = 100 \times \frac{2.5}{6} = 41.67 \text{ cm.}$$

This is the minimum height of the mirror. As the width of the chart is 50 cm (=½ height), then the minimum width of the mirror will be ½x. That is, $y = ½x = 41.67/2 = 20.835$ cm. The minimum size of the mirror will be 41.47 cm × 20.835 cm.

5.1.8 THE MINIMUM SIZE OF A PLANE MIRROR

When a person looks at his image formed by reflection in a plane mirror, it can be shown that the minimum length of mirror in which he can see his whole height is one half of his actual height. This is true for any position of the person.

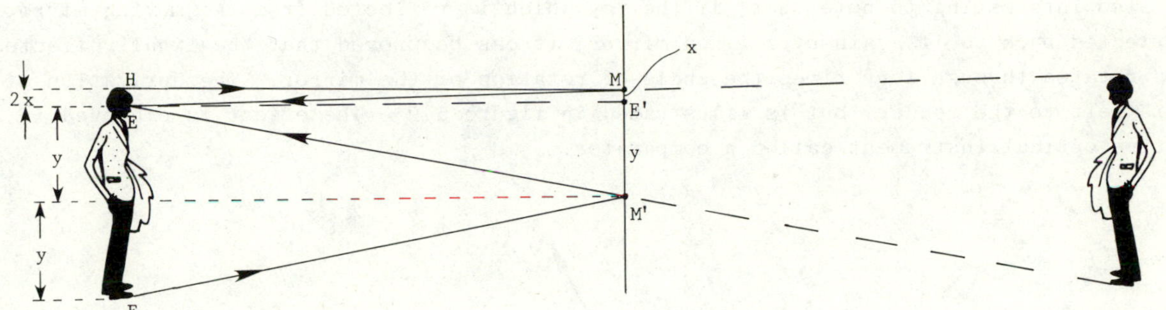

Fig. 5.7 The minimum size of a plane mirror.

In figure 5.7, MM' is the height of the mirror in which the whole image may be seen, and the point E' is level with the eye E of the subject. Now, if the height of the mirror is made up from two dimensions $ME' = x$, and $E'M' = y$ ($\therefore$ $MM' = x + y$), then it follows from the laws of reflection that the height of the subject is made up from two dimensions $HE = 2x$ and $EF = 2y$ ($\therefore$ $HF = 2x + 2y$). Hence, the height of the subject is twice the height of the mirror. Consequently, the minimum size of mirror required is one half of the subject's height.

The next time you look at your face in a plane mirror, draw on the mirror surface with a felt-tip pen, and outline the image. On standing to one side, subsequent measurement will reveal that the height (or width) of the outline is one half of the real height (or width) of your face. This may easily be shown to be true for any distance of your face from the mirror.

5.1.9 THE ROTATION OF A PLANE MIRROR

When an incident ray travelling in a fixed direction is reflected by a plane mirror which is rotating, it can be shown that the reflected ray moves through twice the angle of rotation of the mirror. This fact is derived below, and is illustrated in figure 5.8 . This important effect may be used as a magnifying device for small rotations in applications known as optical levers. Such applications include the mirror galvanometer, the sextant, and the current balance.

Let the mirror rotate through an angle θ. That is, $M_1\hat{A}M_2 = \theta$. Hence, the normal to the mirror will also rotate through θ; i.e. $N_1\hat{A}N_2 = \theta$.

Now, the angle through which the reflected ray moves is $R_1\hat{A}R_2$, and

$$\begin{aligned} R_1\hat{A}R_2 &= R_2\hat{A}I - R_1\hat{A}I \\ &= 2.N_2\hat{A}I - 2.N_1\hat{A}I \\ &= 2(N_2\hat{A}I - N_1\hat{A}I) \\ &= 2.N_2\hat{A}N_1 \\ &= 2\theta . \end{aligned}$$

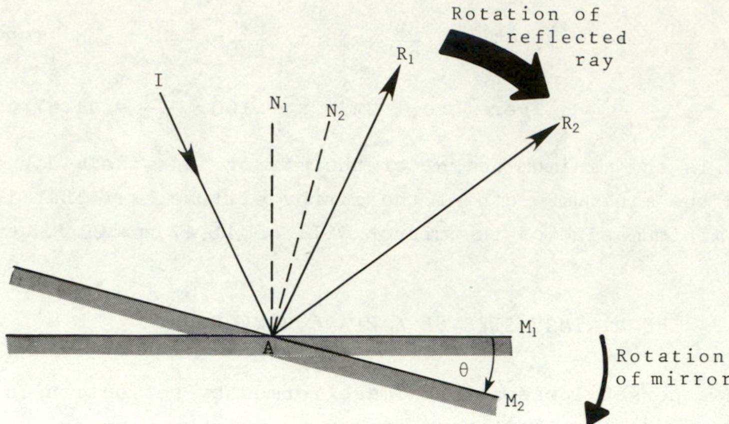

Fig. 5.8 Rotation of a plane mirror.

It is also interesting to note that, if the ray which is reflected from the moving mirror is reflected back to it again by a fixed mirror, it can be proved that the final reflected ray is rotated through four times the angle of rotation of the mirror. The derivation of this is left to the reader, but is illustrated in figure 5.9 . The effect is employed in a sensitive optical instrument called a comparator.

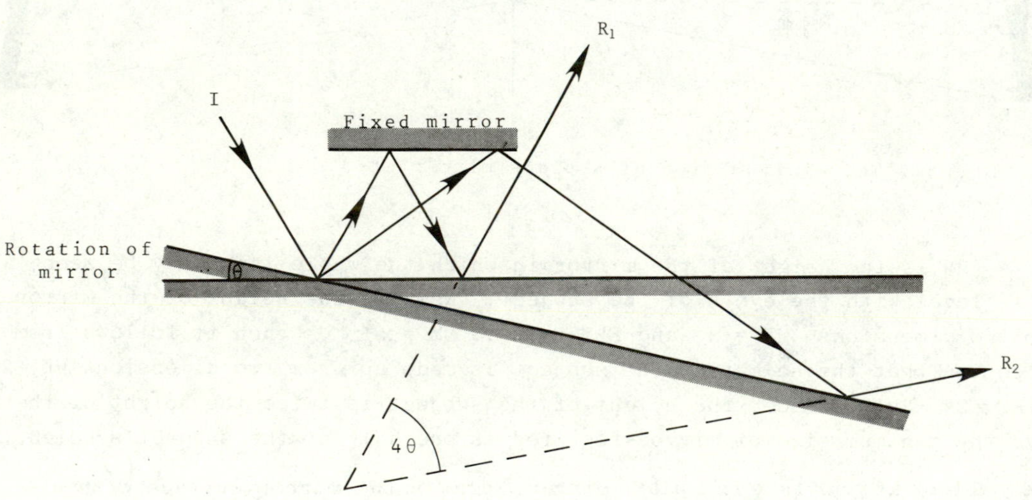

Fig. 5.9 Reflection from a fixed mirror and a rotating mirror.

5.1.10 DEVIATION BY SUCCESSIVE REFLECTIONS AT TWO INCLINED MIRRORS

Figures 5.10 and 5.11 show two plane mirrors inclined at an angle a. A ray of light IRST is reflected in turn at each mirror. The total deviation d of the light is the angle between the incident and reflected rays, measured in the same sense as the two reflections (anticlockwise in this case). It will be shown immediately that the deviation d is independent of the angle of incidence at the first mirror, and is given by $d = 360^0 - 2a$.

In figures 5.10 and 5.11,

$$d = d_1 + d_2$$
$$= 2\alpha + 2\gamma$$
$$= 2(\alpha + \gamma)$$

where α and γ are called glancing angles.

But, $\alpha + \gamma = 180^0 - a$ in triangle ASR.

$\therefore \; d = 2(180^0 - a)$

$$\text{or} \quad d = 360^0 - 2a \qquad (5.1).$$

Hence, the deviation produced depends only on the angle between the two mirrors.

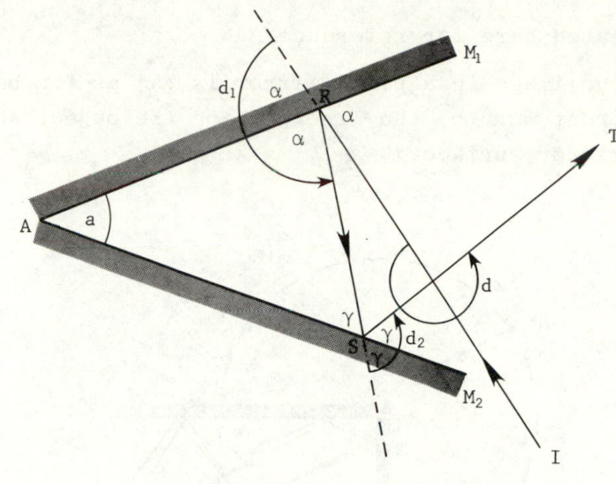

Fig. 5.10 Deviation at successive inclined mirrors.

A special case occurs when $a = 45^0$, figure 5.12. Using equation (5.1) we get

$$d = 360^0 - 2a = 360^0 - 2 \times 45^0$$
$$= 270^0.$$

In this case the light may be considered to be deviated by 90^0 ($=360^0 - 270^0$), and this forms the basis of an optical square used in surveying. This principle is also employed in the pentaprism (see figure 2.30).

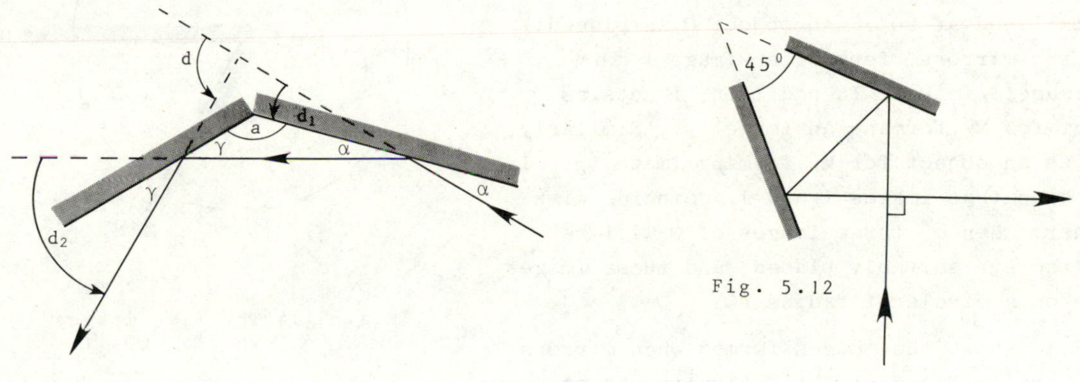

Fig. 5.11

Fig. 5.12

Figure 5.12 illustrates this 90^0 rotation of the incident ray.

5.1.11 IMAGES FORMED BY INCLINED MIRRORS

Figures 5.13 and 5.14 represent an object placed between two inclined plane mirrors. Each mirror forms a direct image, but in addition, other images are also formed. This is because some light reflected by one mirror may be again reflected by the other. It can be seen that the number of images formed depends on the angle between the mirrors.

The most convenient way of constructing the image positions is to use a point object, and to employ the image characteristics (i) (a) and (b) listed in section 5.1.5 . These are repeated here for reference:

the image in a plane mirror is (a) as far behind the mirror as the object is in front, and (b) the line joining the object and the image is perpendicular to the mirror surface.

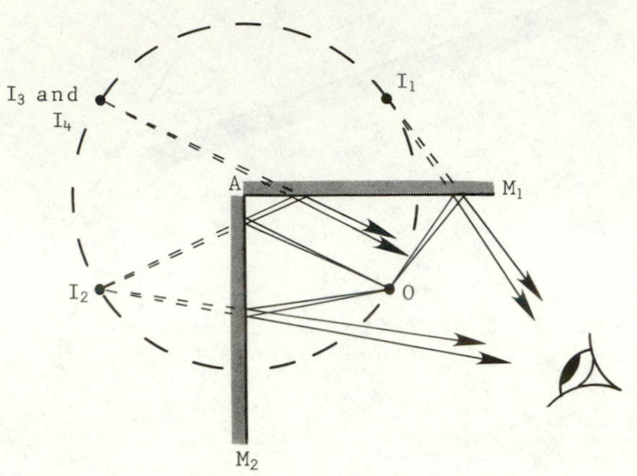

Fig. 5.13 Multiple images: a = 90°.

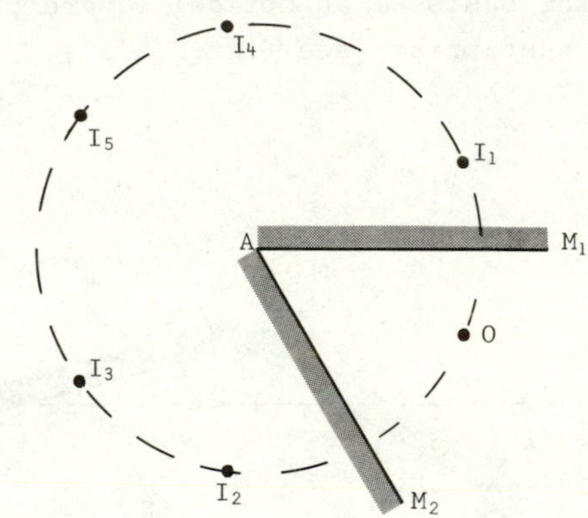

Fig. 5.14 The five images when a = 60°.

Figure 5.13 illustrates two plane mirrors M_1 and M_2 inclined at 90°. An object O is placed between the mirrors, forming an image I_1 in M_1, and an image I_2 in M_2. In addition, I_1 acts as an object for M_2 forming an image I_3. Similarly, I_2 acts as an object for M_1 forming image I_4. It will be seen that images I_3 and I_4 coincide with each other. Hence, three images of O will be seen by the eye suitably placed, and these images will lie on a circle of radius AO.

Figure 5.14 shows the images formed when mirrors M_1 and M_2 are inclined at 60°. Five images are seen, all lying on a circle radius AO.

With two parallel mirrors (a = 0°) the number of images obtained is theoretically infinite, due to repeated reflections from mirror to mirror. However, at each reflection, some light energy is absorbed, and the images become fainter and fainter. Hence, in practice, the number of images depends on the brightness of the object and the reflectance of the mirrors.

5.1.12 MULTIPLE IMAGES IN MIRRORS

With a thick back silvered plane mirror, some light is always reflected at the front surface. The amount of reflected light is greater if the angle of incidence is large (see section 9.10.1). When a brightly illuminated object is viewed at fairly large angles of incidence a series of "ghost" images may be observed in addition to the main bright image. This is shown in figure 5.15 .

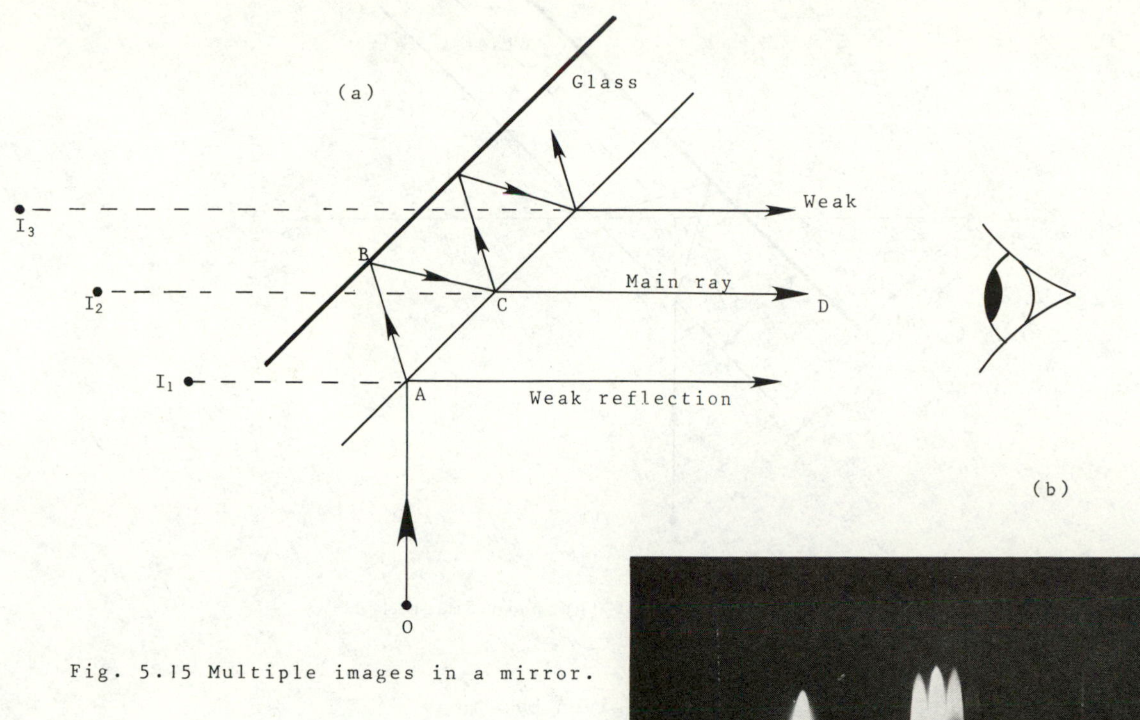

Fig. 5.15 Multiple images in a mirror.

A ray OA from the object is partly (~4%) reflected at the front glass surface. This gives rise to the faint image I_1. The remaining light is refracted along AB. Reflection then occurs at the silver surface at B, followed by refraction into the air along the direction CD. A bright image is observed at I_2. The ray CD comprises about 80% of the energy in the original light ray. Further multiple reflections will occur between the front and back surfaces of the mirror producing other fainter images such as I_3.

The images lie on the normal from O to the mirror with image I_1 as far behind the front surface as the object is in front. Close inspection of figure 5.15(b) will reveal the presence of an extremely faint image I_4. Even in the photograph it is possible to detect the different positions of the images as they are placed laterally with respect to each other, and even to suspect that one can tell that they are at different distances from the observer.

5.1.12.1 The Thick Mirror - Position of the Bright Image

Figure 5.16 shows an object O situated a distance d in front of a plane back silvered mirror. It can be shown that the distance d' of the bright image I_2 (see figure 5.15)

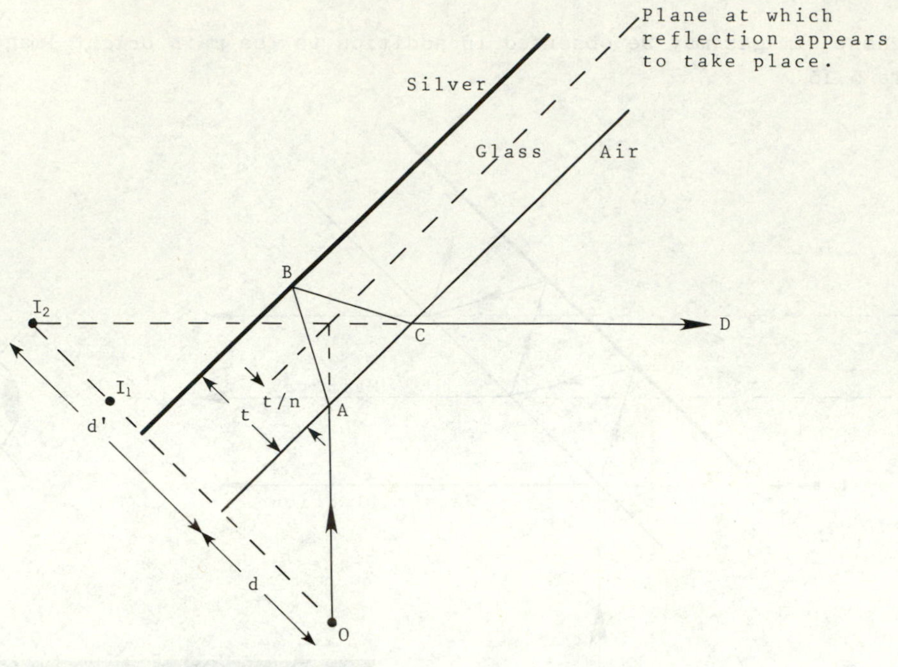

Fig. 5.16 The position of the main image I_2 .

behind the front surface of the mirror is given by:

$$d' = d + \frac{2t}{{}_a n_g} \qquad (5.2).$$

If the refractive index of the glass is ${}_a n_g = 1.5$, and the angles of incidence and of refraction are small, the plane at which the reflection appears to take place is two-thirds the thickness of the glass from the front surface. The derivation of equation (5.2) is included in Appendix 3.

5.2 REFLECTION OF LIGHT AT CURVED SURFACES

Curved mirrors are reputed to have been in use since the time of the ancient Greeks. Euclid, in his writings, referred to both concave and convex mirrors. Curved mirrors may sometimes be used as alternatives to lenses to form images of objects. They are not as widely used as lenses, but find applications as car driving mirrors, and use in reflecting telescopes, cinema projectors, and slide projectors.

The optical properties of spherical mirrors, where each mirror is part of the surface of a sphere, will now be considered. Where necessary, reference will be made to aspherical mirrors, where the surfaces are not parts of spheres. These include the paraboloidal and ellipsoidal surfaces.

5.2.1 TYPES OF SPHERICAL MIRRORS

As previously mentioned, a spherical mirror is formed from part of the surface of a sphere.

The type of mirror produced may be classed as *concave* or *convex* dependent on which surface (outer or inner) of the sphere is coated with the reflecting layer.

Figures 5.17(a) and (b) illustrate a concave (or converging) mirror and a convex (or diverging) mirror, respectively, each mirror being represented by a principal section APB.

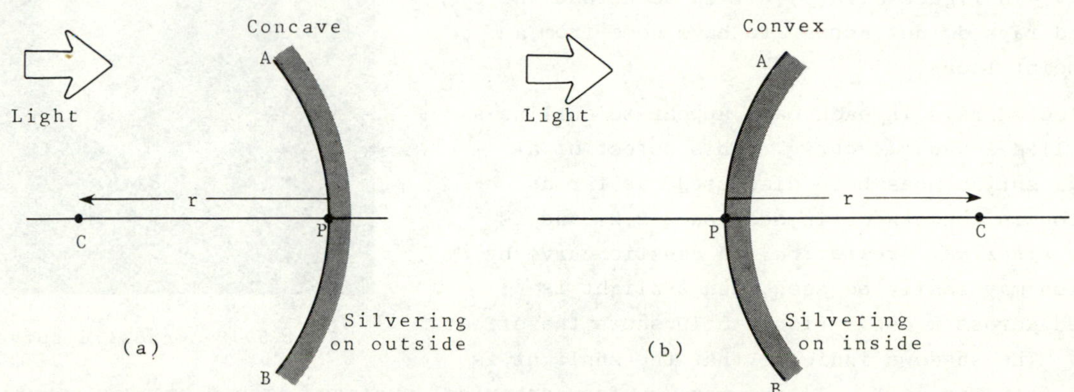

Fig. 5.17 Concave and convex spherical mirrors.

P is the *pole* or *vertex* of the mirror. This is the mid-point of the mirror surface.
C is the *centre of curvature* of the sphere of which the mirror forms a part.
The line PC, produced in both directions, is the *principal axis* of the mirror.
AB is the *diameter* or *aperture* of the mirror.
The distance PC is called the *radius of curvature* of the mirror, symbol r: i.e. PC = r.

5.2.2 REFLECTION OF A PENCIL OF RAYS WHICH IS PARALLEL TO THE PRINCIPAL AXIS

5.2.2.1 *A Wide Aperture Spherical Mirror*

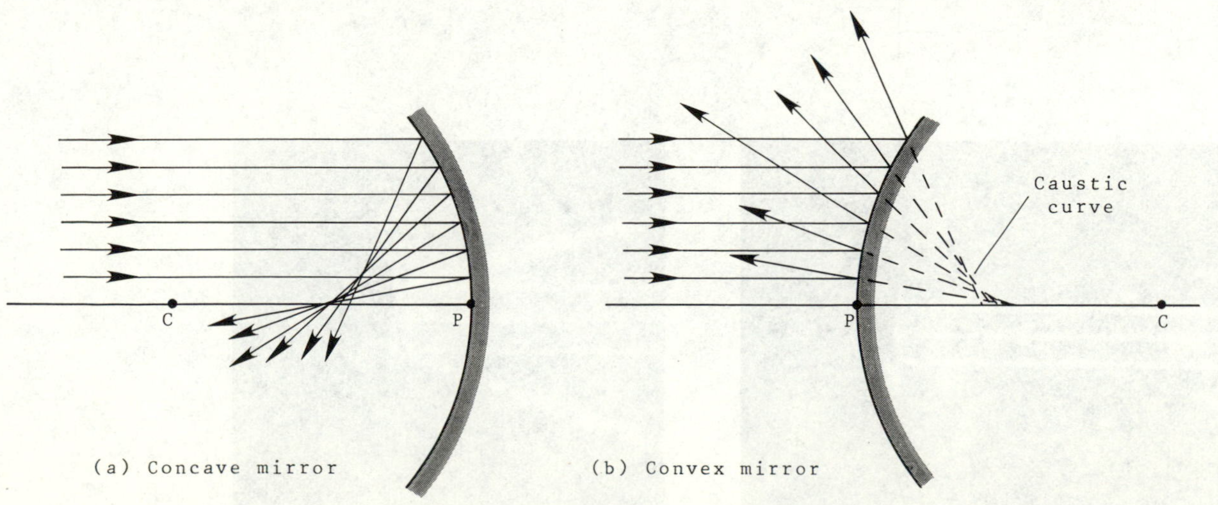

Fig. 5.18 Caustic curves in (a) a spherical concave mirror and (b) a spherical convex mirror.

Fig. 5.19 A caustic curve

Figures 5.18(a) and (b) illustrate the effects of a concave mirror and a convex mirror, respectively, on a wide parallel pencil of rays. Although each individual ray of light obeys the laws of reflection, it is seen in figure 5.15(a) that the reflected rays do not all pass through a single point focus. Similarly, in figure 5.15(b), it is seen that the reflected rays do not appear to have come from a single point focus.

The reflected rays in each case appear to outline a curve called a *caustic curve*. This defect of a spherical surface has been discussed, as far as refraction is concerned, in section 2.2.2, and is known as *spherical aberration*. A caustic curve by reflection may easily be seen when sunlight is reflected across a cup. Figure 5.19 shows the effect clearly. The shadows indicate that the sunlight is incident from the left. A lens made up from spherical surfaces also possesses spherical aberration associated with refraction, and this will be considered in Chapter 7.

5.2.2.2 *Narrow Aperture Spherical Mirrors*

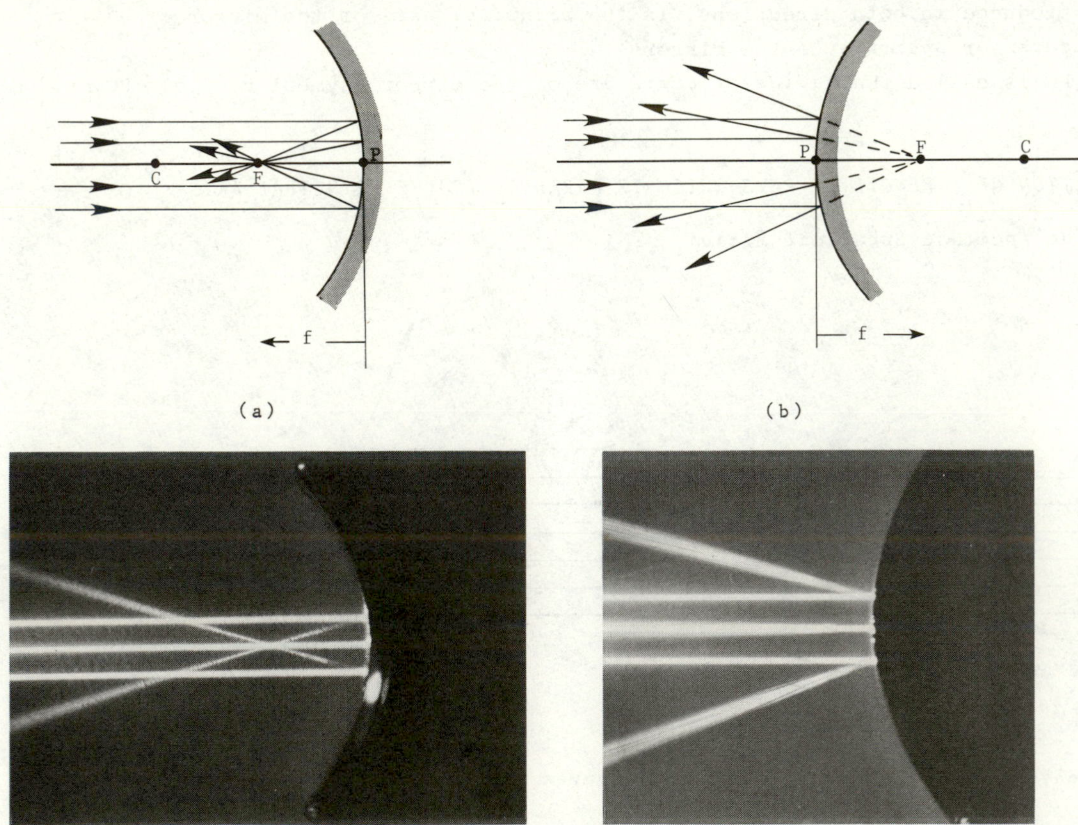

Fig. 5.20 Diagrams and photographs of (a) concave and (b) convex mirrors.

If the incident pencil of rays is limited to a narrow pencil of rays parallel to and close to the principal axis of the mirror the reflected rays all intersect, or appear to intersect, essentially at a single point. This is called the *principal focus* of the mirror and is illustrated in figures 5.20(a) and (b) for a concave and a convex mirror, respectively.

In figure 5.20(a) the principal focus F is a real point because the rays of light actually intersect at that point. A clear image of the distant object would be obtained on a screen placed at F. However, in figure 5.20(b) the principal focus F is a virtual point, since the rays only appear to diverge from it. Hence, no image would be formed on a screen placed at this point.

The distance PF is termed the *focal length*, symbol f, of the mirror.

5.2.3 FORMS OF REFLECTING SURFACE

To converge the light from a distant object point to a single point focus over a wide aperture, the reflecting surface must be concave and paraboloidal, figure 5.21. Reflecting telescopes use paraboloidal mirrors since they do not suffer from spherical aberration. In a searchlight, a paraboloidal mirror is used to form a parallel pencil of reflected rays from the light source.

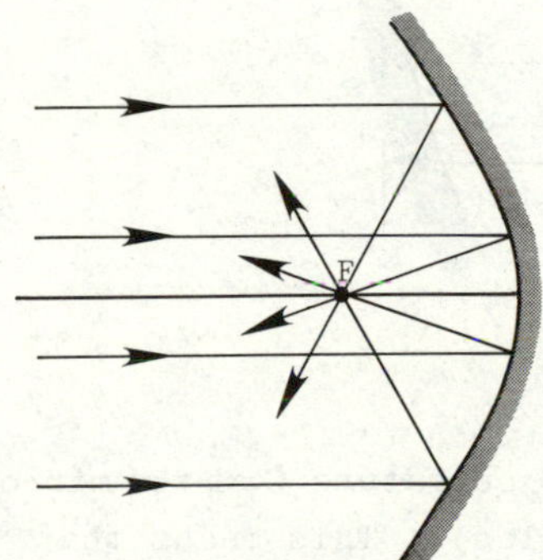

Fig. 5.21
A paraboloidal mirror

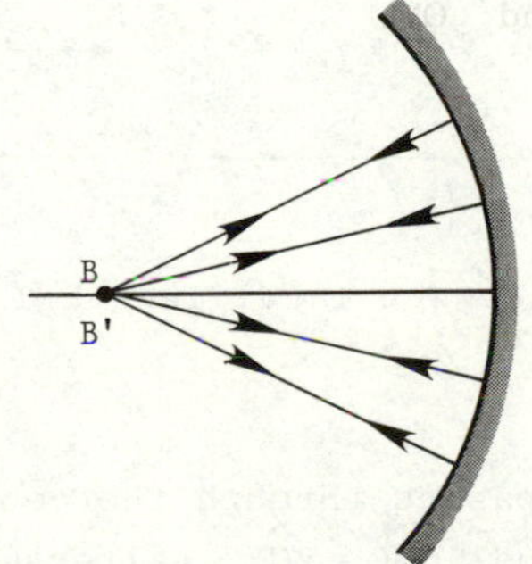

Fig. 5.22
A spherical mirror

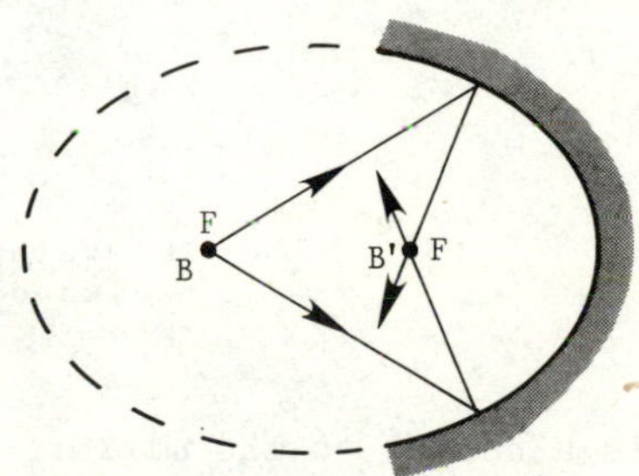

Fig. 5.23
An ellipsoidal mirror

A concave spherical surface must be used if the object and image are at the same distance from the mirror, figure 5.22. In this case they would both be located at the centre of curvature of the mirror.

If the object and image distances are finite and unequal, a concave ellipsoidal mirror will focus all the rays from one focus of the ellipse at the other focus, figure 5.23. Ellipsoidal reflectors are used in cinema projectors to form an image of the light source on the film.

Non-spherical surfaces, such as the paraboloid, ellipsoid, and the plane surface, together with cylindrical and toroidal surfaces met in connection with lens surfaces, are referred to as *aspherical surfaces*. The wide use of spherical surfaces in optics is not because of their superiority in forming images, but because of the relative ease with which they can be manufactured. In the following theory we shall assume that the spherical mirrors are very small parts of spheres, so that the effects of spherical aberration may be ignored.

5.2.4 FOCAL LENGTH AND RADIUS OF CURVATURE

It will now be useful to consider the relationship between the focal length f and the radius of curvature r of a spherical mirror. Figure 5.24 shows one ray of a narrow pencil of rays which is close to and parallel to the principal axis of a concave (converging) spherical mirror. This ray is reflected to pass through the principal focus F.

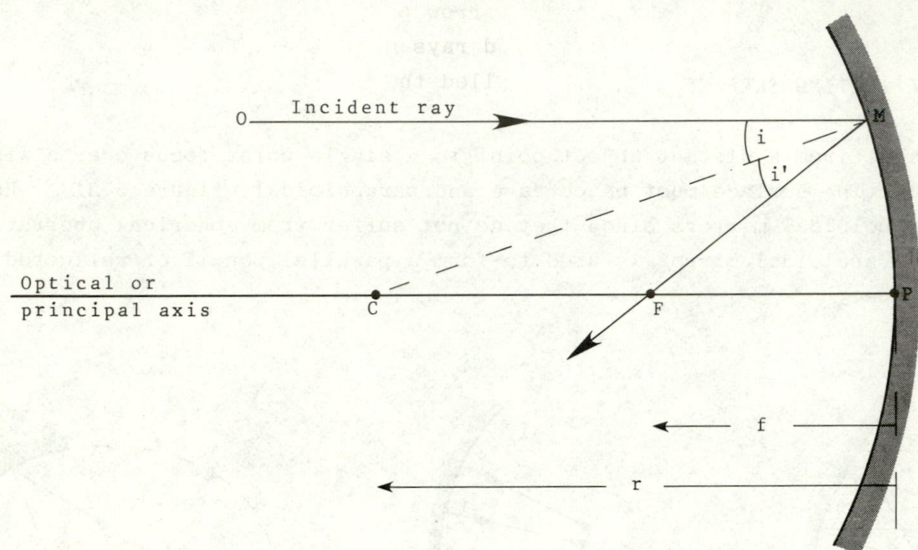

Fig. 5.24 The relationship between focal length f and radius of curvature r: $f = \frac{1}{2}r$.

CM is a normal to the mirror, since it passes through the centre of curvature C of the mirror. Now, $\widehat{OMC} = \widehat{CMF}$ (law of reflection), and $\widehat{OMC} = \widehat{MCF}$ (alternate angles). This means that $\widehat{MCF} = \widehat{CMF}$, and MF = CF.

If M is very near to P (this means only the central part of the mirror is being considered) then MF = PF, very nearly. So, PF = CF from which PC = 2.PF, or r = 2f. The same result may also be obtained for a convex (diverging) mirror. Thus, for both types of spherical mirror,

$$\text{radius of curvature} = 2 \times \text{focal length}$$

$$\text{or} \quad r = 2f \qquad (5.3).$$

5.2.5 FORMATION OF IMAGES BY SPHERICAL MIRRORS

The essential function of a curved mirror is to change the direction of the rays of light and to alter the curvature of the wave fronts. We will make the assumption that a finite planar object perpendicular to the principal axis of a spherical mirror will be imaged, to a first approximation, in a plane which is similarly perpendicular to the axis. This means that each point in the object plane will have a corresponding point in the image plane. This will be true if the aperture of the mirror is restricted somewhat, such that the reflected waves originating from each object point will very closely approximate to spherical waves, figure 5.25 .

When a spherical mirror forms an image, the position, nature, and size of the image relative to the object will depend on the type of mirror and the position of the object. Two methods are available for the determination of these image characteristics. These are:

(a) by scale drawing - called a graphical construction,
(b) by calculation - called an analytical method, using formulae.

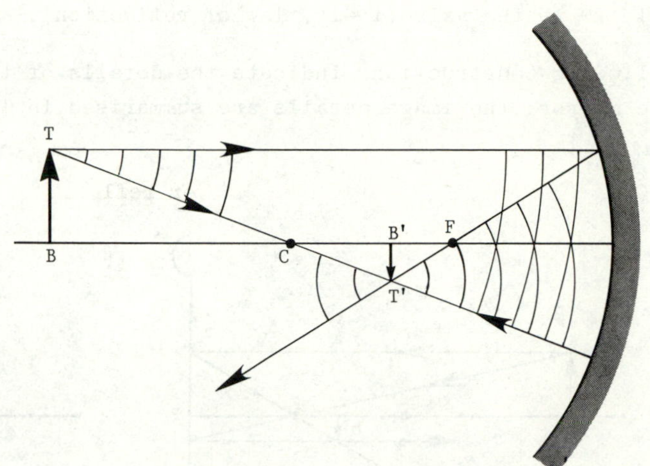

Fig. 5.25 Formation of an image B'T' in a concave spherical mirror.

5.2.5.1 *Graphical Constructions*

This scale drawing method makes use of *construction rays*. These are certain selected rays of light which take definite predictable paths following reflection.

Procedure for a Concave Mirror (refer to figure 5.26).

(a) Using a convenient horizontal scale, draw the principal axis of the mirror and mark in the positions of the mirror, the centre of curvature C, and the principal focus F. The mirror may be represented by a straight line perpendicular to the axis, since only the portion of the mirror near the axis is being considered.

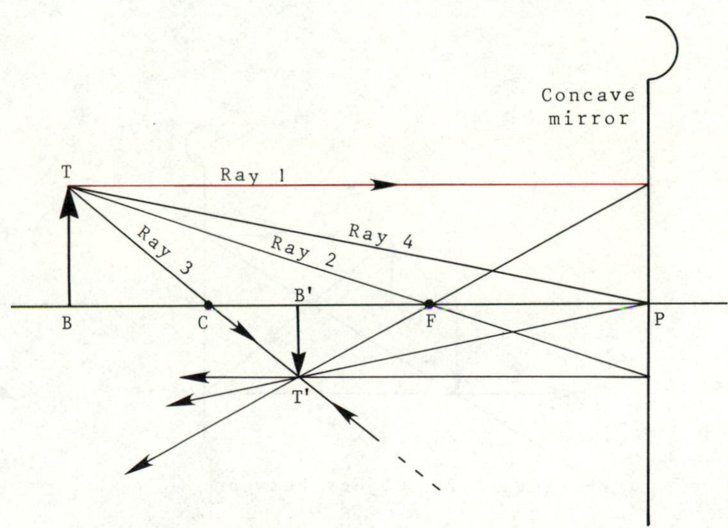

Fig. 5.26 Graphical construction for a concave spherical mirror.

(b) Mark in the top half of the object, using as large a vertical scale as possible, bearing in mind the probable size of the image.

(c) Draw any TWO of the following construction rays from the tip of the object. After reflection, the intersection of these rays indicates the tip of the image.

Construction Rays

1 A ray from T, parallel to the principal axis, is reflected through F.

2 A ray from T, passing through F, after reflection becomes parallel to the principal axis.

3 A ray from T, passing through C, strikes the mirror normally and is reflected back along the same path.

4 A ray from T, meeting the pole at some angle of incidence i, is reflected at the same angle i' to the axis (i = i', law of reflection).

The following constructions indicate the details of the various image formations with a concave mirror: the image details are summarised in Table 5.1.

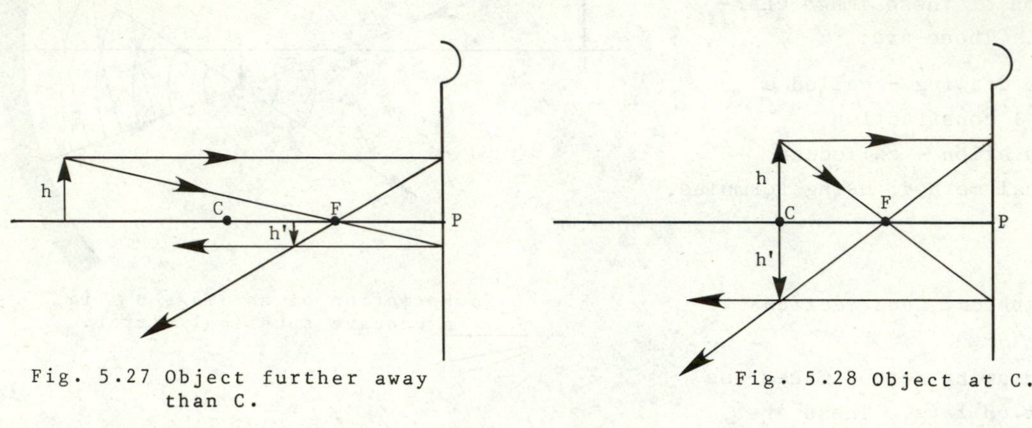

Fig. 5.27 Object further away than C.

Fig. 5.28 Object at C.

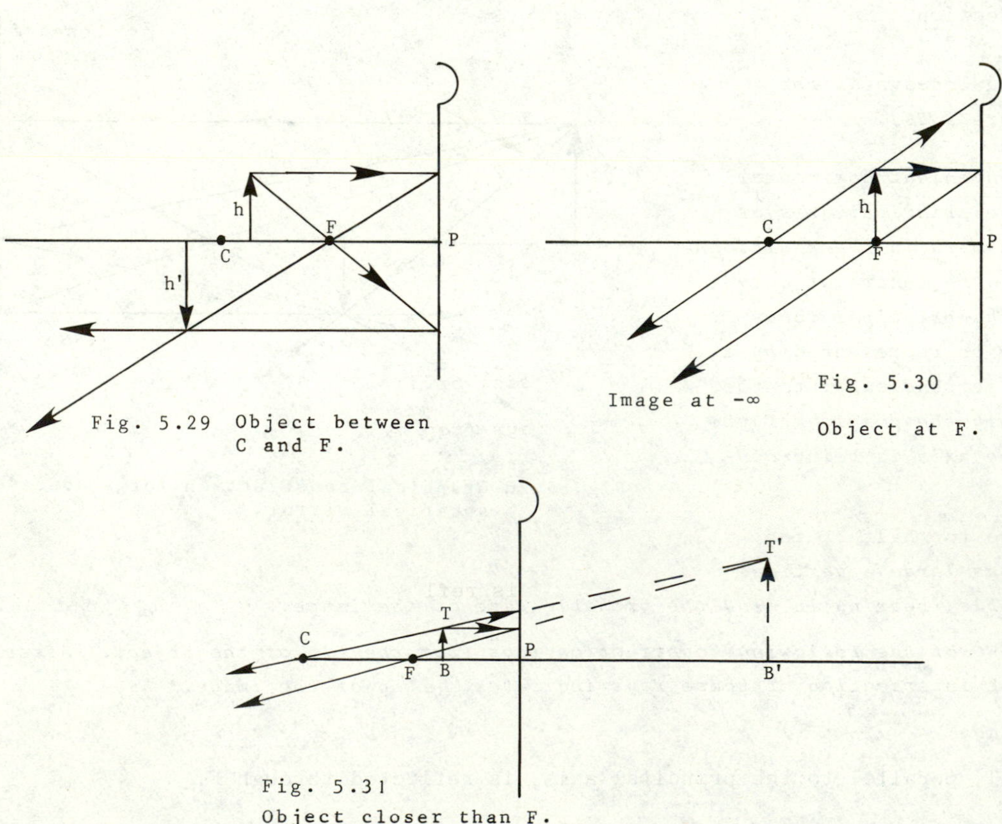

Fig. 5.29 Object between C and F.

Fig. 5.30 Object at F.

Fig. 5.31 Object closer than F.

Table 5.1 Summary of image details.

Real object position	Image position	Nature	Size relative to object
At infinity	At F	Real, inverted	Very small
Between infinity and C	Between F and C	Real, inverted	Smaller
At C	At C	Real, inverted	Same size
Between C and F	Between C and infinity	Real, inverted	Larger
At F	At infinity	Indeterminate	-
Between F and P	Behind mirror	Virtual, upright	Larger

Procedure for a Convex Mirror

(a) Draw the principal axis of the mirror and mark in the position of the mirror, the centre of curvature C, the principal focus F, and the top half of the object, as described for a concave mirror.

(b) Draw any TWO of the following construction rays from the tip of the object. After reflection, these rays will indicate the position of the tip of the image.

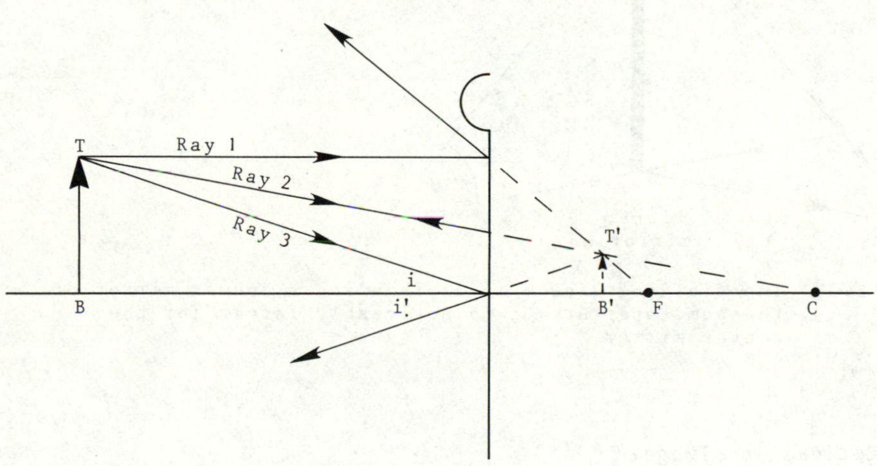

Fig. 5.32 Graphical construction for a convex spherical mirror.

Construction Rays

1 A ray from T, parallel to the axis, is reflected as if coming from F behind the mirror.

2 A ray from T directed towards C meets the mirror normally and is reflected back along the same path.

3 A ray from T, meeting the pole at some angle of incidence i, is reflected at the same angle i' to the axis.

Figure 5.32 shows the details for ANY object position. In all cases, the image will be virtual, upright, and smaller than the object.

A convex mirror is frequently used as a motor car driving mirror. Although the image is diminished, the field of view (the region over which the image is visible to the driver) is greater than that of a plane mirror of the same size. This fact is illustrated in figure 5.33 .

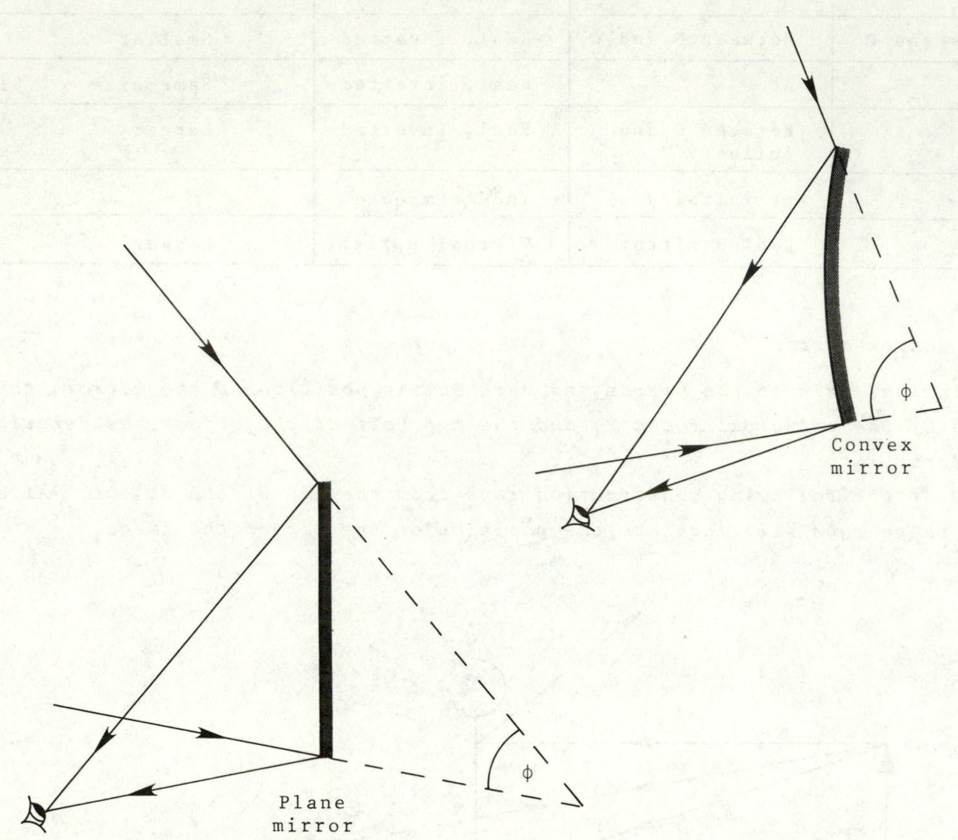

Fig. 5.33 The field of view ϕ in a plane and a convex mirror of the same aperture: ϕ is noticeably larger for the convex mirror.

5.2.5.2 *Calculations for Images*

We can derive formulae for use with spherical mirrors by means of which the position and size of any image may be calculated. The method of calculation is, generally, more accurate than the graphical construction method, but it is useful to check calculation results graphically, even if this is only done roughly. The derivations of the mirror formulae are in Appendix 3. We shall now simply quote the formulae which are:

$$\frac{1}{\ell'} + \frac{1}{\ell} = \frac{2}{r} = \frac{1}{f} \quad (5.4) \qquad \text{and} \qquad m = \frac{\text{height of image}}{\text{height of object}} = \frac{h'}{h} = -\frac{\ell'}{\ell} \quad (5.5),$$

where ℓ represents the distance from the mirror to the object,
ℓ' represents the distance from the mirror to the image,
f represents the focal length of the mirror,
r represents the radius of curvature of the mirror,
h represents the size of the object,

h' represents the image size, and m represents the linear magnification. These quantities are shown in figure 5.34 which is the case of a concave mirror forming a real image.

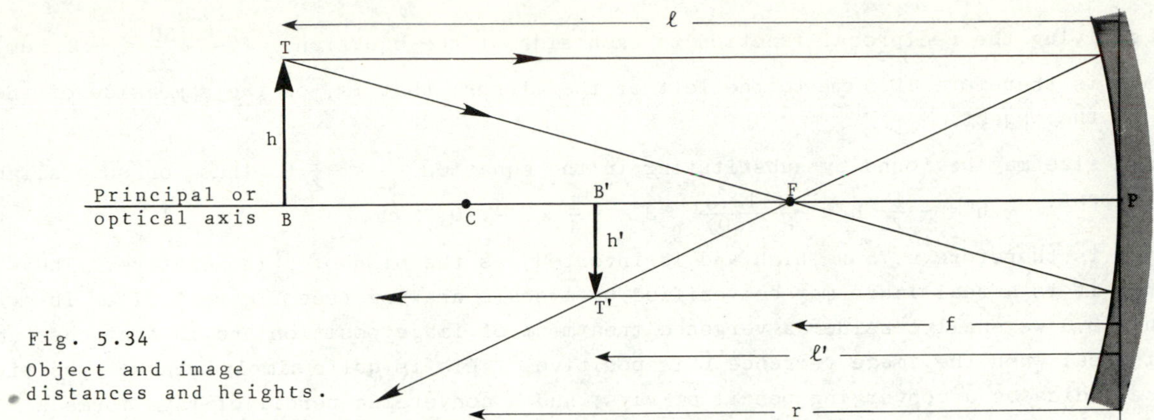

Fig. 5.34
Object and image distances and heights.

In using equations (5.4) and (5.5), careful attention must be paid to the Sign Convention. This has already been discussed in section 2.2.3 in connection with refraction at a single surface, and in Chapter 3 on thin lenses. For reference, the essential features of the Sign Convention as applied to mirrors are written again here.

(a) Distances measured along the axis.

(1) Distances are measured from the pole P of the mirror.
(2) A distance which is measured in the *same* direction as that in which the light is travelling is given a *positive* (+) sign.
(3) A distance which is measured in the *opposite* direction to that in which the light is travelling is given a *negative* (-) sign.

(b) Distances (heights) measured perpendicular to the axis

(1) All heights are measured from the axis of the mirror.
(2) A height which is measured *above* the axis is given a *positive* (+) sign.
(3) A height which is measured *below* the axis is given a *negative* (-) sign.

When drawing a ray diagram the light will customarily be shown travelling from left to right, so that positive distances are to the right and negative distances to the left. With this convention it will be found that

(a) the focal length and radius of a concave mirror have negative values

and (b) the focal length and radius of a convex mirror have positive values.

Worked Example

Calculate the position and size of the image formed when an object 3 cm high is placed 50 cm from a concave spherical mirror of radius of curvature 20 cm.

The data are: $\ell = -50$ cm, $h = +3$ cm, $r = -20$ cm, and $f = \frac{r}{2} = -10$ cm. Using the relationship $\frac{1}{\ell'} + \frac{1}{\ell} = \frac{1}{f}$, we have on rearranging, $\frac{1}{\ell'} = \frac{1}{f} - \frac{1}{\ell} = \frac{1}{-10} - \frac{1}{-50} = \frac{1}{-10} + \frac{1}{50} = \frac{-5+1}{50} = -\frac{4}{50}$,

or, on applying the reciprocal function to each side of the equation, $\ell' = -\frac{50}{4} = -12.5$ cm. The image is therefore 12.5 cm to the left of the mirror; that is, on the same side of the mirror as the object.

The image size may be found by substituting in the equation $\frac{h'}{h} = -\frac{\ell'}{\ell}$. Thus, on some slight rearrangement, $h' = -\frac{\ell'}{\ell} \cdot h = -\frac{(-12.5)}{(-50)} \times 3 = -\frac{1}{4} \times 3 = -0.75$ cm.

The image is therefore 0.75 cm high and is inverted, as the sign of h' is negative. The fact that it is a real image may be verified by a scale drawing (see figure 5.27). In section 5.2.6 we shall consider a vergence treatment of image position and in that case an image is real when the image vergence $\bar{L}'$ is positive. This is quite simply because a positive vergence indicates a converging pencil of rays, and a converging pencil of rays forms a real image.

Worked Example

Calculate the position and size of the image when an object 3 cm high is placed 40 cm from a convex mirror of radius of curvature 25 cm.

The data are: $\ell = -40$ cm, $h = +3$ cm, $r = +25$ cm, and $f = \frac{r}{2} = +12.5$ cm. On substituting these values in the general relationship, we have

$$\frac{1}{\ell'} = \frac{1}{f} - \frac{1}{\ell} = \frac{1}{+12.5} - \frac{1}{-40} = \frac{1}{12.5} + \frac{1}{40} = \frac{16+5}{200} = +\frac{21}{200},$$

$$\text{or} \quad \ell' = +\frac{200}{21} = +9.52 \text{ cm.}$$

The image is thus 9.52 cm to the right of the mirror, or behind the mirror, and is virtual (see figure 5.32). Its size may be calculated as before from

$$h' = -\frac{\ell'}{\ell} \cdot h = -\frac{9.52}{-40} \times (+3) = 0.238 \times 3 = +0.714 \text{ cm.}$$

The image is therefore 0.714 cm high and upright.

Worked Example

A concave and a convex mirror, each of 40 cm radius of curvature, are placed opposite to each other and 50 cm apart on a common axis. An object 2 cm high is placed midway between them. Find the position and size of the image formed when light is reflected first at the convex and then at the concave mirror.

For the convex mirror the data are: $\ell = -25$ cm, $h = +2$ cm, $r = +40$ cm, and $f = \frac{r}{2} = +20$ cm. Then, for the image position,

$$\frac{1}{\ell'} = \frac{1}{f} - \frac{1}{\ell} = \frac{1}{+20} - \frac{1}{-25} = \frac{1}{20} + \frac{1}{25} = \frac{5+4}{100} = +\frac{9}{100}$$

whence, $\ell' = +\frac{100}{9} = +11.11$ cm. The image in the convex mirror is 11.11 cm to the right of, or behind, the mirror, and is virtual.

Its size is given by $h' = -\frac{\ell'}{\ell} \cdot h = -\frac{(+11.11)}{(-25)} \times (+2) = +0.89$ cm, so that the image is upright.

Considering now the reflection at the concave mirror, the image in the convex mirror acts as the object for the concave mirror. That is, in the figure alongside, I_1 is the image in the convex mirror which now acts as the object for the concave mirror shown on the left. The data for the concave mirror are:

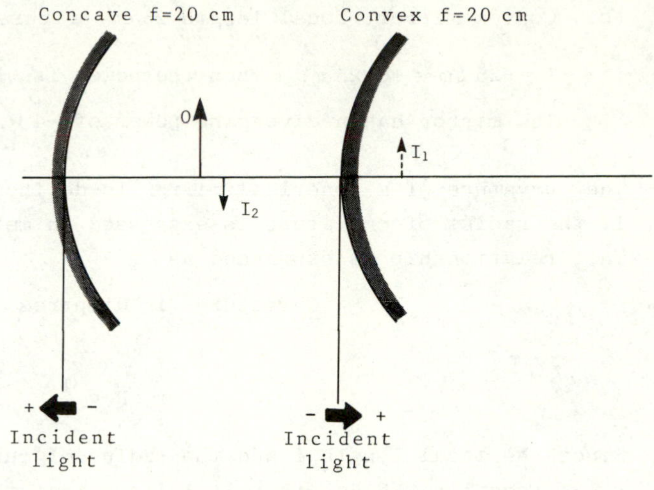

$\ell = -(50 + 11.11) = -61.11$ cm, $h = +0.89$ cm.
$r = -40$ cm, and $f = \frac{r}{2} = -20$ cm.

Then, $\frac{1}{\ell'} = \frac{1}{f} - \frac{1}{\ell} = \frac{1}{-20} - \frac{1}{-61.11}$

$= -\frac{1}{20} + \frac{1}{61.11} = -0.05 + 0.0164$

$= -0.0336$

so that $\ell' = \frac{1}{-0.0336} = -29.73$ cm.

The image I_2 is 29.73 cm to the right of the concave mirror, and is real. A word of caution; note that the light is travelling from left to right for the convex mirror, but in the other direction for the concave mirror. This is indicated on the diagram. This means, of course, that a positive measurement along the axis for one mirror is a negative measurement for the other.

The image size in the concave mirror is $h' = -\frac{\ell'}{\ell} \cdot h = -\frac{(-29.73)}{(-61.11)} \times (+0.89) = -0.43$ cm. The final image is 0.43 cm tall and is inverted.

5.2.6 *POWER AND CURVATURE OF A SPHERICAL MIRROR*

The power of a spherical mirror is defined as $F = -\frac{1}{f}$. If the focal length is expressed in metres, then the power is expressed in dioptres. One dioptre is the power of a surface having a focal length of 1 m. This relationship is expressed as

$$\text{Power (in dioptres)} = -\frac{1}{\text{focal length (in metres)}}$$

$$\text{or} \quad F = -\frac{1}{f} \qquad (5.6).$$

The (-) sign is necessary in this expression so that it results in a concave mirror, which converges light, having a (+) power. This would seem to be a logical choice. Similarly, the power of a convex mirror, which diverges light, turns out to be a (-) value. This is illustrated as follows.

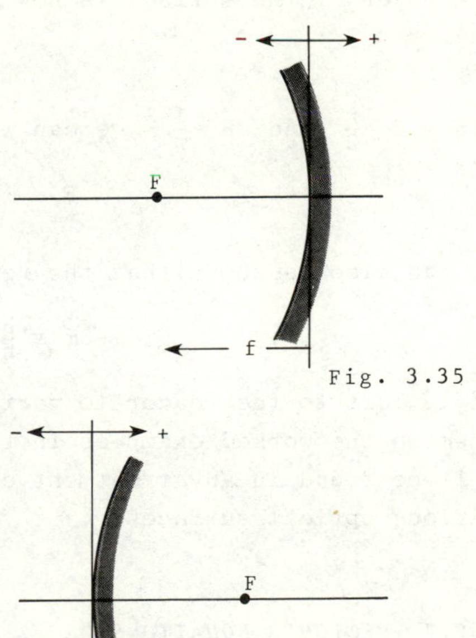

Fig. 3.35

Fig. 3.36

(a) Concave mirror, focal length 20 cm (figure 3.35).

$f = -20$ cm $= -0.20$ m. Using equation (5.6) gives the power $F = -\frac{1}{f} = -\frac{1}{(-0.20)} = +5$ D.

That is, the mirror has a converging power of +5 D.

(b) Convex mirror, focal length 25 cm (figure 3.36).

$f = +25$ cm $= +0.25$ m. Then the power is $F = -\frac{1}{f} = -\frac{1}{(+0.25)} = -4$ D.

The mirror has a diverging power of -4 D.

The curvature of a spherical mirror is defined as the reciprocal of the radius of curvature. If the radius of curvature is expressed in metres, the unit of curvature is the dioptre. This relationship is expressed as

$$\text{Curvature (in dioptres)} = \frac{1}{\text{Radius of curvature (in metres)}}$$

$$\text{or} \quad R = \frac{1}{r} \qquad (5.7).$$

Since the focal length f and the radius of curvature r of a spherical mirror are related by $r = 2f$ (section 2.2.4), then it follows that the power can be written as $F = -2R$. Thus, the power of a spherical reflecting surface depends on the curvature of the surface. The curvature and power of a spherical mirror may be determined optically, or by using the spherometer illustrated in figure 3.28 .

Images can be treated using the vergence relationship derived for refracting surfaces and thin lenses, namely $\bar{L}' - \bar{L} = F$. If the mirror is in a medium of refractive index n_s, then the object vergence is

$$\bar{L} = \frac{n}{\ell} = \frac{n_s}{\ell} \qquad (5.8).$$

The image vergence must take account of the fact that the light reverses its direction on reflection. This is done quite simply by inserting a minus sign so that the expression is

$$\bar{L}' = -\frac{n'}{\ell'} = -\frac{n_s}{\ell'} \qquad (5.9).$$

The power of the surface is now given by the reciprocal of the reduced focal length, so that

$$F = -\frac{1}{f/n_s} = -\frac{n_s}{f} \qquad (5.10).$$

Since $f = \frac{r}{2}$ and $R = \frac{1}{r}$, we can write

$$F = -\frac{n_s}{f} = -\frac{2n_s}{r} = -2n_s R \qquad (5.11).$$

It can also be shown that the relationship for lateral magnification is

$$m = \frac{h'}{h} = \frac{\bar{L}}{\bar{L}'} \qquad (5.12).$$

It is left to the reader to verify these equations and it might be useful exercise to try them on the worked examples in the previous section. The application of these equations will be found in any treatment of reflections occurring in spectacle lenses and the eye's various optical surfaces.

5.2.7 NEWTON'S EQUATIONS

Recall that a single refracting surface and a thin lens both possess two focal lengths f and f'. In the case of the thin lens these were related thus: $f = -f'$. For a single surface the relationship was $f = -\frac{n}{n'} f'$. Now, consider the spherical concave mirror. If an object point is on the principal axis at infinity, then it is imaged at the focal point F.

If we are to be very strict, this point should be labelled F' as it was for refracting surfaces and thin lenses. Conversely, an object point on the axis at F images at infinity. Thus, F and F' are coincident in the case of a spherical mirror. Consequently, we refer to the two focal points as a single focus, and give it the single symbol F. Similar statements apply to the object and image focal lengths f and f'. These are identical, so we may use one or other of the symbols, or both.

Bearing these comments in mind, it is a simple exercise to show that Newton's equations also apply to curved mirrors. That is,

$$m = -\frac{f}{x} = -\frac{x'}{f'} \qquad (5.13) \qquad \text{and} \qquad xx' = ff' \qquad (5.14)$$

where $f = f'$. Figure 3.37 illustrates the extra-focal object and image distances in the three possible cases.

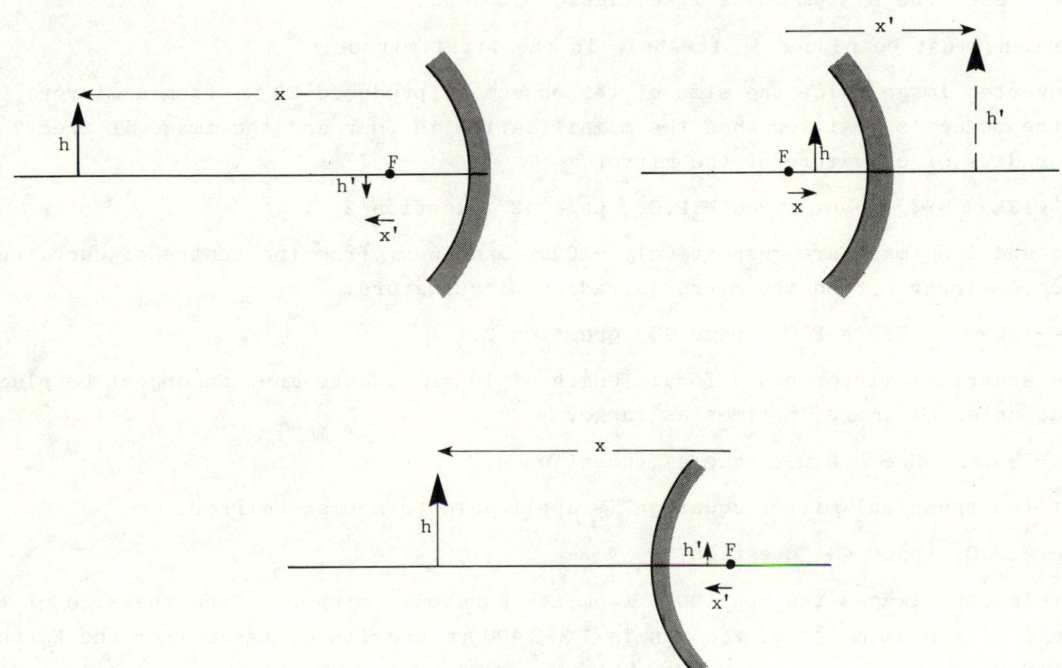

Fig. 3.37 Examples of Newton's extra-focal object and image distances with spherical mirrors.

EXERCISES

1 A sight-testing chart, measuring 120 cm by 80 cm with its longer dimension vertical, is to be viewed monocularly by reflection in a plane mirror. If the observer's eye and the chart are each 3 m in front of the mirror, find the minimum size of mirror which can be used to see the whole chart.
Ans. 60 cm × 40 cm. See W.P.O., page 7, question 1.

2 In the previous question, if the observer's eye is 1.3 m above the floor and the lower edge of the mirror is 1.5 m above the floor, how high is the lower edge of the chart?

Ans. 1.7 m. See W.P.O., page 7, question 2.

3 What will be the rotation of a plane mirror reflecting a spot of light on to a straight scale (a tangent scale) 1 m from the mirror if the spot of light moves through 5 cm?

Ans. 1.43^0. See W.P.O., page 8, question 4.

4 An erect pin is 8 cm from a plane mirror. If the pin is moved 4 cm towards the mirror, how far does its image move? Alternatively, if the mirror is moved 4 cm towards the pin, again find the distance the image moves.

Ans. 4cm. 8 cm. See W.P.O., page 9, question 6.

5 A concave spherical mirror of radius 200 cm converges light from a distant object on to a concave mirror of radius 120 cm. The latter is 40 cm in front of the former. The light comes to a focus and is made parallel by a thin +10 D lens. Where should this lens be placed? The system looks like figure 8.28(c).

Ans. The lens must be placed in the hole in the first mirror.

6 A real inverted image twice the size of the object is produced 20 cm from a mirror. What is the object's position when the magnification is four and the image is erect? Find the radius of curvature of the mirror.

Ans. $r = -13\frac{1}{3}$ cm. $\ell = -5$ cm. See W.P.O., page 42, question 2.

7 An object and its image are respectively -20 cm and +5 cm from the centre of curvature of a concave mirror. Find the mirror's radius of curvature.

Ans. $r = -13\frac{1}{3}$ cm. See W.P.O., page 43, question 3.

8 A concave spherical mirror has a focal length of 10 cm. Where must an object be placed to produce an erect image 1½ times as large?

Ans. $\ell = -3\frac{1}{3}$ cm. See W.P.O., page 44, question 4.

9 Show that the spherical mirror equation is applicable to a plane mirror.

Ans. See W.P.O., page 44, question 5.

10 A solar telescope images the sun 100 m from its parabolic mirror. Find the size of the sun's image if the diameter of the sun is 1 382 400 km and its distance from the Earth is 148 800 000 km.

Ans. 92.9 cm. See W.P.O., page 44, question 6.

11 A thin convex lens is used to provide a virtual object 5 cm from the vertex of a convex mirror of focal length 10 cm. If the real object for the lens is the same size as the virtual object for the mirror and the two objects are separated by 100 cm, find the power of the lens, its position in relation to the mirror, and the relative size, nature, and position of the final image.

Ans. F = +4 D. The lens is 45 cm from the mirror. The final image is inverted, real, and twice the size of the lens' object.

6 THE DISPERSION OF LIGHT AND COLOUR

6.1 INTRODUCTION

The compound nature of the sun's light was first demonstrated by the German astronomer Kepler, and was later more thoroughly investigated by Sir Isaac Newton. As the result of the famous and beautiful series of experiments, Newton was led to the view that ordinary white light is really a mixture of light of many colours. In 1672 he published his first scientific paper and gave an account of his prism experiments.

The splitting of white light into its various coloured components is an effect referred to as *dispersion*. It is a result of the different colours travelling with different velocities in a transparent medium such as glass. For example, the velocity of blue light is less than the velocity of red light in glass. We have stated (section 2.1.3) that the value of the refractive index $_an_g$ of glass with respect to air is given by

$$_an_g = \frac{v_a}{v_g} = \frac{\text{velocity of light in air}}{\text{velocity of light in glass}} .$$

The smaller velocity of the blue light in glass results in a higher refractive index for the glass and a larger degree of refraction when blue light is passed through a prism. (The velocity of light in air is almost independent of colour). Figure 6.1 illustrates the variation of the index of two common types of optical glass with the colour of light.

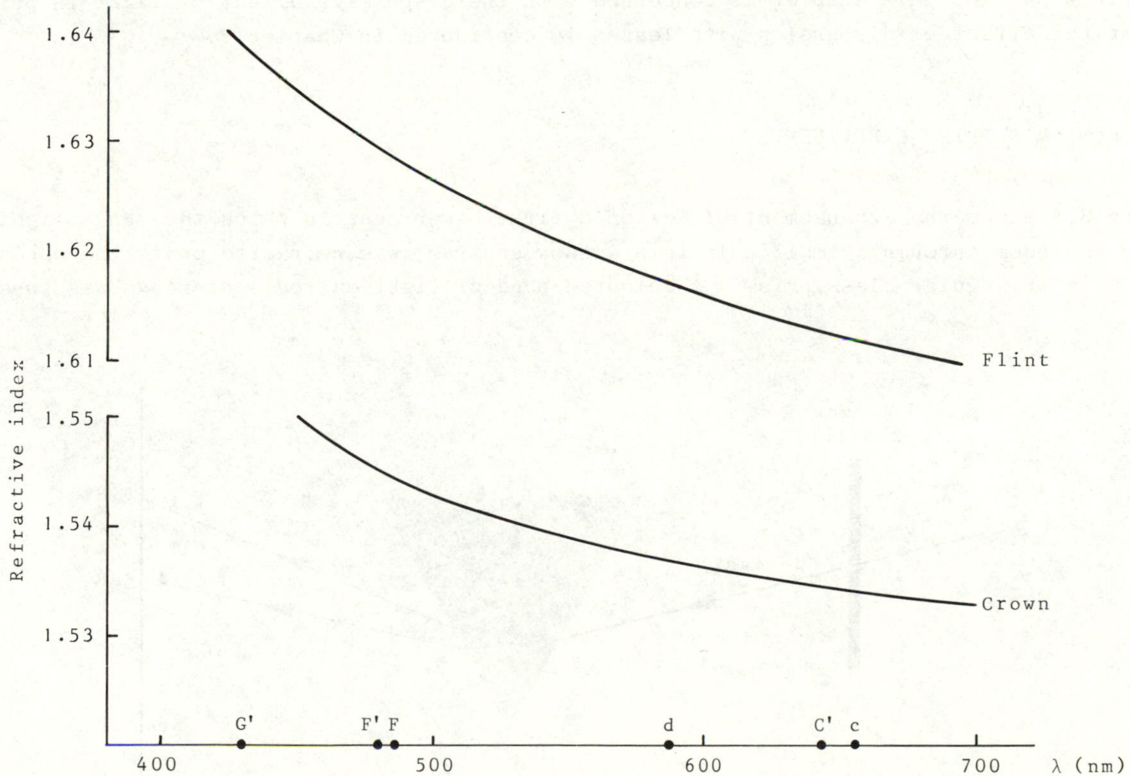

Fig. 6.1 The variation of refractive index with wavelength for a crown and and a flint sample of glass.

Quantitatively, dispersion is defined by the expression dispersion $= -\frac{dn}{d\lambda}$, where n is the refractive index and λ is the wavelength. The minus sign is used to denote the normal case of dispersion in which the refractive index of a substance decreases as the wavelength of the light increases. Substances for which the refractive index increases with increasing wavelength are said to exhibit anomalous dispersion. This effect occurs mainly in the region of an absorption band. Absorption spectra are discussed in section 15.2.5 .

The variation of refractive index with wavelength, in regions of normal dispersion, may be represented by equation (6.1), usually attributed to the French mathematician Augustin Cauchy (1789 - 1857).

$$n = A + \frac{B}{\lambda^2} + \frac{C}{\lambda^4} + \ldots \qquad (6.1),$$

where A, B, C, etcetera are constants for the particular material, and λ is the wavelength of light in vacuum (or approximately in air). Cauchy's equation is one of a number of alternative expressions and gives a tolerable representation of results for many substances.

The property of dispersion is very common. All transparent substances exhibit the effect although the extent to which it is shown will vary. It is now well known that other types of radiation besides visible light are contained in the energy which reaches the Earth from the sun. The most important as far as optometrists and opticians are concerned are those groups of waves immediately next to the visible spectrum in the electromagnetic spectrum (see section 10.6.2) referred to as *ultraviolet* and *infrared*. The nature and biological effects of these invisible radiations will be discussed in section 6.5.7 .

The first part of this chapter is concerned with the dispersive effect of light in prisms, whilst the effect of dispersion with lenses is considered in Chapter 7.

6.2 NEWTON'S PRISM EXPERIMENTS

Figure 6.2 shows the arrangement of Newton's first experiment in which the sun's light, entering the room through a small hole in a window shutter, was refracted on to the opposite wall by a triangular glass prism. A coloured band of light called a *spectrum* was observed,

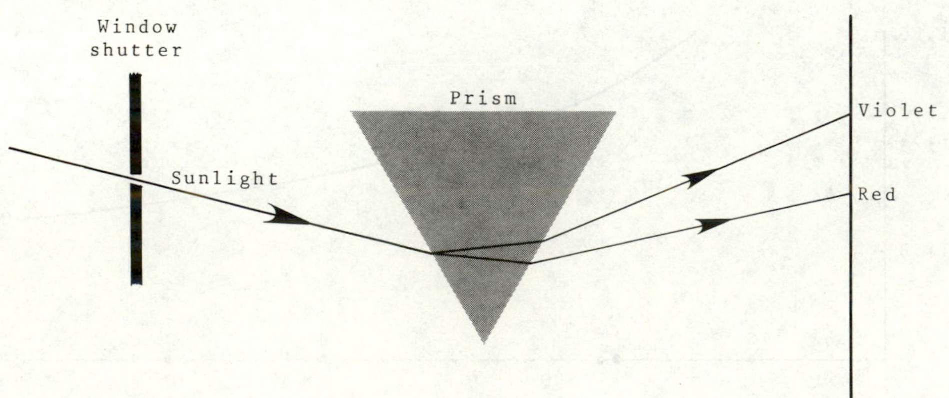

Fig. 6.2 Newton's first experiment: white sunlight shown split into the colours of the rainbow - red, orange, yellow, green, blue, and violet.

containing all the colours of the rainbow in the same order in which they appear in the rainbow. Although these colours are commonly referred to as red, orange, yellow, green, blue, and violet, we should note that this spectrum is a very large gradation of colours from one end to the other.

To check his conclusion about the composite nature of white light Newton recombined the coloured lights by introducing a second prism identical to the first, figure 6.3(a), and also a converging lens, figure 6.3(b).

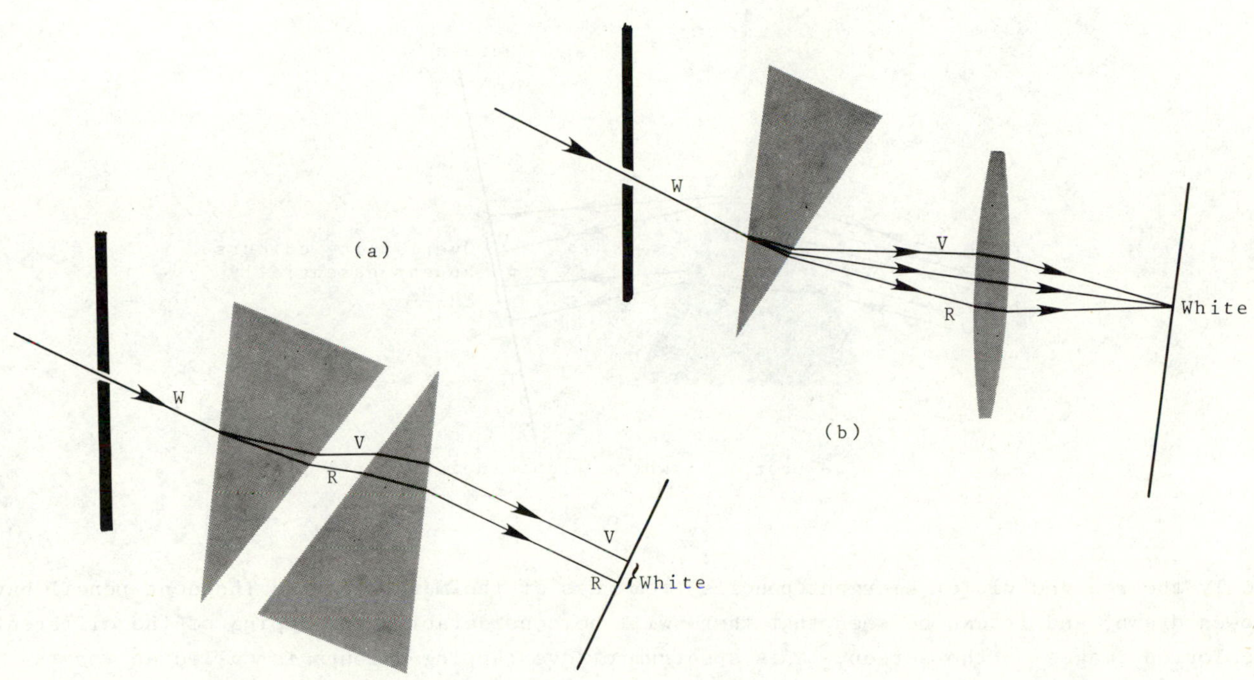

Fig. 6.3 Newton's experiments recombining the spectral colours to form white light.

In each of the above cases the result was white light similar to the original sunlight. Now, having obtained a simple spectrum with a prism, Newton arranged for each of the colours of the spectrum to traverse a second prism in turn, figure 6.4 .

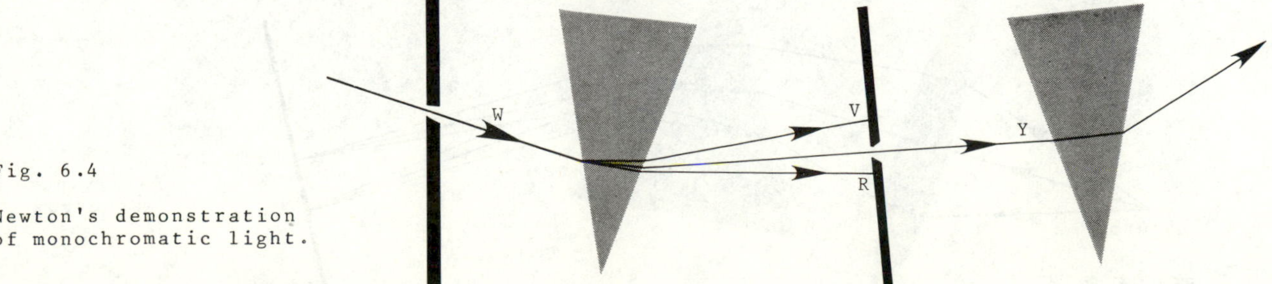

Fig. 6.4

Newton's demonstration of monochromatic light.

In addition to making quantitative measurements of the deviations of the separate colours, he was able to show that no further dispersion was caused by the second prism. Light of only one colour, which is not dispersed into further colours when passed through a prism, is called *monochromatic light*. Newton concluded that the coloured lights were contained in the original white light, and it was not the prism which was introducing the colours. They appear because of their differing deviations through the prism. Light which is made up from a mixture of two or more separate colours is sometimes referred to as *polychromatic light*.

In order that a bright spectrum may be obtained , it is necessary to have a wide pencil of rays of white (polychromatic) light incident on a prism, figure 6.5 .

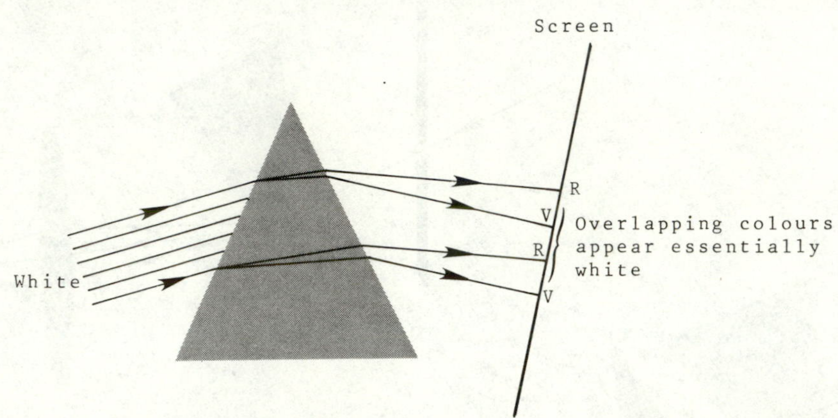

Fig. 6.5 A wide pencil of white light incident on a prism.

Only the red and violet emergent pencils from rays at the margin of the incident pencil have been drawn, and it can be seen that there will be considerable overlapping of the different coloured images on the screen. This spectrum of overlapping colours is called an *impure spectrum*, as the only colour effects are near the edges. Most of Newton's spectra were impure.

In order that a *pure spectrum* is obtained, in which the colours appear separated and unmixed on the screen, the light emerging from the prism must be focused on the screen by a lens. The arrangement illustrated in figure 6.6 will give a reasonably pure spectrum only if the pencil of rays is close to minimum deviation, since only then will a reasonably sharp image of each colour be formed on the screen (see also section 2.1.10.5).

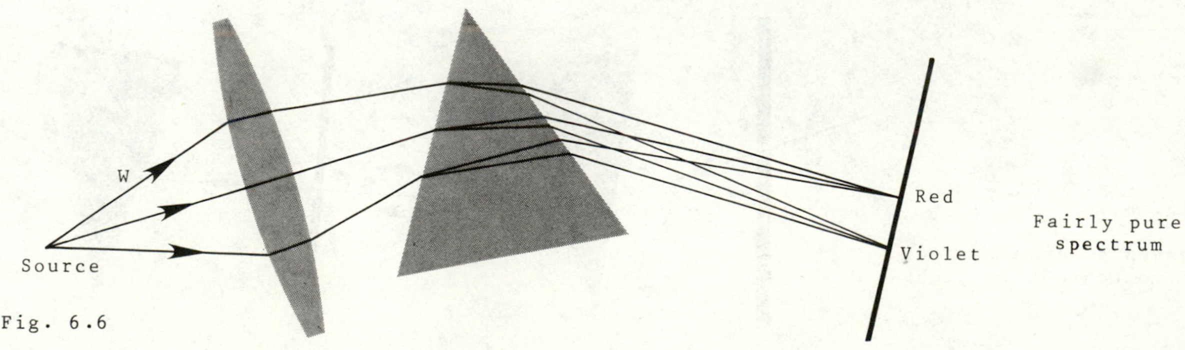

Fig. 6.6

In order that a really pure spectrum is formed the incident pencil is made parallel, and each emergent parallel pencil is brought to a sharp focus by a converging lens following the prism. This is shown in figure 6.7 and forms the basis of the optical system of a spectrometer. This instrument is described in Chapter 8.

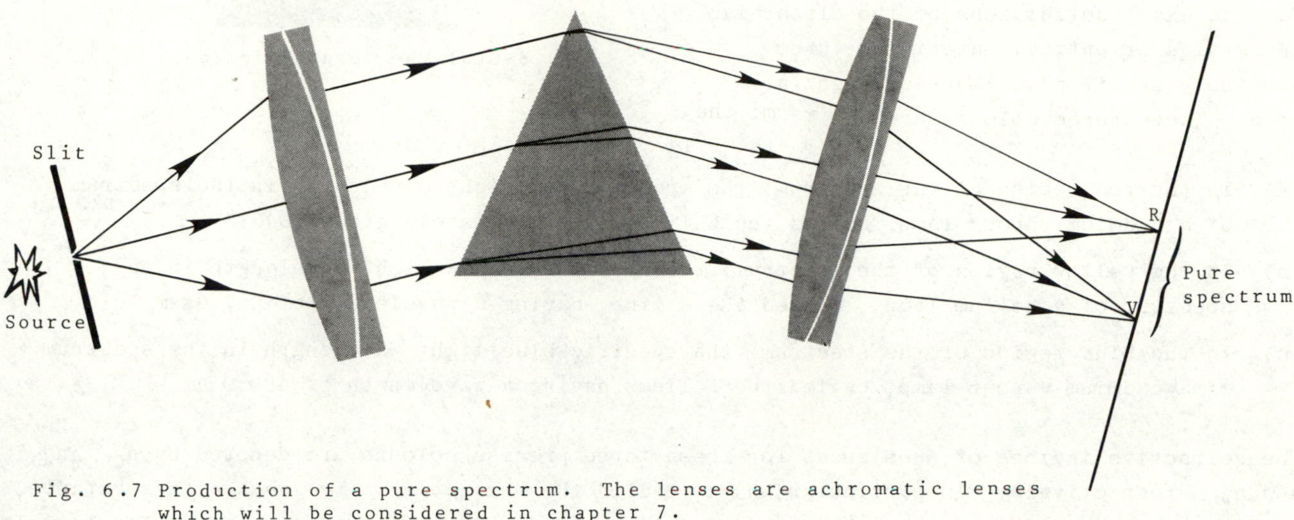

Fig. 6.7 Production of a pure spectrum. The lenses are achromatic lenses which will be considered in chapter 7.

Rays of any one colour emerge as a parallel pencil, but the rays of different colours are deviated through different angles. Hence, the emergent pencils are slightly inclined to one another and will be focused on different parts of the screen. The above system produces the best spectra when the light passes through the prism at or near to minimum deviation.

6.3 DISPERSIVE POWER AND CONSTRINGENCE

We have seen that dispersion occurs in a transparent medium because the refractive index of the medium varies with the colour of the light. Manufacturers of optical glass, from which lenses and prisms are made, supply data regarding the dispersive characteristics of these materials. Single lenses and, of course, single prisms exhibit dispersion, and we shall see later that it is possible to construct composite lenses and prisms for which dispersion effects are negligible.

In figure 6.8(a), a ray of white light is incident on one face of a triangular prism. The angle between the red and blue rays emerging from the prism is called the *angular dispersion*, or sometimes the *dispersion*, of the prism. It is denoted by the symbol α.

It will be seen shortly that for prisms of equal apical angles, but made from materials with different refractive indices, the angular dispersions will be different.

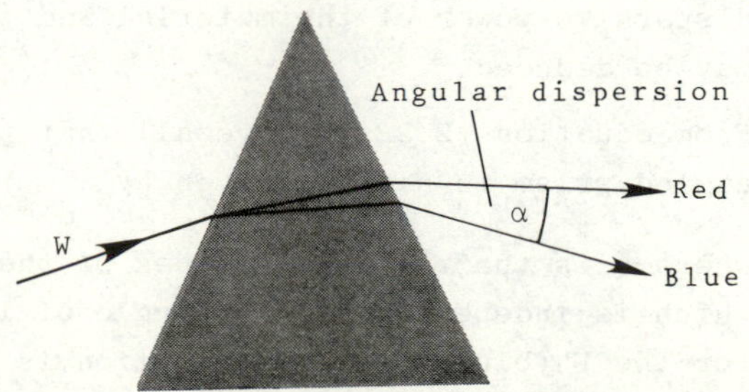

Fig. 6.8 (a) Angular dispersion, α.

In figure 6.8(b) it is convenient to consider also the deviation of a yellow colour of light, which is near the middle of the spectrum. For this reason the yellow ray is called the *mean ray*, and its deviation is referred to as the *mean deviation*.

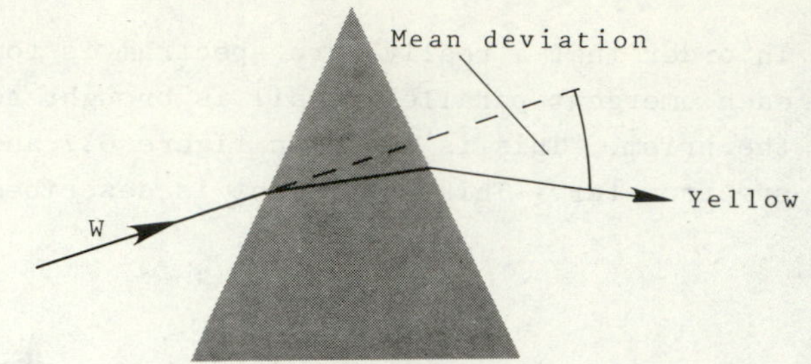

Fig. 6.8(b) The mean deviation

For the exact definitions of the dispersive properties of optical materials, three precisely specified colours of light are used. These three colours are:

(a) In the red region of the spectrum: the specific red light wavelength in the spectrum of a cadmium vapour lamp, called the C' line, having a wavelength of 643.8 nm.

(b) In the yellow region of the spectrum: the specific yellow light wavelength in the spectrum of a helium lamp , called the d line, having a wavelength of 587.6 nm.

(c) In the blue region of the spectrum: the specific blue light wavelength in the spectrum of a cadmium vapour lamp, called the F' line, having a wavelength of 480.0 nm.

The refractive indices of substances for these three precise colours are denoted by $n_{C'}$, n_d, and $n_{F'}$, respectively. In particular, n_d is called the *mean refractive index* of a substance, and is the value understood by the term refractive index unless otherwise stated. For example, the barium crown glass referred to in figure 6.1 has refractive indices

$$n_{C'} = 1.535 \qquad n_d = 1.538 \qquad n_{F'} = 1.545 .$$

Many other spectral colours have been specified for various purposes, symbolised by the subscripts B, b, C, D_1, D_2, F, G, G', etcetera, but these are not important for our present requirements. They will be dealt with in section 15.2.5 . The use of the word line in the definitions of the colours above will also be considered in Chapter 15.

The *angle of dispersion*, or angular dispersion, α is defined as the angle between the direction of the C' (red) and F' (blue) rays in the emergent pencil. The value of α, which depends on the angle of incidence of the polychromatic light, is a measure of the chromatic effect of the prism. It is often referred to as the *chromatic aberration* of the prism.

The *dispersive power* ω of a refracting substance is the angle of dispersion relative to the deviation of the mean (d) ray: that is, by definition,

$$\text{dispersive power } \omega = \frac{\text{angle of dispersion}}{\text{mean deviation}} .$$

If a small angled (thin) prism of material is considered, the relationship between the dispersive power of the material and the refractive indices for the three standard wavelengths may be deduced.

From equation (2.22a) we recall that the deviation of a particular ray of light, by a small angled prism in air, is given by

$$d = (n-1)a$$

where n is the refractive index of the prism, a is the apical angle, and d is the deviation which is independent of the angle of incidence if the latter is small (less than 10^0). Hence, for the F' (blue) ray the deviation is

$$d_{F'} = (n_{F'}-1)a$$

and for the C' (red) ray the deviation is

$$d_{C'} = (n_{C'} - 1)a \ .$$

Thus, the angle of dispersion, or chromatic aberration, of the prism is given by

$$\alpha = d_{F'} - d_{C'} = (n_{F'} - 1)a - (n_{C'} - 1)a \ .$$

That is, abbreviating chromatic aberration to Ch. Ab.,

$$\text{Ch. Ab.} = \alpha = (n_{F'} - n_{C'})a \qquad (6.2).$$

Now, the deviation of the mean d (yellow) ray is

$$d_d = (n_d - 1)a \qquad (6.3),$$

and from equations (6.2) and (6.3) we get the dispersive power

$$\omega = \frac{\text{angle of dispersion}}{\text{mean deviation}} = \frac{\alpha}{d_d} = \frac{(n_{F'} - n_{C'})a}{(n_d - 1)a}$$

or, eliminating the term a in the numerator and denominator,

$$\omega = \frac{n_{F'} - n_{C'}}{n_d - 1} \qquad (6.4).$$

For the barium crown glass for which the figures have already been given the value of ω is $\frac{1.545 - 1.535}{0.538} = 0.01859$.

More commonly used in the optical industry is the reciprocal of the dispersive power, called the *constringence*, or *V-number*, of a substance. This is the value of $1/\omega$ and is denoted by V. That is, constringence V is given by

$$V = \frac{1}{\omega} \qquad (6.5)$$

so that we can write the expression

$$V = \frac{n_d - 1}{n_{F'} - n_{C'}} \qquad (6.5a).$$

The V-number for the barium glass quoted above is $V = \frac{1}{\omega} = \frac{1}{0.01859} = 53.8$. For the two common groups of optical glasses (crown and flint), the crown glasses will have values for V about 60, but for the higher dispersive flint glasses the value will be much lower, being about 30.

Optical glass manufacturers now designate their products with a six-figure number. For example, the above mentioned glass would be specified by the number 538538. The first three digits are the decimal part of the mean refractive index n_d, and the last three signify the constringence V, omitting the decimal point. In another case, the flint glass featured in figure 6.1 would be specified by 617363, indicating that $n_d = 1.617$ and $V = 36.3$.

Now, referring back to equation (6.5a) we should note that the term in the numerator, that is, $n_d - 1$, is referred to as the *refractivity* of the material, whilst the denominator $n_{F'} - n_{C'}$ is called the *mean dispersion* of the material. Thus,

$$\text{refractivity} = n_d - 1 \qquad (6.6)$$

$$\text{and mean dispersion} = n_{F'} - n_{C'} \qquad (6.7).$$

Also included in the specifications of optical materials are quantities of the type $(n_d - n_{C'})$ and $(n_{F'} - n_d)$, which are referred to as *partial dispersions* of a material over the range of wavelengths specified by the subscripts. Hence,

$$\text{partial dispersion (C' to d)} = (n_d - n_{C'}) \qquad (6.8).$$

As an ophthalmic prism is commonly marked with its deviating ability expressed in prism dioptres, it will be useful to state some of the foregoing formulae in prism dioptre notation. We recall (equation (6.3)) that the deviation of the mean ray by a thin prism in air is

$$d_d = (n_d - 1)a\ .$$

If the apical angle (a^0) is expressed in prism dioptres, then the deviation (d) for a given colour of light is a measure of the power (P^{Δ}) of the prism in prism dioptres, for that wavelength. Hence, from equation (6.2) we have Ch. Ab. $= \alpha = (n_{F'} - n_{C'})a$, which becomes

$$\text{Ch. Ab.} = \alpha = (n_{F'} - n_{C'})A \qquad (6.2a),$$

where $A = 100 \tan a^0$ is the apical angle of the prism expressed in prism dioptres, and α is therefore also in prism dioptres.

Also, equation (6.3) may be written

$$P_d = (n_d - 1)A \qquad (6.3a),$$

whence $A = \dfrac{P_d}{n_d - 1}$. Equation (6.2) now becomes

$$\text{Ch. Ab.} = \alpha = (n_{F'} - n_{C'}) \times \frac{P_d}{(n_d - 1)}\ .$$

But from equation (6.5a), $(n_{F'} - n_{C'})/(n_d - 1) = 1/V$, where V is the constringence of the material. Hence, we obtain the expression

$$\text{Ch. Ab.} = \alpha = \frac{P_d}{V}\ .$$

That is, $\text{chromatic aberration of prism} = \alpha = \dfrac{\text{mean prism power (dioptres)}}{\text{constringence}}$.

The subscript d is normally dropped from P_d so that P implies the mean power of the prism, and the expression is then written

$$\text{Ch. Ab.} = \alpha = \frac{P}{V} \qquad (6.9).$$

Thus, a 5^{Δ} prism of spectacle crown glass ($V = 58.6$) has a chromatic aberration of

$$\alpha = \frac{P}{V} = \frac{5}{58.6} = 0.085^{\Delta} = 0^0\ 3' \text{ approximately.}$$

This is the angle between the emergent red (C') and blue (F') rays.

6.4 PRISM COMBINATIONS

A combination of two or more prisms made of different types of glass is generally intended to produce one of the following effects:

(i) dispersion without deviation
or (ii) deviation without dispersion.

These two effects will now be considered.

6.4.1 THE DIRECT VISION PRISM - *dispersion without deviation*

In this effect, the colours of a polychromatic light are separated into a spectrum, but one particular colour, usually the yellow d ray, is undeviated. This forms the basis of the *direct vision prism*. Its main application is in the direct vision spectroscope in which the spectrum of the light from a source is formed in the line of vision of the source.

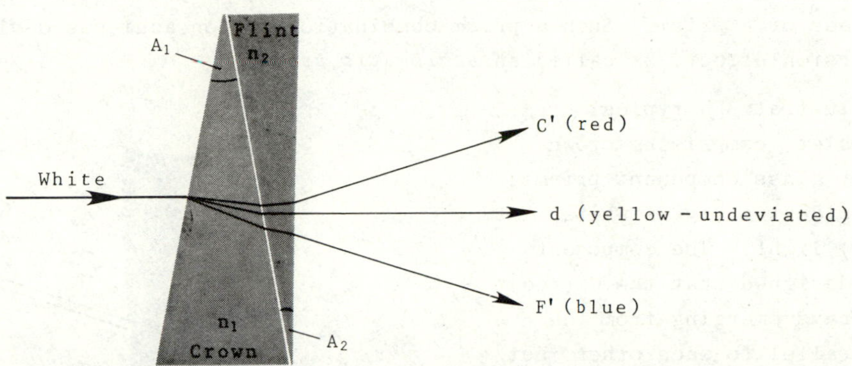

Fig. 6.9 The direct vision prism.

Figure 6.9 shows a simple direct vision prism, which is a combination of two prisms, one of crown glass, the other of flint glass, cemented together. Since the deviation of the d ray must be zero, it follows that the deviation of this ray must be equal and opposite by the separate prism components. For this reason, the second prism is inverted relative to the first. Now, from equation (6.3a) the deviation of the mean (d) ray is given by

$$P_d = (n_d - 1)A$$

using prism dioptre notation. If we now drop the subscript letter d for the mean power and mean index symbols, then we have for the crown prism (component 1)

$$P_1 = (n_1 - 1)A_1$$

and for the flint component prism (component 2)

$$P_2 = (n_2 - 1)A_2 .$$

Thus, the condition for dispersion without any deviation of the d ray is $P_1 = P_2$. That is,

$$(n_1 - 1)A_1 = (n_2 - 1)A_2 \qquad (6.10).$$

The resulting angle of dispersion (chromatic aberration) between the emergent C' (red) and F' (blue) rays is given by applying equation (6.2a) to each component prism, and writing

$$\alpha_{comb} = \alpha_2 - \alpha_1 .$$

The angular dispersion α_2 produced by the flint component is usually larger than α_1 produced by the crown component. So, we have

$$\text{Ch. Ab.} = \alpha_{comb} = \Big[(n_{F'} - n_{C'})A\Big]_2 - \Big[(n_{F'} - n_{C'})A\Big]_1 \qquad (6.11).$$

Usually a larger number of alternate crown and flint glass component prisms is used to produce a large enough dispersion. The direct vision spectroscope is considered in Chapter 8.

6.4.2 THE ACHROMATIC PRISM - *deviation without dispersion*

The image seen through a single prism has coloured edges due to the dispersive effect of the prism material. This affect will be undesirable in certain applications, but it is usually possible to eliminate the dispersion of one prism by that of another prism, whilst at the same time maintaining some required deviation. After all, to produce deviation is one of the main purposes of a prism. Such a prism combination, which achieves deviation but eliminates the dispersion effect, is called an *achromatic prism*.

Figure 6.10 illustrates a typical achromatic doublet, comprising crown glass and flint glass component prisms, on which is incident a ray of white (polychromatic) light. The component prisms are so designed that the C' (red) and F' (blue) rays emerging from the doublet are parallel to each other, but deviated as a whole from the incident direction. The emergent parallel rays will not be seen as separate colours, and so the chromatic aberration is eliminated.

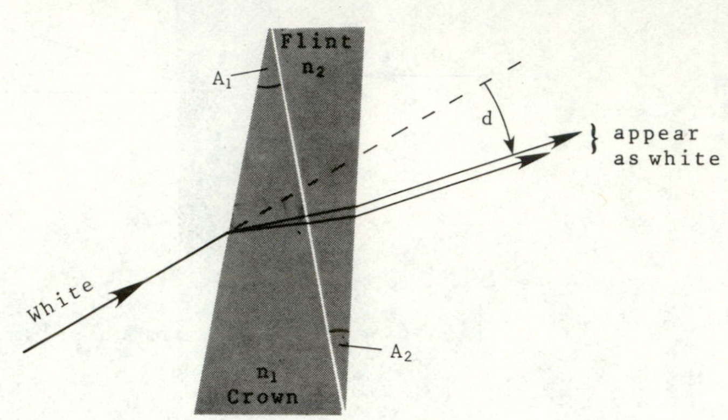

Fig. 6.10 An achromatic prism.

At first glance, the achromatic prism looks similar to the direct vision prism. However, the optical constants of the materials used for the two component prisms in the achromatic prism combination must satisfy the following condition for achromatism; namely

chromatic aberration of crown prism = chromatic aberration of flint prism.

Therefore, using equation (6.9) we have

$$\frac{P_1}{V_1} = \frac{P_2}{V_2} \qquad (6.12),$$

where P_1 and P_2 are the mean powers of the crown and flint components, respectively. Also, as the second prism is inverted relative to the first, the residual prismatic power of the combination is

$$P = P_1 - P_2 \qquad (6.13).$$

Now, from equation (6.12) $P_2 = P_1 \times \frac{V_2}{V_1}$, so substituting for P_2 into equation (6.13) gives

$P = P_1 - P_1\frac{V_2}{V_1} = P_1 \frac{V_1 - V_2}{V_1}$. Rearrangement gives

$$P_1 = \frac{PV_1}{V_1 - V_2} \qquad (6.14).$$

The overall deviation of the light produced by the doublet will be the same for the C' (red), d (yellow), and F' (blue) rays, and is determined by equation (6.13). This may be written in the form

$$P = (n_1 - 1)A_1 - (n_2 - 1)A_2 \qquad (6.15).$$

As usual, the theory is most often more easily grasped when it is seen in applications, so we will now look at three worked examples.

Worked Examples

1 An optical crown glass has the following constants: $n_{C'} = 1.5204$, $n_d = 1.5230$, $n_{F'} = 1.5293$. Determine (a) the refractivity, (b) the mean dispersion C' to F', (c) the constringence, and (d) the dispersive power of this material.

(a) From equation (6.6), refractivity $= n_d - 1 = 1.5230 - 1 = 0.5230$.

(b) From equation (6.7), mean dispersion $= n_{F'} - n_{C'} = 1.5293 - 1.5204 = 0.0089$.

(c) From equation (6.4), dispersive power $\omega = \dfrac{n_{F'} - n_{C'}}{n_d - 1} = \dfrac{0.0089}{0.5230} = 0.01702$.

(d) From equation (6.5), constringence $V = \dfrac{1}{\omega} = \dfrac{1}{0.01702} = 58.75$.

2 Calculate the apical angle of a flint glass prism which when used with a crown glass prism of power 8^{Δ} produces dispersion without deviation. Calculate the chromatic aberration of the combination. The glasses used are:

	$n_{C'}$	n_d	$n_{F'}$
Crown	1.51637	1.51899	1.52496
Flint	1.61749	1.62258	1.63476

The apical angle of the crown prism is given by equation (6.3a)

$$A_1 = \frac{P_1}{n_1 - 1} = \frac{8}{1.51899 - 1} = \frac{8}{0.51899} = 15.41^{\Delta}$$

The condition for dispersion without deviation is given by equation (6.10),

$$(n_1 - 1)A_1 = (n_2 - 1)A_2$$

whence $A_2 = \dfrac{(n_1 - 1)}{(n_2 - 1)} \cdot A_1 = \dfrac{(1.51899 - 1)}{(1.62258 - 1)} \times 15.41 = \dfrac{0.51899}{0.62258} = 12.85^{\Delta}$ or $7^{\circ}19'$.

A_2 is the apical angle of the flint glass prism.

Now, using equation (6.11),

$$\text{Ch. Ab.} = \alpha_{comb} = \alpha_2 - \alpha_1 = \left[(n_{F'} - n_{C'})A\right]_2 - \left[(n_{F'} - n_{C'})A\right]_1$$

$$\therefore \alpha_{comb} = (1.63476 - 1.61749)12.85 - (1.52496 - 1.51637)15.41$$

$$= 0.2219 - 0.1324$$

$$= 0.0895^{\Delta} \text{ or } 0^{\circ}3' \text{ approximately.}$$

3 An achromatic prism is required to have a power of 2^{Δ}. The glasses used in its component prisms are:

	$n_{C'}$	n_d	$n_{F'}$
Crown	1.538	1.541	1.547
Flint	1.644	1.649	1.663

Find the apical angle of each component prism.

Using equation (6.5a), we have

$$\text{for the crown glass} \quad V_1 = \frac{n_d - 1}{n_{F'} - n_{C'}} = \frac{1.541 - 1}{1.547 - 1.538} = 60.11 ,$$

$$\text{and for the flint glass} \quad V_2 = \frac{n_d - 1}{n_{F'} - n_{C'}} = \frac{1.649 - 1}{1.663 - 1.644} = 34.16 .$$

Now, from equation (6.14) the power of the crown component prism is

$$P_1 = \frac{PV_1}{V_2 - V_1} = \frac{2 \times 60.11}{60.11 - 34.16} = 4.63^{\Delta} .$$

And from equation (6.13), $P = P_1 - P_2$, we have

$$\therefore P_2 = P_1 - P = 4.63 - 2 = 2.63^{\Delta}$$

which is the power of the flint glass component. The apical angle of each prism is found by applying equation (6.3a): thus

$$A_1 = \frac{P_1}{n_1 - 1} = \frac{4.63}{1.541 - 1} = 8.56^{\Delta} \text{ or } 4^{\circ}54' \quad \text{and} \quad A_2 = \frac{P_2}{n_2 - 1} = \frac{2.63}{1.649 - 1} = 4.05^{\Delta} \text{ or } 2^{\circ}19' .$$

6.4.2.1 *The Irrationality of Dispersion*

With a simple achromatic prism, such as that depicted in figure 6.10, perfect recombination of all colours in white light is not achieved. The recombination is exact only for those colours used in the calculations. The remaining dispersion among the other wavelengths is small, and for many achromatic prism purposes the small residual colour defect may be safely ignored. However, the residual colour defect is a consequence of an effect known as the *irrationality of dispersion*.

If we consider a line emission spectrum (see Chapter 15) of a typical light source formed at identical distances from separate crown and flint glass prisms of identical apical angle, the result is as shown in figure 6.11 .

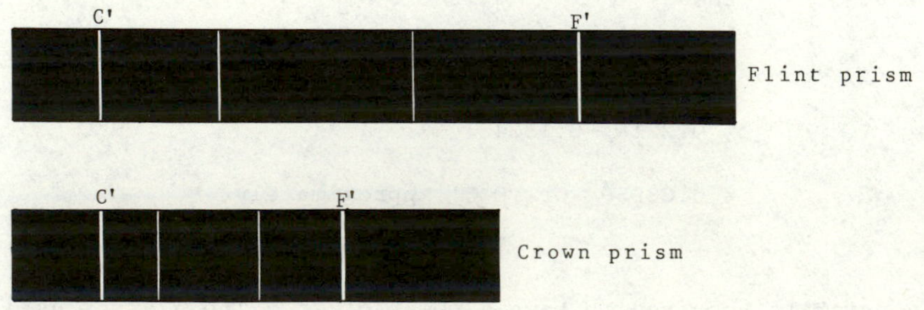

Fig. 6.11 Line spectra for flint and crown glass prisms with the same apical angles.

The spectrum formed by the flint glass prism is wider than that of the crown glass prism due to the greater overall dispersion of the flint glass. In addition, the relative dispersions in the various parts of the spectra are different for the two materials.

Now, if as in figure 6.12, the spectra are made the same width as far as the C' and F' lines are concerned by varying the two apical angles and/or by forming the two spectra at different distances from the prism, it is found that the remaining spectral colours in the light from the source do not coincide in the two spectra.

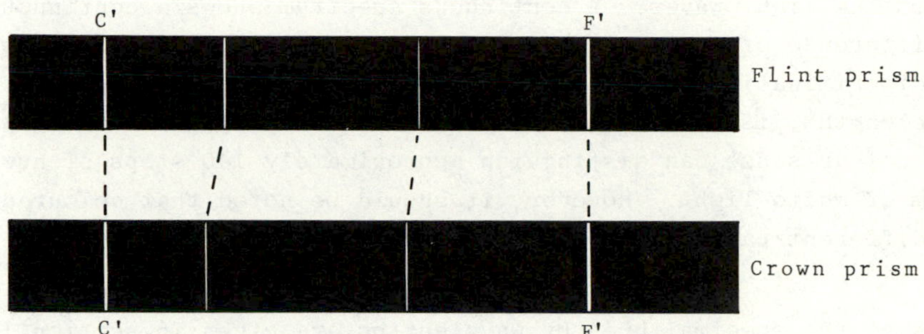

Fig. 6.12 Line spectra for flint and crown glass prisms adjusted to make the spectral lines C' and F' equally spaced: see text for details.

This results in the spectral colours other than the C' (red) and F' (blue) rays not undergoing perfect recombination by the second prism in the achromatic doublet, and is responsible for the formation of a *secondary spectrum*. For better recombination of colours, and the elimination of the secondary spectrum, an achromatic prism containing three or more components is required.

6.5 COLOUR

Colour is subjective. By this we mean that the colours we see - red, yellow, pink, white, green, brown, grey, and many other sensations - are qualities of our mental image, and are entirely different in kind from the physical light rays focused on the retina of the eye. As evidence of this we need only consider the colour perception of a person with normal colour vision in relation to that of a person whose colour vision is defective. In the latter case the person may be unable to distinguish between red and green, and these two colours cannot have the same appearance to him as they do to the normal observer. It is also quite common experience that the colour of an object depends on the nature of the light used to illuminate it, and on the nature of the surface and its surroundings, as well as on the observer.

6.5.1 CHARACTERISTICS OF COLOUR

Before a colour can be measured the variable qualities of colour must be defined. The quantitative measurement of colour, like the measurement of other physiological sensations, is difficult. Three qualities are necessary for the exact specification of a colour, or more strictly, a visual colour sensation, and these are *luminosity*, *hue*, and *saturation*.

The *luminosity* of a colour is the intensity of the luminous sensation which it produces, and this depends on the intensity of the light which enters the eye. The human retina contains two types of light detectors, named rods and cones after their characteristic shapes. The intensity of the light entering the eye must be sufficient to stimulate the cones before colour is perceived. Below this threshold value the rods may detect the light but the vision is achromatic or colourless.

When we say that a colour is blue, or green, or yellow, we are stating its *hue*. This is the most noticeable quality of colour. Hue is determined by the characteristic frequency or vacuum wavelength of the light waves. A continuous spectrum shows a continuous variation of hues. It is the difference of hue which enables us to distinguish the different parts of the spectrum. Colour discrimination, that is, the ability to distinguish between light of nearby frequencies or wavelengths, is at a maximum within the central part of the visual field. A person with normal colour sense can distinguish approximately 140 steps of hue difference across the spectrum of white light. However, it should be noted that colour discrimination is different near different parts of the spectrum, and is maximal at wavelengths of 499 nm and 590 nm.

The broad groupings of the spectral hues by wavelengths are given in section 10.6.2 , from which we see that the normal range of colour vision is from a wavelength of 760 nm at the red end of the spectrum, to 390 nm at the violet end. The visual sensitivity of the eye to the different colours of light is discussed in Chapter 9. Let us note here that the maximum visual effect is obtained with light in the yellow-green region of the spectrum at 555 nm. Visual acuity is also slightly influenced by the spectral composition of the light, some improvement being achieved with monochromatic yellow light.

The third quality of colour, *saturation*, is the variation in depth of colour of a given hue and is not to be confused with luminosity. If red is mixed with white it becomes pink and is said to be *diluted* or *unsaturated*. A saturated hue contains no trace of white. The pure spectral colours produced by a prism (see figure 6.7) are saturated. The hue of an unsaturated colour is that of its dominant frequency (or wavelength); that is, the frequency of the spectral colour which must be diluted with white to match the colour.

The ability of a person to judge the saturation of a colour, that is, freedom from content of a white component, is variable across the various regions of the spectrum. In this respect, the sensitivity is least in the yellow region of the spectrum and increases towards either end.

6.5.2 COLOURED PAPER AND COLOURED GLASS

An object appears coloured by virtue of the light reflected or scattered from it, and involves the process known as *selective absorption*. Much of the colour of nature owes its origin to the polychromatic character of white light, and to the fact that if some of the components of white light are subtracted from it the remainder appears coloured. Consequently, as shown in figure 6.13(a), if we view a piece of blue paper in white light the paper reflects the blue light very well, but absorbs all the other colours. In fact, a smaller percentage of the colours bordering blue in the spectral order, namely green and violet, will also be reflected, but the dominant reflected colour is blue. The surface thus appears blue, although the colour actually comes from the incident white light. If the blue paper is viewed in the yellow light of a sodium lamp, it appears a dull brown. This is because no blue light is emitted from a sodium lamp. Blue paper can only appear blue if the illuminating light itself contains a blue light component.

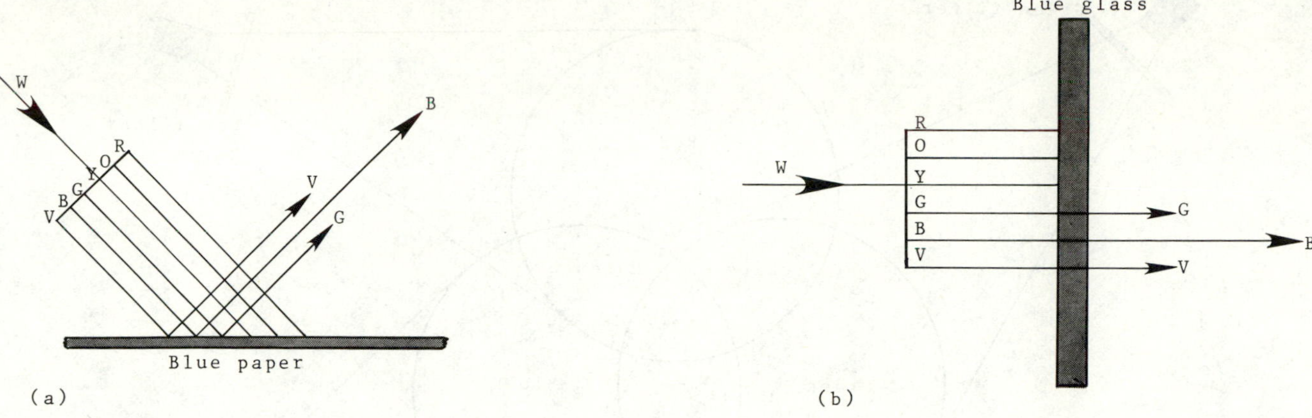

Fig. 6.13 White light incident upon (a) blue paper and (b) blue glass.

Thus, when a coloured surface is viewed in coloured artificial light the appearance of the surface will change. This effect is referred to as *colour distortion*. Colour distortion is observed with many types of modern indoor and outdoor lighting. Certain fluorescent tubes, which emit light having the appearance of white, accentuate blue objects rather than red due to the emission of a disproportionately larger quantity of blue light.

If white light is viewed through a piece of blue glass the glass absorbs light of all the colours except blue, which is transmitted; see figure 6.13(b). The glass itself does not produce the colour; this comes from the incident light. Again, it should be noted that a small percentage of the colours bordering blue in the spectral order will also be transmitted. These bordering colours, green and violet, are indicated with shorter arrows in figure 6.13 .

If a piece of glass is broken into large fragments the appearance of each piece is essentially the same colour as the original. With very small fragments a small part of the incident light is reflected at the surface of each fragment, and this will be white. Hence, finely powdered coloured glass appears white. Also, the small fragments refract the light irregularly without any great absorption of the various colours in the white light. The lack of any real absorption occurs because fragments of powdered glass are very 'thin' and absorption requires a reasonable thickness of material to make it detectable.

6.5.3 COLOUR ADDITION

If beams of red and green light are shone on to a plane white screen, the appearance of the illuminated area is yellow. The yellow sensation is a result of the *additive mixing* of red light with green light. The sensation of yellow may also be achieved with a monochromatic yellow stimulus. We shall refer to the former yellow as *impure yellow* , and the latter as *monochromatic yellow*. When impure yellow light is being observed, the observer is not conscious of the constituent colours in the additive mixture.

Figure 6.14 shows how the effects of additive colour mixing may be demonstrated. The intensities of the light beams may be varied by controlling the power supplies to the light sources. Pieces of glass, coloured red, green, and blue, are placed one in front of each white light source, with the beams of light overlapping on a screen. The area of overlap of the red and

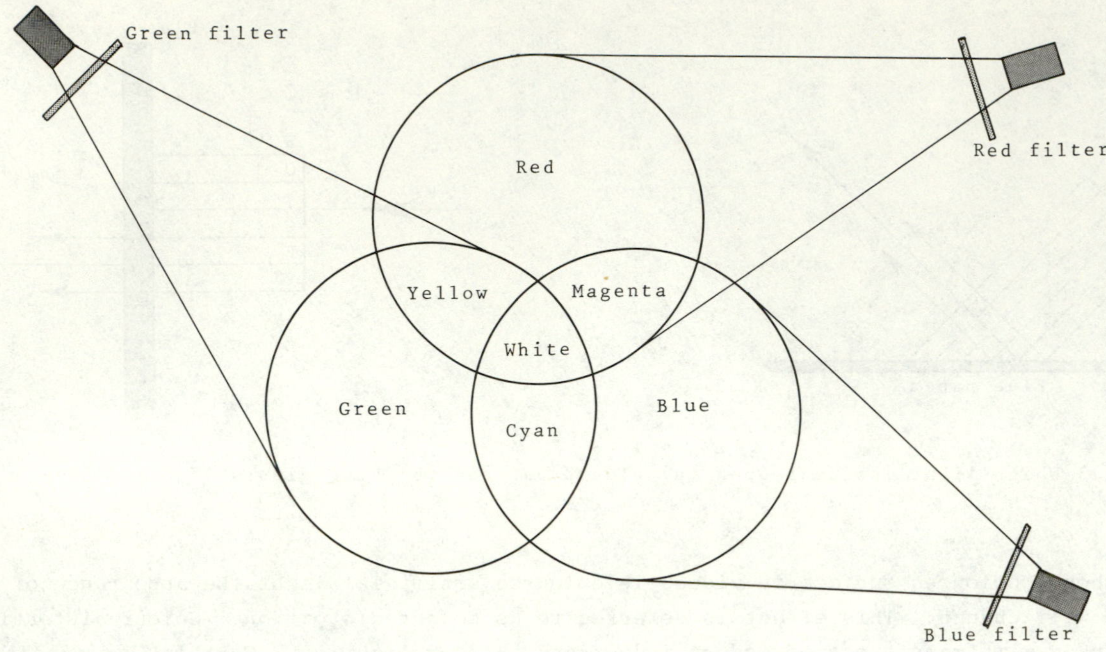

Fig. 6.14 Additive colour mixing.

green patches on the screen appear yellow to an observer. So we may write

red light + green light = yellow light (impure).

Similarly, we observe

red light + blue light = magenta light

and green light + blue light = cyan (turquoise) light.

The area near the centre of the screen where all three light patches overlap has an almost white appearance. It can be made true white by adjustments to the intensities of the light sources. Hence, we have

red light + green light + blue light = white light.

The three colours of light, namely red, green, and blue, are termed *primary colours*. All colours in nature can be matched by a suitable additive mixture of three selected primary stimuli in the red, green, and blue regions of the spectrum. Many sets of suitable primary stimuli can be chosen, although one must be in the red region, one in the green, and the other in the blue region of the spectrum. This is a fact of fundamental importance in the theory of colour vision, and led to the assumption of three colour sensitive photopigments in the retina. Three such pigments were subsequently discovered in the retina.*

It is also possible to achieve the sensation of white by the additive mixing of two colours in suitable proportions. Such pairs of colours are termed *complementary colours*. Using monochromatic light the following are complementary pairs:

red light (650 nm) and blue-green (494 nm); monochromatic yellow (574 nm) and blue (470 nm).

We can also see from figure 6.14 that if we mix impure yellow light, formed from the additive mixing of red and green light, with pure blue light, the result is a white sensation.

* *See Retinal Mechanisms of Color Vision, E.F. MacNichol Jr, in Vision Research, Vol. 4, 1964.*

Hence, impure yellow light + blue light = white. Similarly, from figure 6.14, we have further pairs of complementary colours of light, namely:

cyan light + red light = white and magenta light + green light = white.

6.5.4 *COLOUR SUBTRACTION*

6.5.4.1 *Colour Filters*

The coloured appearance of a glass filter depends on the absorption of some colours and the transmission of others. For example, a pure green filter affects white light incident upon it by absorbing all colours from the incident light except green, which is transmitted and also reflected. This effect may be represented as shown in figure 6.15(a). We should note that the light giving rise to the sensation of green is contained in the incident light and is not 'manufactured' by the filter.

Similarly, a magenta filter appears to be magenta because only the red and blue components of white light are trasmitted and reflected. Seen together, the effect of red and blue light is a magenta sensation; see figure 6.15(b). The remaining colours of the incident light are absorbed by the materials in the filter.

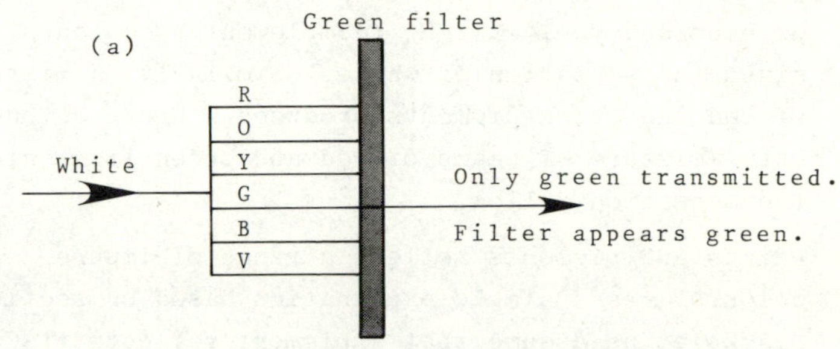

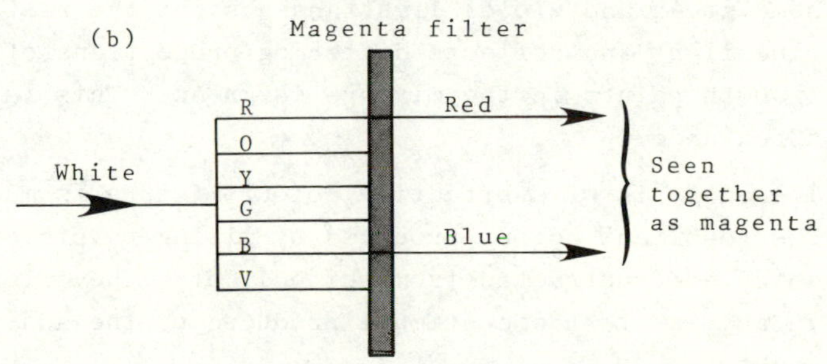

Fig. 6.15 White light incident on (a) a green and (b) a magenta filter.

Figure 6.16 illustrates an impure yellow filter and a magenta filter placed in line with each other in front of a white light source. The yellow glass transmits red, green, and yellow, whilst absorbing the other colours. The magenta filter absorbs the green and transmits the red. Hence, the only colour of light transmitted by the combined filters is red.

These effects illustrate the process known as *colour subtraction*, since colours are absorbed or subtracted from white light when it passes through the glass. If impure yellow, magenta, and cyan filters are placed together, no light is transmitted since no common colour of light is transmitted by all three glasses.

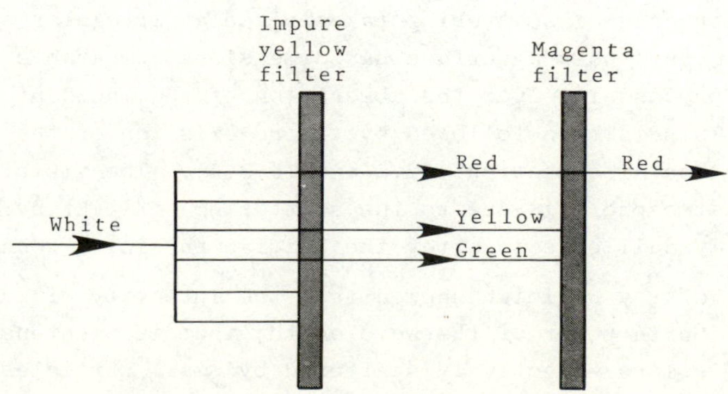

Fig. 6.16 The effect of successive yellow and magenta filters.

Figure 6.17 shows overlapping filters of impure yellow, cyan, and magenta glass and summarises colour subtraction effects with filters.

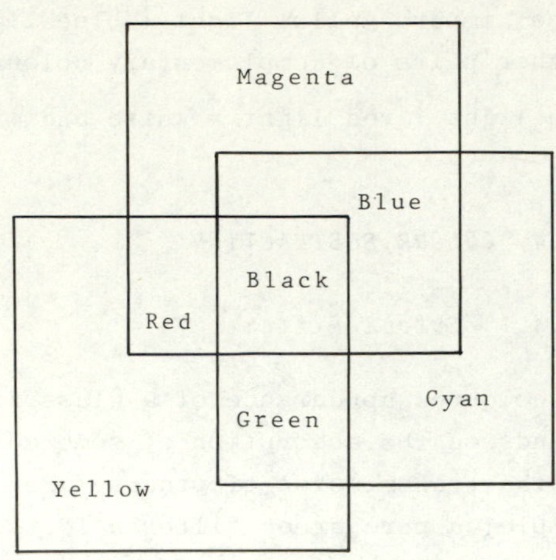

Fig. 6.17 Subtractive colour effects. A white background viewed through overlapping filters.

6.5.4.2 *Paints and Pigments*

The mixing of paints or pigments is a subtractive process. A mixture of blue and yellow paints produces green paint, whereas a mixture of beams of blue and yellow light (complementary colours) gives the sensation of white. Similarly, a mixture of red and green pigments produces a brown colour, but a mixture of beams of red and green light gives the sensation yellow.

Paints and pigments reflect a range of impure colours. As a simple explanation based on section 6.5.2 let us assume that a pigment reflects its own colour of light strongly, together with some light of the immediately neighbouring colours in the spectral order. Thus, in the mixture of blue and yellow paints mentioned above, the blue paint reflects blue mainly, together with some green and violet light and absorbs the rest, whilst the yellow paint absorbs violet and blue light and reflects differing proportions of the rest. The only colour which is not absorbed by both paints in the mixture is green. This is reflected and hence the mixture has a green appearance.

The analysis of subtractive colour mixture is much more involved than that of additive mixture. The resultant colour produced by mixing paints or pigments depends on the range of colours which are subtracted from white light. The colours produced by subtraction do not correspond in any way to those stimuli produced by the additive mixing of similarly coloured beams of light.

6.5.5 *OTHER COLOUR EFFECTS*

Colour produced by interference and diffraction will be discussed in Chapters 13 and 14. Certain other colour effects are due to the *scattering* of light. As was mentioned in section 5.1, the process of scattering takes place at irregular surfaces. More precisely, scattering of light occurs when a surface has dimensions comparable to the wavelength of the incident light. The process involves the absorption of the incident light by an atom or molecule in the surface. This is then followed by the re-emission of the light after a time interval of about 10^{-8} s. (See also sections 12.5 and 15.3.4). The visibility of a beam of sunlight in a dust-laden atmosphere is due to the scattering of light by dust particles. This process is known as the Tyndall effect, after the British physicist John Tyndall (1820 - 1893).

Work by Rayleigh showed that the intensity of scattered light is inversely proportional to the fourth power of the wavelength; that is, intensity $\propto \lambda^{-4}$. Thus, light of shorter wavelengths is more effectively scattered by small particles than that of longer wavelengths. The effect is even more pronounced the finer the scattering particles and accounts for the strong blue colour of a cloudless sky. This is due to the preferred scattering of blue, short wavelength light by molecules in the air. The larger water droplets of clouds do not have the same effect. They scatter and reflect all wavelengths and therefore appear white or grey. From a space satellite the sky is black due to the absence of scattering nuclei.

For similar reasons the smoke from the end of a burning cigarette is coloured blue due to the scattering of light by very small smoke particles. Smoke exhaled from the mouth comprises carbon particles and larger droplets of water, and thus appears white or grey.

It is also interesting to note that the Tyndall phenomenon has made it possible, with suitable illumination, to see small particles such as inflammatory cells and an increased protein content in the aqueous humour of the eye.

6.5.6 THE TRICHROMATIC SYSTEM OF COLORIMETRY

Colorimetry is the science of colour measurement. The appearance of a stimulus is not in itself an adequate definition of colour. Many different combinations, as we have seen, can give rise to the same colour sensation. In order to reduce the number of variable quantities, account has to be taken of the manner in which the eye integrates the various spectral components to produce the final colour sensation. Colour matching experiments show that three variables are necessary and sufficient. The three primary colours in the red, green, and blue regions of the spectrum are chosen and referred to as *reference stimuli*. The instrument used to carry out visual matching is called a colorimeter. Viewing through an eyepiece, the observer sees the colour to be matched in one half of the field of view, and in the other half sees the mixture of the red, green, and blue reference stimuli. The field of view is such as to ensure that the observations are restricted to the eye's most sensitive central region of the visual field. The amounts of the reference stimuli are varied until the mixture matches the test colour.

Since 1931 the C.I.E. system (Commission Internationale de l'Eclairage) has been based on monochromatic reference stimuli of wavelengths 700 nm (red), 546 nm (green), and 436 nm (blue). In practice, these reference stimuli are narrow bands of wavelengths centred at the quoted values.

Let us suppose that a colour stimulus is represented by C, and that by using a colorimeter this is matched by r units of the red reference stimulus (R), mixed with g units of the green reference (G) and b units of the blue reference (B). So we can write

$$C = rR + gG + bB$$

and this is an exact specification of the colour. The quantities r, g, and b may be expressed in terms of the variable areas of the red, green, and blue filters through which the white light is passed in the colorimeter.

For convenience, let us suppose that the colour stimulus C has unit luminance*. This allows a stimulus to be related to hue and saturation only, for which

$$1 = r + g + b$$

where r, g, and b are referred to as the *chromaticity coordinates* of the stimulus. From a preliminary calibration the variable areas of the reference filters can be converted into the chromaticity coordinates, the sum of which must be unity. As $r + g + b = 1$ for any colour,

* For the moment take this to be brightness. See Chapter 9 for a definition.

any two chromaticity coordinates are sufficient to define the chromaticity. It is possible for one of the coordinates to have a negative value for certain colour stimuli, indicating that the colour cannot be matched by a mixture of the primaries without desaturation. To avoid negative values, the chromaticity coordinates r, g, and b are usually transformed into another system based on three imaginary stimuli X, Y, and Z, such that the new coordinates x, y, and z are all positive values. Hence, as above, 1 = x+ y + z and only two coordinates need be specified. Figure 6.18 shows the C.I.E. chromaticity chart, which is commonly used in colorimetry.

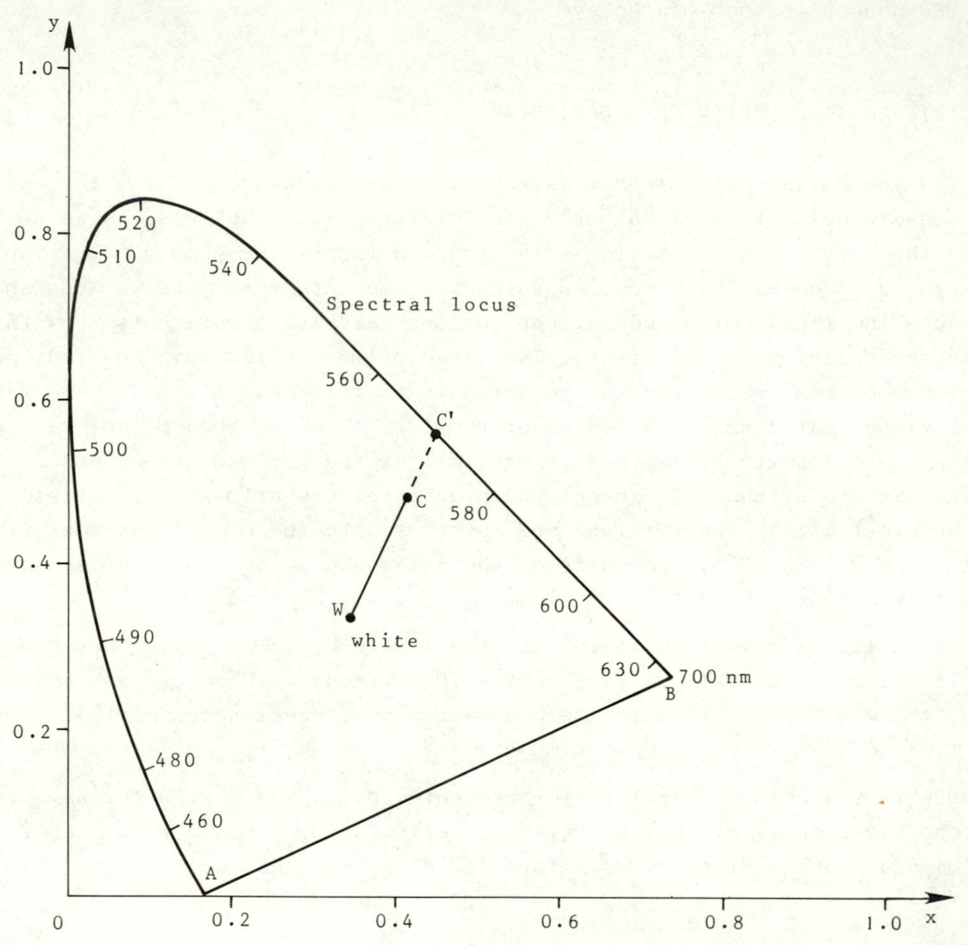

Fig. 6.18 C.I.E. Chromaticity Chart.

Consider a colour represented by the point C (x=0.413, y=0.483). The dominant hue of C is obtained by joining the point W, representing white, to C and extending the line WC to C' on the spectral locus. In the case illustrated this gives a wavelength of 570 nm. The purity, which is the objective stimulus corresponding to saturation, is given by the inverse ratio in which the point C divides the line WC'. As shown, about 1 part of white must be mixed with 2 parts of the dominant hue in order to match the colour represented by C.

Any pair of complementary spectral colours will lie on the spectral locus at the ends of a straight line through the point W.

The ends of the spectral locus are joined by a straight line AB. This line represents the various mixtures of red and violet; that is, the purples (or magentas), which are not spectral colours. The complementaries of the colours at A and B are about 570 nm and 494 nm, respectively. Between these two wavelengths the complementaries are on the line AB and are therefore purples. It follows that there are no spectral complementary colours for spectral colours with wavelengths between 494 nm and 570 nm.

6.5.7 *INFRARED AND ULTRAVIOLET RADIATIONS*

In the period 1800 to 1801, bands of radiations were discovered bordering the red and violet ends of the visible portion of the sun's spectrum. Sir William Herschel, the German-born British astronomer, found that the maximum heating effect of the sun's radiations occurred beyond the red light end of the spectrum. In this way he discovered the invisible *infrared* radiations, having longer wavelengths than light. These rays had been refracted by Herschel's prism less than the red rays. The wavelengths of the infrared (IR) band range from 760 nm to about 10^6 nm (1mm). IR radiations form a very large part of the energy radiated by an electric fire. Filament lamps, although designed to emit visible light, emit much more energy in the IR region as heat.

Ordinary glass is opaque to all but the near IR (near the visible: 760 - 3000 nm), and so for experiments with IR radiation it is necessary to use prisms and lenses made of rock salt.

Infrared radiation is even less scattered by small particles than red light, and photographs have been taken over long distances, and in hazy conditions, using special filters and IR sensitised photographic material.

The energy of IR radiation may be absorbed by whole atoms and molecules, with overdosage of the radiation causing severe skin burns. IR wavelengths below 1500 nm can penetrate the cornea, and larger doses may cause burns of the choroid and retina. Sunblindness is a permanent state caused by IR and intense visible light producing irreversible effects in the visual pigments.

In 1801, Johann Ritter discovered a band of invisible radiations beyond the violet end of the solar spectrum which cause certain minerals to fluoresce. Having shorter wavelengths than light in the approximate range 390 to 1 nm, these rays were termed *ultraviolet*. UV radiation is emitted by hot bodies and also by ionised gases. Absorption by ozone in the Earth's upper atmosphere prevents UV radiation with a wavelength shorter than about 300 nm from reaching the Earth's surface from the sun. Generally speaking, the UV which does reach the the Earth's surface is beneficial, promoting the formation of vitamin D in the skin.

Mild exposure to UV allows the skin to recover from the effects of suntan and sunburn. Longer exposure results in damage to or destruction of the cells in the outer layers of the skin. The eye cannot adapt to UV. Those radiations which cause sunburn can also be responsible for corneal inflammation. This is the effect known as snowblindness, and often no pain is felt until some hours have elapsed. UV can also cause cataract, characterised by the presence of denatured protein in the cells forming the eye lens.

The phenomenon of fluorescence is described in section 15.3.4 . It is used in tubular electric lamps, in which large, and otherwise wasted, quantities of UV from an electric discharge through mercury vapour is converted into visible light approximately resembling daylight. It is also interesting to note the claim of people able to "see" UV down to a wavelength of 300 nm. This is not true vision, being some form of fluorescence effect in the retina.

EXERCISES

1 Describe how a pure spectrum may be produced.

2 Define the terms angular dispersion and mean deviation. If the 30^0 angle of a 30^0-60^0-90^0 prism acts as the apical angle, calculate the angular dispersion and the mean deviation if $n_{C'} = 1.535$, $n_d = 1.538$, and $n_{F'} = 1.545$. Assume a parallel pencil of white light is incident on the face opposite the 60^0 angle.

Ans. $\alpha = 0.13^0$ and $d_d = 18.97^0$.

3 How did Newton come to conclude that white light consisted of the colours seen in the rainbow?

4 A glass prism with an apical angle of 60^0 has a refractive index 1.516 for red light and 1.533 for violet light. A parallel pencil of white light is incident on one refracting face at an angle of incidence giving minimum deviation for red light. Find: (i) the angular dispersion, (ii) the length of the spectrum when focused on a screen by a +2 D achromatic lens.

Ans. $\alpha = 1.52^0$. Length of spectrum = 1.33 cm.

5 Calculate the V-numbers and the dispersive powers for the following glasses:

	$n_{C'}$	n_d	$n_{F'}$
(i)	1.516	1.519	1.524
(ii)	1.609	1.613	1.625

Ans. (i) 64.88 and 0.0154. (ii) 38.31 and 0.0261.

6 What is the dispersive power of the glass in question 2?

Ans. 0.0186.

7 What is meant by the terms dispersion, mean dispersion, and partial dispersion?

8 Calculate the partial dispersions C' to d and d to F' for the glasses in question 5.

Ans. (i) $n_d - n_{C'} = 0.003$, $n_{F'} - n_d = 0.005$.
(ii) $n_d - n_{C'} = 0.004$, $n_{F'} - n_d = 0.012$.

9 If a glass is designated 519604, what does this mean? Calculate the dispersive power for this glass.

Ans. It means $n_d = 1.519$ and $V = 60.4$. $\omega = \frac{1}{V} = 0.016\,56$.

10 A thin prism of apical angle 6^0 has refractive indices

$$n_{C'} = 1.538 \qquad n_d = 1.541 \qquad n_{F'} = 1.547.$$

Calculate it chromatic aberration, mean deviation, and dispersive power.

Ans. $\alpha = 0.054^0$. $d_d = 3.246^0$. $\omega = 0.0166$.

11 A thin +10 D lens is made from the glass in question 9. What is the angular dispersion at a point 6 mm above the optical centre? Hint: use equation (6.9).

Ans. $\alpha = 0.056\,9^0$.

12 Explain why two different glasses must be used in the construction of (i) a direct vision prism, and (ii) an achromatic prism.

13 The direct vision prism discussed in this chapter has a very small resultant dispersion. How do you think this might be increased?

14 A direct vision prism is made from the glasses

	$n_{C'}$	n_d	$n_{F'}$
Crown	1.528	1.531	1.537
Flint	1.630	1.636	1.647 .

If the apical angle of the flint prism is 5^0, what is the apical angle of the crown prism? Calculate the dispersion of the combination.

Ans. $a_1 = 5.99^0$. $\alpha_{comb} = 0.031^0$.

15 Find the resultant angular dispersion in the direct vision prism made from the glasses:

	$n_{C'}$	n_d	$n_{F'}$
Crown	1.523	1.526	1.537
Flint	1.628	1.633	1.644

if the apical angle of the crown prism is 8^Δ.

Ans. $\alpha_{comb} = 0.0424^\Delta$.

16 Explain what is meant by chromatic aberration. How may it be corrected in prisms?

17 An achromatic prism is made from glasses with the following data:

	n_d	$n_{F'} - n_{C'}$
Crown	1.520	0.0087
Flint	1.615	0.0165

If the resultant power is 4^Δ and the minimum edge thickness of each component is 2 mm, find the powers of the components and the maximum and minimum edge thicknesses of the achromatic prism if it is circular and 50 mm in diameter.

Ans. $P_1 = 10.63^\Delta$, $P_2 = 6.63^\Delta$, and the minimum and maximum edge thicknesses are 9.4 mm and 14.2 mm.

18 An achromatic prism is made with a crown component of 10^Δ power. If the data for the glasses are

	Refractivity	Mean dispersion
Crown	0.530	0.008
Flint	0.636	0.017

find the resultant power and the power of the flint prism.

Ans. Resultant power is 4.35^Δ and $P_2 = 5.65^\Delta$.

19 Explain the appearance of a plant with yellow flowers and green leaves seen under (i) red light, (ii) green light, (iii) impure yellow light, and (iv) blue light.

20 Explain the circumstances under which the printing on a poster 'disappears' when viewed through a coloured glass.

21 Why do photographers often use a yellow glass over the camera lens when taking shots of blue sky and clouds?

22 The plant in question 19 is viewed through (i) a green glass, (ii) a red glass, (iii) a yellow glass. Explain its appearance in each case assuming it is illuminated by sunlight.

23 Explain the effects summarised in figure 6.17 .

24 Explain the colour in (i) blue sky, (ii) blue glass, (iii) blue paper.

25 A parallel pencil of white light is incident at 41.041^0 from the glass side of a glass-air plane boundary. If the refractive index of the glass is $n_d = 1.523$, explain what happens to the various colours in the pencil.

26 Explain the terms hue, luminosity, saturation, primary colours, and complementary colours.

27 A pure spectrum is produced on a white screen. Describe its appearance if (i) a red glass is placed between the prism and the screen, and (ii) the screen is replaced by a red cloth.

28 A narrow horizontal strip of white paper is viewed through a prism with a fairly large apical angle, the base-apex line being perpendicular to the length of the paper. Describe the colours seen at the upper and lower edges of the paper.

7 ABERRATIONS

7.0 INTRODUCTION

The derivations of formulae for the refractions at lens surfaces, such as those in section 2.2.3, are based on rays of light making small angles with and travelling on or near the principal axis. Thus, such formulae only apply with reasonable accuracy to paraxial rays of light.

In general, however, rays of light which undergo refraction by a lens originate from a variety of points, some of which may be well away from the lens axis. Furthermore, if the lens has an appreciable aperture any cone of rays from an object point even on the axis will not have the small cone angle necessary for the implementation of paraxial theory. For a lens to produce a clear image of an object, each individual point in the object plane must be imaged at a point on the image plane. This will not occur, in general, with non-paraxial rays. As a consequence of this, the image formed by non-paraxial rays will not be sharp one.

We may also recall (section 6.1) that as a result of the variation of refractive index of an optical medium with wavelength, the focal length of a lens varies with wavelength. Hence, with light which is polychromatic (containing two or more separate colours), a lens will form a number of coloured images which will be in different positions and of different sizes, even if formed by paraxial light rays.

The departure of an image from perfection is known as an *aberration*. Aberrations which are caused by the variation of the refractive index of a material with wavelength are called *chromatic aberrations*, whilst the others, arising even if monochromatic light is considered, are termed *monochromatic aberrations*. This distinction is only one way of classifying the various aberrations.

It is useful to note that aberrations are the consequences of the laws of reflection and refraction at spherical surfaces, and are not due to any fault in the construction of a spherical surfaced lens. The methods of reducing these aberrations to a tolerable level, with the formation of acceptable images, are one of the main problems of geometrical optics. We can be fairly certain that all aberrations cannot be reduced to zero in a system of spherical surfaces. A compromise must be reached depending on the intended use of the lens, with computers nowadays assisting the designer in evaluating a lens or system for its image defects.

Even if it were possible to remove all the image defects which we are going to consider in this chapter, it is evident that the image of an object point will still not be a sharp image point. Instead it would appear as a diffraction pattern, consisting of a bright central maximum intensity surrounded by less bright rings when the aperture is circular. This effect is most pronounced when the aperture of a lens is small. Diffraction effects will be dealt with in Chapter 14.

The final part of this chapter will be concerned with the tracing of non-paraxial rays through systems of spherical surfaces. In order that the designer may compute the course of a ray through an optical system, an accurate trigonometrical method is required. This will be described for a single lens although the procedure can be applied, in turn, to any number of surfaces.

7.1 CLASSIFICATION OF ABERRATIONS

In a book of this nature it is only possible to summarise the main facts concering aberrations and to indicate some means of reducing each effect. The subject is extensive and complicated. It will be found useful to list the aberrations at the commencement and to indicate a number of different ways in which these may be classified.

The seven aberrations to be considered in this chapter are

(i)	*longitudinal chromatic aberration*	*LCA*
(ii)	*transverse chromatic aberration*	*TCA*
(iii)	*spherical aberration*	*S*
(iv)	*coma*	*C*
(v)	*oblique astigmatism*	*A*
(vi)	*curvature of field (Petzval curvature)*	*P*
(vii)	*distortion*	*D.*

The symbols LCA, TCA, S, C, A, P, and D are convenient abbreviations for the aberrations. We have already mentioned in section 7.0 that an aberration may be *chromatic* where the materials used for lenses give rise to colour effects, or *monochromatic* which are largely associated with the form of a lens. Thus, we have:

1(a) due to the material of the lens (chromatic aberrations)

LCA

TCA

1(b) due to the form of a lens (monochromatic aberrations)

S

C — *These are often referred to as Seidel aberrations, after Ludwig von Seidel, who first studied these effects in about 1850.*

A

P

D

Another useful distinction between aberrations occurs when we consider axial and extra-axial or off-axis object points. We can classify the aberrations associated with these two criteria.

2(a) Axial aberrations - axial object points only:

S

LCA.

2(b) Oblique aberrations - off-axis object points only:

TCA

C

A — *Note that spherical aberration and longitudinal chromatic aberration may also still occur.*

P

D

Yet another way of classifying the aberrations is

3(a) Aberrations with lenses of large aperture:

TCA S C D

3(b) Aberrations with lenses of small aperture:

LCA A P.

The aberrations which affect the definition of image points, as distinct from their positions, are LCA, TCA, S, C, A, and these may be subdivided into:

4(a) longitudinal aberrations

LCA S A

4(b) transverse aberrations

TCA C.

We shall now attempt to show how all of these image defects manifest themselves, and what steps may be taken to remedy each defect. Before doing so, it might be useful to note that none of these image defects is noticeable when a single converging lens is held close to the eye and used as a magnifier. In such a situation the effective aperture of the lens is the pupil of the observer's eye and the paraxial approximations nearly hold. Defects due to extended objects are negligible as the field of view is small. Moreover, the virtual images formed by the different colours in white light will all subtend the same angle at the eye so that dispersive effects are not observed. All the defects are observed with real images in a lens, and with the exception of the chromatic aberrations, are present with spherical mirrors also.

7.2 THE CHROMATIC ABERRATIONS

The chromatic aberrations are longitudinal chromatic aberration, LCA, and transvers chromatic aberration, TCA. They both arise due to the dispersive properties of optical materials, and are observed only with polychromatic light. We have seen (section 6.1) how the refractive index of glass depends on the wavelength of light. Let us again note that with the longer wavelength red light, materials exhibit a smaller refractive index than with the shorter wavelength blue light. Since the power and focal length of a lens depend on the refractive index of its material (see equation (3.5b)) it should be clear that the focal length of the lens will be dependent on the wavelength of the light. In particular, the focal length of a lens is smaller for blue light than for red light due to the higher index for blue. This result is shown in figure 7.1(a) for a converging lens where the constituent colours in a beam of white light, which is parallel to the lens axis, are focused at different points on the lens axis.

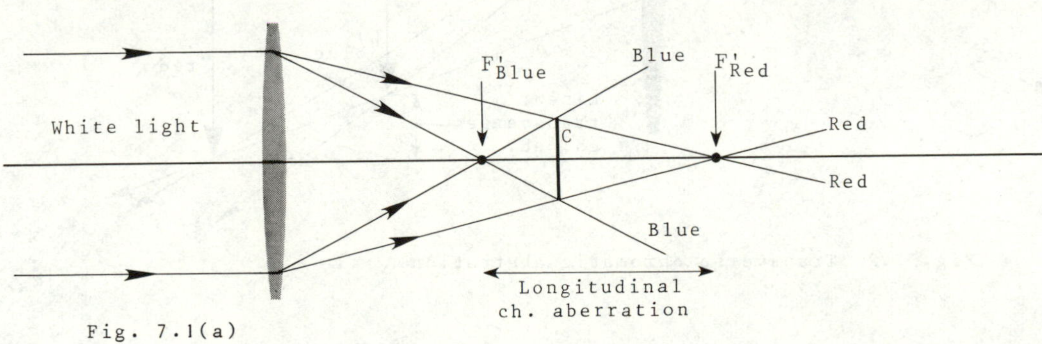

Fig. 7.1(a)

This illustrates longitudinal spherical aberration LCA, and is measured in terms of the axial (longitudinal) distance between two focal points for a given wavelength range. As we described in section 6.3, a commonly used wavelength range for visual use is from red to blue, as specified by the two spectral wavelengths called the C' and F' lines, respectively. Figure 7.1(b) shows the effect of LCA for a diverging lens.

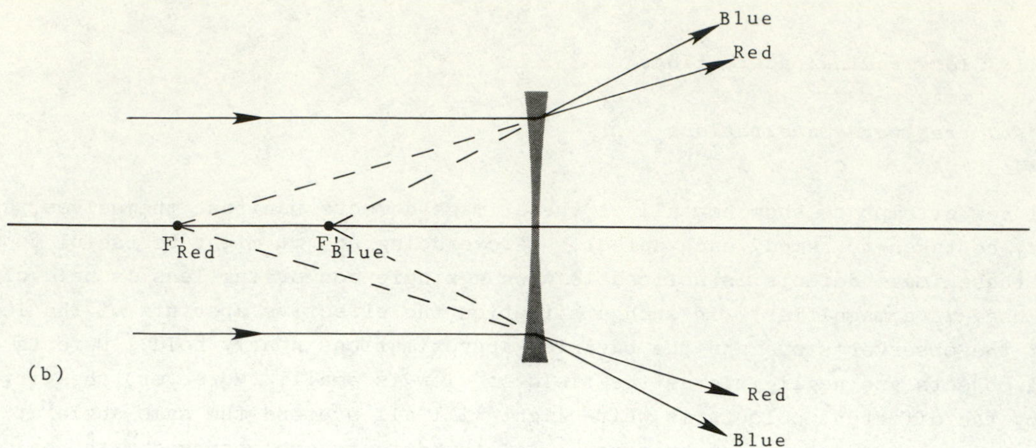

Fig. 7.1 Longitudinal chromatic aberration in positive and negative lenses.

In figure 7.1(a), the position of the best real image is at C, called the *circle of least confusion*. If an image screen is positioned in the refracted pencil of rays, between the lens and C, the image will be blurred with its edges fringed in red. When placed beyond C, the image screen will receive a blurred image with the outline fringed with blue/violet.

In figures 7.1(a) and (b), the magnitude of the longitudinal aberration has been exaggerated for the sake of clarity. If now, as in figure 7.2, we consider rays from an off-axis object point, we can see how the images formed in lights of different wavelengths are of different sizes. Again, only the red and blue images have been drawn and their difference in size has been exaggerated for clarity. Images formed by other colours are located at intermediate points, and are not shown.

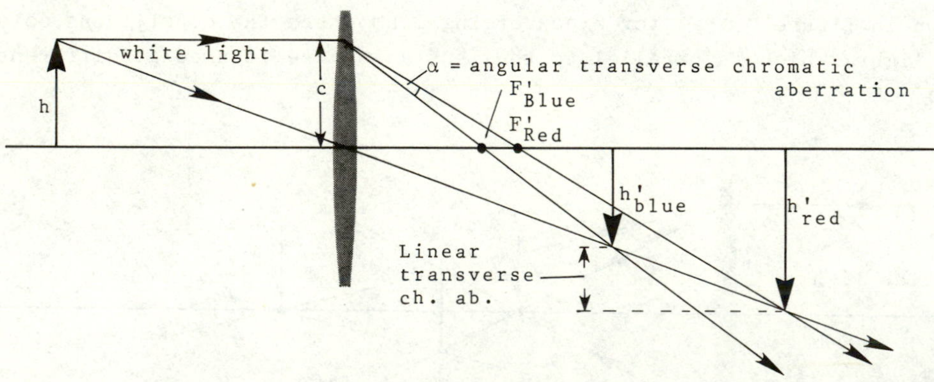

Fig. 7.2 Transverse chromatic aberration.

Thus, there is no single plane in which all the images are simultaneously in focus. This variation of the linear magnification of the various coloured images is referred to as the *transverse chromatic aberration*, TCA, of the image. It may be expressed in the form of:

(i) linear TCA, measured in terms of the difference in image heights for the red (C') and blue (F') colours of light,

or (ii) angular TCA, measured by the angle α between the emergent red and blue rays.

This latter representation of TCA is analogous to the angular dispersion exhibited by a prism with polychromatic light (see equations (6.2), (6.2a), and (6.9)). It is TCA which is responsible for the coloured fringes sometimes seen when viewing an object through the peripheral regions of a spectacle lens.

7.2.1 FORMULAE FOR CHROMATIC ABERRATION

Let us firstly consider the longitudinal chromatic aberration of a single thin lens. When the incident light is parallel to the lens axis the blue and red coloured images will focus at the respective focal lengths $f'_{F'}$ and $f'_{C'}$ from the optical centre. Hence, the LCA is given by

$$\text{LCA} = f'_{C'} - f'_{F'} \qquad (7.1).$$

It may be more convenient to express LCA as the difference in the powers $F_{F'}$ and $F_{C'}$ of the lens for blue and red light respectively. That is,

$$\text{LCA} = F_{F'} - F_{C'} \qquad (7.1\text{a}).$$

If the object is at a finite distance from the lens with the incident vergence constant at the lens, we have for red light $F_{C'} = L'_{C'} - L$, and for blue light $F_{F'} = L'_{F'} - L$. Hence, the longitudinal chromatic aberration in dioptres is given by

$$\text{LCA} = L'_{F'} - L'_{C'} = (F_{F'} + L) - (F_{C'} + L) = F_{F'} - F_{C'}$$

as in equation (7.1a).

Now, for a single thin lens in air, using equation (3.5b):

$$\text{with red (C') light;} \quad \text{Power} = F_{C'} = (n_{C'} - 1)(R_1 - R_2) \qquad \text{(i)},$$

$$\text{and with blue (F') light,} \quad \text{Power} = F_{F'} = (n_{F'} - 1)(R_1 - R_2) \qquad \text{(ii)},$$

$$\text{and with yellow (d) light,} \quad \text{Power} = F_d = (n_d - 1)(R_1 - R_2) \qquad \text{(iii)}.$$

From (i) and (ii) we have

$$\text{LCA} = F_{F'} - F_{C'} = (n_{F'} - 1)(R_1 - R_2) - (n_{C'} - 1)(R_1 - R_2) = (n_{F'} - n_{C'})(R_1 - R_2) \qquad \text{(iv)}.$$

Now dividing (iii) by (iv) gives

$$\frac{F_d}{F_{F'} - F_{C'}} = \frac{(n_d - 1)(R_1 - R_2)}{(n_{F'} - n_{C'})(R_1 - R_2)} = \frac{n_d - 1}{n_{F'} - n_{C'}}.$$

But, from equation (6.5a), $\dfrac{n_d - 1}{n_{F'} - n_{C'}} = V$, the constringence or V-number of the material.

Hence, $\dfrac{F_d}{F_{F'} - F_{C'}} = V$, which on rearranging gives

$$\text{LCA} = F_{F'} - F_{C'} = \frac{F_d}{V} \qquad (7.2).$$

Dropping the subscript d, so that F implies the mean power of the lens, we can write

$$\text{LCA} = \frac{F}{V} \qquad (7.2\text{a}).$$

Thus, the longitudinal chromatic aberration = $\dfrac{\text{mean power of lens}}{\text{V-number of material}}$. Equation (7.2) also applies to a diverging lens. Hence, from equation (7.2a), we can see that a +10 D lens made in glass with V = 50 will have $\dfrac{F}{V} = \dfrac{+10}{50}$ = +0.2 D of LCA.

When the red focus is to the right of the blue focus, as with a converging lens, the LCA is said to be positive. Conversely, a diverging lens will produce negative LCA.

Let us now consider the transverse chromatic aberration with a single thin lens. Referring back to figure 7.2, let the ray of white light be incident at a height c from the optical axis of the lens. We can use Prentice's rule, equation (3.18), to express the prismatic effects of the lens on the red and blue lights separately. Thus, for red (C') light, the prismatic effect of the lens is $P_{C'} = cF_{C'}$, and for blue (F') light the prismatic effect of the lens is $P_{F'} = cF_{F'}$. If we take the angle α as the measure of the angular TCA, then in prism dioptre notation

$$\text{angular TCA} = P_{F'} - P_{C'} .$$

That is, angular TCA $= cF_{F'} - cF_{C'} = c(F_{F'} - F_{C'})$. But from equation (7.2a), $F_{F'} - F_{C'} = \frac{F}{V}$, so we have

$$\text{angular TCA} = \frac{cF}{V}^{\Delta} \qquad (7.3).$$

Thus, the angular transverse chromatic aberration at a point on a lens is given by dividing the mean prismatic effect, cF, at that point by the V-number of the lens material. For example, a +10 D lens made in glass with V = 50 will possess angular TCA at the point 1.5 cm from the optical centre given by $\frac{cF}{V} = \frac{1.5 \times 10}{50} = 0.3^{\Delta}$.

As we have already stated, TCA may alternatively be represented as linear TCA, in which the aberration is taken to be the vertical difference in the image heights $h'_{red} - h'_{blue}$, see figure 7.2 . More precisely, linear TCA $= h'_{C'} - h'_{F'}$, where the subscripts C' and F' refer to the specific red and blue colours.

Now, in general, using equation (3.10), the linear magnification of an image is given by $\frac{h'}{h} = \frac{L}{L'} = \frac{L}{L'}$ for a lens surrounded by air. Thus, since h and L are constant for all colours, but h' and L' vary with colour, we have for the C' and F' images

$$h'_{C'} = \frac{hL}{L'_{C'}} \qquad \text{and} \qquad h'_{F'} = \frac{hL}{L'_{F'}} .$$

Therefore, $\quad \text{linear TCA} = h'_{C'} - h'_{F'} = \frac{hL}{L'_{C'}} - \frac{hL}{L'_{F'}} = hL(\ell'_{C'} - \ell'_{F'}) \qquad (7.4)$

7.2.2 CORRECTION FOR CHROMATIC ABERRATION

We will now show that it is possible to form a combination of two thin lenses such that the focal length is the same for the C' and F' wavelengths. This will result essentially in a correction for chromatic aberration.

7.2.2.1 Two Thin Lenses in Contact

We recall from figure 7.1 that, using white light, the red focus is to the right of the blue focus in the case of a converging lens, but the reverse is true for a diverging lens. This suggests that two thin lenses, one converging and the other diverging, may be combined in contact with one another such that the dispersive effects are cancelled, but the combination still retains some required power. With reference to the red (C') and blue (F') colours, such a combination is called an *achromatic doublet lens*, and is said to be *achromatised* for these two specific wavelengths. There is a direct comparison here between an achromatic doublet lens and an achromatic doublet prism. This latter device was discussed in section 6.4.2 . In order to remove the dispersive effects of individual lenses, but at the same time retain some necessary power, the two components in an achromatic doublet are made of different glasses, usually crown glass and flint glass.

Figure 7.3 shows a typical cemented achromatic doublet with a positive equivalent power for which the contact surfaces have a common curvature. The crown glass component lens will have a larger positive power, but will have the same dispersion as the flint glass component, for which the power is smaller and negative.

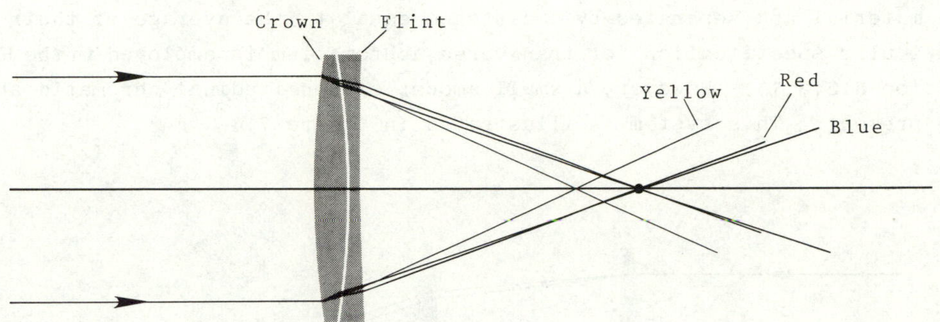

Fig. 7.3 A cemented achromatic doublet.

From equation (7.2a), the longitudinal chromatic aberration is given by LCA $= \frac{F}{V}$, where F is the mean power of the lens and V is the constringence of the material. Since, with respect to the component lenses of an achromatic doublet, the individual chromatic aberrations have to be equal and opposite, we can write

$$(LCA)_{Lens(1)} = -(LCA)_{Lens(2)}.$$

That is,

$$\frac{F_1}{V_1} = -\frac{F_2}{V_2} \qquad (7.5)$$

where F_1 and F_2 are the mean powers for the component lenses, and V_1 and V_2 are the respective constringences of the materials. Also, for two thin lenses in contact the equivalent power is given by equation (3.17)

$$F_E = F_1 + F_2 \qquad (3.17).$$

Rearranging equation (7.5) we get

$$F_2 = -\frac{V_2}{V_1} F_1 .$$

Substituting for F_2 in equation (3.17) we get

$$F_E = F_1 + \left(-\frac{V_2}{V_1} F_1\right) = F_1\left(1 - \frac{V_2}{V_1}\right),$$ from which we have

$$F_1 = \frac{F_E V_2}{V_1 - V_2} \qquad (7.6).$$

Equations (3.17) and (7.6) form the basis of calculations on achromatic doublets. The mean powers of the component lenses, F_1 and F_2 , and the V-numbers of their materials V_1 and V_2 , must satisfy these equations for the doublet to be achromatised for the C' and F' rays (see worked examples following section 7.2.2.2). It must be pointed out that the doublet is not entirely free of chromatic aberration since colours other than the C' and F' rays will be focused separately. In figure 7.3 the yellow focus point is shown to be a short distance from the combined focus point for red and blue. This residual chromatic error is referred to as a *secondary spectrum*. This was discussed in section 6.4.2.1 in connection with achromatic prisms and applies similarly to achromatic lenses.

7.2.2.2 Two Separated Thin Lenses

An alternative method of obtaining an achromatic system is by using two thin lenses made of the same material and separated by a distance equal to the average of their focal lengths. This particular specification for transverse achromatism is employed in the Huygens' eyepiece (see section 8.3.2.1). However, a small amount of longitudinal chromatic aberration will still be present. This system is illustrated in figure 7.4 .

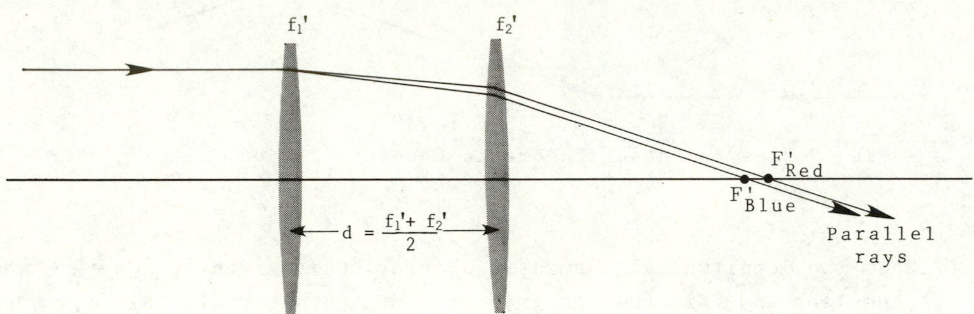

Fig. 7.4 Angular achromatism in a pair of separated thin lenses.

Using equation (4.17) for two separated thin lenses in air, we have

$$\text{equivalent power } F_E = F_1 + F_2 - dF_1F_2 .$$

But, F_1 may be written $F_1 = (n_g-1)(R_1-R_2) = (n_g-1)K_1$, where $K_1 = (R_1-R_2)$ is constant. Applying the constant K_2 to the second lens and noting that the refractive index is the same, we can now write

$$F_E = (n_g-1)K_1 + (n_g-1)K_2 - d(n_g-1)(n_g-1)K_1K_2$$

which simplifies to give

$$F_E = (n_g-1)(K_1+K_2) - d(n_g-1)^2K_1K_2 \ .$$

Now, the value of n_g varies with the colour of the light used, and for achromatism there must be no variation of the power of the system with index. That is, for achromatism we must have $\frac{dF_E}{dn_g} = 0$. Hence, differentiating and equating to zero, we obtain

$$\frac{dF_E}{dn_g} = K_1 + K_2 - 2d(n_g-1)K_1K_2 = 0$$

which is the condition for achromatism. Multiplying each term in the right hand equation by (n_g-1) we have

$$(n_g-1)K_1 + (n_g-1)K_2 - 2d(n_g-1)(n_g-1)K_1K_2 = 0 .$$

This may now be written $F_1 + F_2 - 2dF_1F_2 = 0$ which rearranges to give

$$d = \frac{F_1 + F_2}{2F_1F_2} \qquad (7.7).$$

This is the condition for achromatism with separated lenses. Expressing equation (7.7) in focal length notation, we have

$$d = \frac{\frac{1}{f_1'} + \frac{1}{f_2'}}{\frac{2}{f_1'f_2'}} = \frac{\frac{f_1' + f_2'}{f_1'f_2'}}{\frac{2}{f_1'f_2'}}$$

or $$d = \frac{f_1' + f_2'}{2} \qquad (7.7a).$$

This confirms the statement that two thin lenses, of the same material, separated by the average of their focal lengths, form a system which is achromatic.

Worked Examples

1 An achromatic lens, having a focal length of +20 cm is to be made of crown and flint glass components, cemented in contact, with the following constants

	$n_{C'}$	n_d	$n_{F'}$
Crown	1.520	1.523	1.529
Flint	1.612	1.617	1.629

If the crown glass lens is to be equiconvex, calculate (a) the V-values of the materials used, (b) the mean powers of the component lenses, (c) the radii of curvature of the four surfaces to correct for the C' and F' lines.

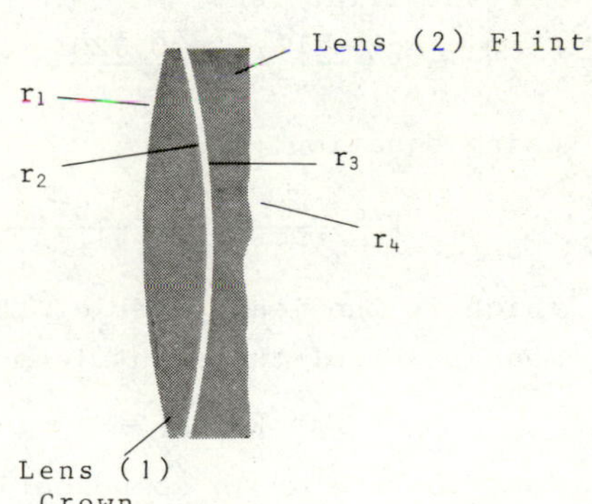

At the moment the form of surface (4) is not known.

For the crown glass $V_1 = \frac{n_d - 1}{n_{F'} - n_{C'}} = \frac{1.523 - 1}{1.529 - 1.520} = 58.1$.

For the flint glass $V_2 = \frac{n_d - 1}{n_{F'} - n_{C'}} = \frac{1.617 - 1}{1.629 - 1.612} = 36.3$.

Using equation (7.6), where F_E = power of doublet $= \frac{1}{+0.20} = +5$ D.

$$F_1 = \frac{F_E V_1}{V_1 - V_2} = \frac{5 \times 58.1}{58.1 - 36.3} = +13.33 \text{ D}.$$

This is the mean power of the crown component. The mean power of the flint component is obtained from equation (3.17):

$$F_2 = F_E - F_1 = +5.00 - (+13.33) = -8.33 \text{ D}.$$

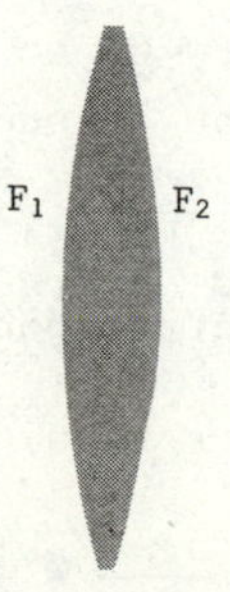

Now, for the crown component as a separate lens, since the form of the crown lens is equiconvex, the power of the first surface (F_1) is given by $F_1 = \frac{+13.33}{2} = +6.67$ D.

For the first surface we have $r_1 = \frac{n_g - 1}{F_1} = \frac{1.523 - 1}{+6.67}$ m = +78.4 mm,

where n_g is the mean refractive index of the crown lens. Hence, $r_2 = -r_1 = -78.4$ mm, and because of the common contact surface between the two components, we have $r_3 = r_2 = -78.4$ mm.

Considering the flint lens as a separate lens we have $F_3 = \frac{n_g - 1}{r_3} = \frac{1.617 - 1}{-0.0784} = -7.87$ D.

But, the mean power of the flint lens is -8.33 D. Therefore, the power of surface (4) is $F_4 = -8.33 - (-7.87) = -0.46$ D. Hence, the radius of curvature of the fourth surface is

$$r_4 = \frac{1 - n_g}{F_4} = \frac{1 - 1.617}{-0.46} = +1.341 \text{ m}.$$

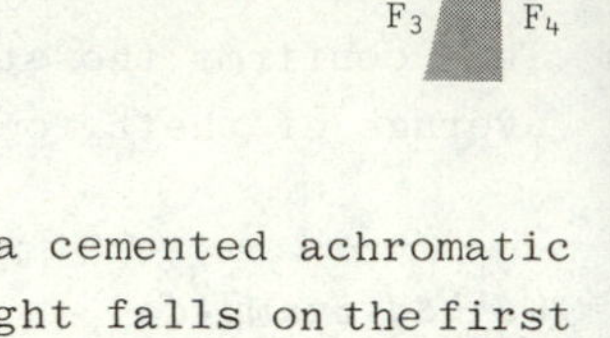

Thus, listing the four surface radii and powers, we have:

For the crown lens $r_1 = +0.0784$ m and $F_1 = +6.67$ D
$r_2 = -0.0784$ m and $F_2 = +6.67$ D

and for the flint lens $r_3 = -0.0784$ m and $F_3 = -7.87$ D
$r_4 = +1.341$ m and $F_4 = -0.46$ D.

2 Calculate the radii of curvature of the surfaces required to produce a cemented achromatic doublet using the glasses specified below, such that when parallel light falls on the first surface of the crown component, which is plane, it leaves the doublet with a divergence of 4 D.

	Refractivity	Mean dispersion (C' to F')
Crown	0.520	0.009
Flint	0.620	0.016

For the doublet $L = 0$, $L' = -4$ D, so $F = F_E = L' - L = -4$ D.

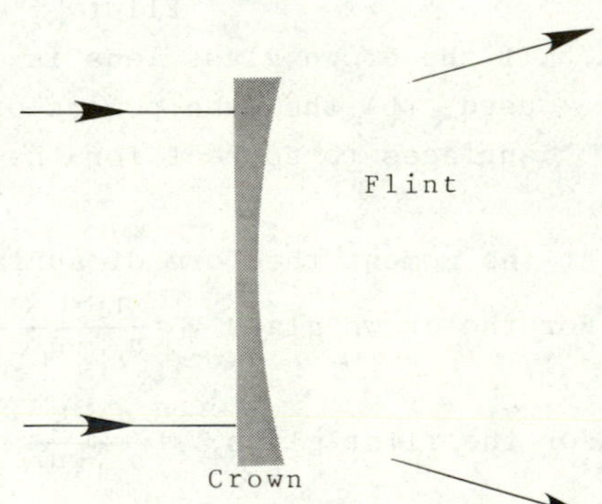

For the crown lens

$$V_1 = \frac{\text{refractivity}}{\text{mean dispersion}} = \frac{n_d - 1}{n_{F'} - n_{C'}} = \frac{0.520}{0.009} = 57.8\,.$$

For the flint lens

$$V = \frac{n_d - 1}{n_{F'} - n_{C'}} = \frac{0.620}{0.016} = 38.8\,.$$

Using equation (7.6)

$$F_1 = \frac{F_E V_1}{V_1 - V_2} = \frac{-4 \times 57.8}{57.8 - 38.8} = -12.17 \text{ D},$$

which is the mean power of the crown lens. The mean power of the flint lens is

$$F_2 = F_E - F_1 = -4 - (-12.17) = +8.17 \text{ D}.$$

Now, for the crown component as a separate lens, the mean power F is given by equation (3.2); that is, $F = F_1 + F_2$ where the subscripts refer to the first and second surfaces of this lens. But $F_1 = 0$, so $F_2 = -12.17$ D. Hence, the radius of the second surface is

$$r_2 = \frac{1 - n_g}{F_2} = \frac{1 - 1.520}{-12.17} = +0.042\,7 \text{ m}.$$

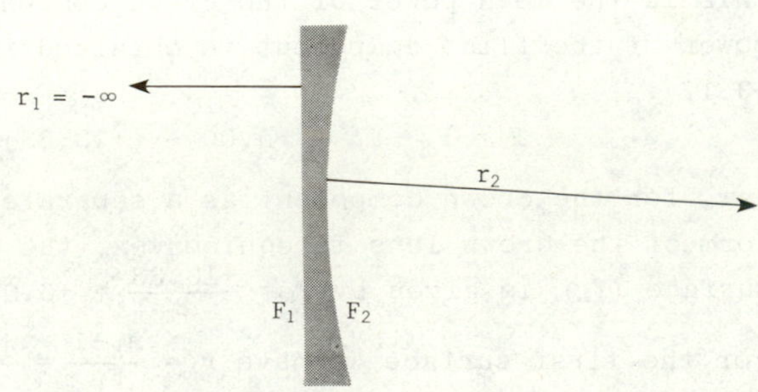

For the flint component as a separate lens, we have

$$r_3 = r_2 = +0.042\,7 \text{ m}$$

whence

$$F_3 = \frac{n_g - 1}{r_3} = \frac{1.620 - 1}{+0.0427} = +14.52 \text{ D}.$$

(Note that F_2 is not the same value as F_3, as the materials have different indices). But the mean power of the flint lens is +8.17 D, therefore $F_4 = +8.17 - F_3 = +8.17 - (+14.52) = -6.35$ D, and we have

$$r_4 = \frac{1-n_g}{F_4} = \frac{1-1.620}{-6.35} = +0.097\,6 \text{ m}.$$

Thus, the four surface radii and powers are:

For the crown component $r_1 = -\infty$ and $F_1 = 0$
$r_2 = +42.7$ mm and $F_2 = -12.17$ D
and for the flint component $r_3 = +42.7$ mm and $F_3 = +14.52$ D
$r_4 = +97.6$ mm and $F_4 = -6.35$ D.

7.3 THE MONOCHROMATIC ABERRATIONS

We have previously derived the Fundamental Paraxial Equation, equation (2.25), and other formulae relating to lenses, such as equations (3.3), (3.4), and (3.5), based on the assumption that sin i could be satisfactorily represented by i alone (in radians). This restriction to paraxial rays means we are limiting ourselves in the use of these equations to a narrow region about the optical axis of the system. The equations referred to above form the basis of what is usually called *First Order* theory. The reason for using this term will shortly become apparent. First order theory is also called *Gauss* or *Gaussian*, after Carl Friedrich Gauss, the German physicist and mathematician who first developed it and published his results in 1841. If we now extend the use of a lens to include rays passing through its peripheral zones, for which larger inclinations of rays with the axis are involved, the statement that $\sin i \simeq i$ is no longer valid. In fact, the sine of an angle is expressed in terms of the angle itself (in radians) by Maclaurin's expansion, which is

$$\sin i = i - \frac{i^3}{3!} + \frac{i^5}{5!} - \frac{i^7}{7!} + \ldots\ldots \qquad (7.8)$$

where 3! is read as "3 factorial" and means $3 \times 2 \times 1$, and $5! = 5 \times 4 \times 3 \times 2 \times 1$, and so on.

For small angles, each term in the above series is much smaller than the preceding term and, for paraxial rays, all terms after the first may be neglected. This is why paraxial theory is sometimes called first order theory, with only the first-order term in i being allowed for. If calculations are attempted using the first two terms of equation (7.8) instead of just the first term, the results are valid for rays of light having a larger range of inclinations with the lens axis than paraxial rays. This improved approximation to the sine of an angle is termed *Third Order* theory, since third order terms in i (that is i^3) have been used in the approximation. This leads to an image position different to that obtained with paraxial rays. Departures from first order theory which consequently occur are embodied in the five *primary aberrations* which are:

spherical aberration, coma, oblique astigmatism, Petzval curvature, and distortion.

These defects of an image are also known as *monochromatic* aberrations because they are present when monochromatic light is considered. They were investigated principally by Ludwig von Seidel in the middle of the 19th century, and are often referred to as *Seidel aberrations*.

Seidel showed that the deviations in the path taken by a ray from that predicted by paraxial theory could be satisfactorily expressed by five terms or coefficients called the *Seidel Terms* or *Seidel Sums*. These are denoted by S_1, S_2, S_3, S_4, and S_5, and respectively represent the contributions made by the five monochromatic aberrations in the order S, C, A, P, and D.

The mathematics of aberration theory is quite involved, but a brief appreciation of the Seidel terms would be useful. Thus, we can say that when a bundle of rays from a small object point fails to unite at a single image point, the emergent wavefronts from the refracting system will not be truly spherical. It has been found convenient to calculate the effects of the primary aberrations in terms of the amount by which the emergent wave differs from the perfect spherical shape. This difference is referred to as the *wavefront aberration* and its form may be deduced from symmetry considerations in each pencil of rays. This results in an aberration function which contains a series of terms which are linked with certain recognisable defects in the image. The first five of these terms are the Seidel terms which we referred to above. If the image formed by a lens is to be free of all defects, the Seidel terms $S_1, S_2, \ldots S_5$ would all have to be equal to zero. This will never be the case since no optical system can be designed to simultaneously satisfy all these conditions. However, the absence of one particular aberration corresponds to one of the Seidel terms being zero. Thus, it may be stated that if $S_1 = 0$, the image is free from spherical aberration, whilst if $S_1 = 0$ and $S_2 = 0$, coma will also be absent. When $S_3 = 0$ and $S_4 = 0$ the image will be free from the effects of oblique astigmatism and Petzval curvature. Finally, for $S_5 = 0$ there would be no distortion in the image.

It can be seen that equation (7.8) contains other terms in i^5, i^7, and so on, and that greater accuracy in the theory will involve other higher order aberrations. We shall restrict the treatment in the remainder of this chapter to the primary aberrations only, for which third order theory gives a reasonably accurate account.

7.3.1 SPHERICAL ABERRATION

The effect known as spherical aberration has been discussed briefly in connection with refractions at single spherical surfaces (see section 2.2.2) and this indicates that it may occur with lenses.

Let us now consider, in figure 7.5(a), a pencil of incident rays parallel to the axis of a thin converging lens. F_p' is the second principal focal point for paraxial rays. The outer zones of the lens transmit non-paraxial and marginal rays which do not intersect at F_p' but, instead, cross the axis to the left of F_p'. This effect is most pronounced for rays passing through the outermost marginal zone, these rays focusing at F_m'.

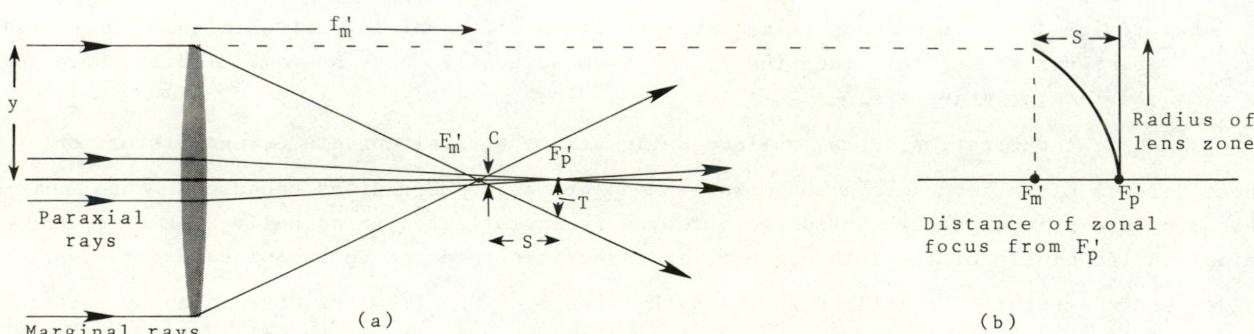

Fig. 7.5 Longitudinal and transverse spherical aberration.

The *Longitudinal Spherical Aberration* is defined as the axial separation of the paraxial and marginal focal points. In figure 7.5(a) this is represented by the distance S. In figure 7.5(b) a graphical illustration shows how the magnitude of longitudinal spherical aberration varies for different zones of the lens. The spherical aberration is said to be positive when, as shown, the marginal rays intersect the axis to the left of the paraxial focus, and the lens is said to be spherically undercorrected. A diverging lens would produce negative spherical aberration with F_m' on the right of F_p'. Note that in this case F_m' and F_p' are both virtual points on the left hand side of the lens. In this case the lens is described as spherically overcorrected.

If an object point off the axis is considered, the same longitudinal separation exists between the image point formed by marginal rays and by paraxial rays.

The magnitude of spherical aberration may be determined by an accurate ray trace procedure, as will be described in section 7.3.6.3 . To give satisfactory results by this method, calculations must be conducted with an accuracy of at least five decimal places. On the other hand, calculations based on third order theory may be undertaken, in which $\sin i \simeq i - \frac{i^3}{3!}$, with i in radians. Results of such calculations show that the longitudinal spherical aberration is dependent on the square of the radius of the lens aperture. That is,

$$S = ay^2 \qquad (7.9)$$

where a is a constant, and y is the radius of the lens aperture.

In this instance, S represents the primary longitudinal spherical aberration. When allowance is made for aberrations of higher order, it is found that S depends on the radius of the lens aperture given by

$$S = ay^2 + by^4 + cy^6 + \ldots.. \qquad (7.9a),$$

where a, b, and c are constants. The second and subsequent terms in equation (7.9a) are due to the inclusion of higher order terms in the expansion of sin i, and are called the secondary, tertiary,.... spherical aberrations.

Sometimes a better measure of spherical aberration may be obtained in terms of the *Transverse Spherical Aberration* of the image. In figure 7.5(a), this is denoted by the distance T, which is the radius of the circular patch of light formed in the plane of the paraxial focus F_p'. It can easily be shown that the primary transverse spherical aberration is dependent on the cube of the radius of the aperture. Thus, from similar triangles in figure 7.5(a),

$$\frac{T}{S} = \frac{y}{f_m'}, \quad \text{or} \quad T = \frac{Sy}{f_m'}. \quad \text{But,} \quad S = ay^2, \text{ therefore } T = \frac{ay^3}{f_m'}$$

$$\text{or} \quad T = ky^3 \qquad (7.10)$$

where k is a constant.

In referring to the "best" image position we should note the position in the refracted pencil of rays where the beam has the smallest diameter. This is the circle of least confusion, and is denoted by C in figure 7.5(a). Using third order theory, it can also be shown that the radius of C depends on the cube of the radius of the lens aperture.

A single lens, having two spherical surfaces, can never form an image which is free from spherical aberration. The defect can be reduced considerably by restricting the transmission of light by the lens to the paraxial region only. This may be achieved by the use of a stop placed near to the lens. However, this simple remedy causes a decrease in the amount of light transmitted to the image, and almost always introduces another defect of the image, namely distortion (see section 7.3.5).

Minimum spherical aberration with a lens may be achieved by suitable choice of the radii of curvature of the surfaces. "Bending" a lens means an alteration in the two surface radii whilst retaining a specified power for the lens, and it has been found that the amount of spherical aberration for a lens of given power and aperture depends on the object distance and the lens form. A good example of this is illustrated in figures 7.6(a) and (b), which shows a lens of given power, in plano-convex form, placed with each surface in turn facing the incident light. The monochromatic incident light is shown from an axial object point at infinity. The focal length, as calculated from equation 3.5(a), is the same whether the plane surface or the convex surface faces the incident light, if the lens is thin.

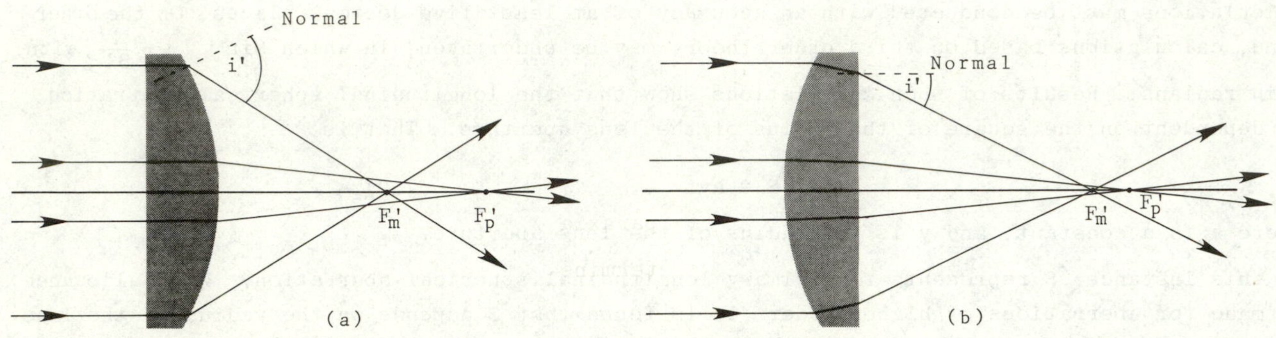

Fig. 7.6 Spherical aberration with a plano-convex lens.

The magnitude of the spherical aberration is much smaller in figure 7.6(b) than in figure 7.6(a). It is evident from these diagrams that, in (a) where the incident light meets the plane face normally, all the deviation of the rays is produced at the curved surface, whilst in (b) some deviation occurs at each surface. Consequently, in (a), the angle i' between the emergent ray and the normal at the point of emergence would be large enough for higher powers in the expansion of $\sin i'$ to be significant. As a general guide, spherical aberration will be minimised if the refraction by the lens is equally shared between the two surfaces.

For an infinitely distant object, the form of the lens made in glass of refractive index 1.5, for which spherical aberration is a minimum, is the biconvex form in which the radii of the surfaces are in the ratio 1:6 . This form of lens is usually referred to as a *crossed lens*, and it is used with the more curved surface (smaller radius) towards the object.

Similarly, if the object and image distances are equal, the best form of lens for minimum spherical aberration is the equiconvex form. A combination of a converging and a diverging lens can also be used to reduce spherical aberration.

It should be carefully noted that the reduction of spherical aberration to a minimum value will only be applicable to one pair of conjugate object and image positions. If spherical aberration is minimised in a system for a selected object position, the defect would appear with increased magnitude for other object positions. It should be mentioned that spherical aberration can be removed by the use of aspherical surfaces, but this is more expensive.

7.3.1.1 *Spherical Aberration and a Single Surface: Aplanatic Points*

In section 2.2.2, we saw that a spherical refracting surface of wide aperture separating two different transparent media formed an image which was affected by spherical aberration. The discussion which then followed showed that a point image of a point object is produced only by paraxial rays. There are, however, three special positions of a point object such that a point image is formed for all rays of light refracted by the surface. Thus, for each of these three object positions, spherical aberration of the image is completely removed. The first case is when the object is located at the vertex of the suface. Consequently, the image is in the same position, and all incident and refracted rays, at all inclinations to the axis, will pass through this point. The second case is when the object point is located at the centre of curvature of the surface, as illustrated in figure 7.7 . All rays of light passing through C will be along normals to the surface and will be undeviated on refraction.

The third case of zero spherical aberration is less obvious than the previous two, and has a number of important applications. It concerns a special pair of object and image points for which a surface or optical system is free of spherical aberration. This pair of conjugate points are called *aplanatic points* and the surface is said to be *aplanatic*. Strictly, the term aplanatic is applied to a surface which produces an image free from the effects of both spherical aberration and coma. Coma will be considered in the next section.

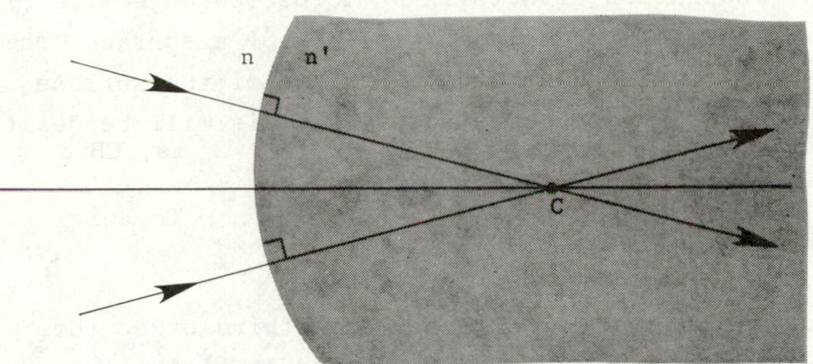

Fig. 7.7 A pencil of rays directed towards the centre of curvature of a spherical surface all pass through C.

In figure 7.8, let B and B' denote the positions of a pair of conjugate points for a spherical refracting surface where the ray directed towards B (virtual object) before refraction becomes the ray passing through B' (real image) after refraction.

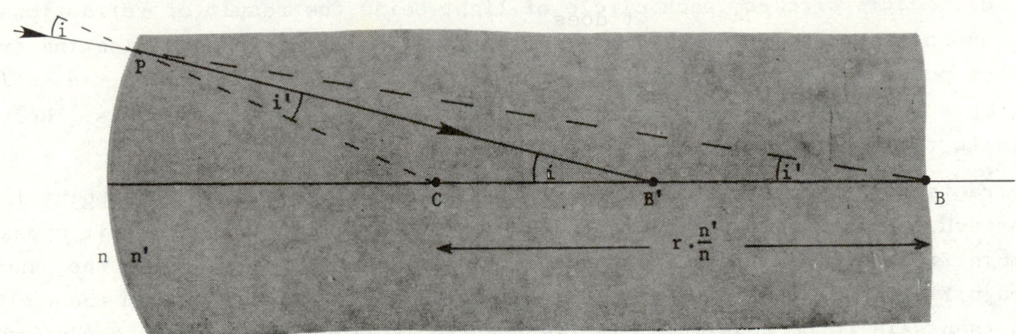

Fig. 7.8 B and B' are aplanatic points.

In particular, let B be the point such that $CB = \frac{n'}{n}.CP$, where CP is the radius of the surface.

That is, $\frac{CB}{CP} = \frac{n'}{n}$ ......(i). Now, in triangle PCB, using the sine rule,

$$\frac{CB}{\sin C\hat{P}B} = \frac{CP}{\sin P\hat{B}C}, \text{ or } \frac{CB}{CP} = \frac{\sin C\hat{P}B}{\sin P\hat{B}C} \quad ....(ii).$$

Now, from (i) and (ii), $\frac{\sin C\hat{P}B}{\sin P\hat{C}B} = \frac{n'}{n}$ and $C\hat{P}B = i$. Thus, it follows by reference to $n \sin i = n' \sin i'$, that $P\hat{B}C = i'$. Now, triangles CPB and CPB' are similar since the angle C is common to both triangles and each triangle contains the angle i'. Therefore $P\hat{B'}C = C\hat{P}B = i$. Hence,

$$\frac{CB'}{CP} = \frac{CP}{CB} = \frac{n}{n'}, \text{ from (i); that is, } CB' = CP.\frac{n}{n'} = \frac{n}{n'}.r = \text{constant} \quad (7.11).$$

Thus, as no assumptions have been made in this deduction, we can say that B' is the same image point for all rays of light directed towards B. That is, the result is independent of the position of P on the surface. B and B' are termed aplanatic points with the image at B' being completely free of spherical aberration. This result is applied to the reduction of aberration effects in a microscope objective of the oil-immersion type. We shall also make use of the aplanatic points of a surface, that is, the points at distances $\frac{n'}{n}.r$ and $\frac{n}{n'}.r$ from the centre of curvature of the surface, in an example of geometrical ray tracing known as Young's construction. This will be dealt with in section 7.3.6.1 .

7.3.2 COMA

The second of Seidel's terms in third order theory refers to the image defect known as coma. Even though a lens may be shaped to minimise spherical aberration for one pair of axial object and image points, so that all the light from the object point essentially passes through a single image point after refraction, it does not necessarily follow that light from an object point adjacent to the first, but off the axis, will similarly pass through a single image point after refraction. Interestingly, coma has been called off-axis spherical aberration. In fact, it can be said that spherical aberration and coma arise from the failure of the lens to image paraxial rays and marginal rays at the same point. Coma differs from spherical aberration, however, in that a point object is imaged not as a circle of least confusion but as a comet-shaped patch, from which is derived the name coma. The comet-shaped patch is made up of a series of *comatic circles*, each circle of light being the result of refractions at a particular zone of the lens. Figure 7.9 illustrates a pencil of rays originating from a very distant object point which is above the axis of the lens. Only rays in the *tangential plane* are shown. This is the plane containing the chief ray and the principal axis. Refractions are confined to three zones of the lens for simplicity.

The rays refracted by the central, axial zone of the lens unite to form the point image T'_p. The rays through the marginal zone unite to form the point image T'_m. Thus, it appears that the image magnification is different for the various zones of the lens. In the example shown, the least magnification is associated with the marginal rays and these form the smallest image. The coma is then said to be negative. If the reverse is true, the coma is described as being positive.

In order to understand why the image is not simply a line $T'_m T'_p$, it is necessary to consider rays incident at the lens in other planes in addition to the tangential plane. It is usual to define the *sagittal plane* as the plane containing the chief ray but being perpendicular to the tangential plane.

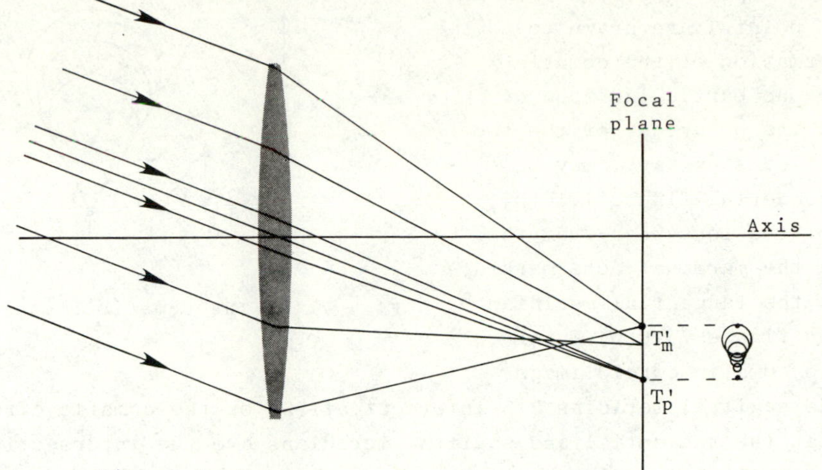

Fig. 7.9 Coma.

T

Lens

(a)

Axis

(1)

(3)

T'_p

Fig. 7.10

1

4 2

(b)

Aperture
2y

3 3

2 4

1

Marginal zone
Plane of lens

(1)

(4) (2) (c) Comatic circle

(3)

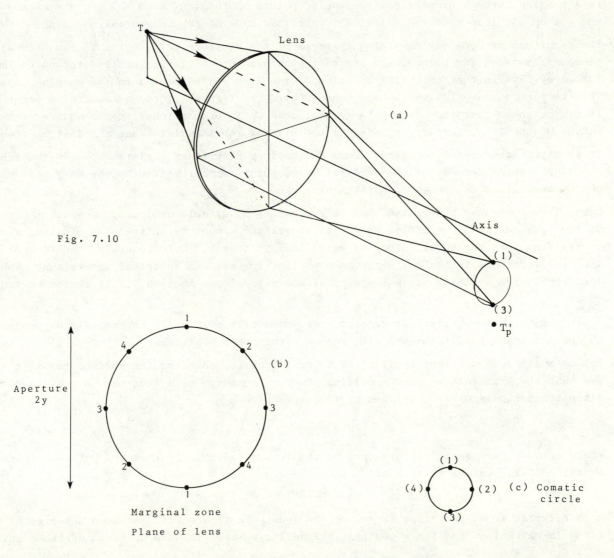

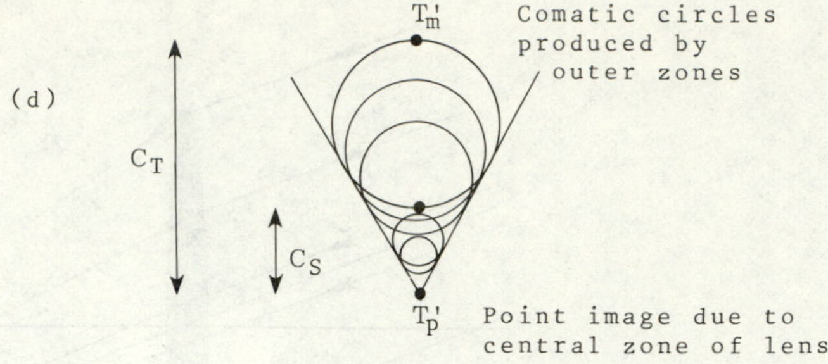

Fig. 7.10 The comatic image of a point.

In figure 7.10(a) two pairs of rays from an above-axis object point T are drawn to illustrate the formation of the comatic circle image, for one particular zone of a converging lens.The behaviour of the two pairs of rays, called skew rays, may be understood by considering figure 7.10(b), which is a view of the lens seen from the object side, with the marginal zone marked. Rays refracted at the tangential meridian of the lens 1,1 in figure 7.10(b) intersect at (1) on the comatic circle image, whilst rays in the sagittal meridian 3,3 intersect at (3) on the comatic circle. see figure 7.10(c). Note that the tangential and sagittal meridians are the intersections of the tangential and sagittal planes with the lens.

Similarly, other points (2) and (4) on the comatic circle are formed by the intersections of rays passing through diametrically opposite points on the lens zone. These are marked 2,2 and 4,4 on the lens zone and are at 45^0 on either side of the tangential meridian.

The magnitude of coma may be stated in terms of the distance $T'_pT'_m$, which is the lateral displacement between the point image due to the paraxial zone of the lens and the point image formed by rays in the tangential meridian of the lens and refracted in the marginal zone. This is termed *tangential coma*, and denoted by C_T in figure 7.10(d). From third order theory it can be shown that the primary tangential coma C_T is proportional to the square of the radius of the lens aperture, y^2, and to the size of the paraxial image h'. That is, $C_T \propto y^2h'$.

It is often more useful to express coma in the form of primary sagittal coma, C_S, which is the displacement from T'_p of the sagittal image point for refraction by the marginal zone. It can be shown that $C_T = 3C_S$ for a particular zone.

Coma, like spherical aberration, may be reduced by a suitable choice of the radii of curvature of the lens surfaces. Although spherical aberration cannot be entirely eliminated, a particular lens form may be selected for which coma is absent. Unfortunately, the form of the lens for zero coma is not quite the same as that for minimum spherical aberration. Thus, a lens form which exhibits minimum spherical aberration (see section 7.3.1) cannot be free from coma.

As in the case of spherical aberration, zero coma will only apply to one pair of conjugate object and image positions, and will reappear for other positions.

Let us again consider the "bending" of a lens, that is, changing the surface radii in such a way that the lens retains some specified power. The shape of a lens may be expressed by the dimensionless parameter σ, defined by the expression

$$\sigma = \frac{r_2 - r_1}{r_2 + r_1} \qquad (7.12)$$

where r_1 and r_2 are radii of curvature of the lens surfaces. In terms of suface curvatures equation (7.12) may be written as

$$\sigma = \frac{R_1 + R_2}{R_1 - R_2} \qquad (7.12a).$$

σ is referred to as the *shape factor* of the lens. In figure 7.11 is shown a series of lens forms in which the lens has a constant refracting power. This is achieved during bending

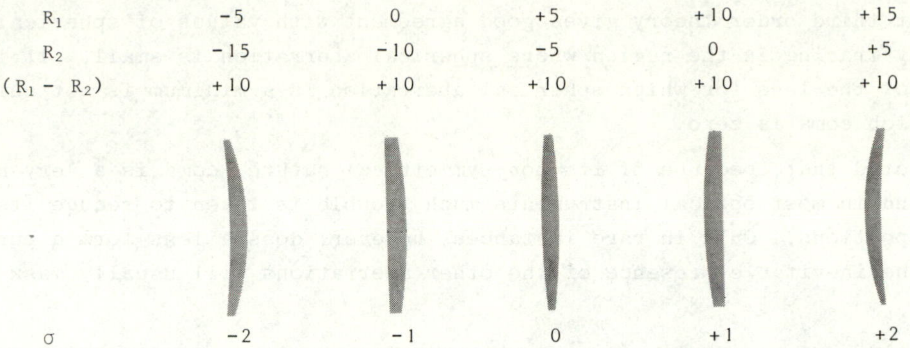

R_1 -5 0 +5 +10 +15
R_2 -15 -10 -5 0 +5
$(R_1 - R_2)$ +10 +10 +10 +10 +10
σ -2 -1 0 +1 +2

Fig. 7.11 The shape factor σ for a series of thin lenses of the same power but with different forms.

if the term $(R_1 - R_2)$ in equation (3.5b) remains constant. It will be seen that the equiconvex form (or equiconcave form) has zero σ, whilst the plano-convex form has $\sigma = \pm 1$. Numerical values of σ greater than 1 indicate meniscus forms.

Figure 7.12 shows how coma varies linearly with the shape factor for a thin lens. It also illustrates the spherical aberration curves resulting from trigonometrical ray tracing, together with the results obtained by application of third order calculations. The numerical values indicated should only be taken as a guide, since they depend on the exact conditions such as aperture, object position and height, refractive index, etcetera which apply, and these are not being stated.

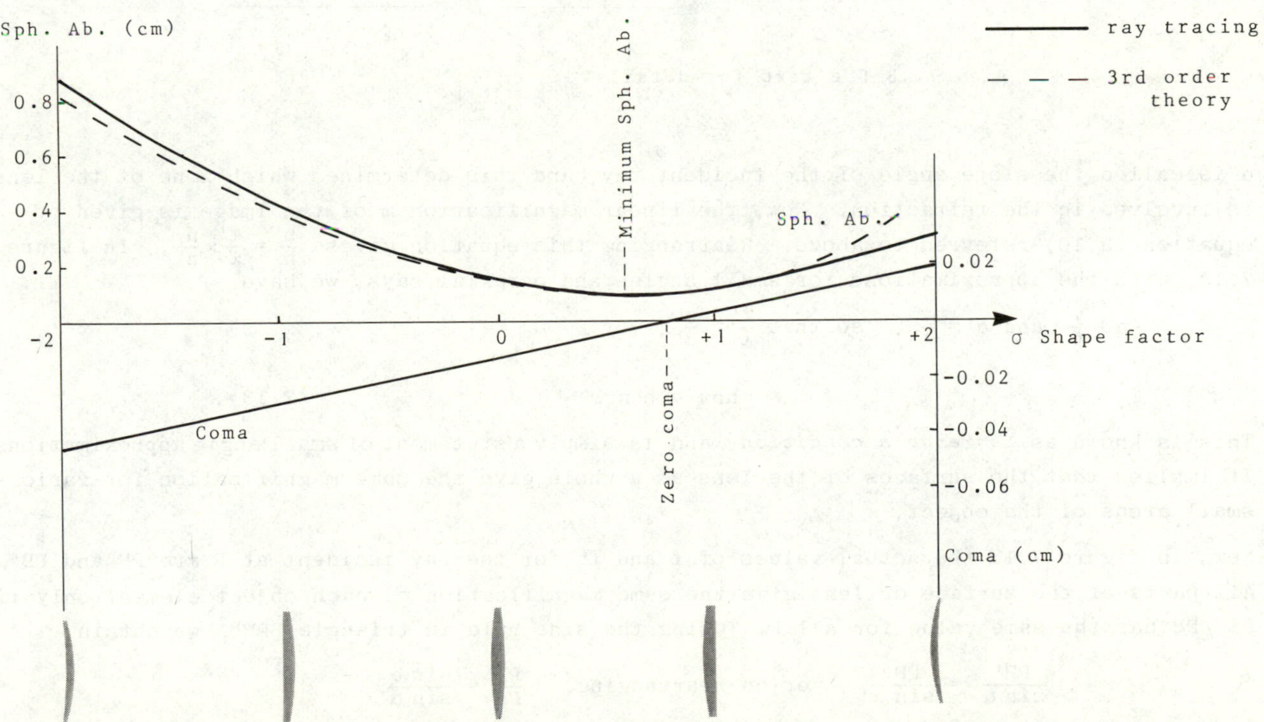

Fig. 7.12 The effect of shape factor on spherical aberration and coma.

It is seen that third order theory gives good agreement with values of spherical aberration obtained by ray tracing in the region where spherical aberration is small. It is also clear that the form of the lens for which spherical aberration is a minimum is not quite the same as that for which coma is zero.

It should be noted that, because of its non-symmetrical nature, coma is a very undesirable image defect and in most optical instruments much trouble is taken to reduce its effects to negligible proportions. Only in rare instances, however, does a lens form a purely comatic image, since the inevitable presence of the other aberrations will usually mask the comatic image shape.

Let us now consider the condition for the absence of coma with a single thin lens. This may be done conveniently by reference to a single refracting surface and equation (2.28); that is $m = \frac{h'}{h} = \frac{\ell'/n'}{\ell/n}$. This equation also applies to a thin lens, with the undashed symbols referring to quantities in the object space and the dashed symbols referring to those in the final image space (see equation (3.10)).

In figure 7.13, let us consider a small element of an object h which is imaged at h', where the refractive indices of the object and image spaces are n and n', respectively.

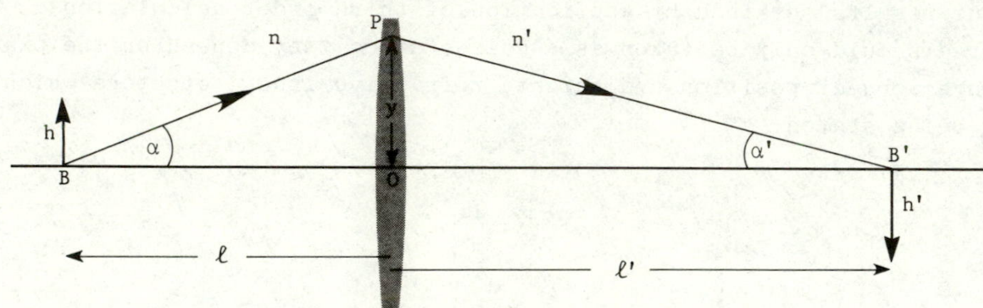

Fig. 7.13 See text for details.

α is called the slope angle of the incident ray, and this determines which zone of the lens is involved in the refraction. Now, the linear magnification m of the image is given by equation (3.10) referred to above. Rearranging this equation gives $\frac{h'}{h} = \frac{\ell'}{\ell} \times \frac{n}{n'}$. In figure 7.13, with the approximations for small angles and paraxial rays, we have

$$\alpha = \frac{y}{\ell} \quad \text{and} \quad \alpha' = \frac{y}{\ell'}, \quad \text{so that} \quad \frac{\ell'}{\ell} = \frac{\alpha}{\alpha'}, \quad \text{or}$$

$$hn\alpha = h'n'\alpha' \qquad (7.13).$$

This is known as *Lagrange's condition*, and is simply a statement of small angle approximations. It implies that the surfaces of the lens as a whole give the same magnification for various small areas of the object.

Now, in figure 7.13 the actual values of ℓ and ℓ' for the ray incident at P are PB and PB'. All parts of the surface or lens give the same magnification of each object element only if PB'/PB has the same value for all P. Using the sine rule in triangle BPB', we obtain

$$\frac{PB'}{\sin\alpha} = \frac{PB}{\sin\alpha'}, \quad \text{or on rearranging,} \quad \frac{PB'}{PB} = \frac{\sin\alpha}{\sin\alpha'}.$$

The constant value of ℓ'/ℓ required is then $\sin\alpha/\sin\alpha'$, from which $\frac{h'}{h} = \frac{n\sin\alpha}{n'\sin\alpha'}$.

Hence, rearranging this relationship,

$$h n \sin\alpha = h' n' \sin\alpha' \qquad (7.14).$$

This is known as *Abbe's Sine Condition*, and is named after the German physicist Ernst Abbe who developed much of the theory of image formation.

If fulfilled, the sine condition indicates that different zones of a lens would give the same magnification $m = \frac{h'}{h}$ of the same elemental area of the object. This is the condition for freedom from coma. It follows from equation (7.14) that, if $\frac{h'}{h}$ is to be constant for all zones of the lens, and n and n' are constants for a given lens, then

$$\frac{\sin\alpha}{\sin\alpha'} = \frac{h'n'}{hn} = \text{a constant} \qquad (7.14a)$$

for all values of α. This is another form of the Abbe Sine Condition.

In the absence of any other aberrations, the sine condition is the true criterion for coma. It is often represented quantitatively by plotting the ratio $\sin\alpha/\sin\alpha'$ against the height y of the incident ray from the lens axis at the point of incidence. In the case of an object at infinity, $\sin\alpha$ is proportional to y and the sine condition equation (7.14a) may be written in the special form

$$\frac{y}{\sin\alpha'} = \text{a constant} \qquad (7.14b).$$

If we now refer to figure 7.14, we see that $y/\sin\alpha'$ represents the sloping distance to the focal point F' from the point H' where the ray deviation appears to occur, and this is the effective focal length f' of that particular zone of the lens.

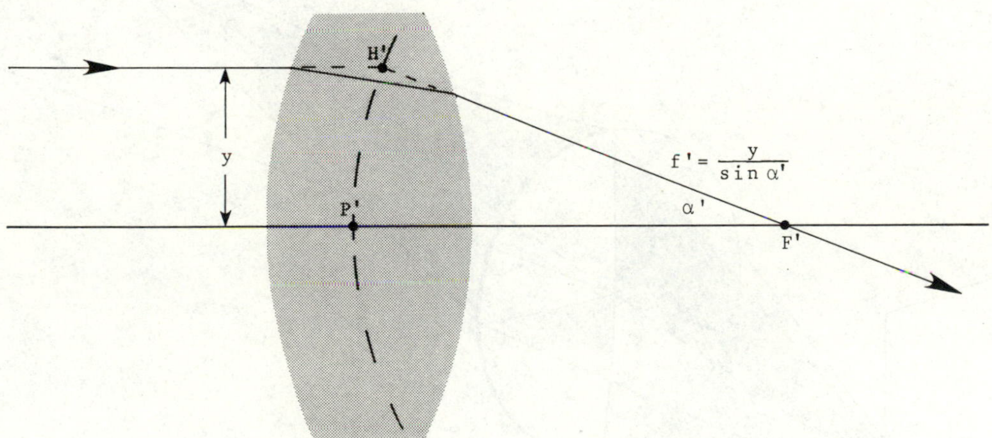

Fig. 7.14 The second principal surface of a lens.

In the absence of coma, f' must be constant for all values of y. Thus, the second principal plane P'H', where the refraction by the lens appears to take place, will be a spherical surface of radius f', and is preferably termed the second principal surface of the lens.

It should be carefully noted that the aplanatic points of a single spherical surface, see section 7.3.1.1, are completely free from spherical aberration and coma, and because of the latter, satisfy the sine condition exactly.

In the lens shown in figure 7.14, light arising from the first focal point defines the first principal surface of the lens with a radius f = PF.

7.3.3 OBLIQUE ASTIGMATISM

The third of Seidel's monochromatic aberrations is oblique astigmatism. This should not be confused with the defect of vision called simply astigmatism. The latter is caused by a lack of spherical curvature in one or more of the refracting components of the eye, usually the cornea. Oblique astigmatism occurs with a lens whose surfaces are parts of spheres and, like coma, affects the image of object points which are not on the lens axis. However, coma results in spreading out the image of an off-axis point object over a plane perpendicular to the lens axis, whereas oblique astigmatism spreads the image in a direction along the axis.

The word astigmatism literally means "not a point" and, although spherical aberration and coma are forms of astigmatism, the term is usually applied to the defective image formed by rays which originate from an object point which an appreciable distance from the lens axis. As the defect is of particular importance with spectacle lenses we shall consider it in some detail. The effect is shown in figure 7.15 where it is assumed that oblique astigmatism is the only aberration present.

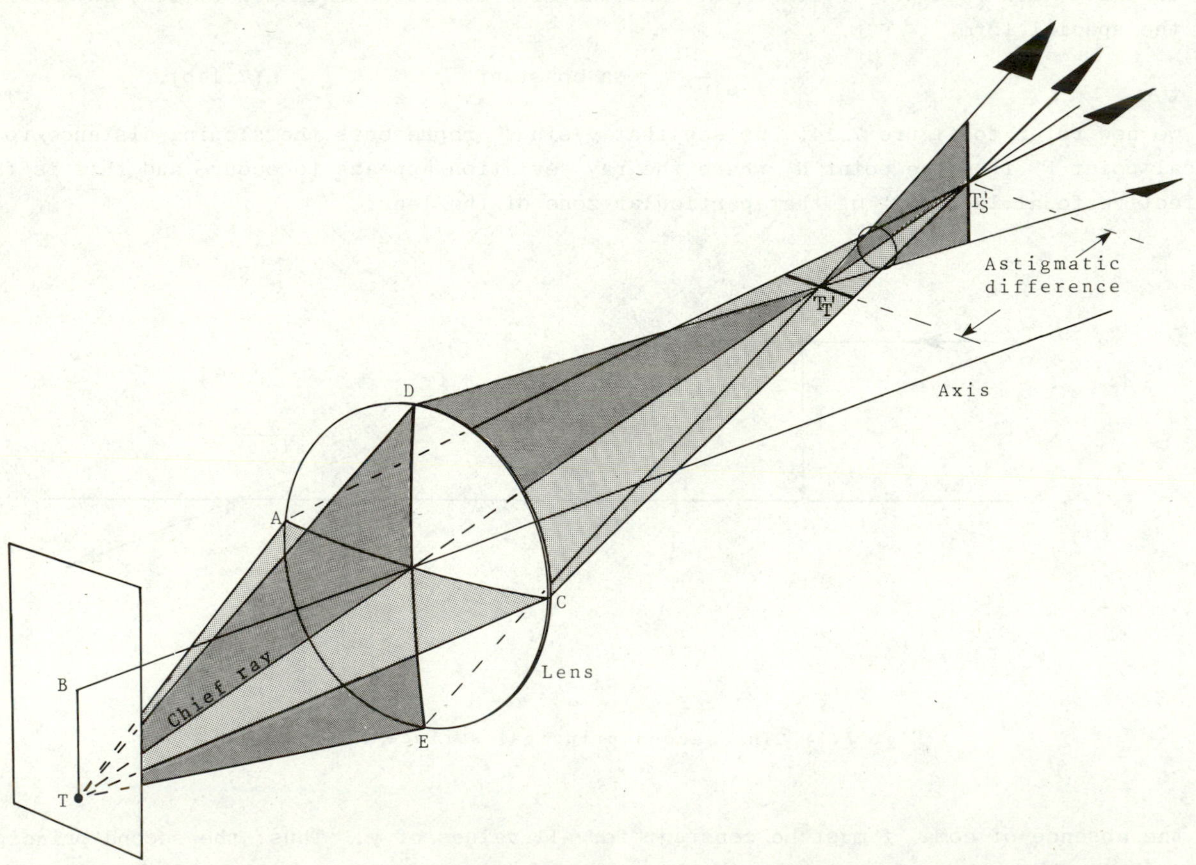

Fig. 7.15 The oblique astigmatic pencil from the object point T.

To assist our description we shall consider the separate effects of the tangential and the sagittal meridians of the lens. In figure 7.15, a pencil of rays from an off-axis object point T is incident on the spherical sufaced lens ADCE. The tangential plane, sometimes called the primary plane, is defined as the plane which contains the chief ray (the ray passing through the centre of the aperture) and the principal axis. The tangential plane in figure

7.15 is therefore represented by the vertical plane TDE, with DE being referred to as the tangential meridian of the lens. The sagittal plane, or secondary plane, is then defined as the plane which contains the chief ray and is perpendicular to the tangential plane. Thus, the plane TAC is the sagittal plane, and AC is the sagittal meridian of the lens. The rays of light in the tangential plane intersect each other, after refraction, at the point T'_T, whilst rays in the sagittal plane intersect, after refraction, at the point T'_S. As the refracted rays in the sagittal plane are still converging as they pass T'_T, the image at this point, called the tangential image, is a line which is perpendicular to the chief ray and the tangential plane. In a similar fashion, the image at T'_S, which is called the sagittal image, is a line perpendicular to both the chief ray and the tangential image. If we traced rays from T refracted by other meridians of the lens than those considered above, it would be found that these rays also pass through the two line images at T'_T and T'_S.

It is instructive to note the appearance of the cross-sectional shape of the refracted astigmatic pencil of rays. Immediately on leaving the lens the cross-section of the pencil is circular, and gradually becomes elliptical with the major axis in the sagittal plane. At the tangential focus the ellipse becomes the line at T'_T, at least when applying third order theory. Beyond this line the pencil changes through elliptical shapes and again becomes circular. This is referred to as the *circle of least confusion* and it represents the optimum position for an image screen. Moving further from the lens the cross-section of the pencil again becomes a line shown at T'_S.

Let us now consider the nature of the astigmatic pencil of rays refracted by the whole lens. In the case of an axial object point B, the incident cone of rays is symmetrical with respect to all the meridians of the lens, and every meridian presents the same diameter d to the object point B. Hence, light is refracted by each meridian in the same way, resulting in a point image B' on the lens axis, as shown in figure 7.16(a). This is, of course, in the absence of all other aberrations. A single point image will not occur, however, when an off-axis object point is considered.

Figure 7.16(b) shows the section of the lens, and the fan of rays from T, which lie in the tangential plane. The letters D and E on the tangential meridian of the lens correspond to their positions as marked in figure 7.15 . The chief ray is now inclined to the lens axis and has a greater thickness of the glass to traverse. Moreover, the diameter which the lens presents to the tangential fan of rays is foreshortened and is denoted by d'. Thus, the centres of the transmitted wavefronts suffer larger retardations and become more curved than for refractions of rays from axial object points. Consequently, the effective power of the lens in the tangential meridian is increased, causing the image T'_T formed by this fan of rays to be closer to the lens than if the power had remained at its axial (sphere) value.

Figure 7.16(c) illustrates the section of the lens, and the fan of rays from T, which lie in the sagittal plane. The letters A and C correspond to their positions marked in figure 7.15 . The off-axis object point T must now be imagined to be below the axis of the lens, with the sagittal image T'_S being formed above the axis. Now, although the chief ray in the sagittal fan of rays is inclined to the principal axis of the lens, the diameter of the sagittal meridian of the lens which is presented to T is the true diameter d, without being foreshortened. Hence, the power of the lens is increased for refractions in the sagittal meridian but not as much as in the case of the tangential meridian. Consequently, the sagittal image T'_S is formed nearer the lens than the axial image point B', but the effect is less than that due to the tangential meridian.

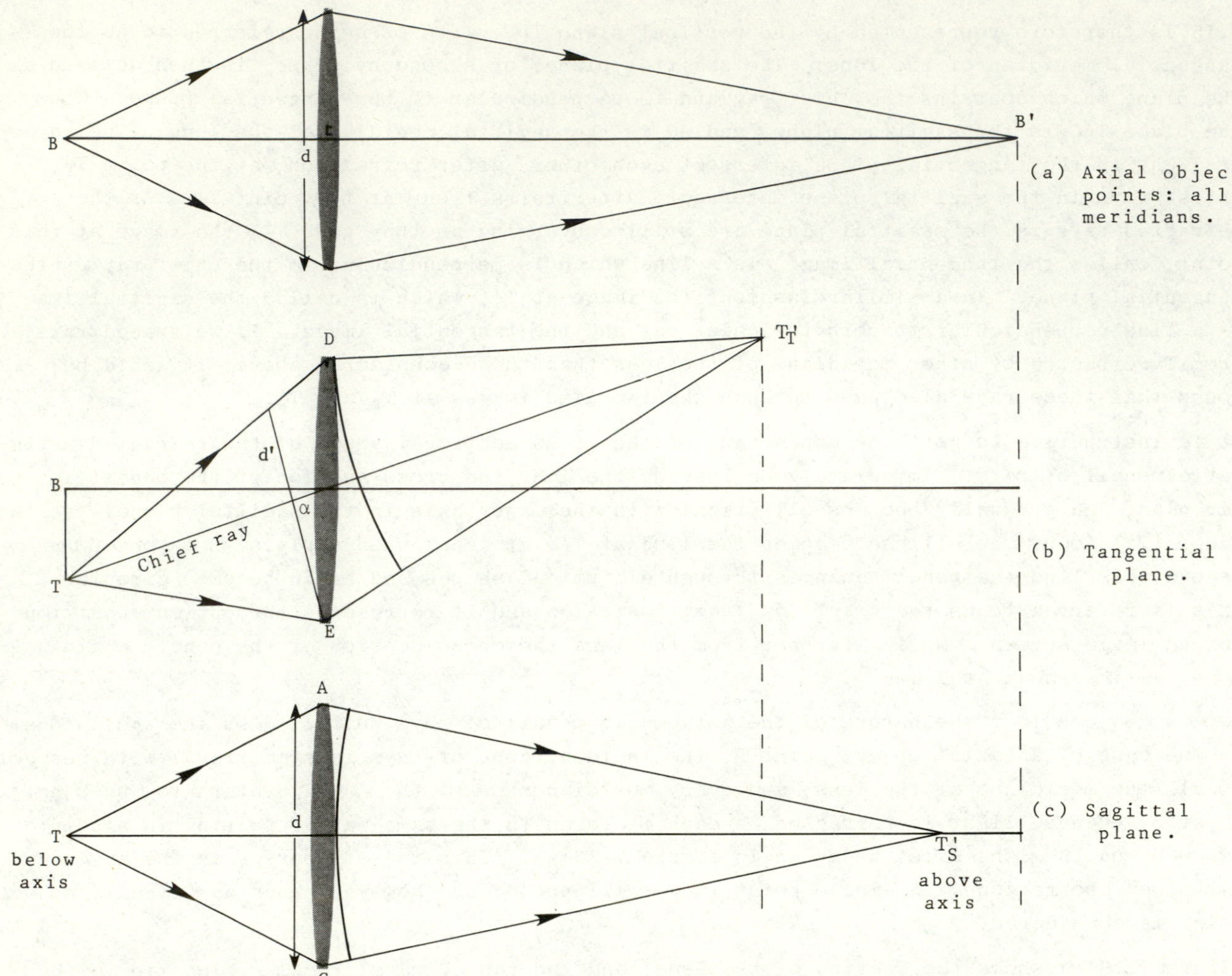

Fig. 7.16 Illustrating the pencils in the tangential and sagittal planes.

7.3.3.1 Variation of Oblique Astigmatism

In figure 7.15, the distance $T'_T T'_S$ between the two focal lines in the astigmatic pencil is referred to as the *astigmatic difference* in the image. When expressed in dioptres, the difference between the emergent vergences in the tangential and sagittal meridians is called the *astigmatic error* for the particular oblquity of the incident light. Obliquity is conveniently measured in terms of the angle α between the chief ray of the incident pencil and the principal axis of the lens. From third order theory, it can be shown that the magnitude of oblique astigmatism in independent of the aperture of the lens, but increases with the square of the point object's distance from the axis, and hence, approximately with the square of the tangent of the angle of obliquity, see equation (7.20). The aberration is considered as positive when the tangential image lies to the left of the sagittal image, and negative when the tangential image lies to the right of the sagittal image.

The method of illustrating the amount of oblique astigmatism produced by a lens is shown in figures 7.17(a) and (b) which depict a fan of rays in the tangential plane for positive and negative lenses, respectively. M and T represent off-axis point objects for which the chief rays have different obliquities.

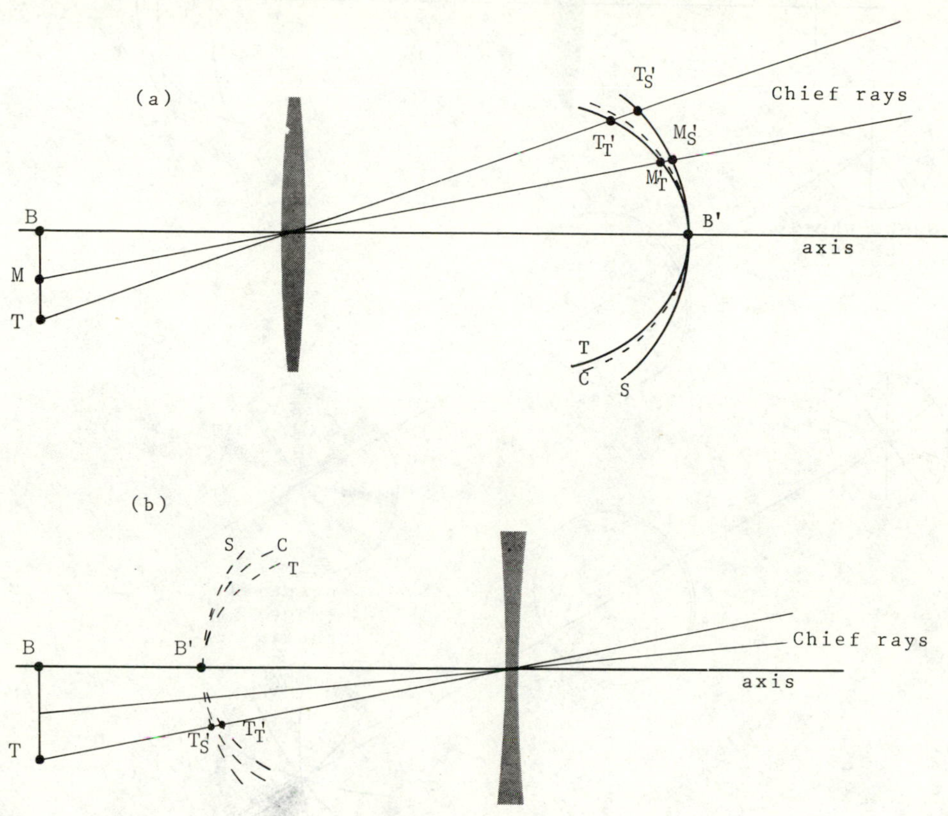

Fig. 7.17 Image shell diagrams

The positions of the line images (T_T', T_S', M_T', and M_S' are the centres of the line images) vary with the obliquity of the chief ray and are more widely separated for object points at larger distances from the axis. For an object of finite size, the centres of the tangential and sagittal images all lie on surfaces or *shells* which are ellipsoidal in shape. These shells are referred to as the T and S shells, respectively. The surface of "best focus" is the locus of the circles of least confusion, C, and this lies between the T and S shells. It is seen , figure 7.17, that the surface of best focus is not a plane, but a curved surface. This additional defect of the image is called curvature of field and will be discussed in conjunction with oblique astigmatism in section 7.3.4 . When the T and S shells are considered in all planes around the principal axis of the lens, visualised by rotating figure 7.17 about the lens axis, the resulting diagram is symmetrical about the axis. It is known as the tea-cup and saucer diagram. Figure 7.18 illustrates this complete pictorial representation of oblique astigmatism for the lens depicted in figure 7.17(a). However, due to the symmetry of the teacup and saucer diagram, it is usually sufficient to represent the shells by the lines T and S as indicated in figure 7.17 . The *astigmatic diagram*, as it is called, is extensively used in spectacle lens design and gives a good indication of the peripheral performance of a lens with respect to oblique astigmatism. On the axis, where the T and S lines come together, the oblique astigmatism is zero, of course.

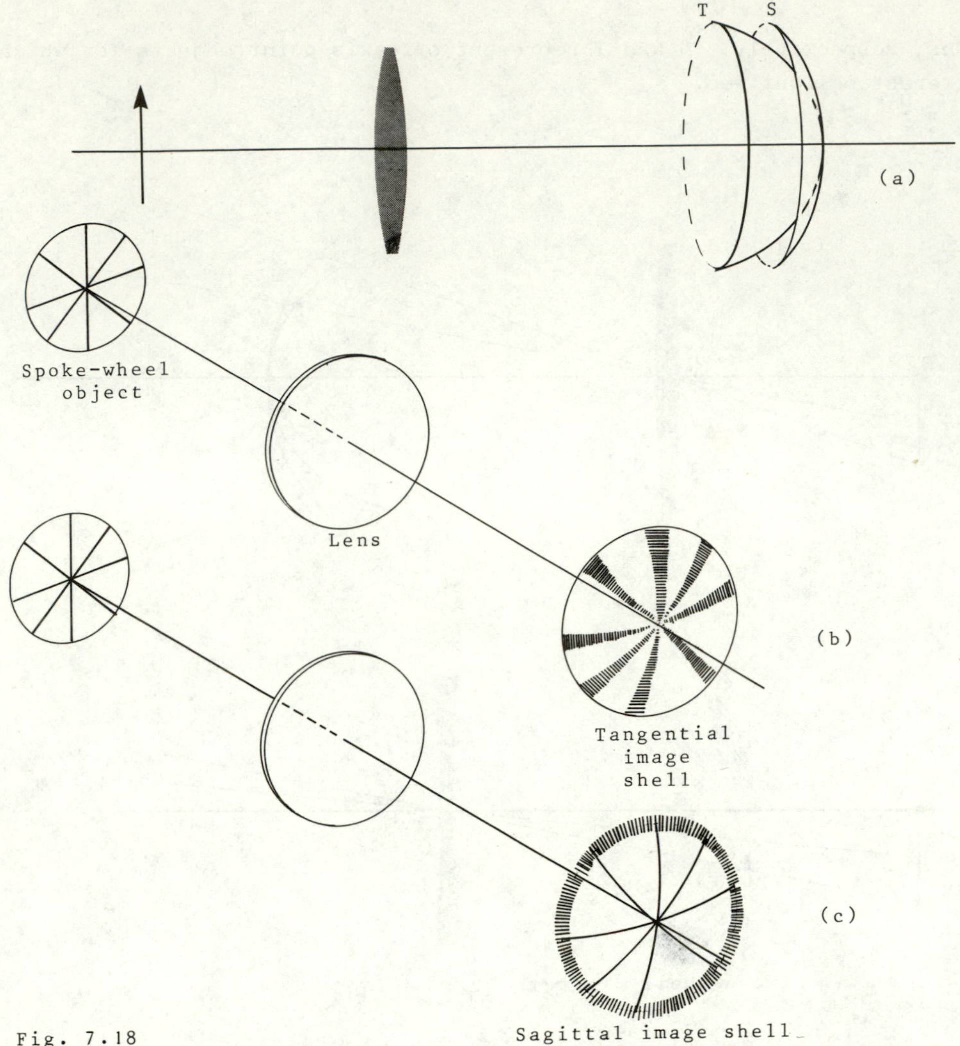

Fig. 7.18

It is evident that, when the form of a lens is altered (known as bending), the form of the refracted astigmatic pencil of rays will vary and, consequently, the amount of oblique astigmatism will change. When the theory of aberrations is applied to a refracting system comprising a spectacle lens and and the eye, the smallness of the aperture of the pupil of the eye renders the effects of spherical aberration and coma negligible. It is extremely important that the effects of oblique astigmatism are reduced, or even eliminated, so that the retinal image will be tolerably sharp when the eye is viewing objects obliquely through the lens. It is also advantageous for the image formation to approximate to the shape of the far point sphere; that is, the locus of the far point of the eye as the eye rotates about its centre of rotation. It is found that the magnitude of oblique astigmatism varies as the form of the lens is altered. Figure 7.19 shows a +4.00 D spherical powered lens in four different forms, together with the astigmatic diagram for each form, the object being at infinity. In each case the index of the lens material is 1.50, and the radii of curvature of the two surfaces are as indicated.

It can be seen that the magnitude of the astigmatic difference TS depends on the form of the lens, and is particularly large for a given obliquity with the equiconvex form (a). At the same inclination to the axis of the chief ray, the magnitude of TS is less with the other forms

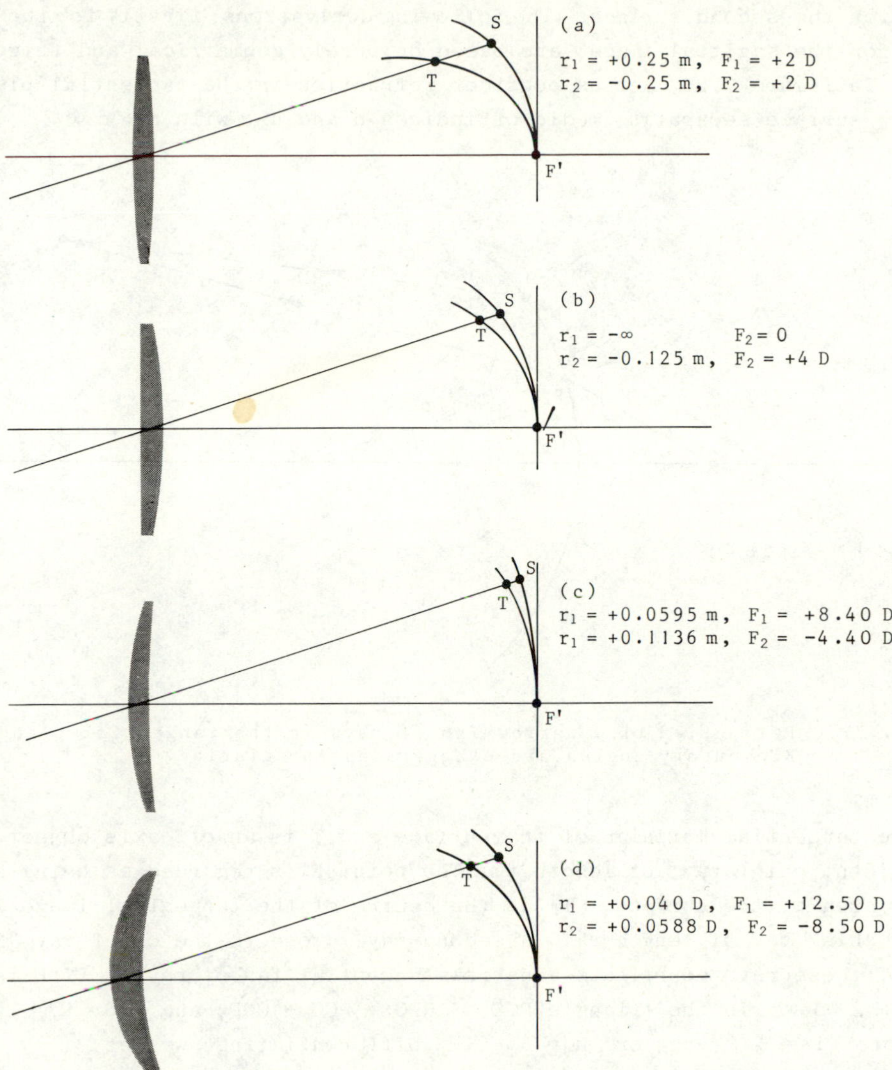

Fig. 719 The effect of lens form on oblique astigmatism.

(b), (c), and (d). Indeed, as the bending of the lens continues, a particular form, called the *best form*, best form is achieved for which the astigmatism is negligible, as in (c). It should be noted that during the process of bending certain lens forms may be achieved for which the T and S shells interchange, that is the oblique astigmatism changes from positive to negative.

In ophthalmic practice, lenses which are free from the effects of astigmatism are known as *stigmatic* or *point focal* lenses, whilst in photography similar lenses are referred to as *anastigmats*.

7.3.3.2 *The Position of the Tangential Image*

The positions of the two focal line images of a point object in an astigmatic pencil may be calculated. It is usual to consider the astigmatic images formed by the first surface of a lens as objects for the second surface. The following derivations, firstly for the tangential image and later for the sagittal image, are based on purely geometrical and trigonometrical considerations. In figure 7.20, let us consider refraction in the tangential plane of a single refracting surface separating media of indices n and n', with $n' > n$.

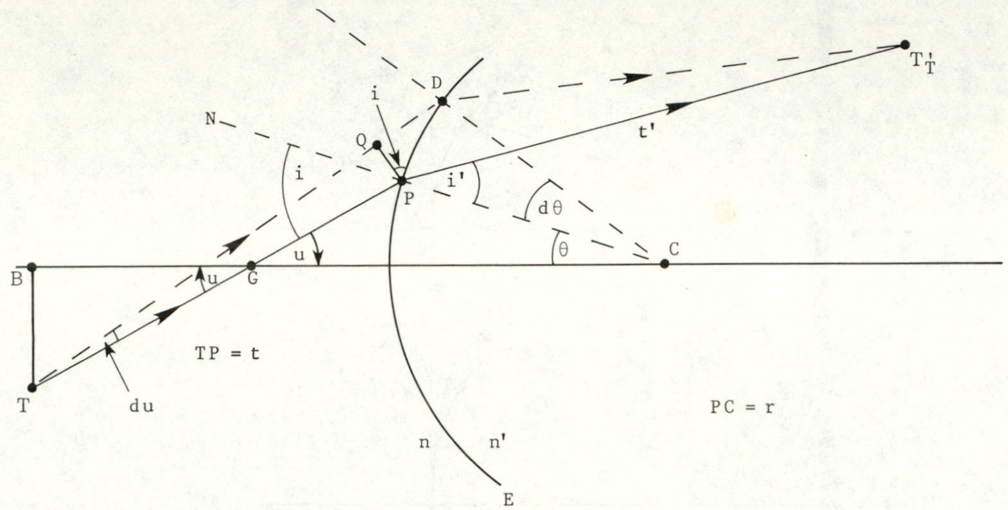

Fig. 7.20 Refraction of a narrow fan of rays in the tangential plane. Elementary angles are exaggerated for clarity.

DE represents the tangential meridian of the surface and T is an off-axis object point. The ray TP is an incident chief ray, of length t. The point P is regarded as being at the centre of an elementary area of the surface. T'_T is the centre of the tangential image line and PT'_T is the refracted chief ray of length t'. A second ray, close to the chief ray, is represented by TD. Together, these rays constitute a narrow tangential fan of rays. If D is very close to P then $PD = r\,d\theta$. Now, in the triangle PCG, $\widehat{NPG} = \widehat{PCG} + \widehat{CGP}$, and $\widehat{NPG} = i$, $\widehat{PCG} = \theta$, but $\widehat{CGP} = -u$, therefore $i = \theta + (-u)$, or $u = \theta - i$. Differentiating, we get

$$du = d\theta - di \qquad \text{(i).}$$

Now, the short line PQ, perpendicular to the incident ray, is given by $PQ = TP\,du = t\,du$. But, in triangle QPD, $\widehat{QPD} = i$, and $PQ = PD\cos i$, with $PD = r\,d\theta$. Hence, $t\,du = PD\cos i = r\cos i\,d\theta$, that is,

$$t\,du = r\cos i\,d\theta \qquad \text{(ii).}$$

Substituting from (i) into (ii) gives $t(d\theta - di) = r\cos i\,d\theta$, so that

$$t\,d\theta - t\,di = r\cos i\,d\theta$$

$$\text{and} \quad di = \frac{t\,d\theta - r\cos i\,d\theta}{t}$$

$$\text{or} \quad di = \left(1 - \frac{r\cos i}{t}\right)d\theta \qquad \text{(iii).}$$

Similarly, for the refracted ray $\quad di' = \left(1 - \dfrac{r\cos i'}{t'}\right)d\theta \qquad \text{(iv).}$

Now, for the refraction at P, Snell's law applies. Therefore, using $n\sin i = n'\sin i'$, and by differentiating, we have $n\cos i\,di = n'\cos i'\,di'$. Substituting for di and di' from (iii) and (iv) we get

$$n \cos i\left(1 - \frac{r \cos i}{t}\right) d\theta = n' \cos i'\left(1 - \frac{r \cos i'}{t'}\right) d\theta$$

On cancelling the $d\theta$ and rearranging we obtain

$$\frac{n'\cos^2 i'}{t'} - \frac{n \cos^2 i}{t} = \frac{n'\cos i' - n \cos i}{r} \qquad (7.15).$$

The term on the right hand side is sometimes referred to as the *oblique power* of the surface for the particular chief ray, and it is always greater than the paraxial power. It can be seen that the oblique power term reduces to $\frac{n'-n}{r}$, the paraxial power of the surface, when the object point T is on the axis. In this case, $i = i' = 0$ and $\cos i = \cos i' = 1$. Also, by writing $t' = \ell'$ and $t = \ell$ we obtain $\frac{n'}{\ell'} - \frac{n}{\ell} = \frac{n'-n}{r}$, the familiar fundamental paraxial equation for a single surface.

7.3.3.3 The Position of the Sagittal Image

Figure 7.21 shows the refraction by the sagittal meridian of the refracting surface, where $n' > n$.

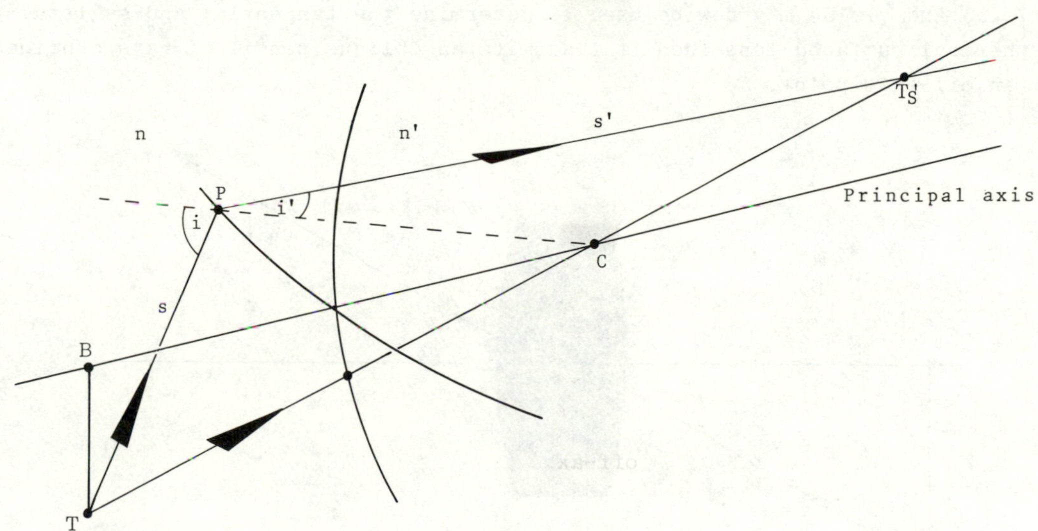

Fig. 7.21 Refraction of a ray TP at a point P in the sagittal meridian.

The ray TP is a chief ray of length s incident on an elementary area around P, with the sagittal image being formed at T'_S, where the distance $PT'_S = s'$. One of the properties of the sagittal image T'_S is that it always lies on the line joining the object point to the centre of curvature of the surface. This is the line TCT'_S.

Now, for any triangle ABC, the area is given by ½ab sin C. For the triangle TPT'_S we have

$$\text{area } \Delta TPT'_S = \text{area } \Delta TPC + \text{area } \Delta PCT'_S .$$

That is, $\quad \tfrac{1}{2}PT.PT'_S.\sin \widehat{TPT'_S} = \tfrac{1}{2}PT.PC.\sin \widehat{TPC} + \tfrac{1}{2}PC.PT'_S.\sin \widehat{CPT'_S}$,

or, using symbols and the sign convention,

$$\tfrac{1}{2}(-s)(s')\sin(180^0-i+i') = \tfrac{1}{2}(-s)(r)\sin(180^0-i) + \tfrac{1}{2}(r)(s')\sin i' .$$

This may be written

$$-ss'\sin(i-i') = -sr\sin i + s'r\sin i' .$$

Expanding sin(i-i') we have

$$-ss'(\sin i\cos i' - \cos i\sin i') = -sr\sin i + s'r\sin i'.$$

Multiplying each term by $\frac{n'}{ss'r}$ gives

$$-\frac{(n'\sin i\cos i' - n'\cos i\sin i')}{r} = -\frac{n'\sin i}{s'} + \frac{n'\sin i'}{s} .$$

But, from Snell's law, $n'\sin i'$ can be replaced by $n\sin i$ whereupon the $\sin i$ term cancels to give

$$\frac{n'}{s'} - \frac{n}{s} = \frac{n'\cos i' - n\cos i}{r} \qquad (7.16).$$

The term on the right hand side is the same oblique power of the surface which we obtained in equation (7.15) for the tangential image.

7.3.3.4 Oblique Astigmatism for Central Refraction

Equations (7.15) and (7.16) may now be used to determine the tangential and sagittal powers of a thin spherical surfaced lens when it transmits an oblique pencil of rays confined to its centre from an off-axis point.

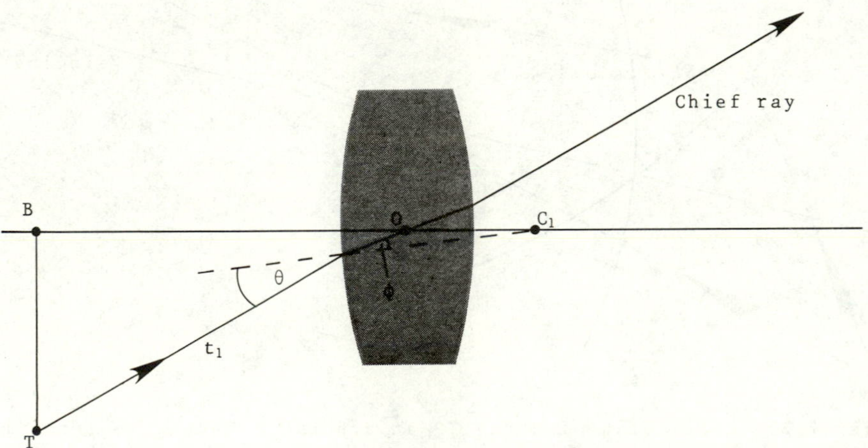

Fig. 7.22 Oblique central refraction.

Figure 7.22 shows the chief ray from an off-axis object point T passing symmetrically through the optical centre O of an equibiconvex lens. The angle of emergence from the second surface of the lens will be equal to the angle of incidence θ at the first surface. Similarly, the angle ϕ inside the lens will be the same at both surfaces. If the lens material has an index n and the lens is in air, then successive applications of the tangential formula, equation (7.15), to each surface gives

$$\frac{n\cos^2\phi}{t_1'} - \frac{\cos^2\theta}{t_1} = \frac{n\cos\phi - \cos\theta}{r} \quad \text{(i)} \qquad \text{and} \qquad \frac{\cos^2\theta}{t_2'} - \frac{n\cos^2\phi}{t_2} = \frac{\cos\theta - n\cos\phi}{r} \quad \text{(ii)}.$$

For a thin lens, the object distance for the second surface equals the image distance for the first surface; that is, $t_2 = t_1'$. So, adding (i) and (ii) and putting $t_2 = t_1'$ gives

$$\frac{\cos^2\theta}{t_2'} - \frac{\cos^2\theta}{t_1} = \frac{(n\cos\phi - \cos\theta)}{r_1} + \frac{(\cos\theta - n\cos\phi)}{r_2}$$

that is

$$\cos^2\theta\left(\frac{1}{t_2'} - \frac{1}{t_1}\right) = (n\cos\phi - \cos\theta)\left(\frac{1}{r_1} - \frac{1}{r_2}\right)$$

$$\text{or}\quad \cos^2\theta\ \frac{1}{t_2'} - \frac{1}{t_1} = \frac{(n\cos\phi - \cos\theta)}{(n-1)f'},$$

where f' is the paraxial focal length. Hence, the oblique tangential power $F_T = \frac{1}{f_T'}$ is given by

$$\frac{1}{t_2'} - \frac{1}{t_1} = \frac{1}{f_T'} = \frac{(n\cos\phi - \cos\theta)}{\cos^2\theta(n-1)}\cdot\frac{1}{f'} \qquad (7.17)$$

$$\text{or}\qquad F_T = \frac{(n\cos\phi - \cos\theta)}{\cos^2\theta(n-1)}\,F \qquad (7.17a),$$

where $F = \frac{1}{f'}$ = paraxial power.

Similarly, two successive applications of the sagittal formula, equation (7.16), for the two surfaces gives

$$\frac{n}{s_1'} - \frac{1}{s_1} = \frac{n\cos\phi - \cos\theta}{r_1} \quad \text{(iii)} \qquad \text{and} \qquad \frac{1}{s_2'} - \frac{n}{s_2} = \frac{\cos\theta - n\cos\phi}{r_2} \quad \text{(iv)}.$$

Adding (iii) and (iv) and noting that, for a thin lens, $s_2 = s_1'$, we get

$$\frac{1}{s_2'} - \frac{1}{s_1} = \frac{n\cos\phi - \cos\theta}{r_1} + \frac{\cos\theta - n\cos\phi}{r_2}$$

$$\text{or}\quad \frac{1}{s_2'} - \frac{1}{s_1} = (n\cos\phi - \cos\theta)\left(\frac{1}{r_1} - \frac{1}{r_2}\right).$$

That is, in the same manner as for the tangential power, and making the same substitution for for the paraxial power F,

$$\frac{1}{s_2'} - \frac{1}{s_1} = \frac{1}{f_S'} = \frac{(n\cos\phi - \cos\theta)}{n-1}\cdot\frac{1}{f'} \qquad (7.18),$$

$$\text{or}\quad F_S = \frac{(n\cos\phi - \cos\theta)}{n-1}\,F \qquad (7.18a).$$

Dividing equation (7.17a) by equation (7.18a), and rearranging, we get

$$F_S = F_T\cos^2\theta \qquad (7.19).$$

We can find an approximate expression for the oblique astigmatic error at some specified θ as follows. Using equations (7.17a) and (7.18a), we can write

$$\begin{aligned}F_T - F_S &= \frac{(n\cos\phi - \cos\theta)F}{\cos^2\theta(n-1)} - \frac{(n\cos\phi - \cos\theta)F}{(n-1)}\\ &= \left(\frac{n\cos\phi - \cos\theta}{n-1}\right)\left(\frac{1}{\cos^2\theta} - 1\right)F\\ &= \frac{(n\cos\phi - \cos\theta)}{n-1}(\sec^2\theta - 1)F\\ &= \frac{(n\cos\phi - \cos\theta)}{n-1}F\tan^2\theta \simeq F\tan^2\theta\end{aligned}$$

since the term in brackets is approximately equal to unity for angles up to about 25°.

That is, the astigmatic error is

$$F_T - F_S = F\tan^2\theta \qquad (7.20).$$

This is a useful expression in ophthalmic practice. For example, suppose a pair of +10.00 D spherical spectacle lenses are used with a poor adjustment so that the patient is looking obliquely through them at an angle of 15^0, when looking downwards for reading, say. Then the oblique astigmatic error will be $10\tan^2 15^0 = 0.72$ D of unwanted astigmatism!

7.3.4 CURVATURE OF FIELD

If third order theory is applied to an optical system, and it is found that the first three Seidel terms are zero, the image will be free from spherical aberration, coma, and oblique astigmatism. In such a situation a point image will be formed for point objects which are off the axis as well as on the axis. We mentioned earlier (see section 2.2.5 and figure 2.44) that an object which is perpendicular to the axis of a refracting system will give rise to an image also perpendicular to the axis only for refraction in the paraxial region. At larger apertures, however, the image of a planar object will be curved. In the absence of astigmatism the image will lie on a curved, paraboloid surface called the *Petzval surface*. This defect of the image is known as *Petzval Field Curvature*, or sometimes as *Curvature of Field*, after the Hungarian mathematician Joseph Petzval (1807 - 1891). We should note that the image is now stigmatic, that is, each object point is imaged as a point, although as stated, the image points lie on a curve. The effect may be appreciated from figure 7.23 which shows a converging lens producing a real image B'T' of a planar object BT. Figure 7.23(b) shows the effect when a diverging lens forms a virtual image.

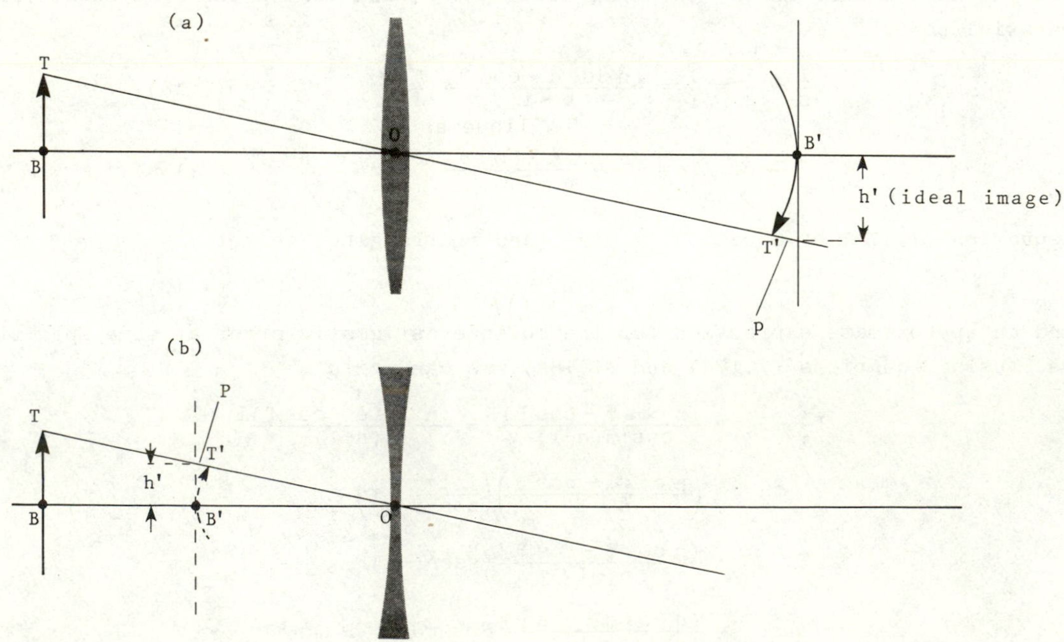

Fig. 7.23 Curvature of field.

In figure 7.23(a) we shall note that, because of the assumed absence of oblique astigmatism, the focal power of the lens will not change from its axial power even when off-axis object points are considered. It can be seen that the tip T of the object is further from the optical centre O of the lens than the base or axial object point B. It follows that the off-axis image point T' will be closer to the optical centre of the lens than the axial image point B', and in the case of a converging lens, the image B'T' will curve inwards towards the object plane.

With a diverging lens, figure 7.23(b), the virtual image curves outwards away from the object plane. For a planar object, the curved image surface in which the image would be formed, if the lens power was constant and equal to its paraxial power, is known as the *Petzval surface* of the lens.

Petzval curvature may be measured quantitatively in terms of the distance p along the chief ray between the Petzval surface and the ideal image plane, where the latter is the plane perpendicular to the axis and through the axial image point B'. This is shown in figure 7.23 for a particular image height h'. The magnitude of p depends on the square of h' except at very large angles to the axis. The curvature of the Petzval surface is dependent only on the focal length(s) of the lens or lenses in a system and on the refractive indices of the components. It is not affected by object distance, aperture size, lens thickness, or separation of the components in a system.

It can be shown that the magnitude of p, measured as shown in figure 7.23 at a height h', is given by

$$p = \frac{h'^2}{2}\sum_{x=1}^{y}\frac{1}{n_x f_x'} \qquad (7.21)$$

where y is the number of thin lenses in the system and n_x and f_x' are the index and focal length of each component lens. This expression is referred to as the *Petzval sum* and it can be seen that a less curved image is formed with lens materials of higher refractive index. As previously mentioned, the Petzval surfaces for converging and diverging lenses curve in opposite senses, and thus, a combination containing converging and diverging components will reduce or eliminate Petzval curvature. For example, let us consider the case of two thin lenses forming a system. It is possible to eliminate Petzval curvature if $p = 0$ in equation (7.21). That is, for zero image curvature, $\frac{1}{n_1 f_1'} + \frac{1}{n_2 f_2'} = 0$, or equivalently,

$$n_1 f_1' + n_2 f_2' = 0 \qquad (7.22).$$

This expression is known as the *Petzval condition* and applies to two thin lenses which are in contact with each other or separated. It is of great importance with camera lens systems because of the need to obtain a flat image over a wide field. This will be dealt with in the next chapter, section 8.1.1.2 . With thick lens systems, the Petzval condition is a useful approximation. In the case of visual instruments, a certain amount of image curvature can be tolerated because the eye can accommodate for it.

In order that we may have a good understanding of Petzval curvature, the defect should be considered at the same time as oblique astigmatism, since the two aberrations are very closely related. In the presence of oblique astigmatism we have previously stated that the lens power for off-axis object points is not the same as the paraxial power, and the difference is larger for rays of light in the tangential plane than for rays in the sagittal plane. Thus, the tangential and sagittal image surfaces both lie on the same side of the Petzval surface, with the sagittal image shell being nearer to it. Figure 7.24 shows that, from the results of third order theory, the distances measured from the Petzval surface along the chief ray to the tangential and sagittal surfaces are in the ratio 3:1 . That is, PT = 3 PS for any given image h'.

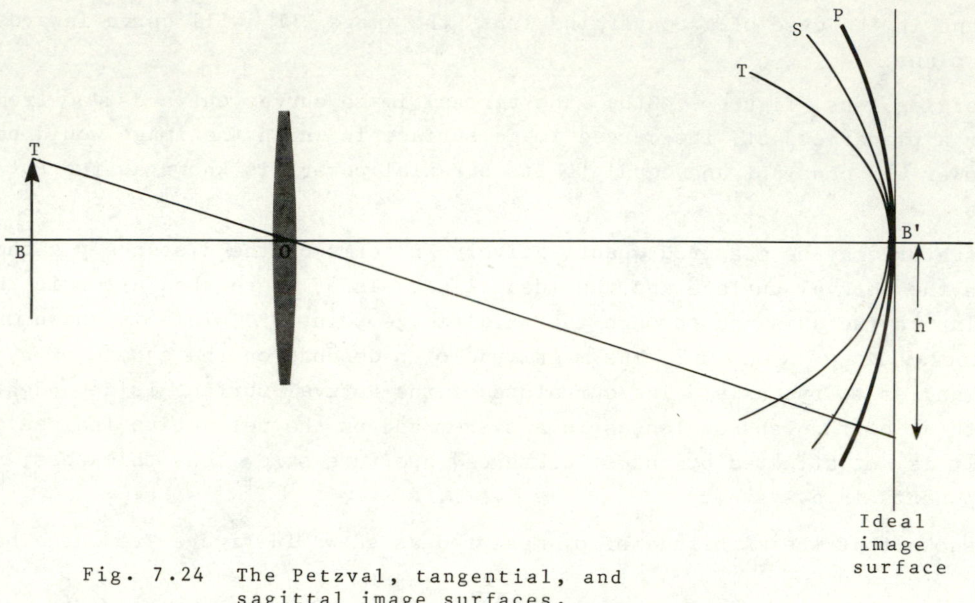

Fig. 7.24 The Petzval, tangential, and sagittal image surfaces.

Thus, astigmatism considered together with curvature of field produces displacements of the tangential and sagittal images along the chief ray from the Petzval surface, and aggravates the curvature of the image. It will readily be appreciated that when coma is also considered, very complicated ray distributions in the refracted pencils will occur. This combined treatment is beyond the scope of this book.

As discussed in section 7.3.3.1 , the shapes of the T and S surfaces may be altered by bending the lens, or by a suitably positioned stop. Attention is drawn to this latter point in section 8.1.1.2 and figure 8.2 , illustrating the importance of the position of the stop with cheap camera lenses. By arranging for the T and S surfaces to curve in the opposite senses, artificial flattening of the image field is obtained. However, the astigmatic difference is increased, resulting in a reduction of definition in the image towards the edges.

Figure 7.25 shows examples of how the astigmatic surfaces T and S vary in relation to the fixed Petzval surface P as the position of the stop in front of a converging lens is altered. The action of the stop is to restrict the rays from each point on the object such that the various chief rays must pass through different parts of the lens. In all the cases shown the distances PT and PS remain in the ratio 3:1, that is, PT = 3 PS. In figure 7.25(a) is represented the usual configuration of the T and S surfaces for a single lens. Introduction of a stop may achieve the situation in figure 7.25(b) in which the T and S surfaces coincide. This corresponds to the elimination of astigmatism, but a curved field remains. Further alterations to the position of the stop may result in the configurations in (c) and (d). With a single lens it is not possible to completely eliminate oblique astigmatism and curvature of field at the same time.

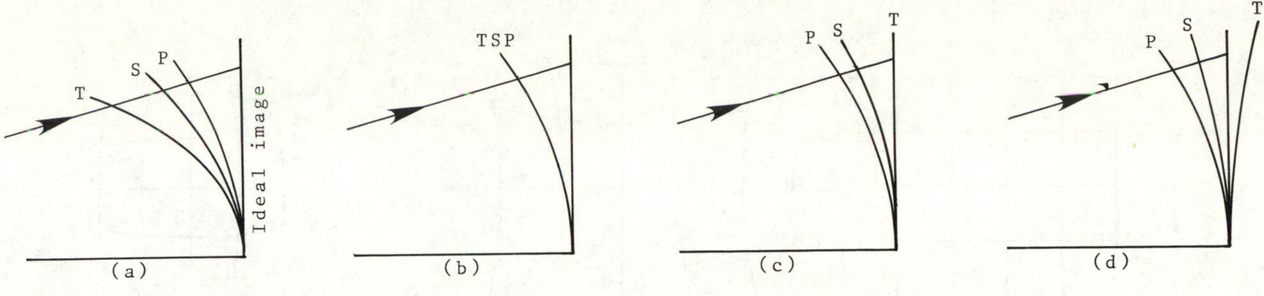

Fig. 7.25 See text for details.

Good quality camera systems are generally anastigmat systems, of which one well known type is shown in figure 8.3. Besides coinciding on the axis of the lens system where the astigmatism is zero, the T and S curves cross one another at an additional point yielding an off-axis angle for which astigmatism is zero. Over the remainder of the field the image is relatively flat, and the astigmatism is quite small. Figure 7.26 shows the ideal image plane and typical T and S curves for an anastigmat system.

We should note that for object points at fairly small angular distances from the axis, astigmatism is less objectionable than coma, while the reverse is true at larger angles. Thus, a telescope objective lens, for which the field of view is quite small, would probably be corrected for coma rather than for astigmatism. As previously mentioned, a wide angle camera lens would include some correction for astigmatism.

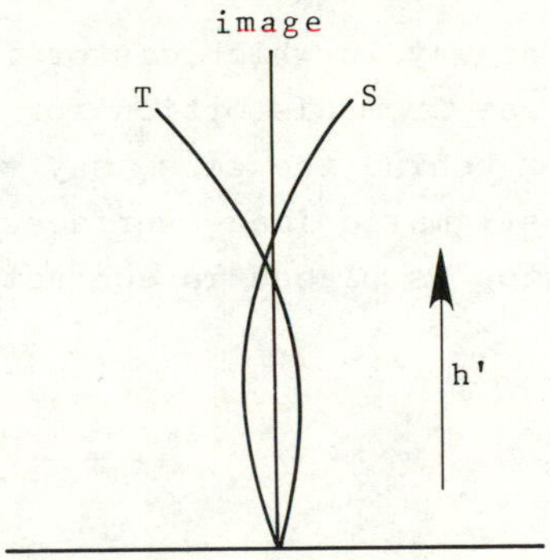

Fig. 7.26

7.3.5 *DISTORTION*

The fifth of Seidel's primary monochromatic aberrations is known as *distortion*. This defect of the image arises not because of a lack of clarity of the image but from a variation of magnification at different distances from the axis of a lens. In order that an image is free from distortion, an optical system must have the same magnification over the whole field. We recall from section 3.1.4.4 that in the paraxial region of a system the linear magnification of an image is given by $\frac{h'}{h} = \frac{n}{\ell} \times \frac{\ell'}{n'}$, or in the simple case where $n = n' = 1$ for a system in air, $\frac{h'}{h} = \frac{\ell'}{\ell}$. The image in a pinhole camera, section 1.4, is formed according to this relationship and is free from distortion at all distances from the axis. It is termed *orthoscopic*.

When a lens image is distorted, the value of the magnification differs from the paraxial value in the outer parts of the field. In some cases the magnification increases with increasing distance from the axis, and the outer parts of the image appear larger than in the case of the ideal image. Alternatively, the magnification may decrease at greater distances from the axis, with the result that the outer parts of the image appear smaller than in the case of the ideal image. These two effects are known as *pincushion distortion* and *barrel distortion*, respectively, for reasons which are evident from figure 7.27.

Figure 7.27(a) shows the ideal undistorted image of a square object, whilst figures 7.27(b) and (c) show pincushion and barrel distortion, respectively.

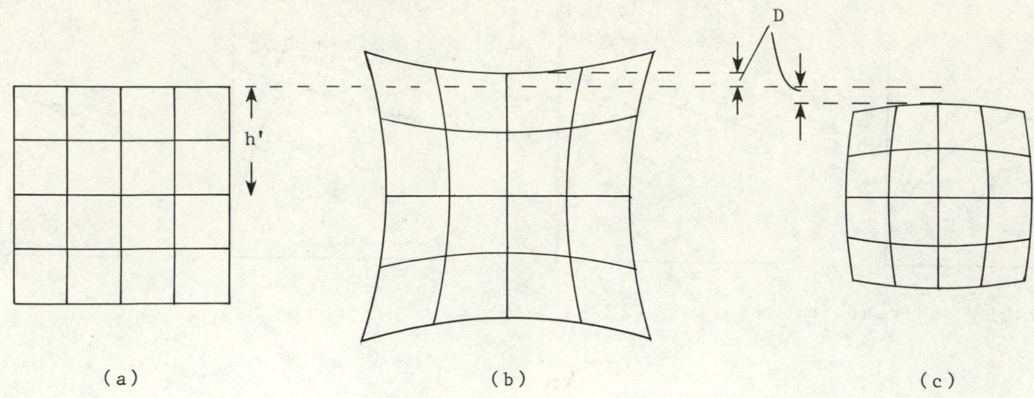

Fig. 7.27 (a) Object. (b) The object imaged with pincushion distortion. (c) The object imaged with barrel distortion.

The magnitude of distortion may be measured by the distance D of an actual off-axis image point from its position on the ideal image as given by paraxial magnification theory. Stated in this way, the value of D is dependent on the cube of h', the ideal image point distance from the axis. As stated, the defect does not affect the clarity of individual image points, but instead displaces them away from or towards the axis of the system.

The way in which distortion arises is shown in figure 7.28. A single thin lens is essentially free from distortion for all object distances. If, however, a stop is introduced in front of or behind the lens, say to reduce spherical aberration, or to modify the shapes of the T and S astigmatic image surfaces, this is invariably accompanied by distortion of the image. If the stop is placed in contact with the lens, as in (a), no distortion occurs.

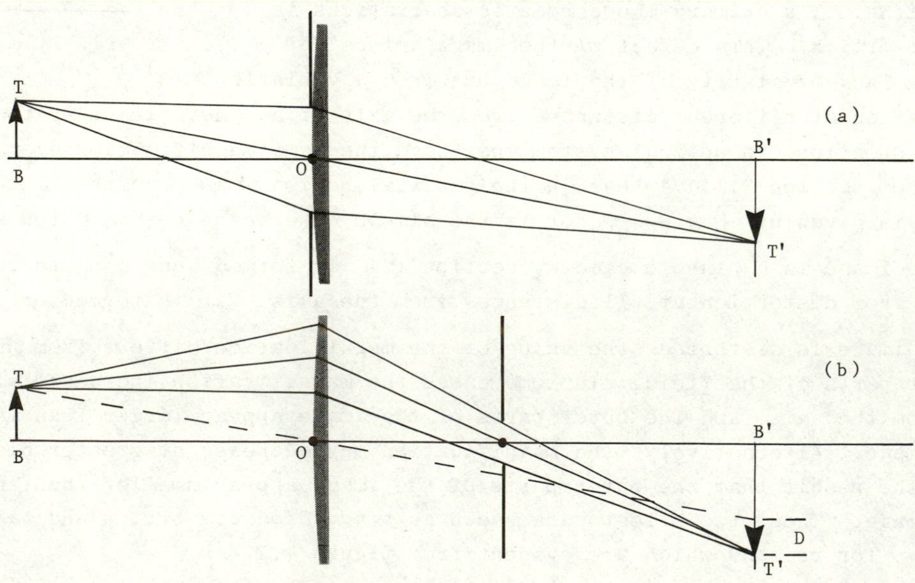

Fig. 7.28(a) and (b).

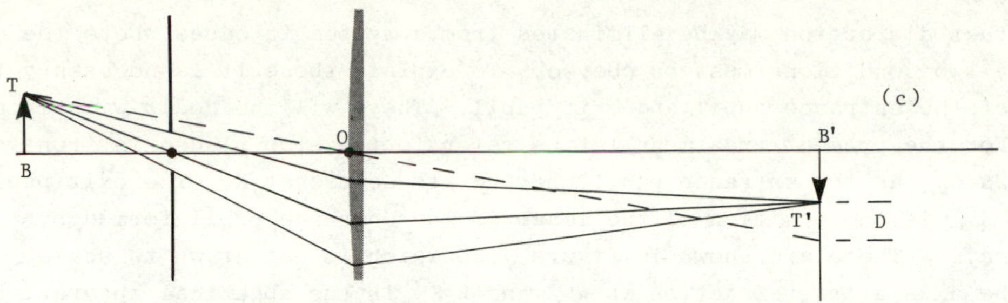

Fig. 7.28 (a) No distortion. (b) Pincushion distortion. (c) Barrel distortion.

It is seen in figure 7.28(b) that, when the stop is placed behind the lens, a different cone of rays is involved in the formation of the tip of the image T' than in figure 7.28(c) where the stop is placed in front of the lens. In the former case, pincushion distortion occurs, and in the latter barrel distortion occurs.

Let us now consider figure 7.28 in more detail. As shown, the chief ray in the pencils from the tip of the object is the one which in each case passes through the centre of the stop. In (a) the chief ray passes through the optical centre of the lens. In (b) the object distance ℓ measured from T along the chief ray is less than it was with the stop at the lens. The corresponding image distance ℓ' measured along the chief ray will consequently be larger and the magnification ℓ'/ℓ will be increased, giving pincushion distortion. In other words, the magnification for an off-axis point will be greater with a rear stop than it would be without it. In (c) the object distance ℓ measured from T along the chief ray is greater than it was with the stop at the lens. Thus ℓ'/ℓ is decreased giving barrel distortion.

Figures 7.28(b) and (c) perhaps suggest a means of reducing the distortion of an image. In figure 7.29 we see a symmetrical doublet lens system with a stop placed between the component lenses. In this way the distortion introduced by the second lens compensates for the distortion produced by the first. Many camera lenses are assembled in this way.

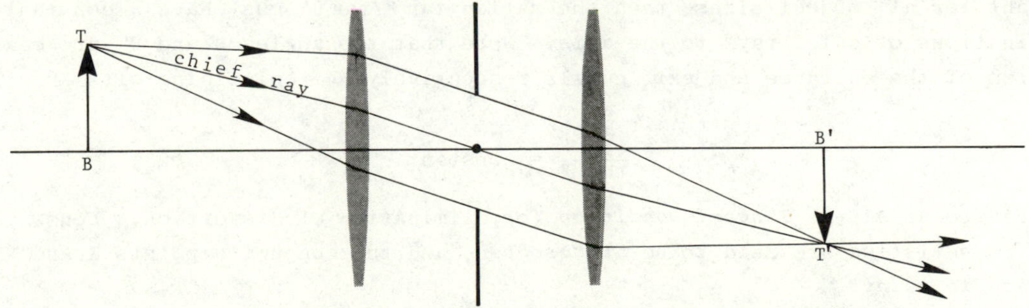

Fig. 7.29 Removal of distortion with a stop placed midway between two identical converging lenses.

If the lens system is perfectly symmetrical, freedom from distortion occurs only for unit magnification. With other magnifications an approximately symmetrical arrangement will still considerably reduce distortion. It is interesting to note the position of the eye's pupil in this connection.

In order that distortion may be eliminated from a system in cases where the magnification is not unity, two conditions must be obeyed. To explain these it is necessary to introduce the concepts of the entrance pupil and exit pupil. These will be dealt with in detail in section 8.4.2 . For the present treatment let us refer to the stop placed in front of a lens, as in figure 7.28(c), as the entrance pupil, having its centre at X. The exit pupil, having its centre at X', is the position of the image of the entrance pupil formed by the system with paraxial rays. These are shown in figure 7.30 which is not drawn to scale. The ray from T crosses the axis after refraction at W', and W'X' is the spherical aberration for this ray.

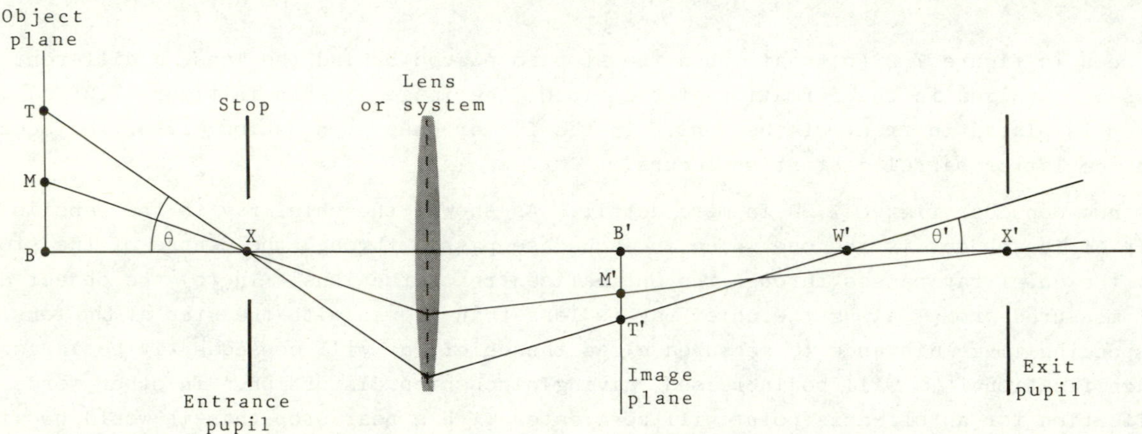

Fig. 7.30 Diagram illustrating Airy's tangent condition for the elimination of distortion.

For the first condition for zero distortion, all rays passing through X must also pass through X'. This means spherical aberration must be absent for the entrance and exit pupil positions. Secondly, the incident and emergent chief rays must all intersect in the plane, shown dashed in the diagram, which is perpendicular to the axis. For the first condition to be satisfied, W' must be imagined to coincide with X'. But for constancy of magnification, that is B'T'/BT is constant for all object sizes, then the ration $\tan\theta/\tan\theta'$ must have a constant value for all inclinations of chief rays to the axis. Note that the angles θ and θ' are measured at the centres of the entrance and exit pupils respectively, W' coinciding with X'. The statement

$$\frac{\tan\theta}{\tan\theta'} = \text{constant} \qquad (7.23)$$

is referred to as *Airy's Tangent Condition* for elimination of distortion. Lens systems for which it is satisfied are said to be *orthoscopic*, and the conjugate points X and X' are known as *orthoscopic points*.

7.3.6 RAY TRACING AT SPHERICAL SURFACES

Many optical systems consist of a series of refracting surfaces, sometimes having large apertures. During the design stages of such systems it will be usual practice to trace rays through the components, surface by surface, where some of the rays may be outside the paraxial region of the system. It is only possible to predict accurately the course of a ray by a trigonometrical computation using various well established procedures. In addition, graphical methods are also available which, although less accurate, are often used in conjunction with the trigonometrical procedure. A graphical method is also referred to as a *geometrical construction*, and besides acting as a check on calculations, it will show clearly the behaviour of a ray at any stage in an optical system.

We shall now consider two geometrical (graphical) constructions and one trigonometrical ray tracing procedure at a single refracting surface. Each of these procedures may be applied to any system by dealing with each surface in turn.

7.3.6.1 Young's Graphical Construction

This method is based on the properties of the aplanatic points of a refracting surface which were discussed in section 7.3.1.1, and is illustrated in figure 7.31(a) for a converging surface and in figure 7.31(b) for a diverging surface.

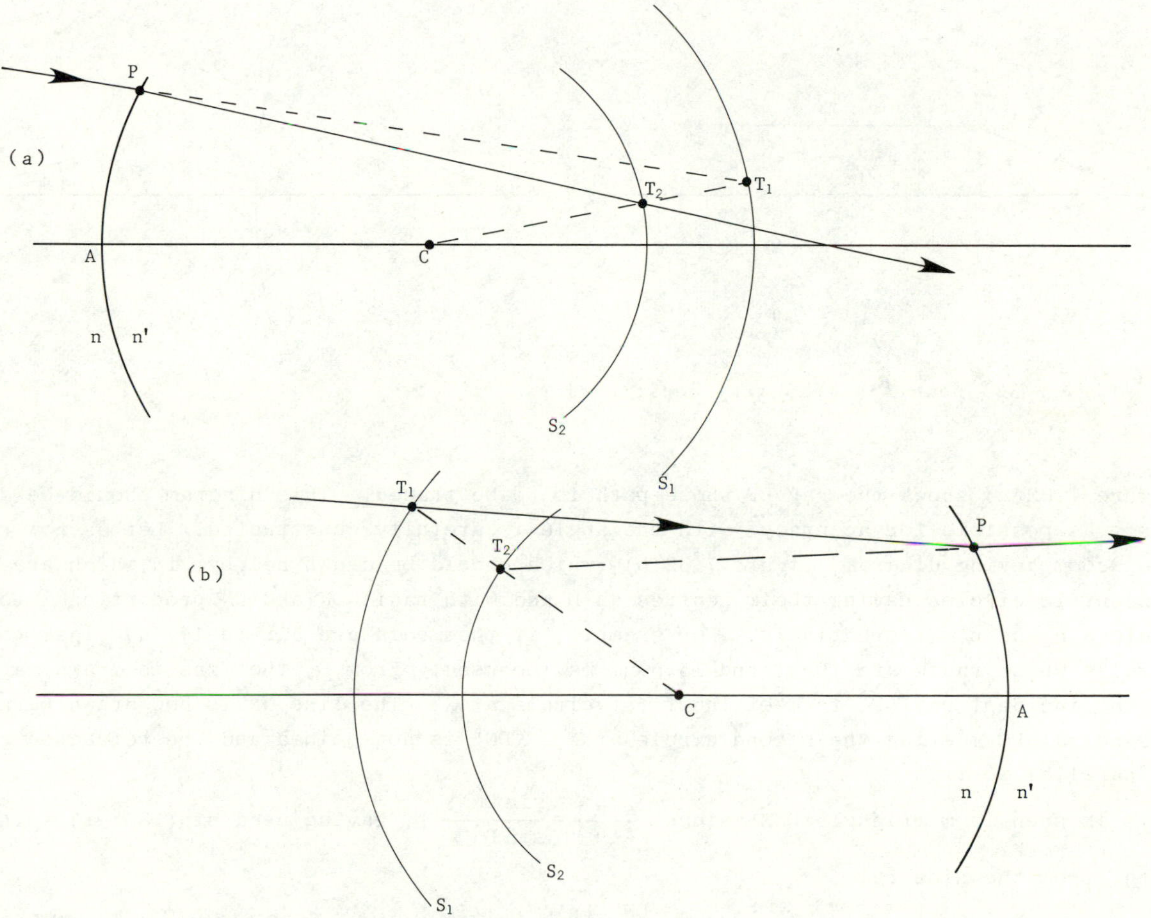

Fig. 7.31 Young's graphical construction.

In each figure, P is the point of incidence of a ray on the surface AP which separates media of indices n and n' (n' > n), and C is the centre of curvature of the surface. AC = r, the radius of curvature. This should be drawn to as large a scale as is convenient. Using C as the centre, two circles are drawn having radii $\frac{n'}{n}\cdot r$ and $\frac{n}{n'}\cdot r$, respectively. These are shown marked S_1 and S_2 in the diagram.

In figure 7.31(a), the incident ray to be traced is directed towards T_1 on the first circle S_1. The refracted ray will pass through T_2, the point of intersection of CT_1 with the second circle S_2. In figure 7.31(b), a similar procedure is followed, and the diagram should be self-explanatory.

7.3.6.2 Dowell's Graphical Construction

This method makes use of Snell's law $n \sin i = n' \sin i'$, and is a generalised version of the method described in section 2.1.4 for refractions at plane surfaces. Figure 7.32 shows how the procedure is applied to a single converging surface where $n' > n$.

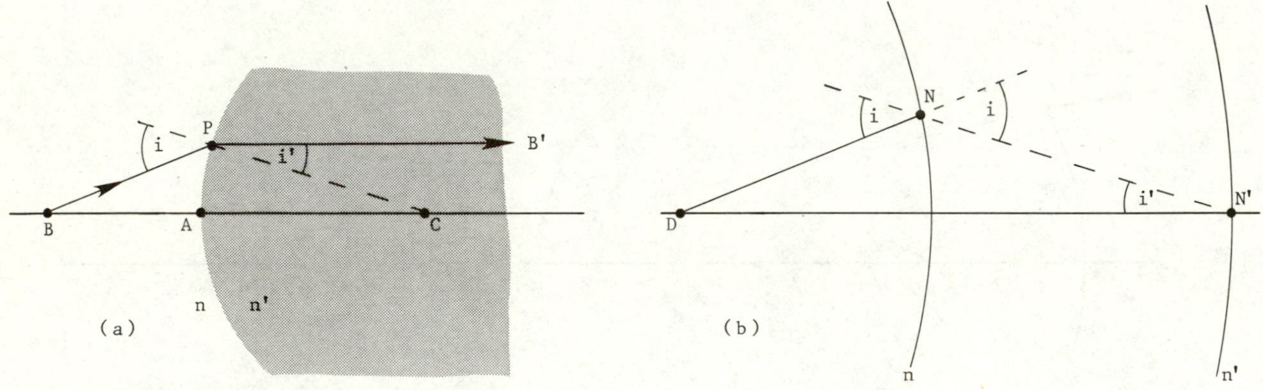

Fig. 7.32 Dowell's graphical construction.

Figure 7.32(a) shows the ray BP whose path is to be traced. This diagram should be drawn as large as possible for accuracy, with the angle i carefully constructed. Let us now refer to the accompanying diagram, figure 7.32(b), which should be drawn nearby, in which are shown concentric circles having their centres at D and with radii DN and DN' proportional to the indices n and n', respectively. For example, if the media are air (n=1) and glass (n'= 1.55), circles whose radii are 10 cm and 15.5 cm may be used. From D, the line is drawn parallel to the incident ray BP, to meet the first circle at N. The line NN' is now drawn parallel to the normal PC meeting the second circle at N'. DN' is now joined and the refracted ray PB' is parallel to it.

This is seen from triangle DNN' since $\frac{\sin i}{\sin i'} = \frac{\sin D\hat{N}N'}{\sin D\hat{N}'N}$, having used $\sin(180^0 - i) = \sin i$.

Then, from the sine rule,

$$\frac{\sin D\hat{N}N'}{\sin D\hat{N}'N} = \frac{DN'}{DN} = \frac{n'}{n}$$

from which $\frac{\sin i}{\sin i'} = \frac{n'}{n}$. Let us note that a ray can easily be traced through parabolic and other aspherical surfaces by the aforementioned methods if the direction of the normal at the point of incidence is known.

7.3.6.3 Trigonometrical Ray Tracing

This procedure is very important for accurate ray tracing with all types of optical systems, including some forms of spectacle lenses. In practice a number of rays in various planes is traced through the system, surface by surface, to give an accurate assessment of the nature of the refracted pencils. The treatment of skew rays is somewhat involved since they do not intersect the axis and is best achieved with a suitable program and a programmable calculator, beyond the scope of this book. However, much information can be obtained by a consideration of ray tracing in a single plane containing the optical axis. Nowadays, simple and relatively inexpensive non-programmable calculators enable these ray tracing calculations to be undertaken easily, and in a systematic manner. If logarithm and trigonometrical tables are used they must allow at least five decimal places.

The equations used in the trigonometrical tracing method may be deduced by reference to figure 7.33 which shows a ray of light from an object point B incident on a spherical refracting surface at the marginal point M. The centre of curvature is at C and the radius of curvature is denoted by r. The refracted ray intersects the axis of the surface at B'.

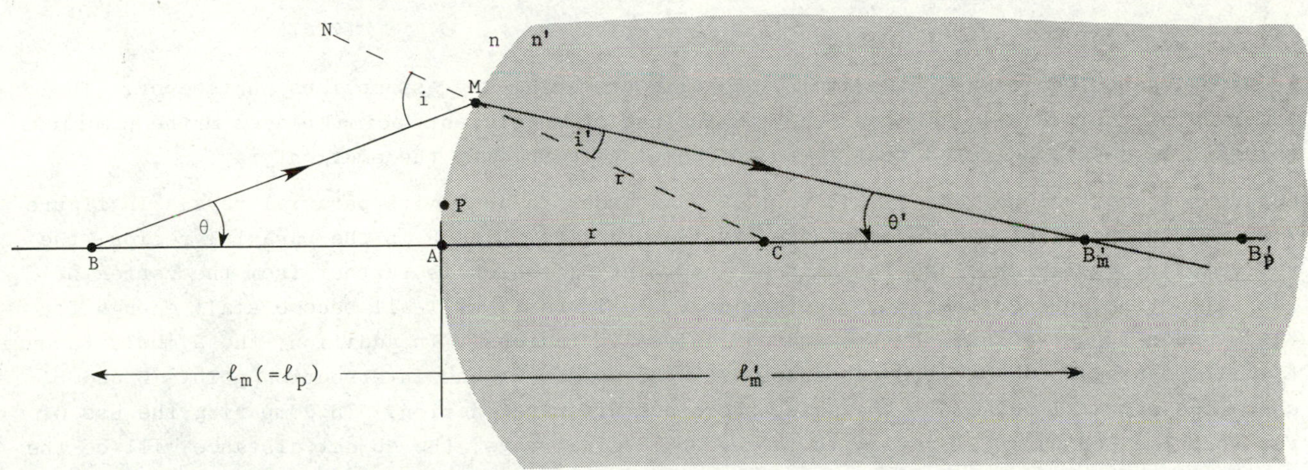

Fig. 7.33 Trigonometrical ray tracing.

The symbols ℓ_m *and* ℓ'_m will be used to denote the distances from the vertex A of the surface of the intersection points B and B'_m of the marginal ray with the axis before and after refraction. Later we shall show how the equations are modified by using the corresponding paraxial distances ℓ_p and ℓ'_p. θ and θ' are the inclinations of the ray with the axis before and after refraction, respectively.

Now in triangle BMC, using the sine rule, we may write $\frac{\sin B\hat{M}C}{BC} = \frac{\sin M\hat{B}C}{MC}$. But, we can see that $\sin B\hat{M}C = \sin(180^0 - B\hat{M}C) = \sin i$, and $\sin M\hat{B}C$ may be written as $-\sin\theta$ since θ is a negative angle. Also, BC = AB + AC and this is written as $-\ell_m + r$, and MC = r. Therefore, we have

$$\frac{\sin i}{-\ell_m + r} = \frac{-\sin\theta}{r} .$$

This may be rearranged to give

$$\sin i = \frac{\ell_m - r}{r} \cdot \sin\theta \qquad (7.24).$$

In addition, applying Snell's law to the refraction we have

$$\sin i' = \frac{n}{n'} \sin i \qquad (7.25).$$

Now, in triangle BMC, $N\hat{M}B = M\hat{B}C + M\hat{C}B$, where $N\hat{M}B = i$ but $M\hat{B}C = -\theta$ as θ is a negative angle.

$$\therefore \quad i = -\theta + M\hat{C}B \qquad (i).$$

Also, in triangle MCB'_m, $M\hat{C}B = C\hat{M}B'_m + M\hat{B}'_mC$, where $C\hat{M}B'_m = i$, and $M\hat{B}'_mC = \theta'$ since θ' is a positive angle.

$$\therefore \quad M\hat{C}B = i' + \theta' \qquad (ii).$$

Substituting $M\hat{C}B$ from (ii) into (i)

$$i = -\theta + i' + \theta'$$

$$\text{or} \quad \theta' = \theta + i - i' \qquad (7.26).$$

Again, in triangle MCB'_m, by the sine rule we have

$$\frac{\sin C\hat{M}B'_m}{CB'} = \frac{\sin M\hat{B}'_mC}{MC}$$

where $C\hat{M}B'_m = i'$, $M\hat{B}'_mC = \theta'$, $CB'_m = \ell'_m - r$, and $MC = r$. Therefore, $\frac{\sin i'}{\ell'_m - r} = \frac{\sin \theta'}{r}$, which gives

$$\ell'_m - r = \frac{\sin i'}{\sin \theta'} \cdot r \qquad (7.27)$$

We have deduced equations (7.24) to (7.27) without making any assumptions whatsoever. The accuracy of their use will depend only on the use of sufficient decimal places in the numerical values. It should be noted that ℓ'_m and θ' completely specify the emergent ray.

Let us now modify equations (7.24) to (7.27) in order to deal with paraxial rays. In figure 7.33, if the point of incidence of the ray is moved from M to P in the paraxial region, the refracted ray will now cross the axis at the point B'_p, which is further from the vertex of the refracting surface than B'_m. The angles θ, θ', i and i' will all become small enough for their sines to be replaced by the angular values in radians. In addition, the symbols ℓ_p and ℓ'_p will be used to denote the distances from the vertex of the intersection points B and B'_p where the paraxial ray meets the axis before and after refraction. In comparing the use of the formulae for paraxial rays with those for marginal rays, the object distance will be the same for both; that is, $\ell_p = \ell_m$. Hence, rewriting equation (7.24) in its paraxial form, we have

$$i = \frac{\ell_p - r}{r} \cdot \theta \qquad (7.28)$$

Similarly, using paraxial notation for equations (7.25) to (7.27) we obtain

$$i' = \frac{n}{n'} i \qquad (7.29)$$

$$\theta' = \theta + i - i' \qquad (7.30)$$

$$\ell'_p - r = \frac{i'}{\theta'} \cdot r \qquad (7.31).$$

It is convenient, from the calculation point of view, to rearrange equations (7.27) and (7.31) as follows: $\ell'_m = r\left(1 + \frac{\sin i'}{\sin \theta'}\right)$ (7.27a) and $\ell'_p = r\left(1 + \frac{i'}{\theta'}\right)$ (7.31a).

The fairly recent introduction of electronic calculators has speeded up the computation procedures at student level. The following example has been worked with the assistance of a Casio fx-31 Scientific Calculator. Before prodceeding with the trigonometrical trace, a few points should be noted.

(i) We shall assume that an incident slope angle θ of about 5^0 will represent the extreme marginal ray.

(ii) The linear nature of the paraxial computing formulae will result in the same intersection length AB_p' for any value of the slope angle θ. So we may choose any conveniently small angle. We chose 0.1^0 in the following example. Note also that the angle may be left in degrees rather than converting to radians.

(iii) In using the computation formulae, the optical sign convention previously discussed in section 2.2.3 is used. In addition, it is necessary to note the adoption of the following:

for rays meeting the surface at points off the axis, such as ray BM in figure 7.33, then measuring from a ray to a radius at that point on the surface, a clockwise direction is taken as a positive angle and an anti-clockwise direction as a negative angle.

(iv) The slope angle θ' and the intersection distance ℓ' of the emergent ray from the first surface are carried through to the second surface where they become θ and ℓ for the second surface. The procedure is then repeated.

Worked Example

A biconvex lens in air is specified by the constants $r_1 = +44$ mm, $r_2 = -247$ mm, $t = 10$ mm, and refractive index 1.5175. Calculate the path taken by a ray originating from an axial object point 220 mm from the first surface and inclined at 5.25^0 to the axis. Also compute the path for a paraxial ray inclined at 0.1^0 to the axis

Marginal ray

First surface

Eq. (7.24)
$$\sin i = \frac{\ell_m - r}{r}\sin\theta = \frac{-220-44}{44}\sin(-5.25^0) = 0.549\,009\,7$$
$$\therefore\ i = 33.299\,102^0$$

Eq. (7.25)
$$\sin i' = \frac{n}{n'}\sin i = \frac{1}{1.5175}\times 0.549\,009\,7 = 0.361\,785\,6$$
$$\therefore\ i' = 21.209\,899^0$$

Eq. (7.26)
$$\theta' = \theta + i - i' = -5.25^0 + 33.299\,102^0 - 21.209\,899^0 = 6.839\,203^0$$

Eq. (7.27a)
$$\ell_m' = r\left(1+\frac{\sin i'}{\sin\theta'}\right) = 44\left(1+\frac{0.364\,185\,5}{0.116\,525\,7}\right) = +177.675\,83\text{ mm.}$$

Second surface

$$\sin i = \frac{\ell_m - r}{r}\sin\theta = \frac{167.675\,83-(-247)}{-247}\sin(6.839\,203^0) = -0.199\,923$$
$$\therefore\ i = -11.532\,458^0$$

$$\sin i' = \frac{n}{n'}\sin i = \frac{1.5175}{1}\times\sin(-11.532\,458^0) = -0.303\,383\,1$$
$$\therefore\ i' = -17.660\,917^0$$

$$\theta' = \theta + i - i' = 6.839\,203^0 + (-11.532\,458^0) - (-17.660\,917^0) = 12.967\,662^0$$

$$\ell_m' = r\left(1+\frac{\sin i'}{\sin\theta'}\right) = -247\left(1+\frac{\sin(-17.660\,917^0)}{\sin(12.967\,662^0)}\right) = +86.936\,173\text{ mm.}$$

Now, ℓ_m(second surface) $= \ell_m'$(first surface) $- t$

$= 167.675\,83$,

and θ(second surface) $= \theta'$(first surface)

$= 6.839\,203^0$.

Paraxial ray

	First surface	Second surface
Eq. (7.28)	$i = \frac{\ell_p - r}{r}\theta$	$i = \frac{\ell_p - r}{r}\theta$
	$= \frac{-220 - 44}{-44} \times (-0.1^0)$	$= \frac{200.299\,11 - (-247)}{-247} \times (-0.104\,612\,9)$
	$= -0.6^0$	$= 0.189\,446\,3^0$
Eq. (7.29)	$i' = \frac{n}{n'}i$	$i' = \frac{n}{n'}i$
	$= \frac{1}{1.5175} \times (-0.6^0)$	$= \frac{1.5175}{1} \times 0.189\,446\,3^0$
	$= -0.395\,387\,1^0$	$= 0.287\,484\,8^0$
Eq. (7.30)	$\theta' = \theta + i - i'$	$\theta' = \theta + i - i'$
	$= 0.1^0 + (-0.6^0) - (-0.395\,387\,1^0)$	$= -0.104\,612\,9^0 + 0.189\,446\,3^0 - 0.287\,484\,8^0$
	$= -0.104\,612\,9^0$	$= -0.202\,651\,4^0$
Eq. (7.31a)	$\ell_p' = r(1 + \frac{i'}{\theta'})$	$\ell_p' = r(1 + \frac{i'}{\theta'})$
	$= 44(1 + \frac{(-0.395\,387\,1^0)}{(-0.104\,612\,9)})$	$= -247(1 + \frac{0.287\,484\,8}{(-0.202\,651\,4)})$
	$= +210.299\,11$ mm.	$= +103.398\,34$ mm.

ℓ_p(second surface) = ℓ_p'(first surface) − t

$= 200.299\,11$ mm

and θ(second surface) = θ'(first surface)

$= -0.104\,612\,9^0$

Hence, at the second surface $\ell_m' = +86.936\,173$ mm and $\ell_p' = +103.398\,34$ mm. Therefore, the spherical aberration $\ell_p' - \ell_m' = +16.462\,167$ mm.

EXERCISES

1 Describe the basic construction of an achromatic doublet lens, and explain why it is necessary to use two different types of glass. Why does an achromatic doublet not focus all colours of light in the same position?

2 Calculate the V-numbers of the following materials.

	$n_{C'}$	n_d	$n_{F'}$
(a)	1.5225	1.5277	1.5335
(b)	1.6340	1.6405	1.6467
(c)	1.6568	1.6625	1.6790

A lens is made of glass (c) above and has a power (for n_d) of +10.00 D. Determine the powers of the lens for $n_{C'}$ and $n_{F'}$.

Ans. 47.97, 50.43, 29.84, +9.914 D, and +10.25 D.

3 An equiconvex thin lens, having surfaces of radii 25 cm, is made of plastics material having a mean index 1.490 and a V-number 54.0. If it forms an image of a distant object calculate (a) the longitudinal chromatic aberration of the image over the range (F'-C'), (b) the angular transverse aberration for a point on the lens 15 mm from the principal axis.

Ans. 0.073 D. 0.109^Δ.

4 Explain what is meant by the chromatic aberration of a lens. A crown glass lens, V = 58.5, is plano-convex in form and has 0.12 D of longitudinal chromatic aberration for the range

C' to F'. The mean index of the glass is 1.5250. An achromatic doublet is to be made using the above lens cemented in contact with an equiconcave lens made of flint glass. The flint glass has V = 34.5 and a mean dispersion $(n_{F'} - n_{C'}) = 0.0175$. Calculate the power of the second lens, and the radii of curvature of its surfaces.

Ans. -4.14 D, -29.17 cm, +29.17 cm.

5 (a) An achromatic doublet consists of converging and diverging component lenses. During manufacture, the diverging component was accidentally made from the same glass as the converging component. What differences would you expect this to make to the paraxial properties of the doublet?

(b) The objective lens in the telescope of a spectrometer consists of a cemented achromatic doublet. Accidentally, it is fixed the wrong way round in the telescope tube. What differences will this make to the paraxial properties of the system? Will it affect the marginal properties of the system?

(c) The components of a telescope objective which consists of an achromatic doublet are moved a short distance apart. Will this affect the quality of an axial image? Explain your answer.

6 The radii of curvature of the surfaces of an equiconvex lens are both 200 mm and the refractive indices of the glass for red and blue light are 1.640 and 1.680, respectively. Find the focal lengths for red and blue light, and determine the diameter of the circle of least confusion if the aperture of the lens is 40 mm.

Ans. 15.63 cm, 14.71 cm, and 1.21 cm.

7 What is meant by the chromatic aberration of a lens? Discuss the importance of this effect in connection with spectacle lenses. A cemented achromatic doublet consists of two components made of the following glasses.

	$n_{C'}$	n_d	$n_{F'}$
Crown glass	1.5015	1.5045	1.5105
Flint glass	1.6225	1.6305	1.6405

If the doublet is required to have a power of +2.00 D, determine the powers of the two components.

Ans. +5.331 D and -3.331 D.

8 A cemented achromatic doublet is to have a focal length of 200 cm. The following materials are available:

	$n_{C'}$	n_d	$n_{F'}$
Hard crown	1.5143	1.5188	1.5228
Dense flint	1.6118	1.6166	1.6288

Determine the powers of the component lenses. If the outer surface of the flint component is to be plane, calculate the radii of curvature of the faces of the component lenses.

Ans. +1.232 D, -0.732 D, $r_1 = +842.1$ mm, $r_2 = -842.3$ mm, $r_3 = -842.3$ mm, $r_4 = \infty$.

9 Explain how, and to what extent, a pair of thin lenses may be assembled to form an achromatic system. An equiconcave lens of glass of index $n_d = 1.600$, of dispersive power 0.0280. and having a focal length of 28 cm is to be combined with a convex lens of glass of index $n_d = 1.500$ and of dispersive power 0.0170 to form an achromatic doublet. Assuming a common contact radius between the two lenses, calculate the radii of curvature of the surfaces of the convex lens. Ans. $r_1 = +113.8$ mm and $r_2 = -335.9$ mm.

10 Using the glasses specified below, find the focal lengths of the crown and flint components of a cemented doublet to give a diverging doublet of focal length 10 cm such that it is achromatised for the C' and F' wavelengths.

	Refractivity	Mean dispersion (C' to F')
Crown	0.5169	0.00853
Flint	0.6226	0.01727

Ans. $F_1 = -24.68$ D and $F_2 = +14.68$ D.

11 Show that a system of two thin lenses made of the same material will be approximately achromatic if the separation of the lenses is equal to the average focal length. Will the system be diverging or converging?

12 Prove that with two lenses of focal lengths f_1' and f_2' spherical aberration will be a minimum if the separation between the lenses is given by $d = f_1' - f_2'$.

13 What is meant by aplanatic points of a surface? Find the positions of the aplanatic points for a positive powered refracting surface, of radius 20 cm, which separates media of refractive indices 1.000 and 1.550.

Ans. 32.9 cm and 51.0 cm from the surface.

14 A spherical surface of radius +9 cm separates two media of refractive indices 1.62 and 1.80. Incident light from the left is converging towards an axial object point 19 cm to the right of the vertex of the surface. Calculate the position of the image point and show that this is free from spherical aberration.

Ans. $\ell' = +17.10$ cm.

15 A narrow pencil of rays from a very distant object is incident on a spherical refracting surface which separates air from glass of refractive index 1.550. If the radius of the surface is +55 mm and the angle of incidence is 40^0, calculate the positions of the tangential and sagittal focal lines.

Ans. $t' = +10.95$ cm and $s' = +13.23$ cm.

16 List the aberrations from which the images formed by lenses may suffer. State and explain which are of particular importance in (a) camera lenses, (b) telescope objectives, and (c) spectacle lenses.

17 Explain what is meant by tangential meridian, sagittal meridian, circle of least confusion, astigmatic pencil, in relation to the defect of images known as oblique astigmatism. With the aid of image shell diagrams, show how the magnitude of oblique astigmatism varies with the form of a lens.

18 A patient is prescribed thin lenses for distance and near vision of powers +10 D and +13 D respectively. Inadvertently, they are both fitted in the same plane, thereby introducing an angle of incidence of 15^0 for near vision. Calculate F_S and F_T if $n_g = 1.5$.

19 A spherical refracting surface of radius 4 cm separates media of indices 1.000 and $\sqrt{3}$. A narrow parallel pencil of rays is incident centrally on the surface at an angle of incidence of 60^0. Show that the positions of the tangential and sagittal images as measured from the point of incidence are given by $3\sqrt{3}$ and $4\sqrt{3}$, respectively.

20 Trace by trigonometrical computation the path taken by a ray which is refracted by a spherical surface for which $n = 1.000$, $n' = 1.5205$, and $r = +5.5$ cm. The ray originates from an axial object point 25 cm in front of the surface and makes an angle of -2^0 with

the principal axis. Find also the point at which a paraxial ray from the same object point crosses the axis after refraction.

Ans. $\ell'_m = +27.225\ 51$ cm, and $\ell'_p = +30.054\ 43$ cm.

21 Show that the angular transverse chromatic aberration produced by a thin lens of power F at a point c cm from the optical centre is given by $\frac{cF}{V}$ prism dioptres.
A bifocal spectacle lens is made from two plano-convex glass lenses cemented together with their plane faces in contact. The distance portion is circular and 52 mm in diameter, whilst the reading segment is the same shape but only 22 mm in diameter. Both portions are worked individually with their optical centres at their geometrical centres and they are cemented with their centres 15 mm apart. Calculate the angular transverse chromatic aberration due to each component individually at a point 10 mm from the optical centre of the distance portion on the line joining the centres of the two components. Assume distance and segment powers of +5.00 D and +3.00 D, respectively, and a V-number of 60. Calculate also the resultant chromatic aberration at the same point.

Ans. Due to distance portion = 0.05 prism dioptres. Due to near portion = 0.025 prism dioptres. Resultant angular dispersion is 0.05 - 0.025 = 0.025 prism dioptres with the deviation in the same direction as that due to the distance portion.

22 In the lens in question 21, find the point at which there is no angular transverse chromatic aberration.

Ans. The point is on the line joining the centres of the components, 5.625 mm from the centre of the distance component. Note that this point is the optical centre of the +8 D near vision portion!

23 The accompanying figure acts as an object for a thin converging lens. The object plane is perpendicular to the principal axis and the central point is on the axis. Assume that the lens suffers only from (a) oblique astigmatism, (b) distortion, (c) spherical aberration, (d) coma. In each case, sketch the image produced by the lens.

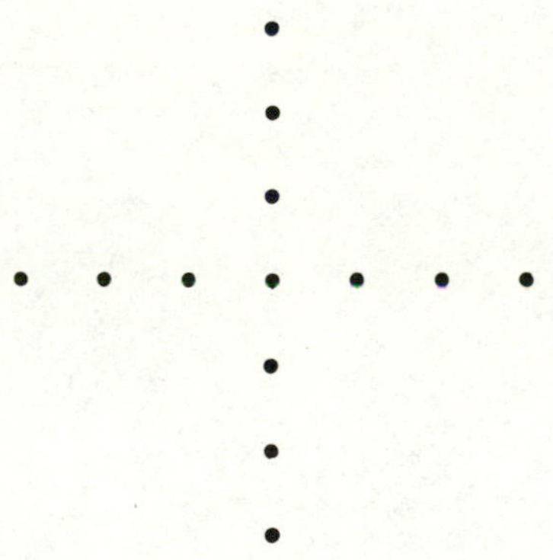

24 Use a trigonometrical ray trace for a plano-convex lens to show the difference in spherical aberration for an object point at 1 000 cm when the lens first has its convex surface and then its plane surface facing the incident light. Choose your own surface powers, centre thickness, and refractive index. Assume angles of incination for the marginal ray and the paraxial ray of 0.1^{0} and 0.001^{0}, respectively. Consider the approximate distances from the principal axis at which these rays will be incident upon the lens, and explain why such small angles have been chosen.

25 A thin lens is made of glass of refractive index 1.65 and is to have a focal length of +20 cm. It is to have minimum spherical aberration for a distant object. Use the crossed lens form (page 238) and calculate the shape factor.

Ans. $\sigma = +0.714$.

26 With the aid of figure 7.12, calculate the radii of curvature of the surfaces if the lens in question 25 is to have zero coma.
Ans. r = +0.144 m and r = −1.3 m.

27 Classify the monochromatic aberrations in terms of those blurring and those not blurring the image.

28 In the presence of a small aperture in a system, which aberrations are unimportant?

29 Consider the following facts about the human eye. (a) The cornea flattens towards its margin. (b) The pupil diameter is about 4 to 5 mm and the pupil is placed in front of the crystalline lens. (c) The image screen, the retina, is curved. (d) the sensitivity of the retina with respect to seeing detail drops off markedly away from the centre. How will these affect the imaging and seeing of images?

8 THE PRINCIPLES OF OPTICAL INSTRUMENTS

8.0 INTRODUCTION

The term optical instrument covers a very wide range of devices. Many of these may be described as devices which magnify. That is to say, they enable the observer's eye to form a retinal image of the object under inspection which covers a greater area of the retina than would be possible without the instrument. For example, microscopes are used for the examination of small objects close to the observer, whilst telescopes are used to examine objects which appear to be small because of their great distance away from the viewer.

Instruments may be classified as non-visual, or visual. In the latter types the instrument is intended to assist the eye. Thus, we shall include in this section a description of the human eye, although this will be somewhat brief but will suffice for the understanding of visual instruments.

The final part of this chapter will be concerned with the effects of stops which restrict the rays transmitted by optical systems.

8.1 NON-VISUAL INSTRUMENTS

8.1.1 THE CAMERA

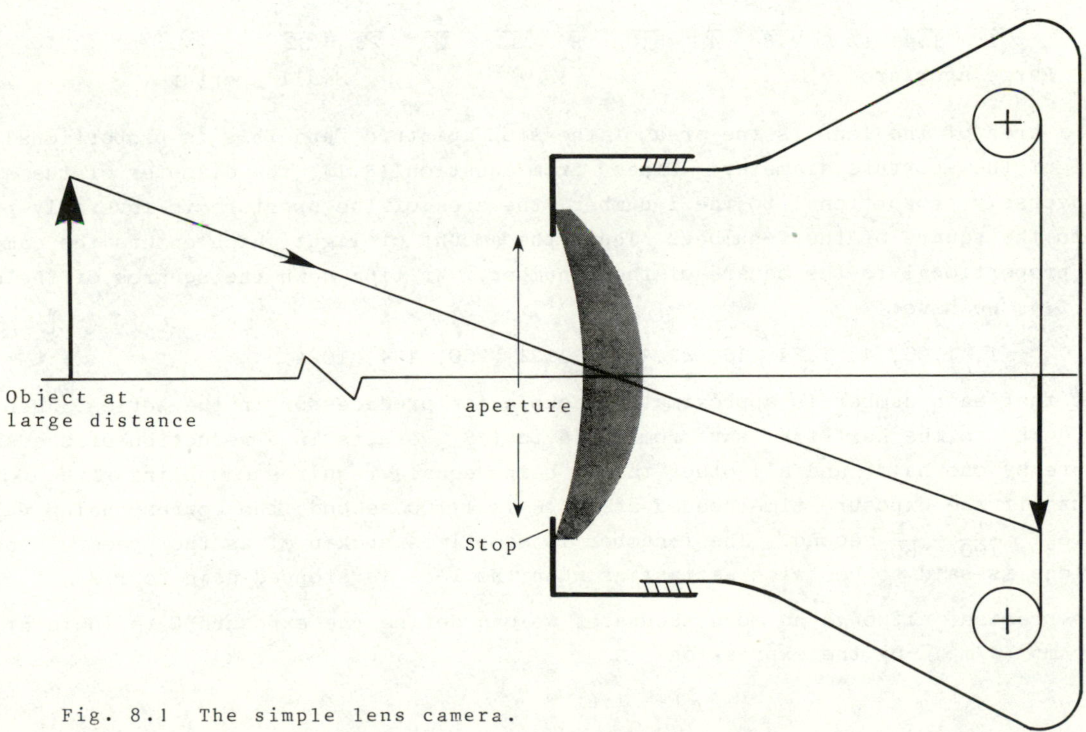

Fig. 8.1 The simple lens camera.

The simple lens camera is shown diagrammatically in figure 8.1 . The principle of the camera is the formation of a real, inverted image on a light sensitive surface by a converging lens. In the cheapest cameras a single positive powered meniscus lens with a fixed stop is employed. Originally called a landscape lens, this device suffers considerably from aberrations, particularly spherical aberration, oblique astigmatism, and chromatic aberration. The inverted nature of the image is no drawback as the image is made upright during the printing process. For various distances of an object from the camera, the image is focused on the film by altering the distance between the lens and the film. This is often achieved by movement of the lens on a screw threaded mounting.

With all but the cheapest cameras the lens assembly is fitted with a stop of variable aperture by means of which the light passing through may be varied. A shutter, operating usually either near the lens assembly or next to the film, in which case it is known as a focal plane shutter, limits the time to which the film is exposed to the light.

8.1.1.1 The f-number

Where the aperture diameter D of the stop is variable, it may be adjusted to be a definite fraction of the focal length of the lens assembly. For example, the aperture setting may be written in the form $\frac{f}{8}$, or f/8, which means that the aperture has a diameter which is one eighth of the focal length f'. This fraction is referred to as the f-number, or f/#, of the aperture, and is given by

$$f/\# = f'/D \qquad (8.1).$$

In many cases the aperture of a camera lens may be set to all or part of the following scale of f-numbers:

$$\frac{f}{1} \quad \frac{f}{1.4} \quad \frac{f}{2} \quad \frac{f}{2.8} \quad \frac{f}{4} \quad \frac{f}{5.6} \quad \frac{f}{8} \quad \frac{f}{11} \quad \frac{f}{16} \quad \frac{f}{22} \quad \frac{f}{32}$$

Large aperture Small aperture

The usable area of the lens is the area of the stop aperture, and this is proportional to the square of the aperture diameter. Since, from equation (8.1), the diameter of the aperture is inversely proportional to the f-number, the area of the aperture is inversely proportional to the square of the f-number. Thus, the amount of light which enters the camera is inversely proportional to the square of the f-number. Writing down the squares of the above series of f/#, we have:

1, 1.96, 4, 7.84, 16, 31.4, 64, 121, 256, 484, 1024 .

We can see that each number is approximately double its predecessor in the series. Hence, "stopping down" to the next f/#, say from f/5.6 to f/8, results in a reduction of the area of the aperture by one half, and all other things being equal, requires a doubling of the exposure time. Thus, if the exposure time needed at f/5.6 is 1/100 second, the corresponding value at f/8 will be $2 \times \frac{1}{100} = \frac{1}{50}$ second. The f-number is sometimes spoken of as the speed of the lens. An f/5.6 lens is said to be twice as fast as when the lens is stopped down to f/8 .

To be more precise, although no more accurate, we can define the exposure E (= incident light energy per m^2 (J/m^2)) by the expression

$$E = I.\Delta t$$

where Δt is the time for which the camera shutter is open, called the exposure time, and I is the intensity (Watts per m^2) in the image plane.

Now, since I is proportional to the area of the stop aperture, we can write $I \propto D^2$ because the area of the stop aperture is $\pi\left(\frac{D}{2}\right)^2$. But from equation (8.1) $D = \frac{f'}{f/\#}$, so we can deduce that

$$I \propto \frac{1}{(f/\#)^2}\ , \quad \text{whence} \quad E \propto \frac{\Delta t}{(f/\#)^2}\ .$$

Thus, repeating the foregoing example, we have $(f/\#)_1 = 5.6$, $(f/\#)_2 = 8$, $\Delta t_1 = \frac{1}{100}$ s. To find Δt_2, we have, for constant exposure E

$$E \propto \left(\frac{\Delta t}{(f/\#)^2}\right)_1 = \left(\frac{\Delta t}{(f/\#)^2}\right)_2$$

$$\text{so,} \qquad \Delta t_2 = \frac{(f/\#)_2^2}{(f/\#)_1^2} \cdot \Delta t_1 = \frac{8^2}{5.6^2} \times 0.01 = 0.02\ \text{s},$$

which is 1/50 second, as before.

Stopping down an aperture has another effect: the apparent distance over which objects appear to be in focus increases. This effect, known as the depth of field, will be considered in section 8.1.1.4 .

8.1.1.2 The Camera Lens

The field of view of a camera lens must be about 45°- 50°. Compare this with the 1°- 2° field of view for a microscope objective lens. Cheap, fixed focus cameras normally utilise a converging meniscus lens together with a stop. The one illustrated in figure 8.2 is a doublet, corrected for chromatic aberration.

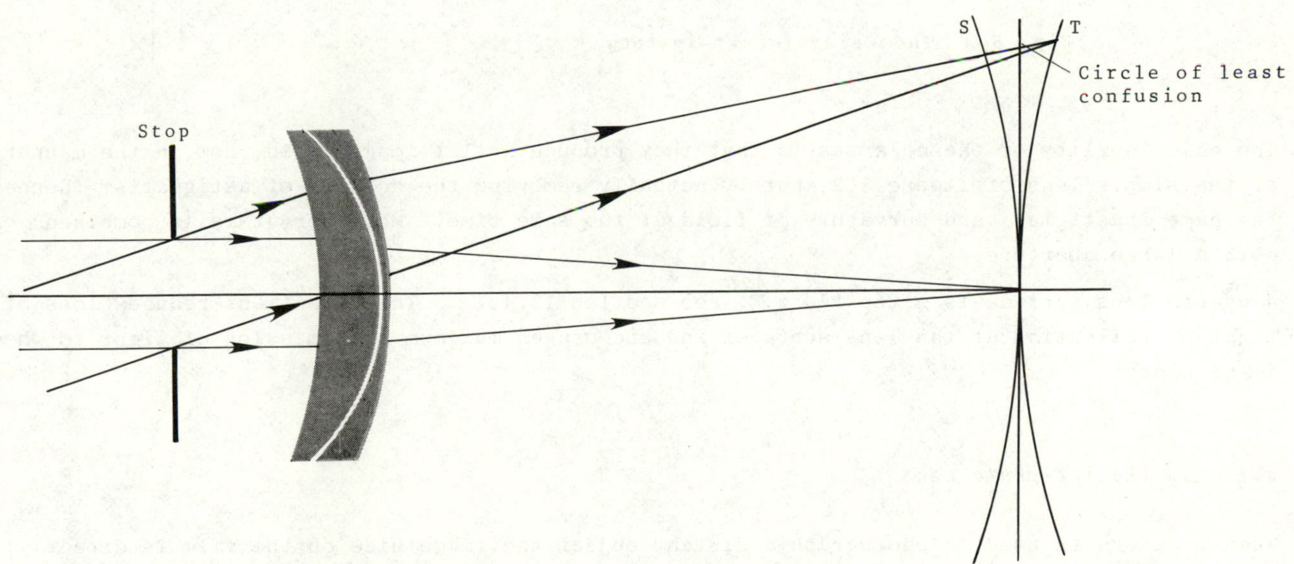

Fig. 8.2 An inexpensive achromatic meniscus camera lens.

Such a lens has considerable oblique astigmatism, but if the stop is suitably placed, the tangential (T) and sagittal (S) astigmatic surfaces are curved in the opposite senses. As the circles of least confusion lie between them, the result is that they lie approximately on a flat surface. For ordinary purposes the flat image field will be sharp enough if the aperture is limited to about f/11 and the field of view to about 40^0.

Good camera lenses are exceedingly complex and come into the category of lens systems. Modern cameras may have stop numbers as small as f/1. In such cases the lenses used are carefully designed and are the most expensive. The combination of wide field and large aperture makes the problem of correcting a photographic lens for aberrations a difficult one. Figure 8.3 shows a modern *Anastigmat* lens, the well known Zeiss Tessar system.

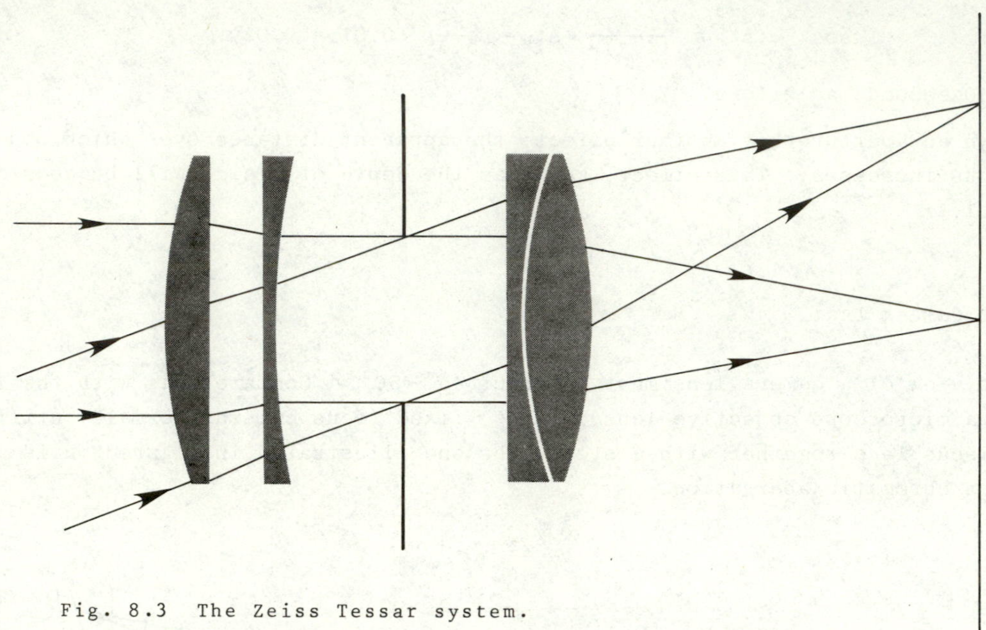

Fig. 8.3 The Zeiss Tessar system.

The chief quality of these lenses is that they produce a flat image field, not in the manner of the simple lens of figure 8.2, but by actually removing the defects of astigmatism (hence the name Anastigmat) and curvature of field at the same time. Moreover, this is combined with a large aperture.

A camera lens surface is often *bloomed*, see section 13.4.1 . This treatment reduces loss of light by reflection at the lens surfaces and encourages maximum transmission of light to the image plane.

8.1.1.3 The Telephoto Lens

When a camera is used to photograph a distant object the image size on the film is directly proportional to the angle θ subtended by the object at the lens, and also to the focal length of the lens. This is shown in figure 8.4 .

If a large image of a distant object is required this necessitates the use of a camera lens with an inconveniently long focal length. This may by avoided by the use of a *telephoto* lens consisting of converging and diverging component lenses. It resembles the Galilean telescope except that the system is not afocal (see section 8.3.4.4).

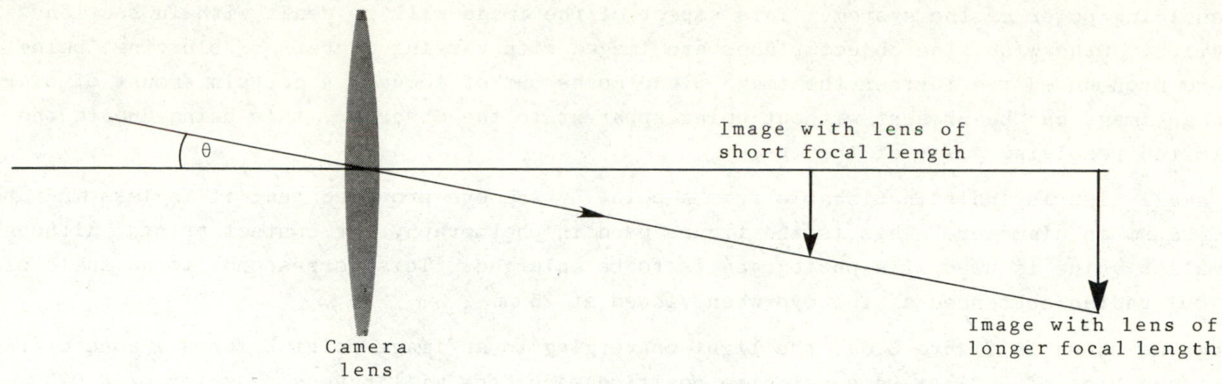

Fig. 8.4 Illustration of image size varying with focal length of lens.

The negative powered component lens L_2, figure 8.5, occupies the normal lens position in the camera with the positive powered component L_1 some distance in front. Figure 8.5 shows how the combination is equivalent to a single positive powered lens of long focal length, shown dashed in the figure, and positioned at the second principal point P' of the system.

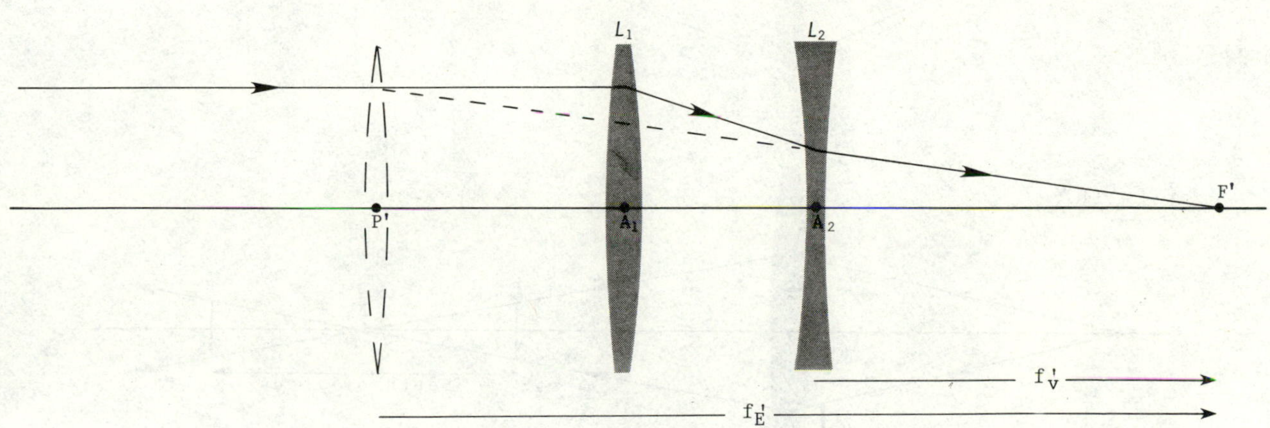

Fig. 8.5 The telephoto lens.

The back vertex focal length f'_V determines the length of the camera, whilst the equivalent focal length is the larger distance P'F' = f'_E. The ratio f'_E/f'_V is called the telephoto magnification. That is,

$$M_T = f'_E/f'_V \quad (8.2).$$

8.1.1.4 Depth of Focus and Depth of Field

The exact distance between a lens and a sharply focused image depends on the distance of the object from the lens. In the case of the camera lens, the photograph is a flat image of a usually three-dimensional object. However, only one plane in the object space is conjugate with the image plane and it is reproduced in the image with a definition dependent on the resolving power of the system. This aspect of the image will be dealt with in section 14.1.3. Otherwise, the object planes are imaged with varying degrees of blurring, being more pronounced the further the image plane moves out of focus. A certain amount of blur in an image can be present without being apparent to the observer, this being due to the limited resolving power of the eye.

A small disc is indistinguishable from a point by the eye provided that it is less than about 0.025 cm in diameter. This is the figure used in photography for contact prints, although a smaller value is used if a photograph is to be enlarged. This corresponds to an angle of 0.001 radian subtended at the eye when viewed at 25 cm.

Now, as shown in figure 8.6a, the light converging to an image point B' forms a cone of rays on each side of B'. At some definite position each cone will have a diameter of 0.025 cm.

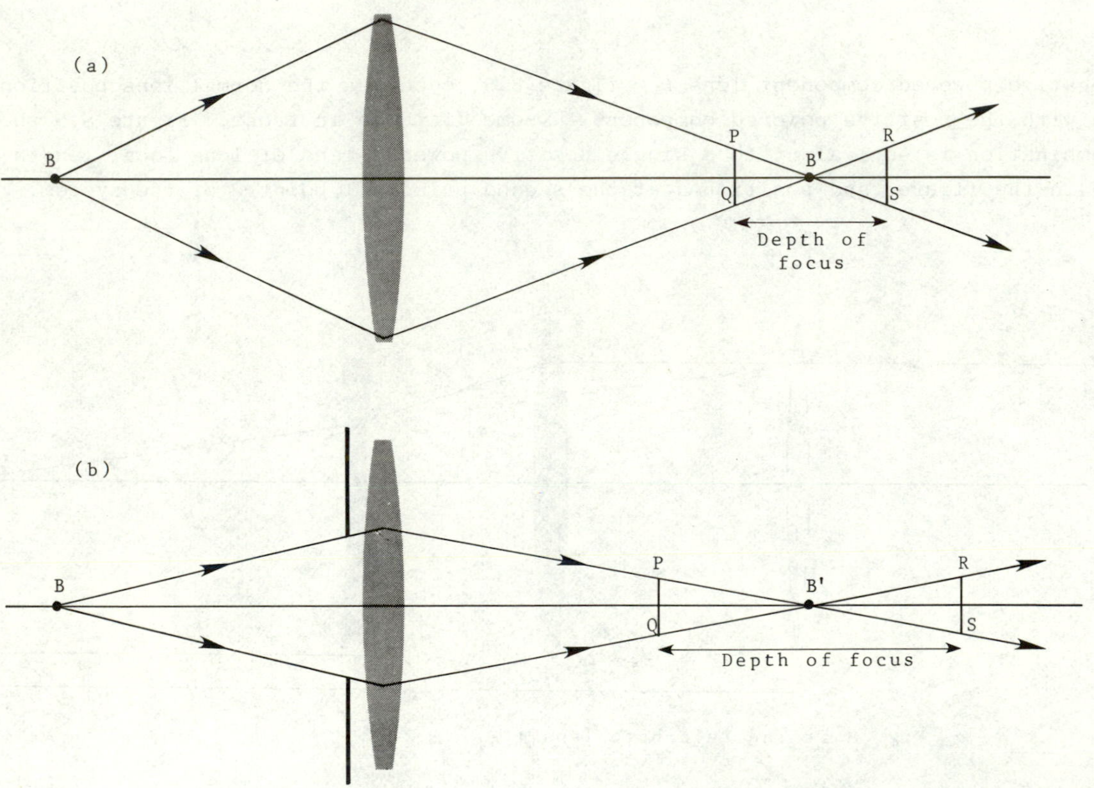

Fig. 8.6 Increased depth of focus with a reduced aperture.

Let these positions be denoted by PQ and RS. Therefore, the image of B will appear as a point anywhere between PQ and RS, and no variation in sharpness will be detectable in this region. In general, when an image screen is placed in the pencil of rays refracted by the lens, a patch of light called a *blur circle* or a *blur disc* is observed. Between PQ and RS therefore, the diameters of the blur circles are smaller than 0.025 cm, and are too small to be distinguishable from points when viewed by the eye.

We can now define the *depth of focus* of a lens, or a lens system, as the total axial range over which the image plane can be moved without noticeable deterioration in the image definition.

The size of the aperture of the lens has a considerable effect on the depth of focus attainable. This point can be appreciated from a consideration of figure 8.6(b). If the aperture of the lens is reduced, without change in the object position, the positions of PQ and RS, where the diameter of the refracted cone of rays is 0.025 cm, will be further apart than when the full lens aperture is used. That is to say, the depth of focus is increased. Thus we conclude that the depth of focus is increased when the aperture of the lens is reduced.

To test for yourself the effect of depth of focus, move a book towards your eye until you are unable to read the print. By inserting a card with a pinhole in it in front of and close to your eye it will now be possible to read the print.

In the same way we can show that objects at different distances from a lens may be tolerably imaged on an image plane which is at a fixed distance from the lens. This will be possible if the cones of light which are converging towards the image points form blur circles less than 0.025 cm diameter at this image plane. This is illustrated in figure 8.7 .

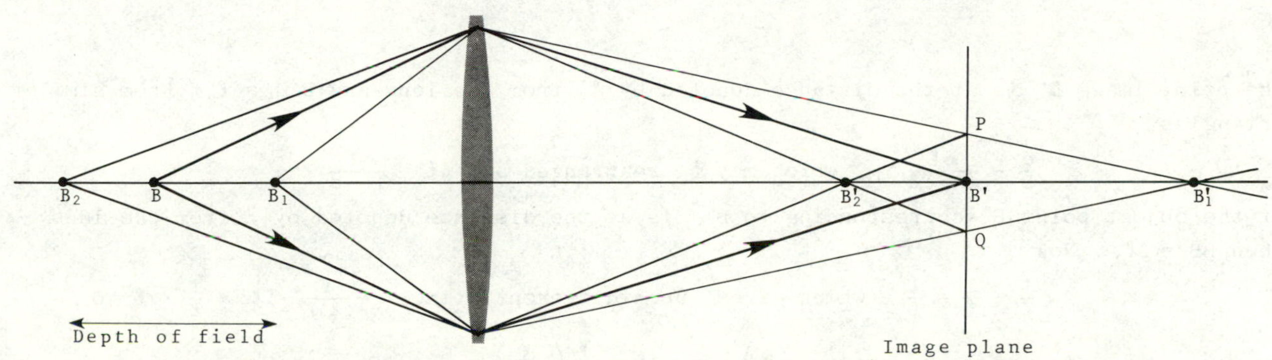

Fig. 8.7 Depth of field.

Let the point object B be focused by the lens so that a point image B' is formed on the image plane. In stating this we are ignoring aberrations and the effects of diffraction. A blur disc PQ, of maximum diameter 0.025 cm, will be formed for the point image B_1' corresponding to the point object at B_1, and thus the image plane will show clearly images of objects in the foreground between B and B_1. The blur disc PQ will also be obtained for the point image B_2' corresponding to a point object B_2, and thus the image plane will also show clearly images of objects in the background between B and B_2. The distance between the axial positions of an object for which it is tolerably in focus on a fixed image plane is termed the *depth of field*. This is the distance B_1B_2 in figure 8.7 .

In general, knowledge of the depth of field of a lens is more important than the depth of focus. Many modern cameras have a simple facility for reading the depth of field when the camera is focused at a particular distance and set to a particular stop size. For example, the Olympus OM-2N camera when focused at 8ft with the aperture at f/4 has a depth of field from about 7ft to 10ft. However, when stopped down to f/8 the depth of field is from about 6ft to 12ft. Note that the depth of field has doubled when the aperture diameter has halved. This doubling of the depth of field as the diameter of the aperture is halved is generally true, and we shall now consider how the depth of field may be calculated.

In figure 8.8 , let D represent the aperture of the lens and d the maximum permissible diameter of the blur circle PQ (0.025 cm).

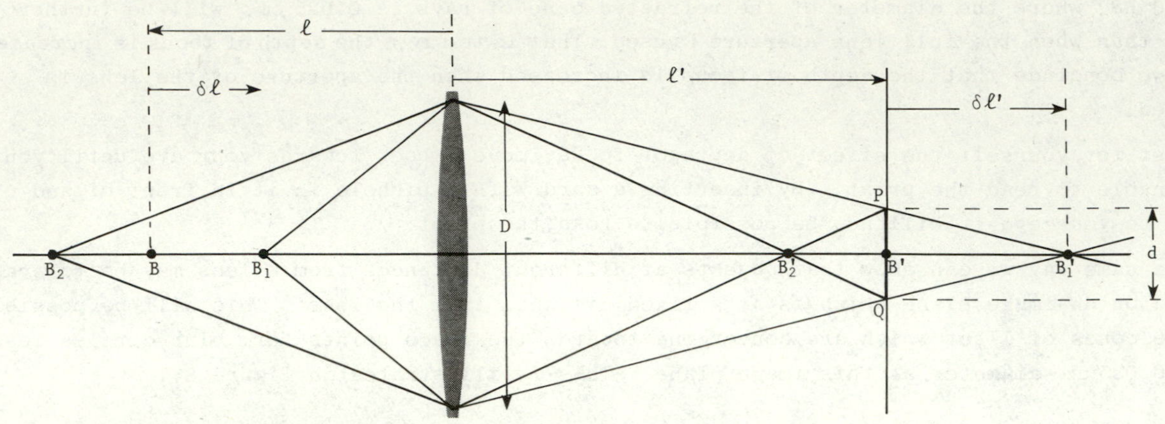

Fig. 8.8 Diagram to demonstrate that depth of field varies inversely as the diameter D of the aperture.

The point image B' is at the distance denoted by ℓ' from the lens and $B'B_1' = \delta\ell'$. From similar triangles

$$\frac{d}{D} = \frac{\delta\ell'}{\ell' + \delta\ell'}, \quad \text{which may be rearranged as} \quad \delta\ell' = \frac{d}{D-d}\cdot\ell'.$$

If the object point B, corresponding to B', is at the distance denoted by ℓ from the lens then $BB_1 = \delta\ell$. Now

$$\frac{1}{\ell'} - \frac{1}{\ell} = \frac{1}{f'}, \quad \text{which gives, upon differentiating,} \quad -\frac{1}{(\ell')^2}\cdot\delta\ell' + \frac{1}{\ell^2}\cdot\delta\ell = 0.$$

$$\therefore \quad \delta\ell = BB_1 = \frac{\ell^2}{(\ell')^2}\cdot\delta\ell' = \left(\frac{\ell}{\ell'}\right)^2 \frac{d}{D-d}\cdot\ell' = \frac{\ell^2}{\ell'}\left(\frac{d}{D-d}\right) \qquad \text{(i).}$$

Also, for the image B_2', the distance $B'B_2'$, or $\delta\ell'$, is again found by similar triangles. Hence, $\frac{d}{D} = \frac{-\delta\ell'}{\ell' + \delta\ell'}$, which rearranges to give $\delta\ell' = \frac{-d}{d+D}\cdot\ell'$. Note that $\delta\ell'$ is negative.

Also again, $-\frac{1}{(\ell')^2}\cdot\delta\ell' + \frac{1}{\ell^2}\cdot\delta\ell = 0$. $\quad\therefore\ \delta\ell = BB_2 = \frac{\ell^2}{(\ell')^2}\cdot\delta\ell' = -\frac{\ell^2}{\ell'}\left(\frac{d}{D+d}\right)$ (ii).

It can be seen from expressions (i) and (ii) that BB_1 is numerically greater than BB_2. Now,

$$\text{Depth of field} = B_1B_2 = BB_1 - BB_2, \text{ since } BB_2 \text{ is negative.}$$

$$= \frac{\ell^2}{\ell'}\left(\frac{d}{D-d}\right) - \left(-\frac{\ell^2}{\ell'}\left(\frac{d}{D+d}\right)\right)$$

$$= \frac{\ell^2}{\ell'}\left(\frac{d}{D-d} + \frac{d}{D+d}\right)$$

$$= \frac{\ell^2}{\ell'}\left(\frac{2dD}{D^2-d^2}\right)$$

Neglecting the term d^2 since it is very small compared with D^2, we have

$$\text{Depth of field} \simeq \frac{2\ell^2 d}{\ell' D} \qquad (8.3).$$

It can be seen from equation (8.3) that the depth of field is approximately inversely proportional to the diameter of the lens aperture D for a given object distance. This shows that, in common with the depth of focus, the depth of field increases as the lens aperture is decreased.

8.1.1.5 The Hyperfocal Distance

Figure 8.9 shows a lens on which is incident a pencil of rays parallel to the principal axis. The rays are brought to a focus at the second principal focus F', a distance f' from the lens.

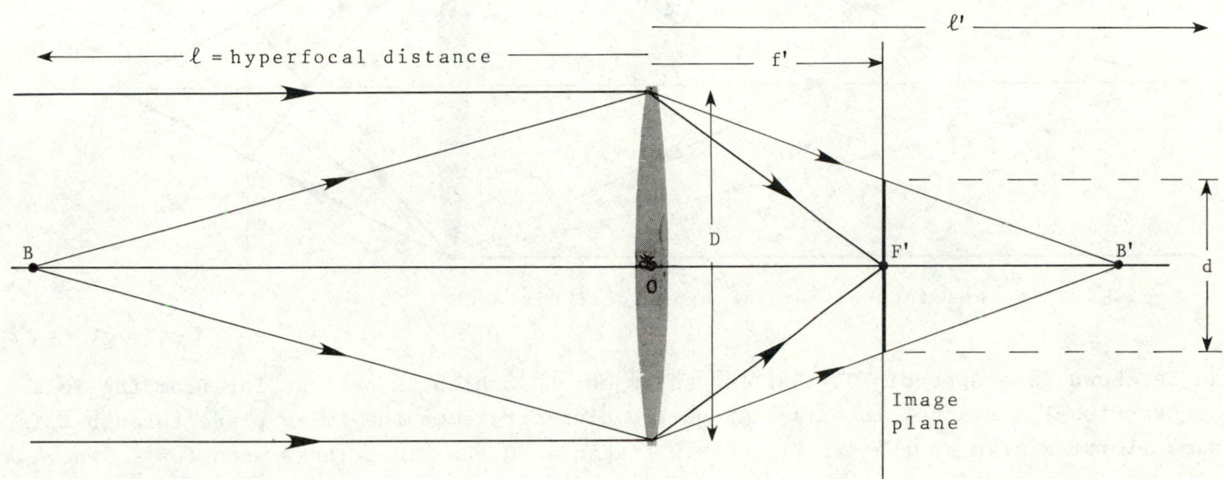

Fig. 8.9 The hyperfocal distance.

The nearest object point B to the lens which can be considered to be in focus on the image plane through F' will be such that the blur circle, formed by the refracted cone of rays from B, has the maximum diameter of 0.025 cm. We recall from section 8.1.1.4 that a small disc is indistinguishable from a point by the eye provided it is smaller than about 0.025 cm in diameter. A more rational way of stating this is to specify the maximum permissible blur circle diameter as a fraction of the focal length. For general photographic purposes, blur circles where the diameters are less than f'/1 000 are subjectively detected as focused points. Now, since $\frac{1}{\ell'}-\frac{1}{\ell}=\frac{1}{f'}$, $\ell = OB = \frac{\ell' f'}{f'-\ell'} = \frac{OB'}{-F'B'}\cdot f'$. But, from similar triangles, $\frac{OB'}{F'B'}=\frac{D}{d}$.

$$\text{Therefore,} \quad \ell = -\frac{D}{d}\cdot f' \qquad (8.4).$$

If the maximum blur circle diameter d is given by d = f'/1 000, then we have $\ell = -\frac{Df'}{f'/1\,000}$, or

$$\ell = -1\,000D \qquad (8.4a).$$

This value of ℓ, which is the distance from the lens of the nearest point which is acceptably in focus on a screen when the lens is focused on infinity, is called the *hyperfocal distance*.

For example, a lens of focal length 5 cm used at an aperture of f/4 has an aperture of $D=\frac{5}{4}$ centimetres. The hyperfocal distance is then -1 000D = -1 000 × 1.25 = - 12 500 cm = - 12.5 m. Thus, objects between infinity and 12.5 m from this lens will be tolerably in focus at the same time in the image plane.

Figure 8.10 shows a lens focused for an axial object point B at the hyperfocal distance, and an image plane placed at the point image B'.

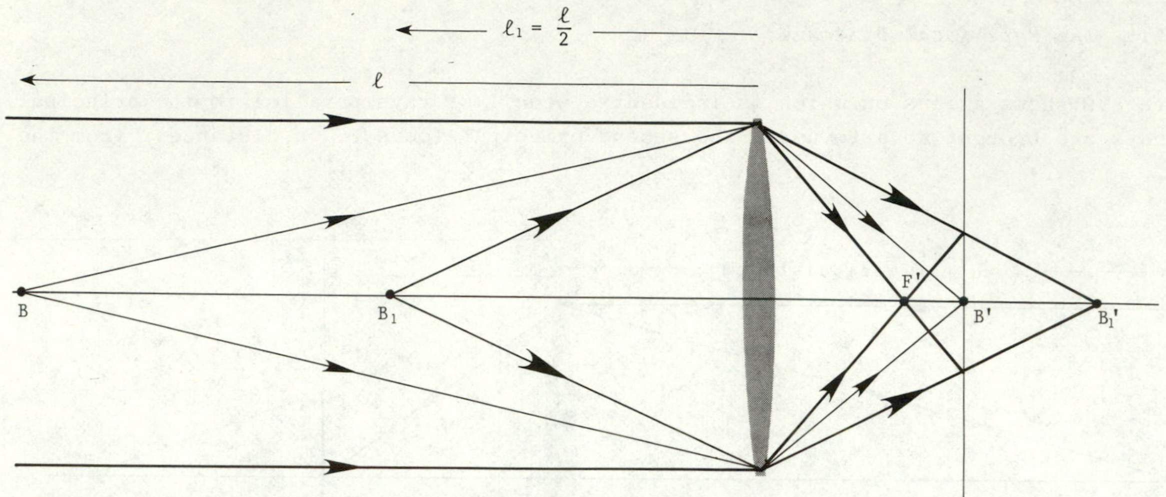

Fig. 8.10 A lens focused for the hyperfocal distance

It can be shown (see Appendix 3) that for an object B_1, which is half as far from the lens as the hyperfocal distance, the diameter of the blur circle on the image plane through B' is the same limiting size as that in figure 8.9 (taken as 0.025 cm). Thus, when $\ell_1 = \frac{\ell}{2}$, where ℓ is the hyperfocal distance and the lens is focused for this distance, all objects lying between infinity and half the hyperfocal distance are tolerably in focus on the image plane. That is, the depth of field for the lens is from infinity to half the hyperfocal distance.

8.1.2 THE PROJECTOR AND THE ENLARGER

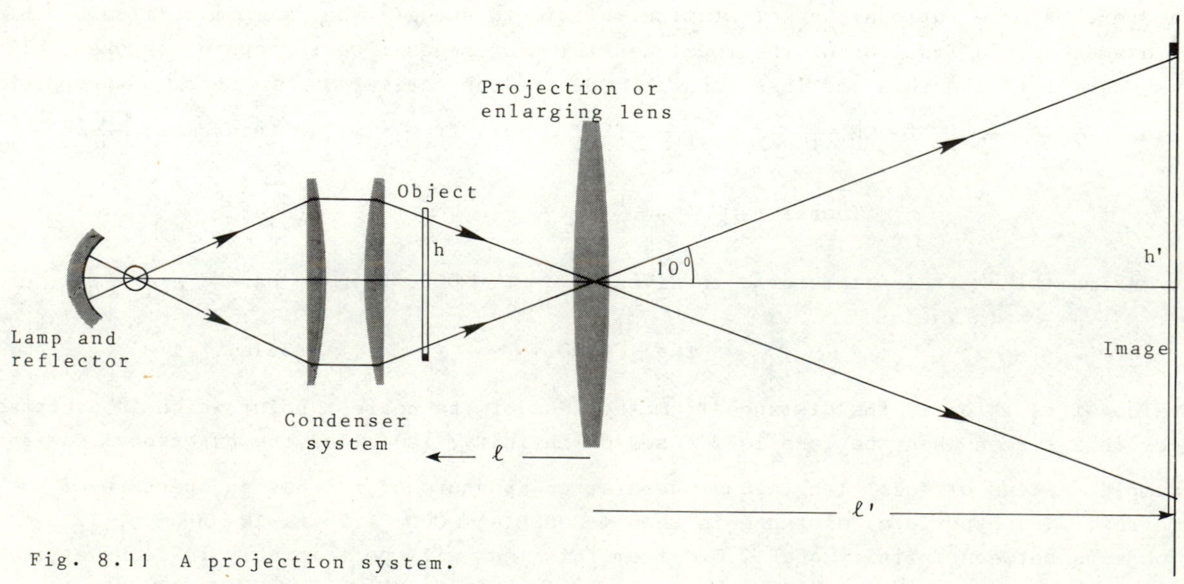

Fig. 8.11 A projection system.

The projector and the photographic enlarger are both essentially the same in principle as the camera. Figure 8.11 shows that the projection or enlarging lens is converging and forms an inverted, real image which in this case is enlarged, ℓ' being greater than ℓ. Recall that $h' = \frac{\ell'}{\ell} \cdot h$. Focusing is achieved by movement of the projection lens. In the case of an enlarger the photographic negative is the object, with photographic paper being mounted in the image plane.

For a projector the object will be a slide or film. As the image on the screen is enlarged, it will not be sufficiently bright unless strong illumination is provided for the film or slide. This is the function of the condenser lens system. Two plano-convex lenses are commonly used for this purpose as it can be shown that they give little spherical aberration. The image of the lamp is formed by the condenser system at or near to the plane of the projection lens, so that no detail of the lamp is formed on the image screen.

A semi-angular field of up to 10^0, see figure 8.11, is usually required for a projection lens, and a flat image field is needed. The condenser lens system plays no part in the final image focusing, and so it is the projection lens which requires careful design for good quality image formation.

A single positive powered projection lens will not form images of sufficient clarity with reasonable magnification. Figure 8.12 illustrates the Projection Ektar lens which is typical of the composite projection lens systems now employed. It is corrected mainly for spherical aberration, coma, and chromatic aberration, and produces the necessary flat field.

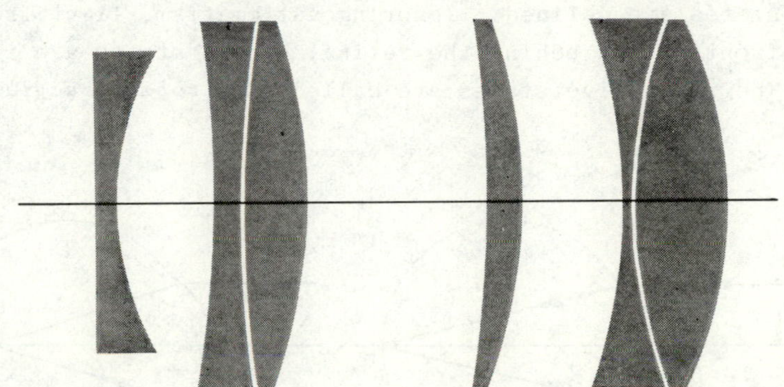

Fig. 8.12 The Ektar projection lens

With losses by reflection of about 4% of the light at each air/glass surface, and taking into account losses by the absorption in the glass, only about 70% of the light from the projector source reaches the image plane. These losses are generally much reduced by the use of anti-reflection coatings, see Chapter 13.

8.2 THE EYE

The human eye is a very complex optical system and a detailed description of its structure and operation is outside the scope of this book. Let it be noted, however, that the eye acts optically in a similar manner to a camera, consisting essentially of a converging lens system which forms an inverted image on the light sensitive retina. The optical constants of a model eye were given in section 4.3.1 , and figure 8.13(a) shows the main optical components in a somewhat simplified manner.

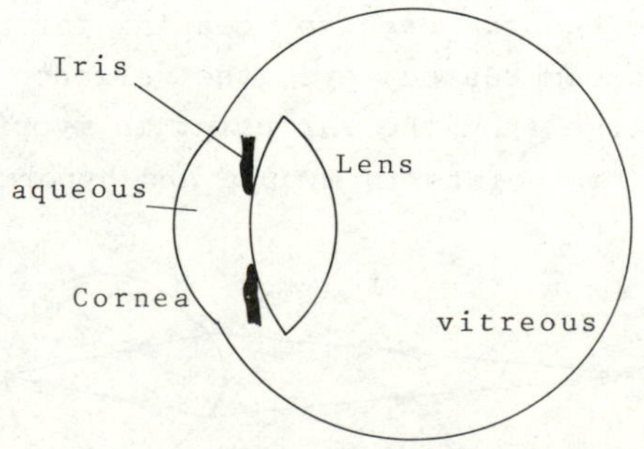

Fig. 8.13(a) A simplified model eye showing the iris.

The converging system of the eye is formed by the cornea, the aqueous humour, the crystalline lens, and the vitreous humour. Most of the refracting power of the system is due to the anterior surface of the cornea. Immediately in front of the lens is the iris, acting as a diaphragm or stop, adjusting its aperture size in response to the intensity of the light falling on it. This prevents overloading of the photosensitive receptors in the retina. The lens itself is very complex, being composed of layers of increasing refractive index towards the centre. The refractive index changes from about 1.37 near the surface to about 1.42 in the nucleus. Muscular action controls the curvature of the lens surfaces, and hence the lens power. In this way focusing for different object distances is achieved, a process known as *accommodation*. In fact, most of the curvature change takes place at the front surface of the lens, so that the distance between the lens and the retina is almost constant during accommodation. It is in this way that the eye differs optically from the camera in its action. The power of the eye cannot be increased indefinitely and the eye cannot form a clear retinal image when the object is closer than a certain distance. Although this distance reduces with age a standard *least distance of distinct vision* is usually taken as 25 cm. The object point for which a focused retinal image is formed when the eye is fully accommodated is called the *near point* of the eye.

When the eye is unaccommodated, that is, when the eye has its weakest power, three focusing states are defined. Ignoring astigmatism, light from a distant object point may focus on, in front of, or behind the retina. These states are illustrated in figures 8.13(b), (c), and (d). The refractive states are called *emmetropia, myopia,* and *hypermetropia (or hyperopia)*.

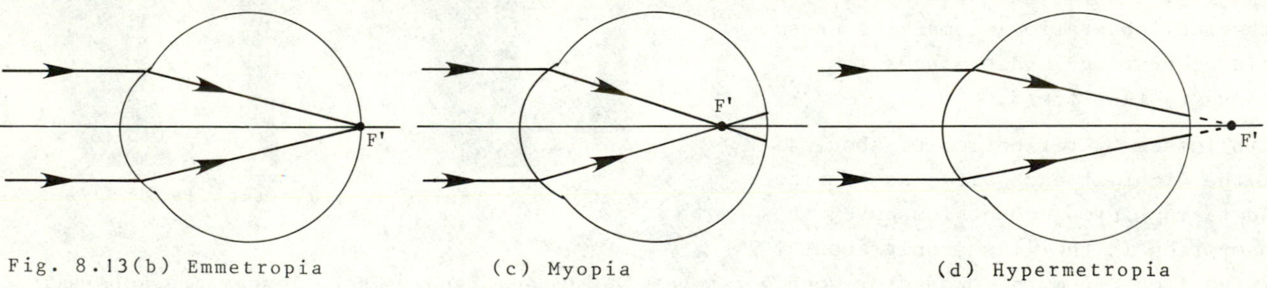

Fig. 8.13(b) Emmetropia (c) Myopia (d) Hypermetropia

The macula is the most sensitive part of the detection layer, the retina, and that point conjugate with the macula in the unaccommodated eye is known as the far point of distinct vision, or the far point of accommodation. It is denoted by M_R, the subscript R being taken from *Punctum Remotum*, meaning far point. Thus, for a clear image on the macula in the relaxed (unaccommodated) eye, the vergence arriving at the eye must be such that the rays appear to diverge from the far point in myopia and converge towards the far point in hypermetropia. The far points in myopia and hypermetropia are shown in figures 8.13(e) and (f).

Fig. 8.13(e) Fig. 8.13(f)

Rays from a distant object point can be made to focus on the macula M' of a myopic or a hypermetropic eye by arranging a diverging and a converging lens, respectively, as shown in figures 8.13(g) and (h). The principle is to make the second focal point of the lens coincide with the far point of the eye.

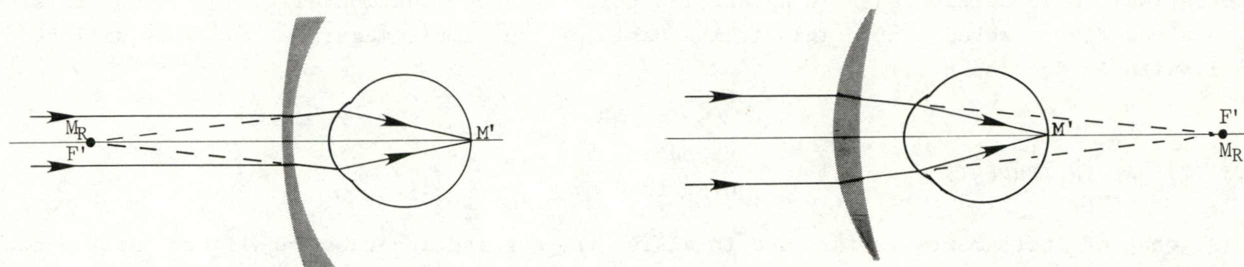

Fig. 8.13(g) A myopic eye focused for distance vision

Fig. 8.13(h) A hypermetropic eye focused for distance vision.

8.2.1 THE VISUAL ANGLE AND RETINAL IMAGE SIZE

Before we undertake a discussion of visual optical instruments it is necessary to understand the main factor on which the size of the retinal image depends. For convenience of drawing, we shall represent the eye in its reduced form (see section 4.3.1) in which all the refraction is deemed to take place at a single curved surface. Figure 8.14(a) shows the eye viewing objects B_1 and B_2. Even though they are at different distances from the eye, and have different sizes, the objects appear to be the same size as each other. This is because they subtend the same angle α at the eye with the result that the retinal images are of the same size, h_2'.

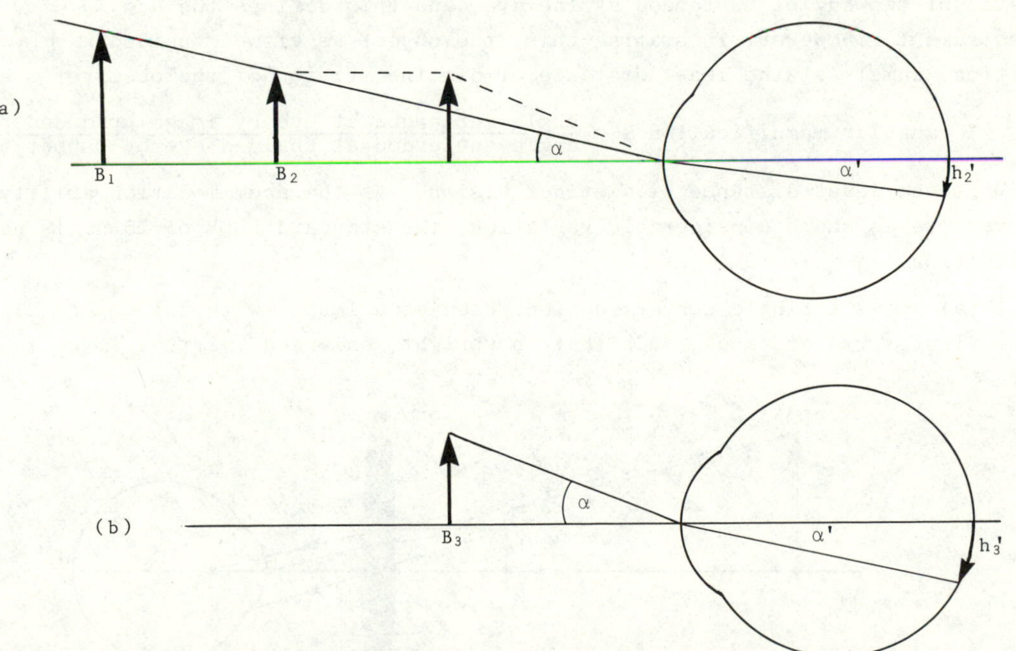

Fig. 8.14 The retinal image size is proportional to the visual angle α.

The angle α is called the *visual angle*, and is related to the angle α' by Snell's law. When the two angles are small they are directly proportional to one another. In order that the object B_2 is made to appear larger it can be positioned at B_3 nearer the eye, figure 8.14(b). This will increase the size of the visual angle α, and the retinal image size also increases from h_2' to h_3'. If the object is moved too close the eye cannot accommodate sufficiently and the retinal image is blurred. In cases like this the accommodation of the eye can be assisted with a positive powered lens. This is the basis of the simple magnifier and this will be dealt with in section 8.3.1 .

8.3 VISUAL INSTRUMENTS

This group of instruments is intended to assist the eye and includes magnifiers, microscopes, telescopes, and oculars.

8.3.1 THE SIMPLE MAGNIFIER

A single positive powered lens can be used to assist the refracting power of the eye. In this way the lens will produce an image which subtends a larger visual angle than that subtended by the object when the latter is placed at the least distance of distinct vision. The lens so used is referred to variously as a magnifying glass, a simple magnifier, a simple microscope, or a loupe (French word for magnifier, often used in optical practice) and is classed as a low vision aid.

The magnifying ability of the loupe may be expressed by the ratio

$$\frac{\text{size of retinal image formed with the aid of the lens}}{\text{size of retinal image of the object viewed directly}}\,.$$

If we limit ourselves to rays of light near to the lens axis the above ratio may be stated as the ratio of two angles subtended at the eye, and this defines the *angular magnification* of the instrument. However, in stating this, the object as viewed unaided is placed at its best position, namely at the least distance of distinct vision of the observer's eye. Hence we have

$$\text{angular magnification } M = \frac{\text{angle subtended at eye by image produced by lens}}{\text{angle subtended at unaided eye by object at LDDV}}$$

where LDDV is the least distance of distinct vision. As the accommodation ability of different observers' eyes shows considerable variation, the standard LDDV of 25 cm is used in the above definition.

Figure 8.15(a) shows a single converging lens used as a loupe, with the object BT positioned within the first focal distance, such that an upright, enlarged, virtual image is seen through the lens.

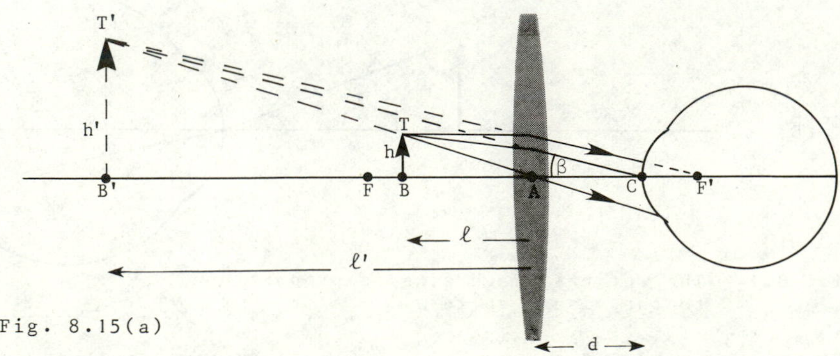

Fig. 8.15(a)

The observer's eye is placed a distance d behind the lens, and β is the visual angle subtended by the image at the centre C of the cornea. Figure 8.15(b) shows the same object placed at the least distance of distinct vision which is given the symbol q (= -0.25 m). At this distance the object subtends an angle α at the unaided eye.

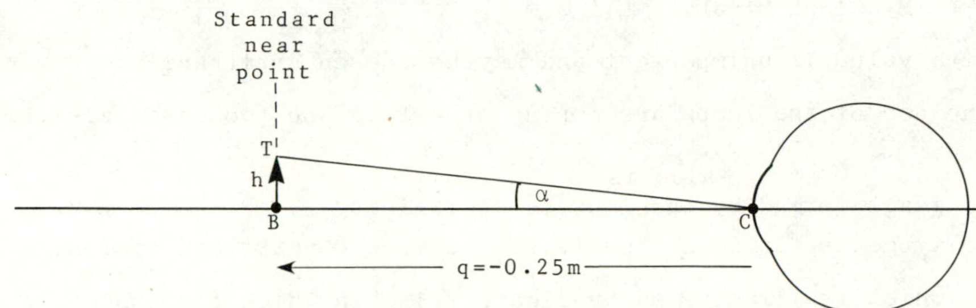

Fig. 8.15(b) See text for details.

Thus, the angular magnification of the loupe is, by definition:

$$M = \frac{\beta}{\alpha} \qquad (8.5).$$

Now, assuming that the angles are sufficiently small to be measured by their tangents, we have

$$M = \frac{\beta}{\alpha} = \frac{\tan\beta}{\tan\alpha} \qquad (8.5a).$$

From figure 8.15(a),

$$\tan\beta = \frac{B'T'}{B'C} = \frac{B'T'}{B'A + AC} = \frac{h'}{(-\ell') + d} = \frac{h'}{d-\ell'} \qquad (i)$$

where h' is the height of the image, and B'A' is denoted by $-\ell'$, as required by the sign convention. Next, from figure 8.15(b),

$$\tan\alpha = \frac{BT}{BC} = \frac{h}{-q} \qquad (ii)$$

where h is the height of the object, and BC is denoted by -q to satisfy the sign convention. So, we have, from (i) and (ii) above:

$$M = \frac{\tan\beta}{\tan\alpha} = \frac{h'}{d-\ell'} \div \frac{h}{-q} = \frac{h' \times (-q)}{h \times (d-\ell')} \qquad (iii).$$

Now, the ratio h'/h represents the linear magnification of the image and from equation (3.10) $\frac{h'}{h} = \frac{L}{\bar{L}'}$, which can be written $\frac{h'}{h} = \frac{L}{L'}$ when the lens is in air. Hence, substituting $\frac{L}{L'}$ for $\frac{h'}{h}$ and letting $\ell' = \frac{1}{L'}$ in (iii) above, we get

$$M = \frac{-Lq}{L'\left(d-\frac{1}{L'}\right)} = \frac{-qL}{dL'-1}$$, which on multiplying by $\frac{-1}{-1}$ (=1) gives

$$M = \frac{qL}{1 - dL'} \qquad (8.6).$$

Equation (8.6) is the general expression for the angular magnification obtained with a loupe. If must be carefully noted that the angular magnification bears no simple relationship to the linear magnification as the latter is independent of the position of the observer's eye. The angular magnification is dependent on the position of the observer's eye, with equation (8.6) containing as it does the term d (see worked example at the end of this section).

It is worth looking at the use of the least distance of distinct vision in the definition of angular magnification. Its choice of value is not too important as can be seen from the following comparison of two loupes. Suppose two loupes of different powers are used to produce an image at some distance ℓ', which is the same for each lens. Then only L will differ and we can compare the angular magnifications thus:

$$\frac{M_1}{M_2} = \frac{qL_1}{1-dL'} \bigg/ \frac{qL_2}{1-dL'} = \frac{L_1}{L_2}.$$

q has disappeared, so its value is unimportant and may be chosen arbitrarily!

Two special cases in the use of the loupe are worthy of note. The two cases are the extremes of angular magnification:

(i) the image formed by the lens is at infinity

and (ii) the image is formed at the least distance of distinct vision.

Consider the first case which is illustrated by figure 8.16, in which the object is placed at the first principal focal point F of the lens, the image then being at infinity.

Using equation (8.6) with $L' = 0$, since the image is at infinity, $q = -0.25$ m, and $L = L' - F = -F$, then

$$M = \frac{qL}{1-dL'} = \frac{-0.25(-F)}{1-0} = 0.25F$$

$$\text{or} \qquad M = \frac{F}{4} \qquad (8.7).$$

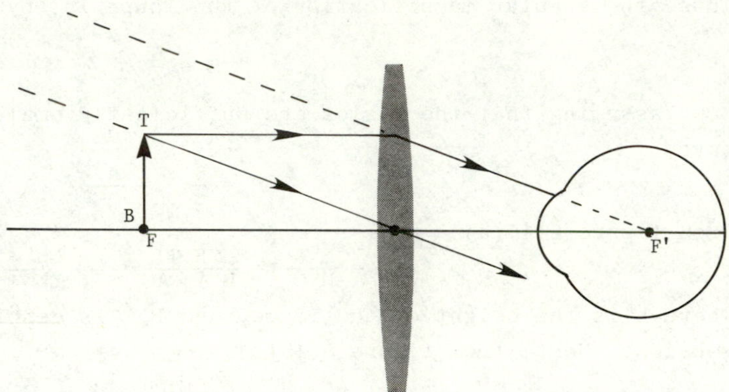

Fig. 8.16 The simple magnifier - object at F, image at infinity.

Hence, we see that when the image is formed at infinity the angular magnification is equal to one-quarter of the lens power. Under these conditions, M is referred to as the *nominal magnification* of the loupe. Because the rays leaving the lens are parallel an emmetropic eye views the image in an unaccommodated state. A magnifier with a power of +12.0 D has a nominal magnification of 3.0. This is usually marked as 3× which means that the retinal image is 3 times larger with the object placed at the first focal distance from the loupe than it would be with the object placed at the standard near point of the unaided eye. The simplest single lens magnifiers are limited by aberrations to about 3×.

Consider now case (ii). The angular magnification will be at its maximum value when the image is formed at the least distance of distinct vision, i.e. $\ell' = -0.25$ m, and the eye is placed close to the lens so that $d = 0$. Using equation (8.6) again with $d = 0$ and $L = L' - F = -4 - F$, where $L' = \frac{1}{\ell'} = \frac{1}{-\frac{1}{4}} = -4$ D.

$$M = \frac{qL}{1-dL'} = \frac{-\frac{1}{4}(-4-F)}{1-0} = 1 + \tfrac{1}{4}F .$$

$$\text{that is,} \quad M = 1 + \frac{F}{4} \qquad (8.8).$$

Equation (8.8) represents the maximum angular magnification of the loupe expressed in terms of the lens power. A lens of power +8.0 D will have a maximum angular magnification of $1 + \frac{8}{4}$ or 3×. Lower angular magnification is obtained when the image is formed further than 0.25 m from the eye, and some users consider it more comfortable to view the image at infinity. When the object is placed within the first focal length the light emerging from the lens is divergent and this requires an accommodative effort on behalf of the subject, unless he just happens to be myopic by the right amount. However, most elderly subjects with subnormal vision will already possess reading spectacles and these will be used in conjunction with the magnifier.

So, to summarise the behaviour of the loupe, we have

$$\text{Angular magnification} \quad M = \frac{qL}{1-dL'}$$

$$\text{Nominal magnification} \quad M = \frac{F}{4}$$

$$\text{Maximum angular magnification} \quad M = 1 + \frac{F}{4}$$

Worked Examples

1 A +8.00 D loupe is held 5 cm in front of the eye, and is used to view an object placed 8 cm in front of the lens. Calculate (i) the linear magnification of the image, (ii) the angular magnification. What nominal magnification would be ascribed to this lens?

For the lens, the given data are:

$\ell = -8\text{ cm} = -0.08\text{ m}$.

$L = \frac{1}{\ell} = \frac{1}{-0.08} = -12.5\text{ D}$.

and $F = +8.00\text{ D}$.

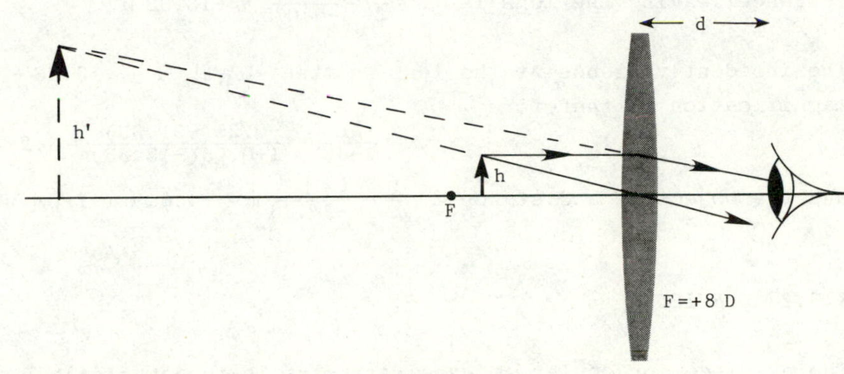

Using $L' = L + F$ gives $L' = -4.5\text{ D}$.

(i) Linear magnification is

$$m = \frac{h'}{h} = \frac{L}{L'} = \frac{-12.5}{-4.5} = 2.78\times.$$

(ii) Angular magnification is

$$M = \frac{qL}{1-dL'} = \frac{-0.25(-12.5)}{1-0.05(-4.5)} = 2.55\times.$$

The nominal magnification is $M = \frac{F}{4} = \frac{8}{4} = 2\times$.

2 An emmetrope exerting 4 D of accommodation views the image of an object clearly when the object is 6 cm in front of a +10.00 D lens. Calculate the distance between the lens and the eye, and determine the angular magnification.

In this problem we can locate the position of the image with respect to the eye by consideration of the accommodative effort involved. When an emmetrope exerts 4 D of accommodation this means that the distance from the eye to the image produced by the loupe is the distance equivalent to a vergence at the eye of -4 D. Thus, the position of the image is $\frac{1}{-4}$ m = -25 cm, or 25 cm in front of the eye.

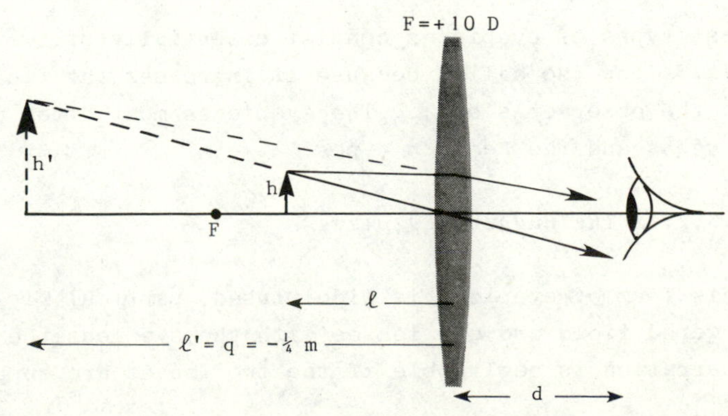

But for the lens, F = +10 D, and $L = \frac{1}{\ell} = \frac{1}{-0.06} = -16.67$ D, whence $L' = L+F = -16.67 + 10 = -6.67$ D. So, the image is a distance $\ell' = \frac{1}{L'} = \frac{1}{-6.67}$ m = − 15 cm from the lens. Since the image is 25 cm to the left of the eye and 15 cm to the left of the lens, the distance between the lens and the eye is (25 − 15) = 10 cm. That is, d = 10 cm. Now, the angular magnification is $M = \frac{qL}{1-dL'} = \frac{-0.25(-16.67)}{1-0.10(-6.67)} = 2.5\times$.

3 A 2.00 D myope uses a +8.00 D lens as a loupe. In order to view the image of an object held in front of the lens, he accommodates 6.00 D. If he holds the lens 5 cm in front of his eye what angular magnification does he obtain, and what is the distance between the lens and the object?

In this problem we can locate the distance of the image from the eye by consideration of the refractive error of the eye and the accommodation involved. For a myope of −2.00 D, exerting 6.00 D of accommodation, he becomes in effect a myope of −8.00 D and this is the vergence of the light at the eye. The image he is viewing is therefore at $\frac{1}{-8}$ m = −12.5 cm from his eye. So, the image distance from the lens is −(12.5 − 5) = −7.5 cm. Thus, the vergence leaving the lens is $L' = \frac{1}{\ell'} = \frac{1}{-0.075} = -13.33$ D.

The incident vergence at the lens is then $L = L' - F = -13.33 - (+8) = -21.33$ D. The angular magnification is therefore

$$M = \frac{qL}{1-dL'} = \frac{-0.25(-21.33)}{1-0.05(-13.33)} = 3.2\times$$

and the object is a distance $\ell = \frac{1}{L} = \frac{1}{-21.33}$ m = −4.69 cm from the lens.

8.3.2 EYEPIECES

The eyepiece, or ocular, of an instrument is fundamentally a magnifier. Its function is to view the image formed by a lens or a lens system preceding it in an optical instrument. Thus, in the microscope (see section 8.3.3) and the telescope (see section 8.3.4), an image of the object under consideration is formed by a lens called the objective, and an eyepiece or ocular is used to view this image. A single lens could serve this purpose but the final image would suffer extensively from the effects of aberrations. Magnification involves a considerable increase in the angle subtended at the observer's eye and those aberrations mainly dependent on field angle, namely oblique astigmatism, curvature of image field, distortion, and transverse chromatic aberration must be carefully considered in the design of an eyepiece.

Most types of eyepieces consist essentially of two lenses referred to respectively as the *field lens* (so called because it increases the field of view) and the *eye lens* (placed next to the observer's eye). The eyepieces most often found on microscopes and telescopes are the *Huygens* and the *Ramsden* types.

8.3.2.1 *The Huygens Eyepiece*

This form of eyepiece is widely used, particularly on microscopes. It consists of positive powered field and eye lenses with the eye lens the stronger component. Transverse chromatic aberration is negligible if the two lenses are separated by a distance equal to the average of their focal lengths. To reduce distortion the lenses are plano-convex with their curved surfaces towards the incident light. Figure 8.17 illustrates the Huygens eyepiece using lenses of focal lengths 3f' and f', where f' is the focal length of the eye lens, separated by a distance 2f'.

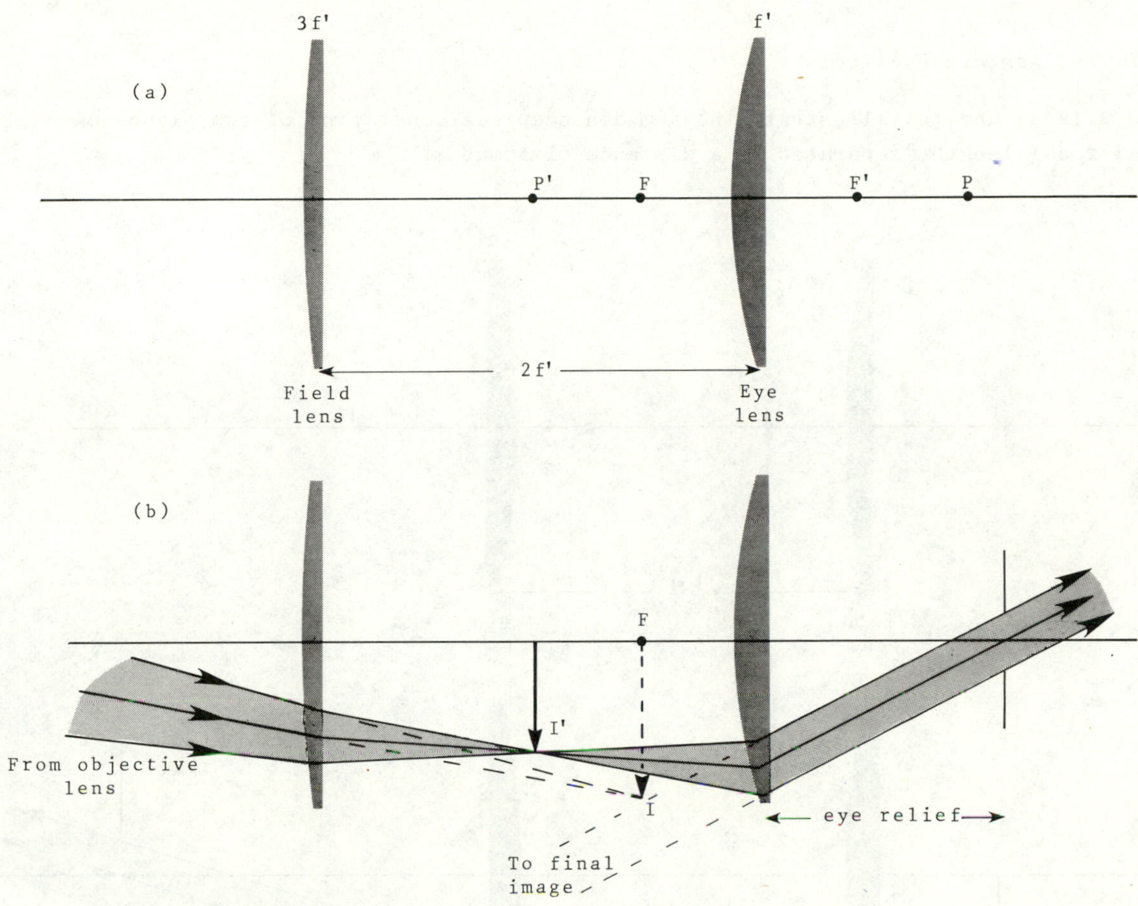

Fig. 8.17 The Huygens eyepiece.

Figure 8.17(a) shows the dispositions of the cardinal points, whilst figure 8.17(b) shows the path of a pencil of rays from the objective lens of the instrument converging towards an image I which serves as a virtual object for the field lens. This lens forms a real image I' which is then usually imaged at infinity by the eye lens to enable comfortable viewing by a relaxed emmetropic eye. The distance from the eye lens to the best position for the observer's eye is known as the eye relief. As this is usually only about 2-3 mm it is often too small for comfort.

The function of the field lens is to deviate inward those rays which would otherwise have missed the eye lens, thus increasing the field of view.

The Huygens eyepiece can only be used to examine an image and cannot be used by itself as a simple magnifying glass. It is called a negative eyepiece but it should be noted that this has no connection with the sign of the power of the eyepiece. It is not suitable for use with cross-wires, but if used, they would be fitted at P', which is also the first focal point of the eye lens. However, the image of the cross-wires placed in this position will be formed by the eye lens alone, whilst the image measured on the cross-wires is formed by the complete eyepiece. Thus the two images would be unequally affected by any residual aberrations, with the image of the cross-wires showing, in particular, strong colour effects at its edges.

8.3.2.2 The Ramsden Eyepiece

Figures 8.18(a) and (b) illustrate the Ramsden eyepiece consisting of two plano-convex lenses of equal focal length, separated by a distance of about ⅔f'.

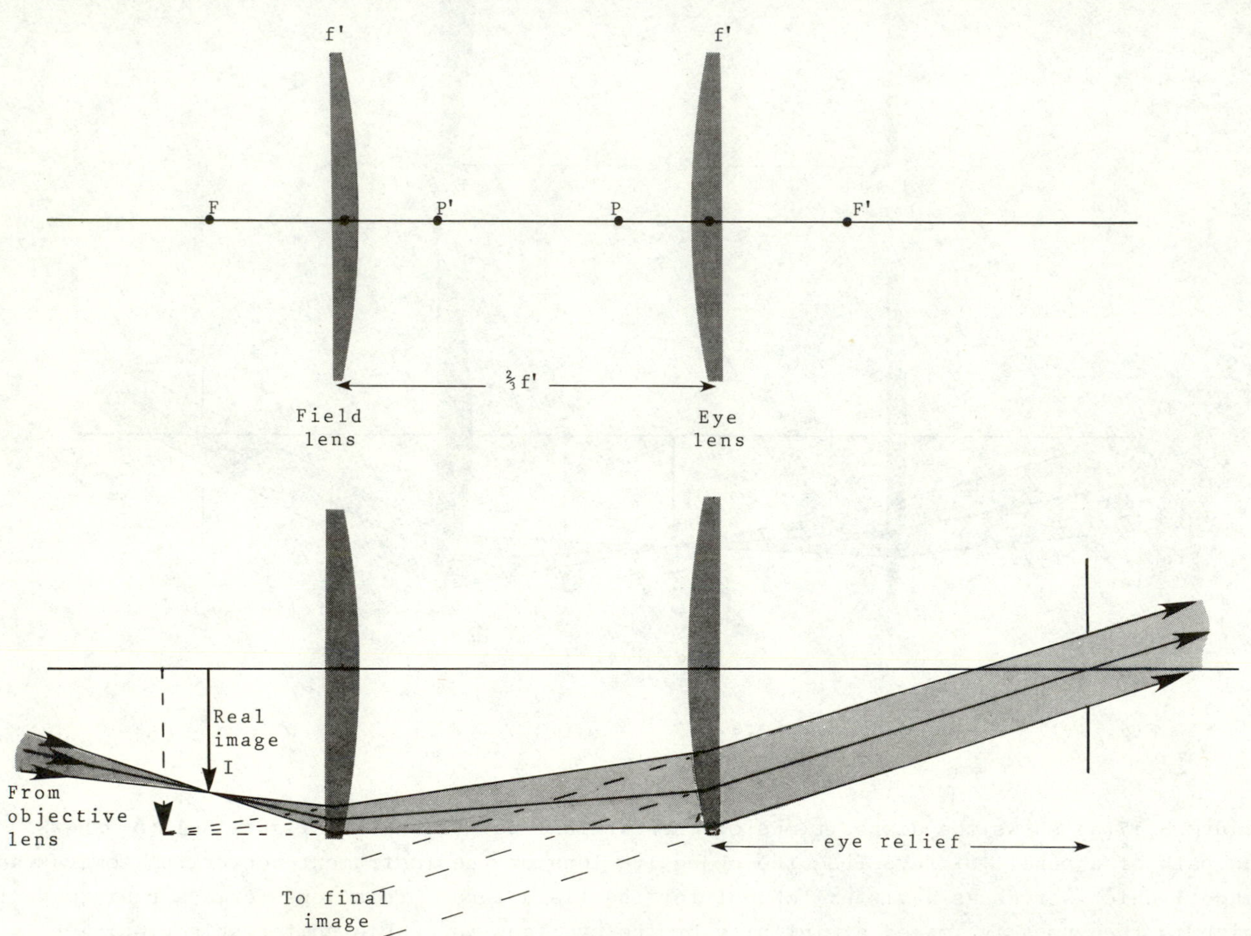

Fig. 8.18 The Ramsden eyepiece.

With this arrangement the first principal focus F is in front of the field lens and this is where the image (or object) to be examined is placed. The eyepiece is called a positive eyepiece. Cross-wires may be mounted to coincide with the plane of the primary image I formed by the objective lens. Cross-wires, which are well reproduced by the eyepiece along with the final image, are essential for quantitative measurements using telescopes or microscopes. The final image is formed at infinity. As with the Huygens type of eyepiece, the function of the field lens is to deviate towards the axis those rays which would otherwise miss the eye lens.

This increases the field of view. In obtaining a real first focal plane, some transverse chromatic aberration remains and must be tolerated, but the Ramsden eyepiece is generally more satisfactory than the Huygens type as regards the reduction of the other aberrations. In addition, the larger eye relief (~12 mm) is more convenient than with the Huygens type.

8.3.3 THE MICROSCOPE

The invention of the microscope* (sometimes called the compound microscope) is generally accredited to the Dutch spectacle-maker Zacharias Janssen, with Galileo publishing the details of his own invention a few years later in 1610. In its simplest form the microscope consists of two positive powered lenses. The lens nearer the object is called the *objective* and is of short focal length, whilst the other lens, having a longer focal length, acts as the *eyepiece*. In reality, both these lenses will contain several components to minimise aberrations and improve the field of view. For simplicity the objective (O-lens) and the eyepiece (ε-lens) will be represented by thin lenses, with the subscripts (o) and (ε) being used to denote points and distances relative to the objective and the eyepiece, respectively. Figure 8.19(a) shows the image formation by the instrument using construction rays only.

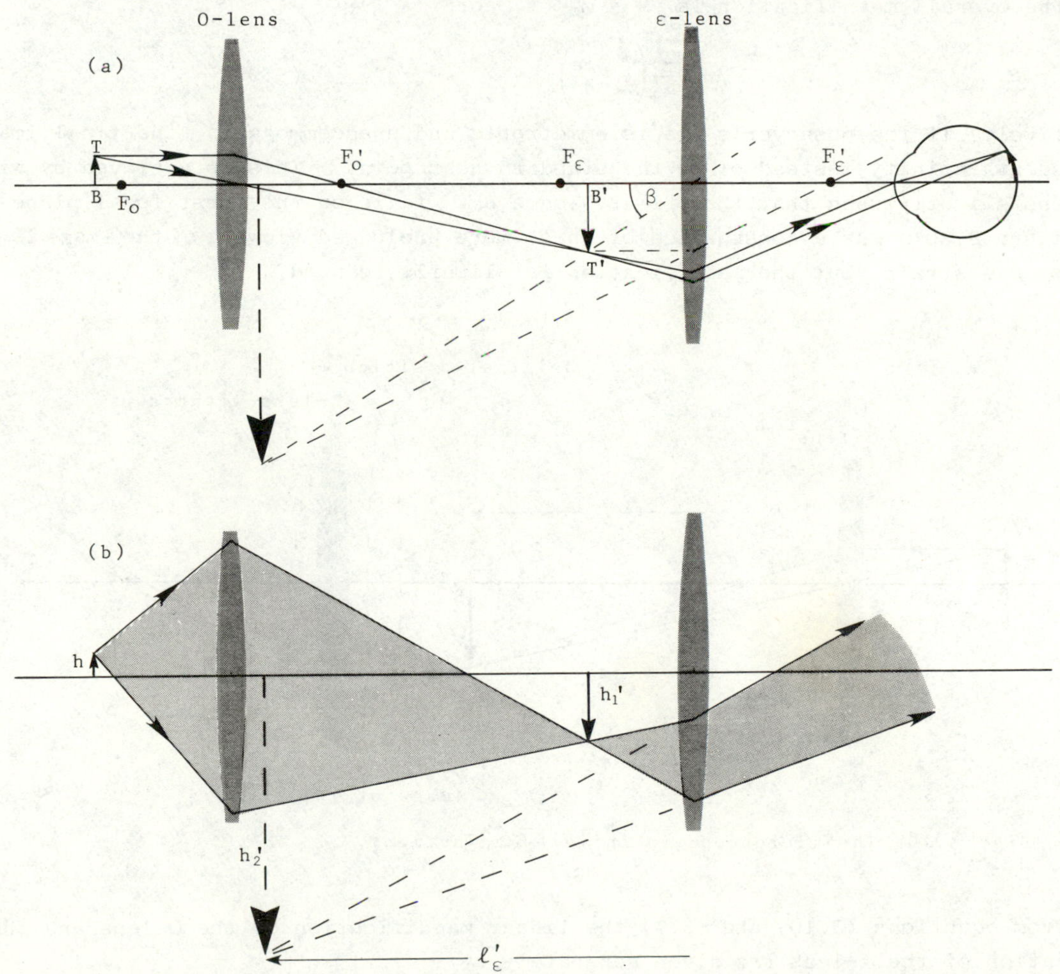

Fig. 8.19 The microscope.

* micro = small; scope = instrument for viewing.

Figure 8.19(b) is the type of diagram that the reader should use to illustrate pictorially the action of the instrument. All the rays from the tip of the object which enter the instrument are included in the shaded pencil. Thus, with the object BT placed just outside the first focal point F_o of the O-lens, a real enlarged primary image B'T' is formed between the lenses, which becomes the object for the ε-lens. Functioning as a magnifier, the ε-lens forms the greatly enlarged virtual final image, which is inverted relative to the original object.

In normal use the final image is formed at the standard near point distance of 25 cm from the ε-lens; that is, $\ell'_\varepsilon = -0.25$ m. This requires the primary image to be formed within the first focal length of the ε-lens, as shown. In this situation the angular magnification will have its maximum value if the eye is placed as close as possible to the ε-lens. As the function of the O-lens is to form an enlarged image that is viewed through the ε-lens, the overall magnification of the microscope is the product of the linear magnification m_o of the O-lens and the angular magnification M_ε of the ε-lens.

From equations (3.10) and (8.8) these are separately given by

$$m_o = \frac{\bar{L}_o}{\bar{L}'_o} = \frac{L_o}{L'_o} \text{ in air} \qquad \text{and} \qquad M_\varepsilon = 1 + \frac{F_\varepsilon}{4}.$$

Hence, the overall magnification is $M = m_o \times M_\varepsilon$, or

$$M = \frac{L_o}{L'_o}\left(1 + \frac{F_\varepsilon}{4}\right) \qquad (8.9).$$

Alternatively, if the observer's eye is emmetropic and unaccommodated, the final image must be formed at infinity instead of at the standard near point. This is achieved by withdrawing the ε-lens slightly such that the primary image B'T' falls on the first focal plane of this lens, figure 8.20 . As a consequence of this, more prolonged viewing of the image is possible with less eye strain, but the magnification is slightly reduced.

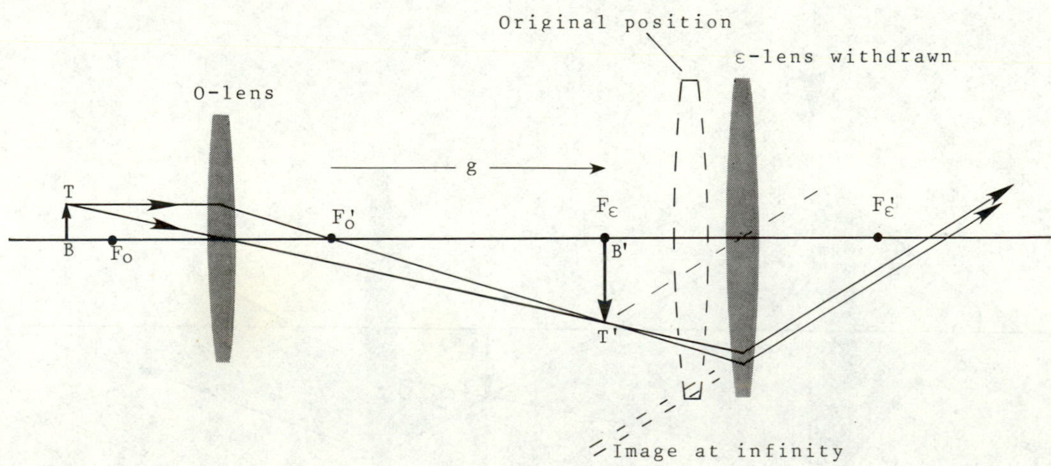

Fig. 8.20 The microscope in infinity adjustment.

Hence, from equations (3.10) and (8.7) the linear magnification of the O-lens and the angular magnification of the ε-lens are given separately by

$$m_o = \frac{L_o}{L'_o} \text{ in air, and } M_\varepsilon = \frac{F_\varepsilon}{4} .$$

Thus, the overall magnification is $M = m_o \times M_\varepsilon$, or

$$M = \frac{L_o}{L_o'} \times \frac{F_\varepsilon}{4} \qquad (8.10).$$

Another way of expressing the magnification of a microscope is as follows. Newton's equation, equation (2.34), states that the linear magnification produced by a lens is $m = \frac{h'}{h} = -\frac{x'}{f'}$, where x' is the distance from the second principal focus to the image. For the objective lens, we have from figure 8.20, $m_o = -\frac{g}{f_o'} = -gF_o$ where g(=x') is the distance from the second focal point of the objective to the first focal point of the eyepiece. Then, since $M_\varepsilon = F_\varepsilon/4$, we have for the overall magnification

$$M = \frac{-gF_o F_\varepsilon}{4} \qquad (8.11).$$

The distance g is known as the optical tube length of the microscope, and is usually about 16 cm.

If, as is sometimes the case, the setting of the microscope is different from the two extreme cases mentioned above, the magnification is calculated from first principles using equation (8.5a). This procedure is illustrated in the second of the worked examples following.

Worked Examples

1 A microscope has an objective of focal length 2 cm and an eyepiece of focal length 4 cm, the lenses being 16 cm apart. Determine the position of the object, and the magnification, when focused for an unaccommodated emmetropic eye.

In solving this problem let us note that the image viewed by an unaccommodated emmetrope is at infinity.

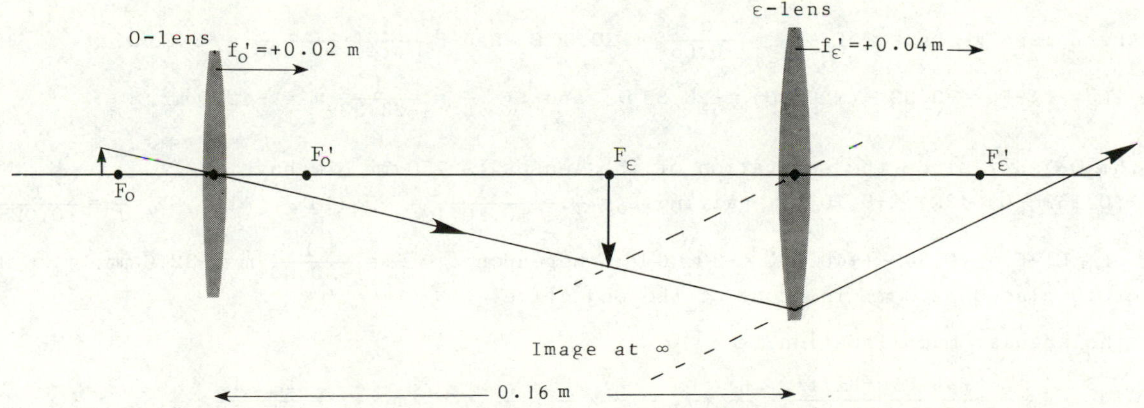

For the ε-lens, the given data are $L_\varepsilon' = 0$ (image at infinity) and $F_\varepsilon = \frac{1}{f_\varepsilon'} = \frac{1}{+0.04} = +25.00$ D.
Then the incident vergence at the ε-lens is $L_\varepsilon = L_\varepsilon' - F_\varepsilon = 0 - 25 = -25$ D, $\ell_\varepsilon = \frac{1}{L_\varepsilon} = \frac{1}{-25}$ m $= -4$ cm.
This confirms that the image formed by the O-lens, acting as an object for the ε-lens, is in the first focal plane of the ε-lens.

Therefore, for the O-lens, since the separation of the lenses is 0.16 m, we have

$$\ell_o' = +(0.16 - 0.04) = +0.12 \text{ m}.$$

Then $L_o' = \frac{1}{\ell_o'} = \frac{1}{+0.12} = +8.33$ D, and $F_o = \frac{1}{f_o'} = \frac{1}{+0.02} = +50.00$ D. We can now find the position of the object:

$$L_o = L_o' - F_o = +8.33 - (+50.00) = -41.67 \text{ D}, \quad \text{whence } \ell_o = \frac{1}{L_o} = \frac{1}{-41.67} \text{ m} = -2.4 \text{ cm}.$$

Hence, the object is 2.4 cm in front of the objective lens.

Now, the magnification is $M = \frac{L_o}{L_o'} \times \frac{F_\varepsilon}{4} = \frac{(-41.67)}{(+8.33)} \times \frac{(+25.00)}{4} = -31.3$. Notice that it is not obligatory to put $-31.3\times$!

2 If the focal lengths of the objective and eyepiece lenses are 25 mm and 50 mm, respectively, and their distance apart is 150 mm, where must an object be placed in order that the image seen by the eye may be 300 mm from the eyepiece lens? What is the angular magnification produced?

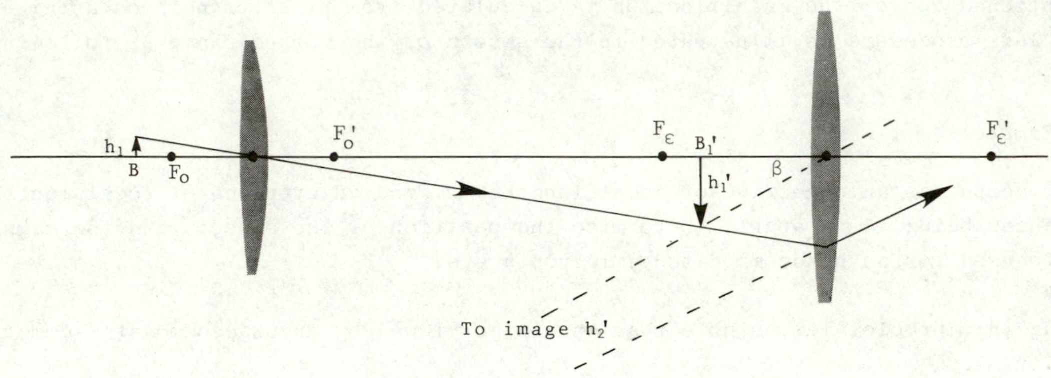

For the ε-lens we have: $F_\varepsilon = \frac{1}{f_\varepsilon'} = \frac{1}{+0.05} = +20.00$ D and $L_\varepsilon' = \frac{1}{\ell_\varepsilon'} = \frac{1}{-0.300} = -3.33$ D.

Then $L_\varepsilon = L_\varepsilon' - F_\varepsilon = -3.33 - (+20.00) = -23.33$ D, and $\ell_\varepsilon = \frac{1}{L_\varepsilon} = \frac{1}{-23.33}$ m $= -42.9$ mm.

For the O-lens, since the separation of the lenses is 150 mm, we have:
$\ell_o' = +(0.150 - 0.0429) = +0.1071$ m, giving $L_o' = \frac{1}{\ell_o'} = \frac{1}{+0.1071} = +9.34$ D. Also, $F_o = \frac{1}{f_o'} = \frac{1}{+0.025} = +40$ D.

Then $L_o = L_o' - F_o = +9.34 - (+40.00) = -30.66$ D, whereupon $\ell_o = \frac{1}{L_o} = \frac{1}{-30.66}$ m $= -32.6$ mm. So the object is placed 32.6 mm in front of the objective.

Now, the angular magnification is

$$M = \frac{\beta}{\alpha} = \frac{\tan\beta}{\tan\alpha} = \frac{-h_2'/\ell_\varepsilon'}{-h_1/q} = \frac{h_2'}{h_1} \times \frac{q}{\ell_\varepsilon'} = \frac{h_2'}{h_2} \times \frac{h_1'}{h_1} \times \frac{q}{\ell_\varepsilon'} = m_\varepsilon \times m_o \times \frac{q}{\ell_\varepsilon'} = \frac{L_\varepsilon}{L_\varepsilon'} \times \frac{L_o}{L_o'} \times \frac{q}{\ell_\varepsilon'}$$

i.e. $M = \frac{(-23.33)}{(-3.33)} \times \frac{(-30.66)}{(+9.34)} \times \frac{(-0.250)}{(-0.300)} = -19.2$. (Note that $h_2 = h_1'$).

8.3.4 THE TELESCOPE

It is usually the Dutch spectacle maker Hans Lippershey who is accredited with the invention of the telescope (*tele = far; scope = instrument for viewing*) in 1608. From this, Galileo developed his version a year or so later and, incidentally, founded modern astronomy. Telescopes may be of the refracting type, using lenses for the formation of images, or of the reflecting type, incorporating a curved mirror. The former may be broadly divided into those employing converging eyepieces and those employing diverging eyepieces, whilst the latter may have several forms, differing in the arrangements for viewing the primary image. Mention will also be made of a system which employs both refracting and reflecting elements, the system being referred to as catadioptric from the words catoptric and dioptric meaning pertaining to reflection and refraction, respectively.

8.3.4.1 The Astronomical Telescope

The optical system of an astronomical refracting telescope closely resembles that of a microscope. In both cases the primary image formed by an objective lens is viewed through a positive powered eyepiece, or ocular. The main difference, however, between these two types of instruments is that the telescope is used to enlarge the retinal or camera image of a distant object.

Figure 8.21(a) illustrates the image formation by a *Keplerian Astronomical Telescope* (invented by Johannes Kepler in 1611), using construction rays only. The reader should compare this with figure 8.21(b) which illustrates pictorially the action of the instrument, all the rays from the tip of a distant object (not shown) which enter the objective being included in the shaded pencil.

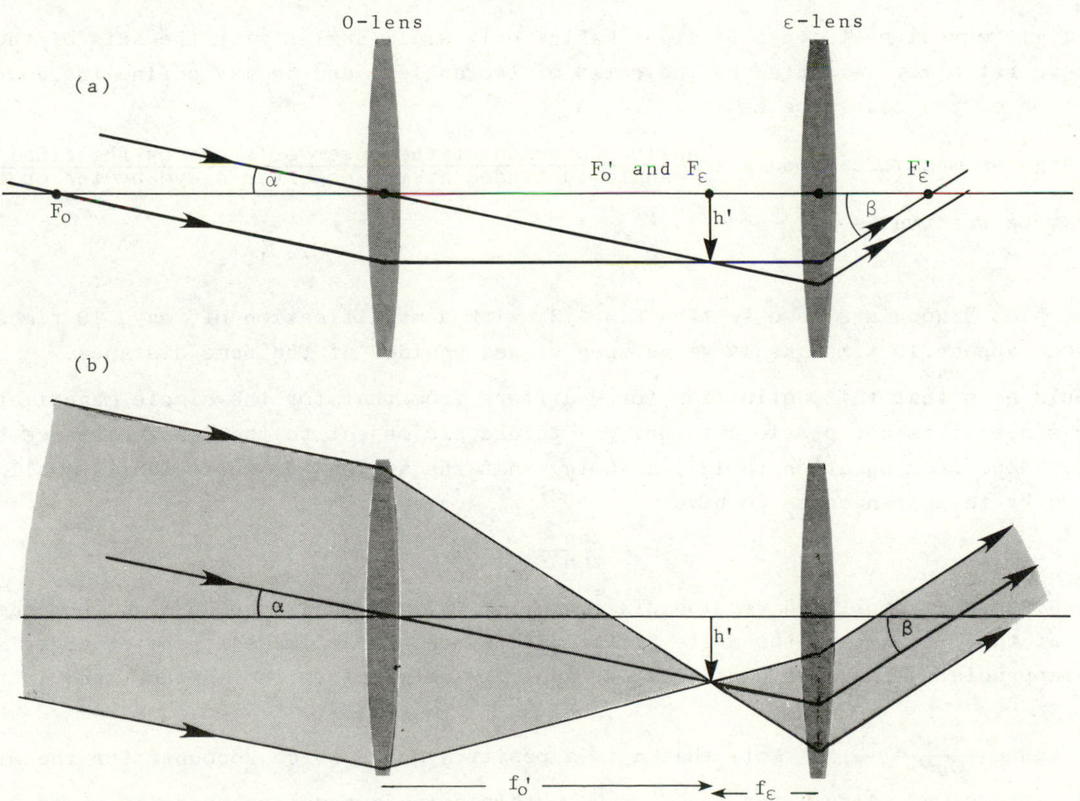

Fig. 8.21 The Keplerian astronomical telescope in its afocal setting.

The rays from the tip of a distant object essentially form a parallel pencil at the objective (the O-lens) of the telescope. This lens forms a real inverted primary image in its second focal plane. In normal useage, the primary image also coincides with the first focal plane of the eyepiece. In figure 8.21 the eyepiece is represented by a single positive lens (the ε-lens), although in reality it will be a lens system. As such, rays emerge from the ε-lens as a parallel pencil, and an emmetropic observer can focus the rays in an unaccommodated state. If the observer is myopic or hypermetropic the ε-lens can be moved in towards or out away from the O-lens respectively to introduce some compensating divergence or convergence.

It can be seen that the distance between the O-lens and the ε-lens, that is the length of the telescope, is the sum of the absolute values of the focal lengths of these lenses. In this configuration the instrument is said to be *afocal*, meaning literally without focus, and this corresponds to both incident and emergent light being pencils of parallel rays.

It is easy to show that in the afocal configuration a telescope has an equivalent power of $F_E = 0$. However, the instrument functions as a magnifying device by ensuring that the angle β subtended by the final image at the eye is larger than the angle α subtended by the object at the O-lens. This angle α is essentially the same angle which would be subtended by the object at the unaided eye.

Notice that the final image is inverted, but whilst the instrument is being used for astronomical purposes this is of no consequence, particularly since most work is photographic.

The magnifying ability of the telescope may be expressed by the ratio:

$$\frac{\text{size of the retinal image formed with the aid of the telescope}}{\text{size of the retinal image with the unaided eye}} .$$

If we limit ourselves to rays of light making only small angles with the axis of the telescope the above ratio may be stated as the ratio of two angles, and we may define the *angular magnification* of the telescope as

$$\text{Angular magnification} \quad M = \frac{\text{angle subtended at the observer's eye by the final image}}{\text{angle subtended at the observer's eye by the object}} .$$

This may be written as

$$M = \frac{\beta}{\alpha} \qquad (8.12).$$

For example, binoculars (see section 8.3.4.3) with a magnification of, say, 10 times, make an object appear 10 times as large as when viewed unaided at the same distance.

We should note that this definition for M differs from that for the simple magnifier, section 8.3.1, since it is not possible to bring a telescopic object to the least distance of distinct vision. Now, from equation (8.12), assuming that the angles are sufficiently small to be measured by their tangents, we have

$$M = \frac{\tan\beta}{\tan\alpha} \qquad (8.12a).$$

Figure 8.22 is a simplified version of figure 8.21(a) and shows the angle α subtended by the object at the O-lens, and the angle β subtended by the final image at the observer's eye. It can be appreciated that the distances O_1R and O_2S are equal to one another, and to the image B'T'.

Now, $\tan\alpha = \frac{T'B'}{O_1B'} = \frac{-h'}{f_o'}$. Note that α is a positive angle which accounts for the minus sign in the expression. Also, $\tan\beta = \frac{O_2S}{O_2F_\varepsilon'} = \frac{h'}{f_\varepsilon'}$, this being a negative quantity since β is a negative angle.

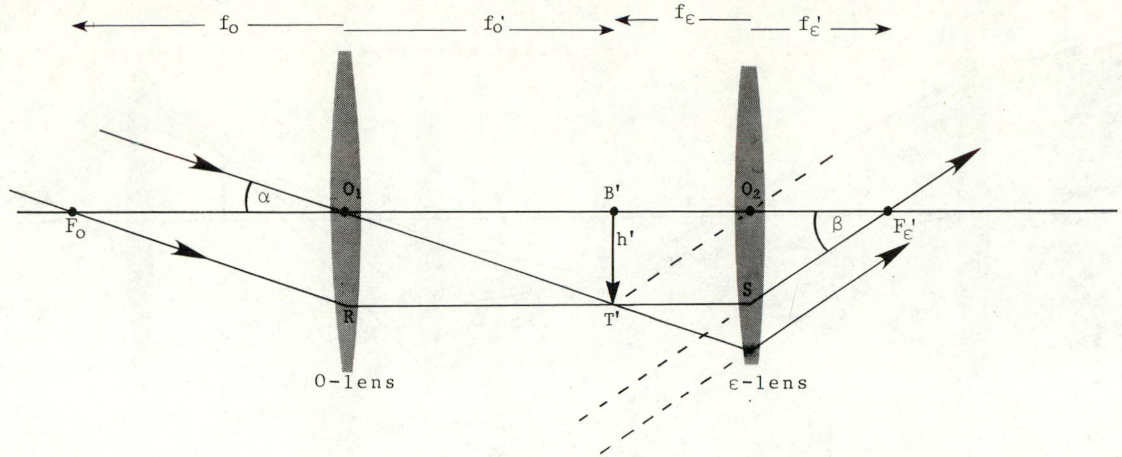

Fig. 8.22 Figure used for derivation of the angular magnification.

$$\text{Hence,}\quad M = \frac{\tan\beta}{\tan\alpha} = \frac{h'}{f_\varepsilon'} \Bigg/ \frac{-h'}{f_o'} = -\frac{f_o'}{f_\varepsilon'} = -\frac{F_\varepsilon}{F_o} \qquad (8.13).$$

Thus, the angular magnification is the ratio of the power of the eyepiece to the power of the objective. The negative sign denotes an inverted image, as is observed. The necessity for an eyepiece of high power and an objective of low power is immediately apparent from equation (8.13).

We have seen already, in figure 8.21, that when set in the afocal configuration, the spacing of the lenses is given by the sum of the absolute values of the focal lengths; that is,

$$\text{the distance between the lenses}\quad d = f_o' + |f_\varepsilon| = f_o' + f_\varepsilon' \qquad (8.14).$$

As already stated, for photographic work with the astronomical refracting telescope, it is no real disadvantage to have the final image inverted relative to the object. However, in cases where the orientation of the image is important, an additional erecting system must be included in the construction of the telescope. The erecting system may be a lens, or lens system, placed between the objective and the eyepiece, or may utilise prisms. In each case the result is referred to as a *terrestrial telescope*, and these will now be considered.

8.3.4.2 The Terrestrial Telescope with Lens Erector

This modified version of the astronomical telescope is shown in figure 8.23 with the lens erector illustrated as a single converging lens. When suitably positioned within the telescope behind the inverted primary image B'T', a further real image B''T'', which is now upright with respect to the object, is formed by this additional lens. The eyepiece then magnifies this second image.

This system has the disadvantage of needing a long tube, since it is inconveniently longer than the basic astronomical telescope. The minimum value of the extra length of the instrument will be four times the focal length of the erecting lens (see equation (3.14)).

In the afocal setting shown, in which the final image is formed at infinity, the angular magnification is determined by multiplying equation (8.13) by the linear magnification produced by the erecting lens.

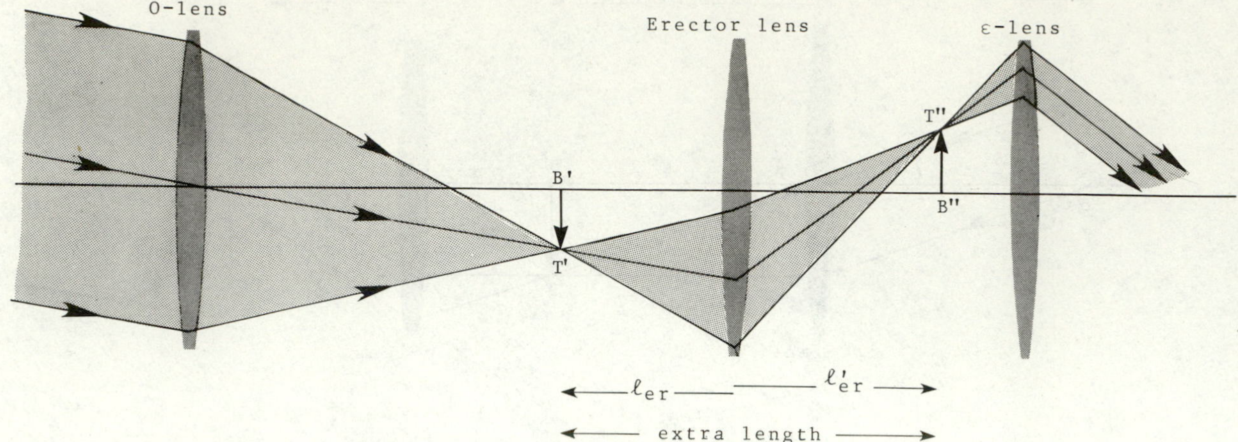

Fig. 8.23 The telescope with an erector lens.

Thus, if the latter is represented by $\frac{B''T''}{B'T'} = \frac{\ell'_{er}}{\ell_{er}}$, we have

$$\text{Angular magnification } M = -\frac{F_\varepsilon}{F_o} \times \frac{\ell'_{er}}{\ell_{er}} \qquad (8.15).$$

8.3.4.3 The Terrestrial Telescope with Prism Erector - the Binocular

The long tube length of the terrestrial telescope with a lens erector is avoided by using a pair of Porro prisms as inverters (see section 2.1.10.6). This achieves the same effect as the lens erector, but in less space. Hence, by forming an upright image, within a short overall instrument length, the two main shortcomings of the astronomical principle are both overcome.

Figure 8.24 shows the arrangement of the Porro prisms. They are mounted with their principal

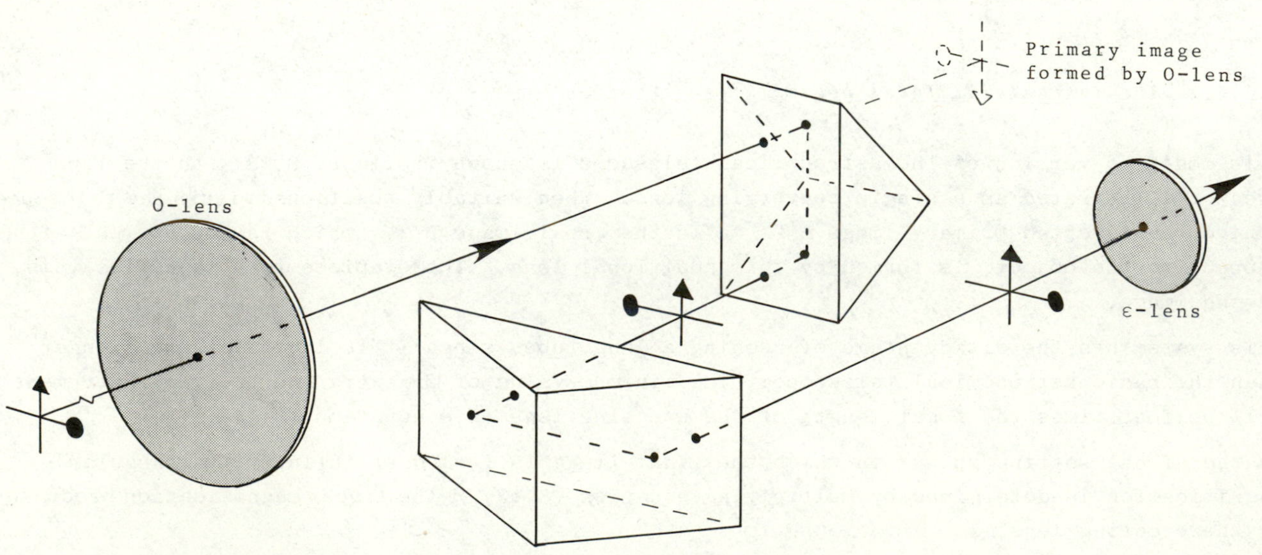

Fig. 8.24 The prism erecting system used in binoculars.

sections orthogonal. The primary image formed by the objective alone is inverted and also reversed left to right. The first prism introduces a top-to-bottom inversion so that the image is rendered upright. However, the left-to-right reversal of the image is not affected by this prism. If the second prism is used as shown, the left-to-right reversal of the image is achieved without affecting the inversion produced by the first prism. Thus, the image viewed by the ε-lens is identically orientated to the object.

Because of the saving in length of the telescope with the prism erector system, a pair of telescopes of this type may be used as *binoculars (binocular telescopes)*. With the two objective lenses set further apart than the two eyepieces, the latter's separation depending on the spacing of the observer's eyes, a greater impression of depth to a scene is achieved with a binocular than with the naked eyes. This is due to the increased stereoscopic effect of the object scene. This aspect of the perceived image is due to the greater disparity of the two images presented to the eyes and its explanation is outside the scope of this text.

Binoculars customarily bear numerical markings of the type 10×50. The first number (10) represents the magnification of the instrument, whilst the second number (50) indicates the diameter of the objective lens expressed in millimetres. The prism erecting system in the binocular does not contribute to the overall magnification of the instrument.

8.3.4.4 The Galilean Telescope

This is an example of a telescope employing a diverging eyepiece. Based on an invention of Galileo, the instrument consists of a converging objective lens and a higher power diverging eyepiece. This instrument overcomes one of the disadvantages of the Keplerian telescope in that the final image produced by the system is enlarged and also upright.

Figure 8.25(a) shows the image formation by the *Galilean telescope* in its afocal setting using construction rays only, whilst figure 8.25(b) illustrates the action of the instrument, all rays from the tip of a distant object (not shown) which enter the objective being included in the shaded pencil.

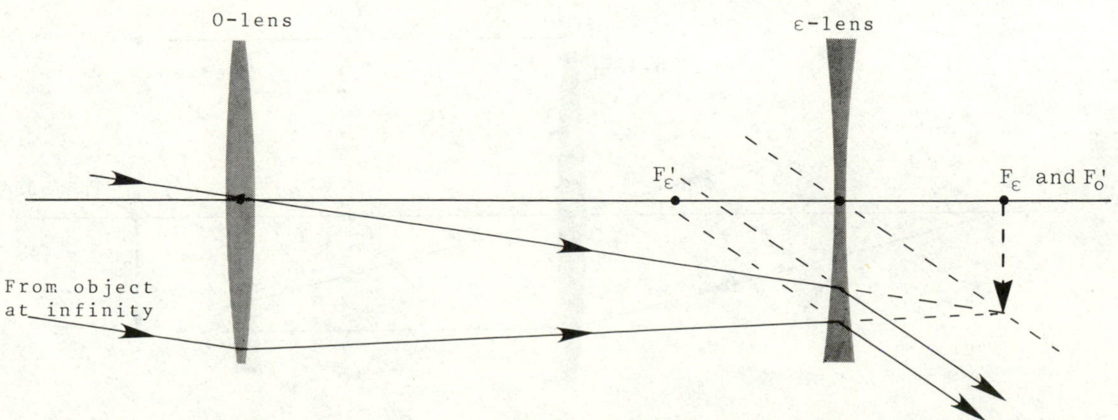

Fig. 8.25(a) The Galilean telescope - construction rays.

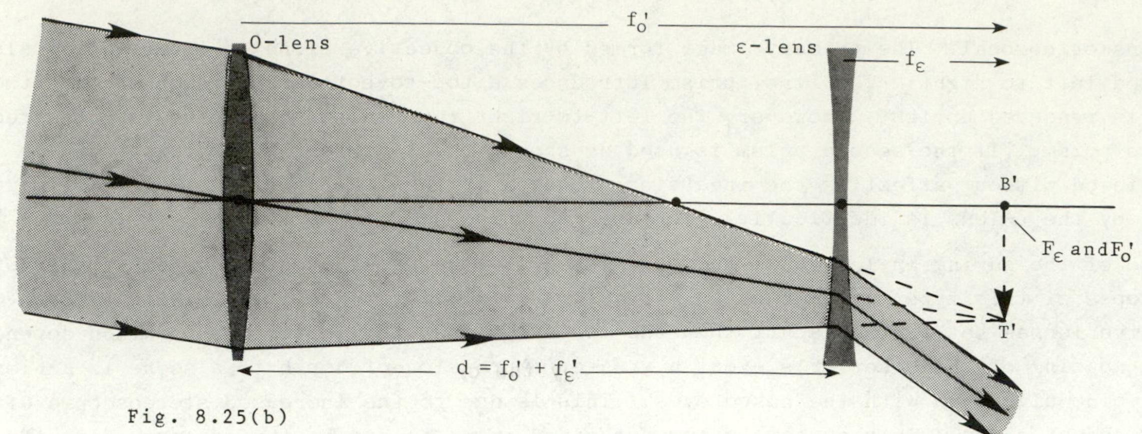

Fig. 8.25(b)

Rays from the tip of the distant object form a parallel pencil at the objective (O-lens). In the absence of any other components, this lens would form an image B'T' in its second focal plane at F_o'. However, with the diverging eyepiece, the ε-lens, positioned as shown, with its first focal plane at $F_ε$ coinciding with the image B'T', the latter acts as a virtual object for the ε-lens. This results in a parallel pencil of rays of reduced diameter emerging from the ε-lens, and corresponds to a virtual upright and enlarged image at infinity.

A disadvantage of this form of telescope is that the pencil of rays emerging from the ε-lens is steeply inclined to the principal axis of the system. This results in a very limited field of view.

In the afocal configuration, the length of the Galilean telescope is equal to the difference in the absolute values of the focal lengths of the objective and eyepiece lenses, and this makes it a shorter device than the Keplerian form. So, we have the distance between the lenses given by

$$d = f_o' - |f_ε'| \qquad (8.16),$$

or $d = f_o' + f_ε'$, allowing for the fact that $f_ε'$ is negative. This is one reason for the use of the Galilean telescope mounted in pairs as opera glasses, where the magnification of about 2 is usually acceptable.

The angular magnification of the Galilean telescope it its afocal configuration may now be derived from equation (8.12a). Figure 8.26 is a simplified version of figure 8.25(a).

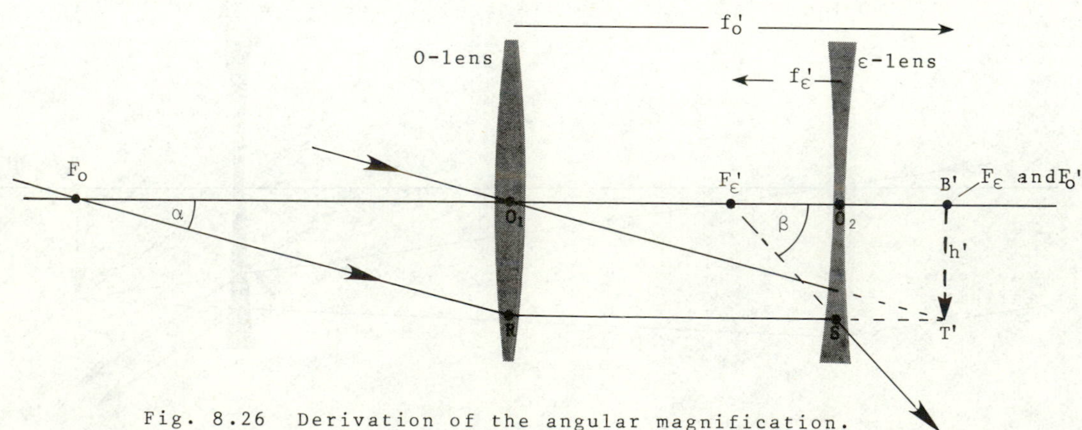

Fig. 8.26 Derivation of the angular magnification.

It shows the angle α subtended by the object at the O-lens, and the angle β subtended by the final image. The former angle is essentially the same angle which would be subtended by the object at the unaided eye. It can be appreciated that the distance O_1R is equal to O_2S, and that they are both equal to the image size B'T'. Hence, the angular magnification is

$$M = \frac{\tan\beta}{\tan\alpha} = \frac{O_2S/O_2F_\varepsilon'}{T'B'/O_1B'} = \frac{h'/f_\varepsilon'}{-h'/f_o'} = -\frac{f_o'}{f_\varepsilon'} = -\frac{F_\varepsilon}{F_o} \qquad (8.13).$$

Note that both α and β are positive angles, and careful consideration of the signs of the distances has been taken to ensure that the tangents of these angles are positive.

The angular magnification is seen to be the same expression that was derived for the Keplerian astronomical telescope. However, since F_ε has a negative value, the sign of the angular magnification will be positive indicating that the image observed is upright.

The Galilean telescope forms the basis of the *telescopic spectacle unit*. This is a low vision aid with which low visual acuity is improved by the magnification. The telescope is adjusted for a finite working distance, and because of limitations in the total length of the system and the need for a reasonable field of view, magnification is usually restricted to about 2×. In a typical case, shown in figure 8.27, the overall length of the telescope is 2.5 cm, with component lens powers $F_o = +20$ D and $F_\varepsilon = -40$ D.

Fig. 8.27 The telescopic spectacle unit.

When prescribed for distance vision the unit can be adapted for near vision by the fitting of a suitable additional positive lens. Figure 8.27 shows such an auxiliary lens for fitting on to the front of the unit.

8.3.4.5 *The Reflecting Telescope*

The term "astronomical telescope" has been used for the type of instrument described in section 8.3.4.1. However, the biggest telescopes used in astronomy make use of a large concave mirror as a means of collecting the light from a distant object. The difficulties encountered in using lenses of large aperture are reasonably obvious. Among the most important ones are the difficulty of making a large lens which is free from defects and strains, and the fact that a lens may be supported only at its rim, with consequent sagging of the bulk of the lens. In contrast, a mirror may be rigidly supported over its entire rear surface. Furthermore, to be free of chromatic aberration in a refracting telescope an objective comprising at least two lenses must be used, whilst a mirror is free of chromatic aberration since the laws of reflection are independent of wavelength. These, and other difficulties, have resulted in the largest refracting telescope in the world, the Yerkes telescope in the USA, having an objective lens diameter about 1 metre only, whilst the largest reflecting instruments are the Hale telescope on Mount Palomar, California, with a mirror diameter about 5 metres or 200 inches, and the new 6 metre reflector at Zelenchukskaya, USSR.

Using a concave paraboloidal mirror as the primary optical component, several arrangements are possible for viewing or recording the image. Figure 8.28 shows some of the mountings used.

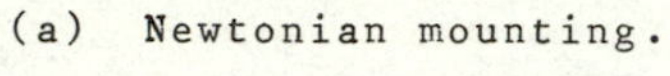

(a) Newtonian mounting.

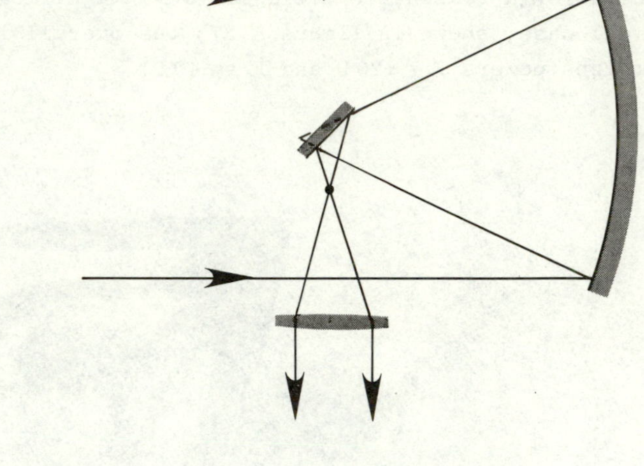

(b) Cassegrain mounting.

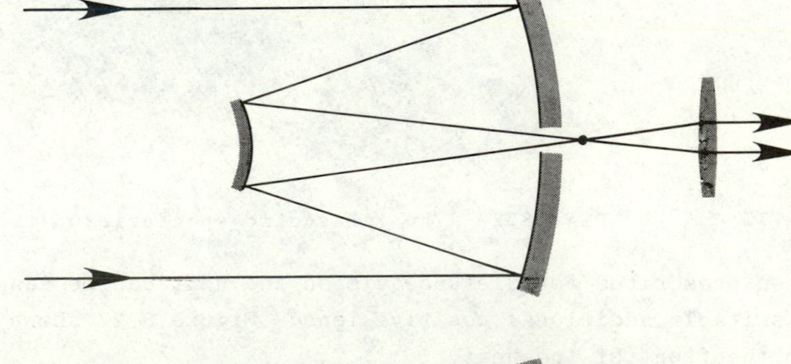

(c) Gregorian mounting.

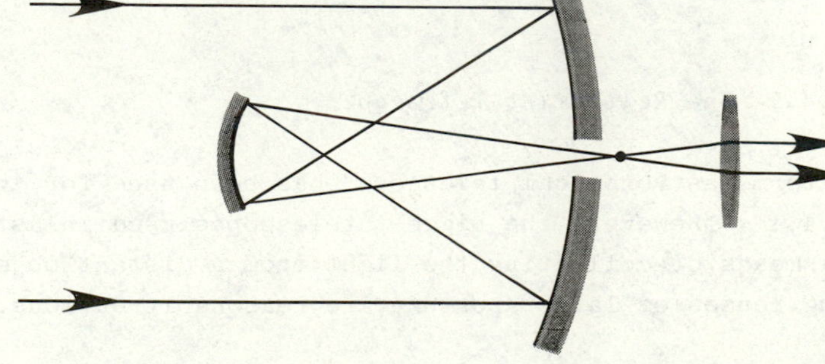

Fig. 8.28 Reflecting telescopes.

In the Newtonian form, figure 8.28(a), a small plane mirror or reflecting prism deviates the light at right angles to the telescope axis. The real image is then viewed from the side with the aid of an eyepiece. With the Cassegrain system, figure 8.28(b), a small convex mirror of hyperboloidal section reflects the light through an opening in the centre of the primary mirror. The image is again viewed with an eyepiece. In the Gregorian arrangement, which is not often used, a small concave mirror of ellipsoidal section is used to reflect the light through an opening in the centre of the primary mirror, figure 8.28(c).

One of the main disadvantages of the paraboloidal mirror is that only a fairly narrow field of view is free from spherical aberration. However, many astronomical applications require a telescope with a wide field of view. Such a system usually incorporates a refracting system and a reflecting system, and is termed a *catadioptric* system. The word catadioptric is derived from *catoptric*, meaning "based on reflection", and *dioptric*, meaning "based on refraction".

8.3.4.6 The Schmidt Catadioptric System

Figure 8.29 shows the general principle of the Schmidt system used as a camera. The unique feature of this system is the refracting corrector plate (lens) which has aspherical surfaces mounted in front of a concave spherical mirror, the centre of curvature of the latter being at C. In such a situation the corrector plate corrects for the spherical aberration of the concave mirror, and axial and off-axial parallel pencils of rays are brought to point foci on a spherical curve through F. The centre of this spherical surface is at C. A film plate may be mounted on this curved surface and f-numbers as low as f/0.5 are attainable, with a wide angle of view.

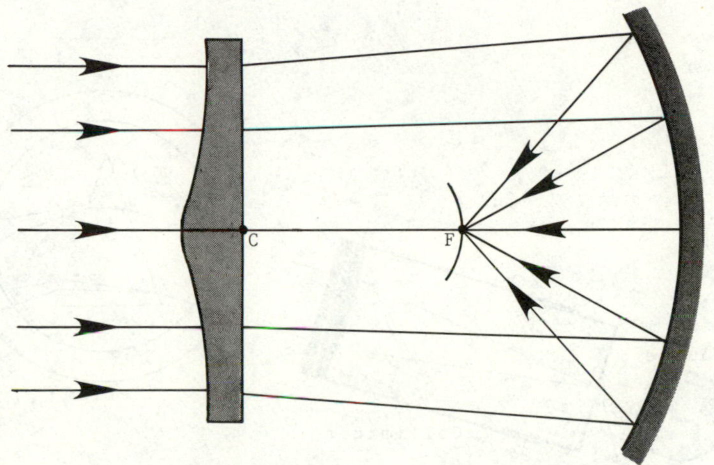

Fig. 8.29 The Schmidt catadioptric telescope.

The most famous Schmidt system is the 122 cm (48 inch) diameter device at the Palomar Observatory. The Schmidt system also has other important applications including TV cameras and satellite tracking.

8.3.5 THE SPECTROMETER

Certain types of optical instrument may be described as analysing instruments. That is, their main function is not to form a detailed image of an object as such, but to determine the composition of a beam of light. The *prism spectrometer* is an instrument of this type and is used to examine sources of light by analysis of their spectra. It also enables an accurate determination of the refractive index of prism materials and liquids to be made. This will be discussed in section 8.3.5.1 . The spectra of light sources are discussed in Chapter 15 and the reader might find it helpful to look at that material in conjunction with this section.

Figures 8.30(a) and (b) show two views of a laboratory spectrometer, whilst figure 8.30(c) illustrates a plan view of its optical action (see also figure 6.7).

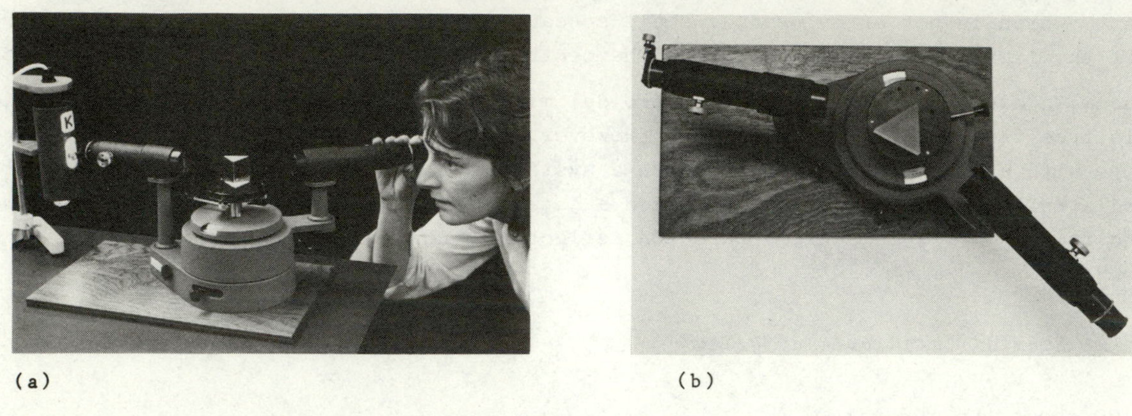

(a) (b)

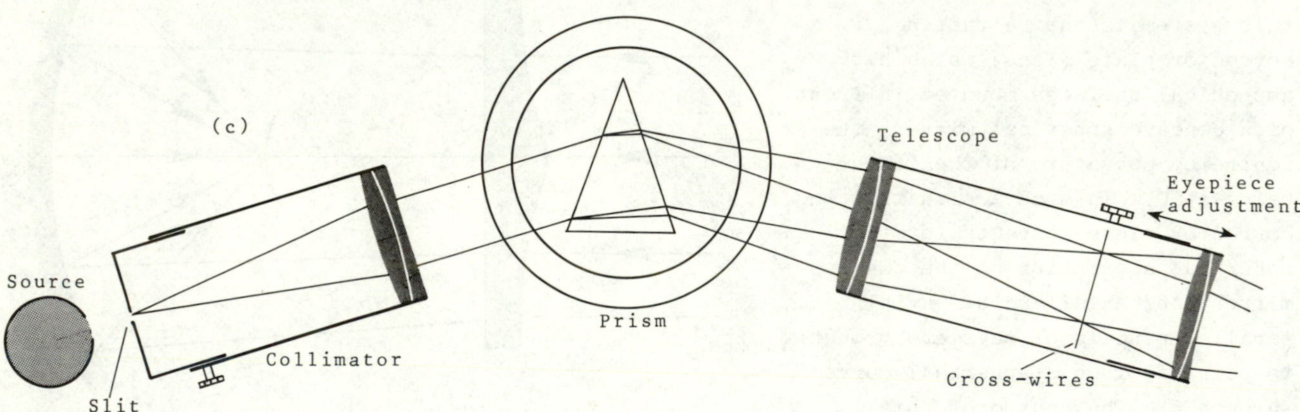

Fig. 8.30 The spectrometer: the spectrum is produced in the telescope objective's focal plane and is viewed through the eyepiece.

The essential components of the spectrometer are:

(i) A collimator, which is arranged to produce a parallel pencil of rays of the light under examination. It consists of an achromatic lens at a suitable distance from an adjustable vertical slit which is illuminated by the light source.

(ii) A dispersing prism (sometimes a diffraction grating is used instead - see Chapter 14) which deviates, as separate pencils of rays, the various wavelengths present in the light.

(iii) A viewing system, usually a Keplerian astronomical telescope, which focuses the light it receives to form a pure spectrum. If the source emits only a few wavelengths the spectrum is really a group of images of the collimator slit formed by each wavelength in the light, and appears as a magnified series of "lines" when viewed through the eyepiece. If the light emitted by the source contains all wavelengths in the visible spectrum, the images form a continuous spectrum.

The telescope also contains a pair of cross-wires, mounted internally in the tube in the second focal plane of the achromatic objective lens.

The prism, which stands on a horizontal table, and the telescope may both be swung independently in a horizontal plane around a vertical axis through the centre of the prism table. The rotations may be measured on a circular vernier scale.

To obtain clear focused views of spectra it is essential that preliminary adjustments of the eyepiece, telescope, and the collimator are made. These are performed as follows with the prism removed from its table.

(i) The eyepiece of the telescope is adjusted to give a clear view of the cross-wires. This should be done by screwing out the eyepiece initially, a move which presents converging rays to the eye and removes the stimulus to accommodate.

(ii) The telescope is pointed at a distant object and the length of the tube adjusted until the object is in focus at the same time as the cross-wires. Thus, the telescope is now set to receive and focus parallel light.

(iii) The collimator is now adjusted to deliver parallel light. This is achieved by aligning the telescope with the collimator, the slit of the latter being illuminated by a light source. The length of the collimator tube is now adjusted, together with the slit width, until a narrow focused image of the slit is observed in the telescope.

8.3.5.1 Measurement of the Refractive Index of a Prism Material

To obtain an accurate value for the refractive index of the material of a prism in air, let us again consider equation (2.20):-

$$n_p = \frac{\sin \frac{1}{2}(a+d_{min})}{\sin \frac{1}{2}a} \qquad (2.20)$$

where a is the apical angle of the prism, d_{min} is the angle of minimum deviation for a particular colour of light, and n_p is the index of the prism material, in air, for the particular colour of light. Since we must determine a and d_{min}, we can see that the experimental determination is in two parts.

(i) *To Measure the Apical Angle (a) of the Prism*

Figure 8.31 shows the prism on its table with the apical angle pointing towards the collimator, the slit of which is illuminated with a light source. Reflected images of the slit are received from the two refracting faces of the prism with the telescope positioned in the positions T_1 and T_2 successively. The telescope positions are noted when the vertical cross-wire is

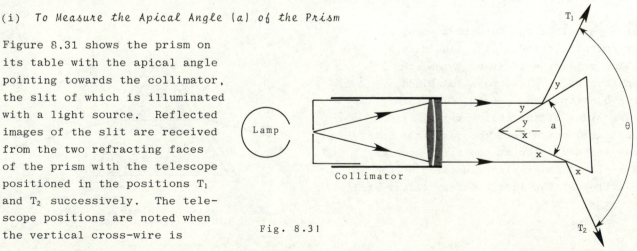

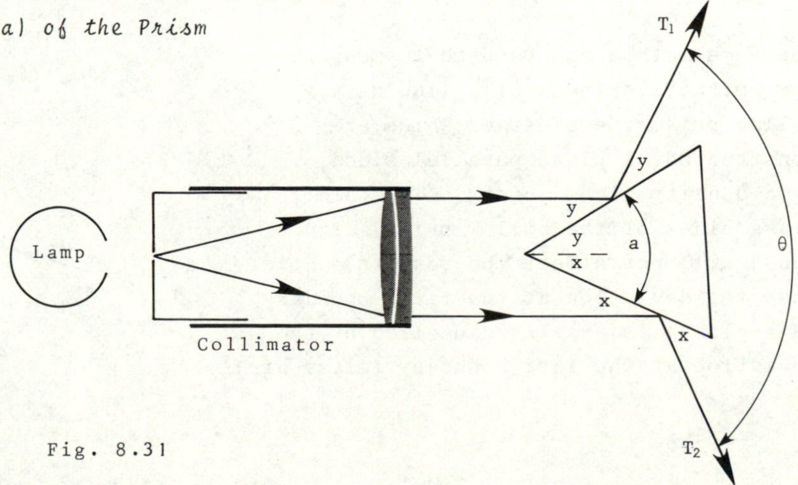

Fig. 8.31

coincident with reflected slit image. The difference in these two readings gives the angle θ in the diagram. It should be clear from figure 8.31 that $\theta = x+x+y+y = 2(x+y)$. But it is also evident that $(x+y) = a$, the apical angle of the prism. Therefore, $\theta = 2a$, or $a = \theta/2$. In this way the apical angle of the prism is determined.

(ii) *To Measure the Angle of Minimum Deviation for a Selected Wavelength*

Figure 8.32 illustrates the prism placed on its table so that the telescope receives light refracted by the prism.

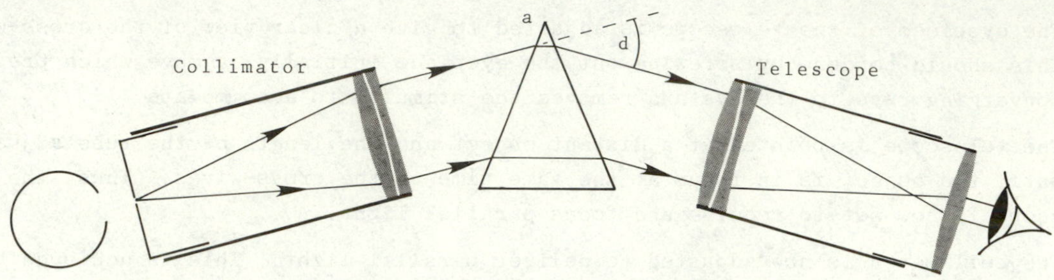

Fig. 8.32 Measurement of the minimum deviation.

The prism table is then rotated to make the selected slit image move across the field of view. By rotating in the appropriate direction and following the slit image in the telescope, the position of minimum deviation may be found. In this position, continuous rotation of the prism in the same direction will cause the slit image to reverse and move back across the field of view. With the vertical cross-wire coinciding with the slit image when in the position of minimum deviation, this position of the telescope is noted. The telescope is then swung back into direct line with the collimator and, with the prism removed, a second reading of the telescope position is noted when the slit image coincides with the vertical cross-wire. The difference in these two positions is equal to the angle of minimum deviation d_{min}. Knowing a and d_{min} the refractive index of the prism may be calculated.

The same method can be used to measure the refractive index of a liquid in a hollow prism the sides of which are constructed of plane parallel sided, good quality glass sheets. The thin glass sides of the hollow prism, figure 8.33, will not affect the results. Note that the deviation at the first boundary (air-glass) is exactly cancelled by the deviation at the last boundary (glass-air)!

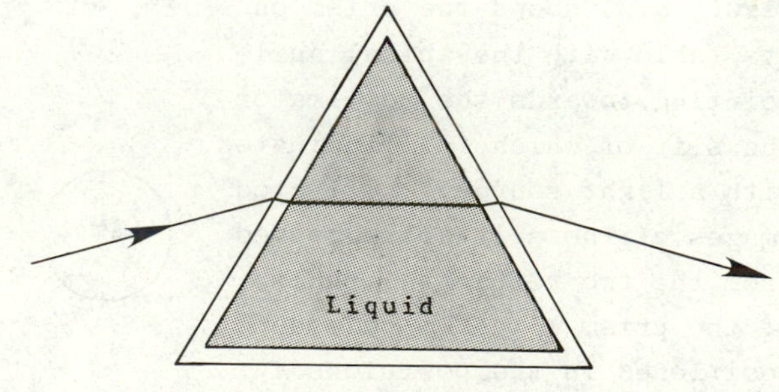

Fig. 8.33 Liquid contained in a thin glass walled prism.

8.3.6 THE DIRECT VISION SPECTROSCOPE

It is sometimes desirable to examine the colours and positions of optical spectra without taking actual measurements of angles. Designed for this purpose, the instrument is called a *direct vision spectroscope*. Its action is based on the direct vision prism, or rather a combination of these to give a better dispersive effect. Figure 6.9 showed a direct vision prism made from two component prisms of differing types of glass. Figure 8.34(a) shows a direct vision spectroscope consisting of several alternate component prisms of crown and flint glasses. It is much less bulky and more compact than a spectrometer, and may be carried in the pocket. The model shown also has an arrangement for illuminating a wavelength scale in the field of view. Figure 8.34(b) illustrates the optical action in a simplified way.

(a)

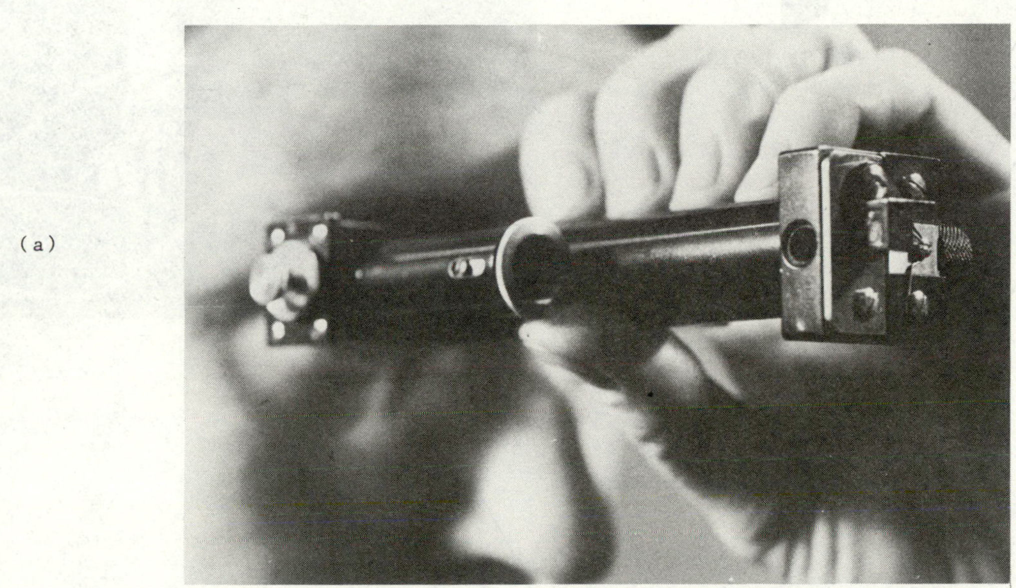

(b)

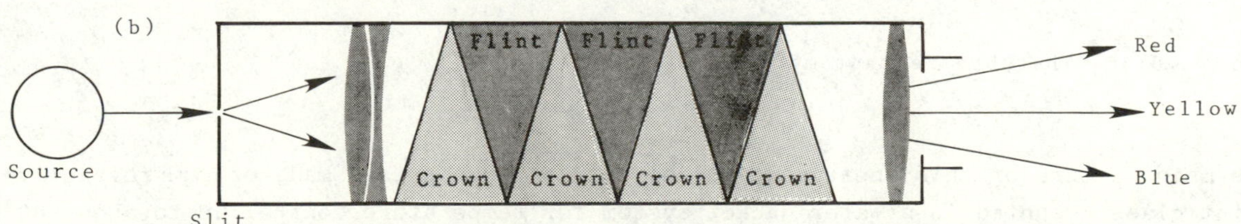

Fig. 8.34 The direct vision spectroscope.

In figure 8.34(b), a narrow pencil of rays enters the spectroscope through a slit and an achromatic lens. It is then refracted and dispersed successively by the component prisms such that dispersion occurs without deviation of the mean yellow ray. A virtual image of the slit in each colour of light from the source is observed in the eyepiece.

8.3.7 THE ABBE REFRACTOMETER

A refractometer is an instrument for the measurement of refractive index. Thus, a prism spectrometer, section 8.3.5, may be classed as a refractometer. It is usual, however, to reserve the name refractometer for those instruments which are calibrated so as to give direct readings of refractive index values. One such instrument is the Abbe refractometer, a very accurate instrument based on the design of Ernst Abbe in 1870, and used for measuring the refractive index of liquids, figure 8.35 .

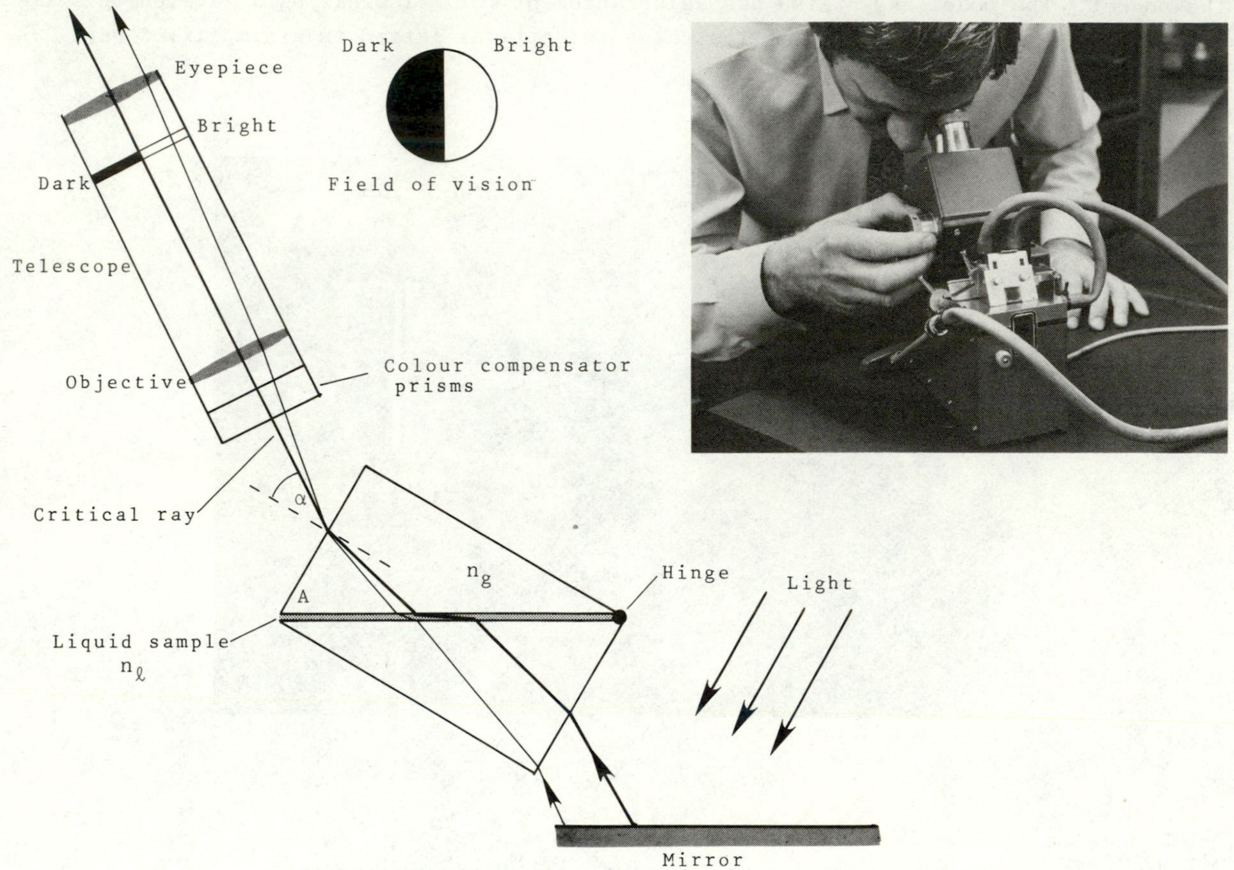

Fig. 8.35 The Abbe Refractometer.

The main feature of this instrument is the pair of Abbe prisms made of high refractive index flint glass, mounted in a water jacket system for temperature control up to about 80°C. The two prisms are hinged, and when clamped together the two hypotenuse faces of the prisms are separated by a space of about 0.1 mm thickness. It is in this space that the liquid under test is held, and a few drops only are required. When the liquid film is illuminated from below, some of the light passes into the liquid at grazing incidence, and this produces a sharp critical boundary. Viewed in the telescope, a dividing line is observed between the bright and the dark parts of the field. When the telescope cross-wires are set on the dividing line a scale attached to the telescope is calibrated to give the refractive index of the liquid directly.

White light may be used to illuminate the liquid film if the instrument is fitted with a dispersion compensator. Without this the view of the boundary as seen in the telescope is indistinct and coloured, unless monochromatic light is used. The compensator often consists of two direct vision prisms that rotate in opposite directions to provide a variable dispersion system.

If α is the minimum angle which the light emerging from the upper prism makes with the normal, and n_ℓ and n_g are the refractive indices of the liquid and the glass, respectively, then it can be shown that

$$n_\ell = \sin A\sqrt{n_g^2 - \sin^2\alpha} - \cos A.\sin\alpha \qquad (8.17).$$

8.4 STOPS IN OPTICAL INSTRUMENTS

In view of our previous assumption that rays of light make only small angles with the principal axes of lenses, the foregoing theory of optical instruments has been limited to the paraxial regions of the lenses. It should be clear that in actual optical systems the amount of light which reaches the image will be dependent on the physical limitations presented by the lens apertures together with any additional diaphragms or stops which may be present.

In Chapter 7 we saw how stops may be used to restrict certain rays of light from reaching the image with the intention of reducing image aberrations. In the present chapter we have already discussed how the presence and diameter of a stop affects the depth of focus and the depth of field of a lens. We have also seen, section 8.1.1.1, that the illuminance of the image depends on the aperture stop diameter. Before we commence a more detailed discussion on stops in optical systems, let us summarise their main effects. These concern:

(i) the aberrations of the image
(ii) the depth of focus and the depth of field
(iii) the field of view
(iv) the illuminance of the image
(v) the resolving power of the instrument.

The effect of a stop on the resolving power of an instrument will be considered in section 14.1.3 .

Before we can determine whether or not a given ray will be transmitted through an optical system, it is necessary to understand the meaning of *iris (or aperture stop)*, *field stop*, *entrance pupil* and *exit pupil*, as applied to optical instruments.

8.4.1 IRIS (APERTURE STOP) AND FIELD STOP

In an optical system, consisting of a number of centred lenses, together with one or more stops, the component which alone determines the amount of light reaching the image is known as the *iris* or the *aperture stop*. We shall use the former term because of its corresponding use in the optical action of the eye. The iris therefore controls the illuminance of the image. It may be a diaphragm or stop mounted in front of or behind the whole system, or it may be located somewhere between the lenses. Alternatively, one of the actual lens apertures may function as the iris. The aperture in the iris is known as the pupil, and it is the size of the pupil which determines the amount of light reaching the image.

Figure 8.36 shows some simple examples of the action of the iris.

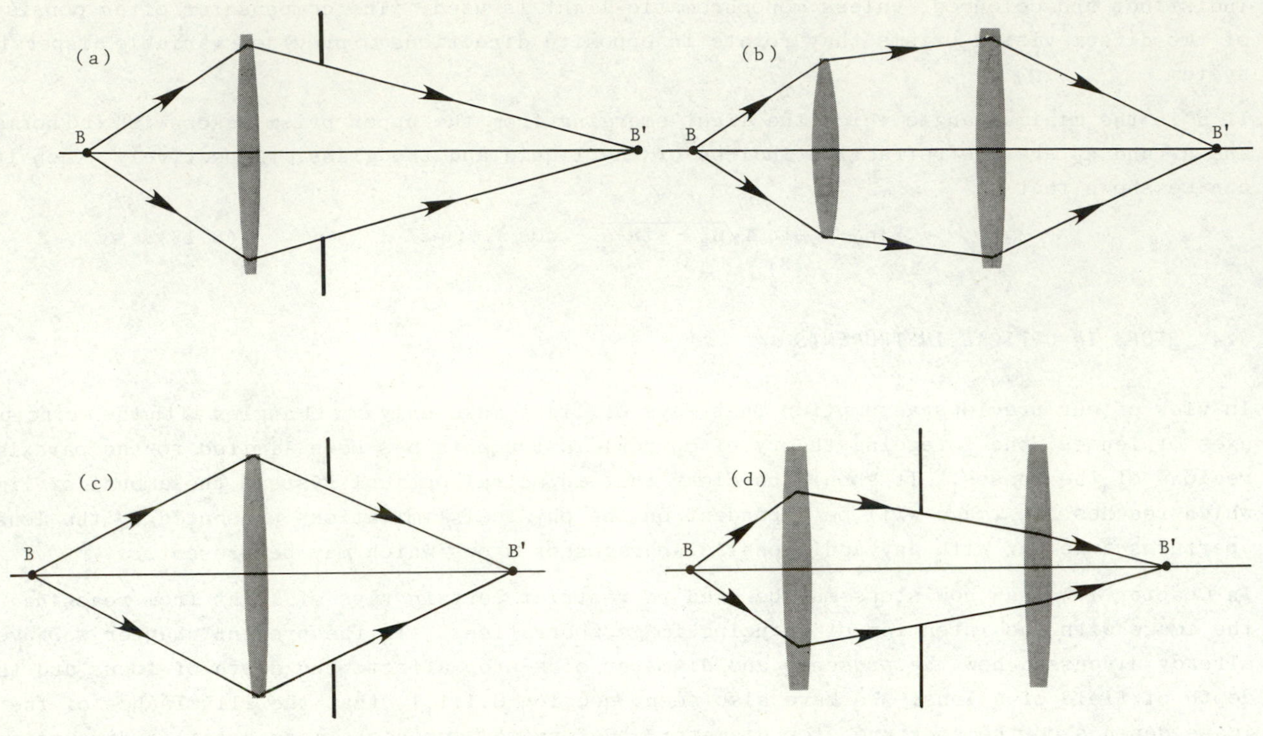

Fig. 8.36 Irises (aperture stops).

In figure 8.36(a), the stop mounted behind the lens is the iris as it is this, rather than the lens aperture, which determines the size of the cone of rays arriving at the image B'. In figure 8.36(b), the diameter of the first lens determines the light reaching the image, whilst in figure 8.36(d) the stop placed between the two lenses is the iris. It should be noted that the same aperture may not always act as the iris if objects at different distances are considered. This will be appreciated from figure 8.36(c). With the stop positioned as in figure 8.36(a) and the object moved further away from the lens, a stage will be reached when the lens rather than the stop will be the limiting factor on light reaching the image. In this situation the lens is acting as the iris. Thus, the iris determines the light gathering capability of a lens system as a whole.

Let us now consider figure 8.37 which shows a single lens forming the image of a distant object. Three separate parallel pencils of rays from three different points on the distant object are shown as being focused in the focal plane of the lens. It is seen that oblique rays can enter the system. However, these are usually intentionally restricted so as to improve the quality of the image. This may be necessary to cut off the indistinct portion of the field, or to prevent reflections from inside the tube from reaching the image plane. In the simple system shown in figure 8.37 the use of a screen of limited size is equivalent to the use of a *field stop*. In the case of a camera, the edge of the film acts as the field stop, whilst in microscopes and telescopes the field stop is positioned in the same plane as the primary image formed by the objective lens. It is often fabricated as part of the eyepiece.

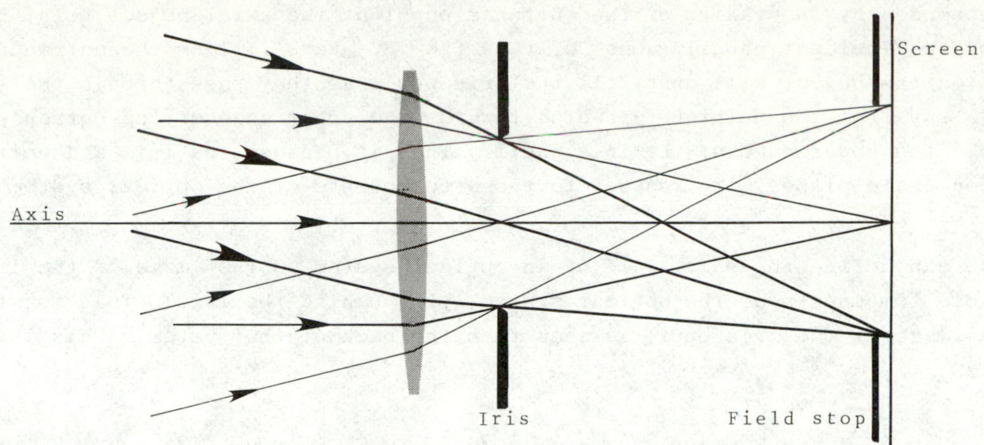

Fig. 8.37 Iris and field stop.

Note that whereas the iris pupil limits the size of the cone of rays entering the system, and therefore limits the energy reaching the image plane, the field stop limits the angle which rays can make with the principal axis.

8.4.2 ENTRANCE AND EXIT PUPILS

Another concept which enables us to determine whether a ray will be transmitted through an optical system is the *entrance pupil*. This may be described as the image of the iris pupil seen through the optical system in front of the iris. In the case of the human eye it is the entrance pupil which is seen as the dark aperture in the coloured iris. The entrance pupil in this case is the image of the pupil formed by the cornea.

Let us consider figure 8.38 which illustrates the most general type of system in which the iris is between the lenses. The image of the iris formed by the front lens only is shown as virtual and enlarged. The "aperture" in this image is the entrance pupil of the system.

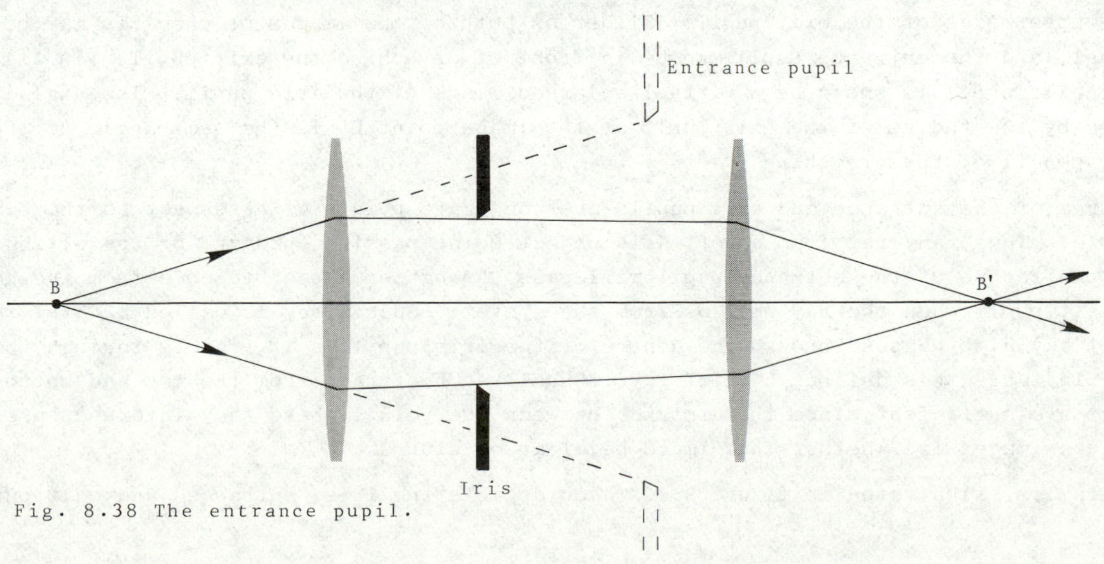

Fig. 8.38 The entrance pupil.

The angle subtended by the radius of the entrance pupil at the axial object point B is the semi-angle of the incident pencil (cone) of rays from B. Rays filling the entrance pupil from a point on the object will just fill the iris pupil as they pass through the system. Notice that a ray from the point B, directed towards the upper edge of the entrance pupil, passes through the upper edge of the iris pupil. This is because the iris and entrance pupils are conjugate planes with respect to those components of the optical system in front of the iris.

Similarly, we can define the *exit pupil* of an optical system as the image of the iris pupil formed by those components of the optical system lying behind the iris. That is, the exit pupil is the image of the iris pupil as seen from the back of the system. This is illustrated in figure 8.39 .

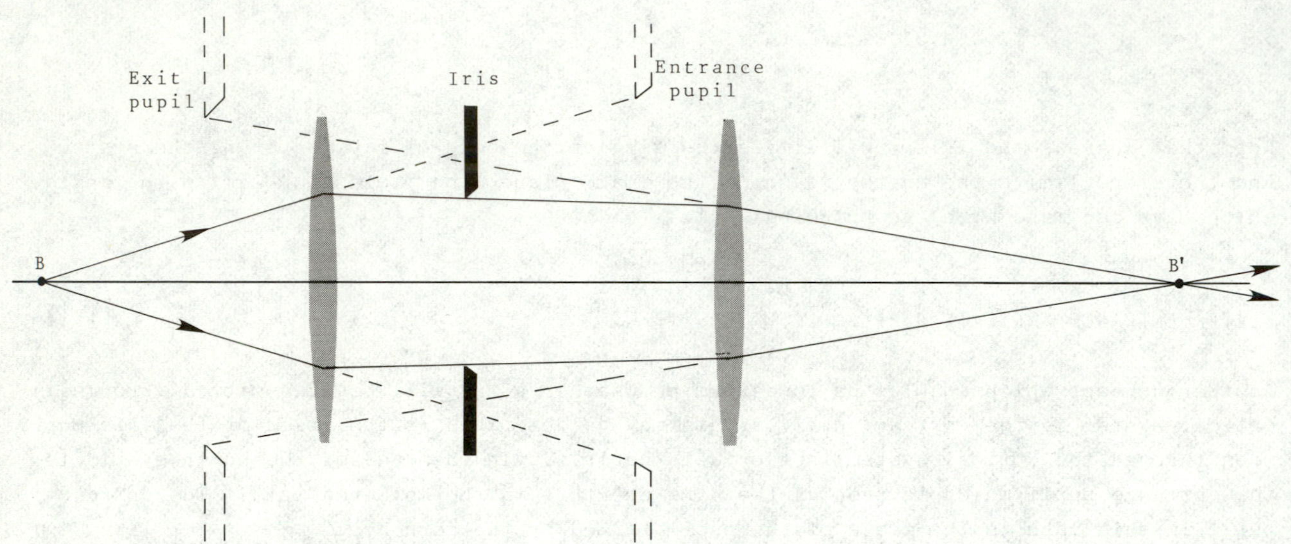

Fig. 8.39 Iris, entrance and exit pupils.

Rays filling the entrance pupil finally emerge from the system filling the exit pupil. A ray just grazing the upper edge of the iris pupil emerges from the system as though coming from the upper edge of the exit pupil. Either or both of the images of the iris may be real or virtual, and the entrance pupil can be in front of or behind the exit pupil. In figure 8.39 the exit pupil is shown as a virtual enlarged image of the iris pupil. The angle subtended by the radius of the exit pupil at the image point B' is the semi-angle of the emergent pencil of rays reaching B'.

The centres of the entrance and exit pupils are conjugate points with respect to the system as a whole. Thus, any ray from an off axis object point passing through, or travelling towards the centre of the entrance pupil will pass through or appear to come from the centre of the exit pupil when the ray emerges from the system. Such a ray is called a *chief ray*. Any chief ray also passes through the centre of the iris pupil as it travels through the system. In fact, this defines a chief ray. Usually, chief rays from the top and bottom of an object are sufficient since the distance between the points where they intersect a screen defines the image size whether that image be clear or blurred.

A chief ray is illustrated in figure 8.40 which depicts the iris, entrance, and exit pupils for a three lens system. In any actual instrument the chief ray rarely passes through the centre of any lens.

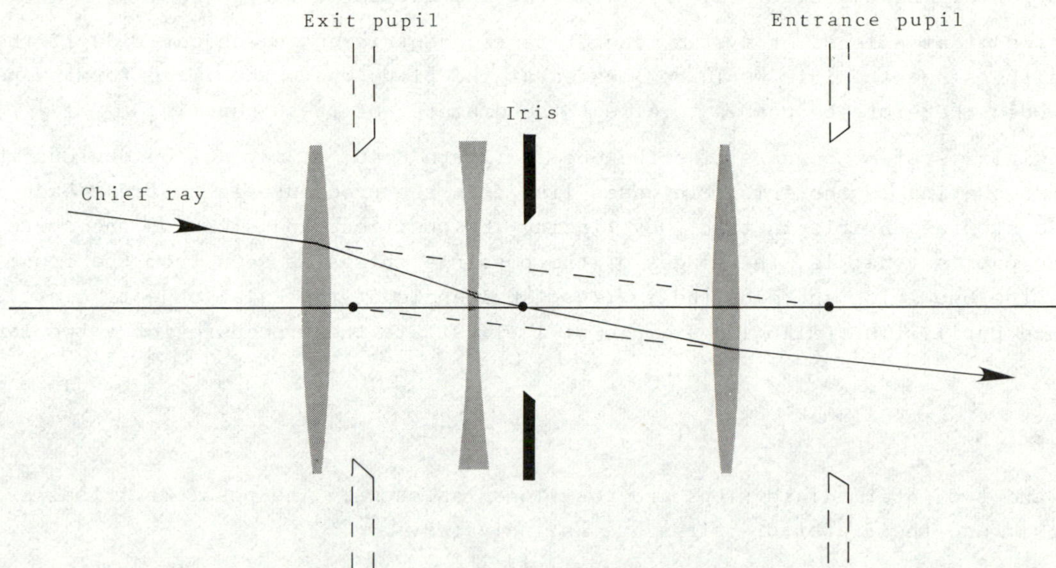

Fig. 8.40 A chief ray, the iris, entrance, and exit pupils.

We should note that in those cases where the iris is fixed in front of the system, or it coincides with the first lens so that there are no lens components in front of it, the entrance pupil coincides with the iris pupil. When the iris is a rear mounted stop, or it coincides with the last lens so that there are no lenses behind it, the exit pupil coincides with the iris.

If the centre of the iris lies at a principal focus the system is said to be *telecentric*. In particular, if the centre of the iris lies at the first, or anterior, focus of a lens system, the chief ray of a pencil traversing the system will emerge parallel to the axis. The exit pupil will be at infinity and the system is said to be telecentric on the image side. This is shown for a single lens in figure 8.41 .

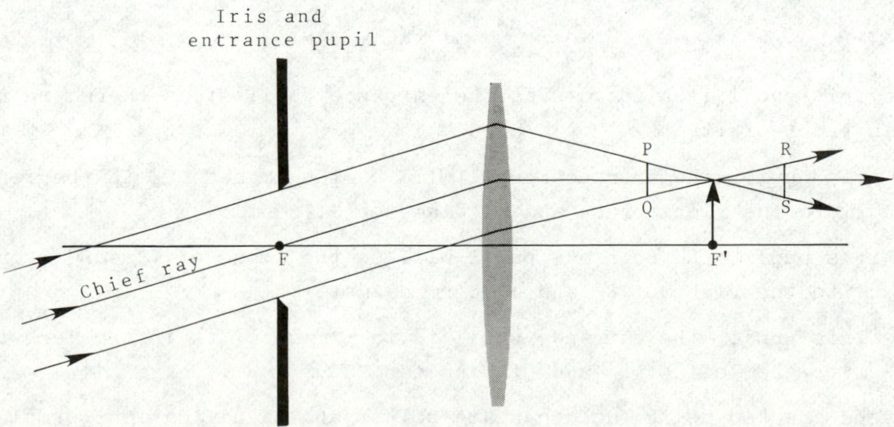

Fig. 8.41 The telecentric principle.

Any focusing error in the image will show the centres of the blur circles, such as PQ and RS, at the same distances from the principal axis. Thus, a measuring scale which is not precisely in the image plane will still give a correct reading of the image size. It is worthwhile noting that blurred image sizes are defined as the distance between the intersections with the screen of the two chief rays from the top and bottom of the object.

It will be seen that a lens, or system, cannot be telecentric on the object side if the object is at infinity, since the iris would be located at the plane where the image forming pencils focus and would therefore be unable to affect the diameter of these pencils.

In a more complex system, perhaps like the one in figure 8.40, it may not be obvious which stop or lens mounting is the iris. In cases like this the procedure is to treat each lens mounting and stop as an iris in turn. By finding the positions and sizes of the corresponding entrance pupils (that is, the images of the possible irises as seen from the front of the system) the one which subtends the *smallest* semi-angle at the axial object point is the true entrance pupil. The following example will illustrate the procedure for a two lens system.

Worked Example

Two lenses and two intermediate stops are positioned as shown. The possible irises are L_1, X, Y, and L_2, and their aperture sizes are as indicated.

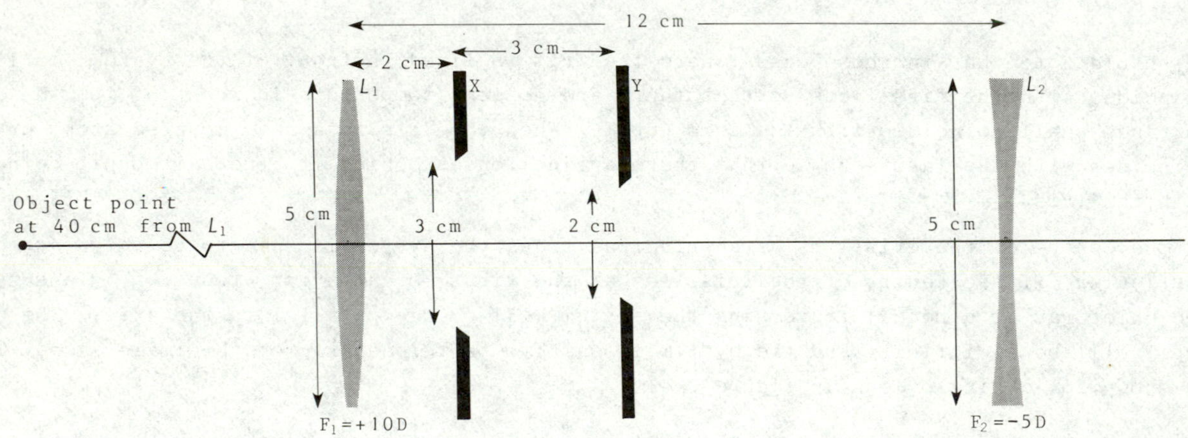

(a) If L_1 is the iris pupil it also acts as the entrance pupil since there are no lens components in front of it.

(b) If X is the iris pupil, the entrance pupil will be the image of X in the lens L_1. This image is 2.5 cm to the right of L_1 and of diameter 3.75 cm.

(c) If Y is the iris pupil, the entrance pupil will be the image of Y seen through L_1. This image is 10 cm to the right of L_1 and of 4 cm diameter.

(d) If L_2 is the iris pupil, the entrance pupil is the image of L_2 in the lens L_1. This image is 60 cm to the left of L_1 and of diameter 25 cm.

These figures may be checked using the conjugate points and magnification formulae for the thin lens L_1.

The corresponding semi-angles at the object point 40 cm to the left of L_1 are:

(a) $\arctan\frac{2.5}{40} = 3.58^0$, (b) $\arctan\frac{1.875}{42.5} = 2.53^0$, (c) $\arctan\frac{2}{50} = 2.29^0$,

and (d) $\arctan\frac{12.5}{20} = 32^0$.

The third of these is the smallest, so that Y is the iris and its image seen from in front of L_1 is the entrance pupil.

A similar procedure may be adopted to find the position of the exit pupil, this being the image of the possible iris pupils, as seen from the back of the system, which subtends the smallest semi-angle at the axial image point B'. The exit pupil is the image of the entrance pupil in the whole system, and as previously mentioned, their centres are conjugate points.

8.4.3 FIELD OF VIEW, ENTRANCE AND EXIT WINDOWS

As stated in section 8.4.1, the field stop in a system limits the extent of the largest object that can be viewed. That is, it limits the field of view. Another way of stating this is to say that the field stop, together with the iris pupil, limits the angle which rays traversing the system can make with the principal axis. Inspection of the simple system in figure 8.37 will illustrate the validity of this. If the field stop were smaller in that case the rays entering via the lens and the iris pupil would be less inclined to the principal axis.

The image of the field stop formed by the optical components in front of it is called the *entrance window* or *entrance port*. Thus, the entrance window is the image of the field stop as seen from the front of the system. The semi-angle subtended at the centre of the entrance pupil by the radius of the entrance window is referred to as the *angular field of view in the object space*. This is sometimes called the *actual field of view*.

The image of the field stop formed by the components behind it is called the *exit window* or *exit port*. Thus, the exit window is the image of the field stop as seen from the back of the system. The semi-angle subtended by the radius of the exit window at the centre of the exit pupil is referred to as the *angular field of view in the image space*. This is sometimes called the *apparent field of view*.

Where the field of view is referred to without qualification, then it is assumed that this is the field of view in object space. As with the entrance and exit pupils, the entrance and exit windows are conjugate planes with respect to the whole system.

The term *field of view* is usually applied to that portion of the object field for which the illuminance of the image is equal to or greater than half its value near the axis. This corresponds to the definition given in section 8.4.2, namely that the angular field of view in the object space is given by the semi-angle subtended at the centre of the entrance pupil by the radius of the entrance window, figure 8.42 .

On some occasions the terms *extreme field* and *full field* are also used. These apply to the object fields over which there is no illuminance at all, and over which the illuminance does not fall appreciably below its axial value.

In figure 8.42 the field stop, the iris and the entrance pupils of a single converging lens system are shown. The semi-angular subtenses of the full, the half, and the extreme fields of view are illustrated.

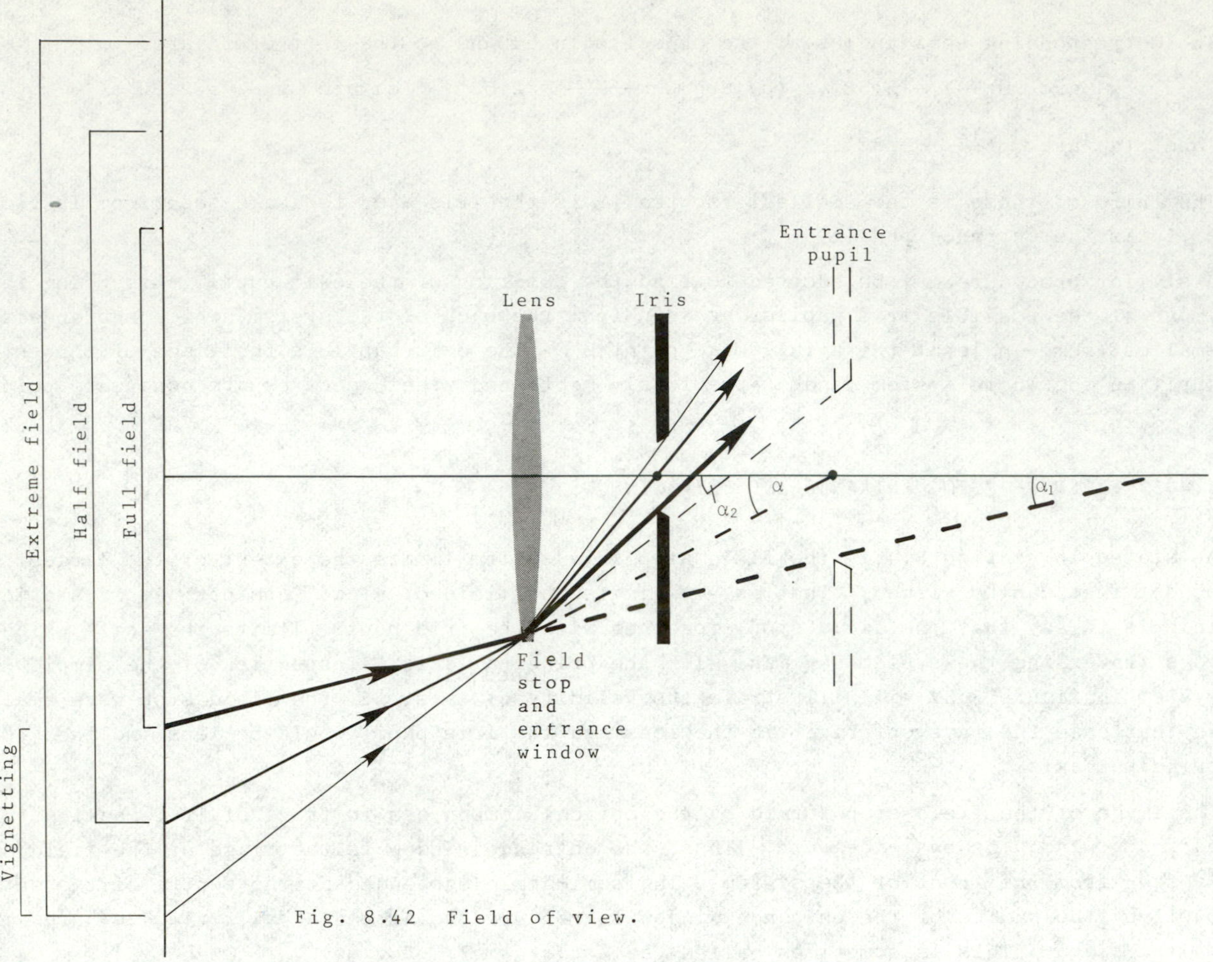

Fig. 8.42 Field of view.

The justification of these names can be derived by inspection of the rays passing through the iris pupil. Notice that a ray from the object plane outside the region designated the extreme field would not pass through the iris pupil - clearly, this is the extreme limit of the field of view. Rays from the full field entirely fill the the iris pupil so that this region of the object field is fully illuminated. At the limit of the half field only half of the iris pupil is filled with rays from the object field, and therefore the light reaching the image plane contains only half the energy of the full field image.

The semi-angles of the full, half, and extreme fields of view are shown as α_1, α, and α_2, respectively. A similar situation exists on the image side with the exit pupil and the exit window. Unless otherwise stated, the term field of view will be used to mean the semi-angular field of half illuminance in the object space.

Near the edge of the full field of view the field will appear restricted with the extremities of the field showing a gradual fall-off in intensity. This is an effect called *vignetting*. In an astronomical telescope this effect is removed by suitably placing the field stop in the focal plane of the objective so that the image of the field stop is in sharp focus and the observer therefore sees a sharp cut-off in the intensity at the edge of the field; see section 8.4.4 .

The foregoing principles regarding entrance and exit pupils, and field of view, will now be applied to magnifying instruments. Since they fall into the classification of visual instruments, these devices are used with the eye behind the eyepiece. Thus, the complete system consists of the instrument itself and the observer's eye. To begin with, we shall consider the instrument itself with particular reference to the entrance and exit pupils, and then finally we shall see briefly what modifications are necessary when the eye is included in the system.

8.4.4 ENTRANCE AND EXIT PUPILS WITH TELESCOPES

In sections 8.3.4.1 and 8.3.4.4 we have seen the basic construction and optical behaviour of the astronomical and Galilean systems, but no reference was made to the presence of stops in these instruments.

Let us again consider the astronomical telescope in the afocal configuration. Figure 8.43 shows this arrangement, and also illustrates the position of the image of the eyepiece formed by the objective. This image, MN is of larger diameter than the objective lens and therefore we may assume that the objective lens acts as the iris. As there are no optical components in front of the iris it also acts as the entrance pupil. Thus, the amount of light entering the system is determined by the diameter D of the objective lens. Consequently, the image ST of the objective lens, of diameter D', seen from the rear of the telescope is the exit pupil. That is, the exit pupil is the image of the objective formed by the eyepiece.

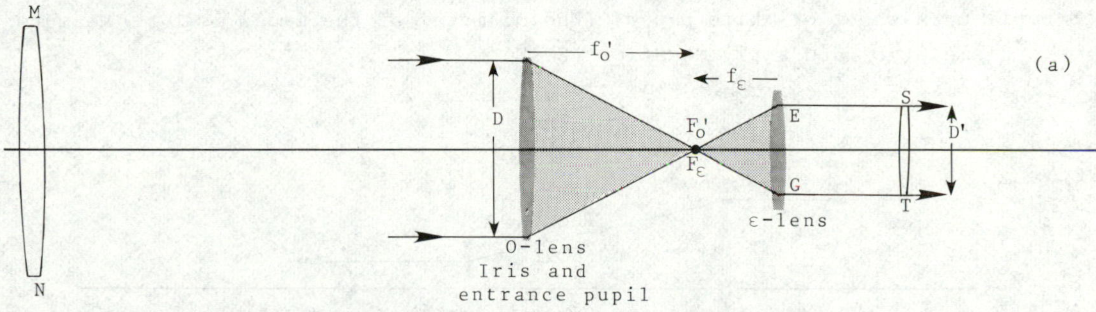

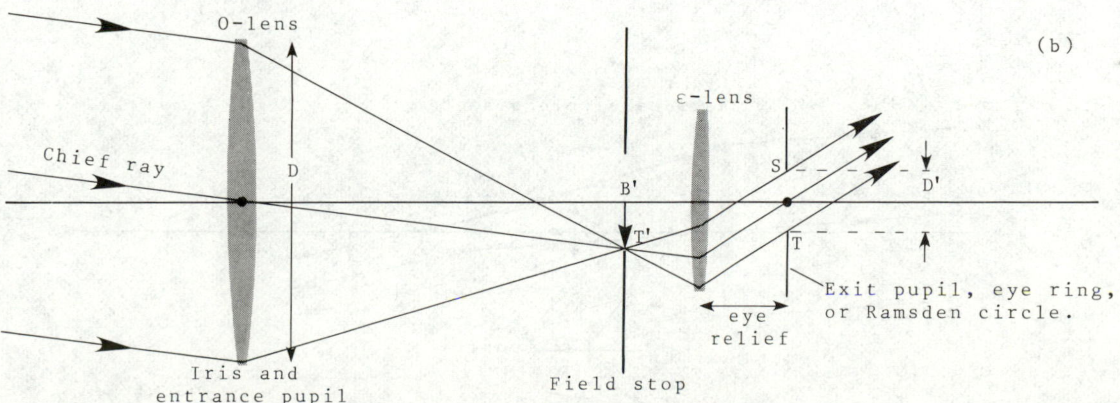

Fig. 8.43 The astronomical telescope - Entrance and exit pupils.

In the astronomical telescope the exit pupil is real and it is convenient to place the entrance pupil of the observer's eye at that point. The exit pupil is often called the *eye ring* or *Ramsden circle*, and its distance from the ε-lens is referred to as the *eye relief*. When the eye is placed at this position the whole field of view is visible at once. If the pupil in the eye, in the plane of the eye ring, is smaller in diameter than the exit pupil of the instrument, this will limit the diameter of the transmitted pencils. In such a case the iris of the eye becomes the iris of the whole system, and its image seen from the front of the instrument becomes the new, smaller diameter entrance pupil. It is important that the diameter of the eye ring should be no bigger than the pupillary diameter of the eye.

In the afocal configuration, figure 8.43(a), the separation of the lenses is given by the sum of the absolute values of the focal lengths of the O-lens and the ε-lens. It can be appreciated that the dimension EG on the ε-lens is the same as the diameter D' of the exit pupil. So, we have in the pair of shaded similar triangles

$$\frac{D}{D'} = \frac{f_o'}{-f_\varepsilon} = \frac{F_\varepsilon}{F_o} \quad , \qquad \text{since } F_o = \frac{1}{f_o'} \quad \text{and} \quad F_\varepsilon = -\frac{1}{f_\varepsilon} \, .$$

But $-\frac{F_\varepsilon}{F_o} = M$, the angular magnification of the instrument in its afocal setting, hence

$$M = -\frac{F_\varepsilon}{F_o} = -\frac{D}{D'} \qquad (8.18).$$

That is, the angular magnification is the ratio of the diameter of the entrance pupil to the diameter of the exit pupil. This leads to a speedy method of checking the angular magnification of an astronomical telescope. Simply point the objective at the daytime sky and image the exit pupil on a sheet of white paper. The diameter of the image is D'. Measure the objective's diameter D, and apply equation (8.18).

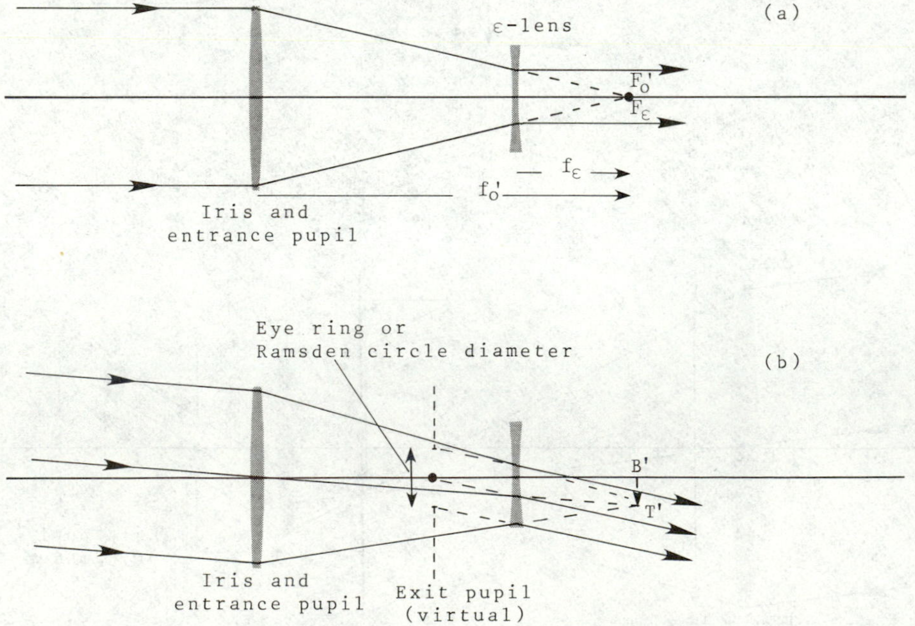

Fig. 8.44 Galilean telescope - entrance and exit pupils.

Figure 8.44 illustrates the principle of the Galilean telescope in its afocal setting (see also figure 8.25). With the diameter of the objective lens acting as the iris pupil and also the entrance pupil, the exit pupil will be virtual since it is the image of the O-lens formed by the negative ε-lens. Moreover, as the primary image formed by the objective is virtual, there is no separate field stop which can be placed in the instrument and the ε-lens acts as the field stop. Unlike the astronomical telescope, the Galilean telescope's field of view therefore suffers from vignetting.

It is not possible to position the observer's eye in the plane of the exit pupil to receive all of the light. It must be placed behind the ε-lens and cannot therefore receive all the light entering the O-lens and emerging from the ε-lens. Due to this restriction the field of view is considerably limited, and this is one of the main disadvantages of the Galilean telescope. Compare this with the astronomical telescope where the eye's entrance pupil is placed at the telescope's exit pupil, the real image of the O-lens, so the eye receives all the light passing through the O-lens without having to move.

With the restriction imposed on the position of the observer's eye, and the fact that its pupillary diameter is only some 3 to 8 mm, then the pupil of the eye becomes the iris pupil of the system. It also acts as the exit pupil, with its image formed by both lenses and the cornea in front of it being the entrance pupil. Strictly, the exit pupil will be the image of the eye's pupil formed by the crystalline lens in the eye, but this is very close to the iris pupil in the eye and we may neglect this slight difference.

8.4.5 FIELD OF VIEW WITH TELESCOPES

Before we discuss the field of view of the astronomical and Galilean telescopes, let us look at a simpler system of a single positive lens placed in front of the observer's eye. This is the loupe arrangement, of course. As shown in figure 8.45, the amount of light passing into the eye is limited by the pupil of the eye, and this will act as the iris of the lens and eye system. It will also act as the exit pupil of the system if we assume the reduced eye model and that the pupil is at the reduced surface for simplicity.

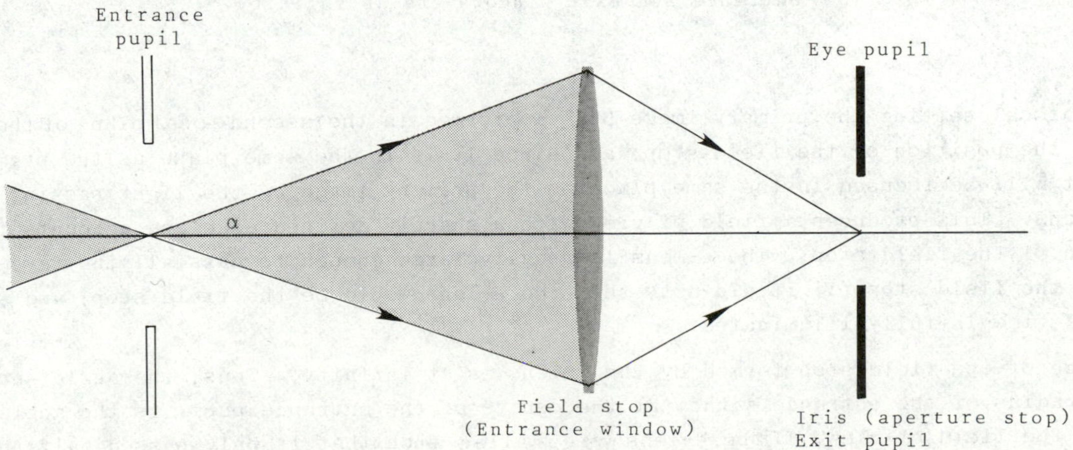

Fig. 8.45 Field of view through a single positive lens.

Since the centres of the entrance and exit pupils of a system are conjugate points, the entrance pupil will be the real image of the exit pupil formed by the lens. This is shown in front of the lens. The shaded cones of rays in front of the lens represent the spaces in which an object must be located in order to be visible in the half field. The lens acts as the field stop, and as there are no refracting components in front of it, it also acts as the entrance window. From the definition of angular field of view (see section 8.4.3) we have:

angular field of view in object space = the semi-angle subtended at the centre of the entrance pupil by the radius of the entrance window

Thus, in figure 8.45, the angular field of view = α .

Now, in the case of an astronomical telescope the aperture of the objective lens is the iris pupil (aperture stop). As there are no optical components in front of it, the objective also acts as the entrance pupil of the instrument. We can now show that the field of view of an astronomical telescope is dependent on the diameter a of the field stop placed in the second focal plane of the objective, figure 8.46 . This figure shows the positions of the entrance and exit pupils, and the field stop.

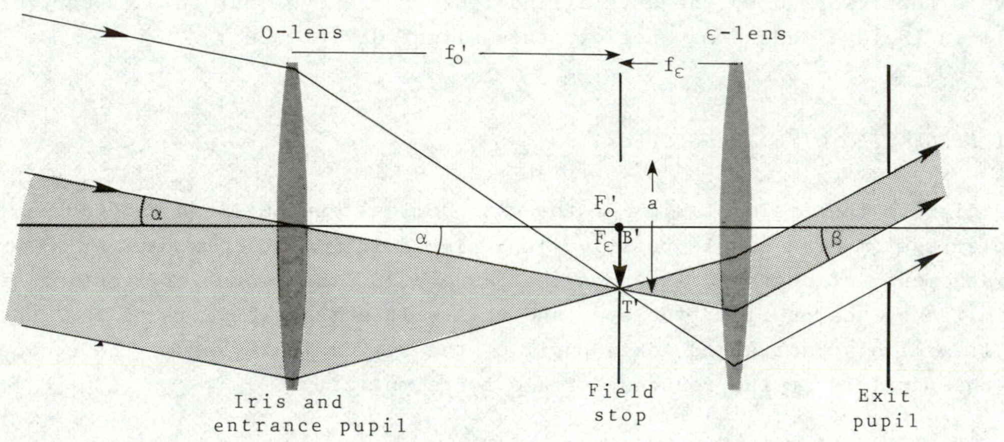

Fig. 8.46 Field of view of the astronomical telescope. Both the entrance and exit windows are at infinity.

In the afocal setting the primary image B'T' is formed in the second focal plane of the O-lens. This is the position of the field stop, and since it is in the same plane as the primary image it will be focused in the same place as the primary image by the light passing through the ε-lens. This produces a field of view with a sharply defined edge and accounts for the position of the field stop. The ε-lens is usually large enough to pass all the rays getting through the field stop (if it did not, then the ε-lens would be the field stop) and so the field of view is fully illuminated.

The image of the field stop formed by the O-lens is at infinity. Thus, the angle subtended by the radius of the entrance window at the centre of the entrance pupil is the angle α, and this is the field of view. If the ε-lens were smaller, such that it only passed half the light to the eye, then α would be the half field of view. This would mean that only half the light passing through the objective lens from an off-axis object point would pass through the system. Such a case would be indicated if the ε-lens were only just large enough to pass the shaded pencil of rays in figure 8.46 .

For numerical purposes it may be more convenient to express the field of view of the telescope as 2α, by taking into account the rays from both sides of the principal axis. Hence, if a is the diameter of the field stop, and this is small compared with the value of the focal length of the objective, then we have

$$2\alpha = \frac{a}{f_o'} \qquad (8.19).$$

For example, suppose the telescope has $f_o' = +40$ cm and $a = 3.8$ cm, then

$$2\alpha = \frac{3.8}{40} = 0.095 \text{ rad} = 5.43^\circ.$$

If the objective is replaced by a lens of focal length 80 cm, doubling the angular magnification, we have

$$2\alpha = \frac{3.8}{80} = 0.0475 \text{ rad} = 2.72^\circ.$$

Thus, we note that as the angular magnification of the telescope doubles the field of view halves. If we rewrite equation (8.19) thus

$$2\alpha = \frac{a}{f_o'} \times \frac{f_\varepsilon'}{f_\varepsilon'} = \frac{a}{f_\varepsilon'} \times \frac{f_\varepsilon'}{f_o'} = \frac{a}{f_\varepsilon'} \times \left(-\frac{1}{M}\right) \qquad (8.19a)$$

we see that the field of view is inversely proportional to the angular magnification. This is generally true, or approximately so, of all magnifying instruments; that is, as the angular magnification doubles the field of view halves, and vice versa.

The corresponding field of view in image space is 2β, and this can be derived simply by making use of equation (8.13), whence

$$2\beta = 2\alpha M = \frac{a}{f_o'} \times \left(-\frac{f_o'}{f_\varepsilon'}\right) = -\frac{a}{f_\varepsilon'} \qquad (8.20).$$

Now, in the case of the Galilean telescope, we can again assume that the aperture of the O-lens is the iris pupil (aperture stop), and once again, since there are no refracting components in front of it, it is also the entrance pupil. The field of view of the telescope is controlled by the aperture of the eye lens.

Referring again to figure 8.44(b), we recall that the primary image B'T' formed by the O-lens is virtual. This image plane is the theoretical location of the field stop which would ensure that the edge of the field of view was sharply defined. However, since the rays of light do not actually intersect in the image plane B'T', a stop cannot be placed there. As a consequence, the ε-lens acts as the field stop, and the resulting field of view has an indistinct margin. The image of the ε-lens formed by the O-lens is the entrance window.

Figure 8.47 illustrates the position of the entrance and exit pupils, together with the entrance and exit windows, for a Galilean telescope in afocal adjustment. The angle α is, as usual, the angular subtense of the radius of the entrance window at the centre of the entrance pupil. We can show that this angle, or more correctly 2α if rays from below the axis are taken into account, represents the maximum half field of view as shown in figure 8.47 .

As previously mentioned, the exit pupil of the Galilean telescope is the virtual image of the objective formed by the eye lens. With the exit pupil being inaccessible to the eye, the latter must be placed as close to the eye lens as possible. The angle β is the angle subtended by the image at the eye. Hence, if a is the aperture diameter of the field stop (the ε-lens), and this is small compared to the focal lengths of the O- and ε-lenses, we then have

$$\text{field of view (of object)} = 2\alpha = \frac{a}{f_O' - |f_\varepsilon'|} = \frac{a}{f_O' + f_\varepsilon'} \qquad (8.21).$$

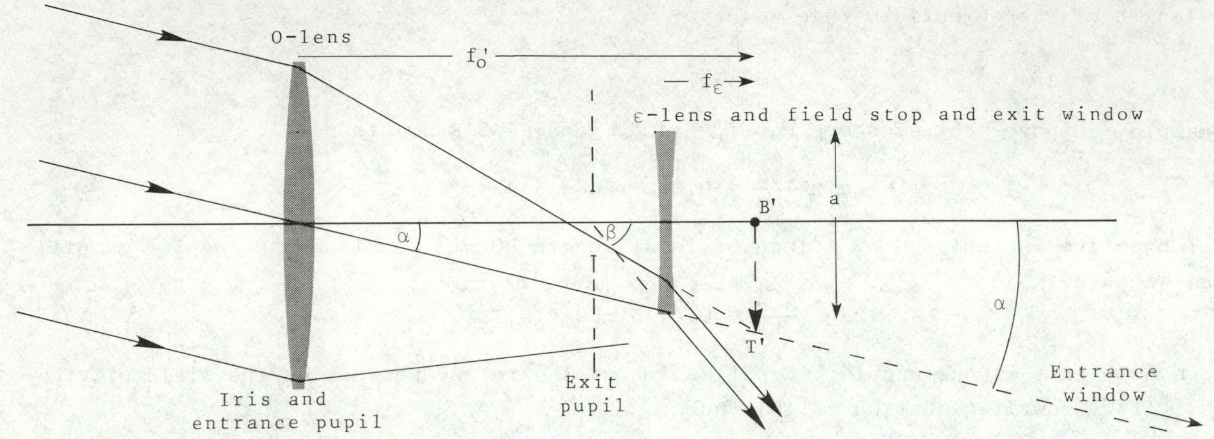

Fig. 8.47 Field of view of the Galilean telescope.

Rewriting equation (8.21), we have

$$2\alpha = \frac{a}{f_O'+f_\varepsilon'} = \frac{a}{f_O'+f_\varepsilon'} \times \frac{1/f_\varepsilon'}{1/f_\varepsilon'} = \frac{a/f_\varepsilon'}{\frac{f_O'}{f_\varepsilon'}+1} = \frac{a}{f_\varepsilon'} \times \frac{1}{-M+1} \simeq -\frac{a}{f_\varepsilon'} \times \frac{1}{M} \qquad (8.21a),$$

the last step having made use of the fact that $M \gg 1$. Once again, the field of view in inversely proportional to the angular magnification.

The field of view in image space is again 2β, where

$$2\beta = 2\alpha\, M = \frac{a}{f_O'+f_\varepsilon'} \times \left(-\frac{f_O'}{f_\varepsilon'}\right) \qquad (8.22).$$

Worked Examples

1 A thin lens of power +8.00 D has a diameter of 42 mm. If the lens is held 10 cm from the eye and a large object is placed 10 cm in front of the lens, what portion of the height of the object can be seen through the lens? Assume the eye's pupil to be located at the reduced eye's surface.

The eye's pupil will be the pupil of the system, and it centre E' will be conjugate with the centre E of the entrance pupil (the image of the eye pupil seen through the lens). The eye pupil will also be the exit pupil of the system. Regarding the entrance pupil as the object and the exit pupil as the image, the object distance $OE = \ell$ from the lens is given by

$$\frac{1}{\ell} = \frac{1}{\ell'} - \frac{1}{f'} = \frac{1}{10} - \frac{1}{12.5} = \frac{5-4}{50} = \frac{1}{50}.$$

Therefore, $\ell = OE = 50$ cm. Now, the lens is the field stop and the entrance window, so the object object field of view is the angle SET. Hence,

$\widehat{SET} = \text{arc tan}\frac{4.2}{50} = 4.80^0.$

The extent of the object which is visible at half-illuminance is QR and is obtainable from similar triangles. Thus,

$$QR = \frac{NE}{OE} \cdot ST = \frac{60}{50} \times 4.2 = 5.04 \text{ cm}$$

QR is known as the linear field of view.

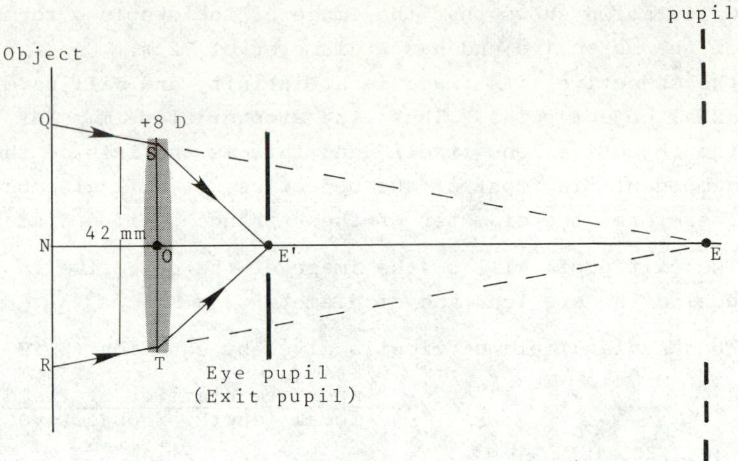

2 An astronomical telescope in the afocal setting consists of an objective lens of +10.00 D, diameter 30 mm, and an eyepiece of +100 D, and diameter 7.2 mm. A field stop of aperture 6.5 mm is positioned in the common focal plane of the two lenses, with its centre on the principal axis of the system. Find the sizes and positions of the entrance and exit pupils, and determine the field of view.

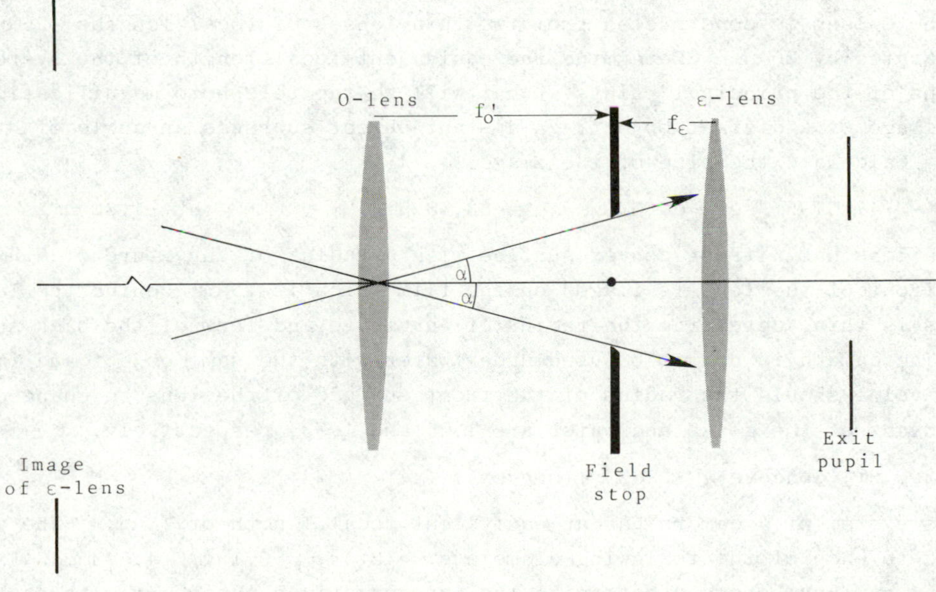

The first step is to establish which stop or lens mount, or image of a stop or lens mount, is acting as the iris (aperture stop) of the telescope. This will be the element or image of an element which subtends the smallest angle at an axial object point at a large distance from the telescope, and will thus be the element or image of an element having the smallest diameter. The possible irises are:

(i) the mounting of the objective, diameter 30 mm,

(ii) the mounting of the eyepiece,

(iii) the stop placed in the common focal plane of the two lenses.

Calculation shows that the image of the eyepiece formed by the objective is 110 cm in front of the objective and has a diameter of 72 mm. As for the stop in the second focal plane of the objective, its image is at infinity and will have no restriction on the light from an axial object point. Thus, the stop or stop image of the least diameter in object space is the objective lens itself, and this is the iris of the telescope. As there are no optical components in front of the objective, it will also be the entrance pupil of the telescope. Therefore, the diameter of the entrance pupil = diameter of objective = 30 mm.

The exit pupil will be the image of the objective in the eye lens. This turns out to be 11 mm behind the eye lens and of diameter 3 mm.

The angular field of view is given by equation (8.19):

$$2\alpha = \frac{\text{aperture diameter of field stop}}{\text{focal length of objective lens}} = \frac{0.65}{10} = 0.065 \text{ radian}$$

$$= 3.72^{0}.$$

EXERCISES

1 Compare the merits of a lens camera and a pinhole camera. Explain the meaning of the f-number of a stop. Explain why an exposure time of 1/50 second at f/11 gives the same activation of the film as an exposure time of 1/100 second at f/8. What are the advantages of using as small an aperture as possible?

2 A telephoto lens is constructed from two thin lenses of focal lengths +25 cm and -12.5 cm, separated by 20 cm. Determine the equivalent focal length of the system and the positions of the principal points. What will be the telephoto magnification when forming an image of a distant object? A distant object subtends an angle of 2.5^{0} at the lens system. Calculate the size of the image.

Ans. $f'_E = +41.67$ cm, $e = -66.67$ cm, $e' = -33.33$ cm, $M_T = 5.0\times$, $h' = -1.82$ cm.

3 A camera lens has a front convex surface with a radius of curvature of 6 cm. An object 1 m in front of the lens is imaged on the film which is 25 cm behind the lens. Assuming the lens is thin, determine the radius of curvature and form of the back surface of the lens. The camera is now to be used under water with the same object and image distances. To what value should the radius of the front surface of the lens be changed? The refractive indices of the glass and water are 1.55 and 1.33, respectively.

Ans. +13.2 cm (concave), +2.32 cm (convex).

4 The lens system of a camera has an equivalent focal length of 10 cm. The aperture may be adjusted to each of the following diameters: 12.5 mm, 9.1 mm, 4.54 mm. Express these apertures as f-numbers and determine the exposure times necessary with the first and the last if the correct exposure time with the 9.1 mm aperture is 1/100 second.

Ans. f/8, f/11, f/22, 1/200 s, 1/25 s.

5 A lens of focal length +20 cm forms an image of an object, the latter being situated 8 m in front of the lens. If the lens aperture is set at f/5.6, and assuming a maximum permissible blur circle diameter of f'/1 000, calculate the depth of field of the lens.
Ans. 3.50 m.

6 A slide projector has a projection lens of focal length 20 cm. What must be the distance

of the screen from the lens in order that it shall project an image which is 100 times the size of the object slide?

Ans. 20.2 m.

7 A projection lens of focal length 15 cm gives an image on a screen 3 m away. The condenser, which may be assumed to be a single thin lens, has a focal length of 8 cm. If the object slide is placed very close to the condenser, determine the positions of the source and the condenser.

Ans. Condenser: 15.79 cm from the projection lens. Source: 16.22 cm from the condenser.

8 The range of distinct vision for a myope extends between the distances of 30 cm and 10 cm. from his eye. Calculate the power of the lenses needed to see distant objects imaged at his far point, assuming the lenses to be in contact with the eyes. With these lenses, what will be the least distance of distinct vision?

Ans. $-3\frac{1}{3}$ D, 15 cm.

9 An object 5 mm long is situated 10 cm from a lens of power +7.50 D. If an observer views the image through the lens and the distance between the lens and his eye is 8 cm, determine the visual angle subtended by the image.

Ans. 2.39°.

10 What must be the position of an object in order that an eye 10 cm behind a +8.50 D lens may see clearly when accommodated for 40 cm. If the diameter of the lens is 38 mm, what length of object can be seen?

Ans. -8.45 cm, 4.28 cm.

11 By means of a diagram show how a single converging lens acts as a simple magnifier. What is the angular magnification produced by a +12.00 D lens 4cm in front of an emmetropic eye which is (a) unaccommodated (b) accommodated 5 D.

Ans. 3×, 3.65×.

12 Distinguish between linear magnification and angular magnification. Explain why the values of linear and angular magnification are, in general, different for a given optical arrangement. An object is placed 4 cm in front of a +10.00 D thin lens. Calculate the linear magnification of the image. The observer's eye, treated as a single refracting surface, is positioned 10 cm behind the above lens. Determine the angular magnification.

Ans. $m = 1.667\times$, and $M = 2.5\times$.

13 Show that the nominal magnification produced by a thin converging lens when used as a magnifier is equal to one quarter of its power. A 3 D myope not wearing his spectacles uses a thin +12.00 D lens as a magnifier. 5 D of accommodation is used to observe the image of an object placed in front of the lens. If the observer's eye is 4.5 cm behind the lens, calculate the angular magnification obtained and the position of the object.

Ans. 3.92×, -4.08 cm.

14 Determine the angular magnification achieved by an uncorrected 4 D hyperope who accommodates 7 D when using a +16.00 D lens as a magnifier held 8.33 cm from his eye. What is the position of the object?

Ans. 3.75×, -5 cm.

15 A Huygens eyepiece consists of two thin lenses of power +10 D asd +20 D respectively,

separated by 7.5 cm. Determine the equivalent focal length of the eyepiece. The eyepiece forms part of an astronomical telescope whose objective lens is 30 cm in front of the first lens of the eyepiece. Determine the position of the exit pupil of the telescope.

Ans. +6.67 cm, 3 cm behind the eyepiece.

16 Compare the Huygens and Ramsden eyepieces, pointing out the advantages and disadvantages of each. Why would one of them be unsuitable for use in a focimeter? An eyepiece consists of two thin converging lenses each of focal length 3 cm, separated by a distance of 2 cm. Find the positions of the cardinal points. If cross-lines are used with this eyepiece, where should they be placed?

Ans. f_v = -0.75 cm, e = +1.5 cm, f_v' = +0.75 cm, e' = -1.5 cm.

17 A compound microscope has an objective lens that produces a linear magnification of 10×. If the observer is emmetropic and unaccommodated, what focal length of eyepiece will produce an overall magnification of 100×?

Ans. +2.5 cm.

18 The focal lengths of the eyepiece and objective of a microscope are 2.5 cm and 1.6 cm respectively, and the separation of these lenses is 22.1 cm. The observer is emmetropic and views the final image without accommodation. Calculate (a) the position of the object, (b) the magnification of the objective, (c) the overall magnification of the microscope.

Ans. -1.74 cm, 11.25×, 112.5× .

19 A microscope is made up from an objective of power +100 D and an eyepiece of power +50 D, the two lenses being separated by 18 cm. How far in front of the objective must an object be placed if its image formed by the microscope is viewed by a myope of 2 D exercising 2 D of accommodation? Assume the observer's eye is close to the eyepiece. What will be the angular magnification in this case?

Ans. -1.07 cm, 204.6× .

20 An astronomical telescope objective has an aperture diameter of 20 cm and a power of +0.80 D. The telescope is fitted with three different eyepieces of focal lengths +25 mm, +12 mm, and +6 mm. Find the afocal magnification and the position and size of the exit pupil with each eyepiece.

Ans.			
Ans.	50×	+2.55 cm	4 mm
	104.2×	+1.21 cm	1.92 mm
	208.3×	+0.60 cm	0.96 mm.

21 With the aid of a diagram, explain the operation of an astronomical telescope. What is meant by the "afocal configuration"? A telescope is made up from a +5.00 D objective and a +20.00 D eyepiece and is used to view a distant object. Determine the angular magnification (a) when used by an unaccommodated emmetrope (b) when used by an unaccommodated myope of 4 D with his eye close to the eyepiece.

Ans. -4×, -4.8× .

22 (a) When a telescope is in the afocal setting, the equivalent power of the system can easily be shown to be zero. How is it possible for an afocal telescope to produce magnification? (b) For an astronomical telescope, the objective power is +1.00 D and the eyepiece is +5.00 D, with the objective diameter 50 mm. Determine the afocal magnification, the separation of the lenses, and the diameter of the exit pupil. Ans. -5×, 1.20 cm, 10 mm.

(c) If this telescope is used to view an object 10 m in front of the objective, find the distance the eyepiece must be withdrawn from the afocal setting to form the final image at infinity. Determine the magnification with this arrangement. Ans. 11.11 cm, -5.56× .
(d) If the system is now used to form an image 25 cm in front of the eyepiece of an object 20 m in front of the objective, find the separation of the lenses and the magnification obtained. Ans. 1.164 m -9.5× .

23 (a) An astronomical telescope consists of a +0.833 D objective and a +12.50 D eyepiece. With the observer's eye close to the eyepiece the telescope is arranged to form an image of the moon which is viewed by an unaccommodated myope of 4 D. If the angular diameter of the moon is 0.5^0, calculate the length of the final image. Ans. 4.32 cm.
(b) If this telescope is used to project a real image of the sun (angular diameter also 0.5^0) on to a screen 1.00 m beyond the eyepiece, determine the separation of the lenses and the diameter of the real image on the screen. Ans. 128.7 cm, 120.4 mm.

24 How would you arrange two thin converging lenses to produce a telescope of magnification 4? Draw a diagram showing an incident axial pencil of rays passing through the system when in the afocal configuration. Draw another diagram to show how the pencil of rays is modified when the eyepiece lens is moved towards the objective a short distance.

25 Using a thin +2.50 D lens and a thin +12.50 D lens an observer sets up an afocal system as a magnifying telescope. Determine the separation of the lenses. Using the same separation and placing his eye at the exit pupil, the observer looks at an object 5 m from the entrance pupil. What accommodative effort is required to see the image clearly if the observer is assumed emmetropic?

Ans. 48 cm, 5 D .

26 A terrestrial telescope has an objective of focal length +25 cm. It is fitted with a thin erecting lens of focal length +2cm placed 2.5 cm beyond the focal point of the objective. The final image is formed by a thin eyepiece lens of focal length +5 cm, and is viewed by an emmetrope without accommodation. If the object is very distant, find the magnification produced by the erecting lens and the magnification by the whole instrment. What is the length of the telescope?

Ans. -4×, +20×, 42.5 cm.

27 By means of a labelled diagram, describe the optical arrangement of a prism spectrometer. Show clearly the paths of red and blue rays from a source of white light when the telescope is in the afocal setting. The telescope cross-lines are set in succession on the red and blue lines in a spectrum, and it is found that the telescope positions differ by $2^0 30'$. If the focal lengths of the objective and the eyepiece of the telescope are 22 cm and 1.9 cm respectively, determine the angle which the images of these lines subtend at the observer's eye.

Ans. $28^0 22'$.

28 The objective of a telescope is a single thin lens of +0.50 D made in glass of dispersive power 0.0175 . The eyepiece is an achromatic doublet of power +20.00 D . If the telescope is focused so that the blue image of a white object is formed at infinity, find the position of the red image. Find the powers of the components of an achromatised objective also of power +0.50 D if a second glass of dispersive power 0.0200 is also available.

Ans. -2.102 cm from the ε-lens, +4.00 D, -3.50 D.

29 A Galilean telescope consists of a thin +16 D lens and a -36 D lens. The system is arranged in the afocal configuration. What is the separation of the lenses and what magnification is produced?

Ans. 2.25×, 3.47 cm.

30 Draw a ray diagram to show how the Galilean telescope forms, at infinity, an image of a distant object. What are the advantages and disadvantages of this type of telescope? Where is the exit pupil? A converging lens of focal length 12 cm is fixed 10 cm in front of a diverging lens of focal length 2.25 cm. An object is situated 3 m from the converging lens. If an emmetropic observer views the image, and his eye is placed close to the diverging lens, what accommodative effort is required and what is the magnification?

Ans. 4.44 D, 5× .

31 A telescope has an objective lens of focal length +15 cm and a negative powered eyepiece. When used by an unaccommodated emmetrope to view a distant object it magnifies 3 times. By what distance must the eyepiece be moved and what will be the resulting magnification when the telescope is used by (a) a hyperope of 4 D (b) a myope of 5 D? Assume the eye is very close to the eyepiece in all cases.

Ans. (a) 0.833 cm, 3.60×. (b) 1.667 cm, 2.25×.

32 a telescopic spectacle unit designed for distance vision is made up afocally from a thin +20 D lens and a thin -40 D lens. What is the separation of the lenses and the magnification produced? If an auxiliary +4 D lens is added to the objective for near vision, what will be the working distance and the new magnification?

Ans. 2.5 cm, 2× . 25 cm, 2× .

33 Define the terms iris, field stop, entrance pupil, and exit pupil as applied to an optical system of centred lenses. Two centred thin positive lenses of focal lengths 8 cm and 2 cm are 5 cm apart, and have diameters 5 cm and 3 cm respectively. A stop of diameter 1 cm is located between the lenses and 2 cm from the first lens. Find (a) the position of the iris (aperture stop) (b) the positions and sizes of the entrance and exit pupils for an axial object point 10 cm in front of the first lens.

Ans. (a) 2 cm behind first lens. (b) 2.67 cm behind first lens, diameter 1.33 cm. 6 cm behind second lens, diameter 2 cm.

34 The focal length of a thin lens is +12 cm and the diameter is 4 cm. If this lens is placed midway between the eye and a large object 10 cm from the eye, and the pupil of the eye has a diameter of 6 mm, over what extent of the object is there no diminution of illuminance?

Ans. 5.7 cm. See W.P.O., 154, question 6.

35 Determine the fully illuminated field of view in the image space for an astronomical telescope for which the data are: angular magnification = 20, objective focal length = +60 cm, diameter 5 cm, eyelens diameter 1 cm, eye-pupil diameter 3 mm. Show where the eye-pupil should be located.

Ans. 6.79^0, 3.15 cm behind the ε-lens. See W.P.O., page 155.

9 PHOTOMETRY

9.0 INTRODUCTION

Photometry may be described as the study and measurement of light in terms of the visual response it produces. However, such are the developments in the methods of measurement that the use of certain non-visual instruments like the photo-electric cell and lightmeter may be included in this study.

Nowadays in the U.K., the Health and Safety at Work Act (1974) puts a duty on employers to provide sufficient lighting of suitable standard for specific purposes, and codes of practice such as the Illuminating Engineering Society Code for Interior Lighting (1977) contain recommendations about lighting levels. Moreover, certain British Standards, such as BS 4274 (1968), are of relevance to photometric measurements for the optical profession. By ensuring that photometric quantities conform to standards, mankind has benefitted in terms of

(i) reduction of eyestrain
(ii) fewer accidents
(iii) better working conditions
(iv) greater production of manufactured goods
(v) improved leisure facilities

and so on. Photometric measurements are particularly important in illumination engineering, photography, and in those industries associated with coloured products such as textiles, dye-stuffs, and paints.

Basic photometric measurement involves the visual response of the observer. Hence, we begin this chapter with a consideration of the *visual sensitivity* of the eye.

9.1 VISUAL SENSITIVITY

The eye is not equally sensitive to the different colours of light in the visible spectrum. Figure 9.1 shows how the visual sensitivity varies with wavelength. The sensitivity curve, sometimes referred to as the *luminosity curve* of an equal energy spectrum, gives the relative brightness, as assessed by the average eye, of the different colours of the spectrum when the incident energies at each wavelength have been reduced to the same mechanical value. For the average light adapted eye at moderate intensities (photopic vision) the maximum visual effect is obtained with light of wavelength 555 nm (yellow-green). From this point the sensitivity decreases towards both ends of the visible spectrum. If the sensitivity for wavelength 555 nm is taken as 1.0, then it is found that the value is 0.5 at wavelengths of about 510 nm and 610 nm. This means that light of these wavelengths is only half as efficient at producing a visual response as light of wavelength 555 nm. The cornea and the crystalline lens of the eye absorb radiations of wavelengths less than about 330 nm, and this value may be taken as the minimum wavelength to reach the retina. The upper wavelength limit to visual sensitivity may be taken as about 760 nm, due mainly to the lack of response of the eye's photosensitive pigments above this value. As stated in Chapter 7, the lower limit of the visible spectrum is

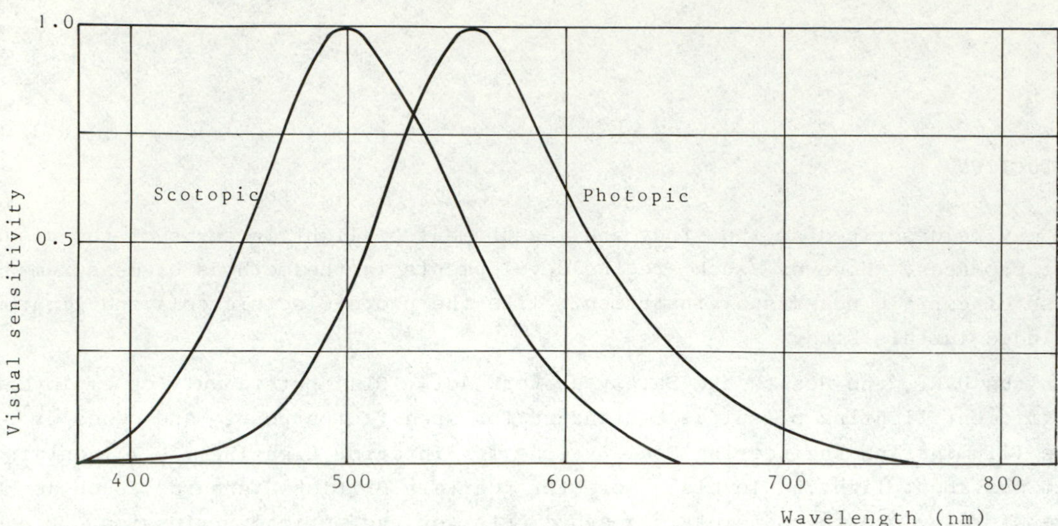

Fig. 9.1 Relative visual sensitivity curve.

about 390 nm.

If the variation of visual sensitivity with wavelength is studied for the average dark adapted eye (scotopic vision) a slightly different sensitivity curve is obtained. In this case the maximum effect is obtained with light of wavelength about 500 nm. There is also a reduction in the range of wavelengths detected. This curve is also shown in figure 9.1. However, it should be noted that although the two sensitivity curves have been drawn with the same maximum height, much lower intensities are needed for scotopic vision.

The shift of the visual sensitivity maximum from 555 nm in photopic vision to 500 nm in scotopic vision is known as the *Purkinje effect*. The two curves in figure 9.1 are associated with the cone (photopic) and rod (scotopic) mechanisms of the retina, respectively. The Purkinje effect and the shortened range of sensitivity at the red end of the spectrum in scotopic vision accounts for the relatively dull appearance of red surfaces under low levels of illumination.

We shall see in a later part of this chapter that the effects of the variable visual sensitivity of the eye render certain photometric comparisons difficult.

9.2 SOLID ANGLES

Light is usually emitted by a source in all directions, and thus it is necessary to consider how the space around a light source is divided up into solid angles. A simple mathematical consideration of plane angle measurement in radians and of solid angle measurement in steradians is included in Appendix 4. For the present we will simply quote the useful result of this consideration which is that:

the total space around a point = 4π steradians

That is, 4π steradians is the total solid angle surrounding a point.

9.3 THE STANDARD SOURCE AND THE CANDELA

Photometry is essentially a science of the comparisons of photometric quantities such as intensities and illuminances, terms which will be defined later in this chapter. It will also become clear that all photometric quantities are related to one another and so it is only necessary to set up one of these quantities as the standard. It has always been most convenient to define the standard in the form of a light source against which other light sources may be compared and calibrated. In designing a standard, it should be clear that the device must give a constant luminous output for an indefinite period of time. This restriction excludes all gas lamps, electric filament lamps, and discharge tubes from being used for standardisation. Although these types of sources have each been used as standards at the time when they were the best sources available, they do not have a constant light output.

The present primary standard source of light is based on the concept of a *black body radiator*. When radiation is incident on a body then, in most cases, some of this radiation is absorbed by the body and the rest is reflected. A black body is a device which absorbs all the radiation incident on it at all temperatures. When at a constant temperature a black body will be radiating the same energy that it receives, the total quantity depending on its temperature. Thus, a black body is also a perfect emitter of radiation, and will emit more energy than any other body at the same temperature. The radiation from a black body is independent of the nature and material of the body and, consequently, is not affected by small variations in the construction of the standard. Moreover, the radiation from a black body closely follows definite laws concerning the total energy emitted and the distribution of this energy amongst the various spectral wavelengths (see Chapter 15).

In basic terms, a small opening in a hollow vessel with a blackened interior will absorb all radiations incident on it by virtue of repeated internal reflections and absorptions. This is shown in figure 9.2 . The small baffle is to avoid a single normal reflection escaping from the opposite side of the enclosure. When such a device is heated the small hole acts as a black body radiator.

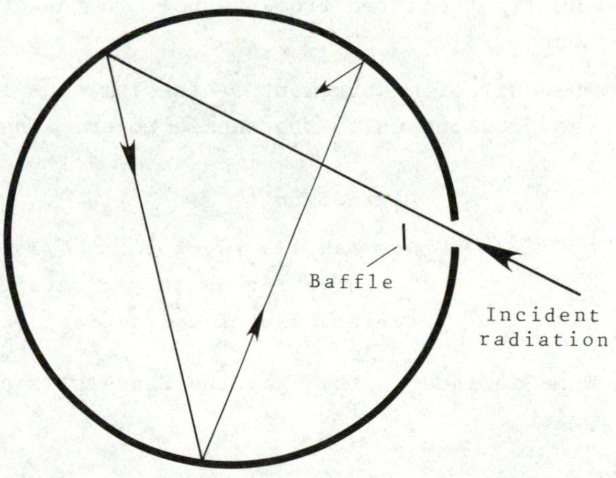

Fig. 9.2

Since 1948 the primary international standard of light has been adopted in terms of the visible radiation from a small hole in the end of a cylinder made of pure fused thorium oxide. This is maintained at the temperature of 1773^{0}C by immersion in pure platinum which is changing from the molten state to the solid state. The main features of this black body standard source are illustrated in figure 9.3 .

The assembly is quite small, the central thorium oxide tube being about 40 mm long and the radiation outlet having a diameter of 1.5 mm.

Platinum is used because it can be obtained in a pure state, has a high freezing point, and does not oxidise in air. Thorium oxide is used because its melting point is higher than

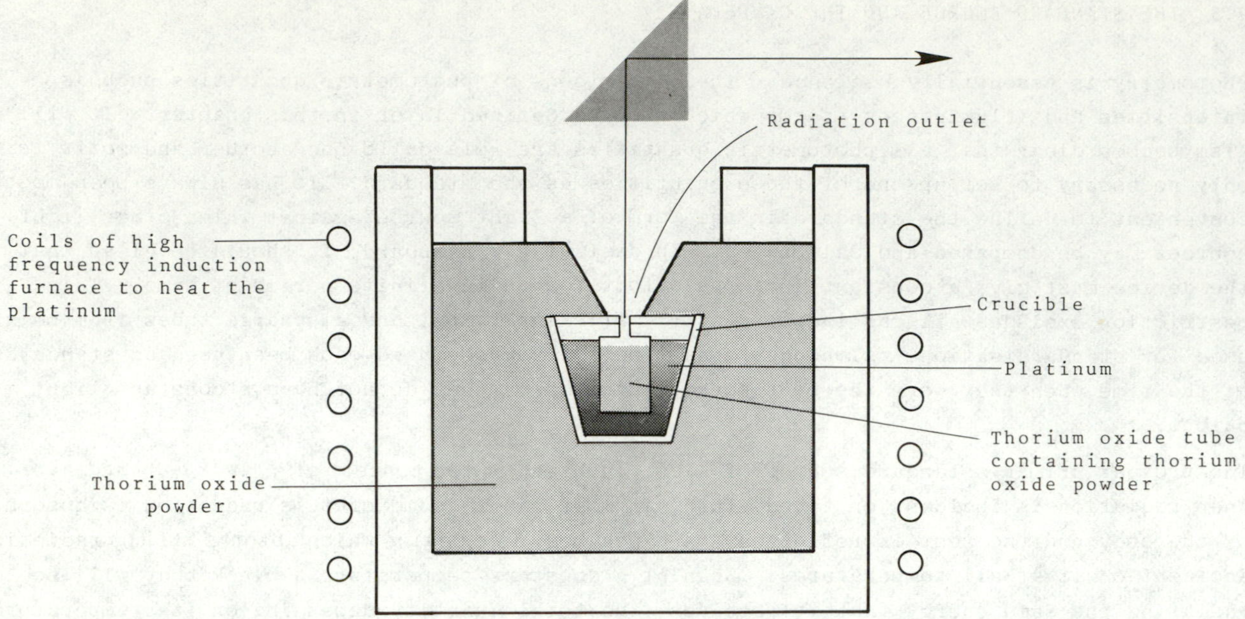

Fig. 9.3 The international black body standard source.

platinum's, and it does not react with or dissolve in the platinum. The platinum is melted by a high frequency electrical induction furnace. When allowed to cool slowly the platinum remains at its freezing temperature whilst it undergoes solidification. During this process the light emitted from the hole in the cylinder defines the intensity of the standard light source.

The unit of measurement of the luminous intensity of a source is the *candela* which has replaced the previous unit, the candle power. The candela is defined as follows:

Definition

One candela (symbol 1 cd) is equal to one sixtieth ($\frac{1}{60}$) of the luminous intensity per square centimetre of a black body at the temperature of solidification of platinum.

When defined in this way the candela is approximately equivalent to the previously used candle power.

9.4 WORKING STANDARDS

Electric filament lamps are much easier to set up and more convenient to use than the primary black body standard source. Consequently, electric lamps are used as everyday working standards for luminous intensity, and are sometimes called sub-standards. They are calibrated periodically with the primary standard in case there is any deterioration in their light output. Many countries maintain a primary standard light source.

9.5 *LUMINOUS FLUX*

The transfer of light from a source is expressed in terms of luminous flux, symbol Φ. To understand the meaning of flux let us consider the formation of an image with a camera. The latent image on a film, which is established chemically in the emulsion by the action of light prior to development, depends on the total light energy reaching the image plane. Thus, a suitable exposure time will help to build up the image of a poorly illuminated object. However, when light energy enters the eye the visual sensation produced depends on the *rate* at which light energy reaches the retina, and not on the total energy. Hence, a dimly lit scene will not continue to grow brighter and brighter as it is observed.

The rate at which light energy flows is called the *luminous flux* (Φ). The word luminous means radiant energy in the visible spectrum, and flux means flowing. Luminous flux is the energy flowing per second, and this could be expressed in joules per second (watts). However, although it is not essential to do so, an independent system of units is employed for photometric quantities.

The unit of luminous flux is called the *lumen*, and this may be defined with reference to the candela, but before we consider the definition it is necessary to explain the significance of a *point source*. Such a light source is one of very small size compared to its distance from a surface or object, and for which the angular subtense is negligible. Much of the theory of photometry assumes point sources.

> *Definition*
>
> *One lumen is the luminous flux emitted into a unit solid angle (1 steradian) from a point source of intensity 1 candela.*

This definition, and the relationship between lumen and candela, is illustrated in figure 9.4 .

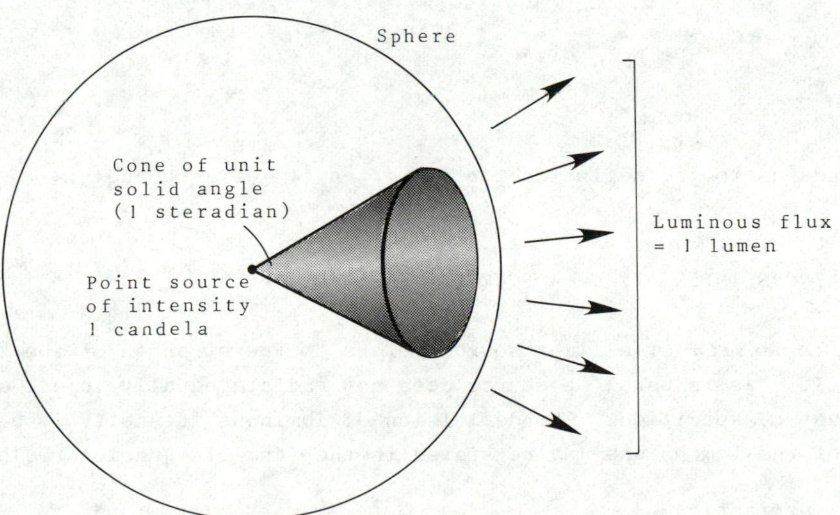

Fig. 9.4 Definition of the lumen.

As explained in Appendix 4, the total space around a point may be represented as 4π steradians. Hence, the total luminous flux emitted in all directions by a point source of intensity 1 candela is 4π lumens. We can write this as:

a 1 candela source emits a total luminous flux of 4π lumens.

For comparison, a 100W light bulb emits about 1200 lumens. In general, if I candela represents the intensity of a source, and Φ lumens of flux are emitted into a solid angle ω steradians, then we can say

luminous flux emitted into a solid angle = intensity of source × solid angle.

That is, $$\Phi = I\omega \quad (9.1).$$

This may be written as

$$I = \frac{\Phi}{\omega} \quad (9.1a).$$

If we specify the direction in which luminous flux is emitted by a source we state the above expression in the form

$$I = \frac{d\Phi}{d\omega} \quad (9.1b)$$

where $d\Phi$ is the flux emitted into the elementary solid angle $d\omega$ in a given direction. This is shown in figure 9.5 .

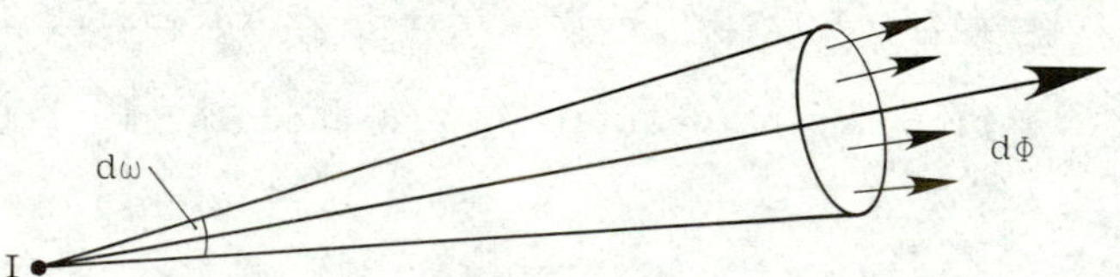

Fig. 9.5 The luminous flux $d\Phi$ emitted the elementary solid angle $d\omega$

This now leads us to the defintion for the *intensity* of a source.

9.6 THE LUMINOUS INTENSITY OF A SOURCE

The *luminous intensity* of a light source refers to the strength of the source in a given direction. If , as is usual, a source does not radiate equally in all directions, then the direction must be specified. The definition of luminous intensity is closely related to the definition of the lumen, and may be stated in the form of equation (9.1b) above.

Definition

The luminous intensity (I) of a source in a specified direction is the luminous flux in lumens emitted per unit solid angle.

That is, writing equation (9.1b) again, $I = \frac{d\Phi}{d\omega}$.

Since a point source of 1 candela emits 1 lumen of flux into 1 steradian of space, then the intensity I of a source will be equal to the ratio

$$\frac{\text{Flux emitted by source into 1 steradian of space}}{\text{Flux emitted by 1 candela into 1 steradian of space}} \text{.}$$

Thus, the unit of luminous intensity of a source is the candela. It is also useful to note that

$$1 \text{ candela} \equiv 1 \text{ lumen/steradian.}$$

Table 9.1 below lists some typical intensities expressed in candela.

Table 9.1

Source	Intensity (cd)
A candle	approx. 1
40W light bulb	30
Electric arc	1500
The sun	10^{27}

9.6.1 THE MEAN SPHERICAL INTENSITY OF A SOURCE

Practical sources of light do not radiate equally in all directions, and are referred to as non-uniform sources. This variation of light output with direction is due, in the simple case of the domestic light bulb, to the design and shape of the filament, the position of the metal cap for fitting the bulb into the socket, and so on. Other practical sources of light will have similar variations in their light output. In commercial use it may be necessary to refer to the average value of the luminous intensity in all directions. This value is referred to as the *mean spherical intensity* or *mean spherical candle power*. (The latter title is still retained in this connection although the "candle power" is a unit of source intensity which is now obsolete).

From equation (9.1a), $I = \frac{\Phi}{\omega}$. But the total solid angle around a point is 4π steradians, hence the mean spherical intensity of a source is equal to the total luminous flux emitted divided by 4π. That is,

$$\text{mean spherical intensity} = \frac{\Phi}{4\pi} \text{ cd} \qquad (9.2).$$

The total luminous flux Φ emitted in all directions by a light source may be measured with an integrating sphere photometer. This is discussed in section 9.11.7 .

If a knowledge of how the intensity of a lamp varies with direction is required this is best represented on a light distribution curve called a *polar curve*. By means of a suitable photometer, see section 9.11.5, the intensity of the lamp may be determined in various directions in a particular plane. Taking the source as the origin, the radius vector in any particular direction is made equal to the intensity in that direction. Figure 9.6 shows a polar curve typical of a 40 W domestic light bulb. This is in fact an extended source rather than a point. It can be seen that, due to the metal cap of the lamp, the intensity in the vertically upward direction is zero. Because of the shape and position of the filament in the bulb, the intensity in a vertically downward direction is about 30 cd, whilst in a horizontal direction it is about 20 cd.

A number of such curves in different planes is required to give a complete light distribution

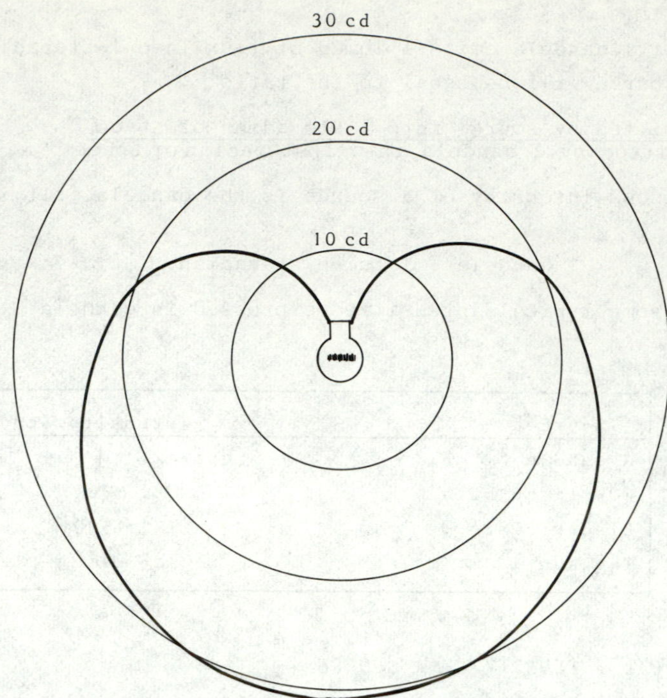

Fig. 9.6 A polar curve distribution of the intensity of a lamp.

around the source. With the aid of a computer, the volume enclosed by the whole series of polar curves may be determined, from which the value of the total luminous flux emitted by the lamp is obtained.

9.7 *ILLUMINANCE*

When luminous flux is incident on a surface, the surface is said to be illuminated. This effect is described by the term *illuminance* which is defined as follows.

> *Definition*
>
> *The illuminance (symbol E) at a point on a surface is the amount of luminous flux falling on unit area of the surface.*

That is, it is the flux per unit area, or flux density. Hence, we may write

$$\text{illuminance} = \frac{\text{amount of incident luminous flux}}{\text{area of the surface}} \quad \left(\frac{\text{Lumens}}{\text{m}^2}\right)$$

or, in symbols,

$$E = \frac{\Phi}{A} \quad \frac{\text{lm}}{\text{m}^2} \qquad (9.3).$$

The basic unit of measurement for illuminance is lumens/square metre which is called the *lux*. That is, $1\ \frac{\text{lm}}{\text{m}^2} = 1$ lux. The abbreviation for lux is lx. Occasionally lumens/square cm is used.

The illuminance at a point on a surface does not depend on the nature of the surface since it is only concerned with incident light. If the illuminance of a surface is due to two or more sources, then the total illuminance is equal to the sum of the illuminances due to each source separately.

It is interesting to note that the eye functions quite well, and can adapt quickly, over a wide range of illuminances corresponding to six or seven orders of magnitude. As the variation between maximum and minimum opening of the pupil represents a factor of only about 10 in area, there are obviously other factors involved in the adaptation process. Table 9.2 lists some typical values of illuminance.

Table 9.2 Typical values of illuminance

Source	Illuminance (lux)
Bright sunlight	10^5
Sight testing chart	500
Interiors (artificial light)	100
Full moonlight	0.2

The Illuminating Engineering Society (I.E.S.) has specified standards of illuminance for many specialised applications. Table 9.3 contains values taken from the I.E.S. Code for Interior Lighting (1977).

Table 9.3

Application	Illuminance (lux)
Operating theatres	up to 50 000
Carpet manufacture (inspection)	1 000
Drawing office	750
Supermarket	500
Library reading area	300
Living room, general	50

9.8 THE TWO FUNDAMENTAL LAWS OF PHOTOMETRY

It can be said that much of photometric measurement is based on two fundamental laws. The first of these states that

the illuminance at a point on a surface is inversely proportional to the square of the distance between the point and the source.

The law applies strictly only in the case of point sources and may be in error for illuminated surfaces near extended sources. This is known as the *Inverse Square Law* and may easily be derived in the following manner.

Imagine a sphere of radius r metres surrounding a uniformly radiating point source which has an intensity I candela, shown in figure 9.7. The total luminous flux emitted by the source is given by equation (9.1):

$$\Phi = I\omega \quad \text{lumens.}$$

But, from section 9.2 we recall that the total solid angle surrounding a point is 4π steradians: that is $\omega = 4\pi$. Therefore, the flux $\Phi = I \times 4\pi = 4\pi I$ lumens. Now, at some particular distance r metres from the source, the flux is distributed normally over the inside surface of the sphere, having a surface area $4\pi r^2$ m^2. That is, $A = 4\pi r^2$. Hence, using equation (9.4) we have

$$\text{illuminance } E = \frac{\Phi}{A} = \frac{4\pi I}{4\pi r^2}$$

which reduces to

$$E = \frac{I}{r^2} \quad \frac{lm}{m^2} \text{ (lux)} \qquad (9.4).$$

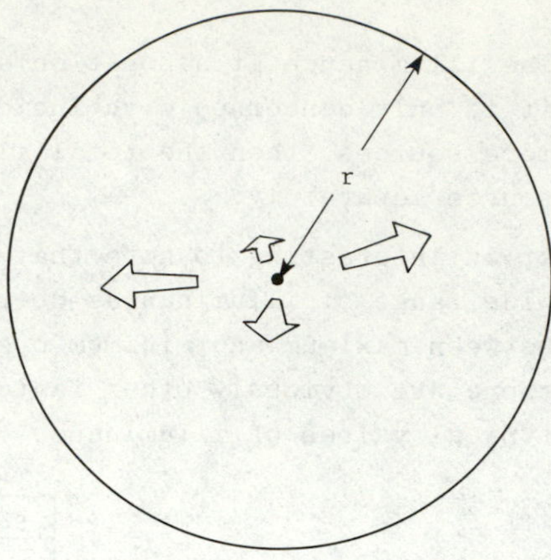

Fig. 9.7 A point source of intensity I candela radiating uniformly into a sphere of radius r metres.

That is, the illuminance E is inversely proportional to r^2. The significance of this is that if we move a surface away from a point source the illuminance drops rapidly. In moving from 1 m to 2 m, then to 3 m from the source, the illuminance reduces in the ratios $1:\frac{1}{4}:\frac{1}{9}$.

The second law is known as the *cosine law* and states that

> *if the normal to an illuminated surface is at an angle θ to the direction of the incident light, the the illuminance is proportional to the cosine of θ.*

This may be appreciated from figure 9.8 .

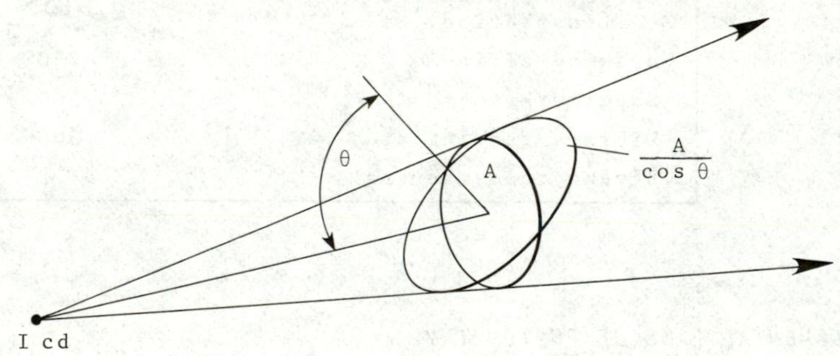

Fig. 9.8 Illuminance $E \propto \cos\theta$.

The luminous flux which would fall on the area A placed to receive it normally falls on the larger area $A/\cos\theta$, which is tilted at an angle θ to A. That is, the angle of incidence at the centre of the tilted area is θ. Thus, the luminous flux per unit area, that is, the illuminance, is reduced by the factor $\cos\theta$. We can appreciate this as follows. Since the illuminance $E = \frac{\text{Flux}}{\text{Area}} = \frac{\Phi}{A}$, the illuminance on the tilted area $A/\cos\theta$ is $E = \frac{\Phi}{A/\cos\theta} = \frac{\Phi}{A}\cos\theta$. But the illuminance for normal incidence on area A is, from equation (9.4), $E = \frac{\Phi}{A} = \frac{I}{r^2}$, so the illuminance on the tilted area is

$$E = \frac{\Phi}{A}\cos\theta = \frac{I}{r^2}\cos\theta .$$

That is, in place of equation (9.4) we now write

$$E = \frac{I}{r^2}\cos\theta \quad \frac{lm}{m^2} \text{ (lux)} \qquad (9.5).$$

Equation (9.5) really combines both the fundamental photometric laws, as we can see when $\theta = 0^0$ (normal incidence). In such a case $\cos\theta = \cos 0^0 = 1$ giving $E = \frac{I}{r^2} \times 1 = \frac{I}{r^2}$. This is equation (9.4) previously derived for normal incidence. The two laws are useful for calculating illuminances and comparing the intensities of sources.

Worked Examples

1 Calculate the solid angle subtended at the centre of a sphere, radius 2 m, by an area of 2.5 m^2 on the surface of the sphere.

From Appendix 4,

$$\text{solid angle } \omega = \frac{\text{size of area on surface of sphere}}{\text{radius}^2}.$$

So, $\omega = \frac{A}{r^2} = \frac{2.5}{2^2} = 0.625$ steradian.

(Note that we may use the S.I. abbreviation for steradian which is sr).

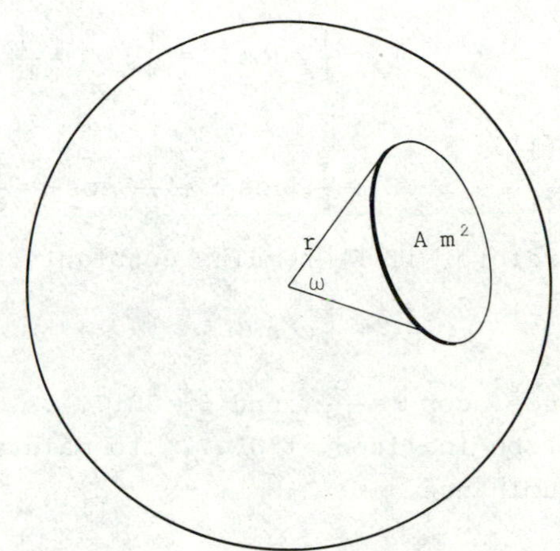

2 A source of light has a mean spherical intensity of 20 cd. How much total flux does it emit?

Using equation (9.2), $\text{MSI} = \frac{\Phi}{4\pi}$, whence

$$\Phi = 4\pi \times \text{MSI} = 4 \times 3.142 \times 20 = 251.3 \text{ lm}.$$

3 A source has an intensity of 250 cd in a particular direction. How much flux is emitted per unit solid angle in that direction?

From section 9.6 we recall that 1 candela ≡ 1 lm/sr. Therefore, 250 cd ≡ 250 lm/sr, whence we see that the source emits 250 lumens of flux into a unit solid angle.

4 A point source of light S, of intensity 100 cd, is suspended 4 m above a horizontal surface. What is the illuminance on the surface

(i) at the point vertically below the source

(ii) at 6 m from this point?

(i) At A, using equation (9.5), with I = 100 cd, $\theta_A = 0^0$ so that $\cos\theta_A = \cos 0^0 = 1$, and $r = SA = 4$ m, then

$$E_A = \frac{I}{r^2}\cos\theta = \frac{100}{4^2} \times 1 = 6.25 \text{ lm/m}^2 \text{(lux)}.$$

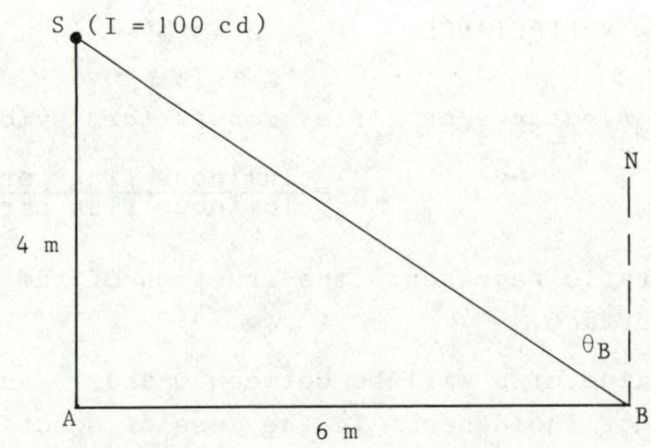

(ii) At B, using equation (9.5) again but with $\theta_B = S\hat{B}N = A\hat{S}B$.

From the diagram at the right,

$$r^2 = SB^2 = AS^2 + AB^2 = 4^2 + 6^2 = 52.$$

Hence, r = 7.211 m. Then $\cos\theta_B = \cos A\hat{S}B = \frac{AS}{SB} = \frac{4}{7.211} = 0.5547$. So, using these values we have

$$E_B = \frac{I}{r^2}\cos\theta_B = \frac{100}{(7.211)^2} \times 0.5547 = 1.07 \text{ lux}.$$

5 Light falls normally on a surface at 4 m from a point source of light. If the surface is moved to a distance of 3 m from the source, at what angle must the surface be inclined in order that the illuminance is the same value?

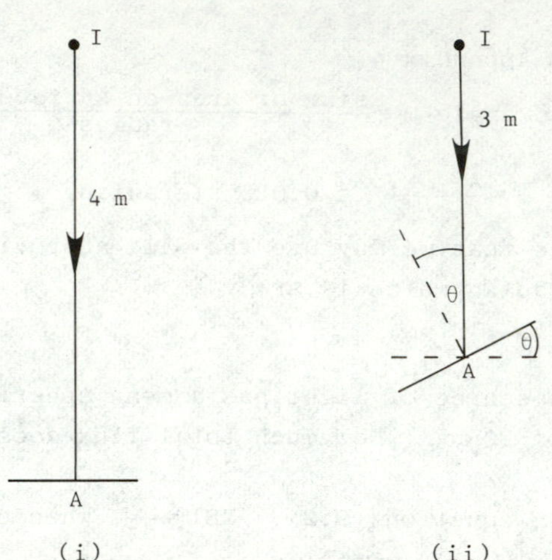

Referring to the diagram, in (i),

$$E_A = \frac{I}{r^2}\cos\theta = \frac{I}{4^2} \times 1 = \frac{I}{16}, \text{ since } \theta = 0^0.$$

In (ii),

$$E_A = \frac{I}{r^2}\cos\theta = \frac{I}{3^2}\cos\theta = \frac{I}{9}\cos\theta.$$

Therefore, if E_A remains constant,

$$\frac{I}{16} = \frac{I}{9}\cos\theta$$

whence $\cos\theta = \frac{9}{16}$, and $\theta = 55.77^0$. So, the surface must be inclined at 55.77^0 to maintain constant illuminance.

9.9 *REFLECTANCE, TRANSMITTANCE, AND OPTICAL DENSITY*

When luminous flux is incident on a surface then, in general, part of the flux is reflected, part is transmitted, and part is absorbed. If ρ represents the fraction of the incident flux which is reflected, τ the transmitted fraction, and α the absorbed fraction, then it follows that $\rho + \tau + \alpha = 1$. If a surface is opaque, that is light cannot pass through the substance, then $\tau = 0$. If a surface is transparent, so that light can pass through the substance, then usually ρ, τ, and α will all have definite values with τ often the largest of the three. No material is perfectly transparent, with good quality optical glass absorbing about 2% of the incident flux per cm of thickness.

9.9.1 *REFLECTANCE*

The *reflectance* or *reflection factor*, symbol ρ, of a surface is the ratio

$$\rho = \frac{\text{luminous flux per unit area reflected by the surface}}{\text{luminous flux per unit area incident on the surface}}.$$

This ratio represents the fraction of the incident flux which is reflected by unit area of the surface.

The value of ρ will be between 0 and 1, and will depend on the nature of the surface and the angle of incidence. In the case of spectacle crown glass the reflection loss at the surface will be at least 4% for normal incidence, and will be greater than this for large angles of incidence. Figure 9.9 shows how the fraction of incident light which is reflected by a plane surface of spectacle crown glass varies for different angles of incidence. Below an angle of incidence of about 50^0 the reflectance is approximately constant, but increases rapidly for angles of incidence larger than this. In Chapter 13 we shall see that this amount of reflected light can be reduced by applying certain thin film coatings to lens surfaces, a process called *antireflection coating*.

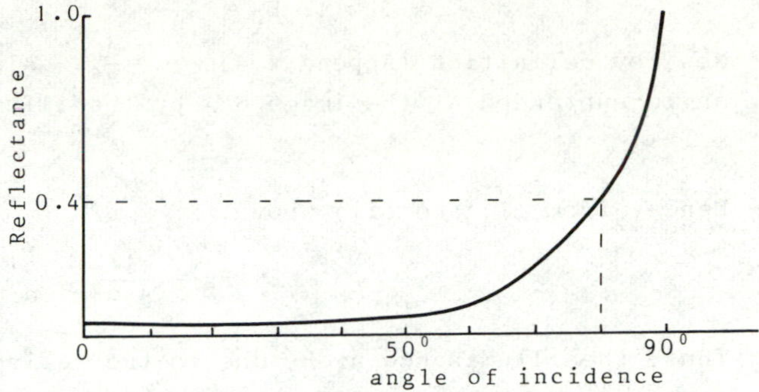

Fig. 9.9

9.9.1.1 Effect of a Reflector on the Illuminance of a Surface

The effect of a mirror, or other reflecting surface, suitably placed near to a light source, is to increase the illuminance on any surface which receives the luminous flux compared to that due to the source alone. Let us consider figure 9.10 . A light source S, of intensity I cd, is placed a distance a in front of a plane mirror M, having a reflectance ρ.

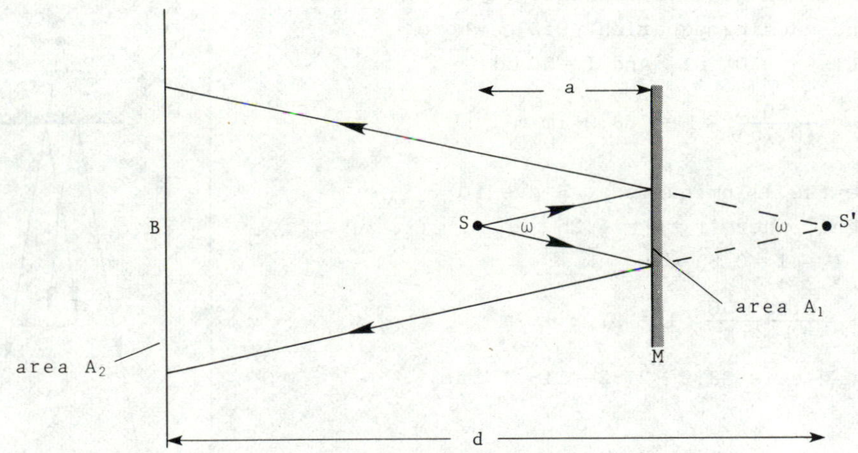

Fig. 9.10 Increased illuminance using a reflector.

The luminous flux Φ contained in the solid angle ω is incident on the area A_1 of the mirror, and is then reflected to form an illuminated patch, of area A_2, on the surface at B. The amount of reflected flux will be ρΦ, and hence, the illuminance on the surface at B, due to the reflected light only is given by equation (9.3):

$$E_B = \frac{\rho\Phi}{A_2} .$$

But from equation (9.1), $\Phi = I\omega$, where I cd is the intensity of the source.

$$\therefore E_B = \frac{\rho I \omega}{A_2} \qquad \text{(i)}$$

Now, by definition (Appendix 4), $\omega = \frac{A_1}{a^2}$. Also, from similar triangles, ω is the solid angle subtended at the image S', by the illuminated area A_2:

$$\therefore \omega = \frac{A_2}{d^2} \qquad \text{(ii)}.$$

Hence, from (i) and (ii) above:

$$E_B = \frac{\rho I A_2}{A_2 d^2} = \frac{\rho I}{d^2} \qquad (9.6)$$

Thus, the illuminance at B, due to the reflected light, is as if from a source of intensity ρI situated at the position of the image in the mirror.

The total illuminance at B will, of course, be due to the reflected light from the mirror together with the light reaching B directly from the source. A similar procedure may be adopted for curved mirrors. In each case the image of the source formed by the mirror will have an intensity ρI cd, where ρ is the reflectance of the mirror.

Worked Example

A small 50 cd source which may be assumed to radiate uniformly in all directions, is placed 75 cm above a horizontal table, and a plane mirror is fixed horizontally 25 cm above the source. If the mirror reflects 85% of the incident light, calculate the illuminance on the table at the point vertically below the source.

Let us firstly determine the illuminance at B by direct light alone. Using equation (9.5), where $r = d = 0.75$ m, $\cos\theta = \cos 0^0 = 1$, and $I = 50$ cd,

$$E_B = \frac{I}{r^2}\cos\theta = \frac{50}{(0.75)^2} \times 1 = 88.9 \text{ lm/m}^2.$$

Now, to determine the illuminance at B due to the reflected light, where $r = d' = 1.25$ m, $\cos\theta = \cos 0^0 = 1$, $I' = \rho I = 0.85 \times 50$.

$$E_B = \frac{I'}{r^2}\cos\theta = \frac{0.85 \times 50}{(1.25)^2} \times 1 = 27.2 \text{ lm/m}^2.$$

The total illuminance is $88.9 + 27.2 = 116.1$ lm/m^2.

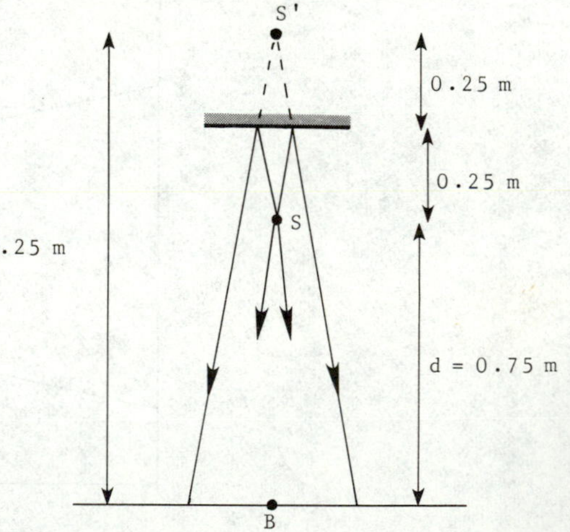

9.9.2 TRANSMITTANCE (*syn. transmission factor*)

The *transmittance* of a transparent body is given by the ratio

$$\tau = \frac{\text{luminous flux per unit area transmitted by the body}}{\text{luminous flux per unit area incident on the surface}}.$$

This ratio represents the fraction of the incident light which is transmitted by unit area of the body. No material is perfectly transparent, such a limiting case meaning $\tau = 1$, and the value of τ will depend on the nature and thickness of the substance. It may also depend on the wavelength of the light used. Some optical materials are available which produce no significant change in the appreciation of colour and display an almost constant transmittance across the whole visible spectrum. Such materials are called *neutral* substances. However, due to their inferior transmission of ultraviolet wavelengths, these materials do offer good protection to sunglare.

Figure 9.11 represents the transmission characteristics of a typical optical material much the same as Chance-Pilkington's SC3 dark neutral tinted glass.

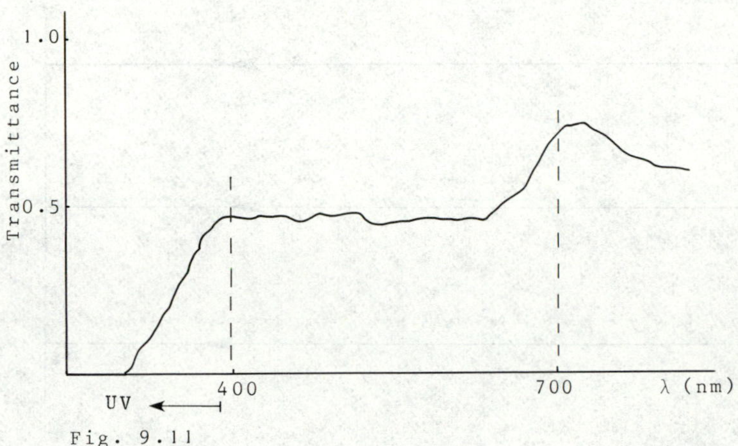

Fig. 9.11

As can be seen, the value of τ is essentially constant at about 0.47 for most of the visible range of wavelengths, but the material is strongly absorbent in the ultraviolet region.

9.9.2.1 *The Effect of a Transmission Filter on the Illuminance of a Surface*

In basic photometric terms, the effect of a filter having a transmittance τ, placed between a source of light of I cd and a surface, is to reduce the effective intensity of the source to a value of τI cd. Thus, if a source of light is fitted with a filter, having an average transmittance of, say 0.62, for the range of wavelengths involved, the real intensity I cd is reduced effectively to an intensity of 0.62I cd. If two of these filters are used together the effect of the second filter is to reduce the intensity from 0.62I to 0.62 × 0.62I cd, that is 0.62^2I.

Hence, we see that three filters in succession will reduce the intensity of the source by a factor of τ^3, and so on. Where two different filters with transmittances τ_1 and τ_2 are placed in front of a source, then the intensity will be reduced to $\tau_1 \tau_2 I$ cd.

We should note here that when the transmittance τ is quoted for a sample we should be a little more explicit. It generally means the ratio of the total transmitted luminous flux to the total incident luminous flux, as defined in section 9.9.2 . This would usually imply that the radiation contained all the visible wavelengths, and is therefore called *total transmittance*.

However, where the transmittance varies for each wavelength τ is usually called the *spectral transmittance*, and is defined for each wavelength. When plotted against wavelength the result is the transmittance curve of figure 9.11 .

The various effects of filters on the illuminance of a surface are summarised in figure 9.12 for normal incidence. The exact positions of the filters between the source and the screen are not important.

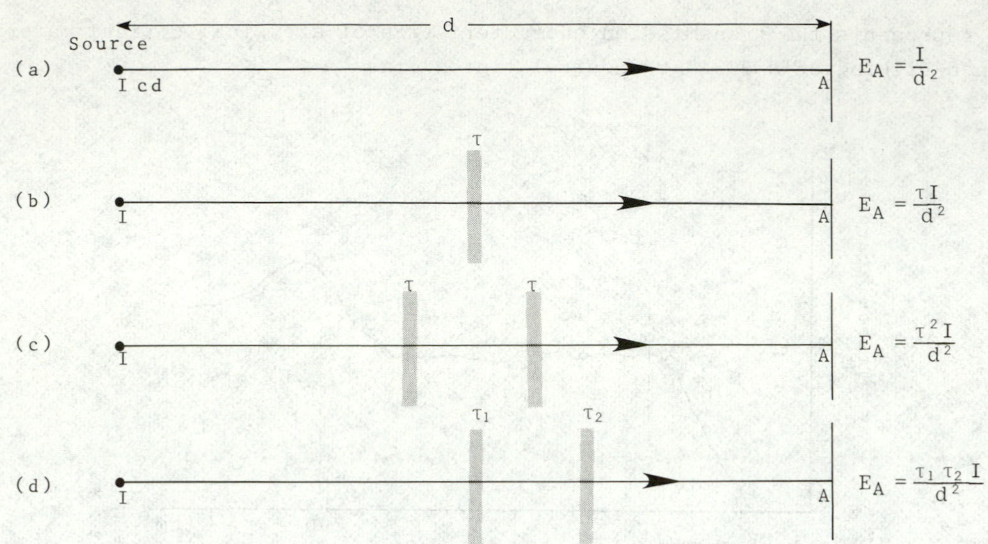

Fig. 9.12 The effect on the illuminance of a surface when filters are interposed between the source and the surface.

Worked Example

A point light source of intensity 200 cd is 2.5 m from a screen. Calculate the illuminance on the screen for normal incidence. If a neutral filter of transmittance 45% is placed between the source and the screen, what is the new value of the illuminance? Where must the source be placed such that, with the above mentioned filter in place, and with normal incidence, the illuminance on the screen is restored to its original value?

Using $E = \frac{I}{d^2}\cos\theta$, with I = 200 cd, $\cos\theta = \cos 0^0 = 1$, and d = 2.5 m, we have as the illuminance on the screen without the filter

$$E_1 = \frac{I}{d^2}\cos\theta = \frac{200}{(2.5)^2} \times 1 = 32 \text{ lux}.$$

When the filter is in position the intensity becomes τI, so we have for the new illuminance

$$E_2 = \frac{\tau I}{d^2}\cos\theta = \frac{0.45 \times 200}{(2.5)^2} \times 1 = 14.4 \text{ lux}.$$

In order that the illuminance is restored to 32 lux, the source must be moved to a new nearer distance d_1. Hence

$$E = \frac{\tau I}{d_1^2}\cos\theta, \text{ and } d_1^2 = \frac{\tau I}{E_1}\cos\theta = \frac{0.45 \times 200}{32} \times 1 = 2.81$$

giving $d_1 = (2.81)^{\frac{1}{2}} = 1.68$ m. That is, with the filter still in place the source has to be moved nearer to the screen to restore the original illuminance.

9.9.2.2 Optical Density

When we wish to consider how the transmittance of a material alters when its thickness changes, we discover that there is no simple linear relationship between these two quantities. Two specific cases of neutral tinted glasses given below show that the relationship between thickness and transmittance is not immediately evident when comparing 2 mm and 3 mm thick samples.

	τ(2 mm)	τ(3 mm)
SC4 (Spectacle Crookes tint)	0.21	0.10
SN2 (Spectacle Neutral tint)	0.58	0.46

These figures are quoted for a wavelength of 550 nm and therefore represent a spectral transmittance value rather than the total transmittance.

It is necessary to consider the transmittance of the material thickness in order to find the relationship between τ and varying thicknesses. In traversing a spectacle tint of 2 mm thickness, some light is lost by reflection at each surface, and some is lost by absorption in the material. This suggests we should define the total or the spectral transmittance of a filter as

τ = *transmittance of 1st surface* $\times$ *transmittance of material* $\times$ *transmittance of 2nd surface*.

Or, in symbols,

$$\tau = T_1 T_m T_2 \qquad (9.7),$$

where T_m is for a given thickness of material, often 2 mm.

It will be shown in section 10.8.2 that the surface transmittance is $T = \frac{4nn'}{(n'+n)^2}$ and for a filter in air and of refractive index 1.523 for a given wavelength, $T_1 = T_2 = 0.957$. Now let us consider the 2 mm thick sample of SC4 material above. Since $\tau = 0.21$, then from equation (9.7) we have

$$T_m = \frac{\tau}{T_1 T_2} = \frac{0.21}{0.957 \times 0.957} = 0.23 .$$

That is, neglecting reflections, 2 mm of this material transmits 0.23 of the incident flux of wavelength 550 nm. Hence, a 3 mm thickness of material will transmit $(0.23)^{\frac{3}{2}} = 0.11$. This is T_m for 3 mm. Hence, the transmittance for the 3 mm sample, including the effect of the surfaces, is

$$\tau = T_1 (T_m)_{3mm} . T_2 = 0.957 \times 0.11 \times 0.957 = 0.10 ,$$

which is the transmittance quoted above.

The slight difficulty in handling the expression for transmittance has led to the adoption of a quantity termed the *optical density* D of a substance, defined by the expression

$$D = \log\frac{1}{\tau} \qquad (9.8).$$

Suppose we wish to consider the effect of increasing the thickness of a filter as we did above. Then, since $\tau = T_1 T_m T_2$, we have

$$D = \log\frac{1}{\tau} = \log\frac{1}{T_1 T_m T_2} = \log\frac{1}{T_1} + \log\frac{1}{T_m} + \log\frac{1}{T_2} = D_1 + D_m + D_2 .$$

That is,

$$D = D_1 + D_m + D_2 \qquad (9.9).$$

We see that the optical density of the sample is simply the sum of the optical densities of the surfaces and the stated material thickness, say 2 mm. Now suppose we wish to find the optical density for a filter 3 mm thick. Then, as above,

$$D = \log\frac{1}{\tau} = \log\frac{1}{T_1 (T_m)^{\frac{3}{2}} T_2} = \log\frac{1}{T_1} + \log\frac{1}{(T_m)^{\frac{3}{2}}} + \log\frac{1}{T_2} = D_1 + \tfrac{3}{2}D_m + D_2 ,$$

since $\log\frac{1}{(T_m)^{\frac{3}{2}}} = \log\left(\frac{1}{T_m}\right)^{\frac{3}{2}} = \frac{3}{2}\log\frac{1}{T_m} = \frac{3}{2}D_m$.

Hence, we see that to find the new optical density we simply multiply the optical density D_m of the material thickness by whatever factor gives the new thickness. For example, if D_m is quoted for 5 mm and we wish to find the optical density when the thickness is 9 mm, the factor will be $\frac{9}{5}$, and so on.

If an absorbing filter is built up from several layers, thereby increasing the thickness, this simple relationship with the thickness is valid only if the spectral characteristics of the separate layers are identical.

Worked Example

At 2 mm thickness, a certain tinted glass has a transmittance for a specified wavelength of 0.47 . Determine the transmittance at 4 mm and 5 mm thicknesses. $n_g = 1.5$.

At 2 mm $\tau = 0.47$ and $T_1 = T_2 = \frac{4nn'}{(n'+n)^2} = \frac{4 \times 1 \times 1.5}{(1.5+1)^2} = 0.96$.

$$\therefore \quad T_m = \frac{\tau}{T_1 T_2} = \frac{0.47}{(0.96)^2} = 0.51 .$$

Then $D_m = \log\frac{1}{T_m} = \log\frac{1}{0.51} = 0.2925$, and $D_1 = D_2 = \log\frac{1}{T_1} = \log\frac{1}{0.96} = 0.0177$.

Hence, the optical density at 4 mm is given by

$$D = D_1 + \tfrac{4}{2}D_m + D_2 = 0.0177 + (2 \times 0.2925) + 0.0177 = 0.6204$$

whence $\frac{1}{\tau}$ = antilog D = antilog 0.6204 = 4.173, and $\tau = 0.24$.

Similarly, for the 5 mm thick sample,

$$D = D_1 + \tfrac{5}{2}D_m + D_2 = 0.0177 + (2.5 \times 0.2925) + 0.0177 = 0.7667$$

and $\frac{1}{\tau}$ = antilog D = antilog 0.7667 = 5.844, and $\tau = 0.17$.

9.10 LUMINANCE AND BRIGHTNESS

In the main, the light sources which have been considered so far have been sufficiently small compared to their distances from surfaces so that they could be treated as point sources. However, in most practical cases, we should be concerned with extended sources of light rather than points. These sources may be self-luminous, like a T.V. screen, or they may be diffusely reflecting or transmitting, as in the cases of the wall of an illuminated room and the glass envelope of a light bulb, respectively. It can be said that there is some similarity between self-luminous surfaces and non-self-luminous surfaces.

For example, when a surface acts as a diffuse reflector of light (section 5.1) the reflected light will be randomly scattered over a wide range of directions, and the surface will have a certain luminous intensity in each direction from which it is viewed.

Figure 9.13 shows an area A of a surface (self-luminous, transmitting, or reflecting) which emits (or transmits or reflects) luminous flux in a specific direction. If the luminous intensity of the surface in the specified direction is I cd, then the luminous intensity of the surface per unit projected area in the direction concerned is referred to a the *luminance* of the surface.

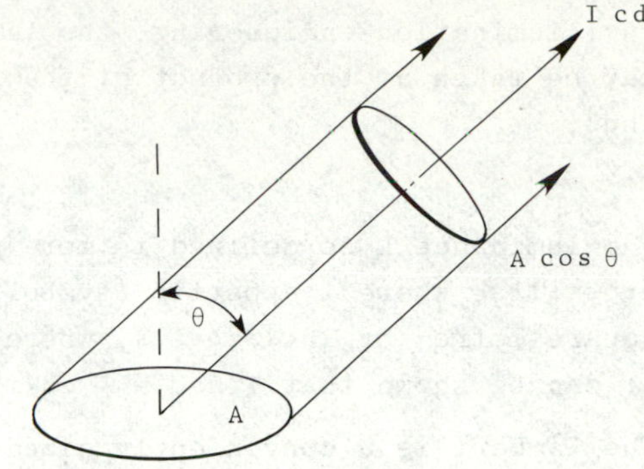

Fig. 9.13 Luminance of a surface.

That is, if the actual area of the surface concerned is A, then the projected area inclined at an angle θ to the normal will be A cos θ, and the luminance L in this direction is given by

$$L = \frac{I}{A \cos \theta} \qquad (9.10).$$

It is important to note that I is not a constant but is θ-dependent. The unit of luminance is the candela per square metre (cd/m^2). It will be seen that the primary standard defined in section 9.3 is really a standard of luminance.

Figures 9.14(a) and (b) show how the projected areas of the radiating surfaces of light sources may vary considerably in different directions. The projected area of a sphere is, however, a disc of constant area.

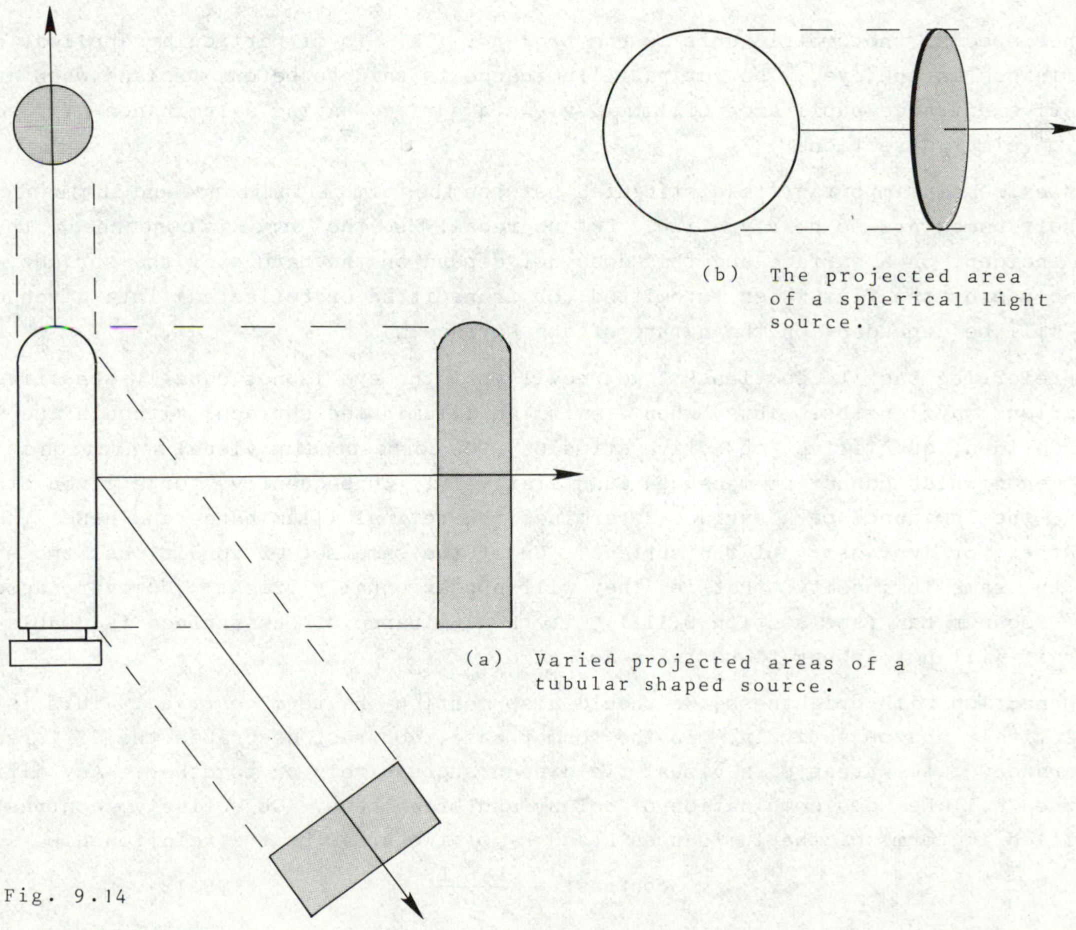

(b) The projected area of a spherical light source.

(a) Varied projected areas of a tubular shaped source.

Fig. 9.14

In illumination engineering, the luminance of a reflecting surface in a particular direction may be taken at the product of the illuminance E and the reflectance ρ in that direction. Thus,

$$L = \rho E \qquad (9.11)$$

The luminance L so defined is more properly called the *luminous exitance* and is expressed in *apostilbs*, where 1 apostilb (symbol asb) corresponds to a reflection of 1 lumen of flux per square metre, or in *lamberts*, where 1 lambert (symbol 1 L) represents 1 lumen per square cm. It can be shown that 1 asb = $\frac{1}{\pi}$ cd/m^2.

The lambert is a conveniently sized unit when dealing with non-self-luminous surfaces such as the ceilings and walls of an indirectly illuminated room or the diffusing globes around incandescent lamps.

Table 9.4 gives typical values for the luminance of a number of extended sources and surfaces.

Table 9.4 Some typical luminance values

Source	Luminance (cd/m^2)
Surface of sun	2×10^9
White paper in sunlight	2.5×10^4
Fluorescent lamp	6×10^3
Moon surface	2.9×10^3
White paper in moonlight	3×10^{-2}

Another special photometric unit is the *troland*. This is of particular application to retinal illuminance in the eye. The retinal illuminance is said to be one troland when an eye, with effective entrance pupil area of 1 mm^2, views a surface having a luminance of 1 cd/m^2 in the line of sight direction.

It is extremely important to distinguish between the terms illuminance and luminance, especially as their names are so nearly alike. Let us recall that the former is concerned with the luminous flux incident on a surface and this does not depend on the nature of the surface. The latter term concerns the flux which is emitted (or transmitted or reflected) in a given direction and this will be dependent on the nature of the surface.

Now, referring back to section 9.1 we recall that the eye is not equally sensitive to light of various wavelengths. Thus, when viewing an illuminated surface, although the luminance is a defined, quantitive, objective stimulus, the corresponding visual sensation is a subjective phenomenon which cannot be measured quantitatively. Consequently, for a given state of adaptation, the luminance of a surface determines the retinal illuminance and hence the subjective *brightness* or *luminosity* of the surface. Under the same set of conditions, two surfaces will have the same luminosity, that is, they will appear equally bright. However, according to the Weber-Fechner Law (see section 9.11.1), if the luminance of one surface is double that of another it will not appear to be twice as bright.

In connection with brightness, we should also mention the term *contrast*. This is used both subjectively and objectively. In the former case, contrast describes the difference in the appearance of two areas of a visual field seen successively or together. Any difference observed will be some combination of colour and brightness. Objectively, contrast may be specified in terms of the luminances L_1 and L_2 of two areas by the relationship

$$\text{contrast} = \frac{L_2 - L_1}{L_1} \qquad (9.12).$$

An area of the visual field which is excessively bright compared to the surroundings is referred to as *glare*, giving discomfort or impairment of vision. It results in a reduction of the sensitivity of the eye and, consequently, a reduction in the perception of contrast. It may be due to lamps which are inadequately shielded, or to the reflection of light by glossy surfaces.

9.10.1 UNIFORMLY DIFFUSING SURFACES

In many instances, observations of diffusely reflecting and self-luminous surfaces indicate that the brightness or luminosity of the surface is independent of the angle at which it is viewed. This will occur if the physical cause of brightness, that is, the luminance of the surface, is the same in all directions. Hence, using equation (9.10), if the luminance L in a given direction is given by $L = \frac{I}{A\cos\theta}$, then $I = LA\cos\theta$. For constancy of luminance we have therefore:

$$I \propto \cos\theta .$$

This means that the luminous intensity of an area of the surface is proportional to $\cos\theta$. This is known as *Lambert's Law of Emission*, and a surface which obeys the law is referred to as a *uniformly diffusing surface*. Thus, an incandescent spherical body, such as the sun, appears like a uniformly bright disc although the central portion is viewed normally and the edges tangentially. A reflecting surface which reflects all the incident light is said to be a *perfect diffuser*.

9.10.2 THE LUMINANCE OF IMAGES

It is often of interest in connection with optical instruments to consider the illuminance and luminance of images formed by a lens. Let us consider, as shown in figure 9.15, a converging lens forming an image of area a' of an object of area a. In the case illustrated the image is an enlarged one.

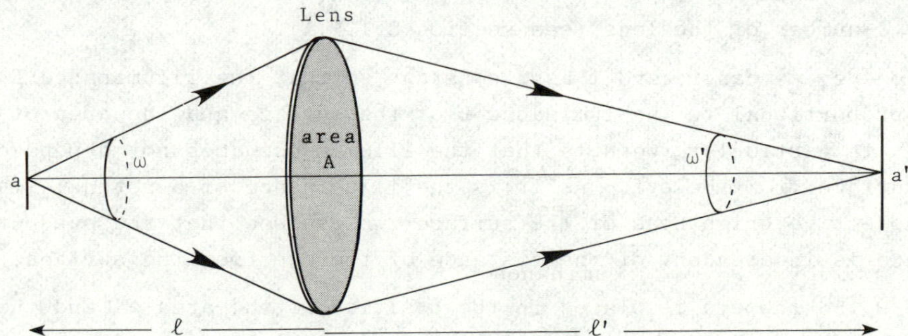

Fig. 9.15 The illuminance of an image.

Let the area of the lens be A, with ω and ω' representing the solid angles of the incident and refracted cones of light rays. Let the luminance of the object area a in the direction of the lens be L.

Then, using equation (9.10): $L = \frac{I}{A \cos\theta}$ with $\theta = 0^0$, then L represents the luminous flux emitted per unit solid angle per unit area in this direction; that is, $L = \frac{I}{a}$ or $I = aL$ (i). Now, $\omega = \frac{A}{\ell^2}$ (ii), and $\Phi = I\omega$ (iii), therefore from (i) and (iii) the flux incident on the lens from the area a is given by

$$\Phi = I\omega = a\omega L \quad \text{lx.}$$

If we now assume that no absorption or reflection losses occur in the system, we can say that all the light incident on the lens will be incident on the image.
Therefore, from equation (9.3):

$$\text{Illuminance of the image} = \frac{\text{Flux incident on it}}{\text{area of image}}.$$

That is, in symbols

$$E = \frac{\Phi}{a'} = \frac{a\omega L}{a'} \quad \text{lx.}$$

But, from similar triangles (taking into account the height and width of the object), we have $\frac{a}{a'} = \frac{\ell^2}{\ell'^2}$. Therefore,

$$\text{Illuminance of image } E = \frac{\omega L \ell^2}{\ell'^2}$$

or $$E = \frac{LA}{\ell'^2}, \quad \text{since } A = \omega\ell^2 \text{ from (ii) above.}$$

This may be written as

$$E = \frac{\pi}{4} L \frac{d^2}{\ell'^2} \qquad (9.13)$$

where d is the lens diameter, and $A = \frac{\pi}{4} d^2$.

Hence, we can see that the illuminance of the image of a surface is proportional to the luminance of the surface and the area of the lens aperture.

In a camera, for distant objects, $\ell'^2 = f'^2$ where f' is the focal length of the lens. The illuminance of the image on the film will be given by

$$E = \frac{\pi}{4} L \left(\frac{d}{f'}\right)^2 \qquad (9.14)$$

The quantity d/f' is called the *aperture ratio* of the lens, and we may recognise it as the reciprocal of the f-number of the lens (see section 8.1.1.1).

In the case of the eye, we can regard ℓ' as a constant so that the illuminance E of the retinal image is proportional to the luminance L of the surface and the area of the entrance pupil of the eye. In particular, we note that the illuminance does not depend on ℓ, the distance of the surface from the eye. As it is the illuminance of a retinal image that determines the subjective brightness of the surface, we can see that the subjective brightness of the surface is independent of the distance of the eye from the surface.

Now, if in figure 9.15, the eye is placed in the position of the area a' and the area a' is no bigger than the area of the entrance pupil of the eye, the luminous flux Φ entering the eye will be contained in a solid angle ω'. Thus the area a' will act like a source of intensity $I' = \frac{\Phi}{\omega'}$, and the luminance along the lens axis will be given by $L' = \frac{I'}{a'} = \frac{\Phi}{\omega' a'}$. Now, since $\Phi = a\omega L$, this may be written as $L' = \frac{a\omega L}{a'\omega'}$, and again, since $a\omega = a'\omega'$, we have

$$L' = L \qquad (9.15).$$

That is, the luminance of the image is equal to the luminance of the object. This treatment has neglected light losses, and when these are taken into account the luminance of the image is less than the luminance of the object.

When comparing the illuminances of two surfaces in terms of their subjective brightness, the surfaces need not be at the same distance from the eye, although it is often convenient for them to be close together. This discussion becomes invalid if the retinal image area becomes so small as to be of similar size to a single receptor in the retina. In this case the brightness decreases as ℓ increases. The reduction in brightness of very distant objects is also affected by atmospheric absorption.

It is interesting to note that if the source is a point, such as a star, the subjective brightness will be dependent on the light entering the eye from the point. In the case of a telescope, provided that all the light passing through the objective lens enters the eye, the brightness is increased in the ratio A_o/A_p, where A_o is the area of the aperture of the objective and A_p is the area of the eye's entrance pupil. It follows that a greater number of stars are visible in a telescope than with the unaided eye. The brightness of the background is not increased as this is not a point source, but the stars themselves have their brightnesses increased greatly.

9.11 PHOTOMETERS

The use of a photometer involves the question of measurement. Since we have defined the standard source intensity as the basis of photometric measurement, and since the illuminance on a surface is dependent on the intensity of the source, we shall begin by outlining the principle by which the intensity of two sources may be compared.

Two sources may be compared indirectly by virtue of the illuminances which they produce on a suitable screen. This screen forms part of the photometer. Photometers may be classed as visual or non-visual. In the former group of instruments the basic principle applicable to their use is that if two adjacent identical white reflecting surfaces appear to be equally bright when illuminated with two sources, then the surfaces will be receiving the same illuminance and the boundary between the surfaces will be difficult to see. Figure 9.16 illustrates the visual photometer principle.

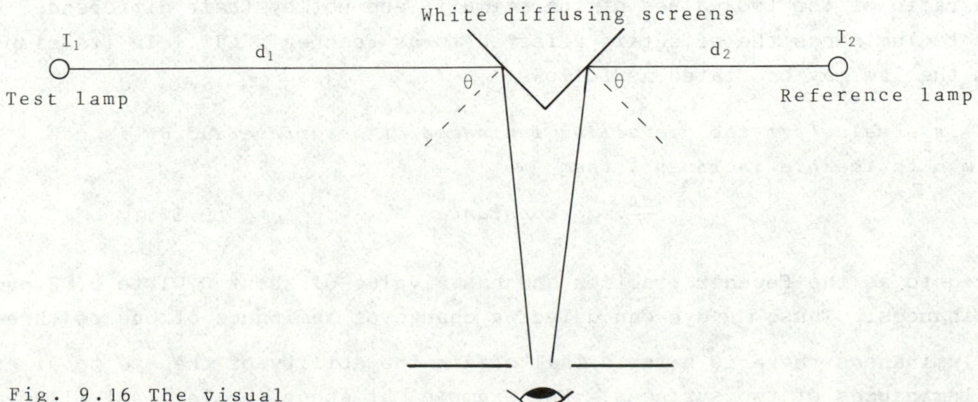

Fig. 9.16 The visual photometer principle.

The distances of the sources are varied until the screens are equally illuminated. With identical screens, this will be when they appear equally bright. The photometer is then said to be balanced. Thus, the illuminance on the left hand screen is $E_L = \frac{I_1}{d_1^2} \cos\theta$. Similarly, the illuminance on the right hand screen is $E_R = \frac{I_2}{d_2^2} \cos\theta$. When balanced, $E_L = E_R$. That is,

$$\frac{I_1}{d_1^2} \cos\theta = \frac{I_2}{d_2^2} \cos\theta$$

$$\text{or} \qquad I_1 = I_2 \left(\frac{d_1}{d_2}\right)^2 \qquad (9.16).$$

The use of the inverse square law is assuming that the two lamps are small enough to be regarded as point sources. If the two surfaces approximate to uniformly diffusing surfaces (see section 9.10.1) this effect applies when the surfaces are viewed from different directions. The ability of the eye to judge the equality of brightness of two surfaces is expressed by the *Weber-Fechner Law*, see section 9.11.1 . Unfortunately, if the colours of the two sources are different the assessment of equal brightness is very difficult.

The non-visual group of photometric devices include photoelectric components using the photo-emissive and photovoltaic effects. These measure directly the illuminance falling on them and can be made very portable. They do not require any visual assessment of brightness to be made by the observer. Devices based on the photovoltaic effect have largely replaced the photoemissive devices, and the former only will be considered in section 9.11.3 .

9.11.1 THE WEBER-FECHNER LAW

As previously stated, the user of a visual photometer is required to assess the equality of brightness of two adjacent illuminated surfaces. It is therefore of value to consider the accuracy with which this can be done. The eye cannot give any quantitative comparison between different luminances; that is, it is not capable of determining, say, that one surface is twice as bright as another. However, the eye can judge with a fair degree of accuracy when two adjacent surfaces appear to be equally bright, provided they appear the same or nearly the same colour.

The work of Weber on various sensations and their stimuli, including hearing, muscular action, and vision, has shown that the assessment of the size of an increase in a stimulus is done in terms of the ratio of the two values of the stimuli, and not by their difference. When applied specifically to luminance the effect is referred to as Fechner's Law. In its simplest mathematical form the law may be stated as follows:

If L is a value for the prevailing luminance of a surface and dL is the minimum noticeable increment, then

$$\frac{dL}{L} = \textit{a constant} \qquad (9.17).$$

$\frac{dL}{L}$ is referred to as the Fechner fraction and has a value of about 0.01 to 0.03 over a wide range of luminances. Thus, the eye can detect a change of luminance of one to three per cent.

At very low luminances there is a rapid fall off in the ability of the eye to discriminate between the luminances of two surfaces. For example, at about the level of illuminance afforded by moonlight $\frac{dL}{L}$ may rise to about 0.5, meaning a 50% difference is required in the luminances of two surfaces before any difference can be discerned.

9.11.2 VISUAL PHOTOMETERS

9.11.2.1 The Grease-Spot Photometer

This is a simple form of photometer and is illustrated in figure 9.17(a) with the detail of the screen and grease-spot in figure 9.17(b).

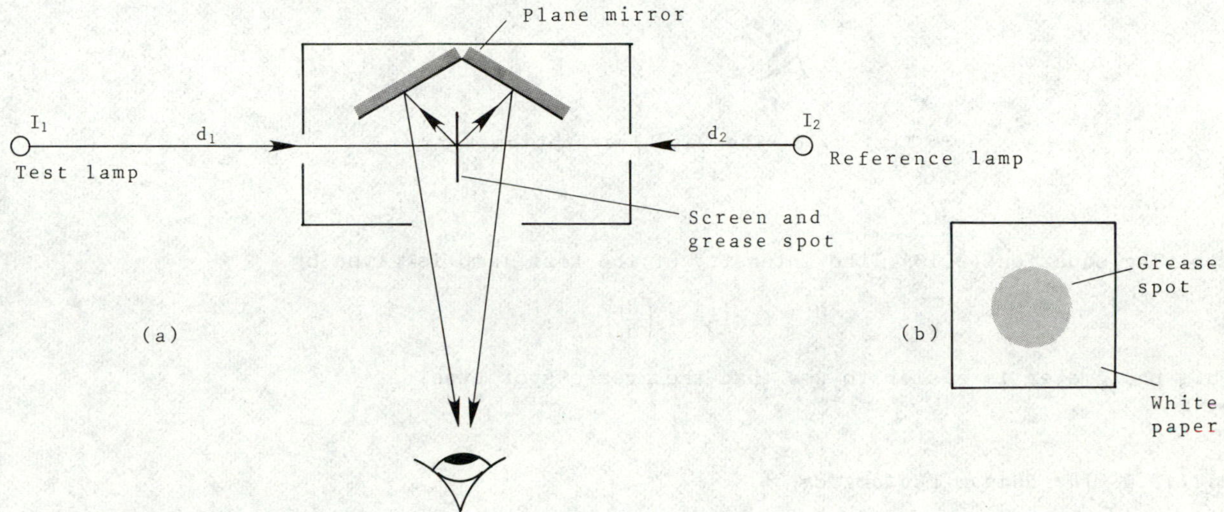

Fig. 9.17 The grease-spot photometer.

The translucent grease-spot or ring is formed by applying oil or wax to the paper screen. When illuminated on each side by the two lamps the intensities of which are to be compared, both sides of the screen may be viewed simultaneously with the aid of two plane mirrors. The grease area transmits more light than it reflects whilst the converse is true for the surrounding paper.

The theory of the grease-spot photometer is beyond the scope of this book but it can be shown that when the contrast between the grease area and the surrounding paper is the same on both sides of the screen, then the illuminances on the two sides are equal. The photometer needs considerable practice to use it effectively. However, it is necessary to adjust the distances of the two lamps from the screen to produce equal contrast between paper and grease area on each side. When this is achieved equation (9.16) applies.

9.11.2.2 The Wax Block Photometer

The essential feature of the wax block photometer is a pair of identical rectangular blocks of paraffin wax (sometimes opal glass is used) placed side by side and separated by a piece of metal foil. In figure 9.18, I_1 and I_2 are the intensities of the two sources to be compared.

Adjustment of the distances d_1 and d_2 is made until the two edges of the wax blocks have the same brightness. When this state of photometric balance is obtained it can be said that both sides of the composite wax block have the same illuminance.

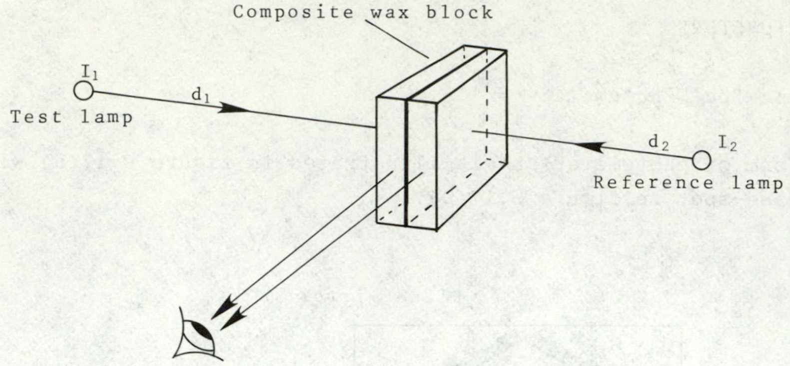

Fig. 9.18 The wax-block photometer.

Applying equation (9.16), the intensity of the test lamp is given by

$$I_1 = I_2\left(\frac{d_1}{d_2}\right)^2 .$$

This photometer is easier to use than the grease-spot type.

9.11.2.3 The Shadow Photometer

Figure 9.19 shows the principal components. Two rods R_1 and R_2 are arranged to partially restrict light from the two sources so that the shadows S_1 and S_2 cast by the rods touch one another. The area S_1 is in the shadow of R_1 formed by the lamp under test, and is therefore illuminated by light from the reference lamp alone. Similarly, the area S_2 of the screen is in the shadow of R_2 formed by the reference lamp and is illuminated by light from the test lamp only. By the adjustment of the distances d_1 and d_2, from the lamps to the screen, the two partial shadow areas S_1 and S_2 can be made to appear equally dark, indicating equality of illuminance.

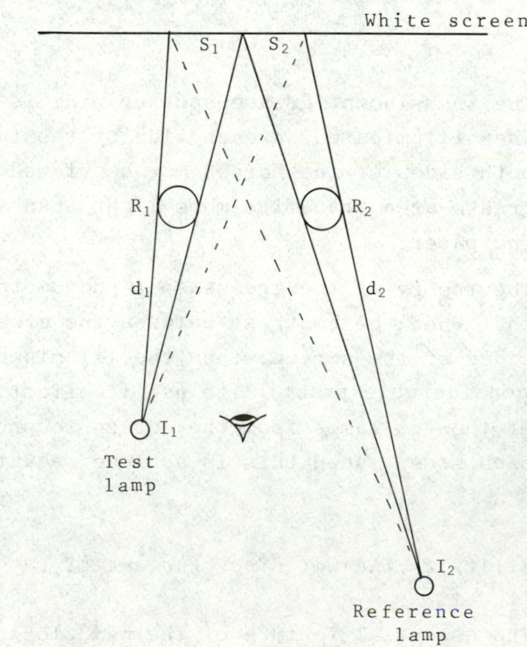

Fig. 9.19 The shadow photometer.

Once again, from equation (9.16), the intensity of the test lamp is

$$I_1 = I_2\left(\frac{d_1}{d_2}\right)^2 .$$

The shadow photometer was invented by Sir Benjamin Thompson (1753 - 1814), and it is often called the Rumford shadow photometer, the name being taken from Thompson's title, the Count von Rumford.

9.11.2.4 The Lummer-Brodhun Photometer

The Lummer-Brodhun photometer is one of the most accurate visual photometers and is similar, in principle, to the grease spot type. It is almost always used when an accurate comparison of the intensities of sources of similar colour is required. The basic form of the photometer is illustrated in figure 9.20(a).

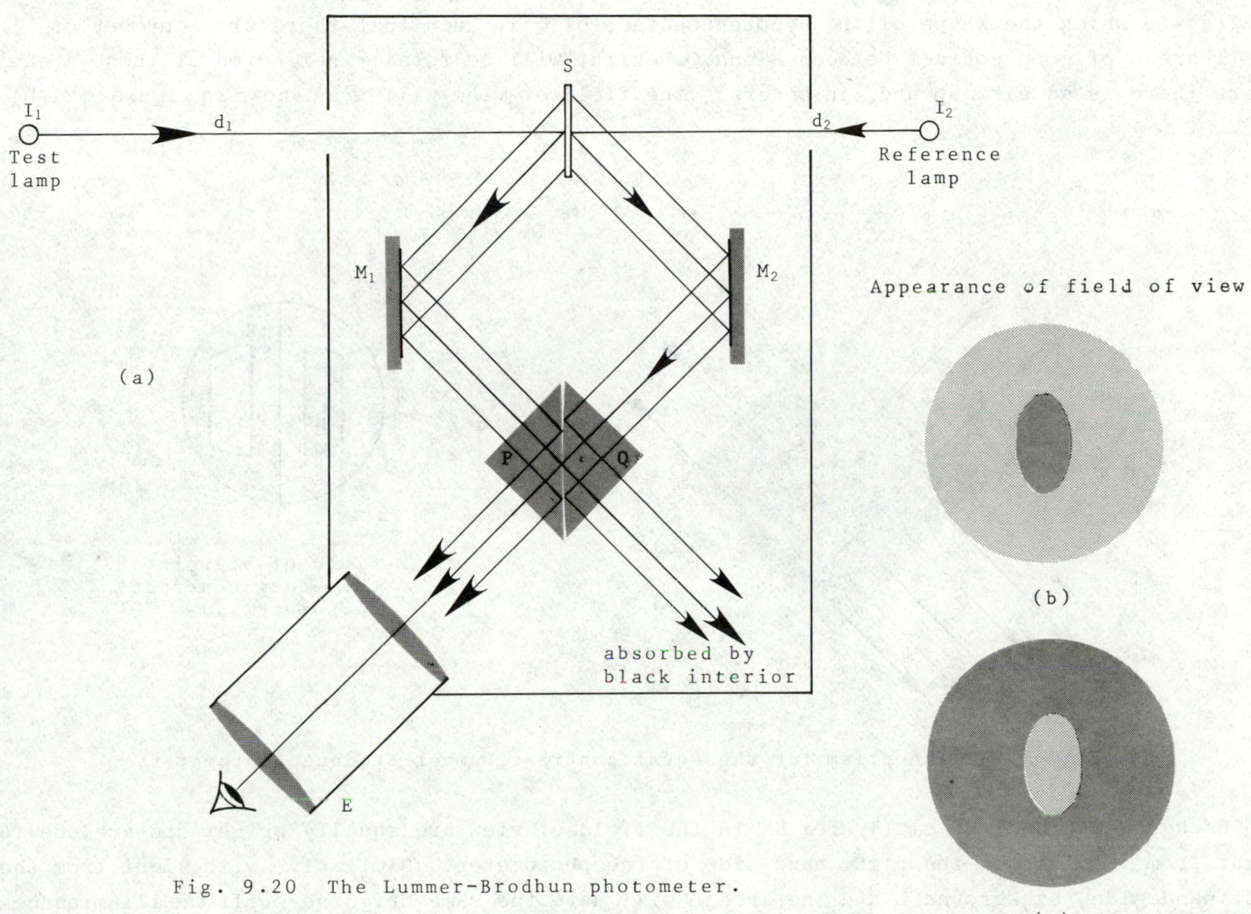

Fig. 9.20 The Lummer-Brodhun photometer.

The unique feature of this instrument is the Lummer-Brodhun composite prism PQ which consists of a pair of 45^0-45^0-90^0 prisms, one of which, Q, has a slightly convex spherical hypotenuse face with a small flat circular area in the middle. This small area is in good optical contact with the hypotenuse face of P, the other parts of the surfaces having an air gap between them. Light from the sources, of intensities I_1 and I_2 candela, is incident normally on the two sides of the white screen S, which is made of a good diffusing material such as plaster of Paris. By means of two plane mirrors M_1 and M_2, light which is diffusely reflected from each side of S reaches the composite prism PQ. The observer views the contact surface of the prism with the aid of an eyepiece E. The field of view consists of a central area which receives light from the lamp on the right hand side, and an outer portion which is illuminated by light from the left hand side. If the illuminances of the two sides of the screen S are not the same there will be a difference in the brightnesses of the central and outer areas of the field of view, and the appearance will be as indicated in figures 9.20(b) or (c). When the sources are adjusted to give equal illuminance at S the field of view will appear evenly bright, and it will be difficult to see the edge of the central area. Equation (9.16) then applies.

The Lummer-Brodhun photometer may be converted to the *equal contrast* form by modification of the composite prism PQ. This is because the eye can judge equality of contrast better than it can judge equality of brightness. The Lummer-Brodhun contrast prism is shown in figure 9.21(a) in which the shape of the hypotenuse face of Q is such that there are a number of small areas of good contact between P and Q. Light will be totally reflected at those areas where there is an air gap and, in general, the field of view will be as shown in figure 9.21(b).

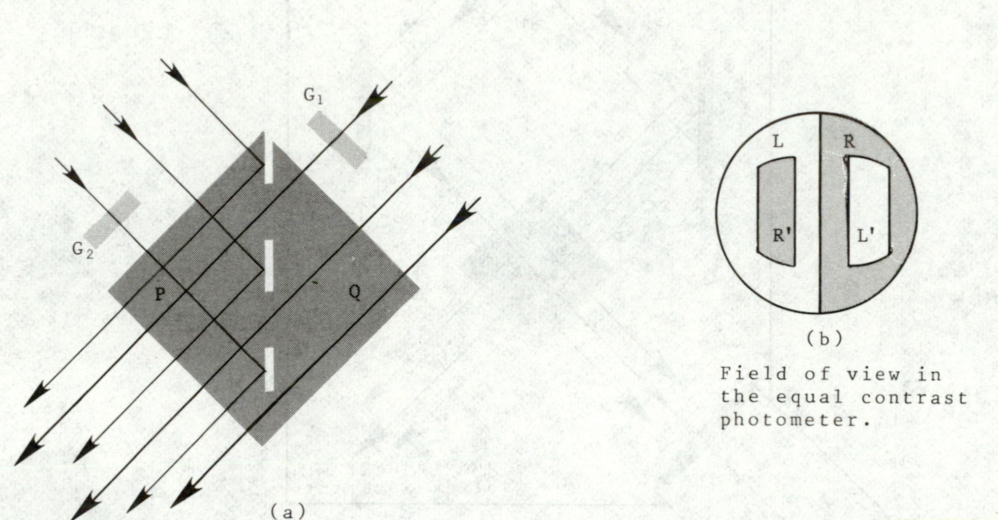

Fig. 9.21 The prism for the equal contrast Lummer-Brodhun photometer.

The background R and the small area R' in the field of view are equally bright and are due to light from the lamp on the right hand side of the photometer. Similarly, with light from the left hand side, background L and the area L' will have the same brightness. If the illuminances of both sides of the screen S of figure 9.20(a) were the same, the composite field of view would be evenly bright. Final modification to the prism assembly PQ is made by fitting two small, thin glass platelets G_1 and G_2, as shown in figure 9.21(a). G_1 has the effect of reducing the brightness of the patch R' compared to the background R by about 8% for the two glass surfaces, whilst G_2 reduces the brightness of the patch L' compared to the background L. In the balanced state, the contrast in each half of the field of view is the same with the backgrounds R and L appearing equally bright.

In practical use, to compensate for any small differences of the reflecting characteristics of the faces of S, and possible unequal reflections by mirrors M_1 and M_2 and by the faces of the prism PQ, the photometer head is rotated through 180^0 (equivalent to interchanging the two lamps) and another balance obtained.

The 8% loss of light due to each thin glass sheet mentioned above is due to the fact that only about 92% of the light is transmitted by each glass, the loss being by reflection.

9.11.2.5 The Flicker Photometer

In all the examples of photometers mentioned so far, we have assumed that the two light sources under comparison are of similar colour. Due to the variable sensitivity of the eye with colour (see section 9.1) it becomes difficult, if not impossible, to assess the equality of brightness for two different coloured areas in a field of view.

Much of the modern lighting in use nowadays, such as sodium and mercury vapour lamps, have colours quite different from that of the standard. To overcome the difficulty of comparisons between sources of this type a *flicker photometer* is employed. There are many forms of this device but the principle is the same for them all. That is, the eye sees alternately, and in rapid succession, the two illuminated surfaces to be compared. It is found that for some particular frequency of this alternation the colour difference disappears and a single combined hue is seen. At this stage the remaining flicker effect will be due to brightness differences, and may be eliminated by adjustment to the lamp positions. As before, when in the balanced position, equation (9.16) applies to the lamps.

One simple form of flicker photometer is shown diagrammatically in figure 9.22(a).

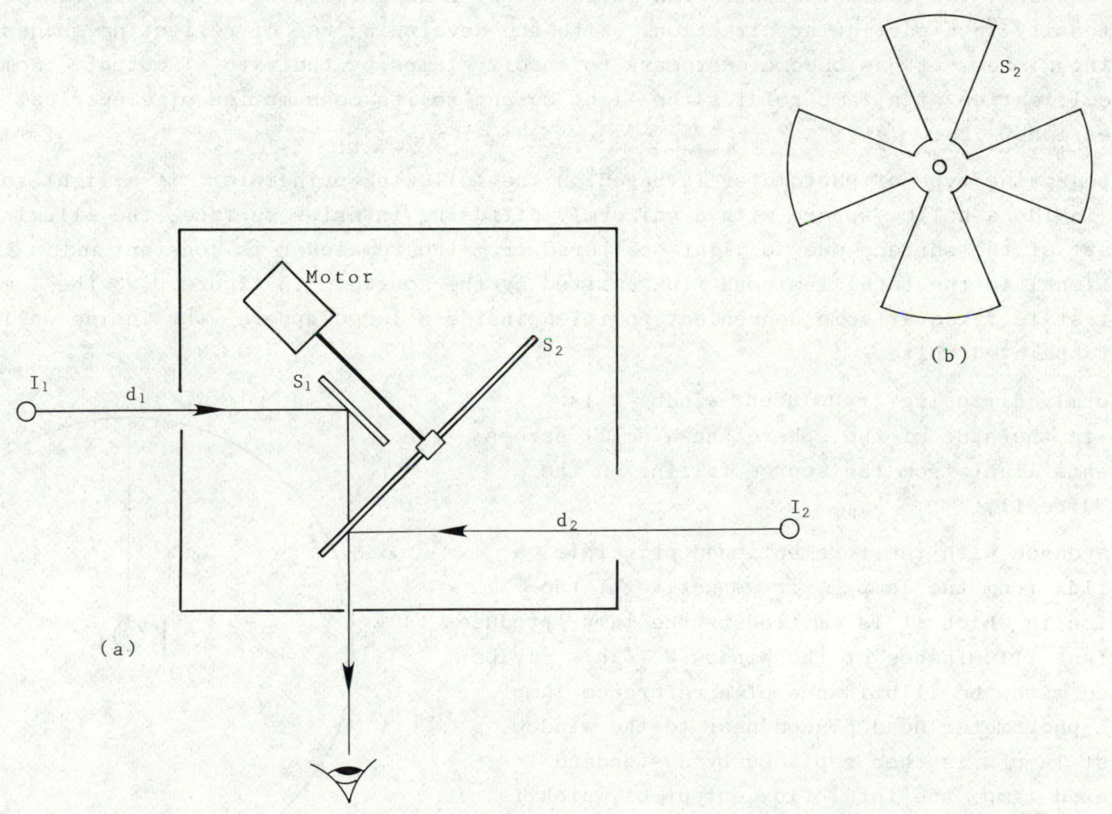

Fig. 9.22 The flicker photometer.

A fixed white screen S_1 is illuminated at 45^0 from the source I_1. A second white screen S_2, having the shape shown in figure 9.22(b), can be rotated by a motor, and is illuminated at 45^0 by light from the source I_2. When S_2 is rotating, the eye, placed as shown, sees first light from S_1 and then light reflected from S_2 in succession.

As described, adjustment of the distances d_1 and d_2, and also the speed of rotation of S_2, reduce the flicker effect to a minimum. In order that satisfactory results may be obtained with this type of photometer, it is necessary that:

(a) The two parts of the field should be visible for equal lengths of time during each rotation of the screen S_2. Thus, the area of the surface "removed" from S_2 must be the same as the area remaining, and the rotation speed should be capable of being controlled at a suitable constant value.

(b) The field of view should be quite small to ensure that the central part of the retina is used.

(c) The illuminances of the two parts of the field should not be too low.

9.11.2.6 The Integrating Photometer

In order to measure the mean spherical intensity (section 9.6.1) of a source, the total luminous flux emitted by the source is determined. This is done using an *Integrating Photometer*. Indeed, the total luminous flux given out by a lamp is often of greater concern than the intensity in a particular direction. With the developing use of reflecting shades and diffusing globes, it has become customary to specify lamps by their total output. Sometimes the specification of a lamp relates the light output to its consumption of electrical power (see section 9.12).

The integrating type of photometer is based on the following principle: if a light source is placed inside a hollow sphere with a uniformly diffusing interior surface, the illuminance of each part of the surface due to light scattered from the remainder is constant and is directly proportional to the total luminous flux emitted by the source. In figure 9.23 the lamp L under test is fixed at some convenient position inside a large sphere, the inside wall of which is painted white.

A uniformly diffusing translucent window W is fitted in the side of the sphere and a small screen S prevents light from the source falling on the window directly.

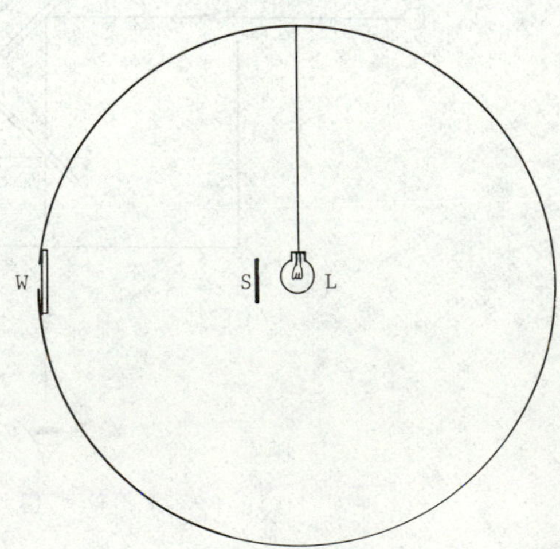

Fig. 9.23 The integrating photometer.

In accordance with the forementioned principle, a given flux from the lamp L, irrespective of the direction in which it is emitted by the lamp, produces a constant illuminance at the window W. This may be compared with the illuminance of a reference lamp using a photometer head placed near to the window. The test lamp L is then replaced by a standard calibrated lamp, the total flux output of which is known, and the illuminance of the window W again compared with the reference lamp.

If the test lamp L produces an illuminance at the window say 2.5 times that produced by the standard calibrated lamp, then the total emitted flux is 2.5 times that emitted by the standard lamp.

9.11.3 NON-VISUAL (PHYSICAL) PHOTOMETRY

Non-visual photometers measure directly the illuminance falling on them. Only one type of non-visual device will be considered in which the action of light on the top surface causes a flow of electrons to occur. This constitutes an electric current which may be indicated by a milliammeter, and is an example of the photoelectric effect. It should be clear that if this effect is to serve our purpose the magnitude of the effect, measured in terms of the electron current which is produced, should be dependent on the colour of the light in much the same way as the visual response of the eye, with a maximum sensitivity at about 555 nm, and no effect at all with the infrared and ultraviolet radiations. In fact, photoelectric devices do not respond like this unless they are fitted with special absorbing filters.

9.11.3.1 *The Photovoltaic Cell*

Most non-visual illuminance photometers use photovoltaic (barrier layer) cells. A typical selenium cell is depicted in figure 9.24 .

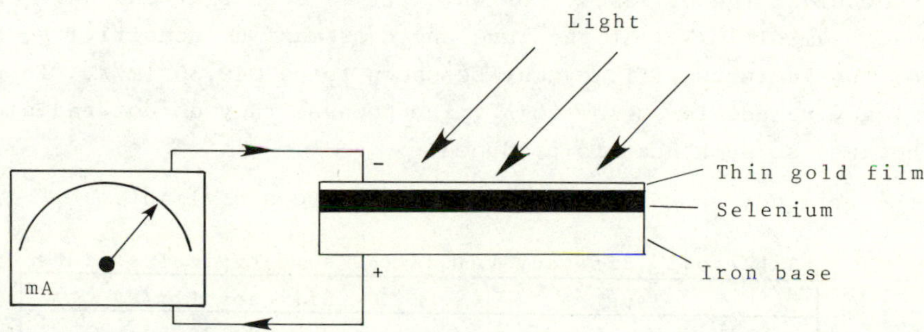

Fig. 9.24 The photovoltaic cell.

A thin layer of the metal selenium (this material is used in several semiconductor devices) is deposited on the iron base plate. The upper surface of the selenium is coated with a very thin film of a metal such as gold. The coating is so thin that it is transparent. When illuminated as shown, electrons are released by the selenium and these are collected by the thin gold coating. A voltage is set up between the top and the base and, consequently, there is a flow of current through the milliammeter. The magnitude of this current is dependent on the illuminance of the selenium surface, although these will not be proportional.

This type of cell forms the basis of a photographic exposure meter which does not require a battery for its operation, and which responds quickly to the conditions. The cell may also be used, instead of a visual photometer, with an integrating photometer. In such cases the milliammeter reading may be calibrated to indicate the total luminous flux radiated by a lamp.

The photovoltaic effect is also used in the solar cell, a device which produces electricity from sunlight. Solar batteries are now under considerable development as alternative sources of energy, and have been used as power sources in satellites and manned space vehicles.

9.12 LUMINOUS EFFICACY OF A SOURCE

As was described in section 9.1, the visual response of the eye differs in different parts of the visible spectrum. That is to say, that for a given amount of radiated power, a red source or a blue source would give less luminous flux than a yellow-green source. From a photometric point of view this means that 1 watt of power corresponds to various numbers of lumens of flux radiated at different wavelengths. In addition, some sources of light waste a large amount of emitted energy as heat, in the form of infrared radiation. It is therefore important to know how effective a light source may be in producing useful luminous flux.

In describing the effectiveness of a source we must be careful to distinguish between the *luminous efficiency* and the *luminous efficacy* of the source. The former is defined by the ratio

$$\text{luminous efficiency} = \frac{\text{luminous flux emitted}}{\text{total flux radiated}} \quad \frac{\text{lumens}}{\text{watts}}$$

and is an indication of the fraction of the total flux radiated which is actually visible. In comparison to this, we can define the luminous efficacy (efficacy = power to produce an effect) by the ratio

$$\text{luminous efficacy} = \frac{\text{luminous flux emitted by lamp}}{\text{power consumed}} \quad \frac{\text{lumens}}{\text{watts}} .$$

We shall briefly consider the efficacy, for which it is seen that the unit of measurement is lumens per watt, symbol lm/W. At the wavelength of maximum sensitivity, 555 nm, the maximum value for the luminous efficacy may be shown to be 649.35 lm/W. In practice, light sources have efficacy values far below this value because they do not radiate all their light at 555 nm, and because so much heat is produced.

Table 9.5 list the efficacy ranges for some typical sources of light.

Table 9.5 Efficacy values for some typical sources

Type	Efficacy (lm/W)
Tungsten filament	8 - 20
Tungsten filament - halogen	17 - 25
Tubular fluorescent	25 - 35
High pressure sodium	75 - 100
Low pressure sodium	115 - 150

As can be seen, sodium lamps are quite efficient as most of the light is emitted in the yellow region, near to the wavelength of maximum visual sensation.

Worked Example

A football pitch 110 m by 87.5 m is illuminated for evening matches by equal banks of 1 000W lamps supported on 16 towers, which are located around the ground to provide approximately uniform illuminance on the pitch. Assuming 35% of the total light emitted reaches the playing area and that an illuminance of 800 lm/m^2 is necessary for TV purposes, calculate the number of lamps on each tower. The luminous efficacy of each lamp may be taken as 25 lm/W.

Let N = number of lamps. The flux emitted by one lamp = 1 000 W × 25 lm/W = 25 000 lm. Therefore, the total flux emitted = 25 000N lm, and the flux reaching the playing area is 35%

of this; that is, $0.35 \times 25\,000\,N = 8\,750\,N$ lm.

Now, $E = 800\ \text{lm/m}^2$; using $E = \dfrac{\text{flux}}{\text{area}}$, $800 = \dfrac{8\,750\,N}{110 \times 87.5}$, giving $N = 880$ lamps.

Since there are 16 towers, each tower will carry $\dfrac{880}{16} = 55$ lamps.

Worked Example

A point source of intensity 40 cd is placed on the axis and 20 cm from a +10 D lens of aperture 4 cm. Find the illuminance on a screen placed 10 cm from the lens, neglecting reflection and absorption losses. What will be the illuminance on the same screen if the aperture of the lens is reduced to 3 cm?

First let us determine the image position B'. $\ell = -20\ \text{cm} = -0.20$ m, so $L = \dfrac{1}{\ell} = \dfrac{1}{-0.20} = -5$ D.
Then $L' = L + F = -5 + (+10) = +5$ D, and $\ell' = \dfrac{1}{L'} = \dfrac{1}{+5}$ m $= +20$ cm.

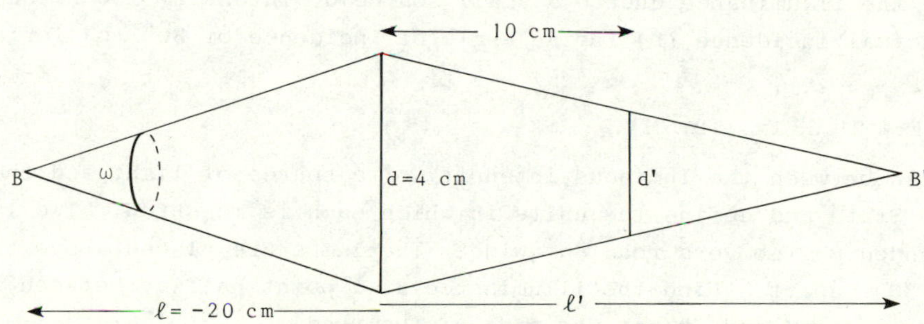

Now, we must find the total flux emitted by the source into the solid angle ω subtended by the lens aperture, where $\omega = \dfrac{\pi \times 2^2}{20^2} = 0.01\pi$ sr. Since the intensity of the source is $I = 40$ cd, then the flux emitted into 1 steradian = 40 lm. Hence, the flux emitted into 0.01π steradian is $40 \times 0.01\pi$ lm.

By similar triangles, $d' = \dfrac{4}{20} \times 10 = 2$ cm. Therefore, the area of the screen illuminated is $\pi \times \left(\dfrac{d'}{2}\right)^2 = \pi \times (1)^2 = \pi\ \text{cm}^2 = \pi \times 10^{-4}\ \text{m}^2$. Hence,

$$\text{the illuminance } E = \frac{\Phi}{A} = \frac{40 \times 0.01\pi}{\pi \times 10^{-4}} = 4\,000\ \text{lx}.$$

If the aperture is reduced to 3 cm the source is emitting flux into $0.005\,625\pi$ sr, giving $40 \times 0.005\,625\pi$ lm. The diameter d' is now 3/2 cm, and the area of the screen illuminated is $\pi \times \left(\dfrac{d'}{2}\right)^2 = \pi \times (0.75 \times 10^{-2})^2 = 5.625 \times 10^{-5}\pi\ \text{m}^2$. Hence, the illuminance is

$$E = \frac{\Phi}{A} = \frac{40 \times 0.005\,625\pi}{5.625 \times 10^{-5}\pi} = 4\,000\ \text{lx}.$$

This case should be compared carefully with the illuminance in the image plane, where reducing the diameter of the lens reduces the illuminance of the image, see equation (9.13).

EXERCISES

1 Define the lumen. A 75 W lamp is rated as producing 15 lumens per watt. Determine the intensity of the lamp in candela, and its rating in watts per candela.

Ans. 89.5 cd, 0.84 W/cd.

2 Calculate the intensity of a light source which emits 6 500 lumens of flux in directions below the horizontal and no flux in any direction above the horizontal.

Ans. 1 035 cd.

2 How does the illuminance of a surface depend upon (i) its distance from the light source and (ii) the angle of incidence of the light on the surface. State the precise conditions under which your statements are valid.

4 The luminous flux incident on the condenser lens of a projector is 12 000 lumens and the average illuminance on the screen 5 m square is 50 lumen/m^2. Determine the fraction of the incident light transmitted to the screen by the optical system.

Ans. 0.104 .

5 Calculate the illuminance due to a small source of intensity 100 cd on a screen 2 m away (a) for normal incidence (b) for an angle of incidence of 30^0 (c) for an angle of incidence of 60^0.

Ans. 25 lx, 21.65 lx, 12.5 lx.

6 Distinguish between the luminous intensity of a source of light and the illuminance of a surface. State and define the units in which each is measured. Two lamps each of 500 cd are suspended 8 m above a road 6 m wide. The lamps are placed above the centre line of the road 30 m apart. Find the illuminance at a point halfway between them (a) in the centre of the road and (b) at the side of the road.

Ans. 1.63 lx, 1.56 lx.

7 Define the unit solid angle, luminous flux, illuminance. Two lamps A and B are fixed 100 cm apart and 120 cm above a working surface. A lightmeter lies on the surface vertically below A. When lamp A alone is switched on the meter reads 52.08 lx. When both lamps are on the meter reads 67.82 lx. Calculate the intensities of A and B.

Ans. A is 75 cd, B is 50 cd.

8 A photometer bench is 2 m long. At one end is placed a source of intensity 16 cd whilst at the other end is placed a source of intensity 25 cd. Determine the position of a screen between the two sources so that each side is equally illuminated.

Ans. 88.9 cm from the source of 16 cd.

9 Lamp A of intensity 50 cd placed 60 cm on one side of a screen matches an 80 cd lamp B placed on the other side. Find the distance of B from the screen. If another 50 cd lamp C is placed 80 cm from the screen on the same side as A, where must B be placed to match A and C together?

Ans. 75.9 cm, 60.7 cm.

10 Two sources of light are arranged to produce equal illuminances on opposite sides of a photometer screen. One source at 50 cm distance has to be moved 5 cm nearer to the screen to restore the balance when a sheet of glass is interposed between it and the screen.

What is the percentage of light transmitted by the glass? How much nearer would this source have to be moved if a second similar sheet of glass were introduced?

Ans. 81%, 4.5 cm.

11 On one side of a photometer screen is placed a 60 cd lamp at a distance of 90 cm. On the opposite side is a 100 cd lamp at a distance of 60 cm. How many identical sheets of glass, each of transmittance 0.85 should be placed between the 100 cd lamp and the screen to give as good a photometric balance as possible?

Ans. 8 .

12 Assuming the maximum value for the luminous efficacy of a source is 640 lm/W, calculate the efficacy of a 50 cd, 60 W source.

Ans. 1.64% .

13 A 200 W compact source hangs vertically over the centre of a bench and 1.75 m above it. If the bench is 1.2 m wide and 2.4 m long determine the illuminance at the centre of the bench and at a corner. The lamp has a luminous rating of 5π lumens/watt and may be assumed to radiate uniformly in all directions.

Ans. 81.6 lx, 40.8 lx .

14 A small lamp of intensity 40 cd is placed 60 cm from a screen. A plane mirror which reflects 70% of the light incident on it is placed 15 cm behind the lamp, parallel to the screen. Find the illuminance of the screen.

Ans. 145.7 lx.

15 A point source of light is placed 5 cm from a large plane mirror of reflectance 0.80 and a light meter, which is facing the source, is moved along the line perpendicular to the mirror and passing through the source. What is the ratio of the illuminances recorded by the meter when it is 15 cm and 10 cm respectively from the mirror?

Ans. 1:3.63 .

16 At 2 mm thickness, a certain tinted glass has a transmittance for a specified wavelength of 0.44 . If the index of the glass is 1.530 determine the transmittance at 1 mm and 3 mm thicknesses.

Ans. 0.63, and 0.31 .

17 The illuminance of a white surface is 55 lux. Assuming the surface is uniformly diffusing and the reflectance is 0.85, find the luminance of the surface in the direction 60^0 to the normal.

Ans. 23.4 lm/m^2.

18 State Lambert's Law of Emission. Explain what is meant by (a) a uniformly diffusing surface (b) a perfect diffusing surface. The luminance of a flat uniformly diffusing surface of area 2.5 mm^2 is 20 cd/mm^2. Determine the luminous intensity of the surface along the normal and along the direction inclined at 50^0 to the normal.

Ans. 50 cd, 32.14 cd.

19 A point source of intensity 48 cd is placed at the principal focus of a concave mirror which subtends 1 steradian at the source. If the reflectance of the mirror is 90%, find the amount of luminous flux in the reflected pencil.

Ans. 43.2 lm.

20 A lamp of intensity 100 cd is used to operate a photocell which requires a minimum luminous flux of 0.1 lm to actuate a relay and switch circuit. The sensitive receiving area of the cell is rectangular measuring 5 cm × 3 cm. Calculate the greatest distance from the cell that the lamp can be placed to operate it.
If a thin sheet of tinted glass of transmittance 0.6 is placed between the cell and the lamp, what is now the maximum distance at which the lamp can be placed?

Ans. 122.5 cm, 94.9 cm.

21 A parallel pencil of light is obtained by placing a small source of light of intensity 20 cd at the first principal focus of a convex lens of focal length 20 cm and diameter 10 cm. The light falls obliquely on a screen at an angle of incidence of 30^0. Find the illuminance of the screen if the lens transmits 80% of the light incident upon it.

Ans. 346.4 lm/m^2.

22 A point source of intensity 5 cd is placed at a perpendicular distance of 100 cm from a screen, and a converging lens of focal length 30 cm is interposed coaxially with the normal to the screen through the source at a distance of 60 cm from the source. Calculate the illuminance at the screen opposite the source (a) without the lens (b) with the lens in position. Assume no losses of light on transmission.

Ans. 5 lx, 125 lx.

23 The following readings were obtained during an experiment in which it was required to find how the intensity I cd of a lamp varies with the electrical power P watts supplied.

Power P (watts)	14.0	22.6	28.4	36.0
Distance of comparison source (metres)	0.635	0.330	0.220	0.170

Assuming that the relationship between I and P is $I = aP^n$, where a and n are constants, find the value of n.

Ans. 2.9 very nearly.

24 A lamp L_1 was mounted on a photometer bench, so that it could be rotated about a vertical axis, while keeping its axis of symmetry horizontal. θ is the angle between the bench axis and the axis of symmetry of the lamp. As L_1 was rotated a reference lamp L_2 was adjusted to different distances d_2 to give photometric balance at the photometer screen. From the following values construct a polar diagram for the lamp L_1. Suggest reasons why the curve is not regular.

θ (degrees)	10	20	50	70	90	110	130	150	180
d_2 (metres)	2.12	1.25	1.10	1.00	0.90	0.75	0.85	1.00	1.00

10 LIGHT WAVES

10.0 INTRODUCTION

Thus far we have considered light from a macroscopic, or large scale, point of view where light was assumed to travel in straight lines within any homogeneous medium. That view was derived from observations such as shadows of straight edges cast by sunlight, figure 10.1, or by beams of sunlight emerging from clouds.

Fig. 10.1 Apparent rectilinear propagation of light. Shadow cast by parallel light (sunlight).

This observation of rectilinear propagation of light led to the geometrical treatment of reflection and refraction. The success of geometrical optics is self-evident in the preceding chapters. However, there is a great number of observed phenomena which cannot be explained using rectilinear propagation and a geometrical optics model. One such common example is the coloured patterns evident in patches of oil on a wet road. Another is the pattern observed in a wedge-shaped film of detergent, figure 10.2 .

Fig. 10.2 The pattern observed in a wedge-shaped soap film.

In order to explain these phenomena, and many more besides, it is necessary to consider light to have a wave-like behaviour. This wave-like behaviour is only exhibited in special circumstances, some of which are important to optometrists and opticians. For example, the development of polarising sunglasses and anti-reflection coatings on lenses depend on the wave theory of light.

It is with these applications and many others in mind that we embark upon the study of physical optics. The term physical optics is reserved for the explanation of various phenomena using the wave theory of light. This chapter is devoted to some general aspects of waves and wave motion prior to dealing with the phenomena arising from the interaction of waves with waves and waves with matter in succeeding chapters.

10.1 ONE-DIMENSIONAL WAVES

10.1.1 *The General Form of a One-dimensional Wave*

Figure 10.3 shows a disturbance moving along a rope.

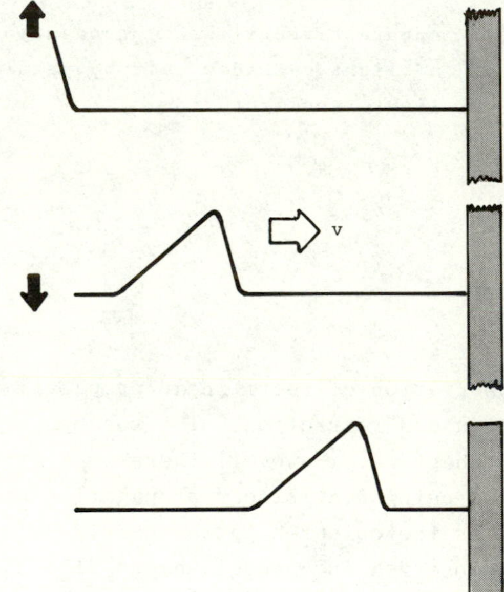

Fig. 10.3 A wave on a rope caused by a vertical displacement.

The disturbance is called a wave. It is the vertical displacement of the rope which moves to the right along the rope. The exact shape of the wave profile is not important at the moment, but what we want to do is to find a general expression to describe a moving wave. Suppose we consider a wave profile which does not change its shape when travelling to the right, as in figure 10.3 . Consider figure 10.4 which shows a wave profile at a time $t = 0$, say, and the same profile at some later time, t. The shape of the wave can be represented as a function of the distance from the origin, x, and the time that has elapsed, t. That is,

$$\psi = f(x,t) \tag{10.1}$$

The figure shows a point on the wave at a distance x from the fixed coordinate system, S_1 .

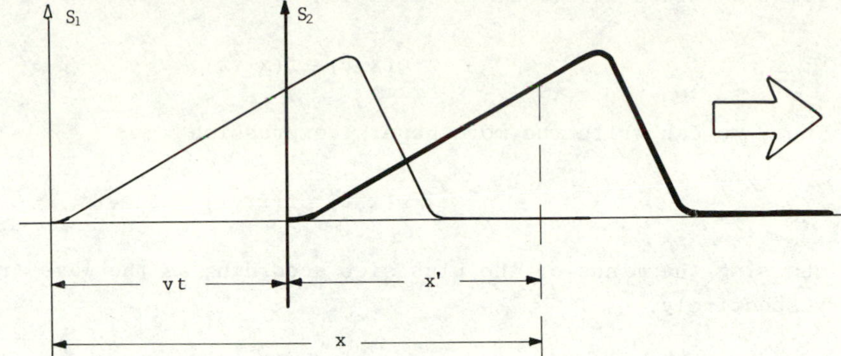

Fig. 10.4 A wave in a moving coordinate system, S_2

Suppose now the wave is travelling at a speed v and we introduce a second coordinate system, S_2, moving with the same speed in the same direction as the wave. Relative to the second system the wave profile is constant so that

$$\psi = f(x') \tag{10.2}$$

where x' is the coordinate in the second system, S_2.
Now, since the second system is moving at the speed of the wave it will have travelled a distance vt in a time t. From figure 10.4 we see that

$$x' = x - vt \tag{10.3}$$

Substituting for x' from equation (10.3) into equation (10.2) we have

$$\psi(x,t) = f(x') = f(x-vt) \tag{10.4}.$$

This represents the general form of a one-dimensional wave function travelling in the positive x-direction.

10.1.2 The Wave Profile after a further time Δt

We have seen that the general form of the travelling wave is

$$\psi(x,t) = f(x-vt) \tag{10.4}$$

Suppose a further time interval Δt elapses during which time x will have increased by v.Δt. If we insert these increments in the wave equation we have

$$\psi(x+v.\Delta t,\ t+\Delta t) = f((x+v.\Delta t) - v(t+\Delta t)) = f(x-vt)$$

which shows that $\psi(x,t) = \psi(x+v.\Delta t,\ t+\Delta t)$, and this indicates that the wave profile remains the same as it travels.

10.1.3 Wave travelling to the left (negative x-direction)

Were we to repeat the above analysis showing the wave travelling to the left instead of to the right we would arrive at the equation

$$\psi(x,t) = f(x+vt) \tag{10.5}$$

Thus, we can write the more general expression

$$\psi(x,t) = f(x\mp vt) \tag{10.6}$$

choosing the minus or the plus sign according as the wave travels to the right or left, respectively.

10.2 SINUSOIDAL WAVES

10.2.1 *Introduction*

The simplest type of wave is that where the function f is a sine or cosine. In fact, it can be shown that many complicated waveforms can be represented as the sum of various sine and cosine waves. We therefore wish to look at some aspects of such waves. It should be mentioned that these waves are also known as simple harmonic, harmonic, sinusoidal, or cosinusoidal waves.

10.2.2 *Sine waves*

Figure 10.5 shows a sinusoidal wave.

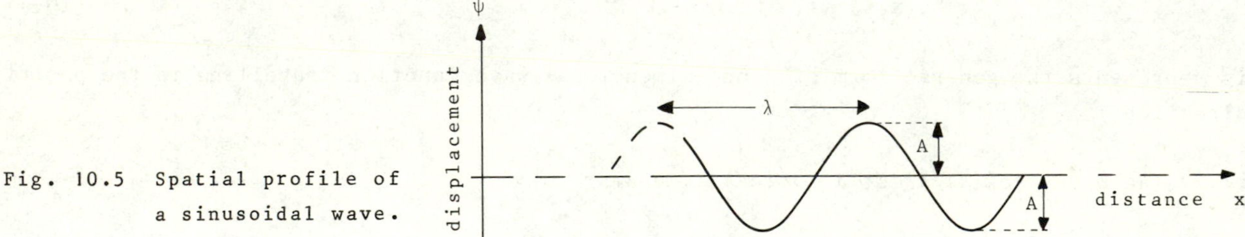

Fig. 10.5 Spatial profile of a sinusoidal wave.

If we take the function $\psi(x,t)$ for time t=0 say, the result is a 'snapshot' of how ψ varies with the distance, x. The function is written

$$\psi(x,0) = A \sin kx \tag{10.7}$$

where k is a positive constant known as the *propagation number* and kx is in radians. Since the sine function varies from -1 to +1 the function $A \sin kx$ varies from -A to +A. A is a positive constant known as the *amplitude* of the wave, see figure 10.5 .

So far the waveform is static. In order to transform it into a moving wave we replace x with x-vt, whence we have

$$\psi(x,t) = A \sin k(x - vt) \tag{10.8}$$

Keeping either t or x constant produces a sine wave varying with distance or in time, respectively. Such waves are periodic: that is, they repeat their profile at regular intervals. When t is held constant the repeat distance is called the *wavelength* or *spatial period* and is denoted by λ. Thus, for $x = x_1$ and $x = x_1 + \lambda$ the function $\psi(x,t)$ will take the same value. For a sine function this is equivalent to adding 2π to the argument. We are saying that, in general,

$$\sin\theta = \sin(\theta \pm 2\pi) \tag{10.9}$$

and applying this to equation (10.8) with x_1 increasing to $x_1 + \lambda$, we have

$$A\sin k(x_1 - vt) = A\sin k((x_1 \pm \lambda) - vt) = A\sin(k(x_1 - vt) \pm k\lambda) \tag{10.10}.$$

Thus, by analogy with equation (10.9)

$$k\lambda = 2\pi \quad \text{or} \quad k = \frac{2\pi}{\lambda} \tag{10.11}.$$

If we now hold x constant, at x_0 say, a sine wave varying in time will result. The repeat interval is called the *temporal period*, τ. This is the time taken for a complete wave to pass a fixed point in space, figure 10.6 .

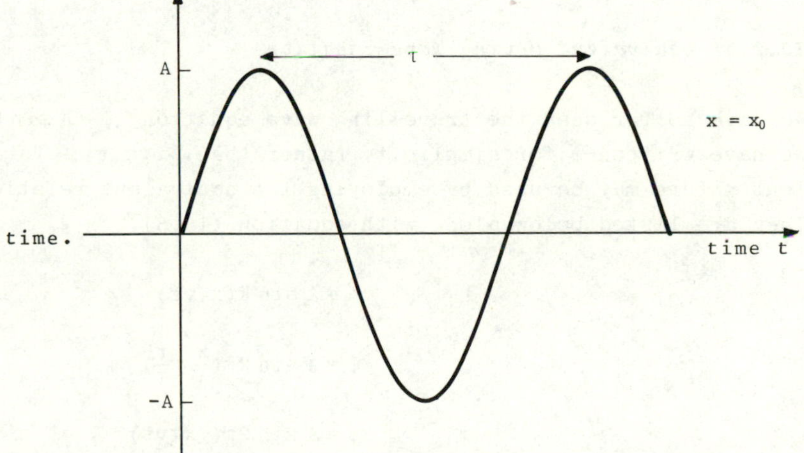

Fig. 10.6 A sine wave varying in time.

Suppose now we consider the value of $\psi(x_0,t)$ with t at two values of t_1 and $t_1+\tau$. Remember τ is the temporal period, the time after which the function takes the same value. Thus,

$$A\sin k(x_0 - vt_1) = A\sin k(x_0 - v(t_1 \pm \tau)) = A\sin(k(x_0 - vt_1) \mp kv\tau)$$

and by analogy with equation (10.9) again we have

$$kv\tau = 2\pi \tag{10.12}.$$

Since $k = 2\pi/\lambda$, we can write $\frac{2\pi}{\lambda}v\tau = 2\pi$ or

$$\tau = \frac{\lambda}{v} \qquad (10.13).$$

Now, τ is the length of time it takes one wavelength to pass a fixed point. Therefore, the number of waves passing the fixed point in unit time is

$$\nu = \frac{1}{\tau} \qquad (10.14)$$

and ν is called the *frequency*, with units of cycles per second or Hertz if τ has units of seconds. Substituting $1/\nu$ for τ, equation (10.13) can be written $\frac{1}{\nu} = \frac{\lambda}{v}$ or, more conventionally

$$v = \nu\lambda \qquad (10.15).$$

The speed v has units of metres per second if the wavelength λ is measured in metres. Two other quantities may be used: they are the *angular frequency*, ω, and the *wave number*, χ. Their relationships with the earlier defined terms are

$$\omega = 2\pi\nu \qquad \text{(radians/s)} \qquad (10.16)$$

$$\text{and} \quad \chi = \frac{1}{\lambda} \qquad (m^{-1}) \qquad (10.17).$$

The wave number is the number of whole wavelengths in one metre.

10.2.3 Equivalent Moving Wave Equations

We have so far seen the travelling wave equation $\psi = A \sin k(x-vt)$. Incidentally, note that we have written ψ for simplicity rather than $\psi(x,t)$. Various other equations of an equivalent nature may be used by employing the equivalent relations developed in section 10.2.2 . They are listed below along with equation (10.8).

$$\psi = A \sin k(x \mp vt) \qquad (10.8)$$

$$\psi = A \sin 2\pi(\frac{x}{\lambda} \mp \frac{t}{\tau}) \qquad (10.18)$$

$$\psi = A \sin 2\pi(\chi x \mp \nu t) \qquad (10.19)$$

$$\psi = A \sin (kx \mp \omega t) \qquad (10.20)$$

$$\psi = A \sin 2\pi\nu(\frac{x}{v} \mp t) \qquad (10.21)$$

It is important to note that all of these waves extend from $x = -\infty$ to $x = +\infty$. Each wave has a single constant wavelength and is said to be *monochromatic*.

10.2.4 Examples

Given the travelling wave function $\psi_1 = 2 \sin 2\pi(0.1x + 6t)$ find the frequency, the wavelength, the temporal period, the amplitude, and the direction of motion.

We simply choose the general form of the equation which matches the form of the given equation. For ψ_1 we can write

$$\psi = A \sin 2\pi(\frac{x}{\lambda} + \frac{t}{\tau}) = \psi_1 = 2 \sin 2\pi(0.1x + 6t) .$$

Then rewriting ψ_1 as $\psi_1 = 2 \sin 2\pi(\frac{x}{10} + \frac{t}{1/6})$, and comparing ψ and ψ_1 we have

$$A \sin 2\pi(\frac{x}{\lambda} + \frac{t}{\tau}) = 2 \sin 2\pi(\frac{x}{10} + \frac{t}{1/6})$$

whence, by identity,

the amplitude $A = 2$ m
the wavelength $\lambda = 10$ m
and the period $\tau = 1/6$ s .

The frequency $\nu = \frac{1}{\tau} = \frac{1}{1/6} = 6$ Hz (Hertz or cycles per second) and the direction of motion is to the left since the sign between the x/λ and t/τ terms is positive.

In this second example you are given the speed of electromagnetic waves in vacuum is 3×10^8 ms^{-1} (metres per second). Find the frequency of yellow-green light of wavelength 555 nm (555×10^{-9} m). Overhead power lines radiate electromagnetic waves at a frequency of 50 Hz. Compare the wavelength with yellow-green light.

Using equation (10.15) $v = \nu\lambda$, we have for the yellow-green light frequency

$$\nu = v/\lambda = (3 \times 10^8)/(555 \times 10^{-9}) \quad ms^{-1}/m$$
$$= 5.41 \times 10^{14} \quad s^{-1} \text{ (Hz)}.$$

The wavelength of the radiation from overhead power lines is given by equation (10.15) rearranged; that is

$$\lambda = \frac{v}{\nu} = \frac{3 \times 10^8}{50} = 0.06 \times 10^8 = 6 \times 10^6 \text{ m}.$$

Comparing wavelengths we have

$$\frac{\text{wavelength from power lines}}{\text{wavelength of yellow-green light}} = \frac{6 \times 10^6}{555 \times 10^{-9}} = 1.08 \times 10^{13} .$$

That is, the wavelength of the radiation from power lines is 10.8 million million times as long as the wavelength of yellow-green light.

10.3 PHASE AND PHASE VELOCITY

10.3.1 *Phase and initial phase*

The argument of the sine function is called the *phase* and is given the symbol ϕ. That is, choosing the form of ψ as $\psi(x,t) = A \sin(kx-\omega t)$ then

$$\phi(x,t) = kx - \omega t \tag{10.22}$$

which is clearly a function of x and t. We can never know the phase of a light wave but we shall find that we can determine the *phase difference* between two waves, and this will be discussed in chapter 11.

Suppose we have two identical sine waves, figure 10.7, propagating in the positive x-direction but one wave leads the other by a constant amount.

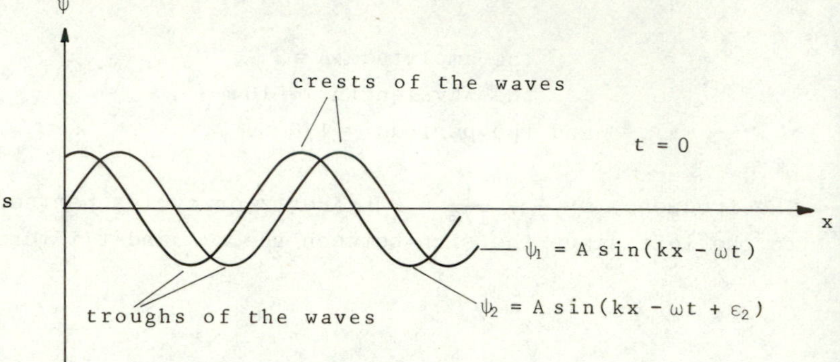

Fig. 10.7 Two identical wave profiles at time t = 0.

The first crest of ψ_1 is seen to be ahead of the first crest of ψ_2. Looking at the phases of ψ_1 and ψ_2 we see that at $x = t = 0$ the phase of ψ_1 is

$$\phi_1(0,0) = k \times 0 - \omega \times 0 = 0$$

and the phase of ψ_2 is

$$\phi_2(0,0) = k \times 0 - \omega \times 0 + \varepsilon_2 = \varepsilon_2 .$$

ε_2 is called the *initial phase* of the function ψ_2. Clearly, the initial phase of ψ_1 is zero. In general, we can write an equation

$$\psi(x,t) = A \sin(kx - \omega t + \varepsilon) \tag{10.23}$$

where ε is the initial phase.

10.3.2 Phase velocity

We have seen that a travelling wave profile moves with a speed $v = \nu\lambda$ (equation (10.15)). This expression can be derived alternatively as follows. Consider the point P on the crest of the wave in figure 10.8 .

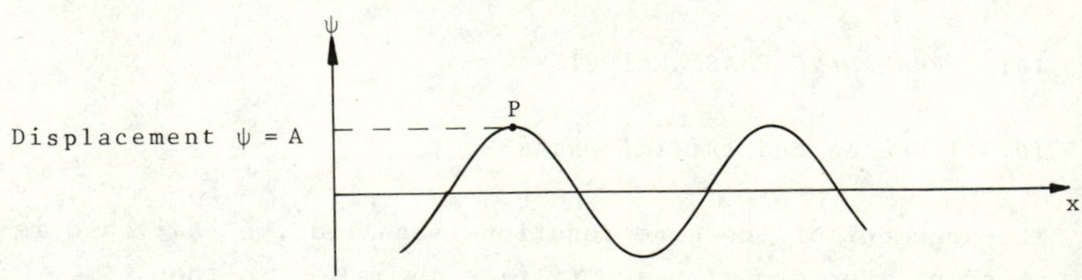

Fig. 10.8 See text for details.

If the function is $\psi = A\sin(kx \mp \omega t + \varepsilon)$ and ψ is constant at $\psi = A$, then we are looking at the progress of the crest marked P. For ψ to be constant the phase must remain constant; that is

$$\phi(x,t) = kx \mp \omega t + \varepsilon = \text{constant} \tag{10.24}$$

Holding x constant we can take the partial derivative of ϕ with respect to t:

$$\left(\frac{\partial\phi}{\partial t}\right)_x = \mp\omega \tag{10.25}$$

Similarly, with t constant

$$\left(\frac{\partial\phi}{\partial x}\right)_t = k \tag{10.26}$$

Using the result from partial differentiation

$$\left(\frac{\partial x}{\partial t}\right)_\phi = \frac{-(\partial\phi/\partial t)_x}{(\partial\phi/\partial x)_t} \tag{10.27}$$

and substituting for $(\partial\phi/\partial t)_x$ and $(\partial\phi/\partial x)_t$ from equations (10.25) and (10.26) we have

$$\left(\frac{\partial x}{\partial t}\right)_\phi = \frac{-(\mp\omega)}{k} = \pm\frac{\omega}{k}\,.$$

Now, $\left(\frac{\partial x}{\partial t}\right)_\phi$ represents the velocity of propagation of the condition of constant phase. That is, in our example, the velocity of the crest of the wave P. So, we can write

$$v = \frac{\pm\omega}{k} \tag{10.28}$$

But $\omega = 2\pi\nu$ and $k = 2\pi/\lambda$, so $v = \pm\frac{\omega}{k} = \pm\frac{2\pi\nu}{2\pi/\lambda} = \pm\nu\lambda$ which is equation (10.15). This is the speed at which the wave profile propagates and is known as the *wave velocity* or the *phase velocity*. When v is positive the wave propagates to the right (increasing x) and when v is negative the wave propagates to the left.

10.4 PROPERTIES OF WAVES

10.4.1 *Wavefronts and rays*

Fig. 10.9(a) Ripple tank photograph of surface waves.

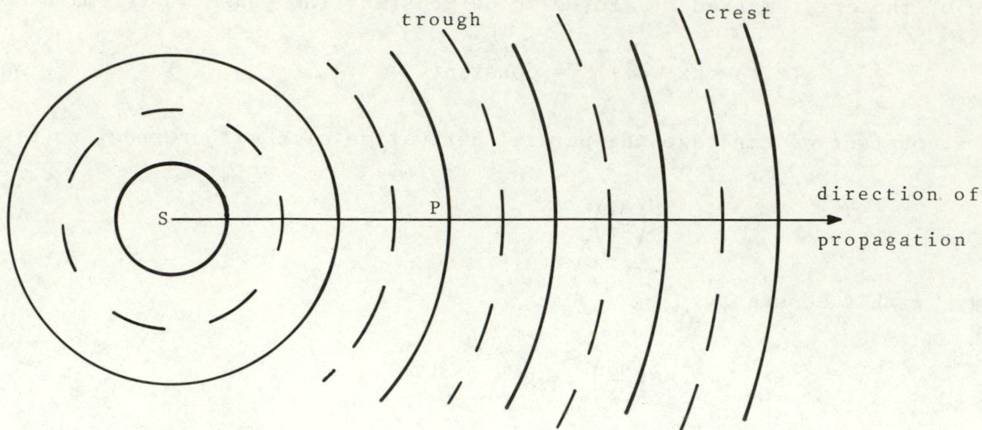

Fig. 10.9(b) Crests and troughs - diagrammatic.

Figure 10.9(b) shows a source S producing circular waves propagating outwards in all directions from S. In the direction of propagation to the right are marked alternate crests and troughs of waves. A line along which all the waves have the same phase, a line marked crest in figure 10.9(b) for example, is called a *wavefront* and the propagation of a wave is regarded as the wavefront advancing at the wave (or phase) velocity.

The straight line in figure 10.9(b) is perpendicular to all wavefronts in the direction of propagation and is called a *ray*. In earlier chapters where we were concerned only with the direction of propagation, rays were drawn almost to the exclusion of wavefronts. Since travelling waves transport energy, a ray can be considered to be the direction in which the wave energy is travelling.

10.4.2 Transverse Waves

Figure 10.9(b) could well represent the waves on the surface of a pool of water. If the source S were a small plunger or dipper oscillating in the vertical meridian it would give rise to waves which would travel outwards from S over the surface of the pool. Imagine now a cork floating in the surface some distance from the source. As the waves reach the cork it will be seen to rise and fall reaching a maximum height +A, the amplitude of the wave motion, above the mean position and a minumum height -A below the mean position. The mean position would be the cork's height when the pool was undisturbed. This motion is illustrated in figure 10.10 where the cork, a black disc in the figure, moves up and down periodically as successive waves sweep by it.

Note that the motion of the cork is transverse to the direction of the waves' motion. Such waves are called, as one might expect, *transverse waves*. Light also has a wave nature although the (electromagnetic) waves in this connection can move through a vacuum. Electromagnetic waves are also transverse waves and the oscillations are disturbances propagating through electric and magnetic fields. In this case the electric field, denoted by E, is perpendicular to the magnetic field, H, and the two fields are perpendicular to the direction of propagation. Figure 10.11 shows the relationship between the electric field disturbance, the magnetic field disturbance, and the direction of propagation.

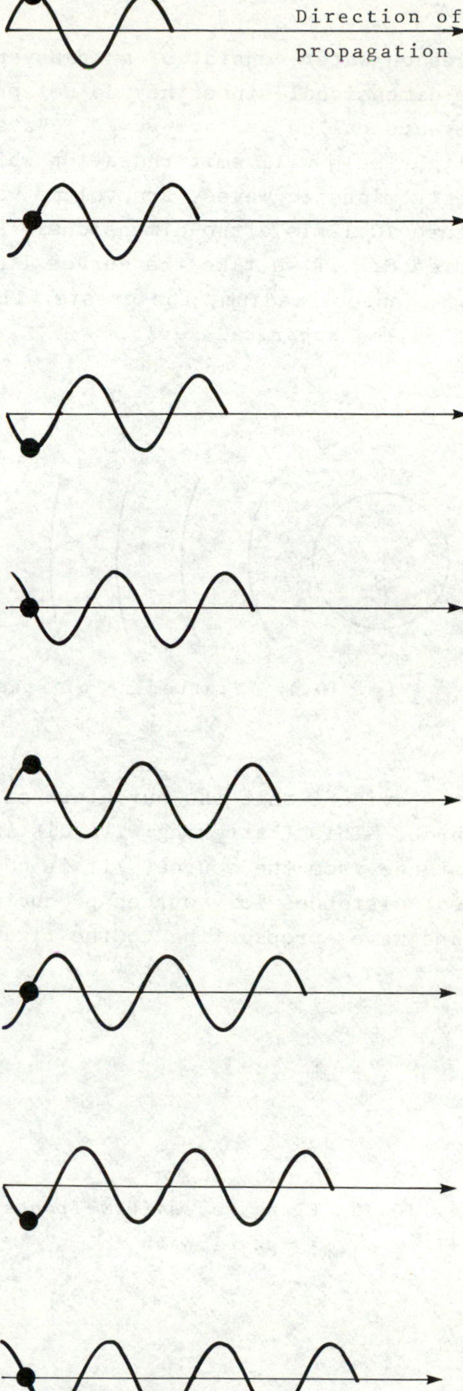

Fig. 10.10 A particle exhibiting transverse motion as waves sweep by.

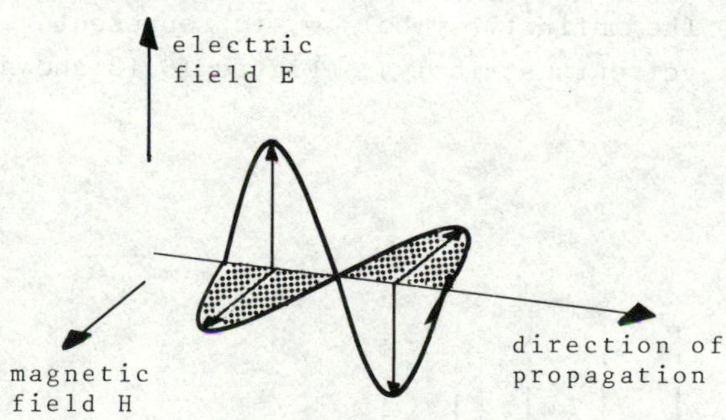

Fig. 10.11 Electromagnetic wave.

The electric field disturbance is the one which will concern us since it can be demonstrated that it has a much greater interaction with matter than the magnetic field disturbance. We shall return to this point in chapter 11 where Wiener's experiment will describe how he was able to come to this conclusion.

10.4.3 Plane and Spherical Waves

Waves on water consist of a transverse wave motion propagating in the surface only. They are two-dimensional since they do not propagate through the bulk of the water. Naturally enough they are called *surface waves*. Waves which propagate in three dimensions are called *volume waves*. A light bulb will emit radiation which propagates in all directions so light waves, which are electromagnetic waves, are volume waves.
Figure 10.12 is a two-dimensional 'slice' through volume waves being emitted from a 'point' source S. If we take the curved lines to be crests of transverse volume waves travelling through a homogeneous medium, the crests will lie on spheres centred on the source, S. Such wavefronts are called *spherical waves*.

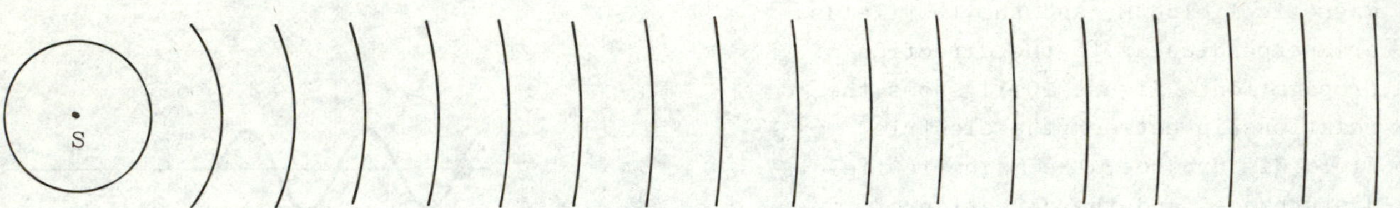

Fig. 10.12 Flattening of spherical waves with distance.

It is evident that the curvature of the wavefronts reduces the further the wavefront is from the source. This flattening with distance results in flat or plane wavefronts at a very great distance from the source. It is conventional to use the infinity symbol, ∞, to represent very great distances for sources producing plane waves (wavefronts strictly). Figure 10.13 shows plane waves propagating to the right.

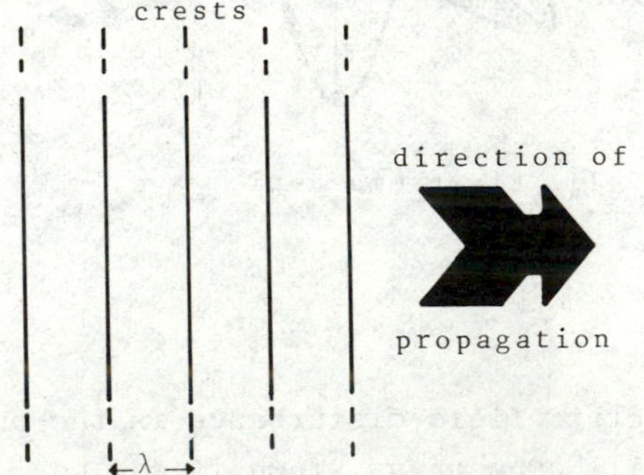

Fig. 10.13 Plane waves (wavefronts); λ = wavelength.

Transverse volume waves have a very important property associated with the direction of the transverse motion of the oscillation. Recall that the oscillation arising at the source of surface waves must be perpendicular to the surface and to the direction of propagation. With volume waves the only constraint is that the oscillation should be perpendicular to the direction of propagation. In particular, a beam of light will in general contain waves which vibrate in random directions. It is possible to pass such a beam through a filter which transmits only light waves vibrating in one meridian. The emergent light vibrates in one plane and is called *plane polarised light*. A beam containing randomly orientated vibrations is said to be unpolarised. Figure 10.14 illustrates plane polarised and unpolarised waves.

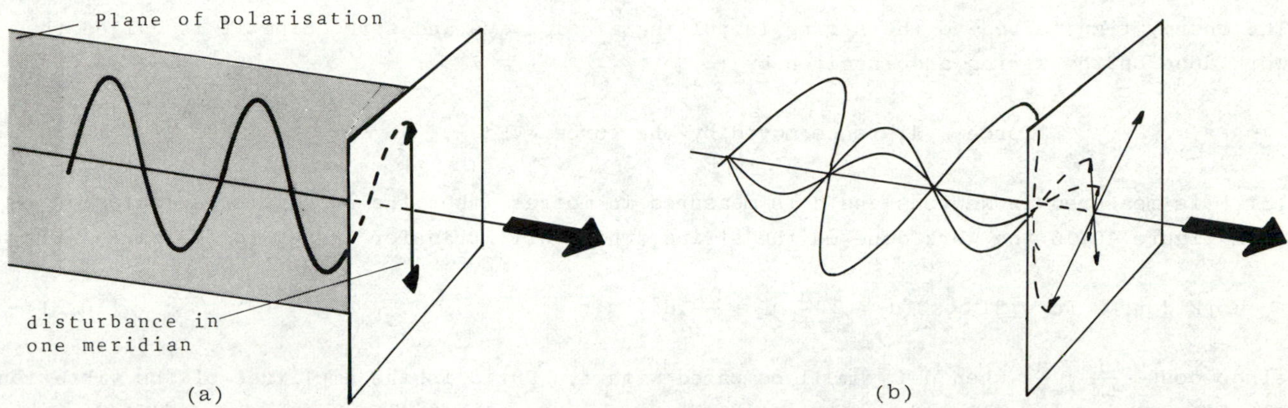

Fig. 10.14 (a) Plane polarised wave. (b) Unpolarised beam of waves.

10.5 AMPLITUDE AND INTENSITY

Travelling waves transmit energy. For example, when light falls on an absorbing medium there is a transfer of energy from the wave to the medium. A simple example of which we are all aware is the absorption of radiated heat, infrared electromagnetic waves, from a central heating radiator. We detect a transfer of energy! Now, there is a simple relationship between the energy content of a wave and its amplitude: *the energy content* $\propto$ *the amplitude*^{2}.

10.5.1 Relationship between the energy content and the wave

Consider a familiar type of vibration; the waves set up on a guitar string when it is plucked. Figure 10.15 shows a string of length 2ℓ.

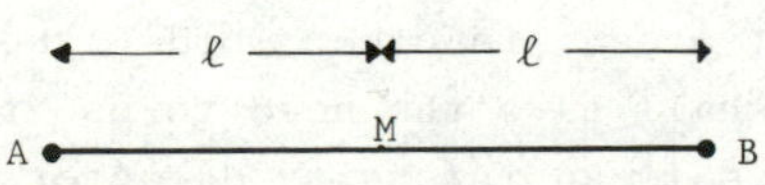

Fig. 10.15 A taut string secured at A and B.

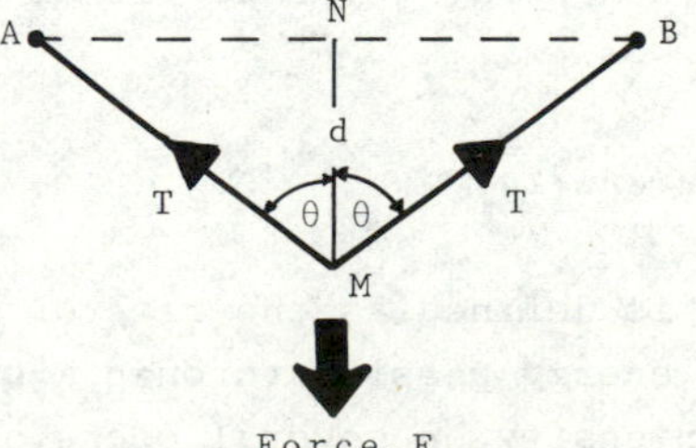

Fig. 10.16 A taut string pulled sideways by a force, F. T is the tension in the string.

If we pull out the string at the mid-point M at right angles to its length, we have the situation shown in figure 10.16 . If we further assume the string is pulled sideways a small distance d then the force F and the tension T will be very nearly constant. Now the force F is countered by the components of the forces in the two halves of the string acting along the direction MN. That is,

$$F = 2T\cos\theta \qquad (10.29).$$

The energy transferred to the string in pulling it sideways and stretching it is called the work done on the string and is given by

$$\text{force} \times \text{distance moved by the force} = Fd \qquad (10.30).$$

(If F is measured in Newtons and d is measured in metres, then the energy is in Joules). From figure 10.16 the work done on the string, which will transfer energy to the wave, will be

$$\text{work done} = Fd = 2T\cos\theta . d \simeq 2T\frac{d}{\ell}.d = \frac{2T}{\ell}.d^2 \qquad (10.31)$$

since $\cos\theta = \frac{d}{AM} \simeq \frac{d}{\ell}$ when d is small compared with ℓ. But d is the amplitude of the vibration when the string is released, so the *energy transferred is proportional to the amplitude squared*, d^2.

10.5.2 Energy content is proportional to amplitude squared - an alternative approach

If we imagine the behaviour of a particle initially at rest floating on the surface of a pool and it is subjected to the wave $\psi(0,t) = A\sin\omega t$, we can observe its oscillation at the point $x = 0$. It will oscillate from the mean (rest) position to +A, back through the zero (rest) position to -A, and back to zero again. Its maximum speed, v_{max}, will occur as it passes through the mean position when all its energy will be kinetic. If the mass of the particle is m then its kinetic energy at the mean position is

$$\tfrac{1}{2}m(v_{max})^2 \qquad (10.32).$$

Its velocity is $\frac{d\psi}{dt} = -\omega A\cos\omega t$ ....(10.33) and, since $\sin\omega t = 0$ when the particle passes through the mean position, $\cos\omega t$ will be ±1. Thus, the energy possessed by the particle will be

$$\tfrac{1}{2}m(v_{max})^2 = \tfrac{1}{2}m(-\omega A)^2 = \tfrac{1}{2}m\omega^2A^2 .$$

That is, the energy is proportional to the amplitude squared, A^2.

10.5.3 Intensity

Intensity is defined as the rate of flow of energy through unit area. In other words, the amount of energy passing through a unit area each second. We shall have the need to use the concept intensity on several occasions, especially since the eye is an intensity detector.

10.6 WAVELENGTH AND FREQUENCY

10.6.1 Wavelength and Refractive Index

Equation (10.15) gave us the fundamental relationship between frequency ν and wavelength λ; $v = \nu\lambda$. When a light wave propagating through one medium is incident on another and passes into that second medium a change of wavelength occurs. In all the cases we shall consider the frequency of the electromagnetic disturbance remains constant. Using this fact we can employ equation (10.15) to find the relationship between the wavelength in vacuum and the wavelength on passing into some other medium.

Let c be the speed of light in vacuum and v the speed of light in a medium of refractive index n. Further, let us denote the wavelength in vacuum by λ. Since for a particular vacuum wavelength a particular refractive index exists for a given medium, defined by $n = \frac{c}{v}$, we can write

$$c = \nu\lambda \quad ....(10.34) \quad \text{and} \quad v = \nu\lambda_m \quad ....(10.35),$$

where λ_m is the wavelength in the medium. Since ν is constant, dividing the equations gives

$$\frac{\lambda_m}{\lambda} = \frac{\nu v}{\nu c} = \frac{v}{c} = \frac{1}{n}, \text{ whence } \lambda_m = \frac{\lambda}{n} \qquad (10.36)$$

That is, the wavelength in a medium of refractive index n (quoted for the wavelength being considered) is the wavelength in vacuum divided by the refractive index.

10.6.2 The Electromagnetic Spectrum

Wavelengths quoted are usually vacuum wavelengths. They form a continuous band from the very long radio waves at one extreme to the gamma rays at the other. The whole range of wavelengths is called the *electromagnetic spectrum*, figure 10.17.

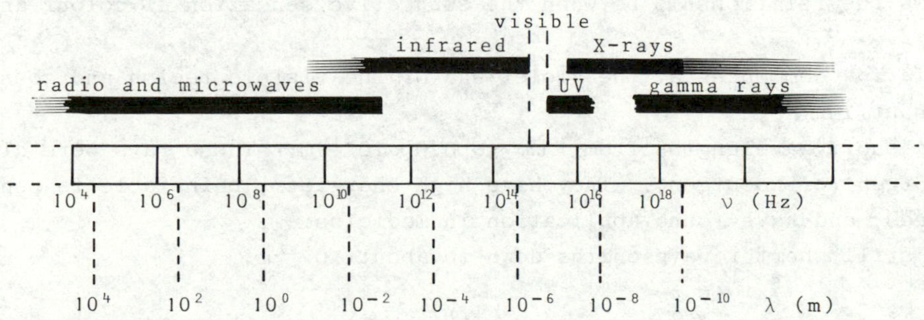

Fig. 10.17 The electromagnetic spectrum.

Radio waves were first generated by H.R. Hertz in 1887. They range from a few Hz up to about 10^9 Hz. The sources include radio, TV, and power lines.

Microwaves have frequencies from 10^9 Hz to about 3×10^{11} Hz which correspond to wavelengths from ⅓ m to about 1 mm. This region is used in communications, radar, radio astronomy, and more recently in microwave ovens.

Infrared radiation, as the name implies, lies in that part of the electromagnetic spectrum just beyond the visible spectrum. The wavelengths of IR (short for infrared) range from 1 mm to about 760 nm.

Visible light is a very narrow band extending from 760 nm to 390 nm. It is generally regarded as the band which can be detected by the human eye. The term light is sometimes applied to those regions immediately adjacent to the visible band, that is the infrared and the ultraviolet regions, although we shall more often than not reserve it for the visible region. Different regions of the visible spectrum give rise to the sensation of colour although it must be said that colour resides in the eye-brain system and is not a property of the frequency of light per se. Light is just another type of electromagnetic radiation.

TABLE 10.1

Light		
Colour	Vacuum wavelength (nm)	Frequency $\times 10^{14}$ Hz
Red	760 - 622	3.94 - 4.82
Orange	622 - 597	4.82 - 5.03
Yellow	597 - 577	5.03 - 5.20
Green	577 - 492	5.20 - 6.10
Blue	492 - 455	6.10 - 6.59
Violet	455 - 390	6.59 - 7.69

Table 10.1 shows the relationship between the subjective sensation of colour and wavelength or frequency.

Ultraviolet radiation commences at the violet end of the visible region and goes down to a wavelength of about 1 nm.

X-rays have very short wavelengths from 1 nm to 6.0×10^{-12} m. These rays were discovered in 1895 by W.C. Röntgen (or Roentgen). They have high energies , being able to penetrate soft biological tissues, and have found application in medicine.

Gamma rays have still shorter wavelengths down to about 10^{-25} m.

10.6.3 Vacuum Wavelengths and Air Wavelengths

The refractive index of air varies slightly with wavelength, temperature, and pressure, and is generally quoted at standard temperature and pressure* as 1.00029. For red light of vacuum

* 25^0C and 760 mm Hg.

wavelength $\lambda = 656.5$ nm the wavelength in air is

$$\lambda_{air} = \frac{\lambda}{n_{air}} = \frac{656.5}{1.000\,29} = 656.3 \text{ nm.}$$

In general, the wavelength of light in air is about 0.2 nm less than the vacuum wavelength. For our purposes we shall take wavelengths in air to be nearly enough equal to vacuum wavelengths.

10.7 WAVE PACKETS OR GROUPS

Thus far all the waves we have considered have been sinusoidal waves of infinite extent. This implies a source vibrating infinitely; but no source of light produces infinitely long, uninterrupted waves! Light or any other electromagnetic radiation is produced when an electrically charged subatomic particle is accelerated or decelerated. We shall have more to say about light sources in chapter 15, but for the moment let us consider qualitatively what happens when an electric charge is accelerated, figure 10.18 .

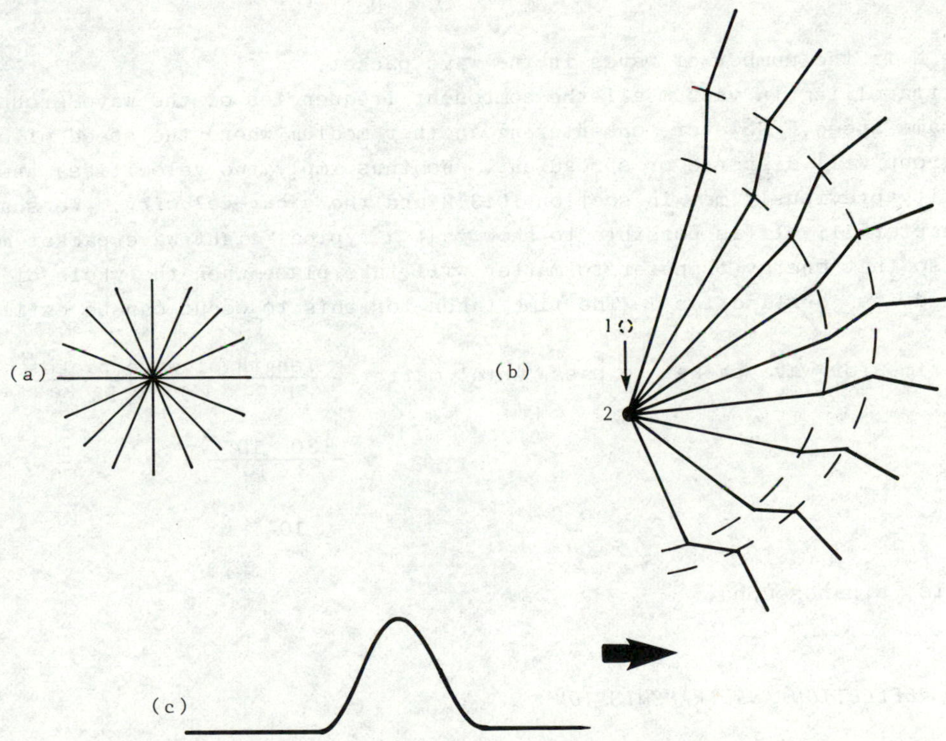

Fig. 10.18 (a) An electrically charged subatomic particle with associated radiating electric field lines. (b) The particle moves rapidly from position 1 to position 2 which causes a 'kink' in the field lines. (c) The kink propagates outwards as a pulse. Several such pulses form a wave.

The pulses do not continue indefinitely and result in a group of waves of finite length, figure 10.19, known as a *wave packet*, *wave group*, or *wave train*. We shall discuss the mathematical representation of such wave packets in more detail in chapter 11 but we should note here that such a wave packet consists of a group of frequencies and not a single frequency. The smaller the number of frequencies in a wave packet the longer will be the wave packet.

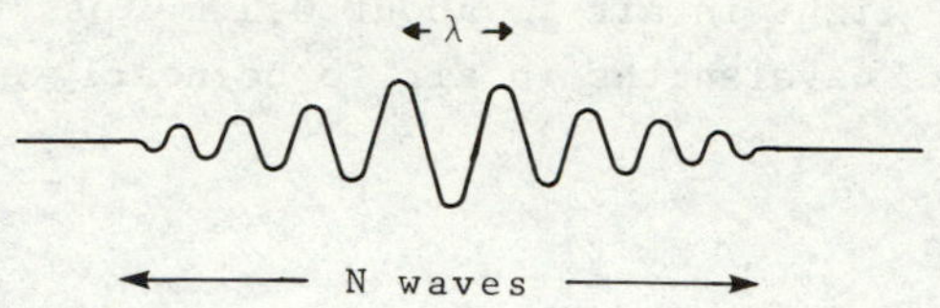

Fig. 10.19 A wave packet.

Different frequencies in a group mean different wavelengths. It is possible to show that the spread of wavelengths $\Delta\lambda$ about the mean wavelength λ is given by

$$\frac{\Delta\lambda}{\lambda} \simeq \frac{1}{N} \qquad (10.37)$$

where N is the number of waves in the wave packet.
When travelling in vacuum all the component frequencies of the wave group or packet travel at the same speed. However, on entering another medium where the speed of each component differs the group will disperse or spread out. We thus imply two velocities: the wave (or phase) velocity previously met in section 10.3.2 and the *group velocity*. We shall discuss the latter in chapter 11. It is possible to show that a typical light wave packet might be about 30cm long so that energy transfer to matter will take place when the whole of the packet has reached its 'destination'. The time taken for this to occur can be estimated thus:

$$\text{time for wave packet to pass 'into' matter} = \frac{\text{length of wave packet}}{\text{speed of wave packet}}$$

$$= \frac{30 \times 10^{-2}}{3 \times 10^{8}} \qquad \frac{\text{m}}{\text{ms}^{-1}}$$

$$= 10^{-9}\ \text{s}$$

that is, a nanosecond.

10.8 REFLECTION AND TRANSMISSION

10.8.1 *Fresnel's Equations - amplitude coefficients*

About one hundred and fifty years ago A.J. Fresnel derived a set of equations which allow the calculation of the amount of light reflected and transmitted at a surface between two optical media. We have briefly mentioned polarisation and intensity in sections 10.4.3 and 10.5.3, respectively, and we shall now make use of these concepts to calculate the fractions of the intensity reflected and transmitted at a surface. The derivation of Fresnel's equations is beyond the scope of this text so we shall merely state them and interpret them.
Figure 10.20 shows a plane surface bounding two media with refractive indices n and n'.

According to our convention in geometrical optics the incident and reflected rays are in the medium of refractive index n, whilst the transmitted (refracted) ray is in the medium with refractive index n'.

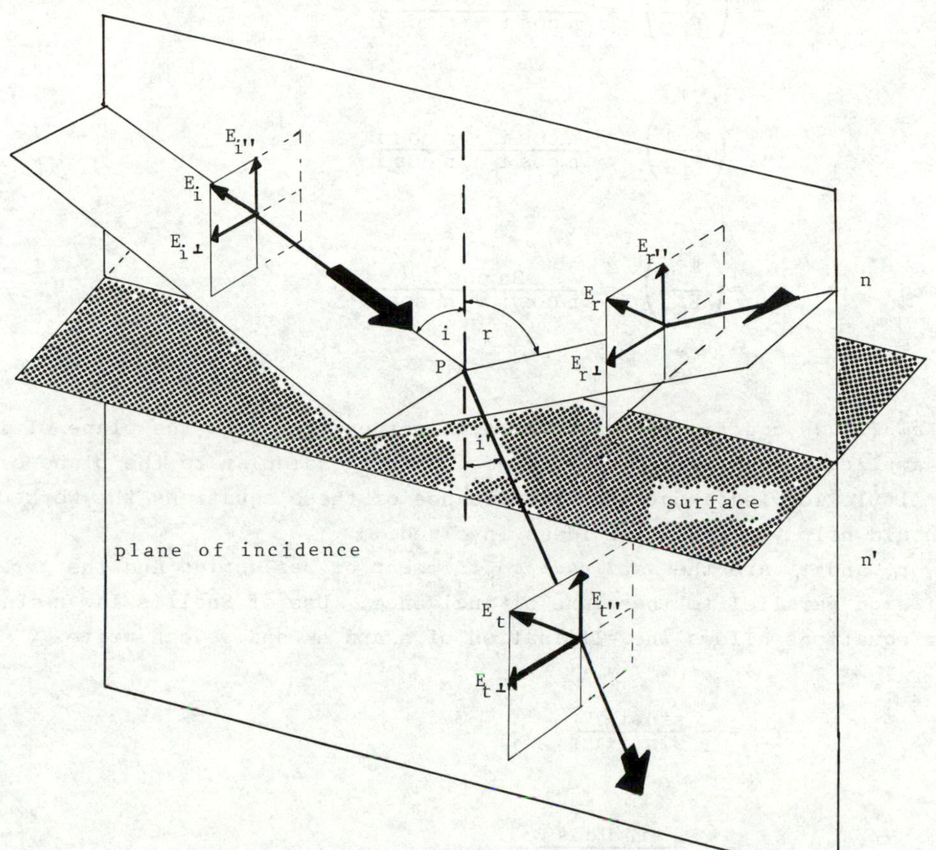

Fig. 10.20 Incident, reflected, and transmitted electric vectors at the point of incidence, P. Note: the the vectors are not drawn proportionately.

The plane of incidence is perpendicular to the surface bounding the two media. Since we are going to be concerned with reflected and transmitted intensities, and since we have shown that intensity is proportional to the amplitude squared, we shall need to consider the incident, the reflected, and the transmitted amplitudes of the waves. In particular, it can be shown that it is the amplitude of the electric wave which is mostly involved in interactions with matter and the magnetic component of the electromagnetic wave is of little consequence. Hence, we shall call the amplitude of the wave the *electric vector*, E. The subscripts i,r, and t refer to incident, reflected, and transmitted.

Look at the incident ray. The plane polarised incident wave signified by its amplitude E_i lies neither in the plane of incidence nor perpendicular to it. However, Fresnel's equations deal with the components of E_i parallel to and perpendicular to the plane of incidence. The subscripts $\parallel$ and $\perp$ are used to indicate components parallel to and perpendicular to the plane of incidence. Hence, $E_{i\parallel}$ and $E_{i\perp}$ are the components of E_i in and perpendicular to the plane of incidence. Now, Fresnel's equations for the fraction of the incident amplitude reflected and transmitted both parallel to and perpendicular to the plane of incidence are:

$$r_\perp = \left(\frac{E_{r\perp}}{E_{i\perp}}\right) = \frac{n\cos i - n'\cos i'}{n\cos i + n'\cos i'} \tag{10.38}$$

$$t_\perp = \left(\frac{E_{t\perp}}{E_{i\perp}}\right) = \frac{2n\cos i}{n\cos i + n'\cos i'} \tag{10.39}$$

$$r_{\parallel} = \left(\frac{E_{r\parallel}}{E_{i\parallel}}\right) = \frac{n'\cos i - n\cos i'}{n\cos i' + n'\cos i} \tag{10.40}$$

and

$$t_{\parallel} = \left(\frac{E_{t\parallel}}{E_{i\parallel}}\right) = \frac{2n\cos i}{n\cos i' + n'\cos i} \tag{10.41}$$

$r_\perp$ is the *amplitude coefficient of reflection* perpendicular to the plane of incidence, whilst $t_\perp$ is the *amplitude coefficient of transmission* perpendicular to the plane of incidence. If you are a little worried about the significance of these equations the worked example in this section should help you grasp the ideas involved.

Similarly, $r_{\parallel}$ and $t_{\parallel}$ are the *amplitude coefficient of reflection* and the *amplitude coefficient of transmission* parallel to the plane of incidence. Use of Snell's law $n\sin i = n'\sin i'$ in these four equations allows the elimination of n and n' and we can write

$$r_\perp = -\frac{\sin(i-i')}{\sin(i+i')} \tag{10.42}$$

$$t_\perp = +\frac{2\sin i'\cos i}{\sin(i+i')} \tag{10.43}$$

$$r_{\parallel} = +\frac{\tan(i-i')}{\tan(i+i')} \tag{10.44}$$

$$t_{\parallel} = +\frac{2\sin i'\cos i}{\sin(i+i')\cos(i-i')} \tag{10.45}$$

Equations (10.42) to (10.45) carry a plus or a minus sign. When particular values of i and i' are considered a minus sign in the result for $r_\perp$, $r_{\parallel}$, $t_\perp$, or $t_{\parallel}$ simply means the electric vector is pointing the opposite way to that indicated in figure 10.20 . This is the same as saying there is a phase change of 180^0 or π radians, which is equivalent to a half-wavelength difference in the medium concerned. For example, suppose $i=48.59^0$ and $i'=30^0$; then equation (10.42) gives

$$r_\perp = -\frac{\sin(i-i')}{\sin(i+i')} = -\frac{\sin(48.59^0 - 30^0)}{\sin(48.59^0 + 30^0)} = -0.3252\ .$$

$r_\perp = -0.3252$ means that the magnitude of the reflected wave's amplitude or electric vector perpendicular to the plane of incidence is, from equation (10.38),

$$E_{r\perp} = r_\perp E_{i\perp} = -0.3252\, E_{i\perp}$$

where the minus sign indicates that $E_{r\perp}$ points in the opposite direction to $E_{i\perp}$: a phase change of 180^0 has taken place.

10.8.2 *Reflectance and Transmittance*

Reflectance is defined as the fraction of the incident intensity reflected whilst *transmittance* is the fraction of the incident intensity transmitted. These can be considered for varying angles of incidence and for component electric vectors parallel to and perpendicular to the plane of incidence, see section 12.6.1 . However, here we shall restrict ourselves to the consideration of light at nearly normal incidence. In that case i and i' are very small which makes the sine approximately equal to the tangent and the cosine very nearly unity. Equating the sine with the tangent in equations (10.42) and (10.44), and making the cosine = 1, gives

$r_{\shortparallel} = -r_\perp$ ....(10.46), and $t_{\shortparallel} = t_\perp$ ....(10.47).

The results we are about to derive for nearly normal incidence will be used when considering *antireflection coatings* on lens surfaces, an application much used in ophthalmic practice nowadays. Recall again that intensity, which we symbolise with I, is proportional to the amplitude (here the electric vector, E) squared: that is

$$I \propto E^2 \qquad (10.48).$$

The reflectance R is defined as $R = \frac{I_r}{I_i} = \frac{E_r^2}{E_i^2}$ (10.49)

Using figure 10.21 it is evident that

$$E_r^2 = E_{r\perp}^2 + E_{r\shortparallel}^2 \qquad (10.50)$$

and $$E_i^2 = E_{i\perp}^2 + E_{i\shortparallel}^2 \qquad (10.51)$$

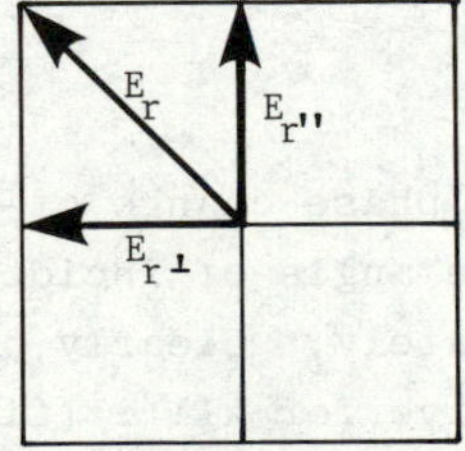

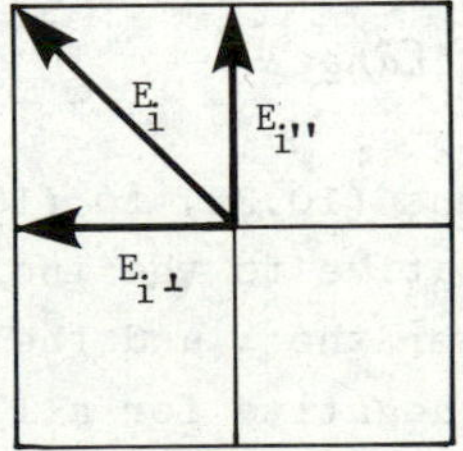

Fig. 10.21 The components of E_r and E_i. Note that they are not to scale and $E_{r\perp}$ does NOT generally equal $E_{r\shortparallel}$, etcetera.

From equation (10.38) and equation (10.40) we can immediately write

$$E_{r\perp} = r_\perp E_{i\perp} \quad \text{..........(10.52)} \quad \text{and} \quad E_{r\parallel} = r_\parallel E_{i\parallel} \quad \text{..........(10.53)}$$

so

$$R = \frac{I_r}{I_i} = \frac{E_r^2}{E_i^2} = \frac{E_{r\perp}^2 + E_{r\parallel}^2}{E_{i\perp}^2 + E_{i\parallel}^2} = \frac{r_\perp^2 E_{i\perp}^2 + r_\parallel^2 E_{i\parallel}^2}{E_{i\perp}^2 + E_{i\parallel}^2} = \frac{r_\perp^2 E_{i\perp}^2 + (-r_\perp)^2 E_{i\parallel}^2}{E_{i\perp}^2 + E_{i\parallel}^2}, \text{ since } r_\parallel = -r_\perp,$$

$$= r_\perp^2 \left(\frac{E_{i\perp}^2 + E_{i\parallel}^2}{E_{i\perp}^2 + E_{i\parallel}^2} \right) = r_\perp^2 = \left(\frac{n - n'}{n + n'} \right)^2 \equiv \left(\frac{n' - n}{n' + n} \right)^2$$

, using equation (10.38) with the cosines = 1 .

The reflectance, then, for near normal incidence is

$$R = r_\perp^2 = r_\parallel^2 = \left(\frac{n' - n}{n' + n} \right)^2 \qquad (10.54)$$

In the last equation you should note that we have used the fact that $r_\parallel = -r_\perp$.

We now define transmittance to be

$$T = \frac{I_t}{I_i} \qquad (10.55)$$

which, using the method above, can be shown to give

$$T = \frac{4nn'}{(n' + n)^2} \qquad (10.56)$$

for normal or nearly normal incidence again. Adding equations (10.54) and (10.56) gives a result we might expect: if no absorption takes place at the surface then the transmittance plus the reflectance equals unity. Thus,

$$R + T = \left(\frac{n' - n}{n' + n} \right)^2 + \frac{4nn'}{(n' + n)^2} = \frac{n'^2 - 2nn' + n^2 + 4nn'}{(n' + n)^2} = \frac{(n' + n)^2}{(n' + n)^2} = 1 \qquad (10.57)$$

Multiplying through by I_i gives $RI_i + TI_i = I_i$ which says that the fractional intensity reflected plus the fractional intensity transmitted equals the incident intensity. Remember this assumes no absorption, of course.

10.8.3 Phase changes

Use of equations (10.38) to (10.41) allows us to plot the phase change of the reflected components relative to the incident components against the angle of incidence for given n and n'. We must consider the $\perp$ and the $\parallel$ reflected vectors separately. Clearly, since from equation (10.42) $r_\perp$ is negative for all i when $n' > n$, $E_{r\perp}$ is negative for all i (figure 10.22(a)) which gives a π phase change for this component when $n' > n$. Figure 10.22(b) shows a zero phase change for $E_{r\parallel}$ upto $i = i_p$, after which the phase change is π.
The angle i_p is of particular interest as we can see from the following. If we consider the case when $i + i' = 90^0$, then equation (10.44)

$$r_\parallel = + \frac{\tan(i - i')}{\tan(i + i')}$$

Fig. 10.22 Phase change for the ⊥ and the ∥ components of the reflected electric vectors for the case when $n' > n$: $n' = 1.5$ and $n = 1$.

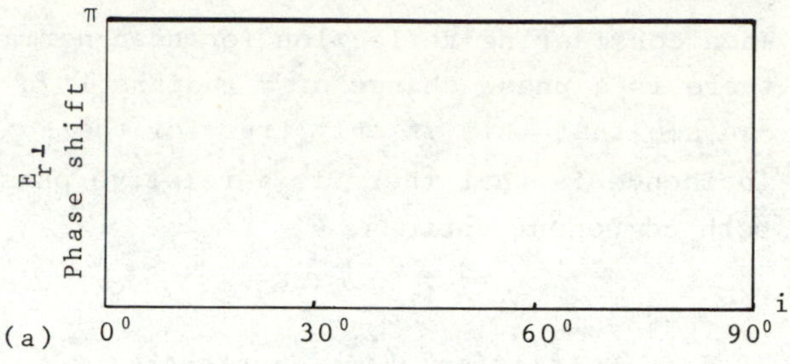

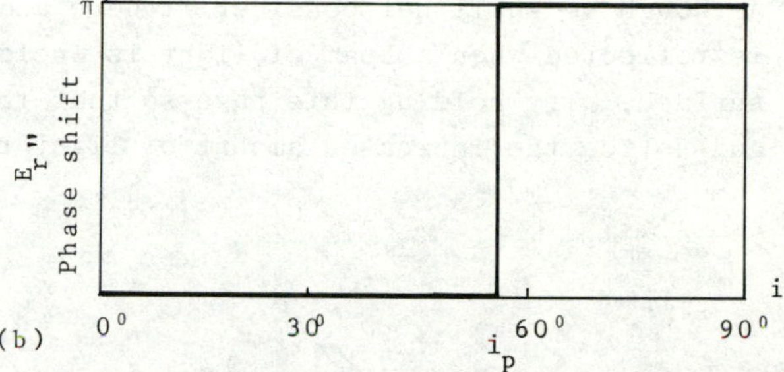

goes to zero since $\tan(i+i') = \tan 90^0 = \infty$. Then the electric vector component parallel to the plane of incidence is

$$E_{r\parallel} = r_{\parallel} E_{i\parallel} = 0 \times E_{i\parallel} = 0 \ .$$

Thus, if we write $i = i_p$, when $i_p + i' = 90^0$ the reflected light consists only of the $E_{r\perp}$ component and is therefore entirely plane polarised perpendicular to the plane of incidence. The angle i_p is called the *polarisation angle* or *Brewster's angle* after its discoverer Sir David Brewster (1781 - 1868), the Scottish physicist.
Since $i' = 90^0 - i_p$, using Snell's law we have

$$n \sin i_p = n' \sin i' = n' \sin(90^0 - i_p) = n' \cos i_p$$

which gives
$$\tan i_p = \frac{n'}{n} \qquad (10.58).$$

This expression, which allows us to calculate the polarisation angle, is known as *Brewster's Law*.
The case considered above where $n' > n$ is commonly called external reflection. We now consider internal reflection where $n' < n$. Figure 10.23 shows the phase changes occurring in this case for $n = 1.5$ and $n' = 1$. i_c is the critical angle, of course.

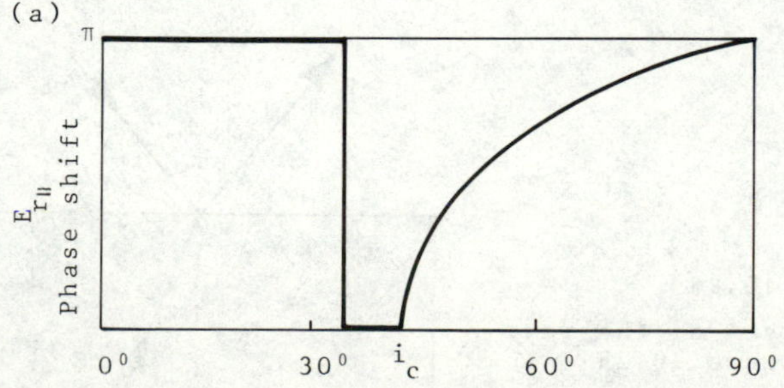

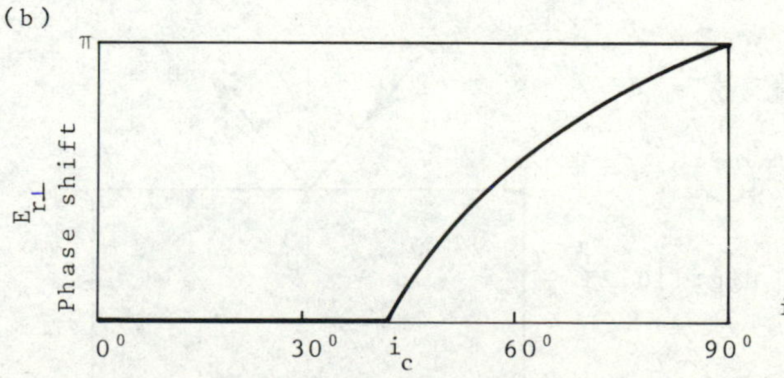

Fig. 10.23 (a) Phase change for $E_{r\parallel}$ and (b) for $E_{r\perp}$ on internal reflection ($n' < n$)

When considering reflection for near normal incident light it is common practice to state that there is a phase change of π radians if n' > n and a zero phase change if n' < n. However, we can see that this is only true for the $E_{r\perp}$ component. What we can say for near normal incidence is that there is a relative phase shift of π radians between the two cases for both component vectors.

10.8.4 Reflection when i approaches 90°

Although we shall not consider it here analytically we should note that almost all the light is reflected when a beam of light is incident at $i \simeq 90^0$. This is even true on a non-transmitting surface. Try holding this page so that the light from a window falls on it at a glancing angle and notice the increased amount of light reflected (figure 10.24).

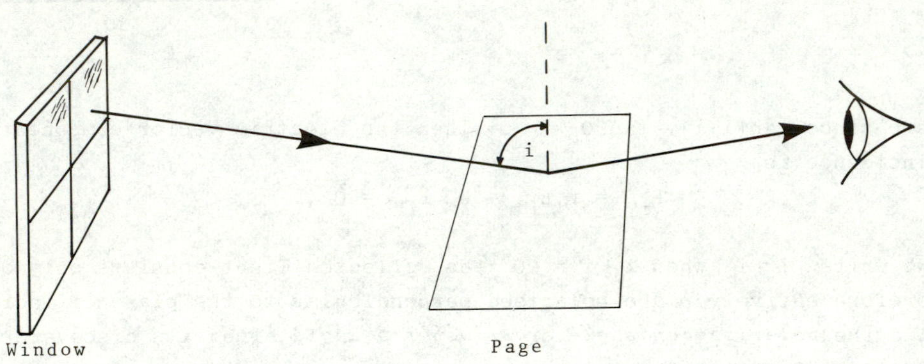

Fig. 10.24 Reflection at a glancing angle: large i.

This topic will be considered in more detail in section 12.6.1 .

10.8.5 STOKES' TREATMENT OF REFLECTION AND REFRACTION

Another way of comparing phase differences between internally and externally reflected waves was presented by Sir George Stokes (1819 - 1903). Suppose a wave of arbitrary polarisation is incident on a plane interface separating two optical media. If the amplitude (the electric vector) is E and the amplitude reflection and transmission coefficients are r and t, respectively, then the reflected and transmitted amplitudes are rE and tE (figure 10.25).

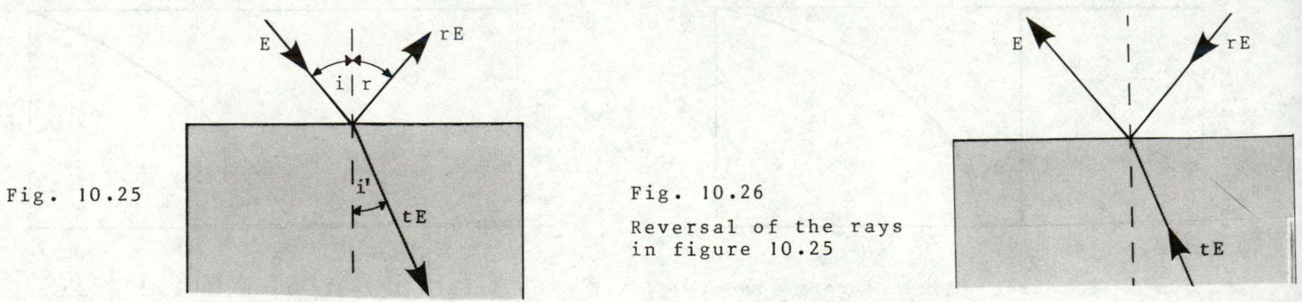

Fig. 10.25

Fig. 10.26
Reversal of the rays in figure 10.25

Using the principle of reversibility we should be able to reverse the directions of all the rays in figure 10.25 as in figure 10.26 . The angles are as before. However, the incident ray from below, amplitude tE, will be partially reflected and partially transmitted. Let these amplitude reflection and transmission coefficients be r' and t' so that figure 10.27 shows the reflected and transmitted parts of the incident wave tE.

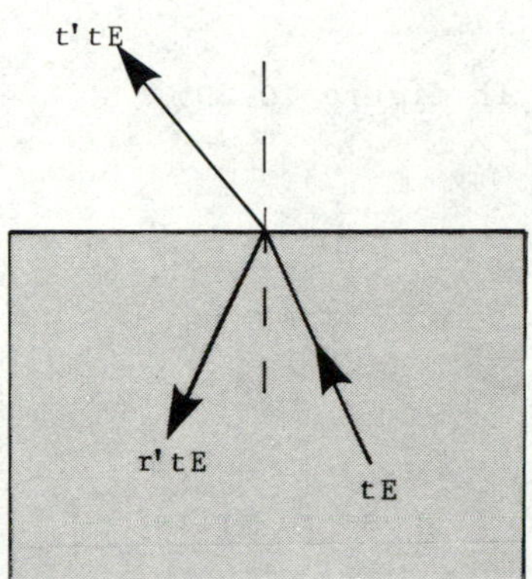

Fig. 10.27

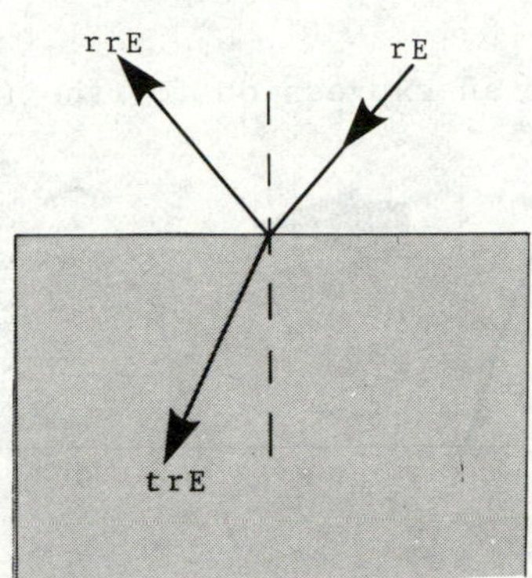

Fig. 10.28

Similarly, figure 10.28 shows the effects for the external, reversed, incident wave rE from figure 10.26 . Now, the sum of the amplitudes in figures 10.27 and 10.28 must equal the amplitudes in figure 10.26 for reversal of the rays to be possible. Hence,

$$t'tE + rrE = E \quad ....(10,59) \quad \text{and} \quad r'tE + trE = 0 \quad ....(10.60)$$

from which Stokes' relations are

$$tt' = 1 - r^2 \quad ....(10.61) \quad \text{and} \quad r' = -r \quad ....(10.62).$$

The negative sign in equation (10.62) indicates a 180^0 or π phase change which means that if there is a π phase change on reflection at a boundary where $n' > n$ then there will be no phase change when $n' < n$. Compare this statement with that at the end of section 10.8.3 ; it is not as rigorous since it does not differentiate between components in and perpendicular to the plane of incidence.

10.9 WORKED PROBLEMS

10.9.1 What is the distance along the wave (in the direction of propagation) between two points having a phase difference of 60^0 if the wave velocity is 3×10^8 ms^{-1} and the frequency is 6×10^{14} Hz? What phase shift occurs at a given point in 10^{-3} seconds and how many waves pass by in that time?

A phase difference of 360^0 is equivalent to one whole wavelength, so 60^0 is equivalent to 1/6 of a wavelength. Now, $\lambda = v/\nu = 3 \times 10^8/(6 \times 10^{14})$ $ms^{-1}/s^{-1} = 5 \times 10^{-7}m$. Hence, the distance required is $\frac{1}{6} \times 5 \times 10^{-7} = 8.3 \times 10^{-8}$ m.

The frequency is the number of waves passing a fixed point in one second, so the number of waves passing by in 10^{-3} s is

$$10^{-3} \times \nu = 10^{-3} \times 6 \times 10^{14} = 6 \times 10^{11} .$$

Since each wavelength is equivalent to a 360^0 or 2π radians phase change the total phase change is $6 \times 10^{11} \times 2\pi$ radians.

10.9.2 Write an expression for the infrared wave shown in figure 10.29.

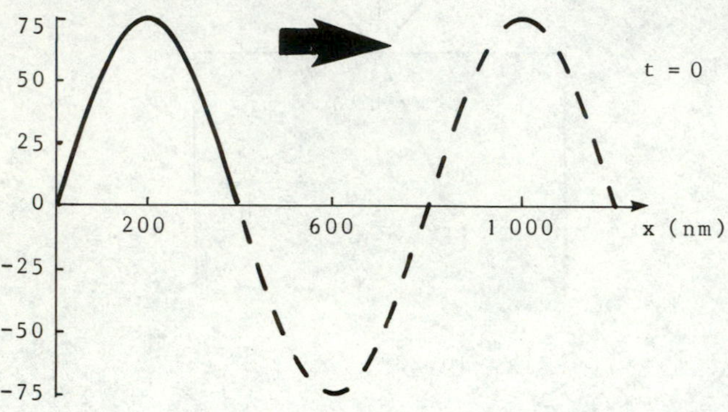

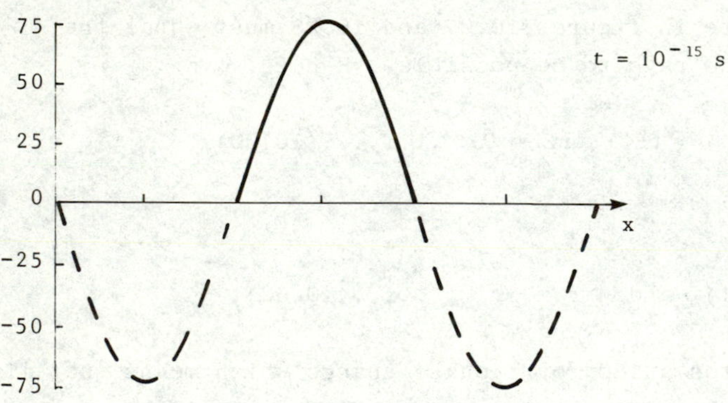

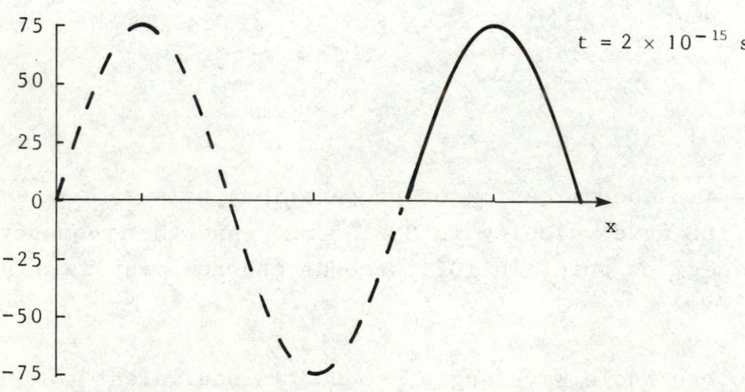

Fig. 10.29 A harmonic wave.

Suppose we write

$$\psi(x,t) = A \sin 2\pi\left(\frac{x}{\lambda} \mp \frac{t}{\tau}\right)$$

for the general expression, which fits this case since $\psi(0,0) = 0$. Clearly, from the diagram, we have

$\lambda = 800 \times 10^{-9}$ m , $A = 75$ units, and the time taken to travel one whole wavelength is 2×10^{-15} s. Thus, the function is

$$\psi(x,t) = 75 \sin 2\pi\left(\frac{x}{800 \times 10^{-9}} - \frac{t}{2 \times 10^{-15}}\right)$$

the negative sign being chosen because the wave is moving in the positive x-direction.

Incidentally, the units of A are volts per metre.

10.9.3 Given the function $\psi(x,t) = 3\sin(kx-\omega t+\pi)$, what is the value of the function at x=0 when t=0, t=τ/4, t=τ/2, t=3τ/4, and t=τ?

Since $\omega = 2\pi\nu = \frac{2\pi}{\tau}$, we can write the wave equation as

$$\psi(x,t) = 3\sin(kx-\frac{2\pi}{\tau}t+\pi)$$

and substituting x = 0 and the various t values we have

$$\psi(0,0) = 3\sin(0-0+\pi) = 0$$

$$\psi(0,\frac{\tau}{4}) = 3\sin(0-(\frac{2\pi}{\tau}\times\frac{\tau}{4})+\pi) = +3$$

$$\psi(0,\frac{\tau}{2}) = 3\sin(0-(\frac{2\pi}{\tau}\times\frac{\tau}{2})+\pi) = 0$$

$$\psi(0,\frac{3\tau}{4}) = 3\sin(0-(\frac{2\pi}{\tau}\times\frac{3\tau}{4})+\pi) = -3$$

$$\psi(0,\tau) = 3\sin(0-(\frac{2\pi}{\tau}\times\tau)+\pi) = 0 \quad .$$

As we should expect in a full temporal period τ the function oscillates from 0, through +3, 0, -3, and back to 0.

10.9.4 A parallel circular beam of light is incident normally on a perfectly absorbing plane surface. If the intensity of the beam is 20 W/cm^2 (Watts per cm^2) and its diameter is $4/\sqrt{\pi}$ cm, how much energy is absorbed by the surface in 2 minutes? (Remember, power equals energy transfer per second: that is, the units are Watts = Joules/second or W = J/s).

The area of the beam is $\pi \times \text{radius}^2 = \pi \times (\frac{2}{\sqrt{\pi}})^2 = 4\ \text{cm}^2$.

So, in one second the beam transfers 4×20 J of energy. Hence, in 2 minutes the energy transferred is

$$4 \times 20 \times 120 = 9600\ \text{J or } 9.6\ \text{kJ}.$$

We shall not be doing very much of this type of problem but it does serve to remind us that light possesses energy and that optics is a branch of physics.

10.9.5 The vacuum wavelength of a light wave is 750 nm. What is its propagation number in a medium of refractive index 1.5?

Firstly, the wavelength in the medium is $\lambda_m = \frac{\lambda}{n} = \frac{750 \times 10^{-9}}{1.5} = 500 \times 10^{-9}$ m.

The propagation number is $k = \frac{2\pi}{\lambda_m} = \frac{6.28}{500 \times 10^{-9}} = 1.26 \times 10^7$ rad/m.

10.9.6 An empty tank is 30 m long. Light from a sodium lamp ($\lambda = 589$ nm) passes through the tank in time t_1 when filled with water of refractive index 1.33 . When filled with carbon disulphide of refractive index 1.63 it takes a time t_2. Find the difference in the transit times. The speed of light in vacuum is $c = 3 \times 10^8$ ms^{-1}.

Refractive index is defined as $n = \frac{\text{speed of light in vacuum}}{\text{speed of light in medium}} = \frac{c}{v}$.

Hence, $v = \frac{c}{n}$. Thus, $t_1 = \frac{\text{distance}}{\text{speed}} = \frac{30}{c/1.33} = \frac{1.33 \times 30}{3 \times 10^8}$

and $t_2 = \frac{30}{c/1.63} = \frac{1.63 \times 30}{3 \times 10^8}$.

So, $t_2 - t_1 = \frac{30}{3 \times 10^8}(1.63 - 1.33) = 3 \times 10^{-8}$ s.

10.9.7 A beam of white light is incident normally on a plane surface separating air from glass. If the refractive indices for particular 'red, green, and violet' light wavelengths are 1.45, 1.50, and 1.55, respectively, find the reflectances for each colour.

The reflectance for normally incident light is given by $R = \left(\frac{n' - n}{n' + n}\right)^2$.

So, for red light $R_{red} = \left(\frac{n' - n}{n' + n}\right)^2 = \left(\frac{1.45 - 1}{1.45 + 1}\right)^2 = 0.034 \equiv 3.4\%$.

For green light $R_{green} = \left(\frac{n' - n}{n' + n}\right)^2 = \left(\frac{1.50 - 1}{1.50 + 1}\right)^2 = 0.040 \equiv 4\%$,

and for violet light $R_{violet} = \left(\frac{n' - n}{n' + n}\right)^2 = \left(\frac{1.55 - 1}{1.55 + 1}\right)^2 = 0.047 \equiv 4.7\%$.

10.9.8 A film of cryolite (Na_3AlF_6) of refractive index 1.31 is deposited on a glass substrate. If three quarters of a wavelength of light with vacuum wavelength 555 nm is to occupy the film, how thick must the film layer be?

The wavelength in the film is $\lambda_f = \frac{\lambda}{n_f} = \frac{555}{1.31} = 423.7$ nm .

Clearly, the film thickness must be $\frac{3}{4}\lambda_f = 317.7$ nm .

EXERCISES

1 State the amplitude, the propagation number, the frequency, the angular frequency, the wavelength, the period, and the direction of propagation of the wave

$$\psi = 10 \sin 2\pi(100x - 3 \times 10^{10} t).$$

Use $c = 3 \times 10^8$ ms^{-1}. From which part of the electromagnetic spectrum does the wave come?

Ans. $A = 10$ Vm^{-1}, $k = 2\pi \times 100$ m^{-1}, $\nu = 3 \times 10^{10}$ s^{-1}, $\omega = 2\pi \times 3 \times 10^{10}$ rad s^{-1}, $\lambda = 1/100$ m, $\tau = 1/\nu = 1/(3 \times 10^{10})$ s, and the wave is propagating in the positive x-direction.

2 A wave propagates in the negative x-direction with an amplitude of 5 Vm^{-1}, a wavelength of 600×10^{-9} m, and a frequency of 5×10^{14} Hz. Write its equation in terms of k and ω.

Ans. $\psi = A \sin(kx + \omega t) = 5 \sin\left(\frac{2\pi}{600 \times 10^{-9}} x + 2\pi \times 5 \times 10^{14} t\right)$.

3 A wave propagating with $\lambda = 300 \times 10^{-9}$ m at a speed of 3×10^8 ms^{-1} has what period?

Ans. $\tau = 10^{-15}$ s.

4 What do you understand by the terms wavelength, amplitude, and phase? Illustrate your answer with reference to the functions $\psi = A \sin \frac{2\pi}{\lambda} x$ and $\psi = A \sin \frac{2\pi}{\tau} t$.

Ans. W.P.O. page 93. N.B. T is used instead of τ in Worked Problems in Optics.

5 State the fundamental equation of wave motion. If the velocity of light in vacuum is 3×10^8 ms^{-1}, find the frequency of the following waves. (i) Red 700 nm, (ii) Orange 600 nm, (iii) Violet 400 nm.

Ans. (i) 4.29×10^{14} Hz. (ii) 5.0×10^{14} Hz. (iii) 7.5×10^{14} Hz. See W.P.O. page 94, qu. 4.

6 Overhead power lines carry alternating current with a frequency of 50 Hz. Find the wavelength of the electromagnetic radiation from the wires.

Ans. 6×10^6 m. See W.P.O. page 95, qu. 5.

7 Given the travelling wave $\psi = 20 \sin 2\pi\left(\frac{x}{6 \times 10^{-7}} - \frac{t}{2 \times 10^{-15}}\right)$, find the frequency, wavelength, and speed.

Ans. 5×10^{14} Hz, 6×10^{-7} m, and 3×10^8 ms^{-1}. See W.P.O. page 95, qu. 6.

8 60 water wave crests pass a fixed post in 15 seconds travelling at a speed of 3 ms^{-1}. Find the frequency and the wavelength.

Ans. 4 Hz and ¾ m.

9 The wave $\psi(x,t) = 0.3 \sin 2\pi\left(\frac{x}{3} - \frac{t}{0.2}\right)$ propagates on the surface of a pond. At time t = 2 s what is the displacement ψ at the points x = 0 and x = 30.75 m?

Ans. 0 and 0.15 m.

10 Sketch the waves $\psi_1 = A \sin(kx - \omega t)$ and $\psi_2 = A \sin(kx - \omega t - \frac{\pi}{2})$ at t = 0. Show the origin and the waves along the positive x-direction.

11 At time t = 0 a wave has the form $\psi(x,0) = 3 \sin \frac{\pi x}{20}$. If the wave is moving in the positive x-direction, write the equation for the disturbance at t = 8 s, when v = 2 ms^{-1}.

Ans. $\psi(x,8) = 3 \sin 2\pi\left(\frac{x}{40} - \frac{8}{20}\right)$.

12 Red light, $\lambda = 660$ nm, enters a glass block whereupon its velocity is ⅔ that of its velocity in vacuum. Find the refractive index of the glass for this wavelength and the wavelength in the glass.

Ans. $n_g = 1.5$ and $\lambda_m = 440$ nm.

13 Describe the electromagnetic spectrum and indicate the regions of different types of waves therein.

14 Explain why we often regard the wavelength in air as being equivalent to the wavelength in vacuum.

15 A ray of monochromatic light is incident at 40^0 on a plane air/glass boundary. If the refractive index of the glass is 1.5 for this wavelength, calculate the amplitude reflection and transmission coefficients, assuming $n_{air} = 1$. At what angle of incidence is $r_{\parallel} = 0$?

Ans. $r_\perp = -0.27777$, $r_{\parallel} = 0.11962$, $t_\perp = 0.72223$, and $t_{\parallel} = 0.43983$. $r_{\parallel} = 0$ when $i = 56.31^0$.

16 Repeat question 15 for a water ($n_w = 1.33$)/glass ($n_g = 1.5$) boundary.

Ans. $r_\perp = -0.09492$, $r_{\parallel} = 0.02508$, $t_\perp = 0.90508$, and $t_{\parallel} = 0.90890$. $r_{\parallel} = 0$ when $i = 48.44^0$.

17 Calculate the reflectance and transmittance for monochromatic light incident normally (a) externally and (b) internally on an air/glass boundary for which $n_g = 1.5$.

Ans. (a) $R = 0.04$, $T = 0.96$. (b) $R = 0.04$, $T = 0.96$.

18 Oil is poured on to a sheet of plastics material, both media having the same refractive index. Show that no light will be reflected from the boundary.

19 Use the Fresnel equations to prove that $t_\perp + (-r_\perp) = 1$.

20 A system of N flat plates of glass, separated by air, transmits a normally incident beam of monochromatic light. Show that the total transmittance is $(1 - R)^{2N}$. If $n_g = 1.5$, find the total transmittance for (a) 2 plates and (b) 10 plates.

Ans. (a) 0.8493. (b) 0.4420 .

21 Repeat the calculation in question 20 with the plates immersed in water of refractive 1.33 .

Ans. (a) 0.9856. (b) 0.9304 .

11 SUPERPOSITION OF WAVES

11.0 INTRODUCTION

In the three immediately succeeding chapters we shall be discussing the phenomena of polarisation, interference, and diffraction. They all have a common basis in that they involve the superposition of waves. That is, they concern the effects of overlapping waves. The way in which the waves overlap determines the intensity distribution in space. From our point of view the superposition gives rise to patterns of distribution of light which may present themselves in certain experimental arrangements. In order to explain the phenomena in these experiments we need to look closely at the results of overlapping waves.
We shall call upon the *principle of superposition* which states that the resultant disturbance in space is the algebraic sum of the constituent disturbances. Figure 11.1 illustrates this with two 'square pulses' travelling in the positive x-direction with speed v.

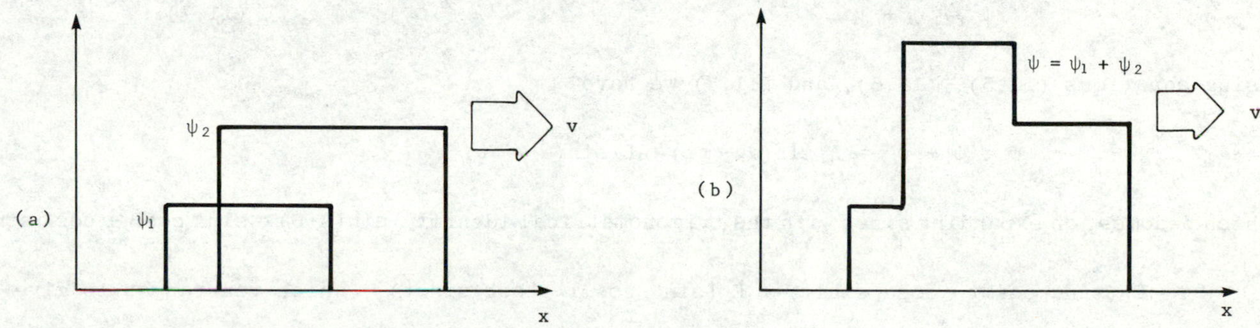

Fig. 11.1 Superposition of two square pulses.

Figure 11.1(a) shows the square pulse waves ψ_1 and ψ_2 occupying the same space. At each point on the x-axis we add the values of ψ_1 and ψ_2 so that $\psi = \psi_1 + \psi_2$. The resultant disturbance is the pulse shown in figure 11.1(b). In all the cases we shall consider the waves will be one-dimensional, or nearly so. Limiting ourselves to this condition simplifies the analysis which none-the-less can be heavy going at times. The essential features will be summarised as we go along and some students may prefer to place most emphasis on the summaries. Failure to follow the mathematics closely will not create too much of a handicap in succeeding chapters.

11.1 ADDITION OF WAVES OF THE SAME FREQUENCY

11.1.1 Algebraic method

Recall that angular frequency ω is related to frequency ν via $\omega = 2\pi\nu$. So, if ω is the same in two wave functions then frequency ν and wavelength λ, $(\lambda = v/\nu)$, will be the same. The following treatment will show that superposing (overlapping) two waves of the same frequency produces a resultant wave of the same frequency but with a different amplitude and phase from either component wave. Let the waves be

$$\psi_1 = E_1 \sin(kx - \omega t + \varepsilon_1) \quad \ldots\ldots(11.1) \quad \text{and} \quad \psi_2 = E_2 \sin(kx - \omega t + \varepsilon_2) \quad \ldots\ldots(11.2).$$

The amplitudes are given the symbol E for *electric* vector since we are dealing with electromagnetic waves later. Note that k and ω are the same in the two equations, but the amplitudes E_1 and E_2 and the initial phases ε_1 and ε_2 are different.

For simplicity let $\alpha_1 = kx + \varepsilon_1$ ......(11.3) and $\alpha_2 = kx + \varepsilon_2$ ......(11.4).

This saves carrying around lengthy terms which are independent of time, t. Thus, the equations are now

$$\psi_1 = E_1 \sin(\alpha_1 - \omega t) \quad \ldots\ldots(11.5)$$

$$\text{and} \quad \psi_2 = E_2 \sin(\alpha_2 - \omega t) \quad \ldots\ldots(11.6)$$

Let the resultant wave when ψ_1 and ψ_2 are added (superposed) be ψ: that is,

$$\psi = \psi_1 + \psi_2 \qquad (11.7)$$

Using equations (11.5), (11.6), and (11.7) we have

$$\psi = \psi_1 + \psi_2 = E_1 \sin(\alpha_1 - \omega t) + E_2 \sin(\alpha_2 - \omega t)$$

which becomes, on expanding sines with the trigonometrical identity $\sin(A-B) = \sin A \cos B - \cos A \sin B$,

$$\psi = E_1 (\sin\alpha_1 \cos\omega t - \cos\alpha_1 \sin\omega t) + E_2 (\sin\alpha_2 \cos\omega t - \cos\alpha_2 \sin\omega t),$$ which rearranges to give

$$\psi = (E_1 \sin\alpha_1 + E_2 \sin\alpha_2)\cos\omega t - (E_1 \cos\alpha_1 + E_2 \cos\alpha_2)\sin\omega t \qquad (11.8).$$

The terms in the brackets of equation (11.8) are independent of time and if we write

$$E_1 \sin\alpha_1 + E_2 \sin\alpha_2 = E \sin\alpha \quad \ldots\ldots(11.9) \quad \text{and} \quad E_1 \cos\alpha_1 + E_2 \cos\alpha_2 = E \cos\alpha \quad \ldots\ldots(11.10)$$

whence, dividing equation (11.9) by equation (11.10) gives

$$\tan\alpha = \frac{E_1 \sin\alpha_1 + E_2 \sin\alpha_2}{E_1 \cos\alpha_1 + E_2 \cos\alpha_2} \qquad (11.11)$$

and squaring equations (11.9) and (11.10) and adding them gives

$$E^2 = E_1^2 + E_2^2 + 2E_1 E_2 \cos(\alpha_2 - \alpha_1) \qquad (11.12)$$

we can see that we can choose E and α in terms of E_1, E_2, α_1 and α_2. Thus, substituting from equations (11.9) and (11.10) into equation (11.8) gives

$$\psi = E \sin\alpha \cos\omega t - E \cos\alpha \sin\omega t = E \sin(\alpha - \omega t) \qquad (11.13),$$

having used the identity $\sin A \cos B - \cos A \sin B = \sin(A-B)$ in the step.

SUMMARY

Equation (11.13) is the equation for $\psi = \psi_1 + \psi_2$. That is, the two waves ψ_1 and ψ_2 superpose (add) to give a wave ψ with a different amplitude E and a different phase $(\alpha - \omega t)$. The frequency is the same since ω is the same. α will, of course, be given by $\alpha = kx + \varepsilon$.

11.1.2 *Intensity of two Overlapping Waves*

Using equation (11.12) we can deduce a most important property of overlapping waves. Because the intensity I of a wave is proportional to the amplitude squared, that is

$$I \propto E^2 \tag{11.14}$$

$$\text{then} \quad E \propto \sqrt{I} \quad \text{or } E = \text{constant} \times \sqrt{I} \tag{11.15}$$

after taking square roots in equation (11.14). Hence, using equations (11.12) and (11.15) we can write

$$I = I_1 + I_2 + 2\sqrt{I_1 I_2}\cos(\alpha_2 - \alpha_1) \tag{11.16}$$

after eliminating the constant arising from equation (11.15), see exercise 14.
This result is most important. We see that the intensity of two overlapping waves of the same frequency is not merely the sum of the separate intensities since there is the additional term $2\sqrt{I_1 I_2}\cos(\alpha_2 - \alpha_1)$ known as the *interference term*. If $(\alpha_2 - \alpha_1) = 0, \pm 2\pi, \pm 4\pi, \ldots$ in the interference term, then $\cos(\alpha_2 - \alpha_1) = 1$ and $2\sqrt{I_1 I_2}\cos(\alpha_2 - \alpha_1)$ is a maximum. When $(\alpha_2 - \alpha_1) = \pm\pi, \pm 3\pi, \ldots$ $\cos(\alpha_2 - \alpha_1) = -1$ and $2\sqrt{I_1 I_2}\cos(\alpha_2 - \alpha_1)$ is a minimum. The first case arises when the crests of ψ_1 and ψ_2 overlap and the two waves are said to be *in phase*. The second case occurs when the crests of ψ_1 superpose over the troughs of ψ_2 and the two waves are said to be *antiphase* or *180^0 out-of-phase*. Figure 11.2 demonstrates these very important special cases.

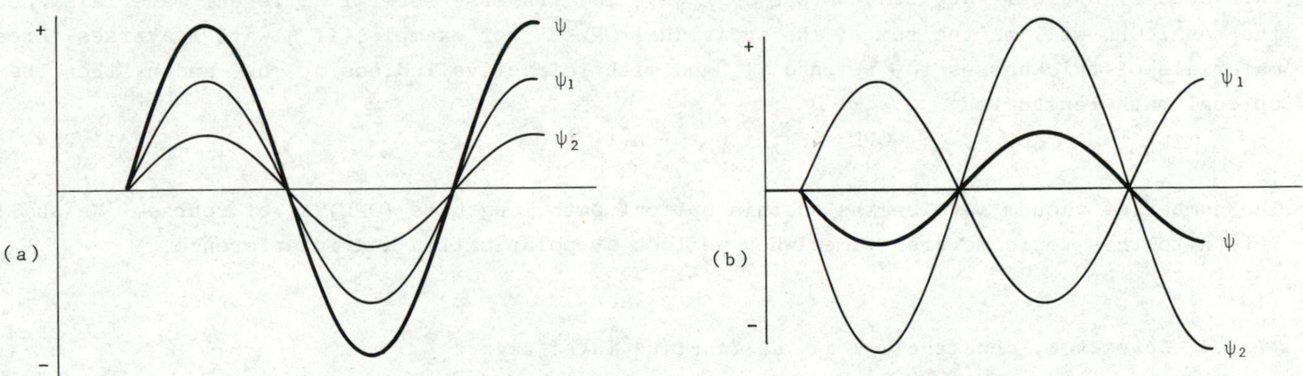

Fig. 11.2 Superposition of two sinusoidal waves ψ_1 and ψ_2 : (a) in phase and (b) antiphase. $\psi = \psi_1 + \psi_2$. Note that this is the sum of disturbances and not intensities.

11.1.3 Phase difference, Optical Path Length, and Optical Path Difference

For the waves ψ_1 and ψ_2 so far considered the *phase difference*, symbolised by δ is given by

$$\delta = \phi_2 - \phi_1 = (kx - \omega t + \varepsilon_2) - (kx - \omega t + \varepsilon_1) = \varepsilon_2 - \varepsilon_1$$

since the phases of ψ_1 and ψ_2 are $\phi_1 = kx - \omega t + \varepsilon_1$ and $\phi_2 = kx - \omega t + \varepsilon_2$, respectively. But suppose the waves are distances x_1 and x_2 from their sources, then the phase difference will be

$$\delta = \phi_2 - \phi_1 = (kx_2 - \omega t + \varepsilon_2) - (kx_1 - \omega t + \varepsilon_1) = k(x_2 - x_1) + (\varepsilon_2 - \varepsilon_1) \qquad (11.17)$$

$$\text{or} \qquad \delta = \frac{2\pi}{\lambda}(x_2 - x_1) + (\varepsilon_2 - \varepsilon_1) \qquad (11.18)$$

since $k = \frac{2\pi}{\lambda}$.

If the waves are in some medium of refractive index n then the wavelength in the medium will be $\lambda_m = \lambda/n$ and the phase difference will be

$$\delta = \frac{2\pi}{\lambda_m}(x_2 - x_1) + (\varepsilon_2 - \varepsilon_1)$$

$$\text{or} \qquad \delta = \frac{2\pi}{\lambda} n(x_2 - x_1) + (\varepsilon_2 - \varepsilon_1) \qquad (11.19)$$

where λ is the vacuum wavelength.

The quantity $n(x_2 - x_1)$ is called the *Optical Path Difference*, which will often be abbreviated to OPD. To look at the physical significance of the OPD suppose we look at the number of wavelengths which can occupy a thickness t of material of refractive index n. The wavelength in the material will be λ/n where λ is the vacuum wavelength. Thus, the number of wavelengths in the thickness t is

$$\frac{t}{\lambda/n} = \frac{nt}{\lambda} \qquad (11.20)$$

The term nt is the 'optical thickness' or the vacuum distance which the same number of vacuum wavelengths would occupy. t is the actual thickness or path length, and nt is the *optical thickness* or *Optical Path Length* (OPL). A wave may traverse several different materials when the total OPL will be the sum of the individual OPLs. For example, if a wave traverses three materials of thicknesses t_1, t_2, and t_3, and with refractive indices n_1, n_2, and n_3 then the optical path length is

$$\text{OPL} = n_1 t_1 + n_2 t_2 + n_3 t_3 \qquad (11.21)$$

The number of vacuum wavelengths in this optical path length is (OPL)/λ, of course. We shall return to this topic several times when we look at polarisation and interference.

11.1.4 Coherence, constructive and destructive Interference

Waves for which $\varepsilon_2 - \varepsilon_1$ is constant are said to be *coherent*. In such cases any OPD will be due to differences in OPLs for the two waves. Coherence will be a necessary condition in many cases we shall consider later.

There is a special case of superposition of two waves which will be of especial interest later. Consider two waves ψ_1 and ψ_2 which have the same amplitude E_1, say, and $x_2 = x_1 + \Delta x$. That is,

$$\psi_1 = E_1 \sin(kx_1 - \omega t) \quad ....(11.22) \quad \text{and} \quad \psi_2 = E_1 \sin(k(x_1 + \Delta x) - \omega t)........(11.23)$$

Adding these waves using the identity $\sin A + \sin B = 2\sin\left(\frac{A+B}{2}\right)\cos\left(\frac{A-B}{2}\right)$ gives

$$\psi = \psi_1 + \psi_2 = 2E_1 \cos\left(\frac{k.\Delta x}{2}\right) \sin\left(k\left(x_1 + \frac{\Delta x}{2}\right) - \omega t\right) \qquad (11.24)$$

Here we have assumed $\varepsilon_1 = \varepsilon_2 = 0$, so the waves are coherent. The amplitude of ψ is $2E_1 \cos(k\frac{\Delta x}{2})$, and when $\Delta x \ll \lambda$ then $\cos(k\frac{\Delta x}{2}) \simeq 1$ so the amplitude is approximately $2E_1$. In fact, if $\Delta x = 0$ then the amplitude of ψ is precisely $2E_1$. This is one of the special cases which is called *total constructive interference*. Figure 11.3 shows constructive interference.

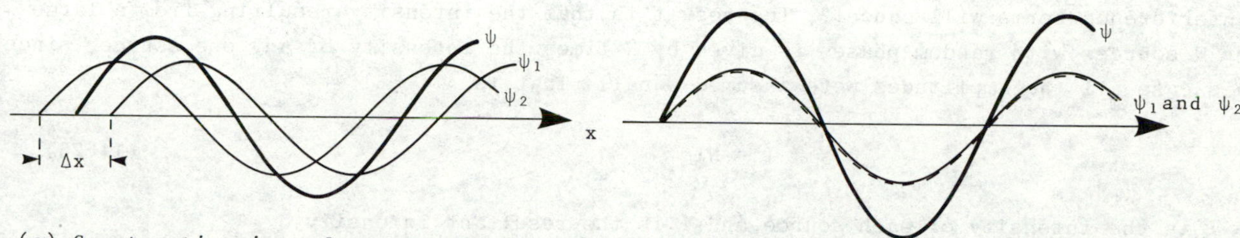

(a) Constructive interference : $\Delta x \ll \lambda$.

(b) Total constructive interference, $\Delta x = 0$.

Fig. 11.3 Cases of constructive interference.

The other special case, *total destructive interference*, occurs when $\Delta x = \lambda/2$ so that the amplitude of ψ is

$$2E_1 \cos(k\frac{\Delta x}{2}) = 2E_1 \cos(\frac{2\pi}{\lambda}\cdot\frac{1}{2}\cdot\frac{\lambda}{2}) = 2E_1 \cos\frac{\pi}{2} = 0 \ , \quad \text{since } \cos\frac{\pi}{2} = 0.$$

Total destructive interference is illustrated below in figure 11.4 .

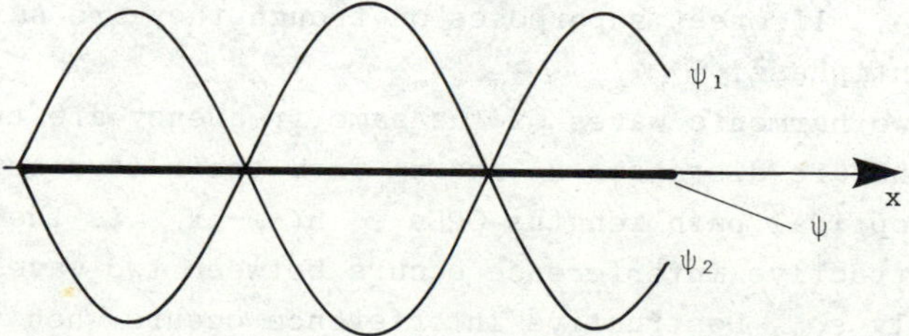

Fig. 11.4

Total destructive interference. Note that a crest of one wave superposes on a trough of the other equal amplitude wave.

11.1.5 Superposition of three or more waves

Section 11.1.1 yielded the result that $\psi = \psi_1 + \psi_2$, where ψ_1 and ψ_2 were harmonic waves with different amplitudes and phases but with the same frequency. The resultant wave ψ was a harmonic wave with the same frequency but with a different amplitude and phase from either component's amplitude and phase. If we were to now take another harmonic wave ψ_3, with the same frequency, and add it to ψ the result would be yet another harmonic wave with yet a different amplitude and phase but retaining the same frequency. By repeated application the superposition of any number of coherent harmonic waves results in a single harmonic wave of the same frequency.

11.1.6 Incoherent sources and intensity

Suppose now we have a very large number N of wave sources where the phase angles are quite random. Such sources are said to be *incoherent*. This means the initial phase ε for each wave may have a randomly chosen value anywhere between 0 and 2π. For simplicity assume all the sources emit waves of the same amplitude. Then the $\cos(\varepsilon_2 - \varepsilon_1)$ term in equation (11.16) will take all values from -1 to +1 resulting in pairs of equal but opposite cosines, so the interference terms will cancel. The result is that the intensity resulting from a large number N sources with random phases is given by N times the intensity of any one source, since in this case all the amplitudes were assumed equal. That is

$$I = NI_1 \tag{11.25}$$

where I_1 is the intensity of each source and I is the resultant intensity.
Light sources other than lasers emit waves from a large number of atomic sources. The waves from these sources have random phases and the resultant intensity is the simple scalar addition of the individual intensities. Since atomic emitters vary in phase rapidly and randomly the resultant wave from these sources does so also. Thus, two or more tungsten lamps, bunsen flames, fluorescent tubes, and so on will be incoherent.

SUMMARY OF SUBSECTIONS 11.1.1 TO 11.1.6

(1) Two or more harmonic waves of the same frequency superpose to produce a resultant harmonic wave of the same frequency.

(2) Two waves of the same frequency for which $(\varepsilon_2 - \varepsilon_1)$ is constant are said to be coherent.

(3) Two waves of the same frequency for which crest superposes on crest are said to be in phase. If crest superposes on trough they are said to be 180^0 (π radians) out of phase or antiphase.

(4) If two harmonic waves of the same frequency are travelling in a medium of refractive index n, and are distances x_1 and x_2 from their sources, then the terms nx_1 and nx_2 are called the optical path lengths OPLs . $n(x_2 - x_1)$ is the optical path difference (OPD).

(5) Constructive interference occurs between two waves when crest falls upon crest, or very nearly so. Destructive interference occurs when the crests of one fall upon the troughs of the other. In both cases the waves have the same frequency.

(6) If a very large number of sources emit waves of the same frequency but with random phases then the resultant intensity is N times the individual intensities: each source assumed here to emit waves of the same amplitude. Where the amplitudes are different one simply sums the individual intensities. Because the phases have no constant relationship the sources are said to be incoherent.

11.1.7 Vector addition of two or more waves

A geometric vector is a mathematical element with two properties: it has a magnitude and a direction relative to some reference axis. Now, the wave function $\psi = E\sin(kx - \omega t + \varepsilon)$ has a magnitude E and an angle $\phi = kx - \omega t + \varepsilon$ associated with it. Perhaps then we can use geometric vectors, or vectors for short, to model our sinusoidal wave function and, more importantly, the addition of wave functions. It turns out that we can.
Firstly, let's rewrite the function for ψ by looking at it when the phase has a value $kx-\omega t+\varepsilon+\pi$. If we do this for all the waves the phase differences will remain the same even though all the phases will be increased by π radians. Then we have

$$\psi = E\sin(kx - \omega t + \varepsilon + \pi) = E\sin(\omega t - (kx + \varepsilon))$$

$$\text{or} \quad \psi = E\sin(\omega t + \alpha) \tag{11.26}$$

where we have written $\alpha = -(kx+\varepsilon)$, and the transformation in equation (11.26) makes use of the identity $\sin(\theta + \pi) = \sin(-\theta)$. Now, $\psi = E\sin(\omega t + \alpha)$ has two identifying characteristics, namely E and α. Suppose we borrow a notation from electrical and electronic engineering and write this concisely as $E\angle\alpha$, then the two waves $E_1\angle\alpha_1$ and $E_2\angle\alpha_2$ represent our familiar forms

$$\psi_1 = E_1\sin(\omega t + \alpha_1) \quad \text{and} \quad \psi_2 = E_2\sin(\omega t + \alpha_2)\ .$$

We can represent ψ_1 by a vector of length E_1 rotating anticlockwise at a rate of ω radian/second so that its projection on the vertical axis is ψ_1, figure 11.5 . Note that E_1 makes a constant angle α_1 with the rotating reference axis, and the latter rotates at ω rad/s, E_1 rotating with it.

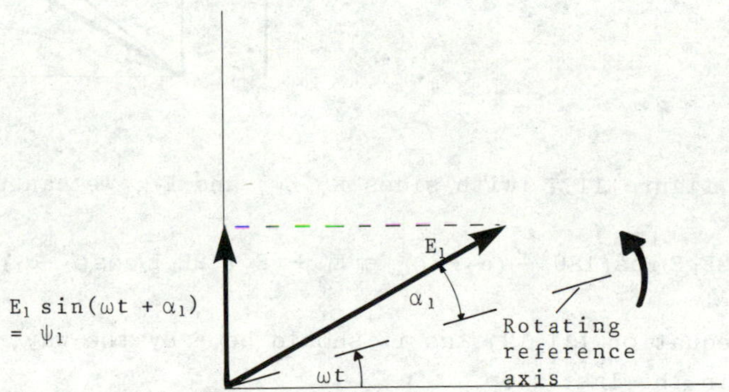

Fig. 11.5 Vector representation of $\psi_1 = E_1\sin(\omega t + \alpha_1)$.

As ωt increases, the projection of E_1 onto the vertical, ψ_1, will increase until $\psi_1 = E_1$ when E_1 is vertical and $\omega t + \alpha_1 = \pi/2$ or 90^0. As E_1 rotates through 2π radians the projection of E_1 on to the vertical will give the successive values of ψ_1. Positive values will be measured upwards and negative values downwards from the horizontal. Now, our purpose is to add two or more harmonic functions of the same frequency. Accordingly, figure 11.6 represents another sine function $\psi_2 = E_2\sin(\omega t + \alpha_2)$.

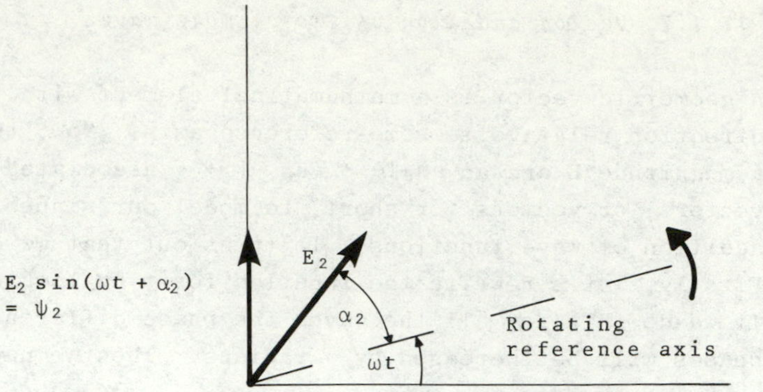

$E_2 \sin(\omega t + \alpha_2)$
$= \psi_2$

E_2 α_2 ωt Rotating reference axis

Fig. 11.6 Vector representation of $\psi_2 = E_2 \sin(\omega t + \alpha_2)$.

In figure 11.7 we add E_1 and E_2 vectorially: note that E is the resultant vector of the addition, α is the angle between E and the rotating reference axis, and $\psi = \psi_1 + \psi_2 = E\sin(\omega t + \alpha)$.

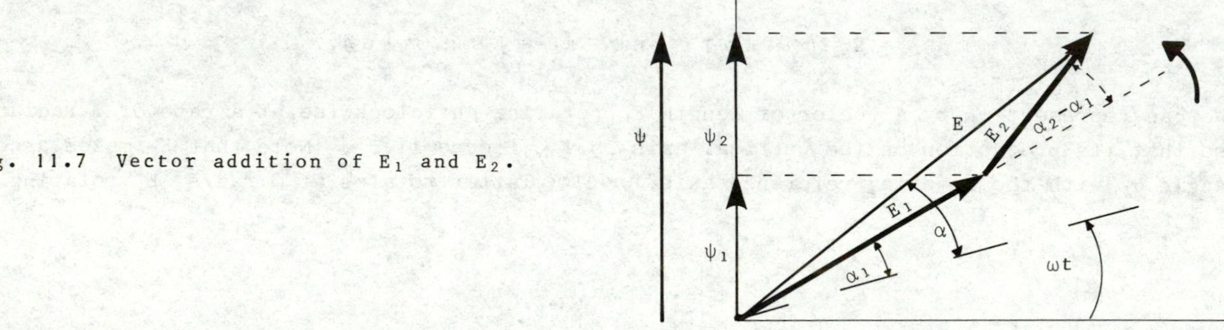

Fig. 11.7 Vector addition of E_1 and E_2.

Using the triangle in figure 11.7 with sides E, E_1, and E_2, we can use the cosine rule to write

$$E^2 = E_1^2 + E_2^2 - 2E_1 E_2\cos(180^0 - (\alpha_2 - \alpha_1)) = E_1^2 + E_2^2 + 2E_1 E_2\cos(\alpha_2 - \alpha_1)$$

which is the same as equation (11.12), as it should be. By the way, we have used the identity $\cos(180^0 - A) = -\cos A$ in the last step.
Note that we are keeping α_1 and α_2 constant which means x_1 and x_2 are constant. The result is the E_1 and E_2 vectors rotate with the rotating reference axis and the phase difference is constant since $\alpha_2 - \alpha_1$ is constant. The function $\psi = \psi_1 + \psi_2$ represents a disturbance at a point in space which varies sinusoidally with time.

Example Find the resultant of adding the sine waves

$$\psi_1 = 20\sin\omega t$$
$$\psi_2 = 10\sin(\omega t + \pi/4)$$
$$\psi_3 = 10\sin(\omega t - \pi/12)$$
and
$$\psi_4 = 15\sin(\omega t + 2\pi/3)$$

We can write these functions in the concise vector notation; thus, we have $20\angle 0$, $10\angle\pi/4$,

$10\angle-\pi/12$, and $15\angle 2\pi/3$. Figure 11.8 shows the vector addition. E and α are measured directly from the diagram. By measurement then, the resultant vector is $E\angle\alpha = 34\angle\pi/6$, which gives

$$\psi = E\sin(\omega t + \alpha) = 34\sin(\omega t + \pi/6) .$$

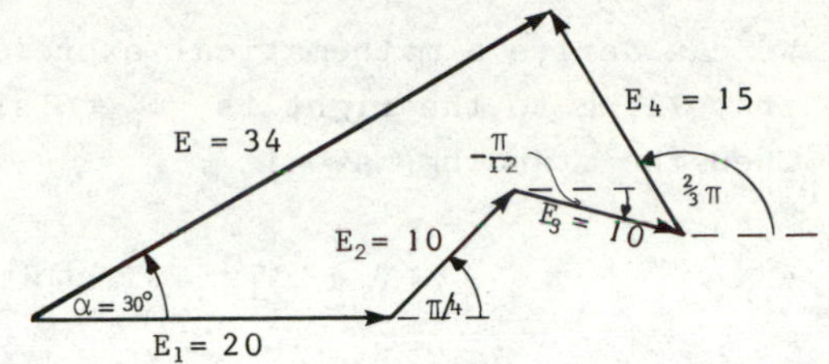

Fig. 11.8 Vector addition of four harmonic waves.

Clearly, the procedure offers an easy method of adding harmonic functions and will be employed when considering antireflection coatings in chapter 13.

11.1.8 *Standing Waves*

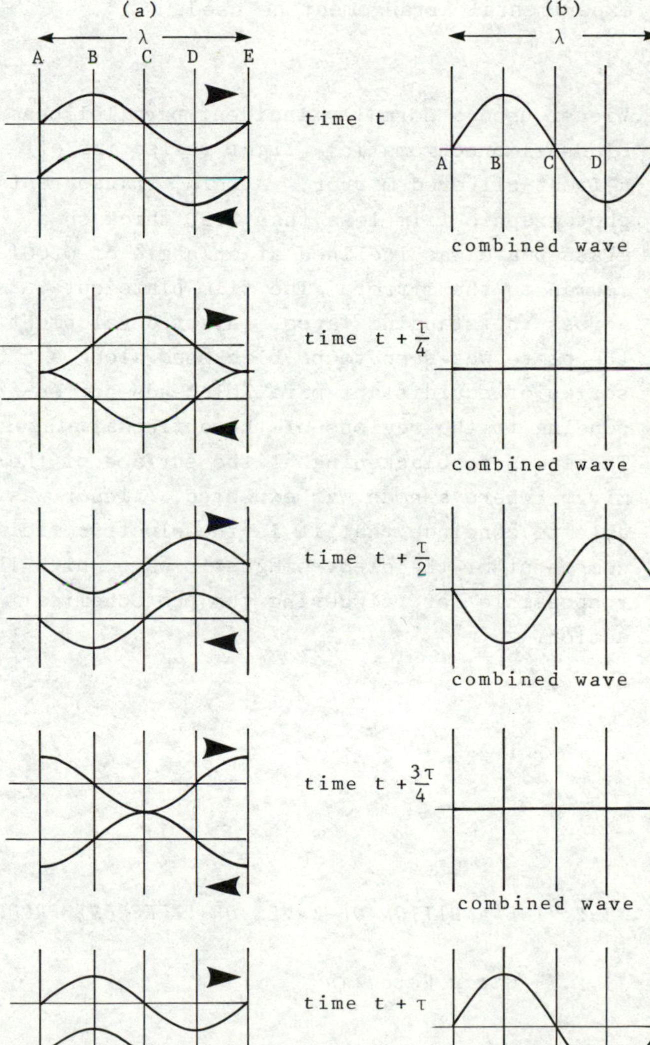

Fig. 11.9 (a) An incident (▶) and reflected wave (◀).
(b) The combined wave: a standing wave.

Figure 11.9 shows two waves of equal amplitude and wavelength travelling in opposite directions. The waves are shown in figure 11.9(a) and the combined wave in figure 11.9(b). The combined wave is a standing or stationary wave. Looking down the (b) diagrams notice that there is zero displacement at the points A, C, and E at all times. These points are called *nodes or stationary points*. However, at the points B and D the displacement oscillates between +2E and -2E where E is the amplitude of the incident and reflected waves. The points B and D are called *antinodes*.

We can derive a mathematical expression for the standing wave. Suppose the incident wave travelling to the right is $\psi_i = E\sin(kx - \omega t)$ and the reflected wave is $\psi_r = E\sin(kx + \omega t)$. Then the standing wave is

$$\psi_s = \psi_i + \psi_r = E\sin(kx - \omega t) + E\sin(kx + \omega t)$$

$$= 2E\sin kx\cos\omega t \qquad (11.27).$$

Notice $\psi_s = 0$ when $\sin kx = 0$ or $\cos\omega t = 0$. Now, $\sin kx = 0$ when $kx \equiv \frac{2\pi}{\lambda} = 0, \pi, 2\pi, 3\pi, \ldots.$ and this is true when $x = 0, \lambda/2, \lambda, 3\lambda/2, \ldots.$, for all times. These values of x are the positions of the nodes so the nodes are a half-wavelength apart.
Further, $\cos\omega t = 0$ when $\omega t \equiv 2\pi\frac{t}{\tau} = \pi/2, 3\pi/2, 5\pi/2, \ldots.$; that is, when $t = \tau/4, 3\tau/4, 5\tau/4, \ldots.$. So, at intervals of $\tau/2$ the combined wave is everywhere zero. Compare these results with those illustrated in figure 11.9.
Standing light waves were first demonstrated by Otto Wiener in 1890. Figure 11.10 shows the experimental arrangement he used.

Wiener used a normally incident parallel beam of quasimonochromatic* light reflecting off a front silvered mirror. A thin, transparent photographic film less than $\lambda/20$ thick on a glass plate was inclined at an angle of 0.001 radian to the mirror. The film plate cut across the standing waves. After development the plate was seen to be blackened along a series of equidistant parallel bands corresponding to the regions of the antinodal planes. There was no blackening at the surface of the mirror where a node was expected. Wiener was able to conclude that it is the electric field component of the electromagnetic wave which is responsible for triggering the photochemical action.

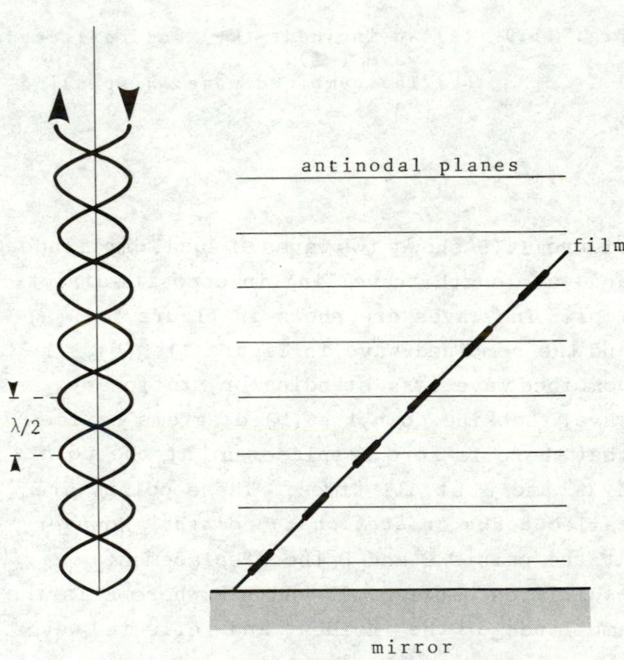

Fig. 11.10 Wiener's experiment

11.2 THE ADDITION OF WAVES OF DIFFERENT FREQUENCY

11.2.1 *Group Velocity*

The most important thing that happens when two sinusoidal waves of different frequencies are added is that the resultant is not itself sinusoidal. One important consequence of this is the fact that wave packets of finite length can be shown to consist of a number of superposed

*See section 11.2.4

sinusoidal waves. Fourier analysis, the mathematical theory which allows one to analyse the sinusoidal components of such a wave packet, is beyond the scope of this text although Appendix 1 has a brief introduction to the methods. We shall, however, use one or two results from the theory. Let us consider the addition of the two waves

$$\psi_1 = E_1 \sin(k_1 x - \omega_1 t) \quad \text{and} \quad \psi_2 = E_1 \sin(k_2 x - \omega_2 t)$$

which have equal amplitudes and zero initial phases. The resultant is

$$\psi = \psi_1 + \psi_2 = E_1(\sin(k_1 x - \omega_1 t) + \sin(k_2 x - \omega_2 t))$$

$$= 2E_1 \cos(\tfrac{1}{2}(k_1 - k_2)x - \tfrac{1}{2}(\omega_1 - \omega_2)t) \times \sin(\tfrac{1}{2}(k_1 + k_2)x - \tfrac{1}{2}(\omega_1 + \omega_2)t) \qquad (11.28)$$

use having been made of the identity $\sin A + \sin B = 2\sin\frac{A+B}{2}\cos\frac{A-B}{2}$.

Writing $\frac{1}{2}(k_1+k_2) = \overline{k}$, $\frac{1}{2}(k_1-k_2) = k_m$, $\frac{1}{2}(\omega_1+\omega_2) = \overline{\omega}$, and $\frac{1}{2}(\omega_1-\omega_2) = \omega_m$ and substituting in equation (11.28) gives

$$\psi = 2E_1 \cos(k_m x - \omega_m t)\sin(\overline{k}x - \overline{\omega}t) \qquad (11.29)$$

where $\overline{k}$ and $\overline{\omega}$ are the average propagation number and the average angular frequency respectively. k_m and ω_m are called the modulation propagation number and the modulation angular frequency. Equation (11.29) represents a travelling wave of angular frequency $\overline{\omega}$ and propagation number $\overline{k}$. The amplitude is given by

$$E = 2E_1 \cos(k_m x - \omega_m t) \qquad (11.30)$$

and, evidently, E is time varying so that the profile is not constant. One can imagine the combined wave travelling to the right but changing its shape. For light ω_1 and ω_2 will be in the range from 7.69×10^{14} Hz for violet light ($\lambda = 390$ nm) to 3.95×10^{14} Hz for red light ($\lambda = 760$ nm). Thus, ω_1 and ω_2 will have closely similar values whence $\overline{\omega} = \frac{1}{2}(\omega_1 + \omega_2)$ will be close to the value of ω_1 or ω_2. However, by comparison, ω_m will be very small since $\omega_m = \frac{1}{2}(\omega_1 - \omega_2)$. Now, the modulation angular frequency is a measure of the rate at which the profile of the group of waves changes shape and the speed at which the profile travels is called the *group velocity*. By analogy with the phase velocity given by equation (10.27) the group velocity is

$$v_g = \frac{\omega_m}{k_m} \qquad (11.31).$$

Suppose now we have more than two sinusoidal waves of different frequencies adding to give a group of waves. The changing profile of these superposed waves is what we have called a wave packet, group, or train. Here, ω_1 and ω_2 will be the extreme angular frequencies and k_1 and k_2 will be the corresponding propagation numbers for the extreme sinusoidal components. It is usual to write $\omega_1 - \omega_2 = \Delta\omega$ and $k_1 - k_2 = \Delta k$ from which we can write the group velocity as

$$v_g = \frac{\Delta\omega}{\Delta k} \qquad (11.32).$$

Example

Suppose a group of waves with frequencies from 4×10^{14}Hz to 5×10^{14}Hz propagates in vacuum. Find the group velocity.

Since the phase velocity is 3×10^8 ms^{-1} for all electromagnetic waves in vacuum we can compute the extreme angular frequencies and wavelengths. Thus,

$$\omega_1 = 2\pi\nu_1 = 2\pi \times 5 \times 10^{14} \text{ rad s}^{-1} \quad \text{and} \quad \omega_2 = 2\pi\nu_2 = 2\pi \times 4 \times 10^{14} \text{ rad s}^{-1}.$$

Then

$$\lambda_1 = v/\nu_1 = (3 \times 10^8)/(5 \times 10^{14}) = 0.6 \times 10^{-6}\text{m and } \lambda_2 = v/\nu_2 = (3 \times 10^8)/(4 \times 10^{14}) = 0.75 \times 10^{-6}\text{m},$$

whence $$k_1 = \frac{2\pi}{\lambda_1} = \frac{2\pi}{0.6 \times 10^{-6}} = \frac{10}{3}\pi \times 10^6 \text{ m}^{-1}$$

and $$k_2 = \frac{2\pi}{\lambda_2} = \frac{2\pi}{0.75 \times 10^{-6}} = \frac{8}{3}\pi \times 10^6 \text{ m}^{-1}.$$

Hence,

$$v_g = \frac{\Delta\omega}{\Delta k} = \frac{\omega_1 - \omega_2}{k_1 - k_2} = \frac{((2\pi \times 5) - (2\pi \times 4)) \times 10^{14}}{(\frac{10}{3} - \frac{8}{3})\pi \times 10^6} = \frac{2\pi \times 10^{14}}{\frac{2}{3}\pi \times 10^6} = 3 \times 10^8 \text{ ms}^{-1}.$$

That is, the profile propagates at the same speed as the components in vacuum.

11.2.2 *Dispersion*

When the wave packet in the last example impinges upon a medium such as glass the wavelengths will change for each component. This means that the propagation numbers will change and the components travel at different speeds in the medium. This results in the dispersal of the components and the consequent 'breaking up' of the group. This is, of course, the familiar splitting up of white light into its constituent colours on passing through a prism, say.

11.2.3 *Bandwidths, Coherence Time, and Coherence Length*

In the example above the *frequency bandwidth* is defined as $\Delta\nu = \nu_1 - \nu_2$, where ν_1 and ν_2 are the extreme frequencies of the group. An important result in Fourier Theory states that the frequency bandwidth is approximately equal to the reciprocal of the temporal extent, Δt, of a pulse or wave packet. That is,

$$\Delta\nu \simeq \frac{1}{\Delta t} \tag{11.33}$$

If a wave packet has a narrow frequency bandwidth equation (11.33) says that it will have a large extent in time, and a large extent in time corresponds to a large length in space. So, the longer the spatial extent of a wavepacket the narrower its frequency bandwidth. The latter is equivalent to saying the longer the wave packet the fewer the component sinusoidal frequencies that add to make the wave packet.

Suppose we were to examine the light emitted by a sodium discharge lamp. The light emitted is often referred to as monochromatic, suggesting that it consists of a sinusoidal wave of a single frequency. However, this is not the case and when the beam is passed through a spectrum analyser it is possible to observe its frequency components. There are a number of narrow frequency ranges containing most of the energy and these regions are separated by regions of darkness. Where the light enters the analyser via a narrow slit images of the slit are formed and they appear as coloured 'lines', known as *spectral lines*. The spectral lines are never perfectly sharp since each slit image is formed by a band of frequencies or wavelengths. Figure 11.11 shows the red line wavelength components from a low pressure cadmium lamp. Note that the figure illustrates the *linewidth* corresponding to the frequency bandwidth. Note also that it is measured between the points where the intensity curve falls to half the intensity maximum: this is borrowed from bandwidth definitions in electronics.

Δt in equation (11.33) is known as the *coherence time*. For waves propagating in vacuum with a speed c (3×10^8 ms^{-1}), the length of the wave packet is known as the *coherence length*, Δx. Waves passing a fixed point in time Δt at a speed c will cover a distance given by speed × time. Thus, the spatial extent of such a wave packet is

$$\Delta x = c.\Delta t \qquad (11.34)$$

This expression will have some practical value when we deal with interference in chapter 13.

Since light from gaseous discharge lamps will always have a spread of frequencies, even if the spread is very narrow, we should strictly talk about quasimonochromatic light. However, we often talk loosely of monochromatic light when we really mean quasimonochromatic light. White light has a frequency range from about 3.95×10^{14} Hz to 7.69×10^{14} Hz, that is, a frequency bandwidth of 3.74×10^{14} Hz. The coherence time Δt is about 2.67×10^{-15} second, given by

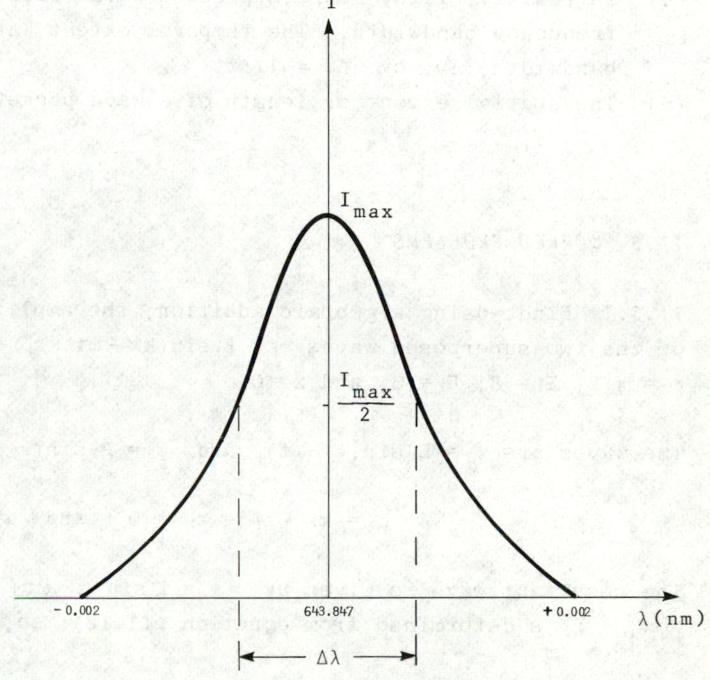

Fig. 11.11 Cadmium red spectral line bandwidth in terms of wavelength. Δλ is called the linewidth.

$$\Delta t \simeq \frac{1}{\Delta \nu} = \frac{1}{3.74 \times 10^{14}} = 2.67 \times 10^{-15} \text{s}.$$

This Δt corresponds to wave packets of length $\Delta x = c.\Delta t = 3 \times 10^8 \times 2.67 \times 10^{-15} = 8 \times 10^{-7}$m, or roughly 0.001 mm. Thus, white light may be thought of as a random succession of short pulses. White light's large number of constituent frequencies can be generated by passing the light through a Fourier analyser such as a prism when the result is the well-known visible spectrum. Ordinary discharge lamps have relatively wide bandwidths leading to coherence lengths of the order of a few millimetres. On the other hand, spectral lines seen with low pressure lamps such as mercury (Hg^{198}, $\lambda_{air} = 546.078$ nm) have bandwidths of about 10^9 Hz. The corresponding coherence length is about 30 cm and the coherence time about 1 ns.

Summary of section 11.2

(1) The addition of two or more harmonic (sinusoidal) waves of different frequencies results in an anharmonic wave: that is, a wave which is not sinusoidal.

(2) The profile of the anharmonic wave propagates at a speed given by the group velocity $v_g = \Delta\omega/\Delta k = (\omega_1 - \omega_2)/(k_1 - k_2)$, where ω_1 and ω_2 are the highest and lowest component angular frequencies in the resultant wave, and k_1 and k_2 are the corresponding propagation numbers.

(3) Dispersion, or spreading of the component frequencies results when the wave passes from vacuum into some optical medium.

(4) In reality, light sources produce waves with a finite range of frequencies known as the frequency bandwidth. The temporal extent, Δt, of a wave packet is related to the frequency bandwidth, $\Delta\nu$, by $\Delta\nu \simeq 1/\Delta t$.

(5) The spatial extent or length of a wave packet is given by $\Delta x = c.\Delta t = c.\frac{1}{\Delta\nu}$

11.3 WORKED PROBLEMS

11.3.1 Find, using algebraic addition, the amplitude and phase resulting from the addition of the two superposed waves $\psi_1 = E_1 \sin(kx - \omega t + \varepsilon_1)$ and $\psi_2 = E_2 \sin(kx - \omega t + \varepsilon_2)$, where $\varepsilon_1 = 0$, $\varepsilon_2 = \pi/2$, $E_1 = 8$, $E_2 = 6$, and $x = 0$.

The waves are $\psi_1 = E_1 \sin(\varepsilon_1 - \omega t)$ and $\psi_2 = E_2 \sin(\varepsilon_2 - \omega t)$. Using equations (11.3) and (11.4)

$$\alpha_1 = kx + \varepsilon_1 = \varepsilon_1 = 0 \quad \text{and} \quad \alpha_2 = kx + \varepsilon_2 = \varepsilon_2 = \pi/2, \quad \text{since } x = 0.$$

The resultant wave is given by $\psi = E \sin(\alpha - \omega t)$, equation (11.13) with $x = 0$.
Now, E is determined from equation (11.12): so,

$$E^2 = E_1^2 + E_2^2 + 2E_1 E_2 \cos(\alpha_2 - \alpha_1) = 8^2 + 6^2 + 2 \times 8 \times 6 \cos(\tfrac{\pi}{2} - 0) = 64 + 36 + (96 \times 0) = 100,$$

whence $E = \sqrt{100} = 10$, taking the positive root.

Next, α is computed from equation (11.11):

$$\tan\alpha = \frac{E_1 \sin\alpha_1 + E_2 \sin\alpha_2}{E_1 \cos\alpha_1 + E_2 \cos\alpha_2} = \frac{8 \sin 0 + 6 \sin(\pi/2)}{8 \cos 0 + 6 \cos(\pi/2)} = \frac{(8 \times 0) + (6 \times 1)}{(8 \times 1) + (6 \times 0)} = \frac{6}{8} = 0.75 .$$

So, $\alpha = \arctan(0.75) = 36.87^0$. Since $1^0 \equiv \pi/180$ radian, $\alpha = 36.87 \times \pi/180 = 0.6435$ radian and the resultant wave is

$$\psi = E \sin(\alpha - \omega t) = 10 \sin(0.6435 - \omega t)$$

or, if x is to vary,

$$\psi = 10 \sin(kx - \omega t + 0.6435) .$$

11.3.2 Two waves $\psi_1 = E_1 \sin(kx - \omega t)$ and $\psi_2 = E_2 \sin(kx - \omega t + \pi)$ are coplanar and overlap. Calculate the resultant's amplitude if $E_1 = 3$ and $E_2 = 2$.

Using equation (11.12) the amplitude squared of the resultant wave is

$$E^2 = E_1^2 + E_2^2 + 2E_1 E_2 \cos(\alpha_2 - \alpha_1) = 3^2 + 2^2 + 2 \times 3 \times 2 \cos\pi , \quad \text{since } (\alpha_2 - \alpha_1) = \pi.$$

Evaluating the expression for E^2 gives $E^2 = 9 + 4 + 2 \times 3 \times 2 \times (-1) = 13 - 12 = 1$. So $E = 1$. This is a case of destructive interference because the phase difference is π radians. Figure 11.12 indicates the result at time $t = 0$.

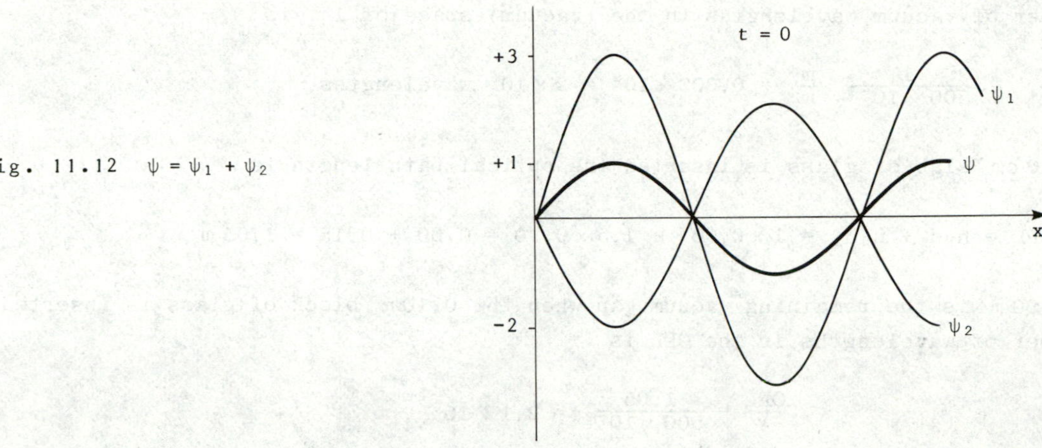

Fig. 11.12 $\psi = \psi_1 + \psi_2$

11.3.3 Show that the optical path length, or more simply the optical path, is equivalent to the length of the path in vacuum which a beam of light of wavelength λ would traverse in the same time.

Let d be the thickness of a material of refractive index n, the latter defined for a vacuum wavelength λ. The speed of light in the medium is $v = c/n$, where c is the speed of light in vacuum. Thus, the time taken by the beam to traverse the thickness d is

$$\text{time} = \frac{\text{distance}}{\text{speed}} = \frac{d}{v} = \frac{d}{c/n} = \frac{nd}{c} .$$

nd is the optical path (length) and a vacuum distance nd will be traversed in a time nd/c by a beam of wavelength λ travelling at a speed c.
If the beam traverses successive distances d_1 , d_2 ,...., in media of indices n_1 , n_2 ,...., then the total time taken is

$$\frac{d_1}{v_1} + \frac{d_2}{v_2} + \ldots = \frac{d_1}{c/n_1} + \frac{d_2}{c/n_2} + \ldots = \frac{n_1 d_1}{c} + \frac{n_2 d_2}{c} + \ldots$$

$$= \frac{1}{c}(n_1 d_1 + n_2 d_2 + \ldots)$$

$$= \frac{1}{c}\sum_{i=1}^{i=q} n_i d_i \quad \text{, for q optical media.}$$

$\sum_{i=1}^{i=q} n_i d_i$ is the optical path length for q media each with refractive index n_i and thickness d_i . In this short-hand notation i takes all the values from 1 to q.

11.3.4 (a) How many vacuum wavelengths of $\lambda = 500$ nm will span a space of 1 m in a vacuum?
(b) How many wavelengths span the gap when the same gap has a 10 cm thick slab of glass ($n_g = 1.5$) inserted in it?
(c) Determine the optical path difference between the two cases.
(d) Verify that OPD/λ is the difference between the answers to (a) and (b).

(a) The number of vacuum wavelengths in the (vacuum) space of 1 m is

$$\frac{1}{500 \times 10^{-9}} \ \frac{\text{m}}{\text{m}} = 0.002 \times 10^{9} = 2 \times 10^{6} \text{ wavelengths.}$$

(b) When a 10 cm slab of glass is inserted the optical path length is

$$\text{OPL} = n_1 d_1 + n_2 d_2 = 1 \times 0.90 + 1.5 \times 0.10 = 0.90 + 0.15 = 1.05 \text{ m},$$

where 0.90 m is the remaining vacuum gap when the 0.10 m block of glass is inserted. The number of wavelengths in the OPL is

$$\frac{\text{OPL}}{\lambda} = \frac{1.05}{500 \times 10^{-9}} = 2.1 \times 10^{6}\ .$$

(c) The Optical Path Difference is the difference between the two optical path lengths. In the first case the OPL is $nd = 1 \times 1 = 1$ m. In the second case it is 1.05 m. So the OPD is $1.05 - 1 = 0.05$ m.

(d) $\dfrac{\text{OPD}}{\lambda} = \dfrac{0.05}{500 \times 10^{-9}} = 10^{5}$. The difference between the answers to (a) and (b) is

$$(2.1 \times 10^{6}) - (2 \times 10^{6}) = 0.1 \times 10^{6} = 10^{5}\ . \quad \text{Q.E.D.}$$

11.3.5 Two waves ψ_1 and ψ_2 both have vacuum wavelengths of 500 nm. The waves arise from the same source and are in phase initially. Both waves travel an actual distance of 1 m but ψ_2 passes through a glass tank with 1 cm thick walls and a 20 cm gap between the walls. The tank is filled with water ($n_w = 1.33$) and the glass has refractive index $n_g = 1.5$. Find the OPD and the phase difference when the waves have travelled the 1 m distance.

Assume air has a refractive index $n_a = 1$ so that the wavelength in air equals the wavelength in vacuum! This is near enough for our purposes. The OPL of $\psi_1 = nd = 1 \times 1 = 1$ m, where d is the one metre distance and n here is $n_a = 1$.

Fig. 11.13

The OPL of ψ_2 is

$$n_1 d_1 + n_2 d_2 + n_3 d_3 + n_4 d_4 + n_5 d_5$$

where, from figure 11.13, $n_1 = n_5 = n_a = 1$, $n_2 = n_4 = n_g = 1.5$, and $n_3 = n_w = 1.33$.

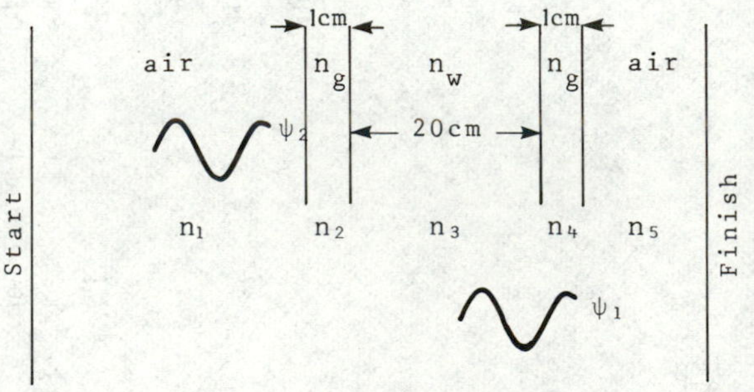

So, the OPL of ψ_2 is

$$n_a(d_1 + d_5) + n_g(d_2 + d_4) + n_w d_3 = 1(1 - 0.22) + 1.5(0.02) + 1.33(0.20)$$

$$= 0.78 + 0.03 + 0.266 = 1.076\text{ m}.$$

Note the air space, $d_1 + d_5$, is 1 m less the space occupied by the tank (= 0.22 m). Hence, the OPD = $OPL_{\psi_2} - OPL_{\psi_1} = 1.076 - 1 = 0.076$ m. The number of vacuum wavelengths which will span this space is $\frac{OPD}{\lambda} = \frac{0.076}{500 \times 10^{-9}} = 1.52 \times 10^5$.

This is the number of wavelengths in the OPL of ψ_2 less the number of wavelengths in the OPL of ψ_1. Each wavelength in the OPD is equivalent to a 2π radians phase difference, so the phase difference δ is given by

$$\delta = 2\pi.\frac{OPD}{\lambda} = 2\pi \times 1.52 \times 10^5 = 9.55 \times 10^5 \text{ radian.}$$

11.3.6 Show that the standing wave $\psi_s(x,t)$ is periodic with time. That is, show that $\psi_s(x,t) = \psi_s(x,t+\tau)$.

The standing wave equation is $\psi_s(x,t) = 2E\sin kx.\cos\omega t$. Then if we increase the time from t to t+τ, we have

$$\begin{aligned}\psi_s(x,t+\tau) &= 2E\sin kx.\cos\omega(t+\tau)\\ &= 2E\sin kx.\cos(\omega t + \omega\tau)\\ &= 2E\sin kx.\cos(\omega t + 2\pi)\\ &= 2E\sin kx.\cos\omega t\\ &= \psi_s(x,t)\end{aligned}$$

since $\omega\tau = 2\pi\nu\tau = 2\pi\frac{1}{\tau}\tau = 2\pi$ and $\cos(\omega t + 2\pi) = \cos\omega t$.

11.3.7 If Wiener's experiment is conducted with red light, $\lambda = 650$ nm, and the photographic film is inclined at 0.5^0 to the mirror, find the distance between the centres of the black bands on the film.

Figure 11.14 shows the geometry where BC is the distance between adjacent antinodes, i.e. $\lambda/2$. AB is the distance required.

$$AB = BC/\sin 0.5^0 = 325 \times 10^{-9}/\sin 0.5^0$$
$$37242.7 \times 10^{-9}\text{ m} \approx 0.037\text{ mm}.$$

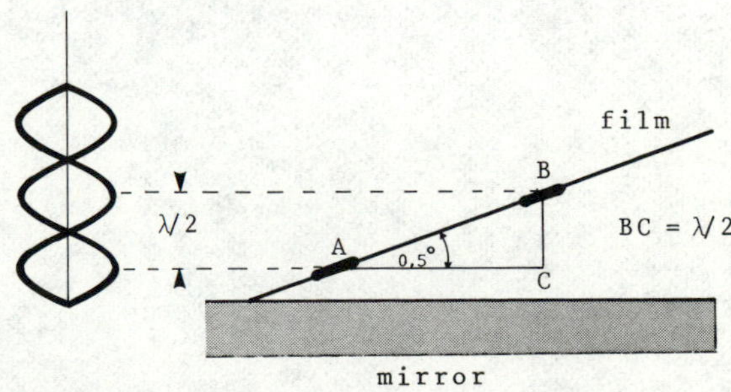

Fig. 11.14

11.3.8 The cadmium red line ($\lambda = 643.847$ nm) has a width $\Delta\lambda = 0.0013$ nm, see figure 11.11. Calculate the coherence length, Δx, if $c = 3 \times 10^8$ ms^{-1}.

The coherence length is given by $\Delta x = c.\Delta t = c.\frac{1}{\Delta\nu}$. We must find the bandwidth $\Delta\nu = \nu_1 - \nu_2$. Recall that ν_1 and ν_2 are the higher and lower extreme frequencies in the band. The wavelengths associated with these frequencies are λ_1 and λ_2 which are equidistant either side of the characteristic wavelength $\lambda = 643.847$ nm. Since $\Delta\lambda = \lambda_2 - \lambda_1 = 0.0013$ nm, $\lambda_1 = \lambda - \Delta\lambda/2$, and $\lambda_2 = \lambda + \Delta\lambda/2$, we can calculate the frequencies.

$$\nu_1 = c/\lambda_1 = 3 \times 10^8/((643.847 - \frac{0.0013}{2}) \times 10^{-9}) = 4.65949 \times 10^{14} \text{ Hz}$$

$$\text{and} \qquad \nu_2 = c/\lambda_2 = 3 \times 10^8/((643.847 + \frac{0.0013}{2}) \times 10^{-9}) = 4.65948 \times 10^{14} \text{ Hz}.$$

Hence, $\Delta\nu = \nu_1 - \nu_2 = (4.65949 - 4.65948) \times 10^{14} = 1 \times 10^9$ Hz. We can now calculate the coherence length: thus

$$\Delta x = c.\frac{1}{\Delta\nu} = \frac{3 \times 10^8}{1 \times 10^9} = 0.3 \text{ m} \equiv 30 \text{ cm}.$$

EXERCISES

1 Add the two coplanar waves $\psi_1 = E\sin(kx - \omega t)$ and $\psi_2 = 3E\cos(kx - \omega t - \frac{\pi}{2})$.

Ans. $\psi = \psi_1 + \psi_2 = 4E\sin(kx - \omega t)$.

2 Add the two coplanar waves $\psi_1 = 3\sin(kx - \omega t)$ and $\psi_2 = 4\sin(kx - \omega t + \frac{\pi}{4})$.

Ans. $\psi = 6.478\sin(kx - \omega t + 0.4518)$.

3 What is the phase difference between the waves ψ_1 and ψ_2 in question 2 ?

Ans. $\phi_2 - \phi_1 = \pi/4$.

4 A light wave, $\lambda = 660$ nm, passes into and out of a glass tank which has walls 5 cm thick, an internal length of 50 cm, and $n_g = 1.6$. If the glass is replaced by a plastics material of refractive index 1.5, what wall thickness will be required to keep the optical path length the same assuming the internal length remains the same? How many waves occupy the wall-interior-wall length?

Ans. Each wall will need to be 5⅓ cm thick. 10^6 waves occupy this OPL.

5 Which two of the following waves will interfere to produce (a) total constructive interference and (b) total destructive interference? $\psi_1 = 4\sin(kx - \omega t)$, $\psi_2 = 4\cos(kx - \omega t + \frac{\pi}{2})$, $\psi_3 = 4\cos(kx - \omega t - \frac{\pi}{2})$, and $\psi_4 = 4\sin(kx - \omega t - \frac{\pi}{4})$.

Ans. (a) ψ_1 and ψ_3. (b) ψ_1 and ψ_2.

6 What is the amplitude of the resultant wave in question 5(a)?

Ans. 8.

7 Two coherent coplanar waves have amplitudes of 2 and 3 units and the same frequency, but a phase difference of $\pi/3$. Find the intensity of the resultant wave relative to that of the individual waves.

Ans. 19 .

8 If the waves in question 7 are incoherent, what then is the resultant intensity relative to the individual waves?

Ans. 13 .

9 Use vector addition to add the two waves $\psi_1 = 10\sin(kx - \omega t)$ and $\psi_2 = 10\sin(kx - \omega t + \frac{\pi}{2})$. Find $E\angle\alpha$ for the resultant wave.

Ans. $10\sqrt{2}\angle 45^0$.

10 Wiener's experiment is undertaken with light of wavelength 650 nm and an angle of 0.001 radian between the film plate and the mirror. Find the distance of the first and second black bands from the line contact between the film plate and the mirror.

Ans. 0.1625 mm and 0.4875 mm.

11 Consider the example in section 11.2.1 and the brief section 11.2.2 . If the group of waves in the example passes into a medium for which the refractive indices are $n_{\nu 1} = 1.51$ and $n_{\nu 2} = 1.49$, calculate the group velocity in the medium using $v_g = \frac{\Delta\omega}{\Delta k} = \frac{\omega_1 - \omega_2}{k_1 - k_2}$, but remember each propagation number will be $\frac{2\pi}{\lambda/n}$, since the wavelength varies inversely with refractive index. Also, calculate the phase (or wave) velocities of the two extreme components and thereby account for dispersion.

Ans. $v_g = 3 \times 10^7\ \text{ms}^{-1}$ and the phase velocities are $\frac{c}{1.49}$ and $\frac{c}{1.51}$.

12 Using the extreme frequencies 5×10^{14} and 4×10^{14} Hz in the wave group of the previous question, calculate the coherence time, Δt, and the coherence length, Δx.

Ans. $\Delta t \simeq 10^{14}$ s. $\Delta x = 3 \times 10^{-6}$ m $\equiv$ 0.003 mm.

13 A 'particle' of light, a photon, is emitted by an atom in about 10^{-8} s. How long is the wave packet?

Ans. 3 m.

14 Substitute $E = \text{constant} \times \sqrt{I}$ from equation (11.15) into equation (11.12) to derive the relationship shown in equation (11.16). Use C for the constant, say, so $E_1 = C\sqrt{I_1}$, $E_2 = C\sqrt{I_2}$, and $E = C\sqrt{I}$. C^2 terms will appear as a common factor!

12 POLARISATION

12.0 INTRODUCTION

In chapters 10 and 11 we dealt with light waves where the electric field or optical disturbance was confined to one plane and that plane was perpendicular to the direction of propagation. The disturbance occurred in the vibration plane or plane of polarisation, figure 10.14(a), and the waves were called transverse, plane polarised waves. Chapter 11 dealt with the superposition of two or more harmonic waves in the same vibration plane and travelling along the same propagation axis, the x-axis. In this chapter, however, we shall be concerned with the interaction of light and matter and the wave resulting from the interaction of two mutually perpendicular plane polarised waves of the same frequency. Whatever the result of the interaction of two such mutually orthogonal waves, and it may not be a plane polarised wave, we shall examine its *state of polarisation*. In particular, we shall be looking for methods of observation of the state of polarisation, methods of production, and applications.

12.1 STATES OF POLARISATION

12.1.1 Plane or Linear Polarisation

We shall consider light propagating along the z-axis and two waves with their electric field vectors E_x and E_y in the x- and y-directions, figure 12.1 .

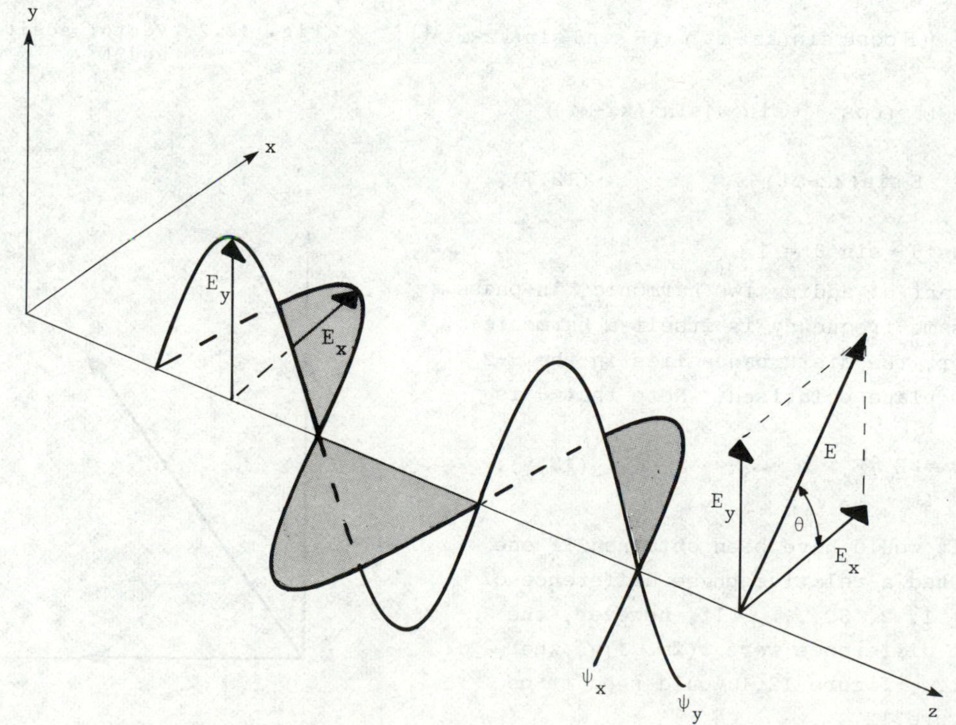

Fig. 12.1 Two superposed linear or plane polarised waves in orthogonal planes.

Let the two waves in figure 12.1 be

$$\psi_x = E_x \sin(kz - \omega t) \tag{12.1}$$

and

$$\psi_y = E_y \sin(kz - \omega t) \tag{12.2}$$

That is, they are of the same frequency, travelling in the positive z-direction, and in phase. Since E_x and E_y can be modelled by vectors we can add them vectorially. Let us call the resultant E, figure 12.2 . Then we can write

$$E_x = E\cos\theta \tag{12.3}$$

and

$$E_y = E\sin\theta \tag{12.4}$$

where θ is the angle E makes with the x-axis. Equations (12.1) and (12.2) now become

$$\psi_x = E_x \sin(kz - \omega t) = E\cos\theta \sin(kz - \omega t) \tag{12.5}$$

and

$$\psi_y = E_y \sin(kz - \omega t) = E\sin\theta \sin(kz - \omega t) \tag{12.6}.$$

ψ_x and ψ_y are the displacements in the x- and y-directions and are vector quantities again. Let the vector resultant from their addition be ψ, figure 12.3 . From this figure we have

$$\psi = (\psi_x^2 + \psi_y^2)^{\frac{1}{2}} = \Big((E\cos\theta \sin(kz - \omega t))^2 + (E\sin\theta \sin(kz-\omega t))^2\Big)^{\frac{1}{2}}$$

$$= (E^2(\cos^2\theta + \sin^2\theta)\sin^2(kz-\omega t))^{\frac{1}{2}}$$

$$= E\sin(kz-\omega t) \tag{12.7},$$

having used $\cos^2\theta + \sin^2\theta = 1$.

So, the resultant of adding two harmonic, in-phase waves of the same frequency is itself a harmonic wave. Moreover, the disturbance lies in the θ-z plane so it is plane polarised. Note that θ is given by

$$\tan\theta = E_y / E_x \tag{12.8}.$$

The same result would have been obtained if one wave, say ψ_x, had a relative phase difference of $\pm 2m\pi$, where $m = 1, 2, 3, \ldots$. If, however, the relative phase difference were $\pm(2m+1)\pi$, where $m = 0, 1, 2, \ldots$, figure 12.1 would need to be redrawn as figure 12.4 .

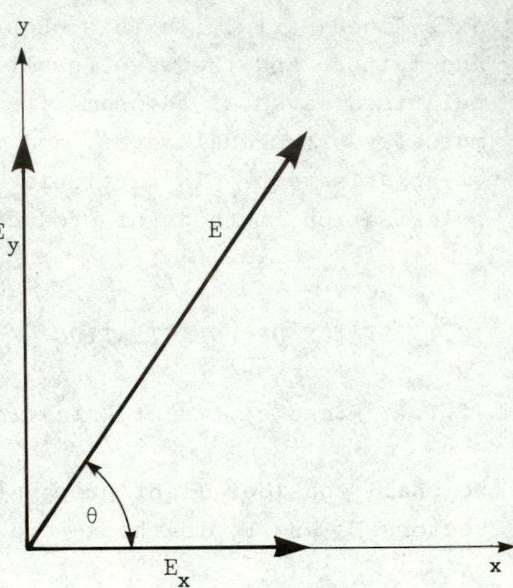

Fig. 12.2 Vector addition of E_x and E_y.

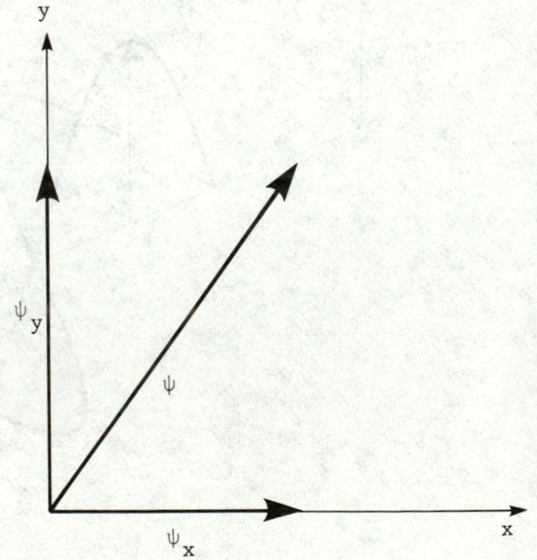

Fig. 12.3 Vector addition of ψ_x and ψ_y.

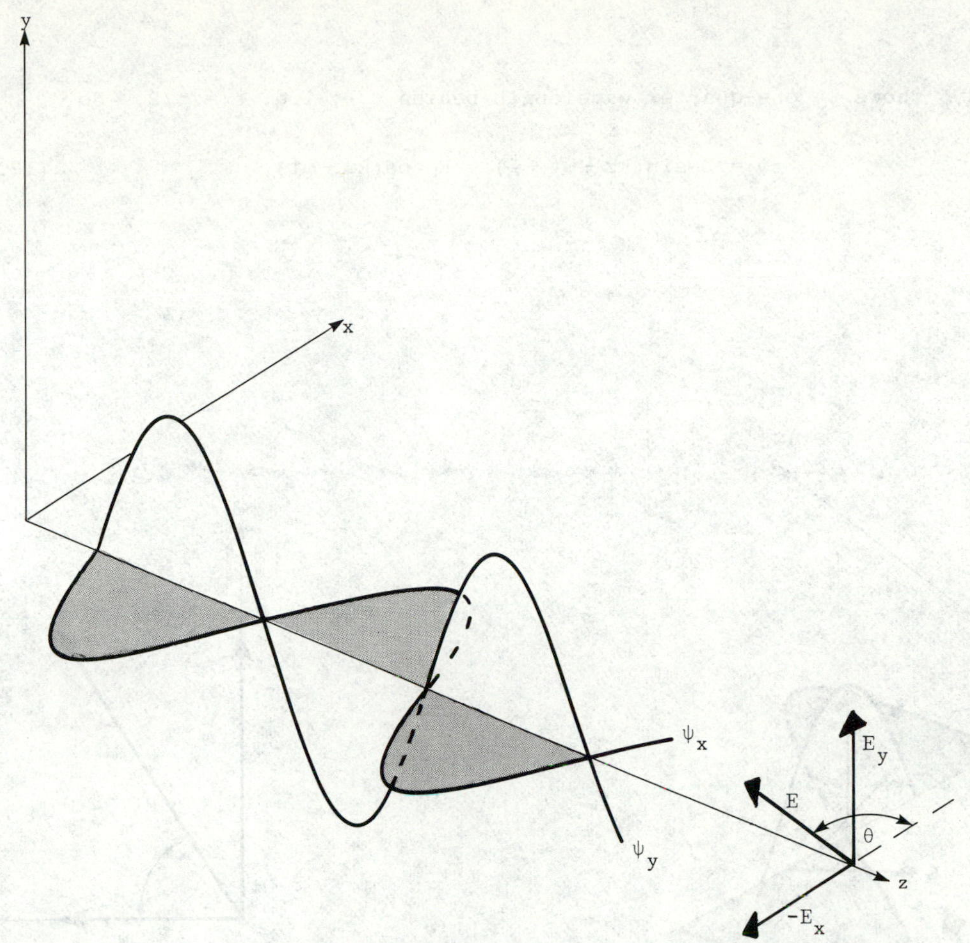

Fig. 12.4 ψ_x with a phase difference π compared with figure 12.1

The resultant disturbance is still $\psi = E\sin(kz - \omega t)$ but now it occupies a plane given by

$$\tan\theta = E_y/(-E_x) \qquad (12.9)$$

Recall that we have been adding two waves to find a single resultant plane polarised wave. The reverse process of taking a single plane polarised wave, ψ, in the θ-z plane, or for simplicity let's call it the θ-plane, and finding its components ψ_x and ψ_y is equally possible and will most certainly be used.

Summarising, adding two orthogonal, in-phase or antiphase harmonic, plane polarised waves of the same frequency results in a single plane polarised wave of the same frequency in some θ-plane determined by equation (12.8) or by equation (12.9).

12.1.2 Circular Polarisation

Consider now another special case of two mutually orthogonal harmonic waves with a relative phase difference of ε and with equal amplitudes E_1. That is,

$$\psi_x = E_1\sin(kz - \omega t + \varepsilon) \;......(12.10) \quad \text{and} \quad \psi_y = E_1\sin(kz - \omega t) \;......(12.11) .$$

Figure 12.5 shows ψ_x one-quarter wavelength behind ψ_y ; i.e. $\varepsilon = \pi/2$. So,

$$\psi_x = E_1 \sin(kz - \omega t + \frac{\pi}{2}) = E_1 \cos(kz - \omega t) \qquad (12.12)$$

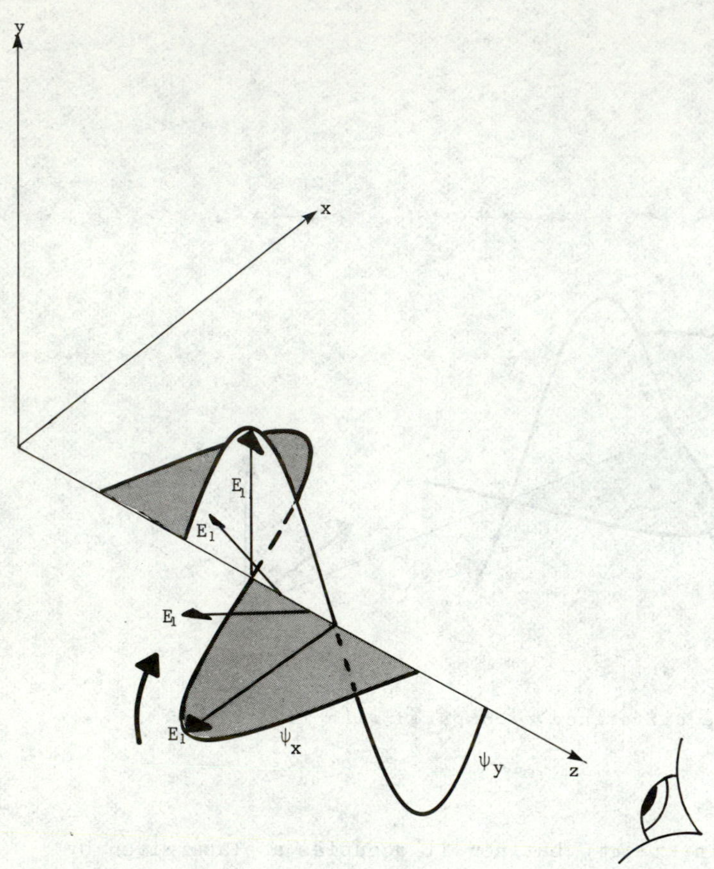

Fig. 12.5 Right circular polarised light.

Fig. 12.6 Resultant disturbance at z = 0 .

The disturbance resulting from adding ψ_x and ψ_y is

$$\psi = (\psi_x^2 + \psi_y^2)^{\frac{1}{2}} = \left((E_1 \cos(kz - \omega t))^2 + (E_1 \sin(kz - \omega t))^2\right)^{\frac{1}{2}}$$

$$= \left(E_1^2 \left(\cos^2(kz - \omega t) + \sin^2(kz - \omega t)\right)\right)^{\frac{1}{2}}$$

$$= (E_1^2)^{\frac{1}{2}} , \quad \text{having used } \cos^2 A + \sin^2 A = 1$$

$$\Rightarrow \quad \psi = E_1 .$$

This is an interesting result: adding two equal amplitude harmonic waves in the manner above has resulted in a disturbance which is constant in magnitude, that is, equal to E_1 . To see exactly how this disturbance propagates along the z-axis let's look at the angle θ which the resultant electric vector makes with the x-axis. Refer to figure 12.6 which shows ψ_x and ψ_y

at $z = 0$, for convenience. Hence,

$$\psi_y = E_1 \sin(-\omega t) \quad \text{.......(12.14)} \quad \text{and} \quad \psi_x = E_1 \cos(-\omega t) \quad \text{.......(12.15)}$$

Now, $\tan\theta = \frac{\psi_y}{\psi_x} = \frac{E_1 \sin(-\omega t)}{E_1 \cos(-\omega t)} = \tan(-\omega t)$ so that $\theta = -\omega t$ ....(12.16)

We see that θ is therefore time dependent, so the resultant $\psi = E_1$ rotates in a clockwise direction as t increases. Seen looking back along the z-axis in figure 12.5 the electric vector rotates clockwise as it sweeps past the observer. Of course, the eye is not really capable of detecting this phenomenon! We are merely employing a little artist's licence to emphasise the effect. This state of polarisation is called *right circular polarisation.*

Inspection of equation (12.16) shows $\theta = 0$ at $t = 0$. Now, $\omega = 2\pi\nu = 2\pi/\tau$, so

$$\theta = -\omega t = -\frac{2\pi}{\tau}t \qquad (12.17)$$

When $t = \tau$ then $\theta = -\frac{2\pi}{\tau}\tau = -2\pi$, so E_1 rotates from 0 to -2π, a full circle, in a time τ. ω is clearly the rate of rotation.

Suppose next that ψ_x is one-quarter wavelength ahead of ψ_y, figure 12.7.

Then $\psi_x = E_1 \sin(kz - \omega t - \frac{\pi}{2})$

$= -E_1 \cos(kz - \omega t)$ ....(12.18)

Again, $\psi = E_1$, but this time at $z = 0$

$$\tan\theta = \frac{\psi_y}{\psi_x} = \frac{E_1 \sin(-\omega t)}{-E_1 \cos(-\omega t)} = -\tan(-\omega t)$$

$$= \tan(\omega t) \qquad (12.19)$$

so $\theta = \omega t$ (12.20).

Compare equations (12.16) and (12.20). The minus sign indicates rotation in the opposite direction so, seen as in figure 12.7, the electric vector E_1 rotates anticlockwise and this is known as *left circular polarised light.*

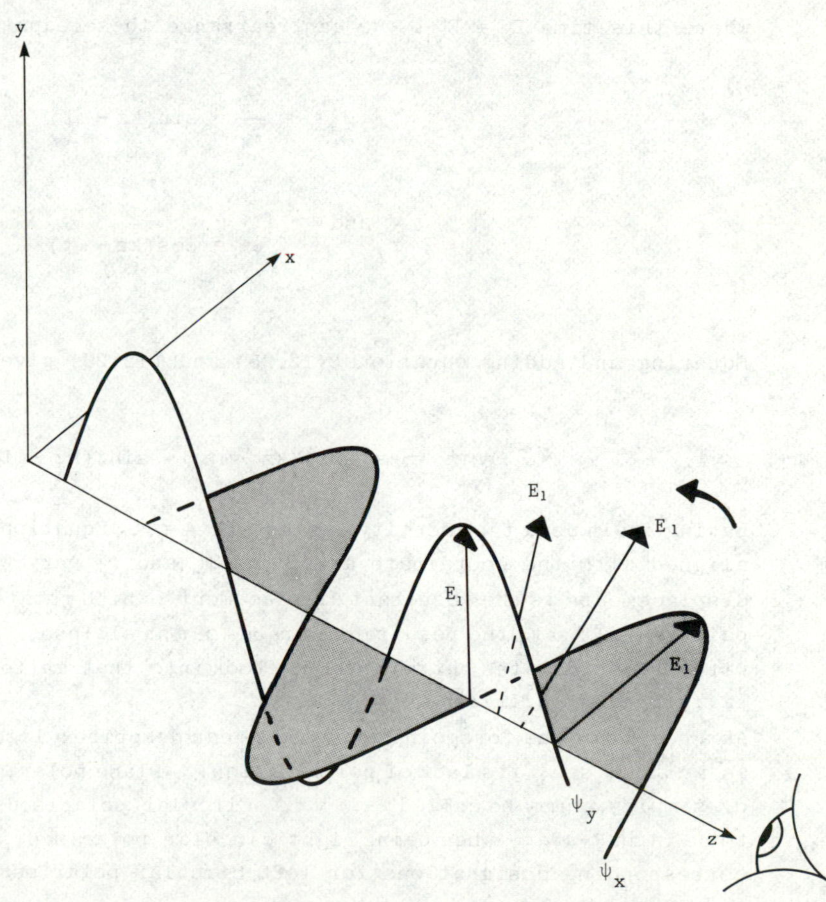

Fig. 12.7 Left circular polarised light.

A plane polarised wave can be synthesised from two oppositely polarised circular waves, providing the amplitudes and frequencies are equal. Adding equations (12.11) and (12.12) produced right circular light, and the addition of equations (12.11) and (12.18) resulted in left circular light. So, if we add the two oppositely polarised circular waves we have

$$E_1 \sin(kz - \omega t) + (-E_1 \cos(kz - \omega t)) + E_1 \sin(kz - \omega t) + E_1 \cos(kz - \omega t)$$

$$= 2E_1 \sin(kz - \omega t) \qquad (12.21)$$

which is a plane polarised wave. The plane wave resides in the y-z-plane since equation (12.21) is evidently given by

$$\psi_y + (-\psi_x) + \psi_y + \psi_x = 2\psi_y \qquad (12.22).$$

12.1.3 *Elliptical Polarisation*

Suppose we now add the waves

$$\psi_x = E_x \sin(kz - \omega t) \;.......(12.22) \quad \text{and} \quad \psi_y = E_y \cos(kz - \omega t) \;.......(12.24)$$

where this time $E_x \neq E_y$. We can rearrange these equations to read

$$\frac{\psi_x}{E_x} = \sin(kz - \omega t) \qquad (12.25)$$

and

$$\frac{\psi_y}{E_y} = \cos(kz - \omega t) \qquad (12.26).$$

Squaring and adding equations (12.25) and (12.26) gives

$$\frac{\psi_y^2}{E_y^2} + \frac{\psi_x^2}{E_x^2} = \cos^2(kz - \omega t) + \sin^2(kz - \omega t) = 1 \qquad (12.27)$$

having again used the identity $\cos^2 A + \sin^2 A = 1$. Equation (12.27) is the equation of an ellipse aligned with the coordinate axes. As ψ_y and ψ_x vary with time, the vector addition of these displacements is the resultant displacement ψ which rotates and varies in magnitude. So, at some point $z = z_0$, say, the resultant ψ traces out an ellipse. In general, the ellipse will not align with the coordinates but we shall not look into that matter. The resultant is then said to be *elliptically polarised*.

As a result of the foregoing analyses we can describe a light wave in terms of its polarisation, or as we shall see, its lack of polarisation. Plane polarised light will be said to be in a *P-state*, or simply, it may be called *P-light*. Circular polarised light may be called *R-light* or be said to be in an *R-state* when being right circular polarised, whilst *L-light* and *L-state* are the corresponding designations for left circular polarised light.

12.1.4 *Natural Light*

An ordinary light source consists of a very large number of atomic emitters randomly orientated.

An excited atom radiates a plane polarised wave packet for about $\Delta t = 10^{-8}$ s and combinations of all emissions of the same frequency give the resultant wave packet. Since the resultant wave packet has a finite length ($\Delta x = c.\Delta t$) natural light consists of wave packets with random polarisations being emitted successively in such a way as to make their polarisation undetectable. A detector such as the eye cannot detect the rapid changes in polarisation resulting from successive wave packets.

Such rapidly varying polarised light is referred to as *natural light* or *unpolarised light*. The latter term is not strictly correct since each individual wave packet is plane polarised but the polarisation varies from wave packet to wave packet too quickly to be discerned.

The term natural light applies to all sources such as the sun, thermal and gaseous discharge lamps, and flames.

12.1.5 Interaction of Electromagnetic Waves and Matter

Matter consists of positively and negatively electrically charged particles. All matter is built from fundamental particles called *protons, electrons,* and *neutrons.* In this simplified picture we imagine atoms, the 'building bricks' of matter, to consist of a nucleus containing positively charged protons and uncharged neutrons with negatively charged electrons 'orbiting' the nucleus. For each positively charged proton in an electrically neutral atom there is one negatively charged electron.

The simplest atom is that of the element hydrogen which is a gas at temperatures above -259.2°C. It consists of one proton and one electron and the electron can occupy several *'energy levels'*. In its lowest energy level the atom is said to be in its *ground state.* Suppose the electron makes a transition from an initial high energy level, E_i, to a final lower energy level, E_f, then an electromagnetic wave packet is emitted with a frequency given by

$$h\nu = E_i - E_f = E \qquad (12.28).$$

That is, $E = h\nu$, a relationship postulated by Albert Einstein in 1905. h is *Planck's constant* and has a value 6.626×10^{-34} Joule seconds. This energy content, E, of a wave packet is transferred to matter when it absorbs the wave packet, so that absorption takes place in discrete amounts. In some materials, notably electrically conducting materials such as metals, an incident wavepacket may be absorbed by 'driving' so-called free electrons in the material. These electrons are driven by the alternating electric field of the wave packet. The driven electrons collide with the lattice atoms of the conductor thereby imparting thermal energy to the material, the energy being lost from the wave packet which is thus absorbed.

In dielectric (non-electrically conducting) materials, such as glass, the incident wave packet can be imagined as exciting electrons into an oscillatory motion. The frequency of the oscillation is that of the incident light. The accelerated electrons in turn radiate light of the same frequency, some forwards to drive adjacent electrons and act as sources of secondary wavelets to transmit the light through the glass, whilst some radiate 'backwards' or reflect the light. The manner in which light interacts with matter can be exploited to advantage to produce polarised light.

12.2 POLARISERS

We are now in a position to look at the methods used to produce, change, and manipulate polarised light. A device which inputs natural light and has an output of some form of polarised light is called a *polariser*. One possible representation of natural or unpolarised light is the superposition of two equal-amplitude orthogonal P-states which are incoherent. A device which separates these, perhaps absorbing one and transmitting the other, would be a linear or plane polariser. Other devices may output circular or elliptical light.

A device which is not one hundred per cent efficient in producing the required polarisation is described as being *leaky* or a *partial polariser*.

Four physical mechanisms can produce polarised light. They are

(i) *dichroism* - the selective absorption of light

(ii) *reflection*

(iii) *scattering*

and (iv) *birefringence* or *double refraction*.

All four mechanisms have one factor in common. The system involved must have some form of asymmetry in order to select one state of polarisation and remove all others from the incident natural light. The asymmetry may be related to the incident or viewing angle or it may be an anisotropy in the polariser itself.

12.2.1 *Detection of P-state light*

We shall need to be able to detect plane polarised light. If natural light is incident on an ideal plane polariser as in figure 12.8, light in a P-state parallel to the polariser's transmission axis will emerge.

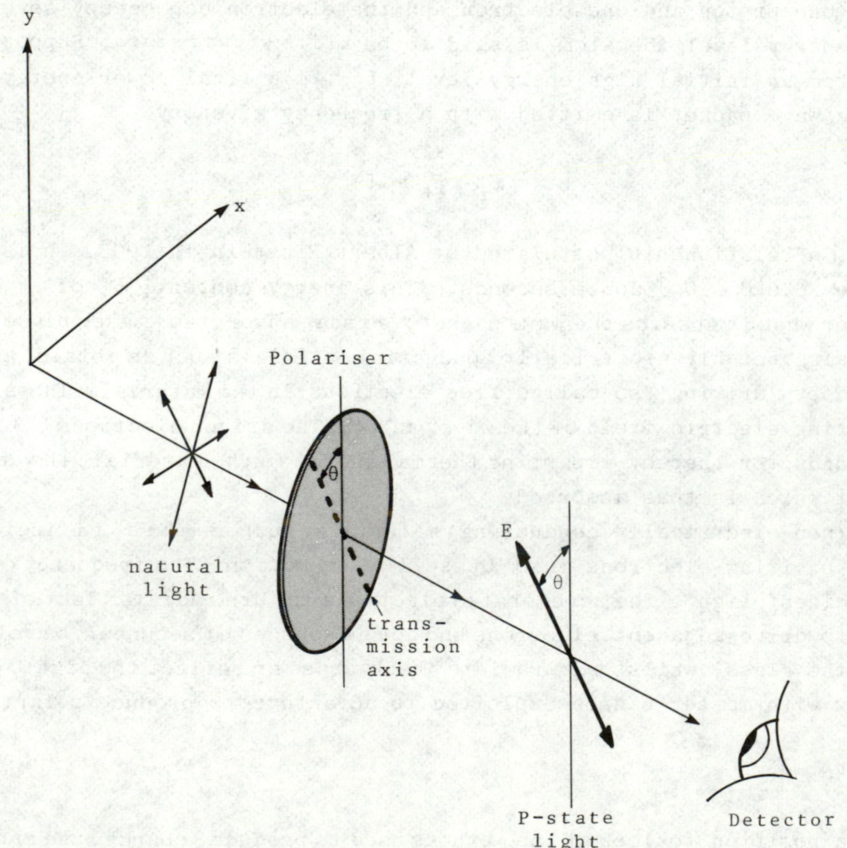

Fig. 12.8 Production of P-state light.

Note that here we are representing natural light by a number of randomly orientated double arrowed lines of unequal lengths. We can imagine these lines as representing twice the amplitude of the component P-state waves. Each component P-state in the incident natural light can itself be resolved parallel and perpendicular to the polariser's transmission axis. Those components parallel to this axis will emerge whilst those perpendicular to it will be extinguished. So, all the emergent light is in a P-state, making an angle θ with the y-axis in figure 12.8 . If we rotate the polariser about the z-axis we rotate the plane of the emergent P-state state. The intensity seen by the detector (e.g. an eye or a photocell) will remain the same since over periods of time much greater than 10^{-8}s the random P-states in natural light will resolve into any two mutually orthogonal vibration planes. Half the energy will be contained in each component when time averaged over relatively long periods, so the intensity emerging will always be half the incident intensity. That is, whatever the orientation of the polariser, half the energy in an unpolarised incident beam will be transmitted in the plane polarised beam which emerges from the polariser. We are here assuming an 'ideal' polariser which does not absorb any energy in the component parallel to its transmission axis and no energy is lost by reflection! Suppose now we add another ideal polariser called an *analyser* in these circumstances, as shown in figure 12.9 .

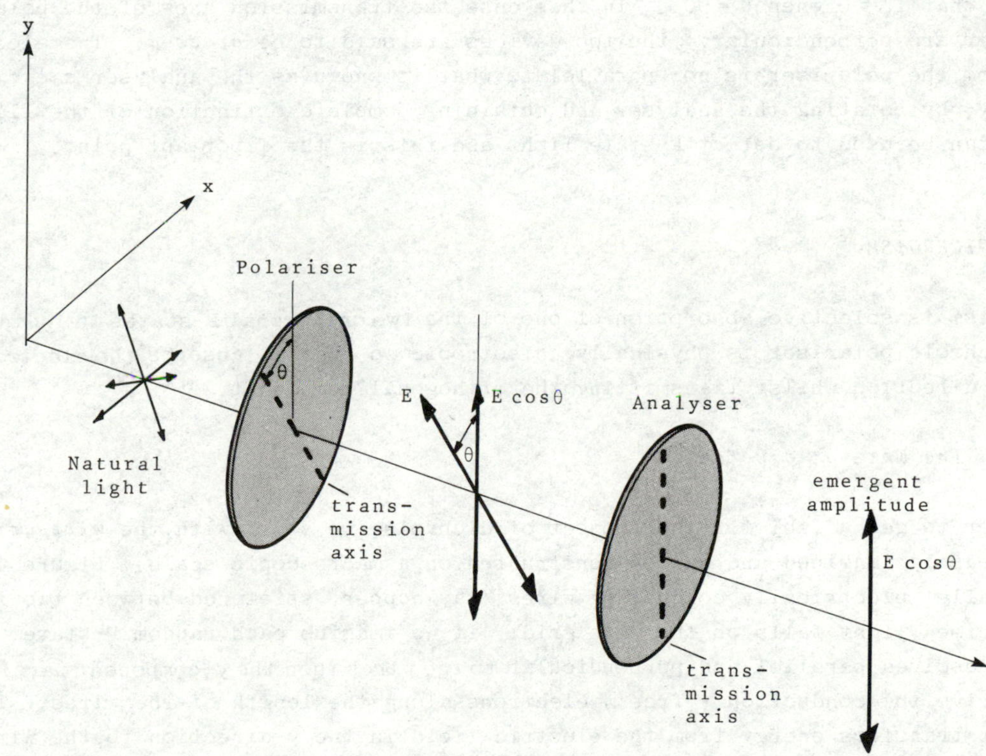

Fig. 12.9 The set-up for detecting P-state light.

Note the direction of the analyser's transmission axis. It will only transmit light with P-states parallel to this axis so the P-state emerging from the polariser must be resolved into components parallel and perpendicular to the analyser's transmission axis. The electric vector component parallel to the analyser's transmission axis is

$$E_\theta = E\cos\theta \qquad (12.29)$$

Now, let the intensity emerging from the polariser be I_0. Then we can write

$$I_0 \propto E^2 \qquad (12.30)$$

since the intensity is proportional to the amplitude squared. Writing I_θ for the intensity emerging from the analyser in figure 12.9, where

$$I_\theta \propto E_\theta^2 = E^2\cos^2\theta \qquad (12.31),$$

and dividing equation (12.31) by equation (12.30) causing the constants of proportionality (not shown) to cancel, then

$$\frac{I_\theta}{I_0} = \frac{E^2\cos^2\theta}{E^2},$$

$$\text{or} \qquad I_\theta = I_0\cos^2\theta \qquad (12.32).$$

Equation (12.32) is known as *Malus' Law* after E. Malus, a French engineer who published it in 1809.

Notice that $I_\theta = 0$ when $\theta = 90^0$. In this case the transmission axes of the polariser and the analyser are perpendicular; the two devices are said to be *crossed*. The electric field emerging from the polariser is now parallel to what is known as the analyser's *extinction axis*. Clearly, by rotating the analyser and obtaining complete extinction of the light we have a system which can be used to detect P-state light and this is the important point.

12.3 DICHROISM

Dichroism is selective absorption of one of the two orthogonal P-states in incident natural light. The dichroic polariser is physically anisotropic so that it absorbs the electric field component in one direction whilst transmitting the orthogonal component.

12.3.1 The Wire-grid Polariser

In order to get a feel for the subject of dichroism we start with the *wire-grid polariser* since it is easily imagined and can be constructed on a macroscopic scale. Figure 12.10 shows a grid of parallel electrically conducting wires, say copper, stretched between two insulating boards. Unpolarised light falls on the wire grid. If we imagine each random P-state in the unpolarised light resolved parallel and perpendicular to the wires then the y-component parallel to the wires will drive the conduction ('free') electrons along the length of the wires. This alternating current transfers energy from the electric field in the y-direction to the wires and the energy appears as heat. The driven electrons collide with the metal's lattice atoms transferring kinetic energy which was initially in the incident beam.

The x-components of each wave packet, perpendicular to the wires, are not able to drive the conduction electrons since they are not free to move very far across the width of the wires. Thus, the x-component of the field is almost unaltered and transmits through the grid. Note that the grid's transmission axis is PERPENDICULAR to the wires.

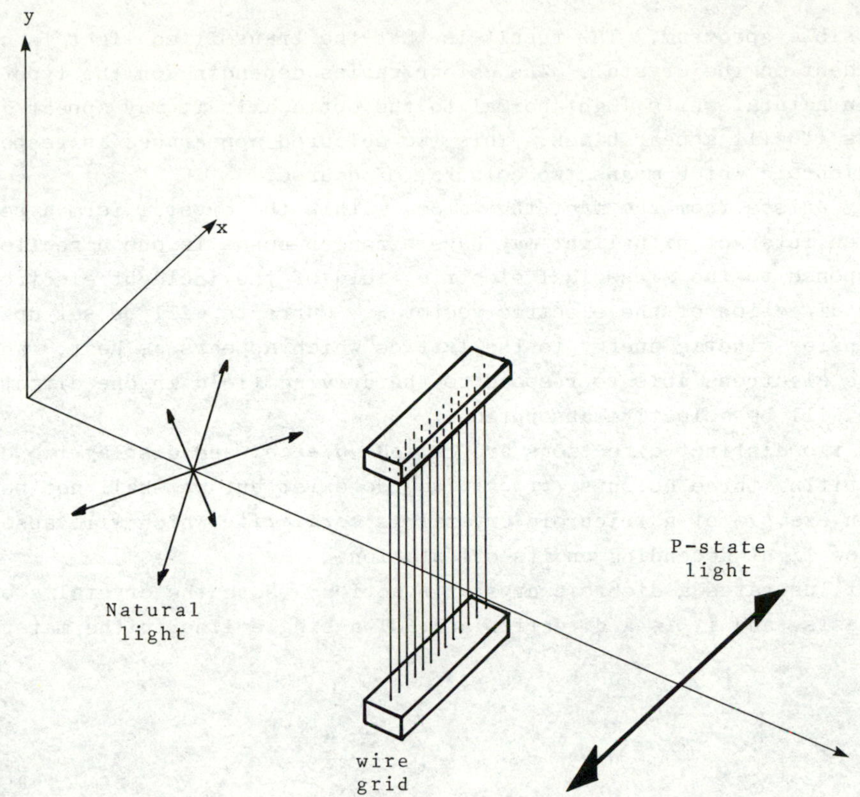

Fig. 12.10 The wire-grid polariser.

It is possible to construct such a grid on a macroscopic scale if we use microwaves, say 1 cm wavelength, and place the wires about 0.5 cm apart. A 'wire-grid' was made in 1960 with 'wires' about 463 nm apart. The 'wires' were in fact made by evaporating gold or aluminium atoms on to a plastics grating replica (see chapter 14). The atoms settled on the edges of the steps in the grating and formed 'wires'.
The wire-grid illustrates the anisotropy which is present in all dichroic polarisers. It is easy to see the physical asymmetry of the wire-grid and equally easy to appreciate the different electrical properties in two mutually orthogonal directions.

12.3.2 Dichroic Crystals

There are certain crystals which are dichroic because of an anisotropy in their structure. Tourmaline is one of these. It is a boron silicate existing in several forms with slightly differing chemical compositions. There is a specific direction within the crystal known as the *optic axis* which is determined by the arrangement of the atoms within the crystal. The electric field component perpendicular to the optic axis is strongly absorbed by the crystal. A plate of tourmaline cut parallel to its optic axis and a centimetre or so thick will effectively extinguish P-states perpendicular to the axis. It thus transmits P-state light parallel to the optic axis. Unfortunately, the transmitted light suffers some absorption and this absorption is not uniform across the visible spectrum. The result is that the transmitted light is coloured when white light is incident on the crystal. The colour varies depending on the type of crystal but viewing the crystal in natural white light normal to the optic axis it may appear green, say, whilst along the axis it will appear black. This two coloured appearance is responsible for calling the crystal dichroic which means two colours, of course.

The anisotropy arises from the fact that atoms within the crystal form a periodic lattice. The electrons which interact with light may have stronger bonds in one direction than in others. Thus, the response to the sinusoidal electric field of the incident electromagnetic wave will depend on the direction of the electric vector E. Currents will be set up and the oscillating electrons transfer kinetic energy to the lattice which appears as heat. When more energy is transferred to electrons able to respond to the driving field in one direction rather than another there will be selective absorption.

Crystals with two distinct directions are called *uniaxial* and display two colours. Biaxial crystals exhibiting three colours (*trichroism*) do exist but we shall not be concerned with them optically. An example of a trichroic crystal is *cordierite* which will absorb either violet, blue, or yellow light depending on its orientation.

Figure 12.11 illustrates a dichroic crystal's action. Note the crystal's optic axis is also its transmission axis, and it is a direction and NOT a single line in the material.

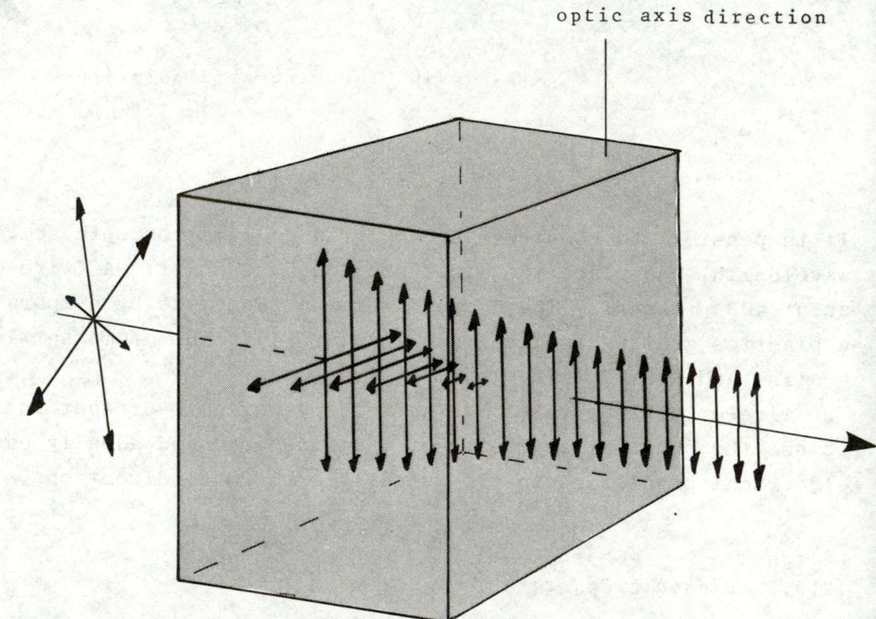

Fig. 12.11
A dichroic crystal's action.

The figure shows gradually reducing electric vectors perpendicular to the optic axis. If the sample is sufficiently thick these vectors will diminish to zero as the energy is transferred from the electric field to the crystal lattice. Some attenuation of the electric vectors parallel to the optic axis occurs but except for specific frequency absorption these vectors are transmitted.

12.3.3 *Polaroid*

Edwin H. Land invented the first dichroic sheet polariser in 1928. It was known commercially as *polaroid J-sheet*. Initially, Land ground herapathite (quinine sulphate periodide) into submicroscopic crystals which were naturally needle shaped. He found that they could be mechanically aligned by extruding a viscous suspension of herapathite needles through a narrow slit. The result was a large flat sheet of dichroic crystal. However, the herapathite needles tended to scatter light and J-sheet was therefore somewhat hazy.

In the late thirties Land invented H-sheet which is based on the wire-grid principle. A sheet of clear polyvinyl alcohol is heated and stretched in one direction during which process the long hydrocarbon molecules become aligned. Dipping the sheet into an iodine-rich solution allows the iodine to attach to the chain-like hydrocarbon molecules thus forming an 'iodine wire-grid'. As a result H-sheet absorbs the E components parallel to the hydrocarbon-iodine chains and transmits P-state light perpendicular to the chains. H-sheet does not present the hazy appearance of its forerunner since the dichroic elements are of molecular size: thus scattering does not present a problem. H-sheet is a good plane polariser across the visible spectrum except at the violet end where some leakage occurs. Holding up crossed polaroids to a bright white natural light source, as in figure 12.12, produces a dark violet colour where they overlap. In less bright incident light the leakage is barely noticeable.

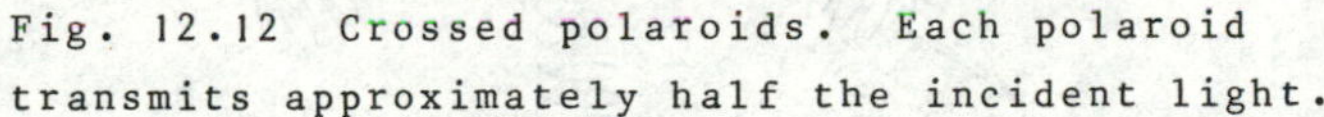
Fig. 12.12 Crossed polaroids. Each polaroid transmits approximately half the incident light.

An ideal H-sheet would transmit 50% of the incident natural light, neglecting reflection, and would appear grey. It would therefore be a neutral filter and would be designated HN-50. In practice about 8% of the light is lost by reflection at the surfaces leaving 92% for transmission. Half of this will be extinguished so 46% emerges in a P-state and the resulting H-sheet might be labelled HN-46. In fact, HN-38, HN-32, and HN-22 are produced commercially, each differing by the amount of iodine included in the sheets. Other forms of polaroid include K-sheet which is heat and humidity resistant and is made from chains of hydrocarbon polyvinylene. HR-sheet, a near infrared polariser, combines the ingredients of H- and K-sheets.

Polaroid vectograph is an interesting material which is not in itself a polariser. It is a laminate of two clear polyvinyl alcohol sheets with their stretch directions orthogonal. When pictures are painted one on each side with an iodine solution, and the result viewed through a plane polariser, the pictures present themselves alternately as the polariser is rotated.

12.4 BIREFRINGENCE

12.4.1 Optical isotropy and anisotropy

Substances such as unstressed glass and plastics have the same optical properties in all directions throughout their bulk. Such substances are said to be optically *isotropic*. We can imagine their electrons being bound to the atomic nuclei by a restoring force which is the same in all directions. Thus, when P-state electromagnetic waves are incident on the atom the electron shows no difference in its reaction no matter what the direction of the electric vector might be. However, in certain crystals the forces binding the electrons to the atomic nuclei do not have this property. They may have a symmetry about an axis which causes electrons to behave differently when acted upon by an electromagnetic wave. To get a feel for this, imagine a 'mechanical' model where a charged particle is 'bound' by springs in the x-, y-, and z-axis directions, figure 12.13(a) and (b).

Imagine a harmonic driving force, a P-state wave, is propagating in the positive z-direction. In figure 12.13(a) it will drive the electron equally well in either the x- or the y-direction if its electric vector is parallel to one or the other. This will not be the case in figure 12.13(b) where the springs are not identical in the x- and y-directions. Imagine now the 'electron' in figure 12.13(a) is displaced along the x- or y-axis and then 'released'. It will oscillate with a characteristic frequency known as the *natural frequency*. The natural frequencies in figure 12.13(b) will be different in the x- and y-directions because of the different spring stiffnesses. It can be shown that if the electron were driven by an oscillating field, an electromagnetic wave, with a frequency less than the natural frequency then the electron would move antiphase with the oscillating field: that is, it would be antiphase with the direction of the electric vector. Conversely, if the electromagnetic wave's frequency were greater than the electron's natural frequency the displacement of the electron would be in-phase with the field.

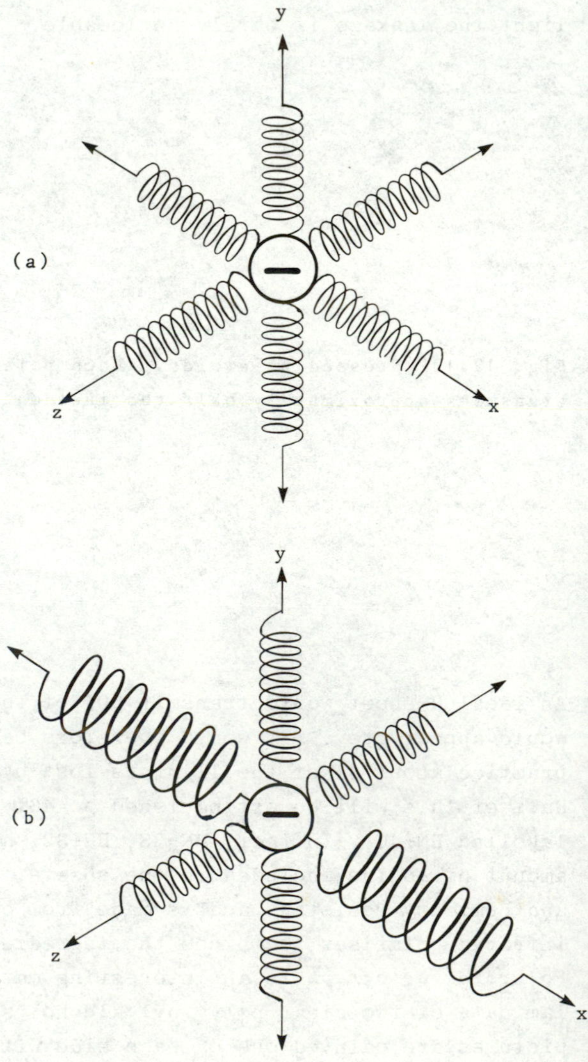

Fig. 12.13 (a) A negatively charged particle 'bound' by springs of equal stiffness. There is no axial symmetry. (b) The spring in the x-direction is here increased in stiffness. There is symmetry about the x-axis.

If these oscillations are damped by interaction with neighbouring oscillators (atoms) then the energy is dissipated within the material in the form of heat (molecular motion). This process is called *absorption*. If there is no damping then the electrons, being accelerated, reradiate at the driving frequency and act as secondary sources. The radiation propagates in a manner analogous to Huygen's secondary wavelets.

If the wave's driving frequency is the same as the natural frequency then the electron will be driven into large amplitude oscillations with a consequently greater chance of interaction with neighbours and a strong absorption of the driving wave. Such frequencies form *absorption bands* for a given material.

Colourless, transparent gases, liquids, and solids have their natural frequencies outside the visible spectrum which is why they are transparent. In particular, glasses have their natural frequencies in the far ultraviolet where they become opaque.

Since light propagates through a transparent substance by driving electrons which reradiate, the speed of the wave, and therefore the refractive index, is determined by the difference between the wave's frequency and the natural frequency of the bound electrons. An anisotropy in the binding forces, as in figure 12.13(b), will result in an anisotropy in the refractive index. For example, if P-state light were to encounter electrons in a crystal which could be represented by figure 12.13(b), its speed of propagation would be governed by the orientation of E. When E is parallel to the x-axis it will propagate at a different speed to when it is parallel to the axis. A material which displays two different speeds of propagation in fixed and orthogonal directions, and therefore displays two refractive indices, is known as *birefringent*.(The word refringence used to be used instead of refraction. It was derived from the Latin *refractus* via the word *frangere*, meaning to break).

Clearly, a birefringent material which absorbs one of the orthogonal P-states is dichroic. So, dichroic crystals are a special case of birefringent crystals. If, as in figure 12.13(b), the binding forces in the y- and z-directions are equal then the x-axis defines the optic axis of a uniaxial crystal. Since a crystal is taken to be an array of these oscillators the optic axis is a direction in the crystal and not a particular discrete line. Birefringent crystals have their two natural frequencies above the range for the visual spectrum and are therefore colourless. Two different refractive indices are present but the absorption is negligible so that both are transmitted.

12.4.2 Calcite

Calcite or calcium carbonate ($CaCO_3$) is a fairly typical birefringent crystal. Marble and lime stone are both made from calcite crystals bonded together. Figure 12.14(a) is view of the atomic arrangement which has a symmetry about any axis perpendicular to the plane of the paper and passing through the centre of a triangular carbonate (CO_3) group. The carbonate groups are all in planes perpendicular to the optic axis. The interaction with light is markedly different when E is either in or normal to these planes. So, this physical asymmetry gives rise to the optical anisotropy.

Calcite crystals will split or cleave smoothly between planes of atoms where the bonding is weakest. The cleavage planes are all normal to three distinct and different directions and a crystal cut in this manner is called a *cleavage form*. A cleavage form crystal of calcite has six faces each of which is a parallelogram with angles very nearly 102^0 and 78^0, as shown in figure 12.14(b). In such a crystal there are only two blunt corners, that is corners where the three angles are all 102^0. The optic axis is in the direction of a line passing through a blunt

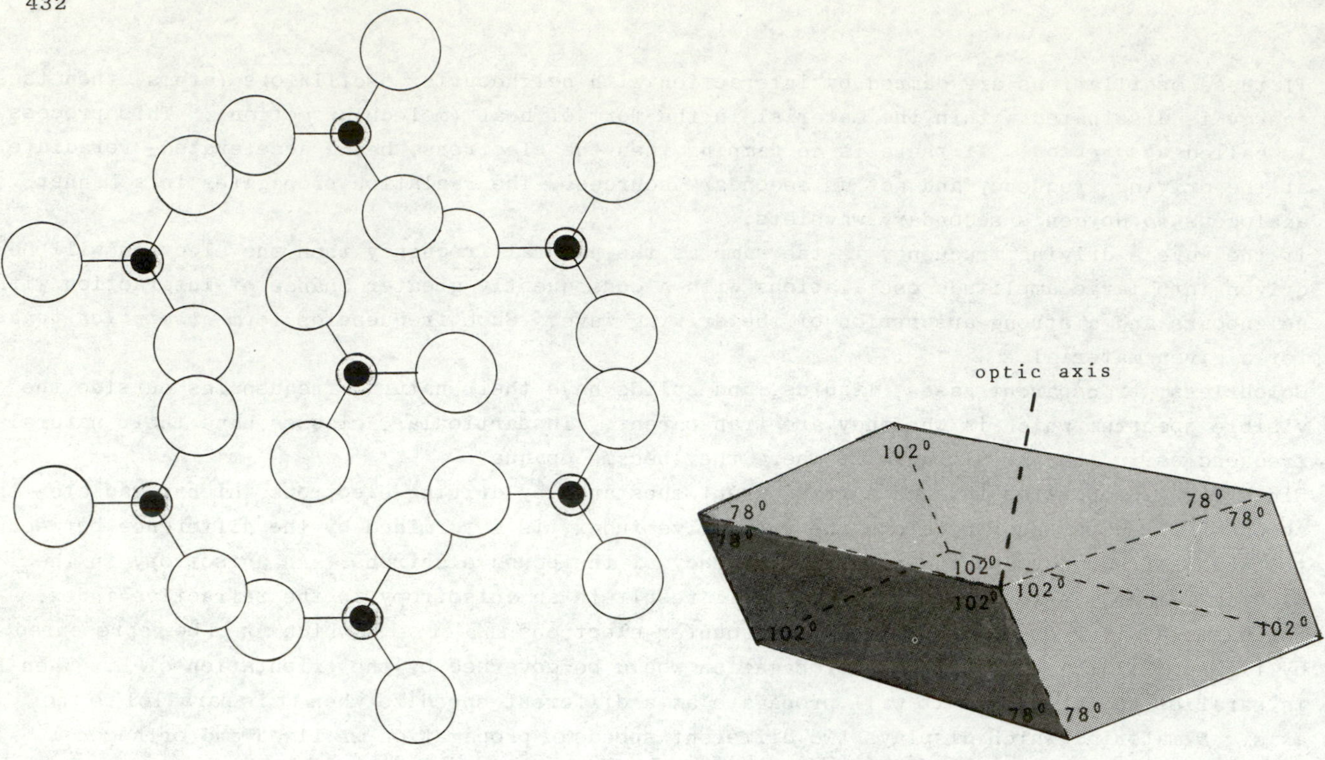

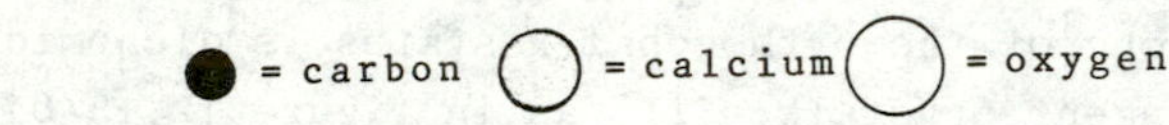

Fig. 12.14a Atomic array looking down the optic axis of calcite.

Fig. 12.14b Calcite cleavage form crystal.

corner so that it makes equal angles with the faces and the edges at that corner.

Erasmus Bartholinus discovered the birefringence property of calcite in 1669. He called it *double refraction*. Figure 12.15 shows the effect he observed.

Fig. 12.15 Double image formed by a calcite crystal.

A narrow beam of natural light incident normally to a cleavage plane of a calcite crystal emerges as two parallel beams displaced laterally. One beam passes through the parallel sided crystal undeviated. That is, it behaves in the expected or ordinary manner. These rays are called *ordinary* or *o-rays*. The other beam is displaced sideways in an extraordinary fashion so Bartholinus named these rays *extraordinary* or *e-rays*. Suppose now we cut a calcite crystal as shown in figure 12.16.

The plane ABCD contains the optic axis: that is, the direction parallel to AC. If we now draw the plane section ABCD as in figure 12.17 we can clearly illustrate what Bartholinus saw.

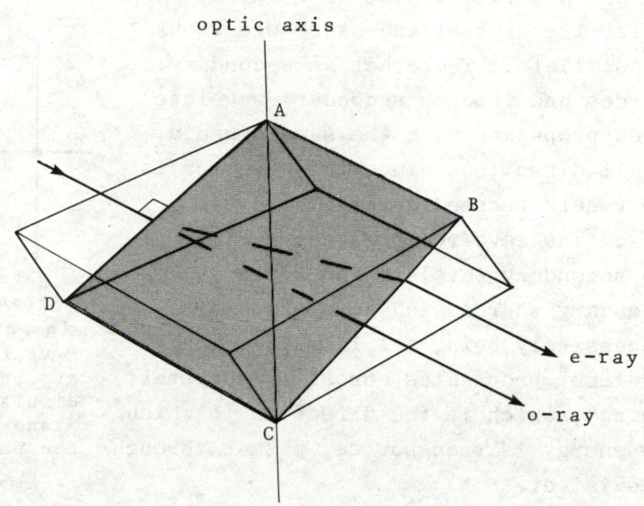

Fig. 12.16 A calcite crystal cut with the optic axis contained as shown.

In figure 12.17 the incident natural light is resolved into components shown ↕ where the electric vector is in the plane ABCD, and · (a dot) where the electric vector is perpendicular to the plane ABCD and therefore perpendicular to the optic axis. If we consider these two component electric vectors separately we can obtain some idea as to why the two beams separate as they do.

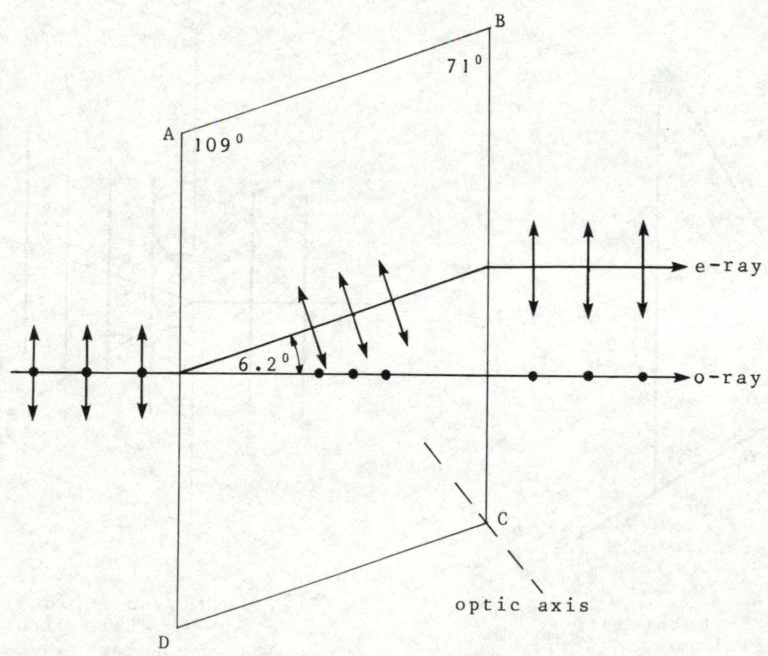

Fig. 12.17 Natural light incident on a calcite crystal with the optic axis contained in the section ABCD.

Consider figure 12.18 which shows the electric vector component perpendicular to the optic axis (and the plane ABCD). The plane waves (a parallel beam of o-waves) strike the crystal and cause electrons to oscillate. These act as secondary sources and produce secondary wavelets which propagate with the same speed, $v_\perp$, in all directions since the E vector is everywhere perpendicular to the optic axis. The wavefront is the envelope of the secondary wavelets and since secondary sources and wavelets are successively being set in motion the wavefront propagates across the crystal. The ray, which is the direction in which the energy is transported, passes through undeviated.

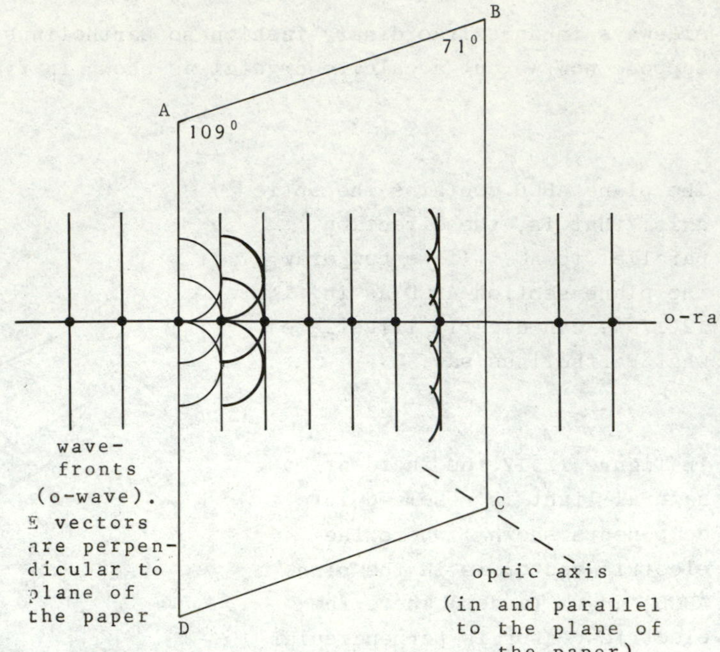

Fig. 12.18 An incident beam in a P-state with E perpendicular to the optic axis.

Now consider figure 12.19 which looks at the E-vector parallel to the plane ABCD. The electric vector has components parallel to and perpendicular to the optic axis as illustrated in figure 12.20. Since the binding forces on the electrons

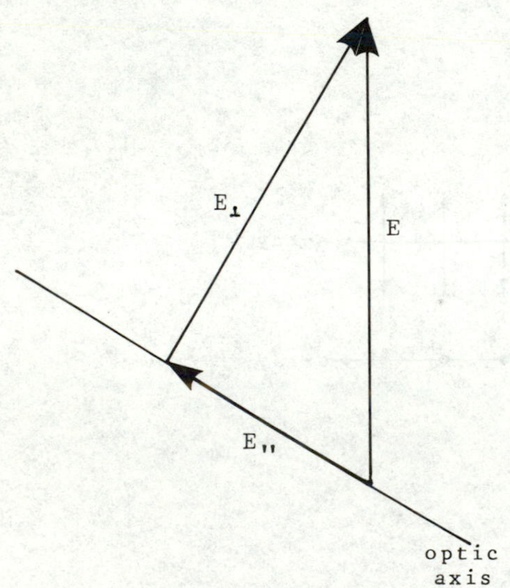

Fig. 12.20 The E-vector with components parallel to and perpendicular to the optic axis.

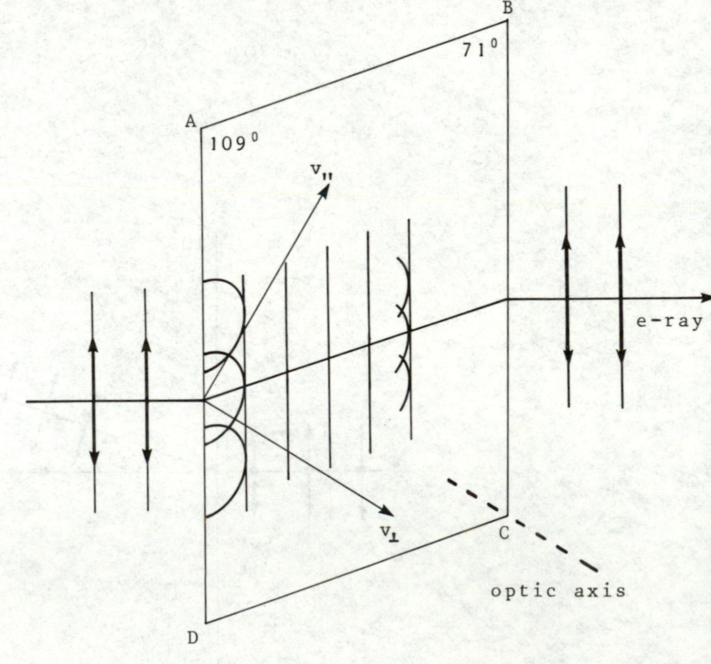

Fig. 12.19 An incident plane wave in a P-state with E parallel to the plane ABCD.

differ in these two directions the secondary wavelets in the crystal will propagate at different speeds shown as $v_{\perp}$ (vee-perpendicular) and $v_{\parallel}$ (vee-parallel) in figure 12.19 . The result is the secondary wavelets are ellipsoids instead of spheres. For calcite and sodium yellow light ($\lambda = 589.3$ nm) the two speeds of propagation are related thus:

$$1.486\, v_{\parallel} = 1.658\, v_{\perp} = c, \quad \text{where } c \simeq 3 \times 10^{8}\ \text{ms}^{-1}.$$

Notice that in figure 12.19 the e-ray is not perpendicular to the wavefront in the crystal - an exception to our 'rule' which applies to spherical wavefronts. The wavefront is again the envelope of the secondary wavelets but because the latter are ellipsoids the wavefront is displaced sideways. This results from the fact that $v_{\parallel} > v_{\perp}$.

12.4.3 Birefringent Crystals

For the o-ray the secondary spherical wavelets are propagating at the speed $v_{\perp}$ in all directions, figure 12.18, but for the e-ray the secondary wavelets propagate at a speed $v_{\perp}$ when the electric vector component is perpendicular to the optic axis, and the propagation direction is parallel to the optic axis, figure 12.19. Where the electric vector component in figure 12.19 is parallel to the optic axis the ellipsoidal wavelet propagates at a speed $v_{\parallel}$ in a direction perpendicular to the optic axis. Note the parallel ($\parallel$) and perpendicular ($\perp$) subscripts attached to the speeds (v) refer to the direction of the electric vector component relative to the optic axis and NOT to the direction of propagation.

We can define the two refractive indices in uniaxial crystals such as calcite as follows:

$$\text{the } \textit{ordinary refractive index} \quad n_o = \frac{c}{v_{\perp}} \qquad (12.33)$$

$$\text{and the } \textit{extraordinary refractive index} \quad n_e = \frac{c}{v_{\parallel}} \qquad (12.34).$$

Table 12.1 lists some ordinary and extraordinary refractive indices for several uniaxial crystal.

Table 12.1 Refractive indices of some uniaxial birefringent crystals ($\lambda = 589.3$ nm).

Crystal	n_o	n_e
Calcite	1.6584	1.4864
Ice	1.309	1.313
Quartz	1.5443	1.5534
Sodium nitrate	1.5854	1.3369
Tourmaline	1.669	1.638

The difference $n_e - n_o$ is called the *birefringence* Δn; i.e. $\Delta n = n_e - n_o$ ....(12.35). Where Δn is negative the crystal is said to be *negatively uniaxial*. Calcite has a birefringence $\Delta n = n_e - n_o = 1.4864 - 1.6584 = -0.172$ and is therefore negatively uniaxial. We see, from equations (12.33) and (12.34) and Table 12.1 it is evident that $v_{\parallel} > v_{\perp}$ in calcite. This then is why the the ellipsoids elongate in the direction shown in figure 12.19 .
Conversely, quartz is *positively uniaxial* since $\Delta n = n_e - n_o = 1.5534 - 1.5443 = 0.0091$.

12.4.4 *Birefringent Polarisers*

Relying on the fact that $n_e \neq n_o$ we can construct numerous crystal arrangements which will act as plane polarisers. Several common arrangements are available commercially and we shall consider three of them: the *Nicol prism*, the *Glan-Foucault prism*, and the *Wollaston prism*.
In 1828 William Nicol invented what is now known as the Nicol prism. The device is made by grinding the ends of a long calcite rhombohedron from 71° to 68°. It is then cut diagonally and cemented back together with Canada balsam, figure 12.21 .

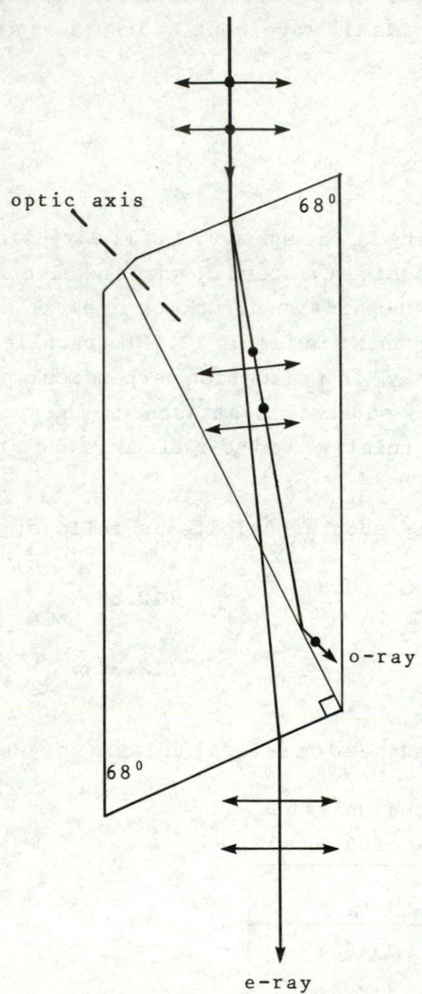

Fig.12.21 The Nicol prism. The flat worked on the upper blunt corner allows one to identify the optic axis.

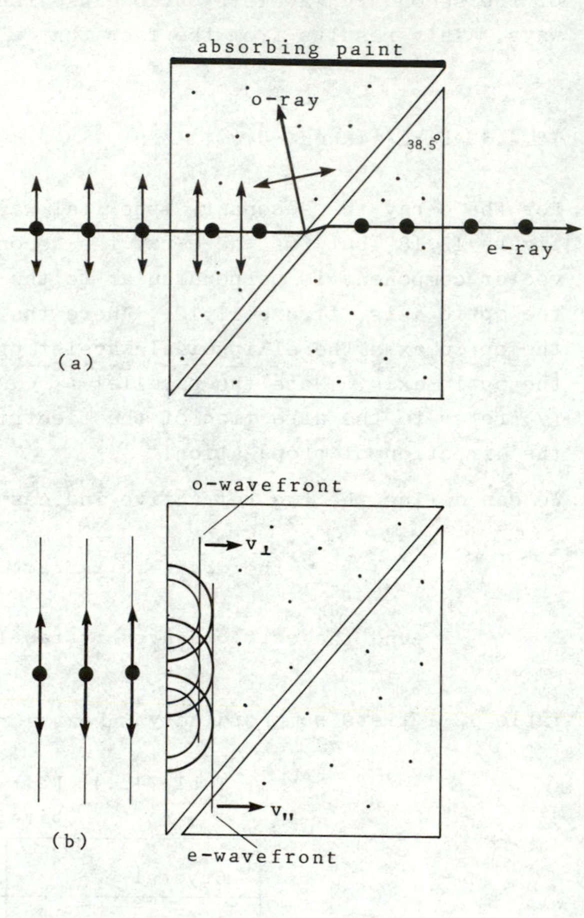

Fig. 12.22 The Glan-Foucault polariser.

Canada balsam is transparent with a refractive index of 1.55 for sodium yellow light. This is nearly midway between n_e and n_o. The result is that the o-ray is totally internally reflected at the calcite-Canada balsam boundary, but the e-ray is transmitted. The o-ray is absorbed by the matt black paint on the sides of the prism.

The Glan-Foucault polariser is constructed only from calcite. Since it is transparent from 230 nm to about 5000 nm the Glan-Foucault device can be used over a wide range of wavelengths.

Figure 12.22 illustrates the arrangement. The optic axis is shown by a multitude of dots, indicating it is perpendicular to the plane of the paper. Since the electric vector marked with a dot (·) is everywhere parallel to the optic axis the e-wave will propagate at a speed $v_{\parallel}$ in all directions. This will therefore result in spherical secondary wavelets for the extraordinary ray and there will be no displacement of this ray. However, because $v_{\perp} < v_{\parallel}$ for calcite the o-wavefront will lag behind the e-wavefront as shown in figure 12.22(b): this is not pertinent to the modus operandi.

Figure 12.22(a) shows the incident light normal to the surface. Both e- and o-rays pass through the first prism segment undeviated. If the angle of incidence at the calcite-air boundary is such that $n_e < 1/\sin i < n_o$ then the o-ray will be totally internally reflected and the e-ray will be transmitted.

The Wollaston prism differs from the Nicol prism and the Glan-Foucault polariser in that it transmits both the e-wave and the o-wave. It is in effect a polarising beam-splitter. Figure 12.23 illustrates how this is achieved.

The electric vector components of the incident beam are propagated at different speeds in the first segment. The waves represented by ↕, parallel to the optic axis in the first segment, are e-waves and lead the o-waves which are represented by · (a dot). Of course, this is because $v_{\parallel} > v_{\perp}$ in calcite. On entering the second prism segment the e-wave becomes the o-wave and vice-versa. This is due to the different orientation of the optic axis in this segment. The o-ray in the second segment bends towards the normal because $n_o > n_e$ in calcite. For the same reason the e-ray in the second segment bends away from the normal.

Wollaston polarising beam-splitters are available with a variety of wedge-angles providing an angular separation of the emergent e- and o-rays from 15^0 to 45^0. The contact surface may be cemented with castor oil or glycerine, or may simply make an optical contact. i.e. it is not cemented.

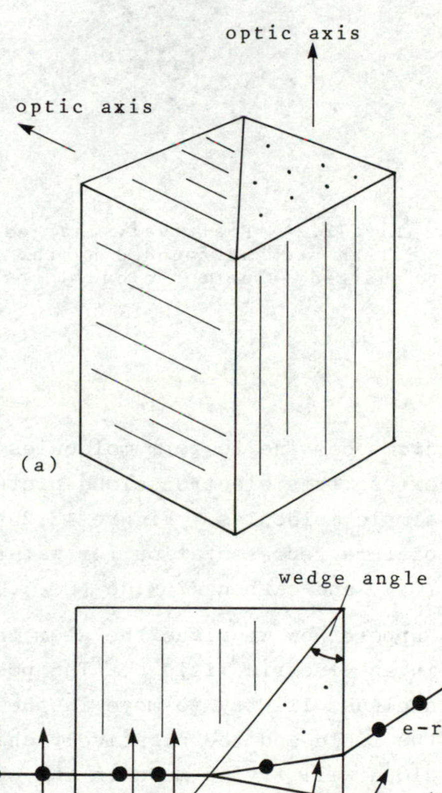

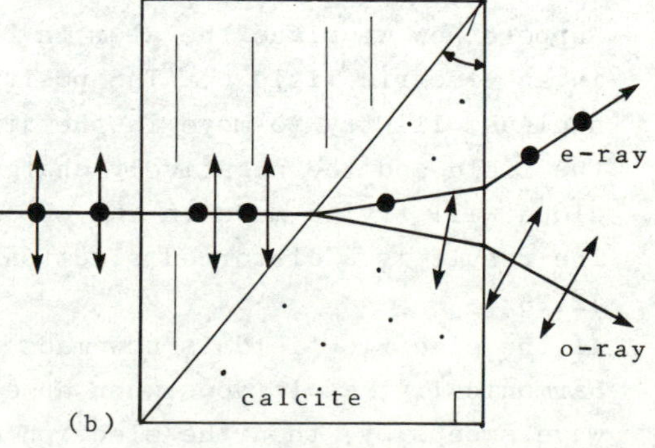

Fig.12.23 The Wollaston polarising beamsplitter.

12.5 SCATTERING

12.5.1 Introduction

So far we have used an electron-mechanical model when discussing the interaction of light and matter. This picturesque model allowed us to imagine different modes of interaction in mutually orthogonal planes. Although we shall not investigate the mathematics of electron-light interactions seriously perhaps we ought to mention another picturesque model - the electron cloud model. Here the electrons surrounding the nucleus are imagined as a cloud. Figure 12.24 illustrates now we might imagine the positively charged nucleus and the negatively charged electron cloud.

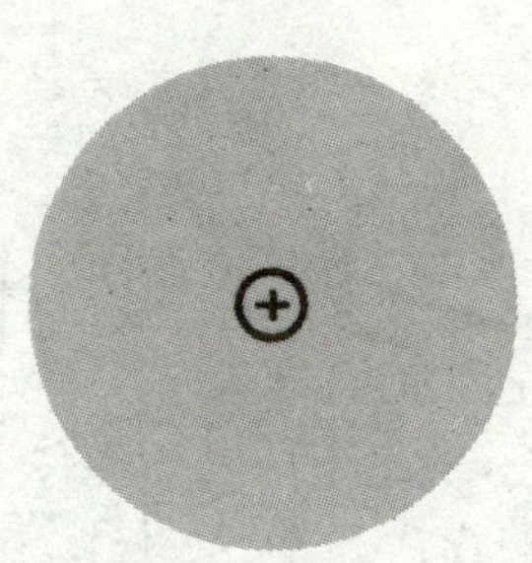

Fig.12.24 Positively charged nucleus of an atom surrounded by the negatively charged 'electron cloud'.

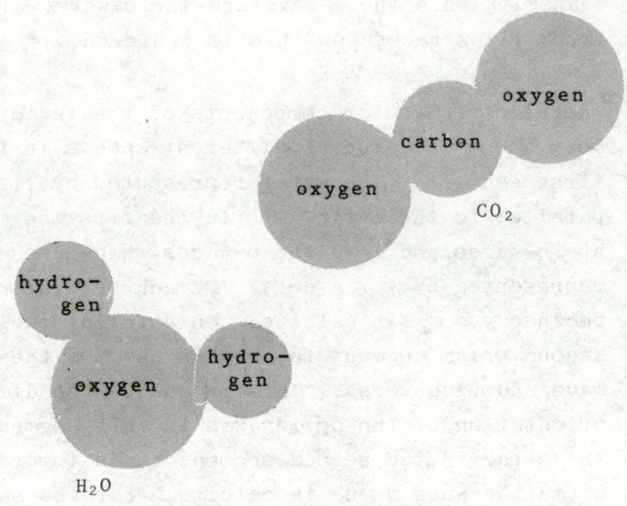

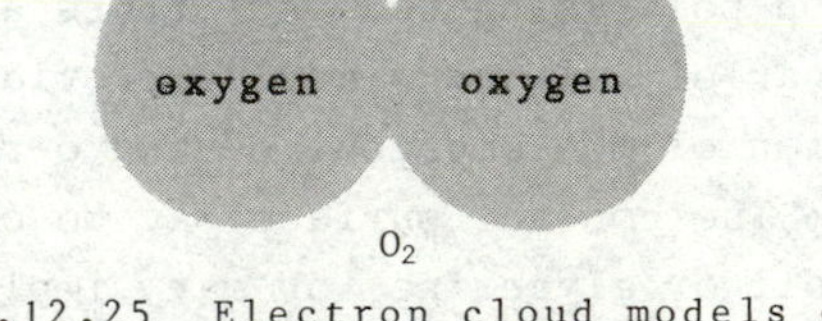

Fig.12.25 Electron cloud models of water, carbon dioxide, and oxygen molecules.

Atoms combine to form molecules and we might extend this electron cloud picture to represent simple molecules. Figure 12.25 shows such a picture representation for water (H_2O), oxygen (O_2), and carbon dioxide (CO_2).

Suppose now we place the atom in figure 12.24 in an electric field ψ. The positively charged nucleus will try to move in the direction of the field and the negatively charged electron cloud will try to move in the opposite direction. The result is a distorted electron cloud, figure 12.26 .

If the electric field is now made to oscillate harmonically, as it does when an electromagnetic wave sweeps by, then the electron cloud will oscillate about the nucleus. The nucleus will move slightly but to a much less extent than the electron cloud since its mass is much greater.

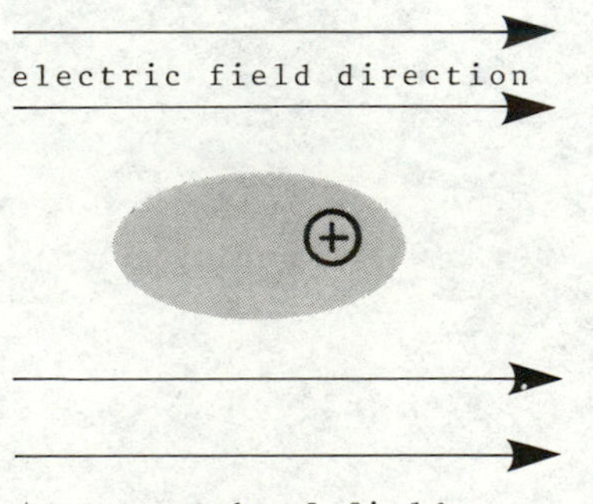

Fig.12.26 Distorted electron cloud in an electric field ψ.

The displacement, x, of a single electron in an electric field oscillating harmonically is given by

$$x = \frac{q/m}{\omega_0^2 - \omega^2} \cdot E \sin\omega t \qquad (12.36)$$

where q is the electric charge on the electron, m is the electron's mass, and ω_0 is the natural frequency of oscillation of the bound electron (remember the electron is 'bound' to a nucleus). The derivation of this equation requires some knowledge of the physics of charged particles and the integration of second order differential equations. Those students who are not familiar with these topics may skip the proof without loss of the qualitative argument which will follow it and proceed immediately to the next statement of equation (12.36).

The force F on a charge q in a time-varying electric field $\psi = E \sin\omega t$ is given by

$$F = q\psi = qE \sin\omega t \; .$$

We are neglecting the magnetic component of the electromagnetic field since its effect on an electron is small compared with that of the electric field. A bound electron will oscillate harmonically when not experiencing a driving force. The force it will experience when oscillating freely will be $-m\omega_0^2 x$, where x is the displacement from the equilibrium position and ω_0 is the natural angular frequency. Using Newton's second law of motion we can equate the sum of the forces to the electron mass times its acceleration, d^2x/dt^2. Thus

$$m\frac{d^2x}{dt^2} = qE \sin\omega t - m\omega_0^2 x \; .$$

Dividing through by m,

$$\frac{d^2x}{dt^2} = \frac{q}{m}E \sin\omega t - \omega_0^2 x \; .$$

There are several methods of dealing with second order differential equations with constant coefficients. Expecting the electron to oscillate harmonically with the same frequency as the driving force $E \sin\omega t$ we guess at $x = x_0 \sin\omega t$ as a solution, where x_0 is the amplitude of the oscillation. Substituting for x in the differential equation gives

$$\frac{d^2x}{dt^2} = \frac{q}{m}E \sin\omega t - \omega_0^2 x_0 \sin\omega t$$

and we can integrate to give

$$\frac{dx}{dt} = \frac{-q/m\, E \cos\omega t}{\omega} + \frac{\omega_0^2}{\omega} x_0 \cos\omega t + C_1 \; , \text{ where } C_1 \text{ is a constant.}$$

$$\text{Now,} \quad \frac{dx}{dt} = \frac{d}{dt}(x_0 \sin\omega t) = \frac{x_0}{\omega}\cos\omega t \; ,$$

so substituting for dx/dt the constant C_1 must be zero since the other terms go to zero when $t = \pi/2\omega, 3\pi/2\omega, \ldots$. Integrating again gives

$$x = \frac{-q/m\, E \sin\omega t}{\omega^2} + \frac{\omega_0^2}{\omega^2} x_0 \sin\omega t + C_2 \; .$$

Substituting for x again we find $C_2 = 0$ since all the other terms are zero when $\sin\omega t = 0$. We now have, on rearranging,

$$x = \frac{\omega_0^2}{\omega^2} x_0 \sin\omega t - \frac{q/m\, E \sin\omega t}{\omega^2} \quad .$$

Substituting $x_0 = x/\sin\omega t$,

$$x = \frac{\omega_0^2}{\omega^2} x - \frac{q/m\, E \sin\omega t}{\omega^2}$$

or $(\omega_0^2 - \omega^2)x = q/m\, E \sin\omega t$ which gives $x = \dfrac{q/m\, E \sin\omega t}{\omega_0^2 - \omega^2}$ (12.36)

Inspection of equation (12.36) shows that when the incident angular frequency ω is close to the natural angular frequency ω_0 the amplitude of oscillation will become large since $\omega_0^2 - \omega^2$ in the denominator will be small. In rare media such as gases one or more electrons may make the transition to a higher energy level, known as an excited state. That is, the wave packet's energy is absorbed. It will subsequently return to the non-excited ground state and emit an electromagnetic wave. This may occur in more than one step so that the radiated wave packets will not have the same frequency as the incident wave packet. If it drops to the ground state in one jump then the reradiated wave packet will have the same frequency as the incident wave. In dense media the atom will probably return to its ground state by losing the energy in the form of joule heat (kinetic energy of molecular oscillation).
At frequencies reasonably below or above ω_0 the electron cloud will oscillate with a smaller amplitude and the incident wave will be reradiated at the same frequency. Where $\omega \simeq \omega_0$ the frequency is called a *resonance frequency*. The nuclei of atoms have resonant frequencies in the infrared whereas the electrons generally have their resonant frequencies in the ultraviolet. At non-resonant frequencies the absorption of the wave packet and its subsequent emission is known as *scattering*.
Scattering is the basis of reflection, refraction, and diffraction (see chapter 14). Some interesting phenomena can be explained on the basis of scattering. For example, on reaching the atmosphere sunlight is scattered in all directions. Lord Rayleigh was the first to deduce that the intensity of the scattered light is proportional to ω^4. Since blue light has a higher frequency than longer wavelengths in the visible spectrum more blue light is scattered and this gives the sky its characteristic blue colour. The red end of the spectrum is largely unscattered, passing through the atmosphere undeviated save for refraction. When the sun is setting the light passes through a greater thickness of atmosphere so more of the short wavelength light scatters sideways with the result that only the sun's red light reaches the observer. Figure 12.27 shows scattering by the atmosphere.

Rayleigh's work was on the scattering of light by particles which are small compared with the wavelength. It is worthwhile noting that the diameter of a hydrogen atom, the smallest atom, is about 10^{-10} m. Compare this with the magnitude of wavelengths in the visible spectrum, 3900×10^{-10} to 7600×10^{-10} m. Naturally enough such scattering is called *Rayleigh scattering*. The molecules in dense transparent media, whether gases, liquids, or solids, will also scatter predominantly blue light. However, in liquids and solids the effect is weak because of the more orderly arrangement of the oscillators which leads to reinforcement of the secondary wavelets in the forward direction and a cancelling of sideways scattering.

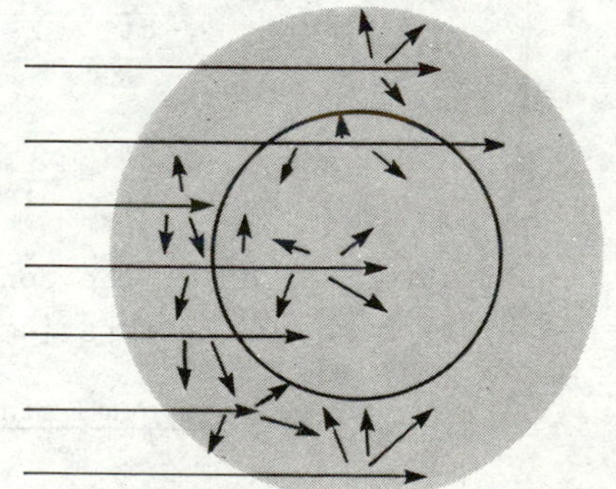

Fig.12.27 Scattering of sunlight in the atmosphere.

Another example of the scattering of blue light will be seen when smoke from the end of a lighted cigarette is viewed against a dark background. The smoke particles are smaller than light wavelengths. However, exhaled smoke has relatively large droplets of water added which can sustain the normal reinforcement of secondary wavelets. This allows normal reflection and refraction of all the wavelengths and the result is that the exhaled smoke appears white. The white appearances of clouds, fog, sugar, salt, paper, and ground glass are all accounted for in a similar manner. In 1908 G. Mie published the results of some rigorous mathematical modelling on scattering from particles of all sizes. Scattering from clouds, paper, and the like is called *Mie scattering*.

12.5.2 Polarisation by Scattering

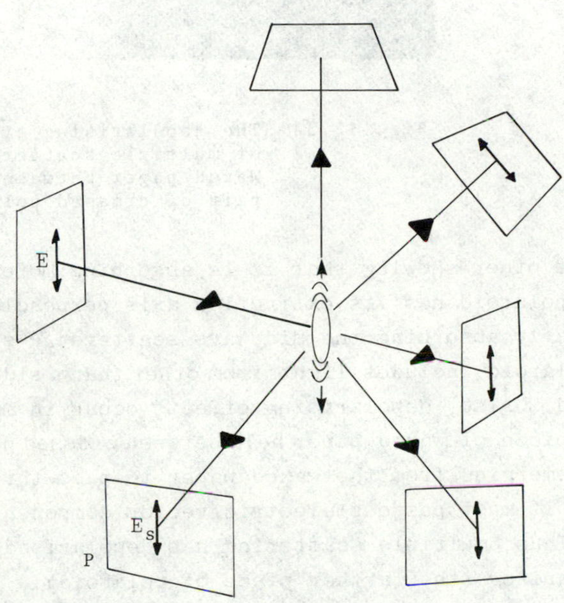

Fig.12.28 Scattering of vertically plane-polarised incident wave by a molecule.

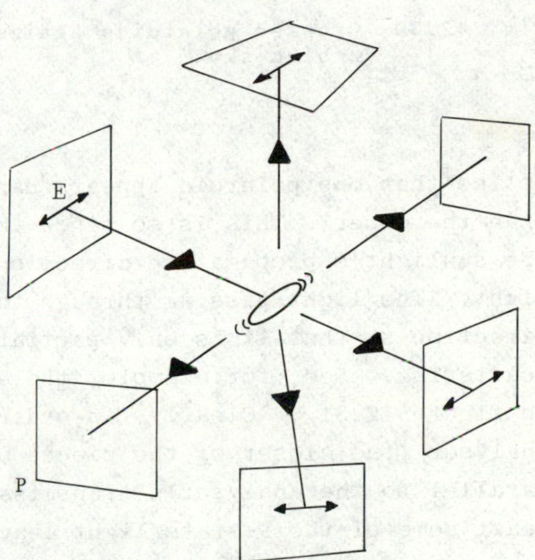

Fig.12.29 Scattering of a horizontally plane-polarised wave by a molecule.

Figure 12.28 shows a plane polarised wave with an electric vector E in the vertical meridian. When it strikes an electron oscillator, in a molecule say, it sets the electron cloud in motion in the same meridian. The reradiated waves are scattered in such a way that the electric vector of the scattered wave, E_s, is coplanar with the incident vector. Suppose now the incident P-state is horizontally orientated as in figure 12.29 . Then, looking at the positions marked P in the diagrams above, we see that only the vertically polarised incident wave has a scattered component in the sideways direction. Since natural light can be considered as a superposition of those two cases, light scattered sideways to the direction of horizontal incident rays will be plane polarised in a vertical meridian. This suggests that blue sky will possess partially polarised light if viewed sideways on to the incident light. Figure 12.30 shows that this is indeed the case.

Fig. 12.30 Crossed polaroids against a blue sky.

Fig. 12.31 The depolarising effect of multiple scattering. Waxed paper between a pair of crossed polaroids.

Notice that one polaroid appears darker than the other showing that it is absorbing more light than the other. This is so since the 'darker' polaroid has its extinction axis perpendicular to the sunlight's propagation direction and is clearly absorbing the sideways scattered P-state light. The light passing through the darker polaroid includes light from other than a sideways-on direction so that it is only partially polarised. Also, depolarising effects occur in multiple scattering. See, for example, the effect of a piece of waxed paper held between crossed polaroids in figure 12.31 . Clearly, in order for light emerging from the waxed paper to pass through the analyser (the closer of the pieces of polaroid) it must possess electric vector components parallel to the analyser's transmission axis. Thus, multiple scattering has depolarised at least some of the P-state light leaving the polariser (the farther piece of polaroid).

12.6 POLARISATION BY REFLECTION

If a plane polarised beam is incident on a plane air-glass boundary, say, and the electric vector is perpendicular to the plane of incidence, then the electron oscillators in the surface molecules will be set in motion as shown in figure 12.32 . The result is that the reradiated reflected and refracted waves also have their electric vectors parallel to the surface. Since natural light can be considered as two orthogonal incoherent equal amplitude waves, and we can choose the component directions arbitrarily, reflected light will have a component E-vector parallel to the surface. Recall that when the angle of incidence is such that $i = i_p$, where i_p is the polarising angle given by $\tan i_p = n'/n$, the reflected wave will be entirely plane-polarised parallel to the surface. This suggests a way of producing P-state light of a reasonable intensity by multiple reflection. Consider figure 12.33 : if a parallel beam of natural light is incident at the polarising angle on a pile of microscope slides, the reflected waves at successive plates will all be polarised parallel to the surfaces. Therefore a relatively high intensity plane polarised reflected beam is produced.

Polarised reflected light offers a convenient way of locating the extinction axis of a plane polariser. For example, if a piece of polaroid is rotated in front of one's eye whilst looking at a reflection from a glossy surface, there will be one obvious position where the intensity of the reflected light is markedly reduced. In this position the extinction axis is parallel to the surface. Clearly, we have also located the transmission axis of the polaroid which is perpendicular to the extinction axis.

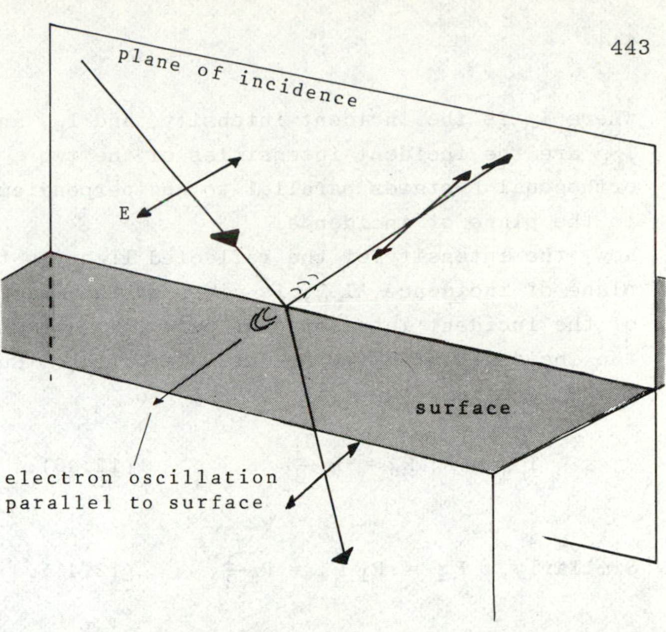

Fig.12.32 Reflection and refraction of a P-state wave with E parallel to the surface.

12.6.1 *Reflectance versus angle of incidence*

Recall equations (10.42) and (10.44) which state

$$r_\perp = -\frac{\sin(i - i')}{\sin(i + i')} \qquad (10.42)$$

and

$$r_{\parallel} = \frac{\tan(i - i')}{\tan(i + i')} \qquad (10.44).$$

$r_\perp$ and $r_{\parallel}$ are the amplitude reflection coefficients. The intensity reflectance, R, is proportional to the square of the amplitude reflection coefficient, see problem 12.10.12 . Thus,

$$R_{\parallel} = r_{\parallel}^2 = \frac{\tan^2(i - i')}{\tan^2(i + i')} \qquad (12.37)$$

and

$$R_\perp = r_\perp^2 = \frac{\sin^2(i - i')}{\sin^2(i + i')} \qquad (12.38).$$

These equations tell us that $R_\perp$ can never be zero but $R_{\parallel} = 0$ when $\tan(i + i') = \infty$, i.e. when $i + i' = 90^0$. Remember that $R_\perp$ is the reflectance due to the electric vector perpendicular to the plane of incidence, which means parallel to the surface.

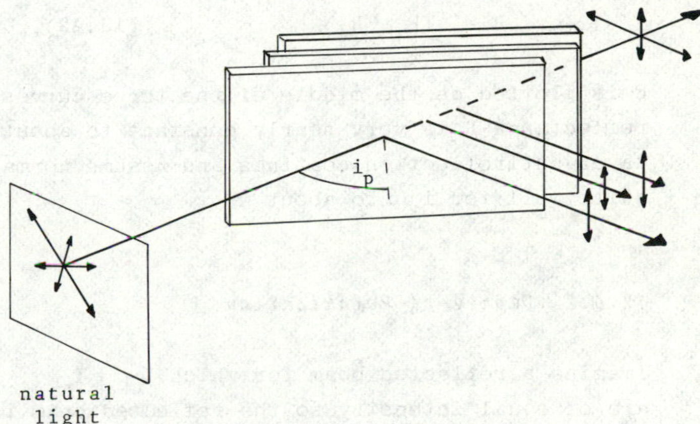

Fig.12.33 Plane-polarised light from a pile-of-plates polariser. Note the unequal amplitudes in the transmitted beam - it is therefore not regarded as natural light.

Figure 12.34 shows a plot of $R_{\parallel}$ and $R_\perp$ against the angle of incidence. Note $R_{\parallel}$ goes to zero at $i = i_p = 56.3^0$ and at glancing incidence, $i = 90^0$, all the light is reflected from a dielectric (non-electrically conducting) surface. This latter condition is easily verified by looking at a window by reflection on a polished table or desk-top. See figure 10.24 which illustrates the idea. Once again, using two equal amplitude, incoherent, orthogonal P-states to represent natural light, since the amplitudes are equal half the intensity will be possessed by each P-state . In the current context this means

$$I_{i\parallel} = I_{i\perp} = I_i/2 \qquad (12.39)$$

where I_i is the incident intensity, and $I_{i\parallel}$ and $I_{i\perp}$ are the incident intensities of the two orthogonal P-states parallel to and perpendicular to the plane of incidence.

Now, the intensity of the reflected light in the plane of incidence, $I_{r\parallel}$, is given by the fraction of the incident intensity reflected, $R_\parallel$, times the incident intensity in this plane, $I_{i\parallel}$. But, $I_{i\parallel} = I_i/2$ from equation (12.39), so

$$I_{r\parallel} = R_\parallel I_{i\parallel} = R_\parallel \frac{I_i}{2} \qquad (12.40).$$

$$\text{Similarly,} \quad I_{r\perp} = R_\perp I_{i\perp} = R_\perp \frac{I_i}{2} \qquad (12.41).$$

The reflectance in natural light is $R = \frac{I_r}{I_i}$ and $I_r = I_{r\parallel} + I_{r\perp}$, so we can write

$$R = \frac{I_r}{I_i} = \frac{I_{r\parallel} + I_{r\perp}}{I_i} = \frac{R_\parallel I_i/2 + R_\perp I_i/2}{I_i}$$

$$\text{or} \quad R = \tfrac{1}{2}(R_\parallel + R_\perp) \qquad (12.42).$$

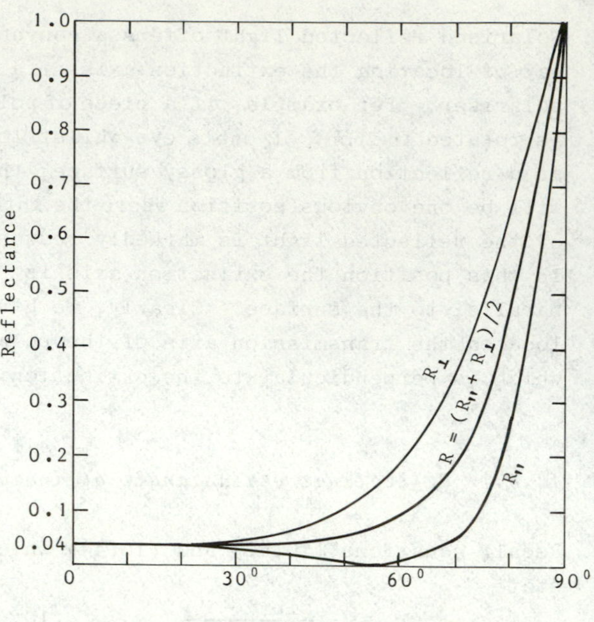

Fig. 12.34 Reflectance v. angle of incidence at an air/glass boundary. $n_g = 1.5$.

R is plotted as the middle of the three curves in figure 12.34 . In that figure notice the reflectance R is very nearly constant to about $i = 40^0$. Therefore when we come to consider single layer antireflection coatings and assume normal incidence for simplicity, the results will hold quite well for i up to about 40^0.

12.6.2 Degree of Polarisation

Imagine a reflected beam for which $I_{r\parallel} = I_{r\perp}$. In this case the two orthogonal reflected waves are of equal intensity so the reflected beam is unpolarised. Suppose now $I_{r\perp} > I_{r\parallel}$, then the intensity of the polarised light in the reflected beam is $I_p = I_{r\perp} - I_{r\parallel}$. Since the intensity of the reflected beam as a whole is $I_r = I_{r\perp} + I_{r\parallel}$ we can express the intensity of the polarised component as a fraction of the total intensity and call it the degree of polarisation, P. So,

$$P = \frac{I_p}{I_r} = \frac{I_{r\perp} - I_{r\parallel}}{I_{r\perp} + I_{r\parallel}} = \frac{R_\perp I_i/2 - R_\parallel I_i/2}{R_\perp I_i/2 + R_\parallel I_i/2}$$

$$\text{or} \quad P = \frac{R_\perp - R_\parallel}{R_\perp + R_\parallel} \qquad (12.43),$$

having used equations (12.40) and (12.41).

When discussing a beam of partially polarised light, whether incident, refracted, or reflected, the degree of polarisation is defined by the relationship

$$P = \frac{I_p}{I_p + I_u} \qquad (12.44)$$

where I_p is the intensity of the polarised light in the beam of total intensity $I_p + I_u$. I_u is the intensity of the unpolarised light in the beam. Imagine now we are looking at a beam of partially polarised light through an analyser, a piece of polaroid, say. Rotating the analyser produces a detected intensity varying from a maximum, I_{max} to a minimum I_{min}. The unpolarised component of the beam will have an intensity $I_u/2$ parallel to the analyser's transmission axis and $I_u/2$ perpendicular to it. The polarised component I_p will be parallel to the transmission axis when the maximum intensity is detected, so the maximum intensity emerging from the analyser will be $I_{max} = I_p + I_u/2$ and the minimum intensity will be $I_{min} = I_u/2$. We can now rewrite equation (12.44) :-

$$P = \frac{I_p}{I_p + I_u} = \frac{(I_p + I_u/2) - I_u/2}{(I_p + I_u/2) + I_u/2} = \frac{I_{max} - I_{min}}{I_{max} + I_{min}}\text{, that is}$$

$$P = \frac{I_{max} - I_{min}}{I_{max} + I_{min}} \qquad (12.45)$$

12.7 RETARDERS

Retarders are optical elements made from birefringent crystals with the optic axis oriented in such a way as to affect the orthogonal P-states differently. In essence, one P-state is caused to lag behind the other and on emerging from the retarder there results a change in the state of polarisation. The name retarders clearly comes from the fact that one orthogonal P-state component is retarded relative to the other.

12.7.1 Wave Plates

If a uniaxial birefringent cuboid crystal is cut with its optic axis parallel to the front and back surfaces, then the electric vector of a normally incident monochromatic P-state wave will have components parallel and perpendicular to the optic axis, figure 12.35. The $E_{\parallel}$ and $E_{\perp}$ components will propagate through the crystal at speeds of $v_{\parallel}$ and $v_{\perp}$, respectively. This will result in their emerging with a phase change which can produce a change in the state of polarisation. Suppose the thickness of the plate is t, then the optical paths of the $E_{\perp}$ and $E_{\parallel}$ components will be $n_o t$ and $n_e t$, respectively, since $n_o = c/v_{\perp}$ and $n_e = c/v_{\parallel}$ from equations (12.33) and (12.34). Then the optical path difference is given by

$$\text{OPD} = (|n_e - n_o|)t \qquad (12.46).$$

If we arrange for the OPD to be a whole number of wavelengths, or an odd number of half-wavelengths, or a whole-number-plus-a-quarter of a wavelength then the plate is called a full-wave, a half-wave, or a quarter-wave plate, respectively. Let us examine each of these.

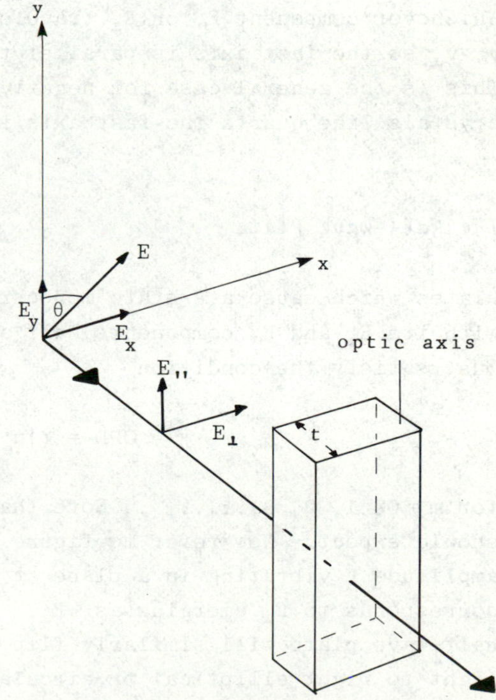

Fig. 12.35 Components of E parallel and perpendicular to the optic axis $E_{\parallel} = E_y$ and $E_{\perp} = E_x$.

The Full-wave Plate

Here the OPD is $$\mathrm{OPD} = (|n_e - n_o|)\,t = m\lambda, \quad m = 1, 2, 3, \ldots \qquad (12.47)$$

where λ is the vacuum wavelength of the monochromatic P-state. In general, the quantity $(|n_e - n_o|)$ changes very little for different wavelengths so the OPD is wavelength dependent. Therefore, a plate of a given thickness t will be a full-wave plate only for a specific wavelength. Such a plate, placed between crossed polaroids with its optic axis at some arbitrary angle θ, will have no effect on the state of polarisation of a monochromatic P-state with amplitude vector E emerging from the polariser. On passing through the full-wave plate the component waves parallel to and perpendicular to the optic axis will have changed relative phase by a multiple of 2π. Thus the orientation of the electric vector of the recombined emergent wave will be unchanged and perpendicular to the analyser's transmission axis. It will therefore be absorbed.
However, consider what happens if the incident beam on the polariser is natural white light. Plane-polarised white light at some angle θ to the optic axis will be incident on the plate. The λ for which the plate represents a full-wave plate will be unchanged in its polarisation state on emerging from the plate and will be absorbed by the analyser, as above. All the other wavelengths will emerge from the plate in some state of elliptical polarisation since they will not satisfy equation (12.47). Components of these parallel to the analyser's transmission axis will emerge and the emergent light will approximate to the complementary colour of the extinguished wavelength.

Fast and Slow Axes

Whichever component $E_\parallel$ or $E_\perp$ travels faster defines the *fast axis* of the plate. In calcite $v_\parallel > v_\perp$ so the fast axis is parallel to the optic axis, whilst the *slow axis* is perpendicular to it. This is the general case for negative uniaxial crystals such as calcite. For positive uniaxial crystals like quartz the fast axis is perpendicular to the optic axis.

The Half-wave Plate

Plates which cause a P-state monochromatic wave, making an angle θ with the optic axis, to emerge with its $E_\perp$ and $E_\parallel$ components half a wavelength out of phase are called half-wave plates. They must satisfy the condition

$$\mathrm{OPD} = (|n_e - n_o|)\,t = (2m+1)\frac{\lambda}{2} \qquad (12.48)$$

for $m = 0, 1, 2, 3, \ldots$. Note that the half-wave plate is again wavelength dependent as we should expect. Now refer to figure 12.36 which shows a monochromatic P-state incident wave of amplitude E vibrating in a plane at an angle θ to the optic axis. The half-wavelength retardation corresponds to $E_\perp$ emerging as $-E_\perp$. This results in E being flipped over through an angle 2θ. half-wave plate will similarly flip elliptical light. It will change left elliptical or circular light to right elliptical or circular light, and vice-versa.

Retarders are most commonly made from quartz or mica. The latter can be made as a half-wave plate with a minimum thickness of 60 µm which corresponds to taking $m = 0$ in equation (12.48). Since $|n_e - n_o|$ is about 0.005 for mica, such a plate would be a half-wave plate for a wavelength

$$\lambda = \frac{2(|n_e - n_o|)t}{2m+1} = 2 \times 0.005 \times 60 \times 10^{-6}$$

$$= 600 \times 10^{-9}\text{m}.$$

A simple half-wave plate can be made by sticking a slightly stretched piece of Sellotape (cellophane tape) onto an uncut lens or a microscope slide. Held between crossed polaroids with its long dimension at 45^0 to the polariser's transmission axis, for the correct λ it will flip over the E-vector from the polariser by $2 \times 45^0 = 90^0$ so that light emerges from the analyser, figure 12.37.

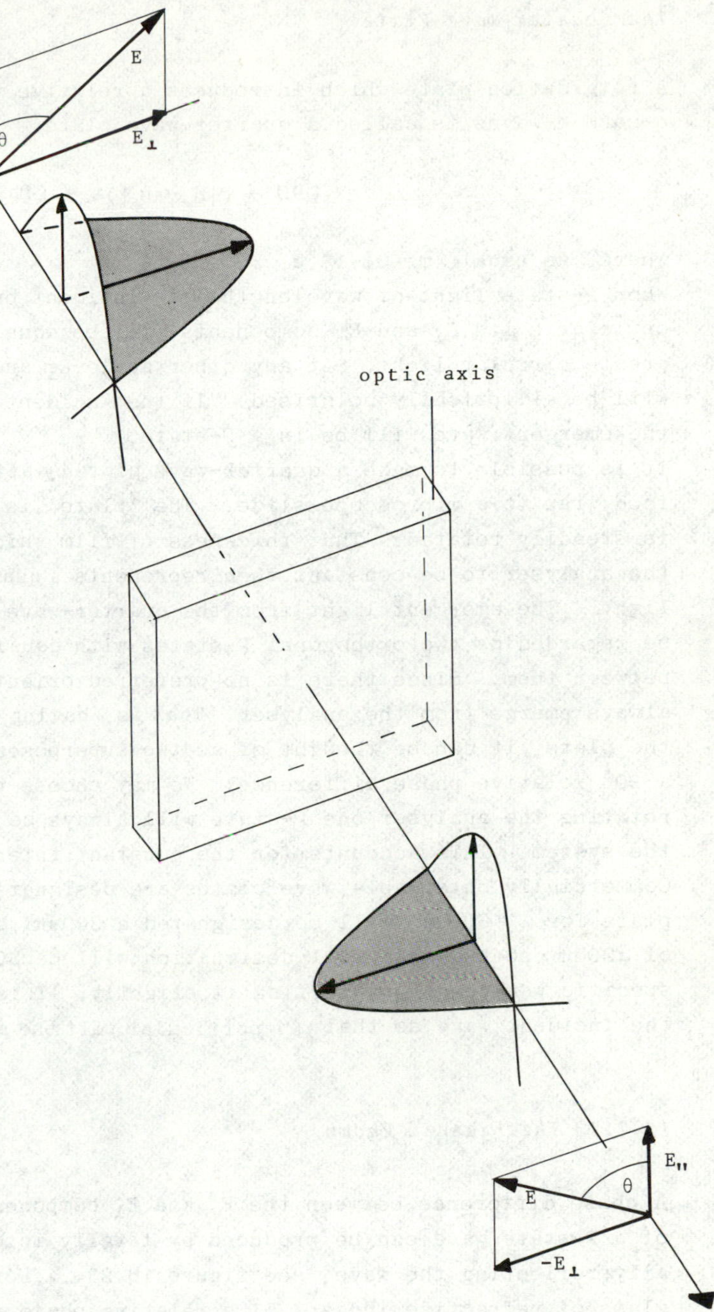

Fig. 12.36 A monochromatic P-state wave incident on a half-wave plate.

Fig. 12.37 Sellotape placed at 45^0 to the transmission axes of crossed polaroids.

If the microscope slide is rotated no light will emerge when the long axis of the sellotape is either parallel to or perpendicular to the polariser's transmission axis. In these cases the P-state incident on the half-wave plate makes an angle 0^0 or 90^0 with the optic axis and cannot be split into two mutually orthogonal components parallel to and perpendicular to the optic axis. Therefore no phase change can occur and the wave emerges from the plate with its E-vector still perpendicular to the analyser's transmission axis, thereafter being extinguished. Any light getting through in this situation is due to depolarisation or local variations in the optic axis orientation.

Stretching the sellotape introduces *stress birefringence* thus producing two different refractive indices along and across the sellotape's length. This type of birefringence occurs in stressed spectacle lenses and in heat toughened lenses. We shall discuss it in some detail in section 12.9.2 .

The Quarter-wave Plate

A retardation plate which introduces a relative retardation of a quarter wavelength between the o- and e-waves is called a quarter-wave plate. It must satisfy the condition

$$\mathrm{OPD} = (|n_e - n_o|)t = (4m+1)\frac{\lambda}{4} \tag{12.49}$$

where, as usual, $m = 0, 1, 2, 3, \ldots$.
When P-state light of wavelength λ is incident on the plate with its E-vector at 45^0 to the optic axis its $E_\perp$ and $E_\parallel$ components will be equal. Producing a $\lambda/4$ relative phase shift will create circular light. At any other angle $E_\perp$ and $E_\parallel$ will not be equal and the emerging light will be elliptically polarised. If the incident wave is elliptically or circularly polarised the emergent wave will be in a P-state.
It is possible to make a quarter-wave plate by attaching successive layers of 'cling-film' kitchen food wrap to a microscope slide. The 'plate' is placed between crossed polaroids and the analyser is steadily rotated. That thickness of film which causes the intensity of the light emerging from the analyser to be constant then represents a quarter-wave plate for the wavelength of the incident light. The emergent light from the quarter-wave plate is circularly polarised and can therefore be regarded as two orthogonal P-states with equal amplitudes with a 90^0 relative phase difference between them. Since there is no preferred orientation for these orthogonal components one will always emerge from the analyser. That is, having created circularly polarised light emerging from the plate, it can be thought of as two superposed orthogonal P-states of equal amplitude and with a 90^0 relative phase difference. We may choose the two orthogonal directions at random so that rotating the analyser one P-state will always be parallel to its transmission axis and emerge from the system. This accounts for the constant intensity mentioned above.
Commercially obtainable wave plates are designated by their linear retardation. Thus, a half-wave plate for $\lambda = 600$ nm will be designated a 300 nm half-wave plate. There will usually be a tolerance of ±20 nm so that the full designation will be 300 ± 20 nm. The wave plate can be 'tuned' to a specific wavelength by tilting it slightly. This effectively alters the thickness traversed by the incident wave so that at a particular tilt the appropriate wave-plate equation will be satisfied.

12.7.2 The Fresnel Rhomb

A phase difference between the $E_\parallel$ and $E_\perp$ components of a P-state wave can be produced by totally internally reflecting the wave, see figure 10.23 . For glass of refractive index 1.51 a relative phase shift of 45^0 occurs when the angle of incidence is 54.6^0. Two internal reflections produce a 90^0 shift. The *Fresnel rhomb* shown in figure 12.38 does just this. If incident P-state light makes an angle of 45^0 with the plane of incidence, the plane of the paper in figure 12.38, its $E_\parallel$ and $E_\perp$ component vectors will be equal. Two internal reflections put them 90^0 out-of-phase and circular light emerges from the rhomb.

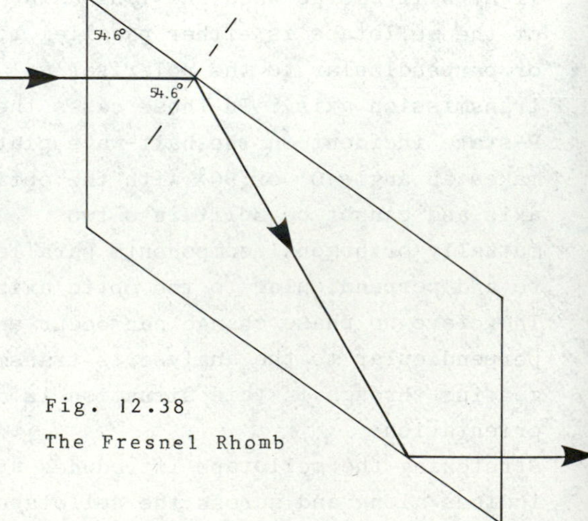

Fig. 12.38
The Fresnel Rhomb

12.7.3 The Babinet Compensator

The devices so far considered in section 12.7 have fixed retarding properties. A device capable of controlling the relative phase shift, the retardance, is called a *compensator*. The Babinet compensator is shown in figure 12.39 . It consists of two quartz prisms with apical angles of about 2.5^0 and the optic axes as indicated. A ray will traverse the thicknesses t_1 and t_2 which can be varied by sliding one prism over the other with a micrometer control. The relative phase difference in the upper prism is

$$\frac{2\pi}{\lambda} \times \text{OPD} = \frac{2\pi}{\lambda}(|n_e - n_o|)t_1 .$$

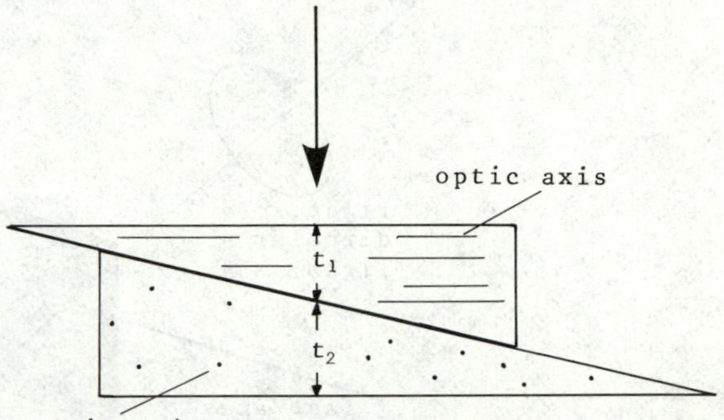

Fig. 12.39 The Babinet compensator.

Since the optic axes are orthogonal the o- and e-rays in the upper prism become the e- and o-rays in the lower prism. This reverses the direction of the phase change in the lower prism which is given by

$$-\frac{2\pi}{\lambda}(|n_e - n_o|)t_2 .$$

The total phase difference is then $$\frac{2\pi}{\lambda}(|n_e - n_o|)(t_1 - t_2) \qquad (12.50).$$

Because $t_1 - t_2$ will vary along the compensator, light is made to enter through a narrow slit with its length perpendicular to the plane of the paper in figure 12.39 .
When the Babinet compensator is placed at 45^0 between crossed polarisers, and in monochromatic light unrestricted by a slit, a series of parallel dark extinction fringes appear across the length of the compensator. These correspond to regions where equation (12.50) satisfies the full-wave plate condition. In terms of the OPD this is

$$(|n_e - n_o|)(t_1 - t_2) = m\lambda \ , \quad m = 1, 2, 3, \ldots .$$

In white light these fringes will be coloured except for the black central fringe where $t_1 - t_2 = 0$.

12.7.4 Circular Polarisers

Figure 12.5 showed that right circular light is produced when ψ_x lags ψ_y by $\pi/2$ radians, or a quarter of a wavelength ($E_x = E_y$ also, of course). Imagine now that right circular light is incident on a quarter-wave plate with its fast axis in the x-direction, figure 12.40 .
On emerging from the quarter-wave plate the x-component will have advanced $\pi/2$ and the ψ_x and ψ_y orthogonal components will then be in-phase. They compound to produce linear or P-state light at 45^0 to the x-axis. If a polariser has its transmission axis at 45^0 the P-state light emerging from the plate will be transmitted.
Suppose instead that left circular light is incident on the quarter-wave plate. The ψ_x component leads the ψ_y component already by $\pi/2$ and will be advanced a further $\pi/2$ on emerging from the

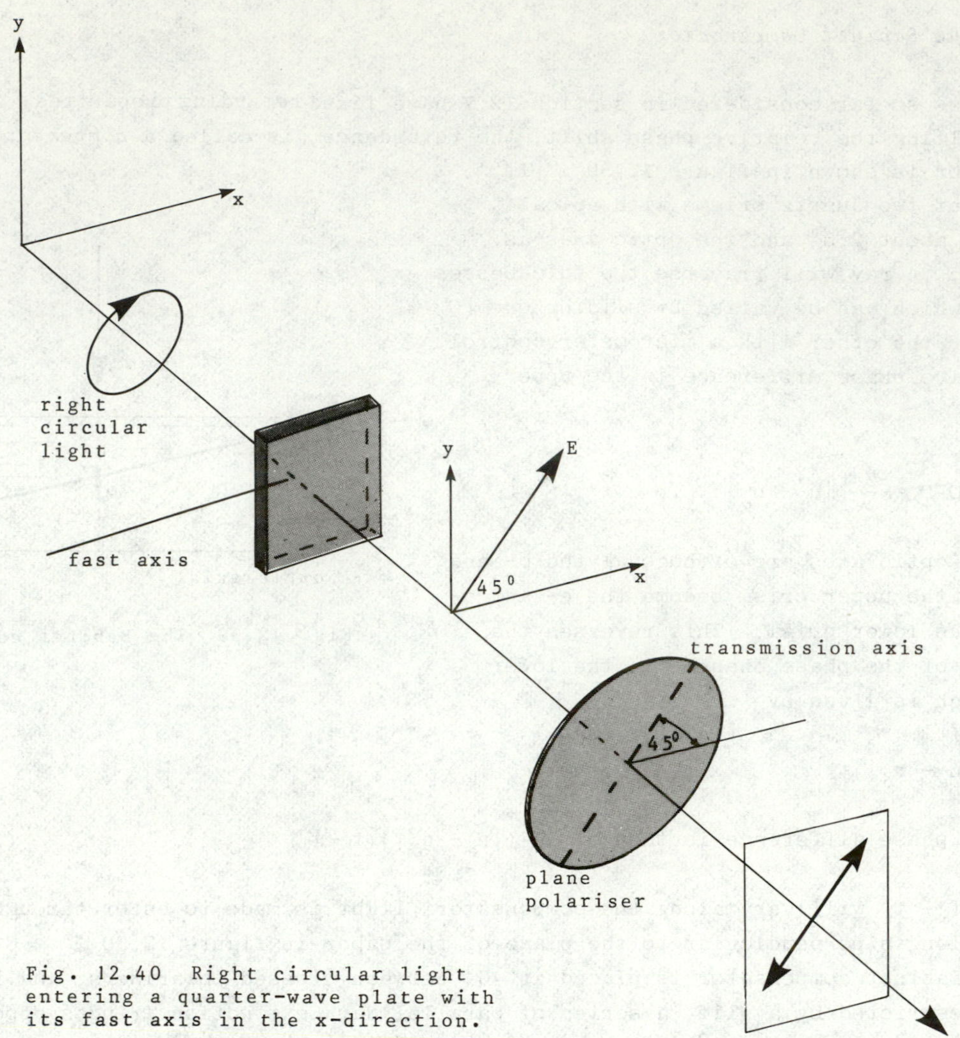

Fig. 12.40 Right circular light entering a quarter-wave plate with its fast axis in the x-direction.

plate since the fast axis is horizontal. Thus ψ_x will lead ψ_y by $\lambda/2$ and the result is P-state light emerges but its vibration plane is now perpendicular to the transmission axis of the polariser in figure 12.40 . Hence, no light will be transmitted by the polariser. We have here the means of checking the handedness of circular light. Right circular light will emerge from the system in figure 12.40, as P-state light, but no light at all will emerge when left circular light is incident on the quarter-wave plate!

12.7.5 Interference Colours

Thus far in this section we have been concerned with monochromatic light passing through polarisers and retarders, the latter having uniform thicknesses. Suppose now instead of a retarder we place a screwed-up piece of cellophane wrapping between crossed polaroids and illuminate the polariser with natural white light. The emergent light is a profusion of variegated colours due to local variations in the OPLs of the ψ_x and ψ_y components. Recall that the OPD is given by $(|n_e - n_o|)t$. Slight variations in the birefringence $n_e - n_o$ associated with the different wavelengths in the white light and differences in the actual path length traversed because of the crumpled nature of the cellophane.

lead to the cellophane acting variously as full-wave, half-wave, and quarter-wave plates for different wavelengths. In between these special cases the optical path lengths will not satisfy any particular condition. The result is that the P-state white light incident on the cellophane will emerge with its constituent colours (wavelengths) in differing states of polarisation. Consider a small localised region of the cellophane where the OPD for the orthogonal components is 900 nm. For orange light of wavelength 600 nm this will represent a half-wave plate since $900 = \frac{3}{2} \times 600$. If the fast and slow axes of the localised region are at 45^0 to the incident P-state white light, and therefore at 45^0 to the P-state constituent orange light, the orange light will be flipped over $2\theta = 2 \times 45^0 = 90^0$ and will emerge from the analyser. Over this localised region most colours will emerge in differing intensities except for violet light ($\lambda = 450$ nm) which will be absorbed by the analyser. That violet will be absorbed can be seen by the fact that $900\text{ nm} = 2 \times 450\text{ nm}$ so that the violet P-state experiences a full-wave plate. As a result, it emerges from the cellophane with its vibration plane still parallel to the polariser's transmission axis and therefore perpendicular to the analyser's transmission axis.
Since one wavelength has been eliminated from the white light the localised region will appear in the complementary colour. This will be happening with different wavelengths in different regions, thereby giving rise to the variegated colours.
Reference will be made to these colours and their production when we consider glass and plastics under stress in subsection 12.9.2 .

12.8 OPTICAL ACTIVITY

Some materials cause the plane of polarisation of P-state light to rotate when it passes through them. For example, a strong solution of household sugar (sucrose) placed between crossed polaroids allows some light to pass through the analyser. If the incident light is monochromatic, rotating the analyser will find some position where the light emerging from the solution is extinguished, figure 12.41. If the analyser must be rotated clockwise to extinguish the light the solution is referred to as *dextrorotatory* or *d-rotatory*. Rotation anticlockwise is referred to as *laevorotatory* or *l-rotatory*. The names are coined from the Latin dexter meaning right and laevus meaning left.

This rotation of the plane of polarisation is not the same as the '2θ flipping over' in half-wave plates. In that case the structure of the crystal was such that two permanent axes, the fast and slow axes, were present in the wave plate. This cannot be the case with the sugar solution since the dissolved molecules are free to move about in the water. The rotational property, known as *optical activity*, is possessed by each molecule and is dependent on the wavelength and the distance travelled by the light through the solution. Travelling a distance twice as long through the solution doubles the rotation of the plane of polarisation. Or, increasing the concentration of the solution causes the electric vector to rotate further.

Fresnel, in 1825, proposed a simple description of the phenomenon. Recall equation (12.21) which represents P-state light resulting from the addition of coherent equal amplitude left and right circular polarised waves. Fresnel reversed the argument and imagined a P-state wave to consist of R-state and L-state circularly polarised components. On meeting a material exhibiting optical activity or *optical rotation* one state is propagated more quickly than the other and the R- and L-states emerge with a relative phase change. When combined they result in a rotated P-state.

Fresnel was able to construct a multiple prism of quartz which separated out the R- and L-states from a P-state. Earlier, in 1811, a French physicist Arago had discovered optical rotation when P-state light was incident on a quartz crystal and propagated along its axis. Then, in 1822, the

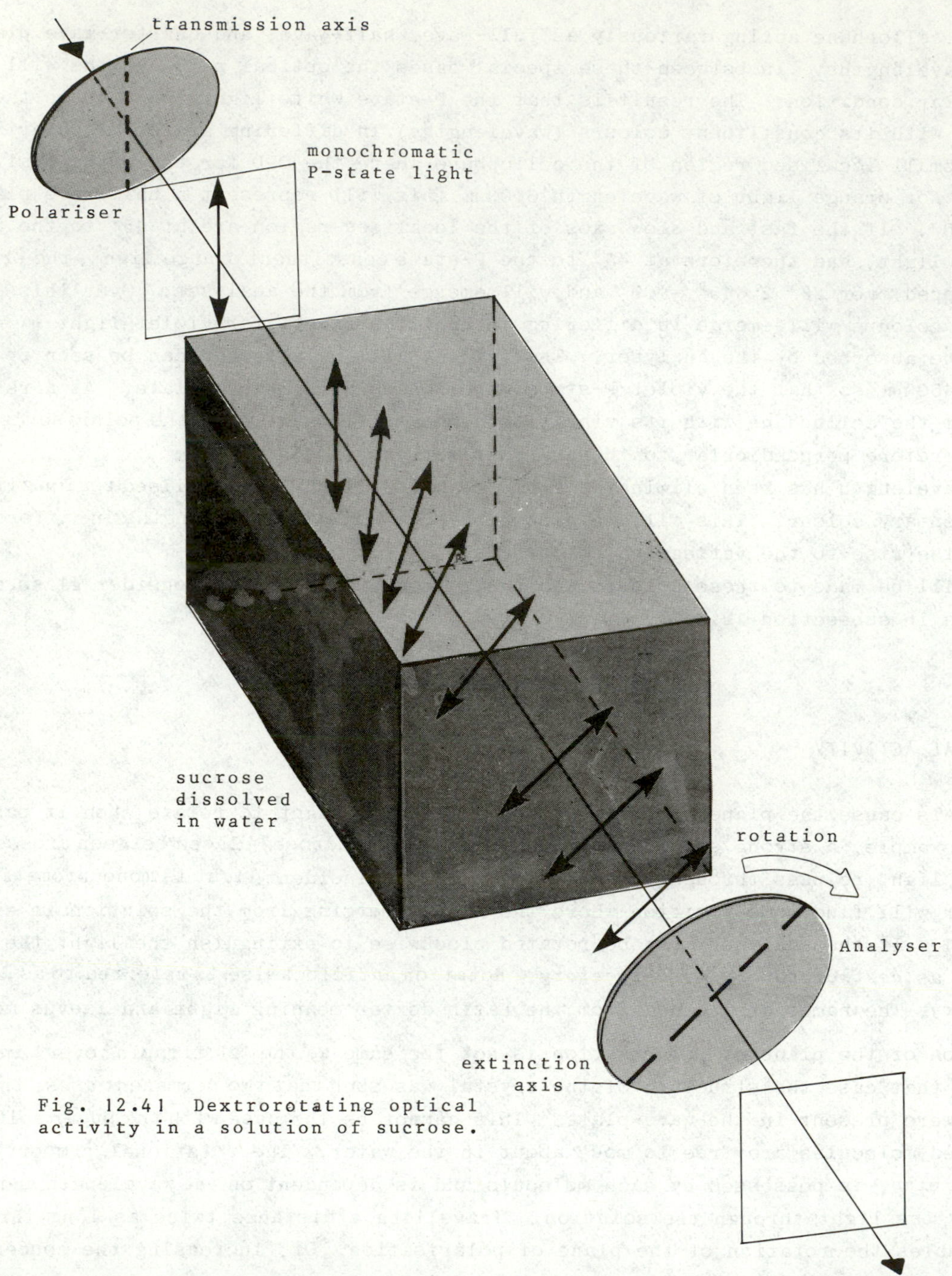

Fig. 12.41 Dextrorotating optical activity in a solution of sucrose.

English astronomer Sir John Herschel had found two different crystallographic quartz structures, one being d-rotatory and the other l-rotatory. Fresnel constructed his multiple prism of alternate d- and l-rotatory segments, all with their optic axes parallel to the composite prism's length, figure 12.42 .

The R-state propagates more quickly in the first segment than in the second and therefore refracts towards the normal in the second. The reverse is true of the L-state. This separation continues at each boundary and the rays emerge as shown.

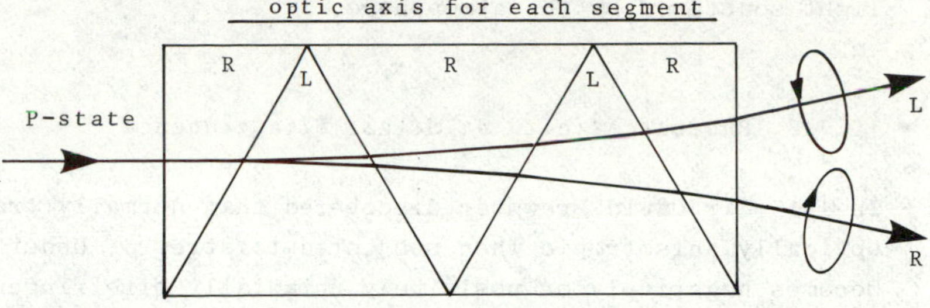

Fig. 12.42 Fresnel multiple prism.

The different speeds of propagation of R- and L-states leads to the definition of refractive indices n_R and n_L for R- and L-state light in a given material. The angle through which the P-state rotates, β, is then given by

$$\beta = \frac{\pi t}{\lambda}(n_L - n_R) \qquad (12.51)$$

where t is the thickness of the material. The term β/t is called the *specific rotatory power* for the material. For quartz it is 21.7^0/mm.

It is interesting to look at the optical activity of naturally occurring biological substances. Natural sucrose is always d-rotatory. No matter whether it is obtained from sugar cane or sugar beet, or wherever it is grown, it is invariably d-rotatory. The same applies to the sugar dextrose which is also called d-glucose to indicate its d-rotatory activity. These sugars can be made in the laboratory and in this case they invariably consist of equal numbers of molecules of l- and d-rotatory activity.

Amino acids, the building bricks of proteins, are all l-rotatory, with one exception. Now there is no reason why d-rotatory amino acids should not exist and they may exist on some other planet. It is not too difficult to imagine Earthmen arriving at some d-rotatory amino acid planet and dying of starvation because they could not digest the fresh-killed d-rotatory 'beef'.

12.9 TECHNOLOGICAL USES OF POLARISED LIGHT

12.9.1 The Uses of Polarised Light caused by Reflection

We have seen that light reflected from a dielectric material's surface is partially polarised, section 12.6. 'Polaroid' sunglasses make use of this effect by having their extinction axes mounted horizontally in order to eliminate the polarised ($E_{r\perp}$) light reflected from horizontal surfaces. They work especially well in reducing the glare from pools of still water since the $E_{r\perp}$ component is entirely extinguished.

A similar effect is achieved in photographing specimens in glass cases. A polarising filter, an analyser, is used over the camera lens to remove the $E_{r\perp}$ component. If the illuminating source is positioned so that $i = i_p$ on the glass cover, then the reflected light will be almost entirely polarised parallel to the surface. There will be some reflected light with $E_{r\parallel}$ vectors since $\tan i_p \neq n'/n$ for all wavelengths since n and n' are wavelength dependent, but this will be insignificant compared to the light which is extinguished.

In photographing white clouds against a blue sky the sky is often darkened by using a yellow coloured filter. The polarising filter above also works quite well by eliminating the polarised light scattered by the atmosphere.

12.9.2 Photoelasticity or Stress Birefringence

In 1816 Sir David Brewster discovered that normally transparent isotropic materials become optically anisotropic when subjected to stress. Under compression or tension the material becomes negatively or positively uniaxially birefringent, respectively. The optic axis is in the direction of the stress and the birefringence is proportional to the stress. Most important if the stress is not uniform over the sample then the birefringence will vary from one localised area to another.

We should here define stress. Stress is defined as force per unit area acting on a body. The forces may place a body under tension or compression, figure 12.43 . When the opposing forces are not collinear they will produce *shear* stress which will deform the body.
There are three orthogonal principal axes parallel to the directions of the principal stresses, σ_1 (the greatest), σ_2, and σ_3 (the least).
When a stressed material such as glass or plastics is placed between crossed polarisers it is convenient to consider two orthogonal electric vector components parallel to two directions of principal stress. These component P-states propagate through the stressed material at different speeds exactly as in wave plates. For a given wavelength the speed of propagation is proportional to the stress in its plane of vibration: i.e. $v_1 \propto \sigma_1$ and $v_2 \propto \sigma_2$. On emerging from the material the two components have undergone a relative retardance proportional to $(v_1 - v_2)t$ where t is the thickness of the stressed material. $(v_1 - v_2)$ is proportional to $(\sigma_1 - \sigma_2)$ so we can say the relative retardance is

$$Ct(\sigma_1 - \sigma_2) \qquad (12.52)$$

where C is a constant for a given material at a given temperature. C is called the *stress-optic coefficient*. Statements made in section 12.7.5 about interference colours apply here. In the case of spectacle lenses mounted in metal frames, t and the principal stresses will vary from one part of the lens to another. Over regions where $Ct(\sigma_1 - \sigma_2)$ is constant the same colour will emerge from the analyser - such coloured bands are called *isochromatics* when the system is illuminated by white light.

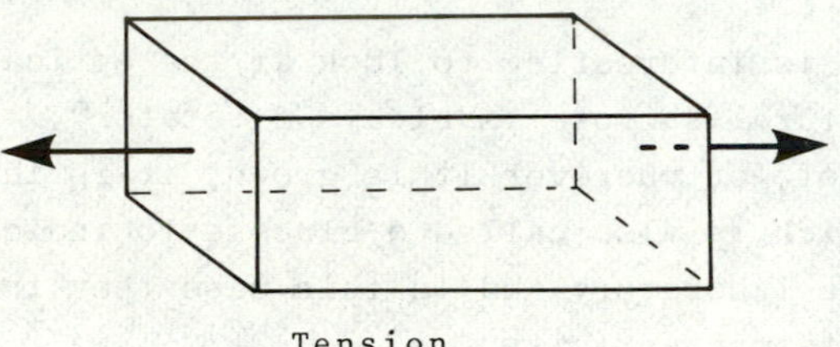

Tension

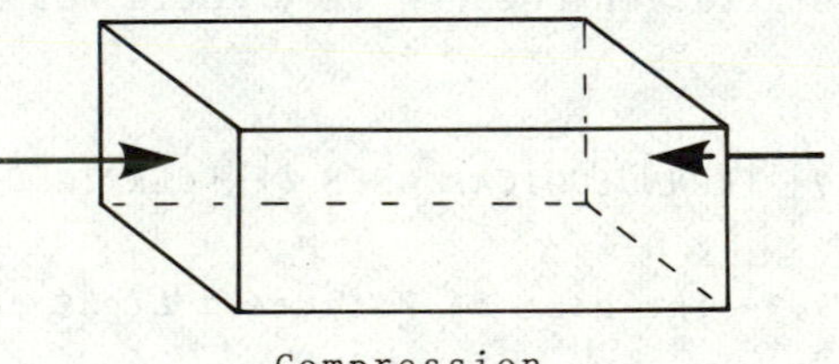

Compression

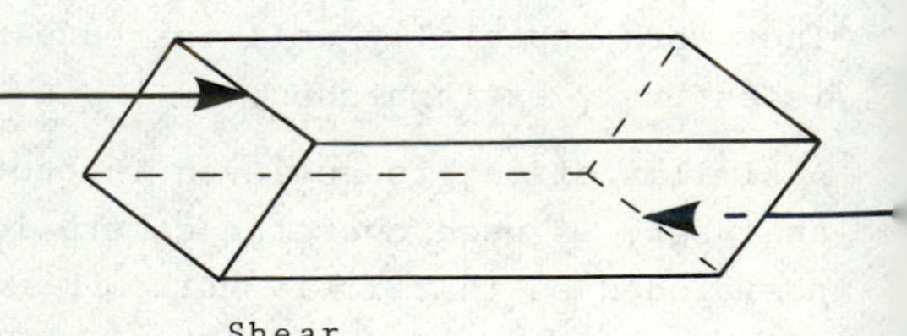

Shear

Fig. 12.43 Types of stress.

If either of the directions of the principal stresses is aligned with the polariser's transmission axis then the P-states leaving the polariser cannot have components parallel to and perpendicular to the two principal stresses. Hence, suc areas on the lens will transmit light through the lens so that its electric vector is not flipped over and thereby suffers extinction by the analyser. The lens will appear black where this applies and these bands are called *isoclinics*. Isoclinics determine all points on the lens where the directions of principal stress are parallel or perpendicular to the polariser's transmission axis. Figure 12.44(a) shows a stressed spectacle lens in a *polariscope*, which is simply a light source with a polariser and an analyser. Figure 12.44(b) is the same lens held between crossed polaroids.

(a)

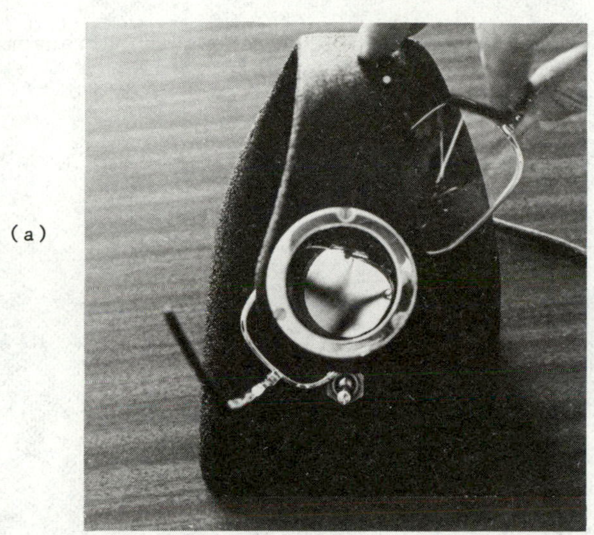

(b)

Fig. 12.44 A stressed spectacle lens seen (a) in a polariscope (a strain-tester) and (b) held between crossed polaroids.

Notice how the irregular pattern of isoclinics indicates a non-uniform stress on the lens. The instrument shown in figure 12.44(a) is known as a *strain tester*.

Strain is defined for a three-dimensional body as the ratio of the change in the body's volume, ΔV, to the volume, V, before a stress was applied. If F is the force applied over an area A, the ratio of stress to strain defines the bulk modulus, κ, for the material under stress. This can be expressed mathematically as

$$\kappa = \frac{\text{stress}}{\text{strain}} = \frac{F/A}{\Delta V/V} \qquad (12.53).$$

Since the denominator is the ratio of two volumes it has no units and the units of κ are those of the stress F/A. A substance that is difficult to compress will have a high bulk modulus. Glass has a bulk modulus about ⅓ that of the metals used in spectacle frames and is therefore three times more compressible. When a lens is stressed energy is transferred to the glass and if this is sufficiently great it will be released by breaking interatomic bonds. The glass fractures, of course. It is for this reason that glass lenses mounted in metal spectacle frames should be tested for undue strain.

Incidentally, heat-toughened spectacle lenses display isoclinics in the form of an approximate cross (convex lenses). Rotating the lens about its optical axis shows little disturbance of the isoclinics and we can deduce a radial distribution of the stress in the lens.

12.9.3 The Kerr Effect

In 1875 the Scottish physicist John Kerr discovered the first electro-optic effect. An isotropic transparent material becomes birefringent when placed in an electric field. The material acts like a uniaxial birefringent crystal with the optic axis in the direction of the applied field. Two indices $n_{\parallel}$ and $n_{\perp}$ are defined for the planes of vibration parallel to and perpendicular to the applied field. The birefringence, Δn, is given by $n_{\parallel} - n_{\perp}$ and it is found to be

$$\Delta n = KE^2\lambda \tag{12.54}$$

where K is the Kerr constant, and E is the field strength. Figure 12.45 shows the arrangement of the Kerr cell.

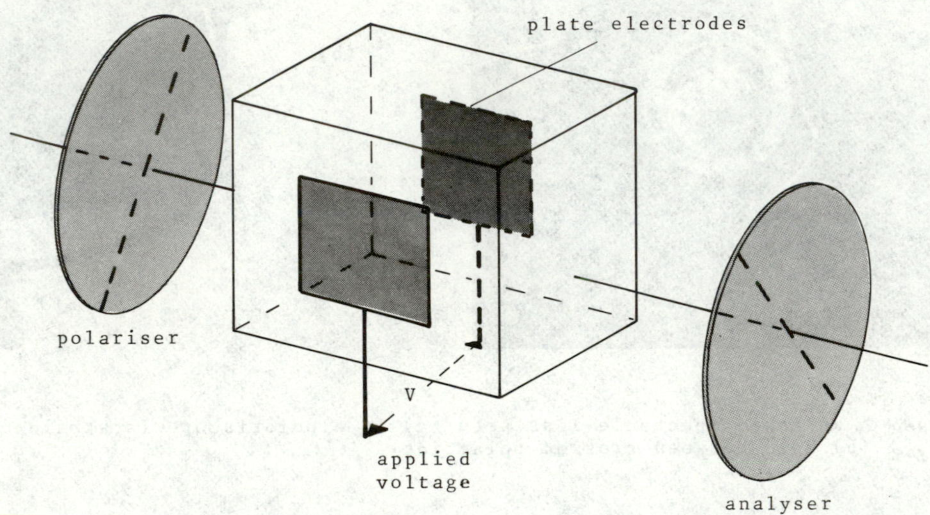

Fig. 12.45 The Kerr cell.

When a polar liquid is placed in the cell the applied electric field partially aligns the molecules thus simulating the anisotropy required for birefringence. The polariser and analyser are placed with their transmission axes at $\pm 45^0$ to the electric field between the electrodes. With zero voltage across the plates no light emerges. Switching on the voltage causes the E-vector from the polariser to emerge from the cell in some new orientation, thus enabling a component to pass through the analyser. The ability to cut off light by switching off the voltage means the cell acts as a switch. It can respond to on-off switching at frequencies as high as 10^{10} Hz so that a Kerr cell can act as a high speed shutter in photography or as a beam chopper in experiments to measure th speed of light.

12.10 WORKED PROBLEMS

12.10.1 A wave ψ has the components $\psi_x = E_1\cos(kz-\omega t)$ and $\psi_y = -E_1\cos(kz-\omega t)$. What is its state of polarisation?

The magnitude of the wave ψ is given by

$$\psi^2 = \psi_x^2 + \psi_y^2 = E_1^2\cos^2(kz-\omega t) + (-E_1^2)\cos^2(kz-\omega t)$$

$$= 2E_1^2\cos^2(kz-\omega t)$$

So, $\psi = \sqrt{2}E_1\cos(kz-\omega t)$ which is a plane polarised wave of amplitude $\sqrt{2}E_1$. The plane of polarisation is given by

$$\tan\theta = \frac{\psi_y}{\psi_x} = \frac{-E_1\cos(kz-\omega t)}{E_1\cos(kz-\omega t)} = -1\ ,$$

so $\theta = 135^0$ measured from the positive x-axis.

12.10.2 Natural light of intensity I_i is incident on three HN-32 sheets of polaroid with their transmission axes parallel. What is the intensity of the emergent light?

Each sheet transmits 64% of the $\frac{1}{2}I_i$ which is parallel to its transmission axis; ($\frac{1}{2}\times 64 = 32$)! None of the $\frac{1}{2}I_i$ intensity parallel to the extinction axes will be transmitted so we need only consider one half of the incident intensity. 0.64 of this will emerge from each polaroid so the transmitted intensity is

$$I_t = (0.64)^3 \times \tfrac{1}{2}I_i = 0.131\ I_i\ .$$

That is, the transmitted intensity is 13.1% of the incident intensity.

12.10.3 Suppose the third polaroid in the last question is rotated through 45^0. What is now the intensity transmitted?

The intensity leaving the second polaroid is $I = (0.64)^2 \times \frac{1}{2}I_i$. This is incident upon the third polaroid and, by Malus' Law, $I\cos^2\theta$ would be the transmitted intensity if the polaroid were ideal.
Now, $\theta = 45^0$, so $\cos^2 45^0 = (1/\sqrt{2})^2 = \frac{1}{2}$. Thus, only one half of the incident intensity $(0.64)^2\times\frac{1}{2}I_i$ would pass through an ideal polariser in the position of the third polaroid. But the third polaroid is not ideal: it transmits only 64% of the intensity of a P-state parallel to its transmission axis. Thus, the emergent intensity is

$$0.64\times\tfrac{1}{2}\times(0.64)^2\times\tfrac{1}{2}I_i = 0.066\ I_i \text{ or } 6.6\% \text{ of the incident intensity.}$$

12.10.4 Two sheets of HN-38 polaroid are held in contact in the familiar crossed position. Let I_1 be the intensity emerging from sheet 1 when natural light of intensity I_i is incident upon it. The intensity emerging from sheet 2 must be $I_2 = 0$. Now insert a third sheet of HN-32 between them with its transmission axis at 45^0 to the other sheets' transmission axes. What is I_2 now?

The intensity emerging from the first sheet parallel to its transmission axis is $0.76I_i/2$. Of this $0.64\cos^2 45^0 = 0.32$ gets through the middle polaroid. The direction of the vibration plane leaving the middle sheet is at 45^0 to the last sheet's transmission axis. Therefore, the fractional intensity getting through the last sheet is $0.76\cos^2 45^0 = 0.38$.

$$\text{So,} \qquad I_2 = 0.38 \times 0.32 \times 0.76 \times I_i/2 = 0.046I_i$$

$$\text{or} \quad 4.6\% \text{ of } I_i .$$

12.10.5 A calcite plate is cut as shown in figure 12.46 with the optic axis perpendicular to the plane of the paper. A ray of natural light, λ=589.3nm, is incident at 30^0 to the normal. The plane of the paper is the plane of incidence. Find the angle between the rays inside the plate.

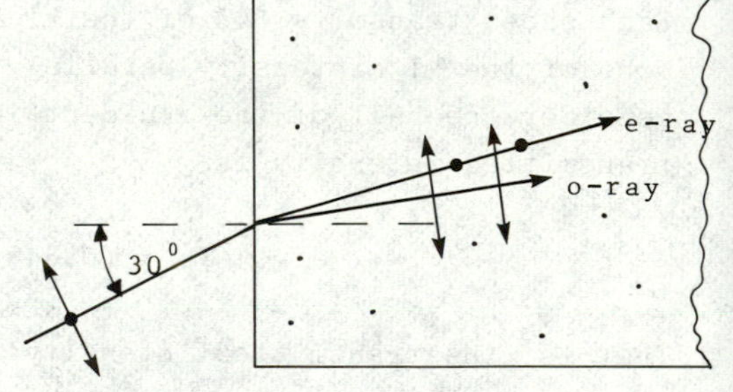

Fig. 12.46

The component E-vector marked · is parallel to the optic axis and will be an e-ray in the calcite. Its angle of refraction, i'_e, is given by Snell's law:

$$n \sin i = n'_e \sin i'_e ,$$

$$\text{So,} \quad \sin i'_e = \frac{n}{n'_e} \sin i = \frac{1}{1.4864} \sin 30^0 = 0.33638$$

$$\Rightarrow \quad i'_e = \text{arc} \sin 0.33638 = 19.66^0 .$$

$$\text{Similarly,} \quad \sin i'_o = \frac{n}{n'_o} \sin i = \frac{1}{1.6584} \sin 30^0 = 0.30150$$

$$\Rightarrow \quad i'_o = \text{arc} \sin 0.30150 = 17.55^0 .$$

So, the angle between the refracted rays is $19.66^0 - 17.55^0 = 2.11^0$.

12.10.6 Natural light is incident on (a) a plane air-glass boundary and (b) a flat sheet of white paper. Suppose the angle of incidence is about 55^0. Examination of the reflected beam from the glass shows a marked degree of polarisation, but this is not the case for the reflected beam off the paper. Explain.

Suppose the glass has a refractive index of 1.5 . Since the beam is incident at $i = 55^0$, the degree of polarisation is given by

$$P = \frac{R_\perp - R_{\shortparallel}}{R_\perp + R_{\shortparallel}} = \frac{(1393 - 1.778) \times 10^{-4}}{(1393 + 1.778) \times 10^{-4}} \simeq \frac{1391}{1395} = 0.997$$

where

$$R_\perp = r_\perp^2 = (-1)^2 \frac{\sin^2(i - i')}{\sin^2(i + i')} = \frac{\sin^2(55^0 - 33.1^0)}{\sin^2(55^0 + 33.1^0)} = 0.1393 = 1393 \times 10^{-4}$$

and

$$R_{\shortparallel} = r_{\shortparallel}^2 = \frac{\tan^2(i - i')}{\tan^2(i + i')} = \frac{\tan^2(55^0 - 33.1^0)}{\tan^2(55^0 + 33.1^0)} = 0.0001778 = 1.778 \times 10^{-4}$$

and $i' = 33.1^0$ from Snell's law. That is, the reflected beam is 99.7% polarised. We should expect this high degree of polarisation since 55^0 is close to the polarisation angle $i = \arctan \frac{1.5}{1} = 56.3^0$.

Reflection from paper involves the excitation of electron oscillators which oscillate in randomly orientated directions (diffuse reflection). Thus, there is no predominance of one P-state in the reflected beam which therefore does not exhibit a noticeable degree of polarisation.

12.10.7 A 50^0 calcite prism is cut with its optic axis as shown in figure 12.47. Sodium light is used in a spectrometer experiment to find n_o and n_e. Two images of the slit are seen and minimum deviation is measured for each. Find n_o and n_e if the angles of minimum deviation are 27.83^0 and 38.99^0. Explain how you would decide which image was formed by the o-rays and which was formed by the e-rays.

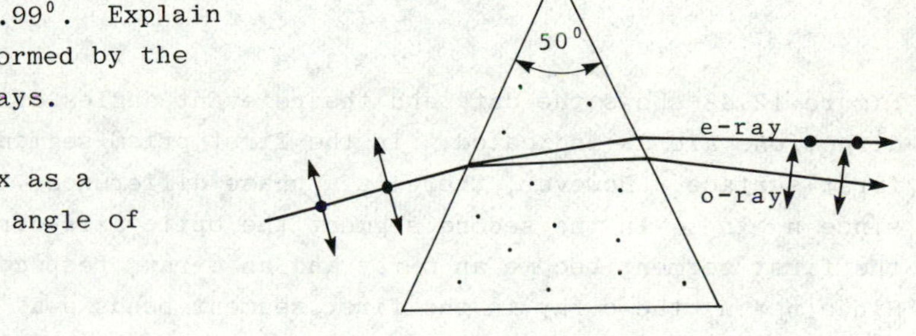

Fig. 12.47

Using the equation for refractive index as a function of the apical angle A and the angle of minimum deviation D:

(i) when $D = 27.83^0$

$$n = \frac{\sin\frac{A+D}{2}}{\sin\frac{A}{2}} = \frac{\sin\frac{50^0 + 27.83^0}{2}}{\sin\frac{50^0}{2}} = \frac{0.62816}{0.42261} = 1.4864$$

(ii) and when $D = 38.99^0$

$$n = \frac{\sin\frac{A+D}{2}}{\sin\frac{A}{2}} = \frac{\sin\frac{50^0 + 38.99^0}{2}}{\sin\frac{50^0}{2}} = \frac{0.70085}{0.42261} = 1.6584 .$$

The o- and e-rays can be determined by using a sheet of polaroid in front of the observer's eye.

With the transmission axis parallel to the optic axis the e-rays will be seen because their E vector is parallel to the optic axis. These will be seen to correspond to the smaller minimum deviation, so the first calculation determined n_e.

12.10.8 A quartz Wollaston beam-splitting polariser is used with a normally incident parallel beam of sodium light for which $n_e = 1.5534$ and $n_o = 1.5443$. If the wedge angle is 45^0 find the angular separation of the emergent e- and o-rays.

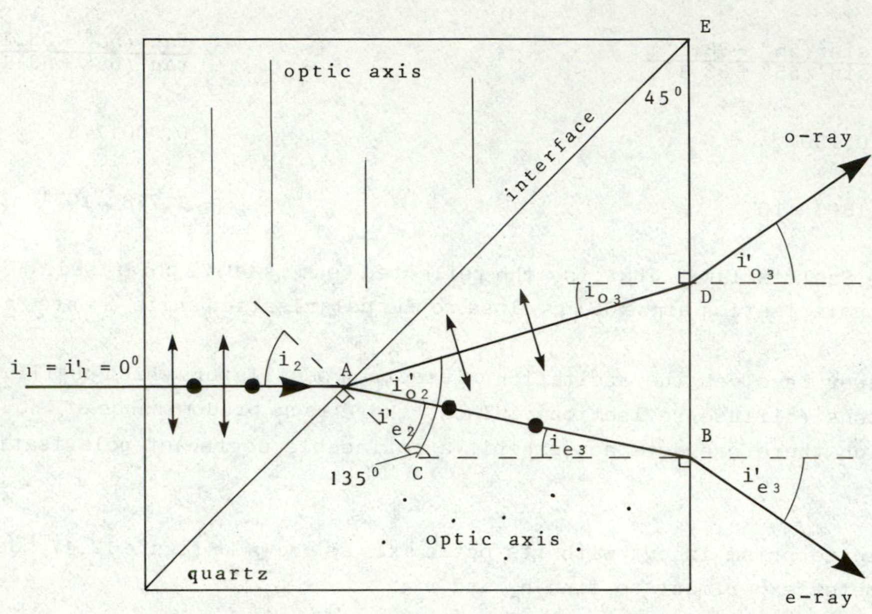

Fig. 12.48

Figure 12.48 shows the data and the relevant angles of incidence and refraction. The optic axis directions are as indicated. In the first prism-segment the o- and e-rays are undeviated at the first surface. However, there is a phase difference, the e-rays (↕) propagating more slowly since $n_e > n_o$. In the second segment the optic axis direction changes and the e-ray and o-ray in the first segment become an o-ray and an e-ray, respectively, in the second segment. Since $n_e > n_o$ the e-ray in the first segment bends away from the normal on entering the second segment, whilst the reverse is true for the other ray.

From the diagram it should be obvious that $i_2 = 45^0$. So, for the o-ray in the first segment whic becomes an e-ray in the second segment

$$n'_{e2} \sin i'_{e2} = n_{o2} \sin i_{o2}$$

$$\text{or} \qquad \sin i'_{e2} = \frac{n_{o2}}{n'_{e2}} \sin i_{o2}$$

$$= \frac{1.5443}{1.5534} \sin 45^0$$

$$= 0.7030$$

$$\text{so} \qquad i'_{e2} = \arcsin 0.7030 = 44.67^0 .$$

From the geometry of triangle ABC in figure 12.48

$$i_{e3} = 180^0 - (135^0 + 44.67^0) = 0.33^0.$$

Then, using Snell's law at the third surface

$$n_3' \sin i_{e3}' = n_{e3} \sin i_{e3} \quad \text{or} \quad \sin i_{e3}' = \frac{n_{e3}}{n_3'} \sin i_{e3} = \frac{1.5534}{1} \sin 0.33^0 = 0.008947$$

so $\quad i_{e3}' = 0.51^0.$

Considering the e-ray in the first segment now:

$$n_{o2}' \sin i_{o2}' = n_{e2} \sin i_{e2} \quad \text{or} \quad \sin i_{o2}' = \frac{n_{e2}}{n_{o2}'} \sin i_{e2} = \frac{1.5534}{1.5443} \sin 45^0 = 0.7113$$

so $\quad i_{o2}' = 45.34^0.$

From the geometry of triangle ADE $\; i_{o3} = 0.34^0$. So,

$$n_3' \sin i_{o3}' = n_{o3} \sin i_{o3} \quad \text{or} \quad \sin i_{o3}' = \frac{n_{o3}}{n_3'} \sin i_{o3} = \frac{1.5443}{1} \sin 0.34^0 = 0.009164$$

whence $\quad i_{o3}' = 0.53^0.$

The separation between the emergent rays is $\; i_{o3}' + i_{e3}' = 0.53^0 + 0.51^0 = 1.04^0.$

12.10.9 Figure 12.49 shows a ray of monochromatic light incident at an angle i_p on extra-dense flint glass for which $n_g = 1.653$. Show that the angle between the transmitted and reflected rays is generally equal to 90^0, and find i_p and i' in this particular case.

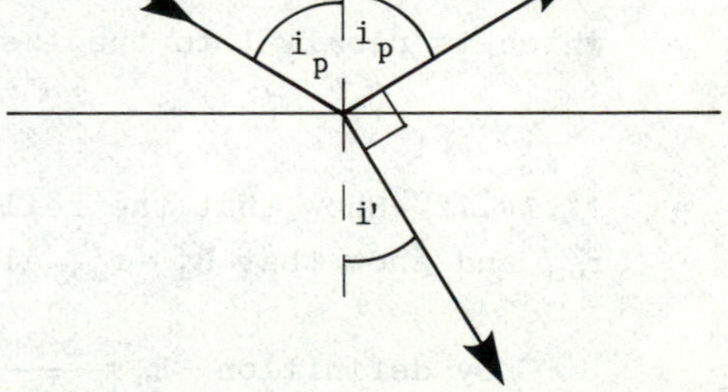

Fig. 12.49

When $i = i_p$ the reflected ray is entirely polarised parallel to the surface so that the amplitude coefficient of the reflected wave parallel to the plane of incidence, $r_{\parallel}$, is zero.

But $\; r_{\parallel} = \dfrac{\tan(i - i')}{\tan(i + i')}$ from equation (10.44) and this will be zero if $i + i' = 90^0$ since $\tan 90^0 = \infty$.

So, $i_p + i' = 90^0$. Also, the angle between the reflected and refracted rays is $\; 180^0 - (i_p + i') = 180^0 - 90^0 = 90^0$, from the diagram, and this is precisely what we were asked to show!

Here, $\; i_p = \arctan \frac{n'}{n} = \frac{1.653}{1} = 58.83^0$, from Brewster's Law. From the foregoing we have

$$i' = 90^0 - i_p = 90^0 - 58.83^0 = 31.17^0.$$

12.10.10 A beam of left circular light meets a quarter-wave plate with its fast axis vertical. What is the state of the emergent light?

If the y-axis is vertical and the wave is propagating in the z-direction then, using our description of left circular light, the ψ_x component leads the ψ_y component by $\lambda/4$. On passing through the quarter-wave plate the ψ_y component, travelling faster than the ψ_x component, emerges advanced by $\lambda/4$ and the two components are now in phase. Since for circular light the amplitude of the waves are equal, we can write

$$\psi_y = E_1 \sin(kz - \omega t) \quad \text{and} \quad \psi_x = E_1 \sin(kz - \omega t)$$

whence $$\psi^2 = \psi_y^2 + \psi_x^2 = 2E_1^2 \sin^2(kz - \omega t) \quad \text{giving} \quad \psi = \sqrt{2}E_1 \sin(kz - \omega t),$$

which is a P-state wave making an angle $\theta = \arctan\frac{\psi_y}{\psi_x} = \arctan 1 = 45^0$ with the x-axis.

12.10.11 A quarter-wave plate made from calcite, $n_e = 1.4864$ and $n_o = 1.6584$, is placed in a beam of normally incident light, $\lambda = 656$ nm, plane polarised at 45^0 to the x-axis. Right circular light emerges. Find the minimum thickness of the plate and the orientation of the optic axis.

The condition for a quarter-wave plate is given by equation (12.49). Rearranging to give the thickness, t,

$$t = \frac{(4m+1)\lambda}{4(|n_e - n_o|)} .$$

For a minimum thickness $m = 0$, so $$t = \frac{\lambda}{4(|n_e - n_o|)} = \frac{656}{4(|1.4864 - 1.6584|)} = 953.49 \text{ nm}.$$

Now refer to figure 12.5 in section 12.1.2 . Right circular light requires ψ_x to lag ψ_y by $\lambda/4$ on emerging from the plate. Or, the other way about, ψ_y must travel faster and emerge leading by λ. So, the fast axis must be vertical (in the y-direction). Since $n_e < n_o$ the extraordinary component will travel faster. But the E-vector for the extraordinary wave is parallel to the optic axis, which is parallel to the fast axis, which we have decided is vertical!

12.10.12 Show that the reflectance $R_{\parallel}$ is equal to the amplitude reflection coefficient squared, $r_{\parallel}^2$, and show that $R_{\perp}^2 = r_{\perp}^2$, also.

By definition $$R_{\parallel} = \frac{I_{r\parallel}}{I_{i\parallel}} = \frac{E_{r\parallel}^2}{E_{i\parallel}^2} = \frac{(r_{\parallel} E_{i\parallel})^2}{E_{i\parallel}^2} = r_{\parallel}^2 .$$

The second part follows by replacing $\parallel$ with $\perp$.

12.10.13 Why may unpolarised light be regarded as the superposition of two orthogonal, equal amplitude, incoherent waves?

Unpolarised light consists of large numbers of randomly oriented vibrations. If we choose any two orthogonal directions perpendicular to the direction of propagation, and resolve each wave's amplitude into components in these directions, with very large numbers the disturbance will on average carry half the energy in each component. In order to be unpolarised these equal energy components must be incoherent. Since they are equal in energy we can imagine equal amplitude incoherent waves.

EXERCISES

1 Explain what is meant by plane or linear polarisation.

2 What condition must be met for two orthogonal P-state waves to superpose and result in another P-state wave?

3 Explain the conditions required to produce circularly polarised light, differentiating between left and right circular light.

4 Add the waves $\psi_x = E_x \sin(kz - \omega t) = 3\sin(kz - \omega t)$ and $\psi_y = E_y \sin(kz - \omega t) = 4\sin(kz - \omega t)$. What is the state of polarisation and the amplitude of the resultant?

Ans. P-state: $\theta = \arctan\frac{4}{3}$ and $E = 5$.

5 Add the waves $\psi_x = 3\sin(kz - \omega t - \pi)$ and $\psi_y = 4\sin(kz - \omega t)$. What is the state of polarisation and the amplitude of the resultant wave?

Ans. P-state: $\theta = \arctan -\frac{4}{3}$ and $E = 5$.

6 What is the state of polarisation when the waves $\psi_x = 7\sin(kz - \omega t + \frac{\pi}{2})$ and $\psi_y = 7\sin(kz - \omega t)$ superpose?

Ans. Right circular light with $\psi = 7$.

7 Explain why natural light may be regarded as the superposition of two incoherent, equal amplitude, orthogonal P-states.

8 Compare the energies of two wave packets (photons) with vacuum wavelengths 1 000 nm and 200 nm. To which regions of the electromagnetic spectrum do these waves belong?

Ans. The ratio of energies for the shorter wavelength photon compared with the longer wavelength photon is 5. The shorter wavelength corresponds to the ultraviolet region and the longer wavelength to the infrared region.

9 A beam of unpolarised light of intensity I_i is incident on two pieces of ideal polaroid held in contact. What is the angle between the transmission axes if the emergent beam's intensity is (a) $I_i/2$, (b) $I_i/4$, (c) $I_i/8$, and (d) 0 ?

Ans. (a) 0. (b) 45^0. (c) 60^0. (d) 90^0.

10 Use the wire-grid polariser to aid the explanation of how polarising sunspectacles work.

11 What is meant by the term birefringence?

12 Natural light of flux density (intensity) I_i is incident on a sheet of polaroid which transmits 32% of the flux. If another sheet of this material (HN-32) is placed with its transmission axis parallel to the first sheet's, what is the emergent intensity?

Ans. 0.21 I_i. See W.P.O. page 134, question 1.

13 What will be the emergent flux density if the analyser is rotated 30^0 in question 12?

Ans. 0.157 I_i. See W.P.O. page 134, question 2.

14 Two perfect linear polarisers (HN-50) are placed with their transmission axes vertical and horizontal, respectively. If natural light, intensity I_i, is incident on the first sheet an intensity $I_i/2$ is transmitted with its plane of polarisation vertical. No light emerges from sheet two, of course. A third sheet is inserted between the other two with its transmission axis at 45^0 to the vertical. What is the intensity of the light emerging from the third sheet now?

Ans. $I_i/8$. See W.P.O. page 134, question 3.

15 Calculate the angle between the o- and e-rays emerging from a calcite Wollaston prism of wedge angle 15^0, for natural light incident normally on the prism. Is calcite negatively or positively birefringent? ($n_o = 1.66$ and $n_e = 1.49$).

Ans. 5.16^0. See W.P.O. page 138, question 10.

16 Calculate the angle between the o- and e-rays emerging from the quartz Wollaston prism (polarising beamsplitter) with a wedge angle of 30^0 when natural light is incident on it normally. ($n_o = 1.5443$ and $n_e = 1.5534$).

Ans. 0.604^0. See W.P.O. page140, question 11.

17 Explain the separation of the e- and o-rays shown in figure 12.17 .

18 Explain how light scattered from air molecules comes to be partially polarised. How would you demonstrate it?

19 Describe the properties of a half-wave plate or retardation plate. A piece of sellotape is stretched slightly along its length and stuck to a microscope slide. It is placed between crossed polaroids at 45^0 to the transmission axes whereupon light passes through the system over the area covered by the sellotape. Explain.

Ans. See W.P.O. page 142, question 12.

20 Find the minimum thickness of a quartz retarder for it to act as a half-wave plate for light of wavelength 590 nm. ($n_e = 1.5534$ and $n_o = 1.5443$).

Ans. 32 418 nm. See W.P.O. page143, question 13.

21 Two polaroid sheets are placed with their transmission axes parallel and placed between them is a piece of sellotape stuck to a microscope slide. If the length of the sellotape is placed at 45^0 to the transmission axes no light emerges from the system over the area covered by the sellotape. Explain.

Ans. See W.P.O. page 144, question 14.

22 Describe the phenomenon known as stress birefringence or photoelasticity. What is its significance in ophthalmic practice?

Ans. See W.P.O. page 145, question 16.

13 INTERFERENCE

13.0 INTRODUCTION

In the last chapter we dealt with the superposition of orthogonal, coherent P-states which arose from the interaction of light waves with matter. In this chapter we shall be concerned with the superposition of coplanar P-states which give rise to the phenomenon of interference.

As a precursor let us remind ourselves of some everyday examples of interference. We are all familiar with the multicoloured patterns seen when water and oil layers form on roads. The light transmitted through these layers is absorbed by the black asphalt and the reflected light from the air-oil and oil-water boundaries interferes by superposition. The effect is quite remarkably beautiful.

Interference is not confined to light waves. The phenomenon is readily witnessed if one drops two stones into a still pool and observes the complicated interaction of the two sets of waves, each set propagating outwards from its source. Figure 13.1(a) is a laboratory ripple tank simulation of this effect.

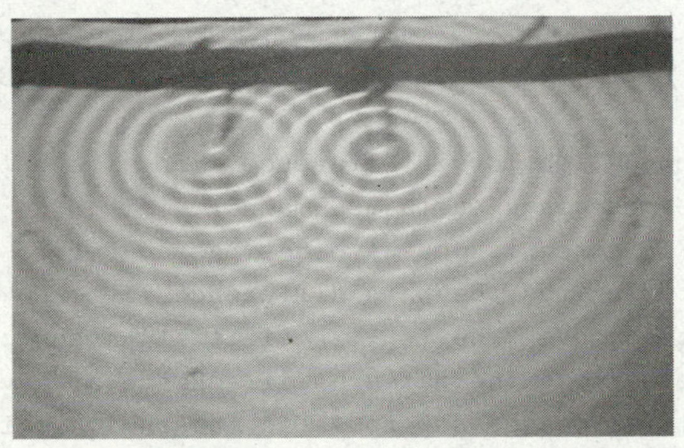

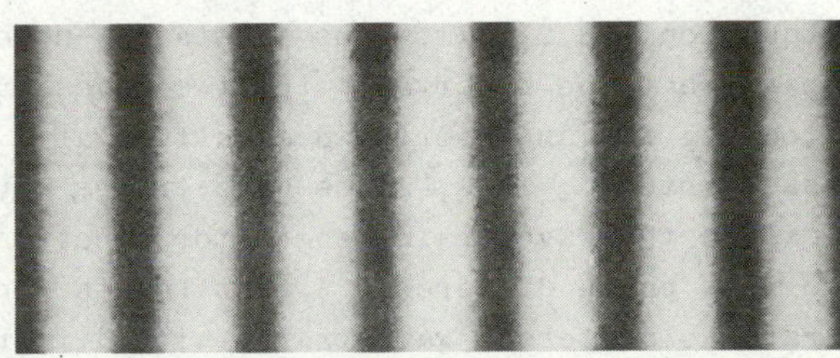

Fig. 13.1 (a) Interference pattern with two dippers in a ripple tank. (b) Interference fringes formed with light.

Another example, an annoying one, is the interference with radio waves. Here a source such as overhead power lines, radiating at 50 Hz, superposes its waves on the incoming modulated carrier wave, the signal. We all know the unpleasant results when the superposed waves are converted to sound and emerge from the loudspeaker as a buzzing noise.

As for light, since the electromagnetic field disturbances in the visible region vary at the very high frequency of about 10^{14} Hz we shall not be concerned with the field displacement, ψ. Instead, we shall observe the intensity distribution which can be detected with a wide variety of detectors such as photoelectric cells, photographic film, and the eye. The sort of intensity patterns we can observe under specially contrived conditions will resemble the interference fringes of figure 13.1(b). The light intensity distribution varies from maxima, I_{max}, to minima, I_{min}, and we shall want to derive mathematical expressions to describe where and when they occur. The alternating dark and light bands are known as *fringes* and they occur in numerous situations. It is worth noting that interference theory is necessary in the explanation of antireflection films and that should be sufficient stimulus for you to get to grips with this material.

13.1 CONDITIONS FOR INTERFERENCE

In chapter 11, section 11.1.2, we examined the intensity resulting from the interaction of two monochromatic, coplanar waves of equal frequency. The result was given by equation (11.16),

$$I = I_1 + I_2 + 2\sqrt{I_1 I_2}\cos(\alpha_2 - \alpha_1)$$

where I_1 and I_2 were the intensities of the two waves. α_1 and α_2 were 'shorthand carrier terms' for $(kx + \varepsilon_1)$ and $(kx + \varepsilon_2)$, respectively. The argument of the cosine function is the phase difference between the two waves. Suppose now the phases of the two waves are $\phi_1 = kz_1 - \omega t + \varepsilon_1$ and $\phi_2 = kz_2 - \omega t + \varepsilon_2$, then the phase difference is δ or $\phi_2 - \phi_1$: i.e.

$$\delta = k(z_2 - z_1) + (\varepsilon_2 - \varepsilon_1) \qquad (13.1)$$

where the waves are coplanar, of the same frequency, and travelling in the positive z-direction instead of the x-direction as they were in chapter 11.

Since $k = 2\pi/\lambda$ we can write equation (13.1) as

$$\delta = \frac{2\pi}{\lambda}(z_2 - z_1) + (\varepsilon_2 - \varepsilon_1) \qquad (13.2).$$

Equation (13.2) allows us to state a necessary condition for interference. For the intensity pattern to be observable the two waves must superpose in a sustained manner. It is no good looking for an interference pattern at a region in space if one wave has passed before the other has arrived! Thus, since wave-trains have finite lengths the path difference $(z_2 - z_1)$ must not exceed the wave-train length for superposition to occur. The $(\varepsilon_2 - \varepsilon_1)$ initial phases component of the phase difference δ must remain constant otherwise the pattern will change at a very rapid rate as different wave-trains with differing initial phases superpose in the region being observe If $\varepsilon_2 - \varepsilon_1$ is constant the sources are said to be *coherent*. If two or more sources are incoheren the pattern will change about every 10^{-8} seconds, and this will be too fast to observe. Thus, f incoherent light no interference patterns are visible.

13.2 INTENSITY MAXIMA AND MINIMA: VISIBILITY

Let us return now to the intensity of two interfering coplanar waves. We can now rewrite equati (11.16) as

$$I = I_1 + I_2 + 2\sqrt{I_1 I_2}\cos\delta \qquad (13.3).$$

The intensity I will be a maximum I_{max} when $\cos\delta = 1$, and this occurs when $\delta = 0, \pm 2\pi, \pm 4\pi, \ldots$. This is equivalent to saying the optical path difference is an integer number of wavelengths; i.e. OPD $= m\lambda$, $m = 0, \pm 1, \pm 2, \pm 3, \ldots\ldots$, if $\varepsilon_2 - \varepsilon_1 = 0$. In this case the phase difference between the two waves is an integer multiple of 2π and the disturbances are in-phase. This is equivalent to saying a crest of one wave is superposed on the crest of the other, the condition being known as *total constructive interference*. When $\cos\delta = -1$ the intensity will be a minimum I_{min}. This occurs when $\delta = \pm\pi, \pm 3\pi, \pm 5\pi, \ldots$, and is referred to as *total destructive interference*. Again, we can state the equivalent OPD condition for minima: OPD $= (m + \frac{1}{2})\lambda$, $m = 0, \pm 1, \pm 2,$

Once again, this is if $\varepsilon_2 - \varepsilon_1 = 0$. It corresponds to a crest of one wave overlapping a trough of the other and the waves are said to be antiphase. For conditions between total constructive and total destructive interference there are varying degrees of constructive and destructive interference. However, in all the cases we shall consider only in-phase and antiphase conditions will be important. This will greatly simplify any mathematical equations we shall evolve. A further simplification occurs if the two waves have equal amplitudes. This allows us to write $I_1 = I_2 = I_0$, say. Equation (13.3) then becomes

$$I = I_0 + I_0 + 2\sqrt{I_0 I_0}\cos\delta$$

$$\text{or} \qquad I = 2I_0(1 + \cos\delta) \qquad (13.4)$$

In this case $I_{max} = 4I_0$ and $I_{min} = 0$, when $\cos\delta = 1$ and $\cos\delta = -1$, respectively.

Many of the cases we shall consider will have $I_{max} = 4I_0$ and $I_{min} = 0$ although a few will not be so. None-the-less, whatever the values of I_{max} and I_{min} there is a useful parameter called the *visibility* which can be applied to the interference fringe pattern. Defined by Michelson, the visibility is given by

$$V = \frac{I_{max} - I_{min}}{I_{max} + I_{min}} \qquad (13.5).$$

When $I_1 = I_2 = I_0$ we have seen that $I_{max} = 4I_0$ and $I_{min} = 0$. In this case the visibility is

$V = \frac{4I_0 - 0}{4I_0 + 0} = 1$, and this is the maximum value attainable. Clearly, V will equal zero when I_{max} equals I_{min}; that is, when there is a uniform intensity with no fringes. So, it is advisable to conjure up experimental situations where $I_1 = I_2$ for maximum visibility.

13.3 WAVEFRONT SPLITTING INTERFEROMETERS

In order to produce interference fringes with maximum visibility recall that we need coherent sources, path differences much less than wave-train lengths, and equal amplitude (or intensity) waves. The use of waves with the same frequency also simplifies the ensuing pattern. One way of achieving all these conditions is to have a single monochromatic source and split the wavefronts into two by passing them through two slits, for example. This was the arrangement used by Thomas Young, the English physicist and polymath. Incidentally, it was Young in 1817 who proposed that light waves were transverse rather than longitudinal.

Young's experiment is shown in figure 13.2 where monochromatic plane waves illuminate the slit S which acts as a primary source from which cylindrical waves propagate. The waves from S illuminate the slits S_1 and S_2 which act as coherent secondary sources. Each wavefront is 'split', and since a wavefront is the locus of points on a disturbance where the phase is everywhere the same, the phase difference is therefore constant for waves arising from S_1 and S_2. If the distances SS_2 and SS_1 are equal the phase difference for the secondary waves arising from S_1 and S_2 will be zero. However, this is not essential. As long as $\varepsilon_2 - \varepsilon_1$ is constant the fringe pattern will be sustained. If the slits S_1 and S_2 are symmetrically placed, as in figure 13.2, a fringe pattern will be evident on the screen. Maxima will occur along the lines joining overlapping crests. Midway between the overlapping crests will be a trough superposed on a crest. Along lines joining trough-crest superpositions minima will occur. Moving the screen to the right will cause the fringe separations to increase since the lines joining the superposing crests diverge from left to right.

Now consider figure 13.3 from which we can quantify the observations. a, the distance between the slits S_1 and S_2, is small compared to the distances between each set of slits and the slits and the screen. The path difference between the waves from S_1 and S_2 to a point P on the screen is very nearly given by the distance S_1A. That is, $S_1P - S_2P \simeq S_1A$. If the system is immersed in a medium of refractive index n then the optical path difference is

$$OPD = n.S_1A \ .$$

But, $S_1A = a\sin\theta$, from triangle S_1S_2A, so the OPD is now given by

$$OPD = na\sin\theta \ .$$

For small θ, $\sin\theta \simeq \tan\theta \simeq y/s$, from triangle POP_0, so now the OPD is

$$OPD = nay/s.$$

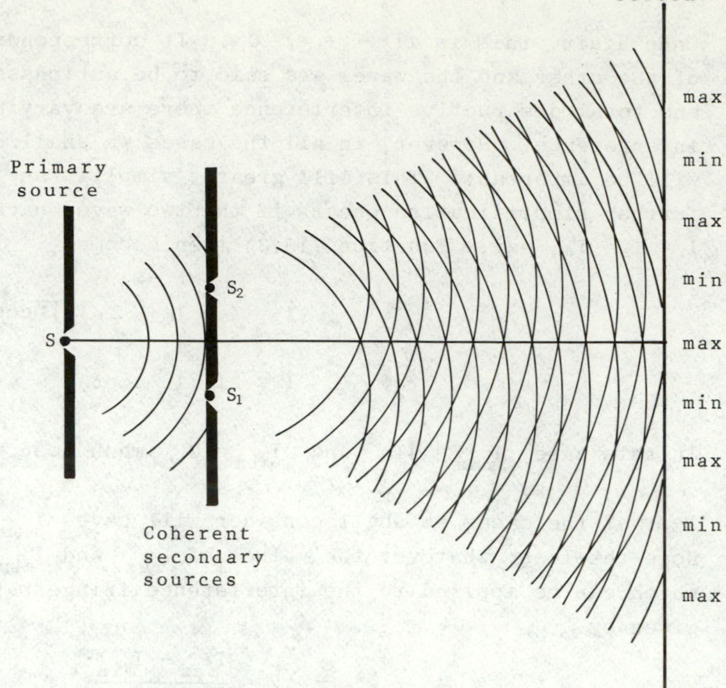

Fig. 13.2 Young's experiment - maxima and minima on a screen.

13.3.1 Bright Fringes or Maxima

In section 13.2 we stated that an intensity maximum will occur when the OPD is a whole number of wavelengths: i.e. $OPD = m\lambda$, $m = 0, \pm1, \pm2, \ldots$. Thus, for an intensity maximum at P on the screen we equate the OPD to $m\lambda$. This gives

$$\frac{nay}{s} = m\lambda \ , \quad m = 0, \pm1, \pm2, \ldots$$

Rearranging, we have $y = \dfrac{sm\lambda}{na}$.

It is convenient to call the central bright fringe the zeroth fringe, or zeroth order fringe, when $m = 0$, and similarly for the 1st fringes at each side of the central fringe when $m = \pm1$, and so on. So, putting y_m for y, the mth bright fringe occurs at a distance y_m from P_0 given by

$$y_m = \frac{sm\lambda}{na} \qquad (13.6)$$

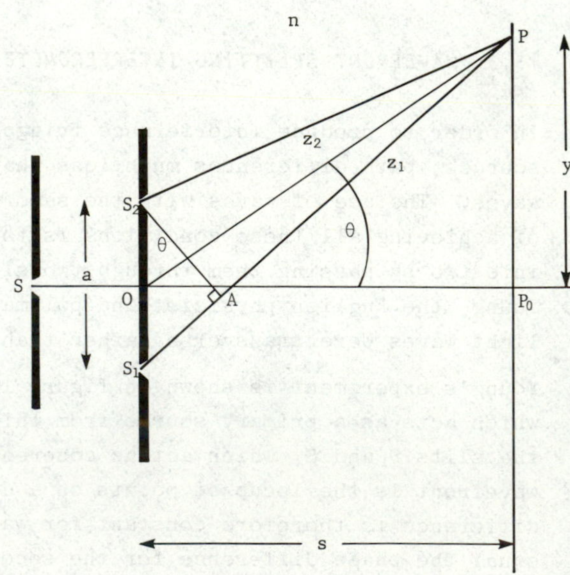

Fig. 13.3 The geometry of Young's experiment.

Example

The slits S_1 and S_2 in Young's experiment are 1 mm apart (centre to centre) and the screen placed such that $s = 5$ m. If the light illuminating S has wavelength $\lambda = 589.3$ nm and the system is in air, $n = 1.000\,29$, find the distance of the first bright fringe from the central bright fringe.

Here $m = 1$, $a = 10^{-3}$m, $\lambda = 589.3 \times 10^{-9}$m, $n = 1.000\,29$, and $s = 5$ m. Thus,

$$y_1 = \frac{sm\lambda}{na} = \frac{5 \times 1 \times 589.3 \times 10^{-9}}{1.000\,29 \times 10^{-3}} = 2.946 \times 10^{-3}\ \text{m} \equiv 2.946\ \text{mm}.$$

If the system were immersed in water, $n = 1.33$, the result would be $y_1 = 2.215$ mm. We see that the fringe pattern reduces in size.

The distance between consecutive bright fringes, called the *fringe separation* or *fringe width*, can easily be found as follows. Let Δy be the distance between the $(m+1)$th and the mth bright fringes; then

$$\Delta y = y_{m+1} - y_m = \frac{s(m+1)\lambda}{na} - \frac{sm\lambda}{na} = \frac{s\lambda}{na}.$$

That is,

$$\Delta y = \frac{s\lambda}{na} \qquad (13.7).$$

Note that Δy is independent of m so all the bright fringes are equidistant apart, at least within the small angle constraints of our equations.

The angular position of the mth bright fringe, θ_m, can be obtained from $\theta = y/s$ by writing

$$\theta_m = y_m/s = \frac{m\lambda}{na} \qquad (13.8),$$

having used equation (13.6).

13.3.2 *Dark Fringes or Minima*

Intensity minima, or dark fringes, will occur when the OPD is equal to an odd number of half-wavelengths so that total destructive interference results. Thus,

$$\text{OPD} = (m + \tfrac{1}{2})\lambda, \quad m = 0, \pm 1, \pm 2, \ldots..$$

or

$$\frac{nay_m}{s} = (m + \tfrac{1}{2})\lambda$$

which rearranges to

$$y_m = \frac{s(m + \frac{1}{2})\lambda}{na} \qquad (13.9).$$

It should now be apparent that we have to hand the means of measuring the wavelength of light from quasimonochromatic sources. Equation (13.7) can be used as the following example will illustrate.

Example

Helium yellow light illuminates two slits, separated by 2.644 mm, in a Young's experimental set-up. If 21 bright fringes occupy 20 mm on a screen 4.5 m away, calculate the wavelength. Assume the refractive index of air is 1.

21 fringes have 20 fringe separation spaces covering a distance of 20 mm, so the fringe separatic must be 1 mm, or 10^{-3} m. Using equation (13.7) rearranged

$$\lambda = \frac{na.\Delta y}{s} = \frac{1 \times 2.644 \times 10^{-3} \times 10^{-3}}{4.5} = 587.6 \times 10^{-9}\ \text{m}.$$

13.3.3 Intensity Distribution on the Screen

Recall equation (13.4) which gave the intensity at a point P where two equal intensity coplanar waves of the same wavelength overlap; that is, $I = 2I_0(1 + \cos\delta)$. Using the trigonometrical identity $\cos^2 A = \frac{1}{2}(1 + \cos 2A)$ we can rewrite equation (13.4) as

$$I = 4I_0 \cos^2\frac{\delta}{2} \qquad (13.10).$$

Now, if $\varepsilon_2 - \varepsilon_1 = 0$ in Young's experiment, $\delta = k(z_2 - z_1)$ or $k(z_1 - z_2)$ if $z_1 > z_2$ as in figure 13 Then the intensity distribution on the screen is

$$I = 4I_0 \cos^2\frac{\delta}{2} = 4I_0 \cos^2(\tfrac{1}{2} \times k(z_1 - z_2)) = 4I_0 \cos^2(\tfrac{1}{2} \times \frac{2\pi}{\lambda}(nay/s))$$

having used equation (13.2) for δ with $(\varepsilon_2 - \varepsilon_1) = 0$ and $k(z_1 - z_2) = \frac{2\pi}{\lambda}(nay/s)$.

Simplifying this result slightly, we have $$I = 4I_0 \cos^2\left(\frac{nay\pi}{s\lambda}\right) \qquad (13.11).$$

Figure 13.4 is a plot of I against y on the screen of figure 13.3 .

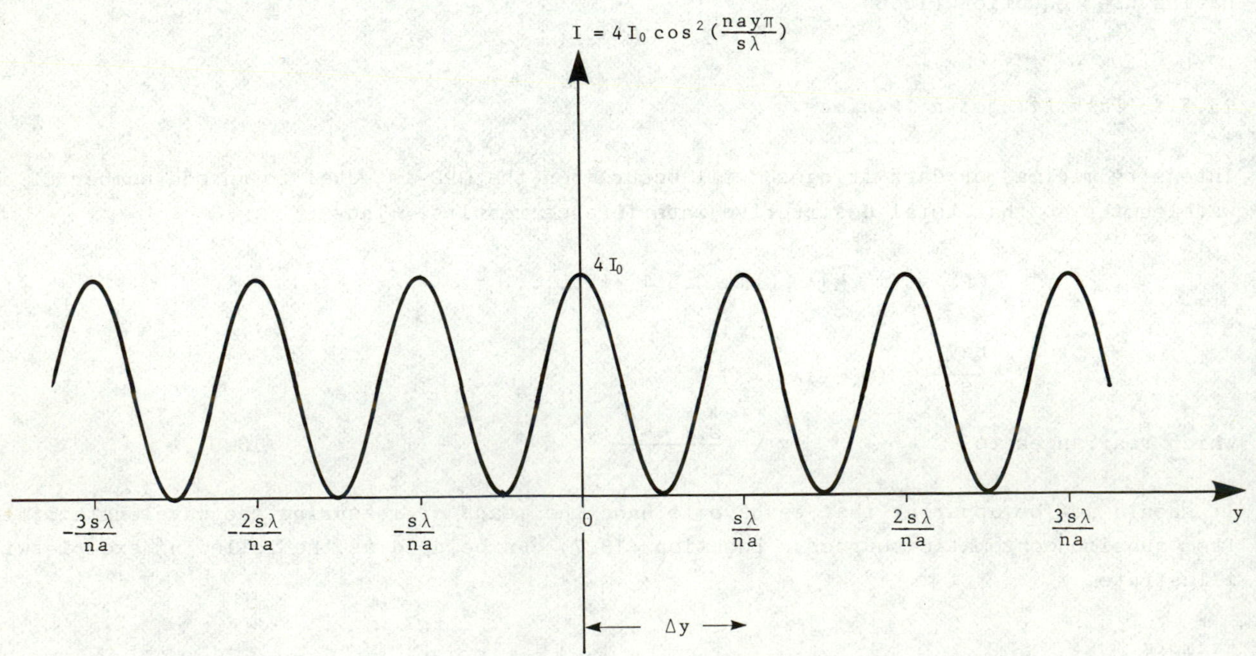

Fig. 13.4 Intensity distribution in Young's two-slit experiment.

If you punch two pinholes about 1 mm apart in a thin sheet of cardboard and look at a sodium street lamp about 100 m away you can obtain a direct view of the fringe pattern when the card is held very close to the eye. The pattern of the fringes will be perpendicular to the line joining the centres of the holes.

13.3.4 Other Wavefront-splitting Configurations

The Young's experiment configuration forms the basis of several wavefront-splitting interferometers the most common of which are *Fresnel's double mirror*, *Fresnel's biprism*, and *Lloyd's mirror*. The mathematical and physical principles are as for the two-slit configuration.

Fresnel's Double Mirror

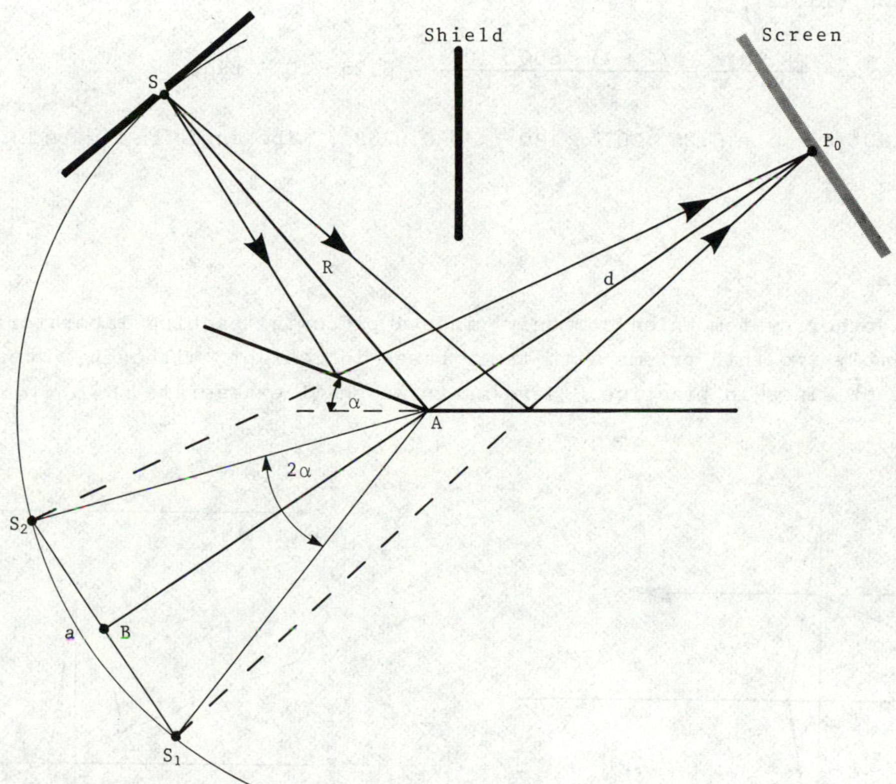

Fig. 13.5 Fresnel's double mirror.

The portions of the primary wave from S are reflected from the two mirrors so that there are two secondary coherent sources S_1 and S_2. The arrangement is then identical to the two-slit geometry so that $\Delta y = \frac{s\lambda}{na}$, where $s = BP_0$, the perpendicular to the screen.

Now, from the figure, $s = BA + d \simeq R + d$, since $BA \simeq R$.

Hence, $\Delta y = \frac{(R+d)\lambda}{na}$. We can then substitute for the angle α between the mirrors. If we imagine the line SA is an incident ray, first on the left mirror and then on the right mirror, we can sa that this is the same effect as rotating a single mirror through an angle α. In doing so the reflected ray rotates through 2α. S_2A and S_1A correspond to the reflected rays in the two positions so $S_2\hat{A}S_1 = 2\alpha$. From triangle S_2AB, $a/2 = S_2A.\sin\alpha \simeq S_2A.\alpha = R\alpha$, since $S_2A = R$. Thus, $a = 2R\alpha$ and

$$\Delta y = \frac{(R+d)\lambda}{2Rn\alpha} \qquad (13.12).$$

To obtain some idea of the scale of things, especially the angle α, let us look at an example.

Example

A Fresnel double mirror is used with the source slit at 1 m from the mirror intersection A. When the screen is 4 m distant and $\lambda = 500$ nm the fringe width (separation) is 2 mm. Find the angle between the mirrors.

Rearranging equation (13.12),

$$\alpha = \frac{(R+d)\lambda}{2Rn.\Delta y} = \frac{(1+4)\times 500\times 10^{-9}}{2\times 1\times 1\times 2\times 10^{-3}} = 6.25\times 10^{-4} \text{ rad}.$$

Since 1 radian $\equiv 180^0/\pi$, $\alpha = 6.25\times 10^{-4}\times 180^0/\pi = 0.0358^0$. The angle in figure 13.5 is obviously exaggerated.

The Fresnel Biprism

Fresnel invented another system which commonly finds a place in teaching laboratories: the Fresne biprism. It is simply two thin prisms with their bases in contact, although, of course, it is mad from a solid piece of glass in practice. Here again we shall exaggerate the angles for the purpos of clarity.

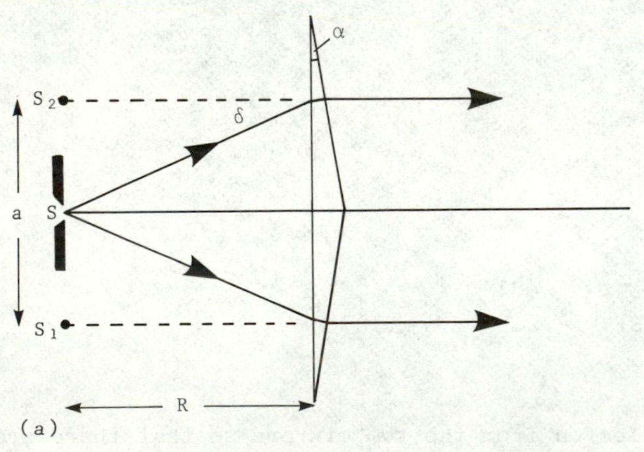

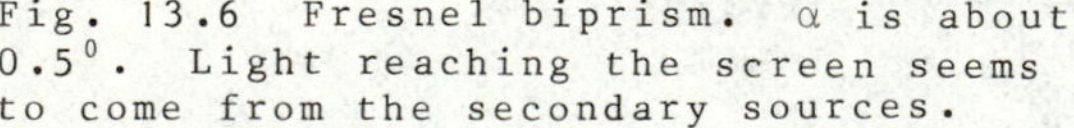

Fig. 13.6 Fresnel biprism. α is about 0.5^0. Light reaching the screen seems to come from the secondary sources.

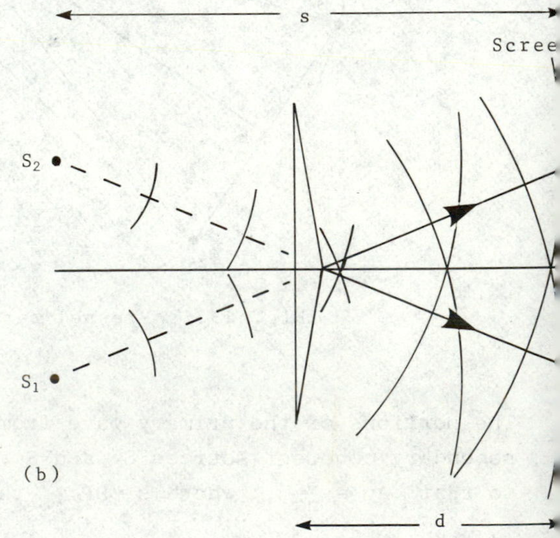

The usual equation for the fringe separation holds, only this time we can find the distance a between the secondary sources S_1 and S_2 in terms of the apical angle α of each prism.

If the deviation of each thin prism is δ then, from the thin prism equation,

$$\delta = \left(\frac{n_g}{n} - 1\right)\alpha$$

where n_g is the refractive index of the glass from which the prism is made and n is the surrounding medium's refractive index. From figure 13.6(a) we can write

$$\frac{a}{2} = R\delta \quad \text{or,} \quad a = 2R\delta = 2R\left(\frac{n_g}{n} - 1\right)\alpha \ .$$

Substituting this expression for a in equation (13.7), and putting $s = R + d$,

$$\Delta y = \frac{s\lambda}{na} = \frac{(R+d)\lambda}{n}\left(\frac{1}{2R\left(\frac{n_g}{n} - 1\right)\alpha}\right)$$

$$\text{or} \qquad \Delta y = \frac{(R+d)\lambda}{2nR\left(\frac{n_g}{n} - 1\right)\alpha} \qquad (13.13).$$

Lloyd's Mirror

Figure 13.7 depicts two portions of a wave from the source slit S, one reaching the screen directly and the other by reflection in the mirror. There is one important difference in this arrangement: the angle of incidence on the mirror is close to 90^0 so both the $E_{r\perp}$ and $E_{r\parallel}$ reflected amplitudes will suffer a π phase change, see figure 10.22 . Hence, at any point P on the screen the optical path length for the reflected wave must include a term $\lambda/2$ corresponding to the π phase change.

Recall the OPD in Young's experiment was nay/s and this was equated to $m\lambda$ for bright fringes. The OPD must here include $\lambda/2$ so we have

$$\frac{nay}{s} - \frac{\lambda}{2} = m\lambda$$

$$\text{or} \qquad \frac{nay}{s} = (m + \tfrac{1}{2})\lambda \qquad (13.14).$$

Thus, at P_0 in figure 13.7 where $y = 0$, the OPD = $\tfrac{1}{2}\lambda$ and a dark fringe will exist. We can see this in another way by using equation (13.11) with the π phase change included in the argument of the $\cos^2$ function:

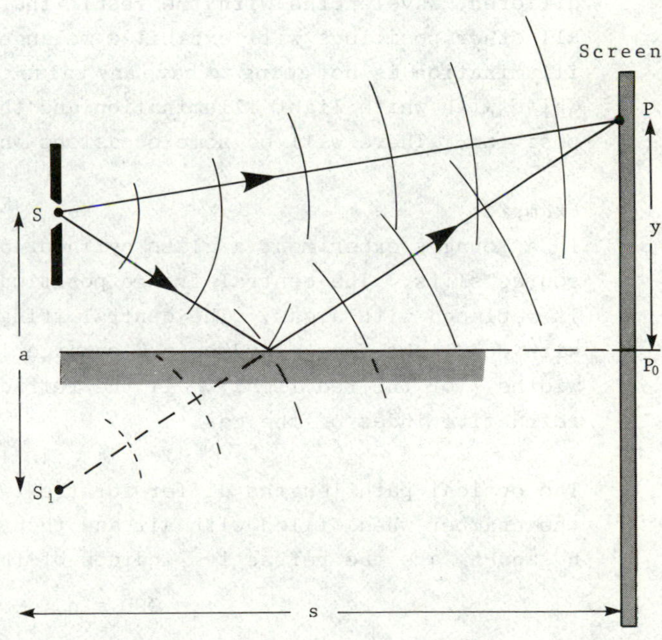

Fig. 13.7 Lloyd's mirror.

$$I = 4I_0 \cos^2 \tfrac{1}{2}\left(\frac{2\pi}{\lambda}\left(\frac{nay}{s}\right) - \pi\right)$$

$$= 4I_0 \cos^2\left(\frac{nay\pi}{s\lambda} - \frac{\pi}{2}\right) = 4I_0 \sin^2\left(\frac{nay\pi}{s\lambda}\right) \qquad (13.15),$$

having used the identity $\cos(A - \frac{\pi}{2}) = \sin A$, in the last step. When $y = 0$ the right hand side of equation (13.15) is zero, so a dark fringe exists at P_0.

Example

Lloyd's mirror is used with sodium light ($\lambda = 589.3$ nm) and the slit placed 3 mm above the reflecti surface and 3 m from the screen. Find the position of the first bright fringe above the level c the reflecting surface.

We know that a dark fringe or minimum occurs at the level of the reflecting surface and the distanc between fringes is $\Delta y = \frac{s\lambda}{na}$, as before. So, the first dark fringe is

$$\Delta y = \frac{s\lambda}{na} = \frac{3 \times 589.3 \times 10^{-9}}{1 \times 6 \times 10^{-3}} = 2.95 \times 10^{-4}\ \text{m} \equiv 0.295\ \text{mm}$$

above the level of the reflecting surface. The figure 6 in the denominator occurs because $S_1S =$ is $2 \times 3 = 6$ mm. The first maximum, or bright fringe, is $\frac{1}{2}\Delta y$ above P_0, of course.

13.3.5 Fringes with White Light

We have consistently referred to quasimonochromatic sources in all the foregoing arrangements. Suppose instead the source slit S were illuminated by white light, what then would be the patte on the screen? Consider the equations $y_m = \frac{sm\lambda}{na}$ and $\Delta y = \frac{s\lambda}{na}$. Clearly, both are wavelength dependent. Therefore, for values of m other than zero the bright fringes will occur at differe heights above and below the central fringe position. The fringe widths will be different for different wavelengths with the result that, except for the central fringe which will appear whi all other positions will exhibit a mélange of colour. This all sounds as though white light illumination is not going to have any value. This is not the case, however, since the central fringe is white with white light illumination and this therefore enables us to locate the central fringe position. There will be some occasions when we want to know the location of the central fringe

Example

In a Young's experiment a glass cylinder 50 mm long is placed in front of one of the secondary source slits. The central fringe position is located with white light. The air is evacuated a is replaced with a gas. The central fringe is now seen to be displaced. Illuminating the syst with red light from a hydrogen discharge lamp, $\lambda = 656.3$ nm, the central fringe had moved 42 fri widths (for the red light). If the refractive index of air is 1.000 28 for $\lambda = 656.3$ nm, find t refractive index of the gas.

The optical path lengths differ for the two cases by the difference in the optical path lengths the chamber when filled with air and then with gas. If ℓ is the internal length of the chamber an n_a and n_g are the refractive indices of the air and the gas, respectively, then the OPD is

$$\text{OPD} = n_g\ell - n_a\ell = (n_g - n_a)\ell\ .$$

This OPD must equal 42 wavelengths, so

$$42\lambda = (n_g - n_a)\ell$$

$$\text{or} \qquad n_g = \frac{42\lambda}{\ell} + n_a = \frac{42 \times 656.3 \times 10^{-9}}{50 \times 10^{-3}} + 1.000\,28 = 1.000\,83\ .$$

In the last example white light was used to locate the central fringe in two conditions, one with air in the chamber and one with gas in it. The only fringe which appears white is the central fringe and this occurs where the OPD is zero. Suppose the upper slit has the cylindrical chamber placed in front of it, then the optical path length increases from the upper slit to the screen when the gas is in the cylinder. To equate the optical path lengths from each slit to the screen more waves must fill the distance from the lower slit to the screen, and this must mean the central white fringe moves upwards.

13.4 AMPLITUDE SPLITTING BY THIN FILMS

In section 13.3 interference was arranged by splitting wavefronts and arranging for the two parts to superpose. The two secondary waves had electric vectors (amplitudes) very nearly the same as the primary wave from which they originated. Now we are going to look at cases where the entire primary wavefront is split into two and each part has a much lower amplitude than the primary. The effect is produced by reflection of an incident wave at the two surfaces of a thin film. The film may be a soap film, an oil slick, or an air film between two glass plates. Because the amplitude reflection coefficients are small the amplitudes of the two reflected waves are small relative to the primary wave. For this reason we shall ignore the multiple reflections and consider only the first two shown as E_{1r} and E_{2r} in figure 13.8(a). These waves have nearly the same amplitude so the visibility will be quite high.

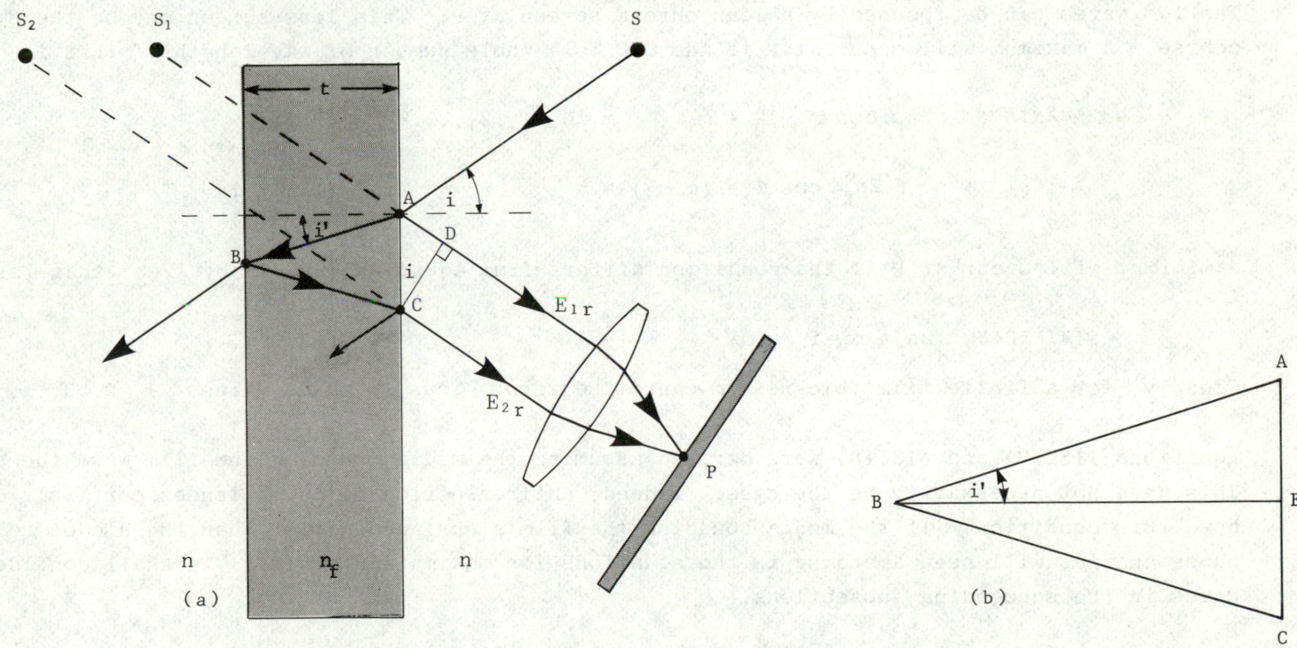

Fig. 13.8 Amplitude splitting by a thin film. S is a monochromatic point source.

The reflected waves can be thought of as originating from the virtual sources S_1 and S_2. The optical path difference for the two waves is

$$n_f(AB + BC) - n.AD$$

where n_f is the refractive index of the film and n is the refractive index of the surrounding medium.

Now, $AB = BC = t/\cos i'$, see figure 13.8(b), and $AD = AC \sin i = 2.AE \frac{n_f}{n} \sin i' = 2t \tan i' . \frac{n_f}{n} \sin i'$, since $AC = 2.AE$, $\sin i = \frac{n_f}{n} \sin i'$ from Snell's law, and $AE = t \tan i'$.

The OPD is therefore

$$n_f\left(\frac{2t}{\cos i'}\right) - n.2t.\tan i' . \frac{n_f}{n} . \sin i'$$

$$= \frac{2n_f t}{\cos i'}\left(1 - \sin^2 i'\right)$$

$$= \frac{2n_f t}{\cos i'} . \cos^2 i'$$

$$= 2n_f t \cos i' \qquad (13.16).$$

There is a further hidden optical path difference of $\lambda/2$ due to the π relative phase change caused by one internal and one external reflection. So, the OPD $= 2n_f t \cos i' - \lambda/2$.

The two waves can be focused by a lens onto a screen at P. This lens-screen may be the eye, of course. A maximum will occur at P if the OPD is a whole number of wavelengths: that is

MAXIMUM $\quad 2n_f t \cos i' - \frac{\lambda}{2} = m\lambda , \quad m = 0, 1, 2, \ldots.$

or $\quad 2n_f t \cos i' = (m + \tfrac{1}{2})\lambda \qquad (13.17).$

A mimimum will occur at P if the condition differs from equation (13.17) by $\lambda/2$; that is

MINIMUM $\quad 2n_f t \cos i' = m\lambda \qquad (13.18).$

Clearly, for a finite film thickness m cannot be zero. Thus, m takes values 1, 2, 3,.... .

Equations (13.17) and (13.18) were derived assuming the media bounding the film were the same. This need not necessarily be the case. Indeed, antireflection films on lenses certainly do not have this condition. If the media bounding the film's surfaces differ then the hidden relative phase changes will need amending in the equations for maxima and minima. We shall consider these cases in the succeeding subsections.

Now consider figure 13.9 which differs from the last figure in that the source is no longer a point. With an extended source it is possible to focus rays which have differing angles of incidence on the film. All rays with an equal angle of incidence or inclination will focus at the same point on the screen providing they pass through the lens, and the fringes produced are therefore called *fringes of equal inclination*.

The thickness of the film influences the appearance of interference fringes. As the film becomes

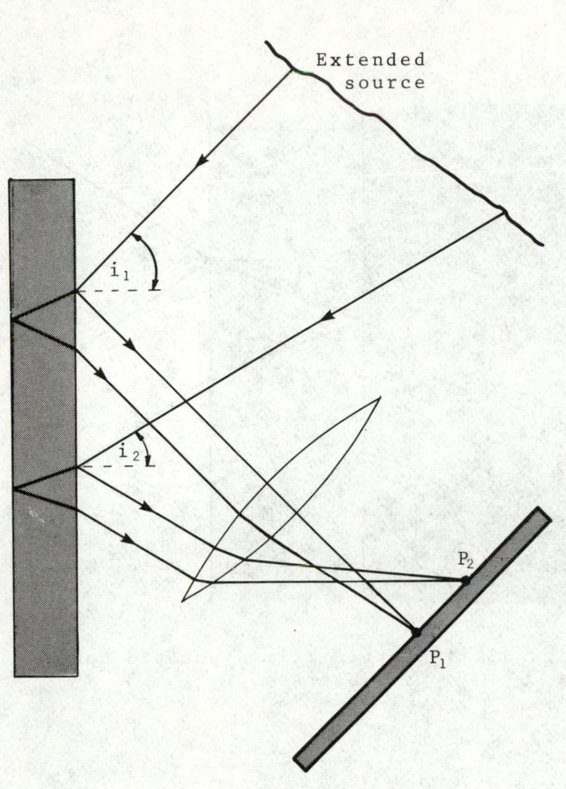

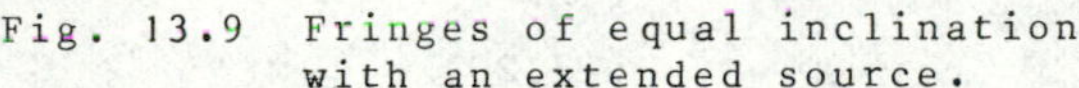

Fig. 13.9 Fringes of equal inclination with an extended source.

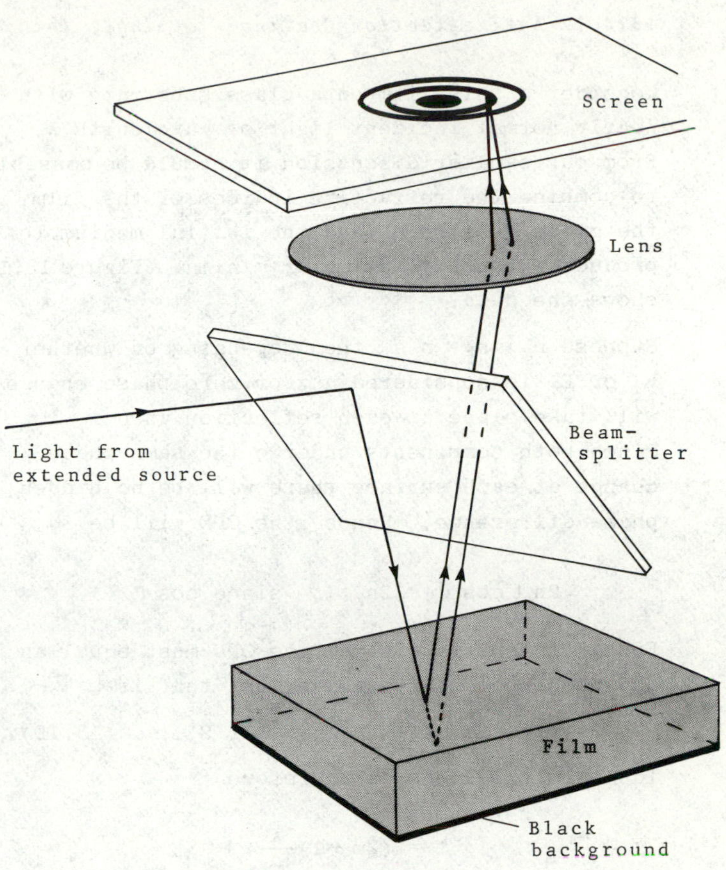

Fig. 13.10 Haidinger fringes.

thicker the separation of the two reflected rays, one from each surface, increases and if only one of the rays enters the lens the interference pattern will disappear. Reducing the angle of incidence will reduce the separation of the rays so that at near normal incidence the fringes are most easily observed. Equal-inclination fringes at nearly normal incidence seen in thick films or plates are known as *Haidinger fringes*. When the source is extended they consist of concentric circular bands centred on a perpendicular through the optical centre of the lens to the film or plate, figure 13.10 .

Example

Red light from a hydrogen discharge lamp, $\lambda = 656.28$ nm , is incident at 30^0 on a thin film of refractive index 1.5 . What is the minimum thickness of film if an intensity maximum is to be observed?

Maxima occur when $2n_f t \cos i' = (m + \frac{1}{2})\lambda$, and t will be a minimum when $m = 0$. Rearranging the equation, with $m = 0$, gives $t = \frac{1}{2}\lambda/(2n_f \cos i')$. We can find i' from Snell's law:

$$i' = \arcsin\left(\frac{n}{n'}\sin i\right) = \arcsin\left(\frac{1}{1.5}\sin 30^0\right) = 19.47^0 .$$

$$\text{So,}\quad t = \frac{\lambda/2}{2n_f \cos i'} = \frac{656.28/2}{2 \times 1.5 \times \cos 19.47^0} = 1650 \text{ nm} .$$

13.4.1 Antireflection Coating - a single thin deposited film

Consider a thin film on a glass substrate with nearly normal incident light of wavelength λ. From our earlier discussion it should be possible to combine the refractive indices of the film, the glass substrate, and the initial medium to produce intensity maxima or minima. Figure 13.11 shows the data.

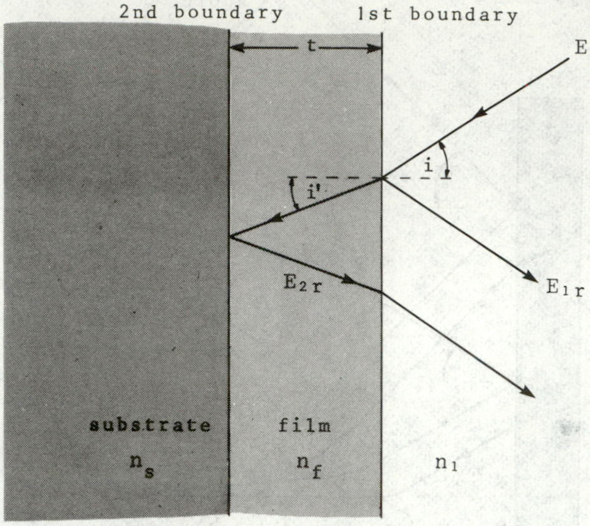

Suppose $n_s > n_f > n_1$, then depending on whether $E_\perp$ or $E_\parallel$ is considered a π or zero phase change will take place at each reflection when $i \simeq i' \simeq 0^0$. Since both components undergo the same phase change at each surface there will be no hidden phase difference. Hence, the OPD will be

$$2n_f t \cos i' = 2n_f t \ , \quad \text{since } \cos i' \simeq 1.$$

For an intensity minimum the OPD must equal an odd number of half-wavelengths: that is

$$2n_f t = (2m+1)\frac{\lambda}{2} \ , \quad m = 0,\ 1,\ 2, \ldots. \quad (13.19).$$

The film thickness is therefore

$$t = (2m+1)\frac{\lambda}{4n_f}$$

Fig. 13.11 A thin film deposited on a substrate.

and this will be a minimum when $m = 0$. So, $t = \dfrac{\lambda}{4n_f}$ (13.20)

will be the minimum thickness for an intensity minimum in light of wavelength λ. This minimum will be a zero if the amplitudes of the completely destructively interfering waves are equal. Recall equation (10.54) which states that the fraction of the intensity reflected, R, for normal incidence is given by $R = ((n'-n)/(n'+n))^2$. We already have the condition for the film's thickness, equation (13.20), and the expression for R will allow to find the relationship between n_s, n_f, and n_1 in the antireflection film.

Now, the amplitude coefficient of reflection r is proportional to the square root of the intensity reflectance R since the amplitude $\propto \sqrt{(\text{intensity})}$. That is, for normal incidence, $r \propto \sqrt{R}$. In problem 13.8.5 we show that $r = -R^{\frac{1}{2}} = -(n'-n)/(n'+n)$. Thus, if r_1 and r_2 are the amplitude coefficients of reflection for normally incident light at the first and second boundaries given by

$$r_1 = \frac{E_{1r}}{E_i} = -\left(\frac{n_1' - n_1}{n_1' + n_1}\right) \quad \text{and} \quad r_2 = \frac{E_{2r}}{E_i} = -\left(\frac{n_2' - n_2}{n_2' + n_2}\right) \ ,$$

neglecting the slight reduction in E_i reaching the second boundary, then we can write

$$E_{1r} = r_1 E_i = -\left(\frac{n_1' - n_1}{n_1' + n_1}\right) E_i = -\left(\frac{n_f - n_1}{n_f + n_1}\right) E_i \quad \text{and} \quad E_{2r} = r_2 E_i = -\left(\frac{n_2' - n_2}{n_2' + n_2}\right) E_i = -\left(\frac{n_s - n_f}{n_s + n_f}\right) E_i \ .$$

These two reflected waves leave the first boundary in an antiphase condition because the wave in the film traverses the film twice. It therefore introduces an OPD of $2 \times \lambda_f/4 = \lambda_f/2$, or a half film-wavelength, which makes the waves antiphase. Note that $\lambda_f = \lambda/n_f$. If we now equate E_{1r} and E_{2r}

we will have total destructive interference of the reflected light and a zero minimum reflection. So,

$$-\left(\frac{n_f - n_1}{n_f + n_1}\right) E_i = -\left(\frac{n_s - n_f}{n_s + n_f}\right) E_i .$$

Dividing through by $-E_i$ and after some algebraic manipulation, we have

$$n_f = \sqrt{n_1 n_s} \qquad (13.21).$$

Equations (13.20) and (13.21) are called *the path* and *the amplitude conditions*, respectively, for zero intensity minimum reflection. This is the principle of single layer (film) antireflection coating or blooming on lens surfaces. We should emphasise again the approximations used in the above analysis: the incident wave's amplitude on the second boundary is strictly somewhat reduced by loss due to the reflection at the first boundary, and we have also neglected multiple reflections.

When n_1 is air, as it is with spectacle lenses, the amplitude condition becomes

$$n_f = \sqrt{n_s} \qquad (13.22)$$

If the amplitude and path conditions are satisfied for a given wavelength, zero reflection for normal incidence will occur whether the light is incident from the air or the substrate side. This means that both surfaces of a lens should be antireflection coated. Further, since it is not possible to destroy energy, the light which is not reflected is now transmitted. This means that the antireflection film actually prevents light being reflected whilst allowing more to be transmitted through the system. It also means that our model of light being reflected at the two surfaces and then totally destructively interfering is not correct. In fact, the interference must take place between the electron oscillators in the boundaries, preventing them from reradiating light backwards! None-the-less, the equations for the path and amplitude conditions do work and antireflection coating is now widely used on both glass and plastics substrates.

With spectacle lenses λ is chosen to be 555 nm (yellow-green) because it is this wavelength to which the eye is most sensitive. The single layer material used on glass spectacle lenses is magnesium fluoride (MgF_2), $n_f = 1.38$, so the actual thickness of the film given by equation (13.20) is

$$t = \frac{\lambda}{4n_f} = \frac{555}{4 \times 1.38} = 100 \text{ nm, very nearly.}$$

Since the film thickness is computed for only one wavelength the OPD must be increasingly in error as λ is further from 555 nm. Thus, more and more light is reflected with λ approaching the red and blue ends of the visible spectrum. This explains the characteristic red-blue (purple) appearance of the reflection. This colouration of the surface is often called a *bloom*, a name taken from the wax bloom appearing on plums and grapes. Several substances are available for use as thin films on glass. Table 13.1 lists some of them together with their refractive indices.

Table 13.1 Some materials used for thin films

Material	Refractive index
Cerium oxide	2.2
Cerium fluoride	1.63
Cryolite	1.31
Germanium	4.0 (infrared)
Magnesium fluoride	1.38
Titanium dioxide	2.40
Zirconium dioxide	2.1
Zinc sulphide	2.35

Although equation (13.20) was derived using $\cos i' = 1$, i.e. $i = i' = 0$, since the cosine function does not vary greatly for small angles (it is 0.9397 at 20^0) the antireflection film works quite well for a wide range of angles of incidence about the normal.

Finally, we should mention that the minimum thickness single layer antireflection films are ofte referred to as quarter-wavelength films. This refers specifically to the quarter-wavelength in the film and not to a quarter vacuum wavelength. Since we write $\frac{\lambda}{4n_f}$ as $\lambda_f/4$, the path conditio is simply

$$t = \frac{\lambda}{4n_f} = \frac{\lambda_f}{4}$$

from which we see that t is one-quarter of the wavelength in the film.

13.4.2 Increased reflection with a Single Film

If a single high refractive index film is deposited on glass we can arrange for the reflected wa to produce total constructive interference by putting them in-phase. In this case we cannot ign the drop in amplitude which takes place at each reflection and, strictly, we should not ignore multiple reflections. However, to simplify matters we shall ignore multiple reflections in order arrive at some idea of the principles involved.

Figure 13.12 shows an incident ray with wave amplitude E_i. There will be considerable loss by the first reflection so the transmitted amplitude E_{1t} will be less than E_i. Similarly, $E_{2r} < E_{1t}$ and $E_{2rt} < E_{2r}$, so we must calcultate their relationships before we are able to find how much light is reflected. Also, we must note that for nearly normal incidence there will be a π relative phase change at the air-film boundary compared with at the film-glass boundary. Recall equation (10.56),

$$T = \frac{4nn'}{(n' + n)^2}$$

for the fractional intensity transmitted. Then the fractional amplitude transmitted, t_1, at the first boundary will be

$$t_1 = (T)^{\frac{1}{2}} = \frac{2\sqrt{nn'}}{n' + n}$$

so that $E_{1t} = t_1 E_i = \frac{2\sqrt{nn'}}{n' + n} \cdot E_i = \frac{2\sqrt{1 \times n_f}}{n_f + 1} \cdot E_i$

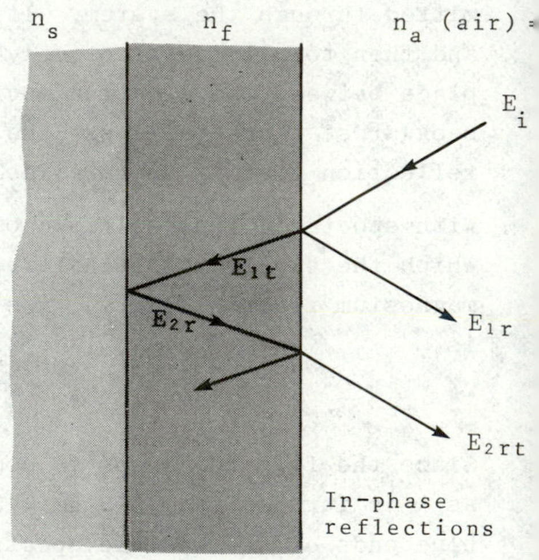

Fig. 13.12 Increased reflection with a thin film: $n_f > n_s$.

The fraction of this reflected at the second boundary will give E_{2r}. Again, since the intensity reflectance R is given by

$$R = \left(\frac{n' - n}{n' + n}\right)^2$$

the amplitude reflection coefficient is the square root of this; that is,

$$r_2 = (R)^{\frac{1}{2}} = -\left(\frac{n' - n}{n' + n}\right) = -\left(\frac{n_s - n_f}{n_s + n_f}\right)$$

the negative root being taken to indicate a hidden π phase change when $n_s > n_f$. So, E_{2r} is given by

$$E_{2r} = r_2 E_{1t} = -\frac{n_s - n_f}{n_s + n_f} \cdot \frac{2\sqrt{(n_f)}}{n_f + 1} \cdot E_i \ .$$

Now, a fraction t_1 of this amplitude will be transmitted from the film into the air and emerge with amplitude E_{2rt}, given by

$$E_{2rt} = t_1 E_{2r} = \frac{2\sqrt{(n_f)}}{n_f + 1}\left(-\frac{n_s - n_f}{n_s + n_f}\right) \cdot \frac{2\sqrt{(n_f)}}{n_f + 1} \cdot E_i \ .$$

Now, the amplitude of the first reflected wave, E_{1r}, is simply $E_{1r} = r_1 E_i = -\frac{n_f - 1}{n_f + 1} \cdot E_i$.

For these reflected amplitudes to be in-phase the film thickness must be given by equation (13.17) and the thickness will be a minimum when $m = 0$. So,

$$t = \frac{\lambda/2}{2n_f \cos i'} = \frac{\lambda_f}{4} \quad \text{, when } \cos i' \simeq 1 \text{ again.}$$

This twice traversed quarter-wavelength path injects a π-phase change so the amplitude of the second reflection now becomes $-E_{2rt}$, the minus sign indicating the π phase change. Given a quarter-wavelength film then, we add the amplitudes E_{1r} and $-E_{2rt}$ and express the sum as a fraction of the incident amplitude E_i. Squaring this gives the fractional intensity reflected, the reflectance, for normal incidence. Thus,

$$\text{fractional amplitude reflected} = \frac{E_{1r} + (-E_{2rt})}{E_i}$$

$$= \frac{\left(-\frac{n_f - 1}{n_f + 1} \cdot E_i\right) + \left(-\frac{2\sqrt{(n_f)}}{n_f + 1} \cdot \left(-\frac{n_s - n_f}{n_s + n_f}\right) \cdot \frac{2\sqrt{(n_f)}}{n_f + 1}\right) \cdot E_i}{E_i} \ .$$

The E_i cancels and simplifying slightly gives

$$\left(-\frac{n_f - 1}{n_f + 1}\right) + \left(\frac{4n_f}{(n_f + 1)^2} \cdot \frac{n_s - n_f}{n_s + n_f}\right) \ .$$

Suppose the substrate is ophthalmic crown glass, $n_s = 1.523$, and the film is zinc sulphide, $n_f = 2.35$, then the fractional amplitude reflected is

$$\left(-\frac{2.35 - 1}{2.35 + 1}\right) + \left(\frac{4 \times 2.35}{(2.35 + 1)^2} \cdot \frac{1.523 - 2.35}{1.523 + 2.35}\right)$$

$$= -0.582 \ .$$

Squaring this gives the reflectance which is 0.339 or 33.9%. This same maximum reflectance will occur for $t = 3\lambda_f/4$, $t = 5\lambda_f/4$, and so on. However, in increasing the thickness of the film losses will occur by absorption.

13.4.3 Fringes of Equal Thickness

So far in section 13.4 we have dealt with films with parallel surfaces. We now consider films where the film thickness varies and so determines the fringe pattern. A soap film, figure 13.13, will thin at the top as gravity causes it to sink downwards. The patterns produced are dependent on thickness variations, and the alternating dark and light fringes correspond to regions of constant thickness. In effect, such fringes are a contour map of the film thickness and they can be used to determine the surface features of lenses and prisms. When placed in contact with an optical flat* any irregularities in a lens surface will show up as an irregular fringe pattern. Here it is the air between the surfaces which forms the film.

Fig. 13.13 A wedge-shaped soap film. Notice how the fringe width reduces as the film thickness increases towards the bottom.

When viewed at nearly normal incidence such fringes of *equal film thickness* are called *Fizeau fringes*. The simplest geometrical film of varying thickness is a wedge which can be formed between two optical flats by putting a piece of paper between them at one end, figure 13.14 .

Since one reflection is at a glass-film boundary and the other is at a film-glass interface, there will be a hidden π phase change equivalent to an optical path difference of $\lambda/2$. Equation (13.17) applies, and if the incident light is nearly normal ($\cos i' \simeq 1$) the condition for maxima is

$$2n_f t = (m+\tfrac{1}{2})\lambda \;, \quad m = 0, 1, 2, \ldots \quad .$$

But, from figure 13.14, $t = x\tan\alpha \simeq x\alpha$, where the wedge angle α is small and in radians. If the mth maximum occurs at x_m, then $t_m = x_m\alpha$ is the wedge thickness there.

We can now substitute for t_m in the condition for maxima to give

$$2n_f x_m \alpha = (m+\tfrac{1}{2})\lambda \qquad (13.23)$$

or $$x_m = \frac{(m+\frac{1}{2})\lambda}{2n_f\alpha} \qquad (13.24).$$

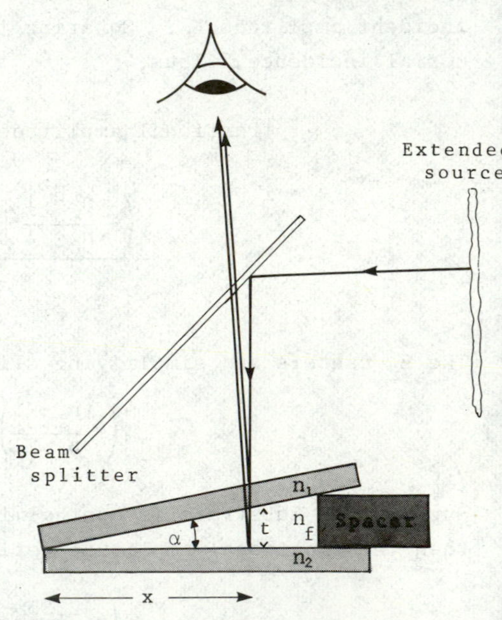

Fig. 13.14 A wedge-shaped film between optical flats.

The fringe width or separation between maxima, Δx, is found by subtracting x_m from x_{m+1} : thus,

$$\Delta x = x_{m+1} - x_m = \frac{(m+1+\frac{1}{2})\lambda}{2n_f\alpha} - \frac{(m+\frac{1}{2})\lambda}{2n_f\alpha}$$

or $$\Delta x = \frac{\lambda}{2n_f\alpha} \qquad (13.25).$$

Wide fringes therefore mean large λ or small α, or both! If λ is constant, that is monochromatic light is used, variations in fringe width depend on variations in α which, in turn, means t is

* *Optical flat - a surface which does not deviate from a plane by more than $\lambda/4$.*

varying. This is the principle referred to for checking surfaces held against an optical flat. You can witness these effects by pressing together two well-cleaned microscope slides when irregular coloured fringes will present themselves in normal room illumination. Pressing the slides together changes the air film thickness and the fringe pattern is seen to change. If a pointed instrument is used to press the slides together the fringe pattern produced consists of concentric rings. Figure 13.15(a) shows concentric rings fortuitously obtained without pressing the slides with a pointed instrument.

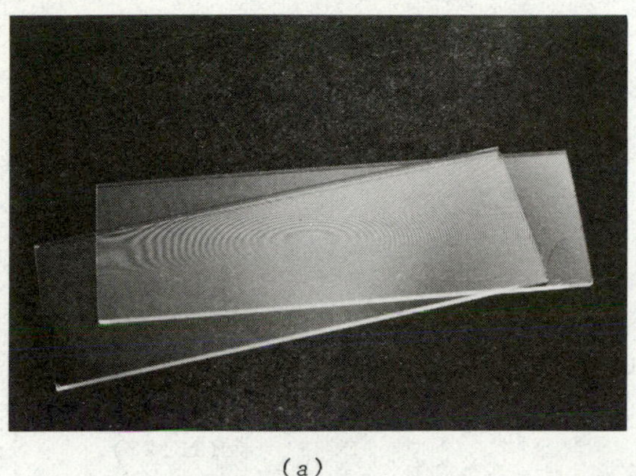

(a)

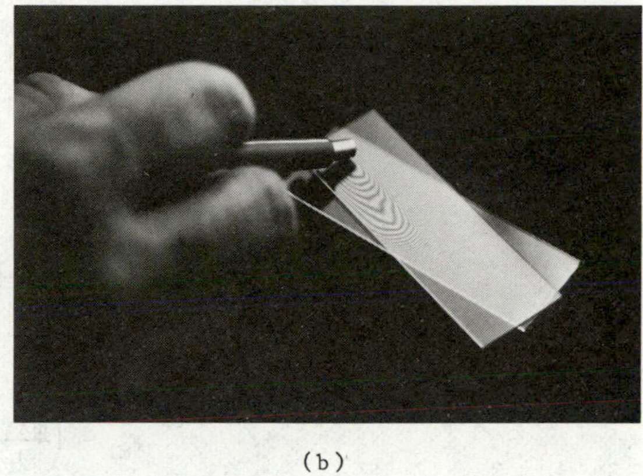

(b)

Fig. 13.15 (a) Newton's rings. (b) Fringes obtained by pressing two slides together. A sodium discharge lamp was used for the illumination.

These ring shaped fringes are known as *Newton's rings* and may be quantitatively examined using the arrangement in figure 13.16 .

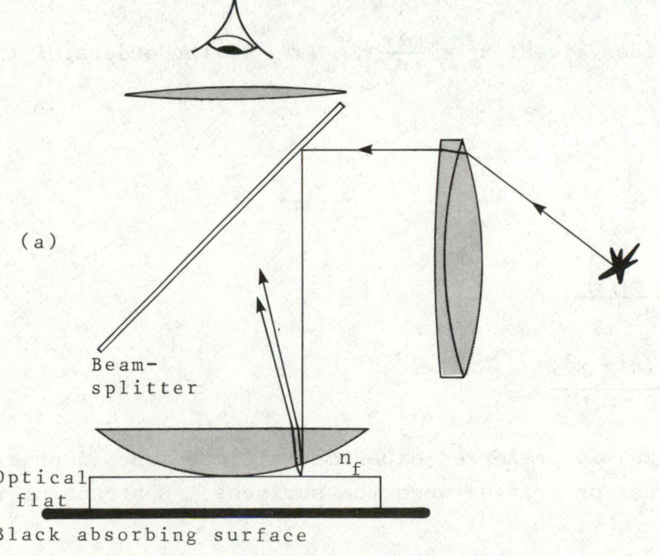

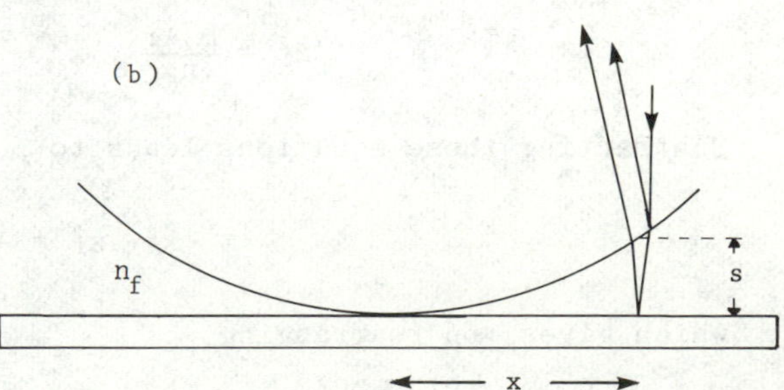

Fig. 13.16 a) The arrangement for Newton's rings.
b) The sagitta s at a distance x from the vertex of the lens surface.

When the light is very near normal incidence on the optical flat the ray reflected at the flat has traversed the distance s twice. s is the sagitta at the semi-chord diameter x of the lower lens surface. It is a simple matter to show that $s \simeq x^2/2r$, where r is the radius of curvature of the lens surface. If the gap between the two surfaces is filled with a medium of refractive index n_f then the optical path difference between the two reflected waves, one reflected at the optical flat and the other at the lower lens surface, is $2n_f s$. However, once again we must not forget the hidden π relative phase change which one reflection will suffer relative to the other. Thus, for a maximum the condition will be

$$2n_f s - \frac{\lambda}{2} = m\lambda \ , \quad m = 0, 1, 2, \ldots .$$

Putting $s = x_m^2/2r$ and rearranging gives

$$\frac{2n_f x_m^2}{2r} = (m + \tfrac{1}{2})\lambda$$

$$\text{or} \qquad x_m = \left(\frac{(m + \frac{1}{2})\lambda r}{n_f}\right)^{\frac{1}{2}} \qquad (13.26)$$

where x_m indicates the radius of the mth bright fringe. In similar vein, the radius of the mth dark fringe is

$$x_m = \left(\frac{m\lambda r}{n_f}\right)^{\frac{1}{2}} \qquad (13.27).$$

If the two surfaces make a good contact the central fringe will be a minimum. This is understandable since the sag s goes to zero at the vertex of the lens surface. Note here that counting starts at zero, so that when m = 20, say, this is the 21st ring in natural numbers.

Example

In an experiment the Newton's rings apparatus is illuminated with the green light from a mercury lamp, $\lambda = 546.1$ nm. If the diameters of the 10th and 20th dark rings are 2.10 mm and 2.96 mm, respectively, calculate the radius of the convex surface. Assume $n_f = 1$.

Using equation (13.27) and squaring both sides gives $x_m^2 = \frac{m\lambda r}{n_f}$. Now, if the orders of the rings are designated m_1 and m_2, we can write

$$x_2^2 = \frac{m_2\lambda r}{n_f} \qquad \text{and} \qquad x_1^2 = \frac{m_1\lambda r}{n_f}$$

Subtracting these equations leads to

$$x_2^2 - x_1^2 = \frac{(m_2 - m_1)\lambda r}{n_f}$$

which gives, on rearranging,

$$r = \frac{n_f (x_2^2 - x_1^2)}{(m_2 - m_1)\lambda} \qquad (13.28)$$

This method of finding r via equation (13.28) is preferred experimentally because in practice the sagitta s may be increased by a piece of dust or grit between the surfaces. Subtracting the equations eliminates this error*.

* *See Worked Problems in Optics, A.H.Tunnacliffe*

Now, with all the measurements in millimetres,

$$r = \frac{n_f(x_2^2 - x_1^2)}{(m_2 - m_1)\lambda} = \frac{1 \times ((1.48)^2 - (1.05)^2)}{(20-10) \times 546.1 \times 10^{-6}} = 1.992 \times 10^2 \text{ mm} \equiv 19.92 \text{ cm},$$

where $\lambda = 546.1 \times 10^{-6}$ mm, $x_2 = \frac{2.96}{2} = 1.48$ mm , and $x_1 = \frac{2.10}{2} = 1.05$ mm.

It is interesting to note that if n_f is intermediate between the indices of the lens and the plate then equations (13.26) and (13.27) are reversed and a bright centre occurs in the pattern.

13.4.4 *Interference Fringe Types*

When focusing on fringes it is necessary to have some idea of where they are located and what type they are. Fringes are classified as *real* and *virtual*, figure 13.17, and *localised* or *non-localised*. Real fringes can be focused on a screen but virtual fringes arise from diverging light which must be passed through a converging system to be focused.

Non-localised fringes are usually produced by small sources, are real, and exist in an extended region of space. The fringes in Young's experimental arrangement are non-localised. Moving the screen to the right in figure 13.2 will intersect successive real fringes. Localised fringes are visible only over a given surface, even if they are at infinity (parallel interfering rays), and are generally produced by extended sources.

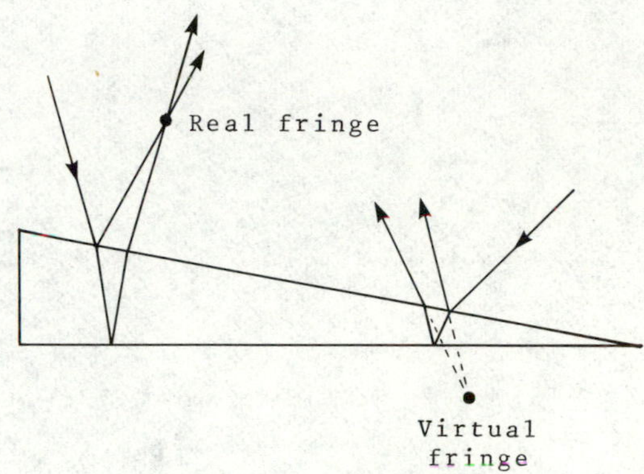

Fig. 13.17 Location of real and virtual fringes.

13.4.5 *Amplitude-splitting Interferometers*

In the amplitude-splitting thin films we have considered the two interfering beams travelled much the same path except where the optical path difference occurred. There is a group of instruments, known as *amplitude-splitting interferometers*, in which the two beams are sent along different paths before being recombined. Mirrors are used to produce these different paths.

The Michelson Interferometer is representative of a group of instruments which amplitude-split the beam into two parts. Figure 13.18 shows the arrangement. Light from an extended source is divided into two waves by a beamsplitter, B, which may be a glass plate, a half-silvered mirror which is semi-transparent because the metallic layer is too thin to be opaque, or a thin stretched plastic film known as a pellicle. One wave reflects from the front-silvered mirror M_2 and part of this reflected wave passes through the beamsplitter to reach the detector. The other wave reflects off the mirror M_1 and again part of this reflected wave is reflected at the beamsplitter to reach the detector. A compensator plate, C, is placed between the mirror M_1 and the beamsplitter to make the optical paths of the two waves identical when the mirrors are equidistant from the beamsplitter. When M_1 and M_2 are exactly perpendicular then M_2 and M_1', the image of M_1 by reflection in the beamsplitter, are parallel and the light reaching the detector comes effectively from two virtual sources S_1 and S_2 which generate circular fringes. If M_2 and M_1' are close together and slightly inclined to each other they act as a thin wedge-shaped film and produce straight parallel fringes. If we assume the beamsplitter is a glass plate there will be a phase difference of π due to one reflection being internal and the other external.

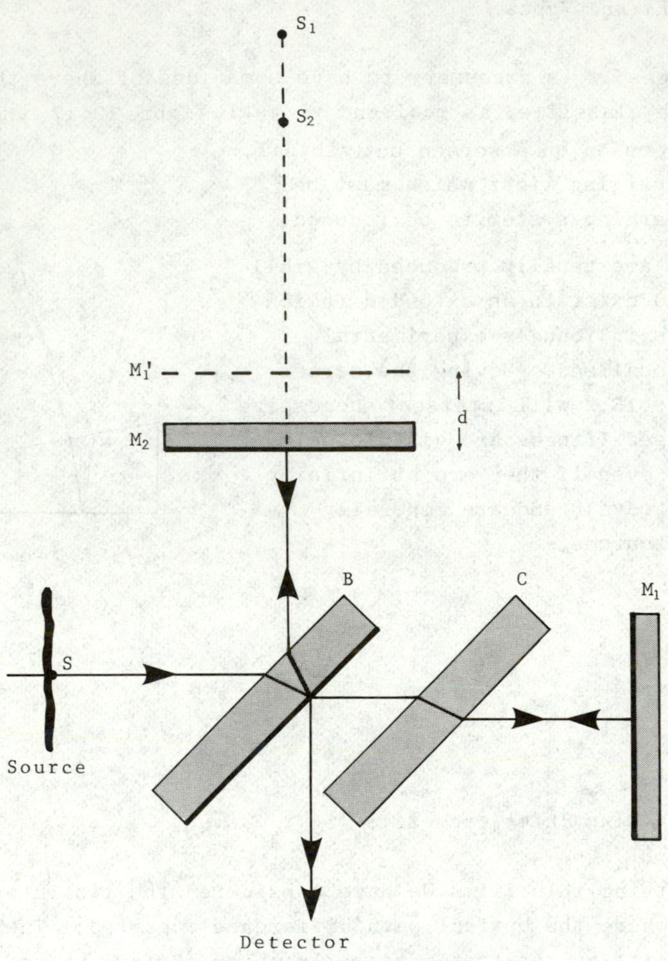

Fig. 13.18 The Michelson interferometer.

One of the two mirrors is moveable by means of a screw which is calibrated to read the lateral displacement of the mirror. The other mirror and the compensating plate are mounted so that they can be tilted into precise positions. It is necessary to make the compensating plate exactly parallel to the beamsplitter and the fine adjustment of the mirrors is essential to determine perpendicularity, or departures from it.

Figure 13.19 shows the geometric paths of the two rays from the virtual sources S_1 and S_2 as seen by the detector.

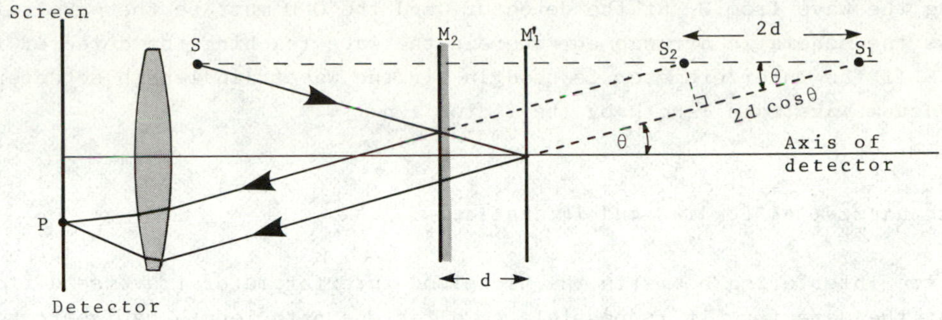

Fig. 13.19 The detector's view in a Michelson interferometer.

The path difference of the two waves is then $2d\cos\theta$, and if the apparatus is immersed in a medium of refractive index n the OPD is $2nd\cos\theta$. If the beamsplitter is an uncoated glass plate we have already mentioned there will be a hidden π relative phase change. So, the condition for an intensity minimum at P, on the screen, will be

$$2nd\cos\theta = m\lambda \qquad (13.29).$$

This condition will hold for all points on the source equidistant from the detector's axis. Simply imagine rotating figure 13.19 about the axis of the detector. By symmetry, the condition will hold for all points mapped out by P as it rotates about the axis. Therefore the resultant fringe pattern will be circular and centred on the axis. Equation (13.29) shows that θ is dependent on λ, and it is useful to rewrite the equation as

$$2nd\cos\theta_m = m\lambda \qquad (13.30)$$

where θ_m is the angular radius of the mth dark fringe for wavelength λ. Clearly, a polychromatic source will generate a fringe pattern for each wavelength and a monochromatic source will be preferable for an easily distinguished pattern. Also, it is important to realise that the OPD must not exceed the coherence length (the wavetrain length) if interference is to occur. Since the wavetrain length increases as the wave disturbance approaches monochromaticity, the distance $2nd\cos\theta$ can be larger with quasimonochromatic sources. In fact, it is possible to use the Michelson interferometer to measure wavetrain lengths. Starting with $d = 0$, the mirror M_1 is gradually screwed out until the fringe pattern disappears. In this situation the amplitude-divided wave from M_1 has travelled an optical path distance 2nd greater than its corresponding amplitude-divided wave from M_2. Here we assume we are observing along the axis and $\theta = 0$, so that $\cos\theta = 1$. Since the pattern has disappeared the wave from M_2 must not be

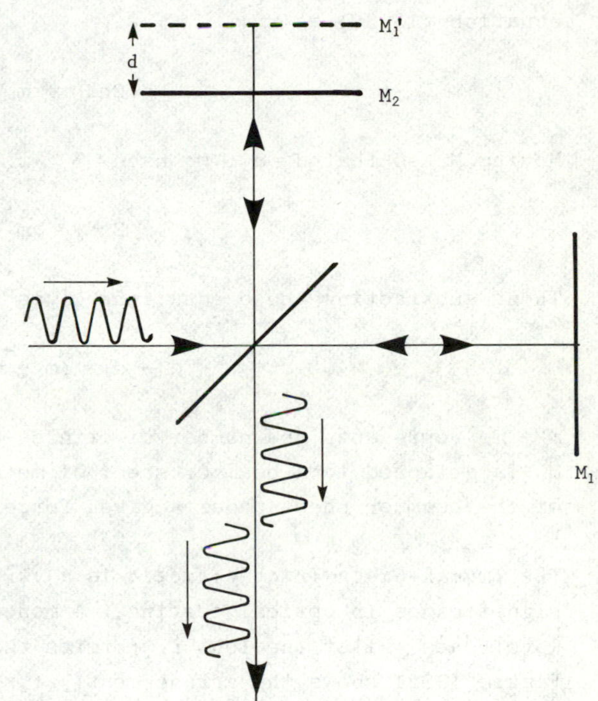

Fig. 13.20 Determination of wave-packet length.

overlapping the wave from M_1 at the detector, and the OPD must be the wavetrain length. Figure 13.20 shows the schematic arrangement wherein the wave reaching the detector first is from the mirror M_2. If the interferometer is used in air the wavetrain length so determined will be short than the vacuum wavetrain length by the factor $1/n_{air}$.

13.4.6 Measurement of Lengths and Thicknesses

Since the two interfering beams in the Michelson interferometer traverse different paths before arriving at the detector, it is possible to alter one path length by inserting some refracting material in the form of a thin sheet or a cell containing a liquid or a gas. Accurate measuremen of the change in OPD is possible even to a tenth of a fringe width, or better. If we observe the fringe pattern at the centre ($\cos\theta = 1$) and count the number of fringes in monochromatic light crossing the centre of the field of view, this will enable us to state the change in optical path length in one arm of the instrument in terms of the wavelength.

For example, with the mirror M_1 in a position such that $d = d_1$, the central dark fringe is given b equation (13.29) as

$$2nd_1 = m_1\lambda \qquad (\cos\theta = 1) .$$

Moving M_1 so that $d = d_2$, we have

$$2nd_2 = m_2\lambda .$$

Then, subtracting these equations gives

$$d_1 - d_2 = (m_1 - m_2)\frac{\lambda}{2n} \qquad (13.31).$$

$m_1 - m_2$ represents the number of fringes crossing the centre of the field of view. If the mirror M_1 is attached to a beam or sheet of metal this method can be used to measure the amount of bendi of the beam or sheet under a given force.

The Twyman-Green Interferometer is a variation of the Michelson interferometer which has great significance in optical testing. A monochromatic point source and a lens L_1 provide plane incide wavefronts whilst the lens L_2 permits the whole fringe pattern to enter the eye (the detector). Figure 13.21 shows the arrangement.

Continuous wave lasers are now superceding the illumination arrangement shown in figure 13.21 si they provide long coherence lengths and allow short photographic exposure times due to their hig intensity. The long coherence length allows for long path differences which means that thicker objects can be placed in one arm of the interferometer.

Arranged to examine a lens the spherical mirror M_2 has its centre of curvature at the focal poin of the lens. The mirror M_1 is tilted to produce Fizeau fringes of equal thickness which show as parallel bands, of course. If the lens is free of aberrations and defects the returning wavefro will be plane and the fringe pattern will remain as parallel bands. However, aberrations or othe defects such as inhomogeneous materials, scratches, bubbles, and the like, will distort the fring pattern and the result can be photographed.

If M_2 is replaced by a plane mirror, prisms and optical flats can be tested. The optical enginee

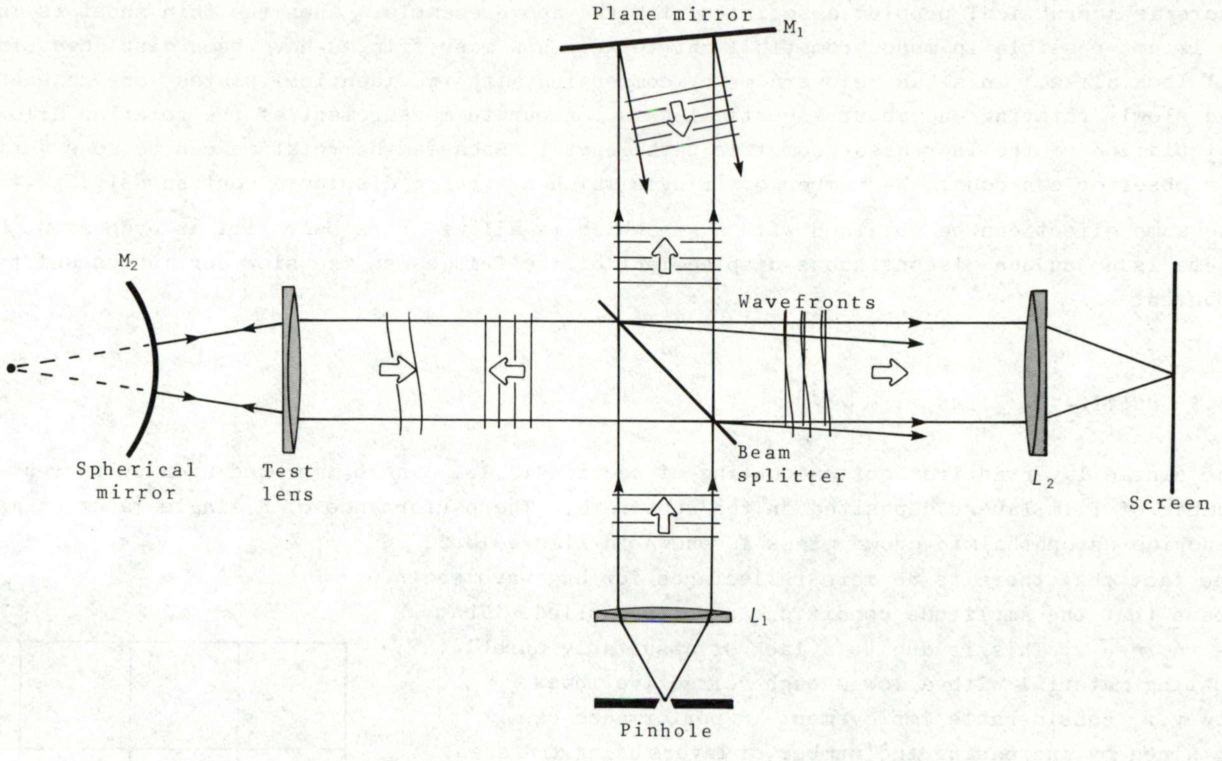

Fig. 13.21 The Twyman-Green interferometer.

can mark a surface showing where further working is required to remove 'high' or 'low' spots.

13.4.7 Interferometric Measurement of Refractive Index

If a thickness t of a substance of refractive index n_s is placed in one arm of the Michelson interferometer the optical path length of that arm is increased by $n_s t - nt = (n_s - n)t$, where n is the refractive index of the medium in which the instrument is immersed. If $n = 1$ (air), then the optical path difference is $2(n_s - 1)t$, since the sample is traversed twice, and if we equate this to $(m_1 - m_2)\lambda$, where $m_1 - m_2$ is the number of fringes which are displaced on inserting the material, then we can calculate n_s.

As an example suppose a thin sheet of glass of refractive index n_s is inserted in one arm of a Michelson interferometer which is illuminated by mercury light, $\lambda = 546.1$ nm. If 94 fringes are displaced when the sheet is inserted, find its refractive index if the thickness is 0.0513 mm.

The OPD = $2(n_s - 1)t = (m_1 - m_2)\lambda$, which rearranges to give

$$n_s = \frac{(m_1 - m_2)\lambda}{2t} + 1 = \frac{94 \times 546.1 \times 10^{-9}}{2 \times 0.0513 \times 10^{-3}} + 1$$

$$= 1.5$$

There is a practical problem associated with the above example. When the thin sheet is inserted it is not possible in monochromatic light to know how many fringes have been displaced since they all look alike. This can be overcome by commencing with two identical plates, one in each arm, and slowly rotating one about a vertical axis. Accurate measurement of the rotation allows the calculation of the increased geometric path length. Because the rotation can be done very slowly the observer can count the number of fringes which are being displaced continuously.

The same effect can be obtained with a gas which is allowed to slowly fill an evacuated chamber. There is no sudden discontinuous displacement of the fringes so the slow continuous shift can be counted.

13.5 MULTILAYER FILMS

The single layer antireflection coating of section 13.4.1 can be improved upon by increasing the number of film layers deposited on the substrate. The performance of a single layer of magnesium fluoride on ophthalmic crown glass is shown in figure 13.22 . The fact that there is no zero reflectance for any wavelength means that the amplitude condition is not satisfied. That is $n_f \neq \sqrt{n_s}$. This is due to a lack of a suitably durable coating material with a low enough refractive index. However, considerable improvement in performance can be obtained by increasing the number of layers using different refractive indices for the successive layers. Such systems are called *multilayer films* and are widely used in camera lenses and more recently on spectacle lenses. Not only does more light reach the image plane to increase the contrast in the image, but multiple internal reflections are markedly reduced in intensity. The reduction in internal reflections leads to a drop in 'flare spots', those unwanted secondary images formed when light from bright sources reaches the image plane after suffering internal reflection in the lens system. The magnitude of the problem with camera lenses is easily illustrated if one considers a lens system with 10 surfaces, say, each uncoated. The transmittance at each surface for normally incident light is

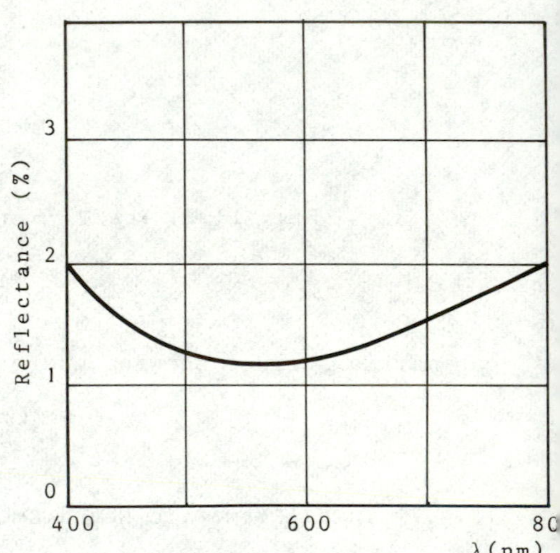

Fig. 13.22 Reflectance from a coat glass surface (n_g = 1.52

$$T = \frac{4nn'}{(n'+n)^2} = \frac{4 \times 1 \times 1.5}{(1.5+1)^2} = 0.96,$$ having used equation (10.56) and assumed the refractive index of the glass to be 1.5 . Since successive surfaces only receive 0.96 of the light inciden at the preceeding surface, the emergent light has an intensity of only $(0.96)^{10} = 0.665$ or 66.5% the intensity incident at the first surface. A single layer coating of magnesium fluoride on ea surface would increase the transmittance to a little less than 0.99, and a similar calculation f ten surfaces now shows something like 90% of the light is transmitted.

Multilayer films can improve upon this considerably. The principle is easily understood even if the practicalities of the calculation are difficult. We shall not pursue the practicalities too far in this text! Figure 13.23 illustrates the problem for a two layer film. Figure 13.23(a) shows the three reflected waves' amplitudes E_1, E_2, and E_3 from the three interfaces. The relati phases of these waves can be adjusted by altering the optical path length which is done by carefully controlling the thicknesses of the films. Hidden phase changes are taken into account, of course. Figure 13.23(b) shows one way in which the phases can be arranged to add to a zero resultant reflection vector. Another way is to have two of the reflection vectors in-phase but make the pair antiphase with the third, figure 13.24 .

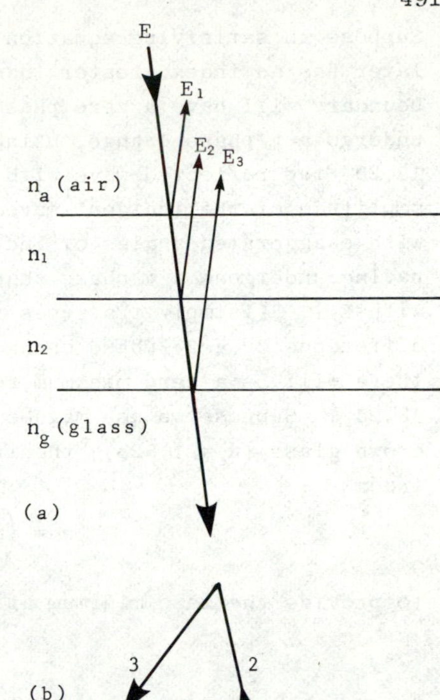

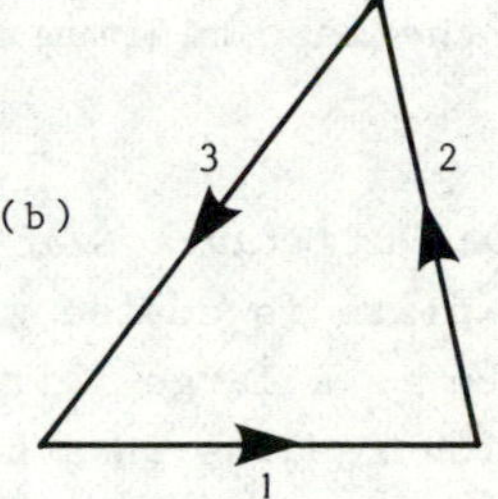

Fig. 13.23
(a) Near normal incident light and reflections from each boundary.
(b) Zero resultant reflected vector.

The performance for a double film using glass ($n_g = 1.5$), cerium oxide ($n_2 = 2.2$), and magnesium fluoride ($n_1 = 1.38$) is shown in figure 13.25(a). Note that the reflectance at $\lambda = 500$ nm is zero and the film optical thicknesses are not simple quarter-waves.

If the optical thicknesses of the two layers are chosen to be $\lambda/4$, calculation for a substrate of refractive index n_s gives the condition

$$\frac{n_s}{n_a} = \left(\frac{n_2}{n_1}\right)^2 \qquad (13.32)$$

for a zero reflectance of the wavelength chosen, where $n_a = 1$ (air). This kind of film is referred to as a *double-quarter, single minimum* coating. The double-quarter clearly refers to the two films having $\lambda/4$ optical thicknesses, and the single minimum means there is zero reflectance for only one λ.

Clearly, $n_2 > n_1$ in equation (13.32) and it is now common practice to designate a glass-high index-low index-air system as gHLa. H layers are commonly zirconium dioxide ($n = 2.1$), zinc sulphide ($n = 2.35$), titanium dioxide ($n = 2.4$) or cerium oxide ($n = 2.2$), whilst L layers are usually magnesium fluoride ($n = 1.38$) or cerium fluoride ($n = 1.63$).

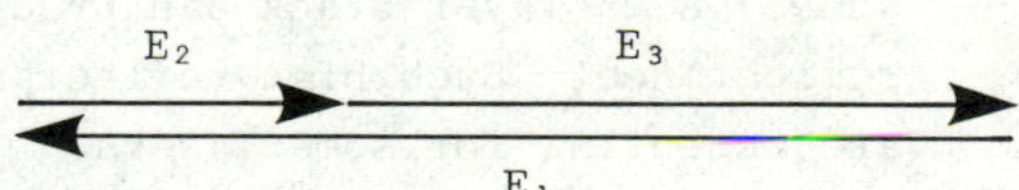

Fig. 13.24 Three reflection vectors adding to zero.

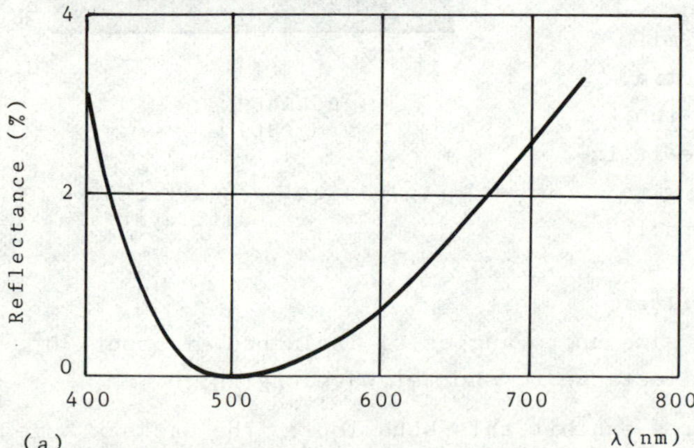

(b)

Air

MgF_2 $\quad n_1 t_1 = 0.326\lambda$

CeO_2 $\quad n_2 t_2 = 0.063\lambda$

Glass

Fig. 13.25 (a) Reflectance for a double film: $n_g = 1.5$. (b) Film data.

Suppose in satisfying equation (13.32) and the double-quarter, single minimum requirement the H layer has an index greater than that of the glass substrate. Then the reflection at the H-g boundary will have a zero phase change whilst the reflections at the a-L and L-H boundaries will undergo a π phase change, using the convention discussed in problem 13.8.5 . Referring to figur 13.23, the reflected wave with amplitude E_1 will have undergone a π phase change relative to E, t amplitude of the incident wave. Remember the rays are incident near the normal and they are sho with exaggerated angles of incidence and reflection, and ignoring refraction, in this figure. E having undergone a π phase change and also traversed a quarter wave film twice, will be antiphas with E_1 . E_3 simply traverses two quarter-wave films twice, which is equivalent to a λ path difference or a 2π phase change. This means E_3 is also antiphase with E_1 . Thus, if $E_1 = E_2 + E_3$ there will be a zero minimum reflectance for the wavelength chosen. The condition satisfies figu 13.24 . Suppose we use magnesium fluoride for the outer film, and the substrate is ophthalmic crown glass ($n_s = 1.523$), then from equation (13.32) we would need to choose the inner film with index

$$n_2 = \left(n_1^2 \cdot \frac{n_s}{n_a}\right)^{1/2} = \left((1.38)^2 \times \frac{1.523}{1}\right)^{1/2} = 1.70$$

to provide the zero minimum at the wavelength chosen for a double-quarter, zero minimum coating.

Quarter-wave Multilayer Stacks

If a glass plate is coated with alternate H and L layers there is a large refractivity ($n' - n$) at each surface which results in a high reflected intensity. The *multilayer stack* shown in figure 13.26 is designated gHLHLHLHLa or $g(HL)^4a$. Adding a further H layer, so that the stack becomes $g(HL)^4Ha$, increases the reflectance further. If the optical thicknesses are all a quarter-wavelength then the emergent waves are all in-phase which results in a very high reflectance. A 25 layer stack can produce better than 99.8% reflectance. Such high reflectance multilayer stacks are essential for some lasers.

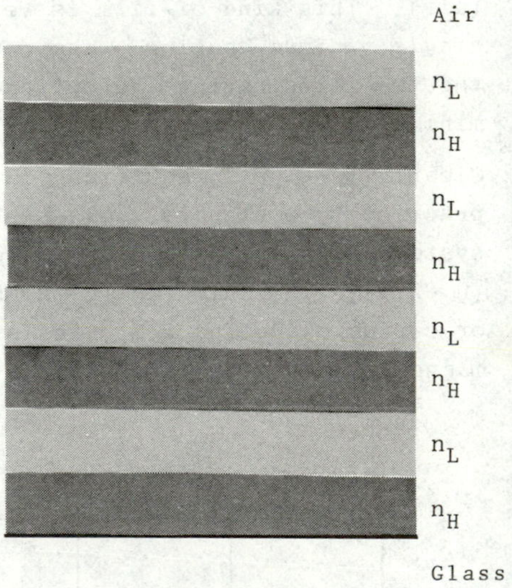

gHLHLHLHLa
or $g(HL)^4a$

Fig. 13.26 A quarter-wave multilayer stack.

Interference Filters

Multilayer stacks can also be designed to transmit particular bands in the UV, visible, or IR spectra. Accordingly, if a filter effectively increases the (short wavelength) *high* frequency transmittance it is known as a *high-pass filter*, the analogy being with radio frequency filters in electronics. Conversely, a stack with an increased (long wavelength) low frequency transmittance is called a *low-pass filter*. The response of these filters remains much the same up to angles of incidence of about 30^0. Increasing i beyond 30^0 generally moves the reflectance to shorter wavelengths.

Several natural periodic or stack-like structures exhibit this behaviour: the backs of some beet peacock feathers, and the wing scales of butterflies are examples. The lustre of human hair in light has the same origin; the light is reflected by minute transparent scales around each hair.

Multilayer systems which transmit a very narrow band of frequencies or wavelengths are called *band-pass filters*. Figure 13.27 illustrates the transmittance for a yellow-green band-pass filter. The bandwidth may be as little as 1 nm. The transmissions at each end of the spectrum can be removed by conventional absorbing filters and the result is the narrow transmitted band shown at 555 nm.

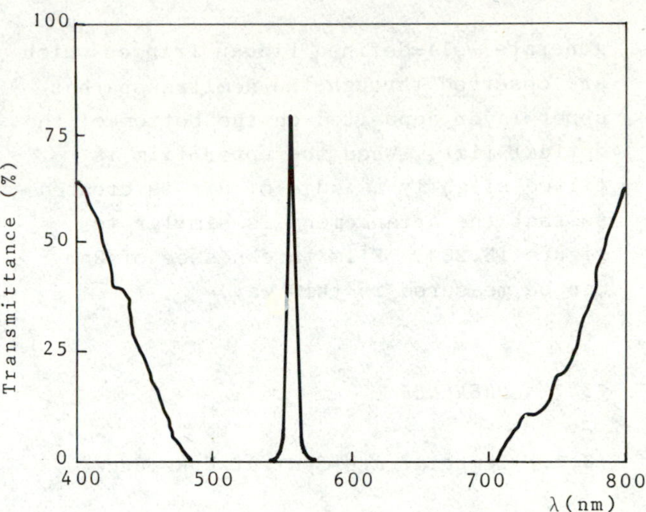

Fig. 13.27 Band-pass filter transitance.

13.6 THIN FILM MEASUREMENTS

Return to figure 13.14 for a moment and recall the condition for maxima in a wedge-shaped film:

$$2n_f t = (m + \tfrac{1}{2})\lambda$$

for normal incidence. Suppose now we write t_m for the film thickness at the mth bright fringe. We can find the change in film thickness between the mth and the (m + 1)th fringes, and therefore between any two fringes, quite simply.

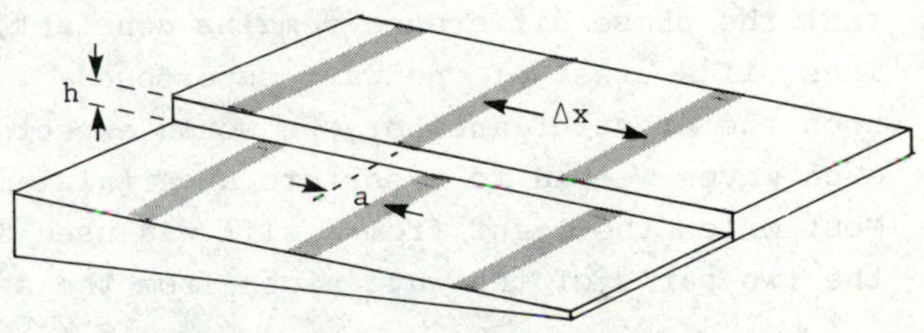

Fig. 13.28 Fringes formed by a stepped wedged-shaped film.

We write

$$t_m = \frac{(m + \frac{1}{2})\lambda}{2n_f} = (m + \tfrac{1}{2})\frac{\lambda_f}{2}$$

where $\lambda_f = \lambda/n_f$, and in similar fashion

$$t_{m+1} = (m + 1 + \tfrac{1}{2})\frac{\lambda_f}{2} \quad .$$

Subtracting these equations gives the change in film thickness, Δt, between successive fringes. Thus,

$$\Delta t = t_{m+1} - t_m = (m + 1 + \tfrac{1}{2})\frac{\lambda_f}{2} - (m + \tfrac{1}{2})\frac{\lambda_f}{2} = \frac{\lambda_f}{2} \qquad (13.33).$$

We see that $\Delta t = \lambda_f/2$. That is, the film changes in thickness by one-half film-wavelength between fringes.

Now consider a wedge with a step in it, figure 13.28. In effect, we have two wedges with the same wedge angle but there is a sudden thickness change at the step. If the fringe width is Δx, then the step height is given by

$$h = \frac{a}{\Delta x}\cdot\frac{\lambda_f}{2} \qquad (13.34)$$

by inspection of the relationship between a and Δx in figure 13.28, and using equation (13.33). Figure 13.29 shows a system for measuring the thickness of a film deposited on a glass substrate. The film being measured is coated with a layer of opaque silver. This silver layer accurately contours the surface of the film and is about 60 - 80 nm thick. The two opposing silver layers generate well-defined Fizeau fringes which are observed through the semitransparent upper layer deposited on the bottom of the optical flat. When the upper film is tilted slightly a wedge of air is created so that the arrangement is similar to figure 13.28 . Film thicknesses of 2nm can be measured in this way.

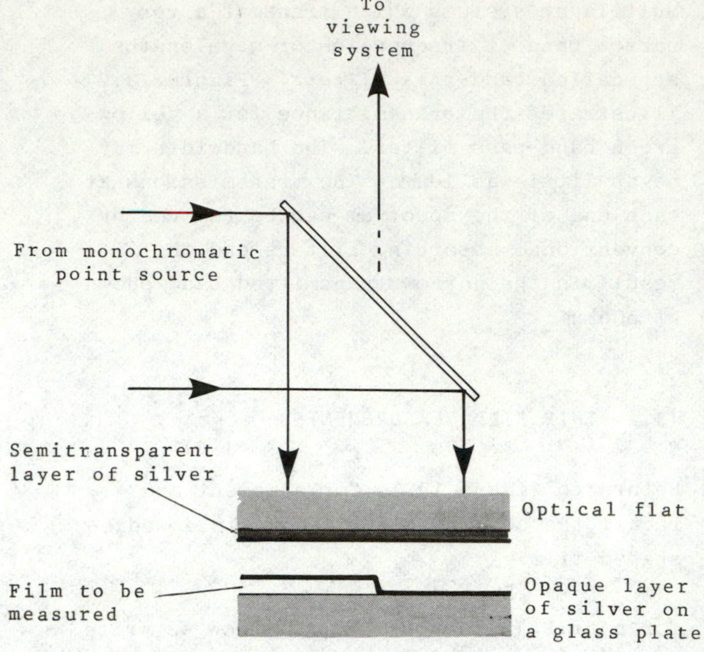

Fig. 13.29 Measurement of a thin film's thickness.

13.7 COHERENCE

13.7.1 *Temporal and Spatial Coherence*

In order to produce interference fringes we have assumed point or extended sources, mainly monochromatic to prevent overlapping of fringes for different wavelengths, and coplanar, or very nearly coplanar plane polarised waves. Further, the waves have been coherent which means, you will recall, that the phase difference remains constant. Thus, if a crest of one wave superposed upon the crest of another, we would expect the next crests to superpose, and so on. To produce such waves we had to resort to a certain amount of ingenuity in the experimental arrangements. Most often the light from a slit was used and the wave was sheared into two components. Since the two parts of the wave arose from the same source they were therefore coherent.

Because the very best source emits a finite range of wavelengths or frequencies the concept of monochromatic waves is an idealisation. A true monochromatic wave is infinitely long and has onl[y] one frequency. Figure 13.30 shows a wave extending to infinity in both directions and its single frequency.

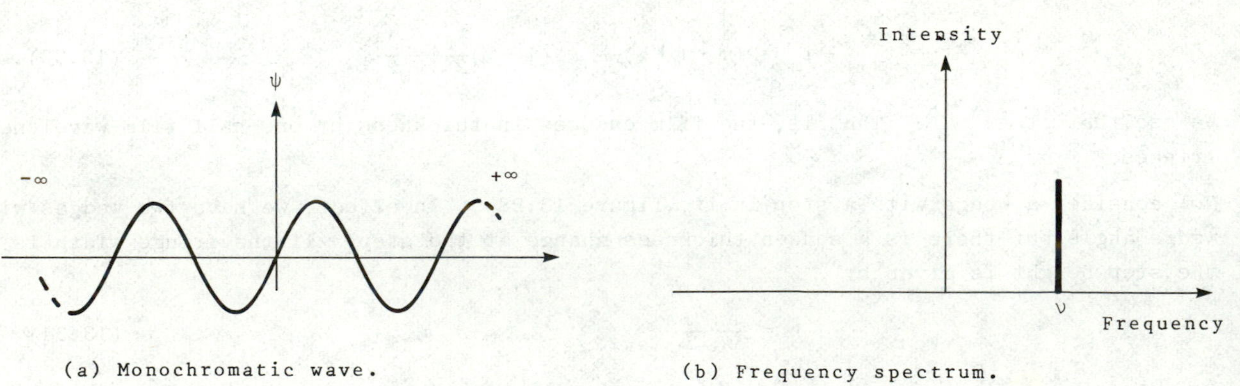

(a) Monochromatic wave.

(b) Frequency spectrum.

Fig. 13.30 (a) A monochromatic wave which is necessarily infinitely long. (b) The single frequen[cy] component of the monochromatic wave: the bandwidth, $\Delta\nu$, is zero.

Such waves do not exist in nature. To exist the source would need to have been switched on for an infinitely long time. We therefore picture true electromagnetic waves as finite wavetrains, or wave packets, with a spread of frequencies. Figure 13.31 shows a wave train and the sort of frequency spectrum it might have. The frequency $\bar{\nu}$ is known as the *characteristic or mean frequency*, and is that frequency most representative of the wavetrain.

If the intensity at the dominant frequency is scaled to unity then the width of the frequency spectrum at half that intensity is called the *bandwidth*. Equation (11.33), $\Delta\nu \simeq 1/\Delta t$, related bandwidth to the coherence time Δt. The latter is the time over which the phase of an emission from a source is fairly constant. Related to coherence time we defined the coherence length or wavetrain length in equation (11.34) as $\Delta x = c.\Delta t$. For a true monochromatic wave $\Delta t = \infty$ and since this is not possible the best approach to monochromaticity is called quasimonochromatic light. Even the emission from a laser is quasimonochromatic. For example, $\Delta\nu = 1.3 \times 10^9$ Hz for a typical teaching laboratory helium-neon laser.

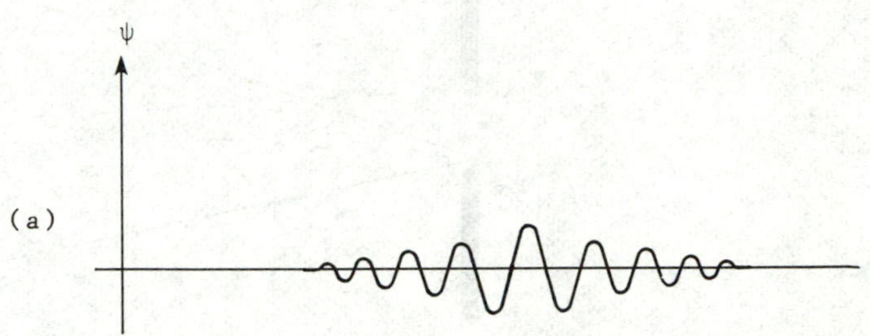

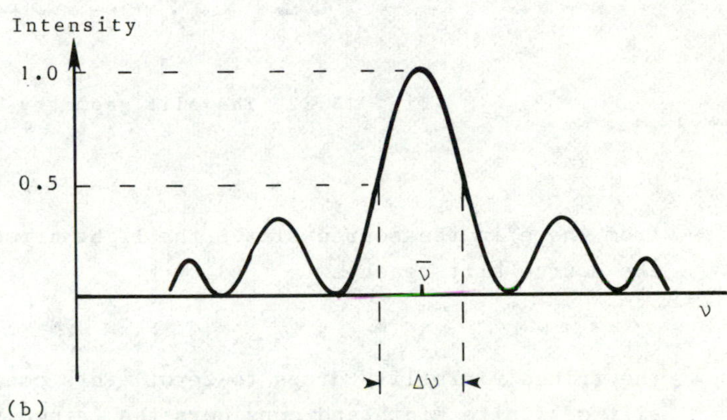

Fig. 13.31 (a) A finite wavetrain.
(b) Frequency spectrum for the finite wavetrain.

When the bandwidth increases the coherence time and the coherence length decrease and this is spoken of as a decrease in *temporal* or *longitudinal coherence*. The occurrence of fringes then, in a system where the two waves travel different paths and are then brought together, is dependent on bandwidth. Bandwidth in turn describes a property of the source which means we must choose the source critically if we wish to produce fringes. Clearly, we must choose sources with exceedingly small $\Delta\nu$, and therefore large Δt and Δx.

The foregoing discussion of temporal coherence helps to explain the requirements in a Michelson interferometer; that is, fringe visibility will be high when $\Delta\nu$ is near zero, and the separation, 2d, of the secondary sources S_1 and S_2 in figure 13.19 is much less than the coherence length.

The Young's two slit experiment used a primary source slit S in front of two secondary source slits S_1 and S_2. No mention has yet been made about the slit width which is desirable for the maximum visibility of fringes. In figure 13.32, if $r_1 = r_2$ and $SA = SS_1$, the path difference AS_2 influences the interference pattern. If the path difference AS_2 to the secondary source slits exceeds the coherence length then no fringes can form. This is the same as saying that the wavetrain from S_1 will arrive at P, on the screen, before the wavetrain from S_2 so that they cannot overlap and form fringes. AS_2 will obviously increase as the distance a between the secondary source slits increases. There will therefore be a maximum value for a after which no fringes will form for a given coherence length. AS_2 also depends on the spatial extent of the source slit S since AS_2 will vary depending

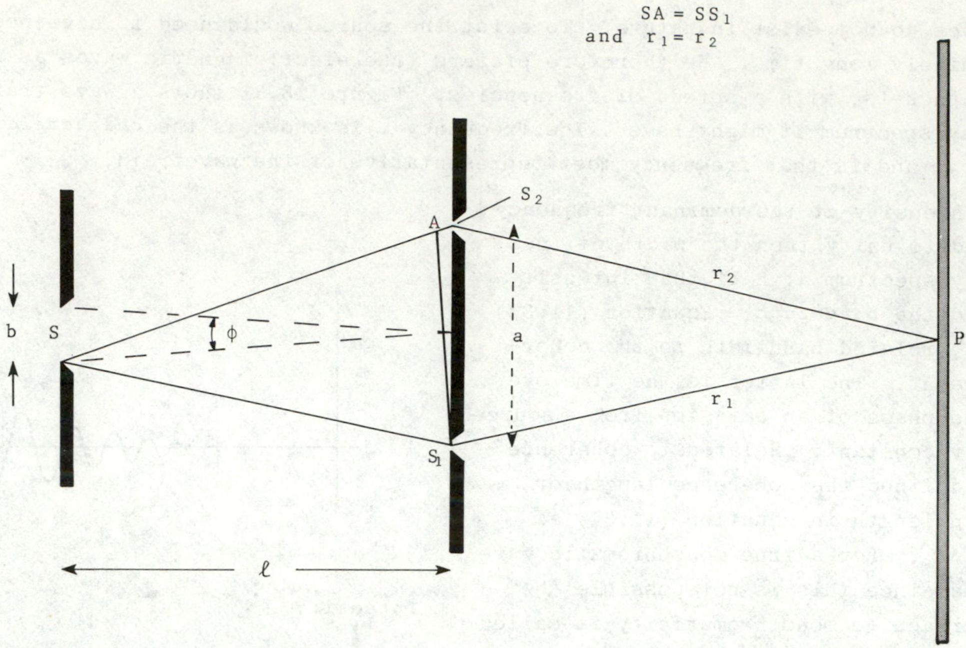

Fig. 13.32 The slit geometry in Young's experiment.

from where in the source slit S the light arises. It can be shown that when the angular width of the source slit reaches

$$\phi = \frac{\lambda}{a} \tag{13.35}$$

the fringe visibility drops to zero. This condition, which we shall not derive, assumes a source slit of finite width and considers the fringe patterns arising from different incoherent 'line sources' within the slit. That is, the slit is divided up into elementary strips. This limiting angular width is the limit of what is known as *spatial coherence*. If b is the source slit width and it is a distance ℓ from the secondary source slits, then the angular width of S is $\phi = b/\ell$. Combining this expression for ϕ with equation (13.35) gives the condition for good fringe visibility

$$\frac{b}{\ell} = \frac{\lambda}{a} \tag{13.35a}.$$

Slits are not the only type of sources. A common shape of source is a circular disc which may be nothing more than an aperture in a mask over an extended source, or it may be a natural shape like the sun. It can be shown that interference fringes will disappear when the angular diameter of the circular disc is given by

$$\phi = 1.22\frac{\lambda}{a} \tag{13.36}.$$

So, the diameter of the circular source must not exceed $\ell\phi$, or $1.22\frac{\ell\lambda}{a}$.

The visibility of the fringes, $V = (I_{max} - I_{min})/(I_{max} + I_{min})$, in an interferometric system is a measure of the *degree of coherence*. When $I_{min} = 0$ the visibility takes its maximum value of 1 and the two interfering waves are coherent. When $V = 0$ the waves are incoherent, and for $0 < V < 1$ the waves are said to be partially coherent.

13.7.2 Examples of the Use of Coherence

(1) A Michelson interferometer is illuminated with cadmium red light ($\lambda = 643.847$ nm). Initially $d = 0$ in figure 13.19 . After one mirror has been moved through 15.94 cm the fringes disappear. Calculate the linewidth, $\Delta\lambda$, of the cadmium red line. (See figure 11.11 for an illustration of the linewidth).

The OPD = 2d when the instrument is in air, if we assume $n_{air} = 1$. Fringes disappear when the coherence length (wavetrain length) is equal to the OPD. Thus, the coherence length is

$$\Delta x = 2d = 2 \times 15.94 = 31.88 \text{ cm} .$$

We can find the linewidth $\Delta\lambda$ which corresponds to the bandwidth $\Delta\nu$ as follows. Using $c = \nu\lambda$, rewritten as $\lambda = c/\nu$, and differentiating λ with respect to ν, we have

$$\Delta\lambda = -\frac{c}{\nu^2} \cdot \Delta\nu .$$

The minus sign merely indicates that λ decreases as ν increases, so we drop the sign. Since $\nu = c/\lambda$ we can write the last expression

$$\Delta\lambda = \frac{c}{\nu^2} \cdot \Delta\nu = \frac{\lambda^2}{c} \cdot \Delta\nu \qquad (13.37).$$

Using $\Delta\nu \simeq \frac{1}{\Delta t}$ in the last expression gives

$$\Delta\lambda = \frac{\lambda^2}{c} \cdot \Delta\nu = \frac{\lambda^2}{c} \cdot \frac{1}{\Delta t} = \frac{\lambda^2}{\Delta x} \qquad (13.38)$$

since the coherence length $\Delta x = c.\Delta t$. Now, we have measured Δx above so we can calculate $\Delta\lambda$:

$$\Delta\lambda = \frac{\lambda^2}{\Delta x} = \frac{(643.847 \times 10^{-9})^2}{31.88 \times 10^{-2}} = 0.0013 \text{ nm}.$$

$\Delta\lambda$ is the 'spread of wavelengths' around the characteristic or dominant wavelength 643.847 nm. It would be profitable for you to look again at problem 11.3.8 .

(2) In the Young's two-slit experiment the primary source slit S is 0.2mm wide and 1 m from the plane of the two slits. At what separation a of the two slits do the fringes first vanish if the source emits quasimonochromatic light of wavelength 589.3 nm?

Equation (13.35) gives the required condition: $\phi = \lambda/a$, where $\phi = b/\ell$ in figure 13.32 . ϕ is the angular width of the slit seen from the two slits, b is the width of the slit S, and $\ell = 1$ m in this problem. Thus, the width between the two slits is

$$a = \frac{\lambda}{\phi} = \frac{\lambda}{b/\ell} = \frac{\lambda\ell}{b} = \frac{589.3 \times 10^{-9} \times 1}{0.2 \times 10^{-3}} = 2.95 \times 10^{-3} \text{ m} ,$$

$$\text{or,} \quad a = 2.95 \text{ mm}.$$

For fringes to be observed with a high visibility the width between the two slits should be less than 2.95 mm. We should note that a can be increased by increasing ℓ or reducing b.

(3) If the source slit in the last problem were replaced by a uniformly illuminated circular aperture where b is now the aperture's diameter, find a where the fringes first disappear. Take b = 0.1 mm this time.

The condition required is now equation (13.36). Hence,

$$a = 1.22\frac{\lambda}{\phi} = 1.22\frac{\lambda}{b/\ell} = 1.22\frac{\lambda\ell}{b} = \frac{1.22 \times 589.3 \times 10^{-9} \times 1}{0.1 \times 10^{-3}}$$

$$= 7.19 \times 10^{-3} \text{ m}$$

or $\quad a = 7.19$ mm.

(4) Figure 13.33 shows an arrangement known as the *Michelson Stellar Interferometer*.

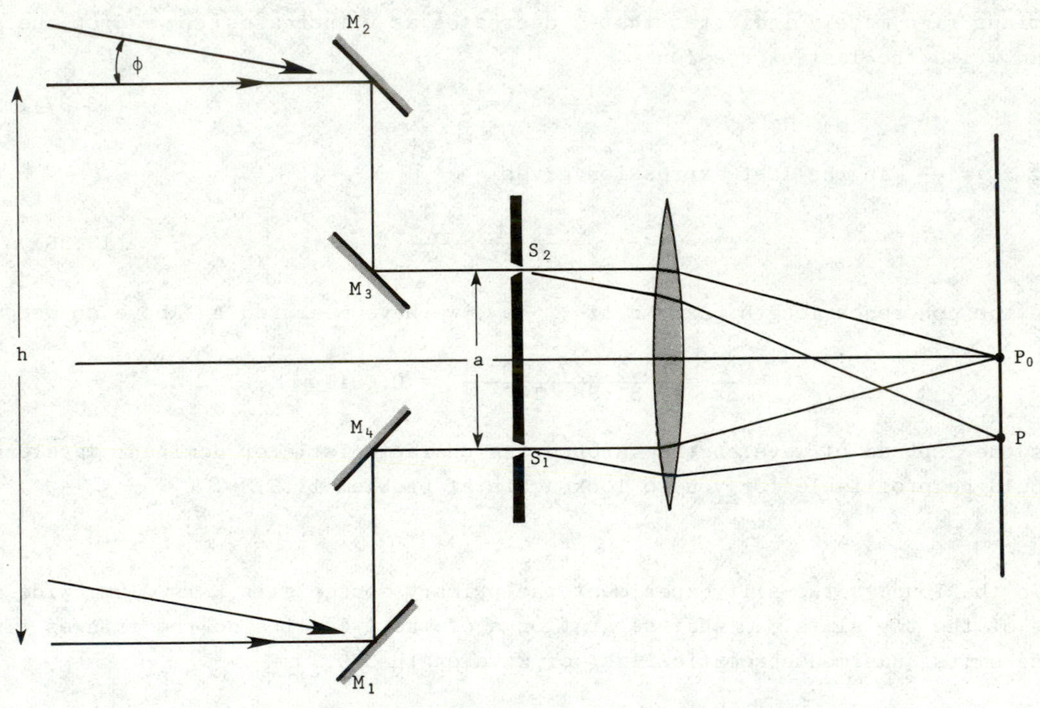

Fig. 13.33 The Michelson stellar interferometer.

The mirrors M_1 and M_2 are moveable. When directed towards a star the mirrors are adjusted until the fringes disappear, then h is measured. Find the angular diameter of the star if h = 307.34 cm when $\lambda = 570$ nm.

We are really repeating problem 3 where the distance between the slits is now effectively h. Therefore the angular diameter of the star is

$$\phi = 1.22\frac{\lambda}{h} = \frac{1.22 \times 570 \times 10^{-9}}{307.34 \times 10^{-2}} = 2.26 \times 10^{-7} \text{ rad} \equiv 1.296 \times 10^{-5} \text{ degree.}$$

If the two slits are illuminated not by light from a single star but by a double star, which is effectively a two point sources system, then it can be shown that the angular separation of the point sources is

$$\phi = \frac{\lambda}{2h} \tag{13.39}$$

in the Michelson stellar interferometer when the fringes disappear. This relationship has been used to find the angular separation of two stars in a double star system. The stellar interferometer can also differentiate between a single and two star system. Turning the mirrors M_1 and M_2 about an axis perpendicular to the line joining them gives the same result for a single star because of the latter's circular symmetry. This is not the case with a double star.

13.8 WORKED PROBLEMS

13.8.1 Find the refractive index of a single layer antireflection film used on glass of refractive index 1.7 if no light of wavelength 550 nm is reflected on normal incidence. What is the reflectance for $\lambda = 400$ nm?

Equation (13.22) immediately gives

$$n_f = \sqrt{n_s} = \sqrt{1.7} = 1.3038,$$

assuming the refractive indices to be for $\lambda = 550$ nm. The optical thickness of the film is

$$n_f t = \frac{\lambda}{4} = \frac{550}{4} = 137.5 \text{ nm}.$$

At $\lambda = 400$ nm the optical thickness 137.5 nm traversed twice by a wavelength of 400 nm represents a fraction $2 \times 137.5/400 = 0.6875$ of a wavelength. This in turn is equivalent to a phase change of $2\pi \times 0.6875 \equiv 247.5^0$, and this is the phase difference between the waves reflected from the first and second interfaces of the coated material. Assuming n_f and n_s are the same at 400 nm for simplicity, the amplitude reflection coefficients at the air-film and the film-substrate boundaries are

$$r_1 = -\frac{n_f - 1}{n_f + 1} = -\frac{1.3038 - 1}{1.3038 + 1} = -0.1319$$

and

$$r_2 = -\frac{n_s - n_f}{n_s + n_f} = -\frac{1.7 - 1.3038}{1.7 + 1.3038} = -0.1319,$$

respectively. They should be equal, of course! We can now write $E_{1r} = r_1 E_i$ and $E_{2r} = r_2 E_i$. However, if we arbitrarily put $E_i = 1$, then

$$E_{1r} = r_1 \quad \text{and} \quad E_{2r} = r_2$$

and the resultant reflected amplitude E_r will be the vector addition of r_1 and r_2. We shall have

$$E_r = r E_i = r, \quad \text{since } E_i = 1.$$

We now find the resultant amplitude reflection coefficient r by adding r_1 and r_2, vectorially.

Thus, r is as shown in figure 13.34 . Using the cosine rule

$$r^2 = r_1^2 + r_2^2 - 2r_1 r_2 \cos 67.5^0$$

$$= 2r_1^2(1 - \cos 67.5^0) \quad , \quad \text{since } r_1 = r_2 \, ,$$

$$= 2 \times (0.1319)^2 (1 - 0.38268)$$

$$= 0.0215 \; .$$

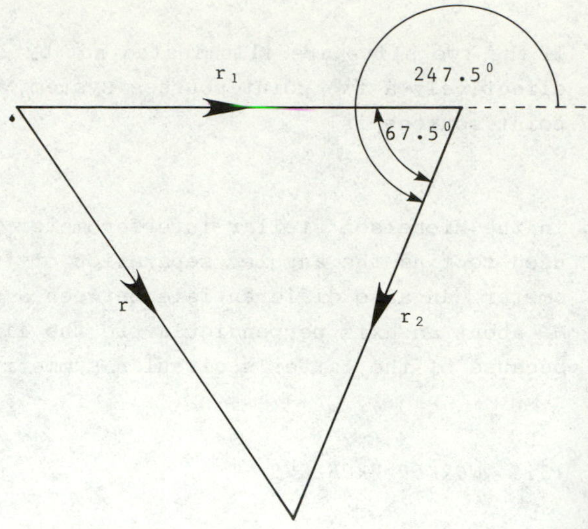

Fig. 13.34 Vector addition of r_1 and r_2.

But the reflectance $R = r^2 = 0.0215 \equiv 2.15\%$, so 2.15% of 400 nm wavelength light is reflected.

A similar calculation for $\lambda = 500$ nm produces a reflectance of 0.17%, showing that more light is reflected the further the wavelength departs from 550 nm , in this case. For red light, $\lambda = 750$ nm , the reflectance is 1.2%. The mixture of a preponderance of red and violet giving the purple 'bloom' characteristic of single layer antireflection coatings is clearly evident.

13.8.2 Figure 13.35 shows the Jamin interferometer with two identical chambers placed in the path of the interfering beams. The inner lengths of the chambers are 25 cm and a mercury source, $\lambda = 546.1$ nm, illuminates the system. If both chambers are evacuated and air is allowed to slowly fill one of them, the observer sees 133 fringes sweep by a cross-line hair in the focal plane of the telescope objective. Find the refractive index of the air in the chamber.

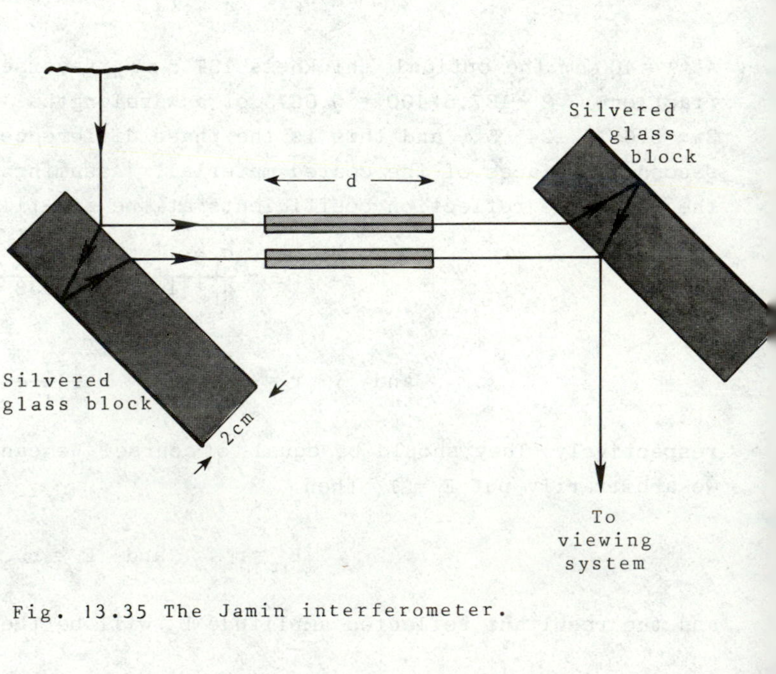

Silvered glass block

d

Silvered glass block

2 cm

To viewing system

Fig. 13.35 The Jamin interferometer.

The optical path difference between the two states, one with the air-filled chamber and the other with the evacuated chamber, is given by

$$nd - 1.d = (n - 1)d$$

where n is the refractive index of air, 1 is the refractive index of the vacuum, and d is the internal chamber length. Each fringe sweeping by represents a change in OPD of 1 wavelength. Suppose m fringes sweep by, then we can write

$$OPD = (n-1)d = m\lambda$$

which rearranges to give

$$n = \frac{m\lambda + d}{d} = \frac{m\lambda}{d} + 1$$

$$= \frac{133 \times 546.1 \times 10^{-9}}{0.25} + 1$$

$$= 2.91 \times 10^{-4} + 1$$

$$= 1.000291$$

13.8.3 A thin wedge of transparent liquid is formed between two flat glass plates. The spacer is a hair 0.1 mm in diameter placed 60 mm from the apex of the wedge and lying at the centre of a dark fringe. If there are 462 dark fringes from the apex to the spacer, calculate the refractive index of the film when sodium light, $\lambda = 589.3$ nm, is used for the illumination.

The condition for dark fringes or minima differs from the equation for maxima by $\lambda/2$. Equation (13.23) gave the condition for maxima: $2n_f x_m \alpha = (m + \frac{1}{2})\lambda$. Thus, the condition for minima is

$$2n_f x_m \alpha = m\lambda\,, \qquad m = 0,\ 1,\ 2,\ldots. \qquad (13.40).$$

The zeroth dark fringe occurs when $x_m = 0$, so that there is a dark fringe at the wedge apex. The fringe separation, the distance between the centres of two adjacent dark fringes, is

$$\Delta x = x_{m+1} - x_m = \frac{(m+1)\lambda}{2n_f \alpha} - \frac{m\lambda}{2n_f \alpha} = \frac{\lambda}{2n_f \alpha} \qquad (13.41)$$

just as for bright fringes, which it ought to be, of course!

The thickness of the film at the mth dark fringe is

$$t_m = x_m \alpha \qquad (13.42).$$

Since there are 462 dark fringes occupying 60 mm along the wedge, we must have 461 fringe widths. So,

$$\Delta x = \frac{60 \times 10^{-3}}{461} = 0.13 \times 10^{-3}\,\text{m}$$

is the fringe width or separation. Using equation (13.42) now to find the wedge angle,

$$\alpha = \frac{t_m}{x_m} = \frac{0.1}{60} = 1.66667 \times 10^{-3} \text{ rad.}$$

Note that we start counting the fringes from zero, so $m = 461$. Now, from equation (13.40)

$$n_f = \frac{m\lambda}{2x_m \alpha} = \frac{461 \times 589.3 \times 10^{-9}}{2 \times 60 \times 10^{-3} \times 1.66667 \times 10^{-3}}$$

$$= 1.358\ .$$

Alternatively, having calculated Δx and α we could have used equation (13.41).

13.8.4 Figure 13.36 shows the arrangement for a Mach-Zehnder interferometer. It consists of two mirrors, M_1 and M_2, and two beamsplitters, B_1 and B_2. Explain how it can produce Haidinger and Fizeau fringes, and suggest what advantage it might have over the Jamin interferometer.

When all the mirrors and beamsplitters are parallel it will produce circular Haidinger fringes of equal inclination. Tilting one mirror slightly will create Fizeau fringes of equal thickness, as in the wedge-shaped film. Since the beams follow markedly different paths, before being brought together to create the interference pattern, one can insert large test objects in one path.

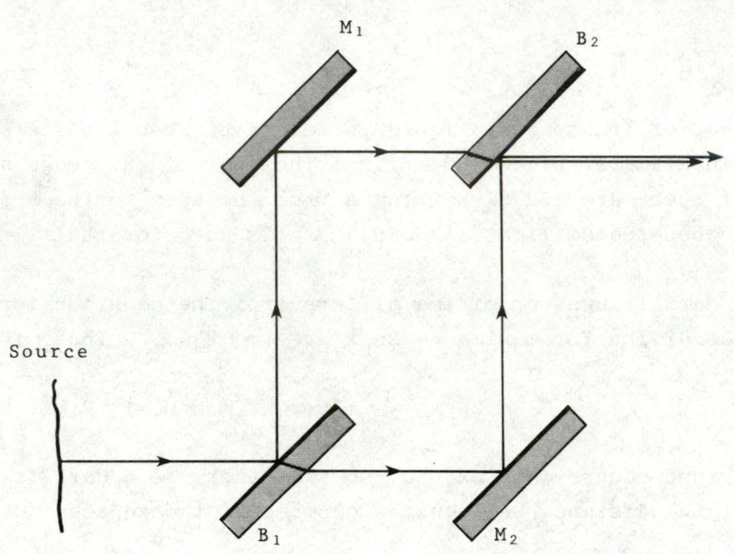

Fig. 13.36 The Mach-Zehnder interferometer.

13.8.5 When discussing reflection of normally incident light from a dielectric (non-conducting) surface such as glass, we often state that there is a π phase change if $n' > n$ and no phase change if $n' < n$. Strictly, we should consider the $E_{r\perp}$ and $E_{r\parallel}$ components separately for near normal incidence but we should note that for normal incidence the plane of incidence can be in any plane perpendicular to the surface. This means there is no preferred direction and we tend to lump the two together and talk of E_r. We then define the amplitude coefficient of reflection as $r = E_r/E_i$ for normal or near normal incidence. Explain why this is permissible.

Firstly, refer to figures 10.22 and 10.23. We see that for $i = 0^0$, or indeed for $i < i_p$, the polarisation angle, the $E_{r\perp}$ and $E_{r\parallel}$ components undergo π and zero phase change for $n' > n$, whilst the reverse is true for $n' < n$. When dealing with thin films such as antireflection coating it is the relative phase change at the two boundaries which counts when considering OPDs. Take the sing layer antireflection film as an example. Here $n_3 > n_2 > n_1$, so $E_{r\perp}$ undergoes a π phase change at each surface and the relative phase change due to the reflections above is zero. The $E_{r\parallel}$ compone undergoes zero phase change at each surface so the relative phase change is again zero by reflecti alone. In this case, any phase difference between the two reflected beams travelling backwards i therefore introduced by the optical path difference only. Also, from equation (10.46) $r_\parallel = -r_\perp$ for normal incidence. But if the incident ray is normal to the surface we can draw the plane of incidence in any plane perpendicular to the surface, so there is no preferred direction. This also means that $|E_{r\perp}| = |E_{r\parallel}|$ and half the intensity is associated with each component. Thus, we may consider either for convenience. If we choose the $\perp$ component, then from equation (10.38) with $\cos i \simeq \cos i' \simeq 1$

$$r_\perp = \frac{n - n'}{n + n'} = -\frac{n' - n}{n' + n} .$$

Writing r for $r_\perp$, we can drop the $\perp$ subscript since there is no preferred direction and the treatment in section 13.4.1 follows. Because $r_\perp = r_\parallel$ for normally incident light the calculation is valid for both components. It follows that we can define $r = E_r/E_i$ and we can speak of a π phase change at an $n' > n$ boundary and a zero phase change at an $n' < n$ boundary.

EXERCISES

1 What are the conditions necessary to produce sustained interference between two waves? In a Young's double slit arrangement the screen is 1.0 m away from the double slits which are 1.8 mm apart. If light of 546 nm wavelength is used, find the separation of the fringes on the screen.

Ans. 0.303 mm. See W.P.O. page 110, question 1.

2 In question 1, a glass chamber filled with air is placed before the upper slit S_2. The air is then replaced with a gas, whereupon the fringe pattern is seen to be displaced by 14 bright fringes. If the chamber's internal length is 16 mm, find the refractive index of the gas, n_g, given the refractive index of air is 1.000 275. In which direction is the fringe pattern moved?

Ans. 1.000 753 . N.B. the pattern moves upwards and not downwards as stated in W.P.O. page 111, question 2.

3 In Young's experiment with light of wavelength 587.5 nm , it was found that the separation of the first and eleventh dark fringes was 4.44 mm at a distance of 2 m from the slits. (i) Find the separation of the slits. (ii) What would be the separation of the first and eleventh bright fringes if the apparatus were immersed in water of refractive index 1.33? (iii) How would you find the central bright fringe?

Ans. (i) 2.65 mm . (ii) 3.33 mm . See W.P.O. page 112, question 3.

4 Explain how interference fringes are produced with a Fresnel biprism. If the apical angles are 0.01 radian, the prism is 10 cm from the slit, and the screen is 90 cm from the prism, calculate the fringe width for light of wavelength 600 nm . The prism glass has refractive index 1.5 .

Ans. 0.6 mm. See W.P.O. page 113, question 4.

5 An expanded laser beam impinges directly on a Fresnel biprism. Show that the fringe width is independent of the position of the screen.

Ans. See W.P.O. page 115, question 5.

6 Explain how interference fringes are produced by thin films.

Ans. See W.P.O. page 115, question 6.

7 State the conditions for dark and bright fringes in a thin film of refractive index n_f. A thin transparent film of refractive index 1.432 is to generate a minimum in reflected light of wavelength 500 nm, under normal incidence. Find the minimum thickness of film to achieve this.

Ans. 174.6 nm. See W.P.O. page 117, question 7.

8 Starting with $r_1 = -\left(\frac{n_f - n_1}{n_f + n_1}\right)$ and $r_2 = -\left(\frac{n_s - n_f}{n_s + n_f}\right)$, show that a film can be made anti-reflecting if $n_f = \sqrt{(n_1 n_s)}$.

Ans. See W.P.O. page 117, question 8.

9 Magnesium fluoride, refractive index 1.38, is used as an antireflection film on the surface of an ophthalmic crown glass lens, refractive index 1.523 . Show that this film cannot reduce the reflected light to zero and find what percentage ($\lambda = 555$ nm) is reflected.

Ans. 1.2%. See W.P.O. page 118, question 9.

10 Suppose a thin wedge-shaped film is formed between two flat glass plates. Describe the pattern formed when monochromatic light is incident normally on the wedge. If the wedge spacer is a hair, find its diameter if it lies in the position of the 345th bright fringe when the wedge is illuminated by sodium light ($\lambda = 589.3$ nm) under normal incidence.

Ans. 0.102 mm. See W.P.O. page 119, question 10.

11 Newton's rings are formed by reflection at the film between the convex surface of a lens and the flat glass plate upon which it rests. When sodium light ($\lambda = 589.3$ nm) is used the diameter of the second dark ring is 0.238 cm and that of the twenty-second ring is 0.788 cm. What is the radius of curvature of the convex surface?

Ans. 120 cm. See W.P.O. page 120, question 11.

12 Suppose, in the Newton's rings experiment, some dust separates the lens and the plate by an unknown distance Δs. Show that the previous method in question 11 still works.

Ans. See W.P.O. page 122, question 12.

13 Describe the Michelson interferometer and its underlying principles. The device is set up to show circular fringes. Suppose the mirror M_1 is initially a distance d further from the beam-splitter than is mirror M_2. As d is reduced fringes sweep towards the centre of the field of view. If 800 bright fringes pass by when d is reduced by 1.2×10^{-4} m, find the wavelength of the monochromatic light used.

Ans. 600 nm. See W.P.O. page 122, question 13.

14 A Michelson interferometer is set to give maximum visibility of fringes for a sodium source emitting a doublet with wavelengths 589.0 nm and 589.6 nm. The mirror is moved until the fringes disappear. Find the movement of the mirror.

Ans. 0.1447 mm. See W.P.O. page 124, question 14.

14 DIFFRACTION

14.0 INTRODUCTION

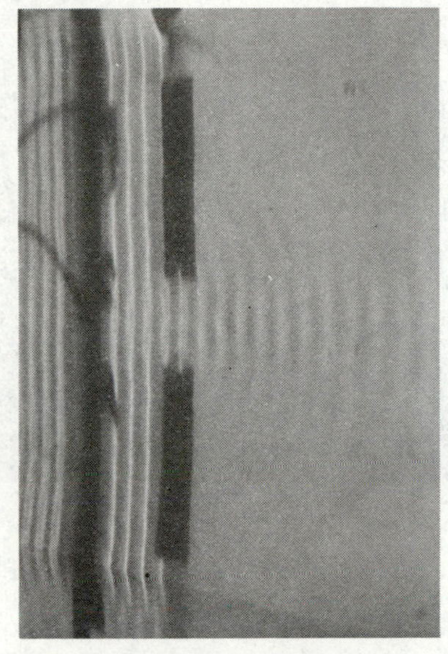

(a)

When we considered apertures and obstacles in the path of light from extended and point sources in chapter 1, the light patches and shadows were constructed geometrically: light was assumed to travel in straight lines. However, if we investigate the formation of shadows and images from a wave-theory point of view we can demonstrate departures from rectilinear propagation. Francesco Maria Grimaldi (1618 - 1663) first investigated such departures and he named the phenomenon diffractio, from which we derive the modern term *diffraction*.

The effect occurs with all wave propagation whenever a wavefront meets some form of obstruction. For example, Grimaldi, and somewhat later Robert Hooke (1635 - 1703), noticed bands of light within the shadow of a rod illuminated by a small source. Since light wavelengths are very short the effect mostly goes unnoticed, but water waves in a laboratory ripple tank easily demonstrate the phenomenon on a macroscopic scale. Figure 14.1(a) shows plane wavefronts incident on a narrow slit. The secondary wavelets arising in the slit are seen to propagate into what would be the geometrical 'shadow'. The effect is most noticeable

(b) 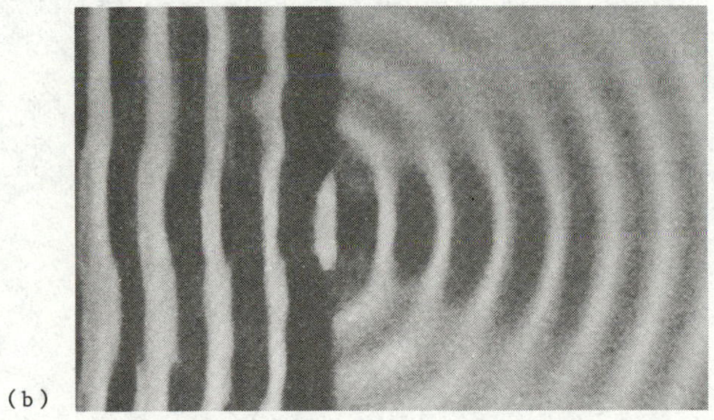

(c) 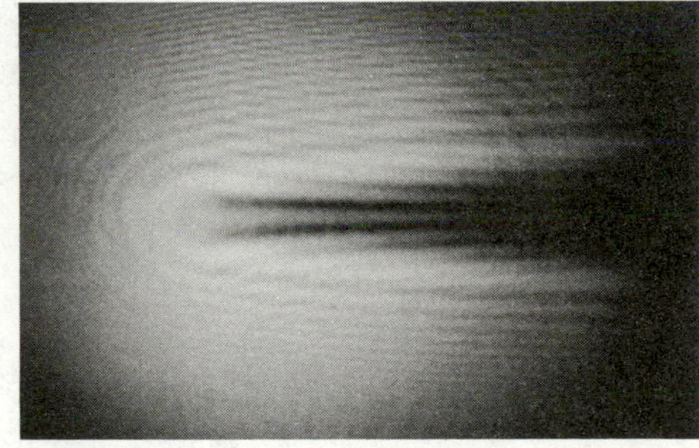

Fig. 14.1 (a) A ripple tank demonstration of diffraction through a narrow slit. (b) A close-up of (a) when the slit-width is about equal to the wavelength. (c) The shadow of a needle point in a helium-neon laser beam.

when the wavelength approaches the aperture size, figure 14.1(b). Figure 14.1(c) is an example with light. It is the shadow of a needle point in a helium-neon laser beam.

Diffraction results from the superposition of secondary wavelets propagating into the geometrical shadow. These waves interfere and under suitable conditions produce what are really interference effects. However, where these effects are caused by waves diffracted by an obstacle, be it a slit, a circular aperture, a rod, or the rim of a lens, the results are called diffraction patterns when imaged on a screen of some sort. Optical instruments invariably involve lenses and apertures in the image forming system and, in the limit, the ultimate sharpness of an image depends on diffraction. For example, figure 14.2 is the image of a point source formed by a lens in front of which was a 0.5 mm diameter circular aperture. The source was a helium-neon laser which is the best approximation to an idealised point source.

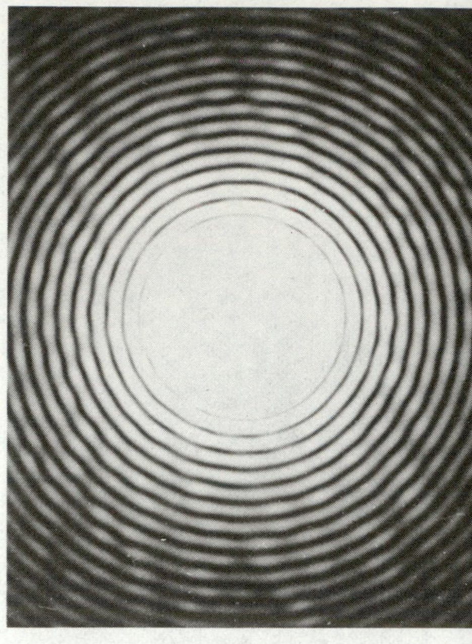

Fig. 14.2 The image of a point source!

Our approach to an analysis of the diffraction effects resulting from the superposition of the diffracted waves will depend on the Huygens principle of secondary wavelets. Huygens assumed each point on a wavefront was the source of a secondary wavelet of the same frequency. The shape of the wavefront at some later time is taken as the envelope of the secondary wavelets, but only a small portion of the secondary wavelet is employed in forming the envelope. Fresnel modified Huygens principle by introducing the concept of interference between secondary wavelets, and this allows us to derive quantitative conditions for diffraction. The addition of the idea of interference to the Huygens principle leads to the Huygens-Fresnel principle which may stated

Every point on a wavefront acts as a source of a spherical secondary wavelet of the same frequency. The magnitude of the displacement at any point beyond is the superposition of all the secondary wavelets at that point, having regard for their amplitudes and their relative phases.

Diffraction phenomena are divided into two classes:

Fraunhofer diffraction patterns are observed in the image plane of a source as at P in figure 14.

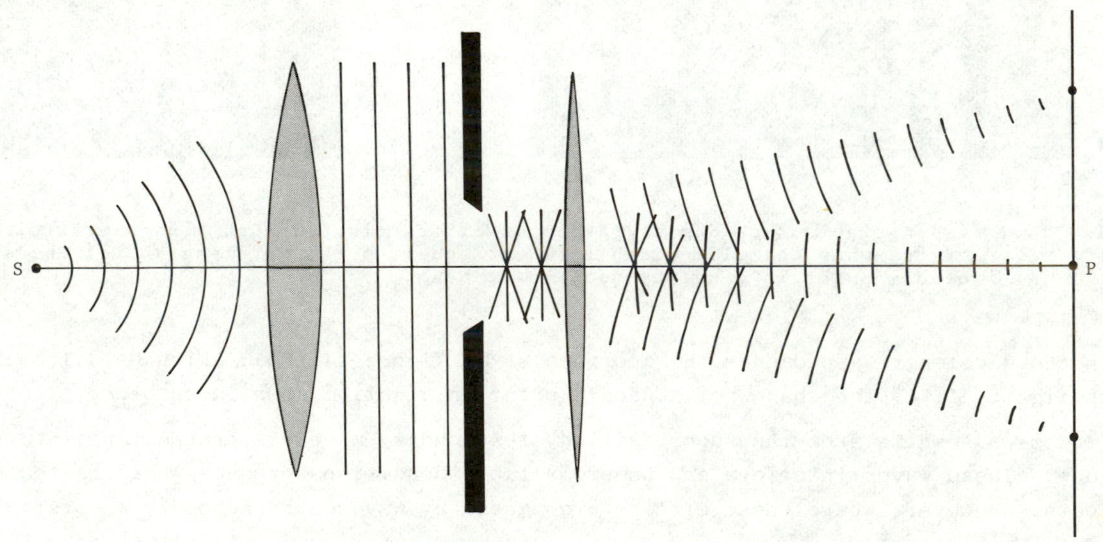

Fig. 14.3 Fraunhofer diffraction.

Such patterns occurring in the image plane are of the utmost importance in optical instruments, including the eye. Fraunhofer diffraction is sometimes referred to as *far field diffraction*. *Fresnel* diffraction or *near field diffraction* patterns occur anywhere in the diffracted beam other than in the image plane. Moving the screen to the left or right in figure 14.3 would therefore display a Fresnel diffraction pattern.

Although we shall not be overly concerned with the process giving rise to diffraction, we can imagine that electromagnetic waves interact with a physical obstruction. This interaction drives electrons into oscillations at the frequency of the incident wave and these act as secondary sources. If the electrons are at the edge of a large aperture, and the point of observation is far away, the secondary wavelets do not show marked superposition effects. However, for small apertures or a point of observation close to the aperture the sources of secondary wavelets are close together and appreciable effects might be expected. In fact, this turns out to be the case. Narrowing a slit, for example, is seen to cause the Fraunhofer diffraction pattern to spread out on the screen and make it more discernible.

14.1 FRAUNHOFER DIFFRACTION AT SINGLE APERTURES

14.1.1 Fraunhofer Diffraction at a Single Slit

Figure 14.4 shows a plane monochromatic wave incident normally on a relatively long, narrow slit. Some light is undeviated and some is diffracted both upwards and downwards, but we consider only a beam diffracted upwards at the moment. Suppose the diffracted rays shown make an angle θ with the normal to the slit. The path difference between the upper and lower diffracted rays is labelled $b\sin\theta$, where b is the width of the slit. Using the Huygens-Fresnel principle, we divide the slit into N equal strips parallel to the slit and regard each of these as giving rise to secondary wavelets of the same amplitude, E'. This is a reasonable assumption if the slit is uniformly illuminated. Each strip is then b/N wide, and this results in a path difference between adjacent strips of $\frac{b}{N}\sin\theta$, as shown in figure 14.5 .

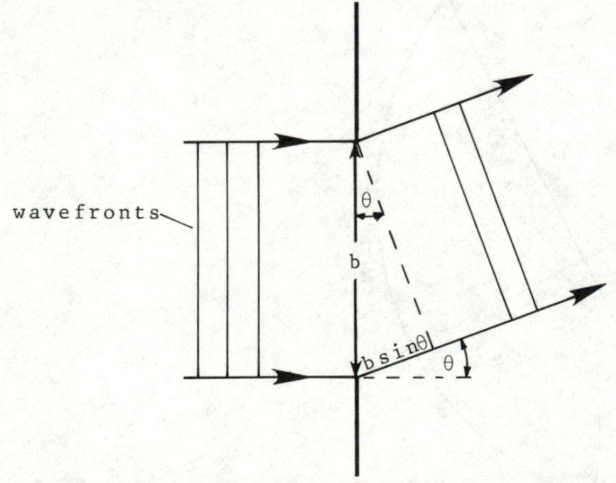

Fig. 14.4 The diffracted rays in the θ direction from a single slit of width b.

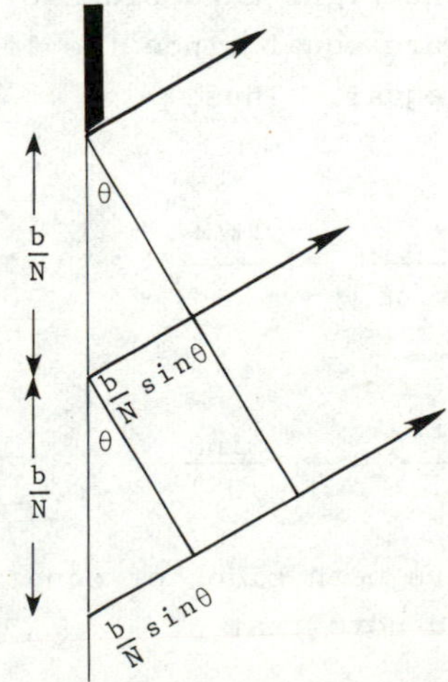

Fig. 14.5 Path difference b/N sinθ between adjacent strips in a slit of width b.

In the θ direction each secondary wavelet has a phase difference $\frac{2\pi}{\lambda}.\frac{b}{N}\sin\theta$ from its adjacent strip, and the extreme strips have a phase difference $\frac{2\pi}{\lambda}b\sin\theta$, which we shall call 2β: that is $2\beta = \frac{2\pi}{\lambda}b\sin\theta$. At some point to the right, a great distance from the slit, the electric field will be the sum of the disturbances from each of the secondary wavelets, and we can find its magnitude by adding the secondary wavelets vectorially. Figure 14.6 illustrates this with an equal phase difference between successive vectors.

E_θ is the resultant amplitude of the electric field at a point P in the image plane in the direction θ measured from the normal to the slit. Suppose now the number of strips increases until $N \to \infty$, then figure 14.6 smooths out as in figure 14.7 . Here, the circular arc ADB represents E_0, the amplitude of the field in the forward or zero direction, and the angle 2β is the phase difference between the first and last elementary strip secondary wavelets. To see why the arc length ADB represents the amplitude in the $\theta = 0$ direction, we can imagine what happens in figure 14.6 when $\theta = 0$. The phase difference between each secondary amplitude E' will be zero, so all the E' vectors lie along the horizontal and E_θ ($=E_0$) is equal to their algebraic sum.

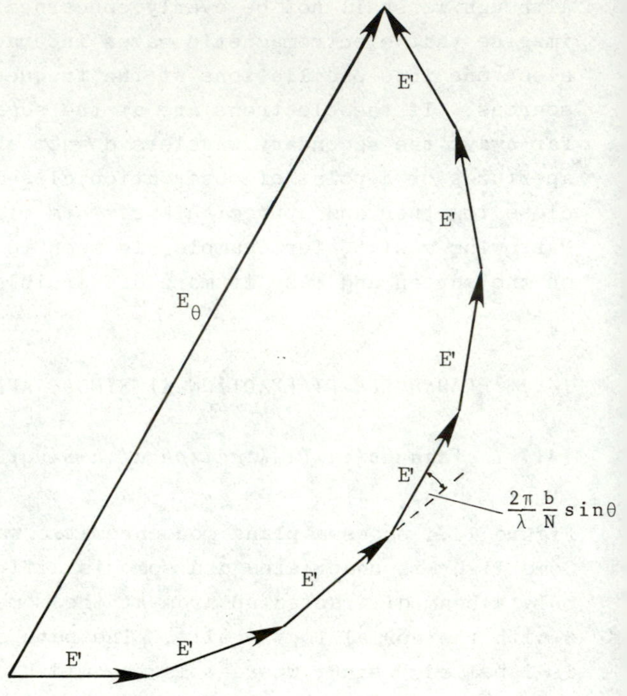

Fig. 14.6 N vectors with a phase difference $\frac{2\pi}{\lambda}\frac{b}{N}\sin\theta$ between each.

From figure 14.7,

$$OA = \frac{AC}{\sin\widehat{AOC}} = \frac{E_\theta/2}{\sin\beta}$$

and

$$OA = \frac{\text{arc } AD}{\beta} = \frac{E_0/2}{\beta}.$$

Clearly, the right hand sides of these equations are equal since the left hand sides are equal. Thus,

$$\frac{E_\theta/2}{\sin\beta} = \frac{E_0/2}{\beta}$$

or

$$\frac{E_\theta}{E_0} = \frac{\sin\beta}{\beta} \qquad (14.1).$$

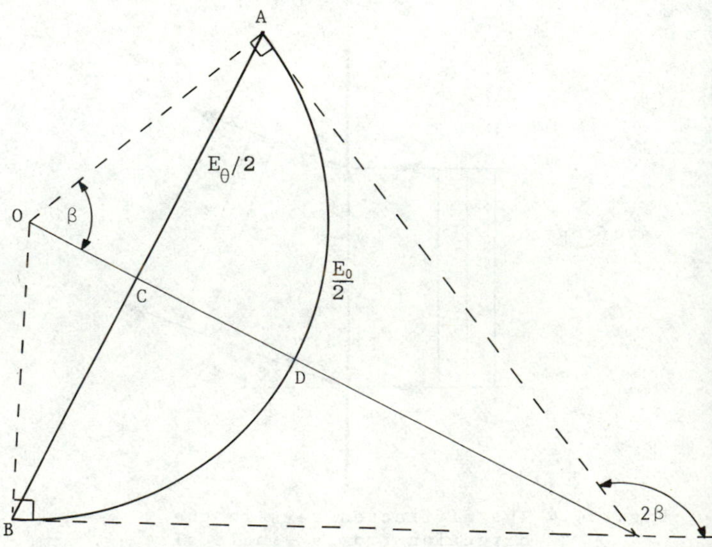

Fig. 14.7 The number of elementary strips increased to infinity.

If we square both sides of equation (14.1), and note that

$$\frac{I_\theta}{I_0} = \frac{E_\theta^2}{E_0^2}$$

we can write

$$\frac{I_\theta}{I_0} = \left(\frac{\sin\beta}{\beta}\right)^2 \qquad (14.2)$$

Equation (14.2) tells us that the intensity on the screen in the image plane in a direction θ is $(\sin\beta/\beta)^2$ of the intensity I_0 in the $\theta = 0$ direction. Plotting I_θ/I_0 against β gives figure 14.8 .

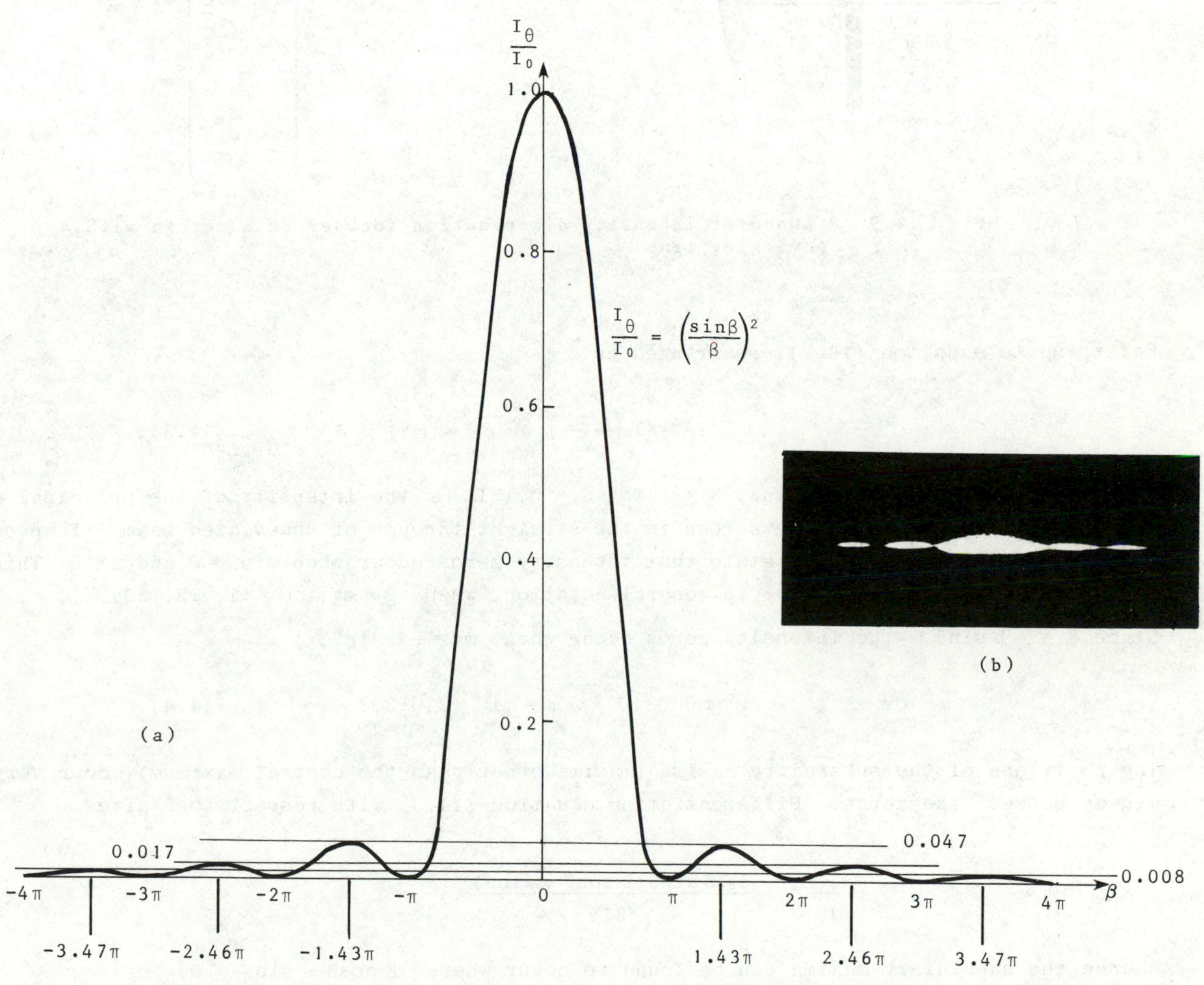

Fig. 14.8 (a) The intensity distribution in the Fraunhofer diffraction pattern of a single slit.
(b) A photograph of the single slit pattern using a helium-neon laser for the source.

Both parts of this figure illustrate how most of the light goes into the undeviated beam to form the central maximum intensity. Using a lens following the slit the Fraunhofer pattern can be imaged at a convenient distance, as shown in figure 14.9 .

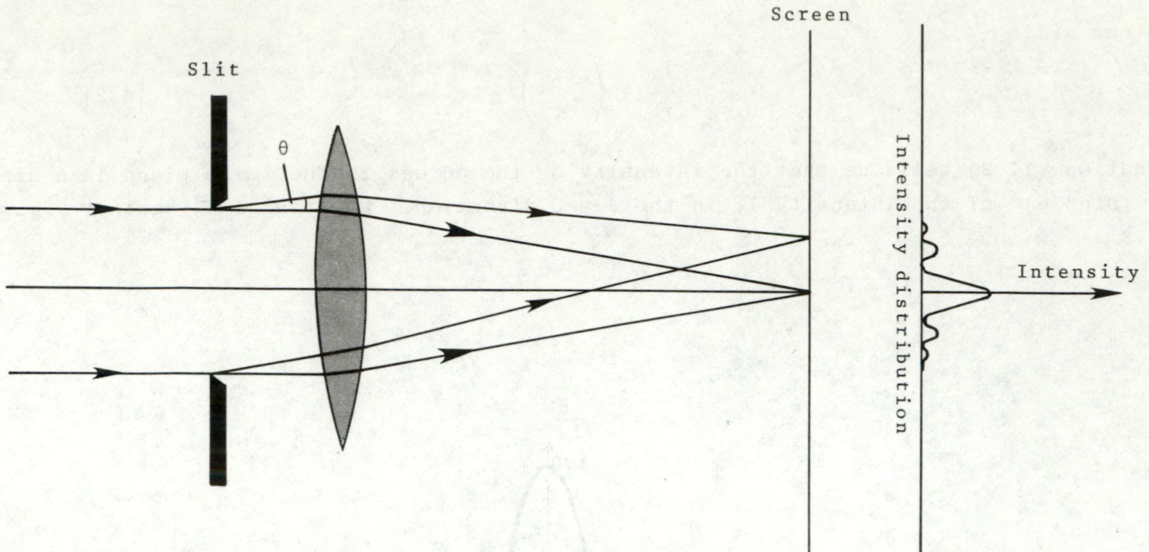

Fig. 14.9 Fraunhofer intensity distribution focused on a screen with a converging lens.

Referring to equation (14.2), rearranged as

$$I_\theta = I_0\left(\frac{\sin\beta}{\beta}\right)^2 \qquad (14.3),$$

$I_\theta = I_0$ when $\beta = 0$, since $(\sin\beta)/\beta = 1$ for $\beta = 0$. I_0 is the intensity of the principal maximum. That is, the intensity on the screen in the straight-through or undeviated beam. Inspection of equation (14.3) allows us to state that intensity zeros occur when $\sin\beta = 0$ and $\beta \neq 0$. This occurs when $\beta = \pm\pi, \pm2\pi, \pm3\pi, \ldots$, or in general notation, when $\beta = m\pi$, $m = \pm1, \pm2, \pm3, \ldots$. Since $\beta = \frac{\pi}{\lambda} b\sin\theta$, the intensity zeros occur when $m\pi = \frac{\pi}{\lambda} b\sin\theta$,

$$\text{or} \qquad b\sin\theta = m\lambda, \quad m = \pm1, \pm2, \pm3, \ldots \qquad (14.4).$$

The positions of the subsidiary maxima (maxima other than the central maximum) occur very nearly midway between the zeros. Differentiating equation (14.3) with respect to β gives

$$\frac{dI_\theta}{d\beta} = \frac{I_0 . 2\sin\beta(\beta\cos\beta - \sin\beta)}{\beta^3},$$

whence the subsidiary maxima can be found to occur where $\beta\cos\beta - \sin\beta = 0$, or

$$\tan\beta = \beta \qquad (14.5).$$

The first three solutions of this equation are $\beta = \pm1.4303\pi, \pm2.4590\pi, \pm3.4707\pi$. To a close approximation, we see that the subsidiary maxima occur very nearly midway between the zero minima at $\beta = \pm\frac{3}{2}\pi, \pm\frac{5}{2}\pi, \pm\frac{7}{2}\pi$. Using these values of β, the values of the subsidiary maxima are approximately $0.045I_0$, $0.016I_0$, and $0.008I_0$, respectively. It is clear that the values of the subsidiary maxima fall rapidly from the central value of I_0. You should compare these approximate values with the more precise values given in figure 14.8. They compare very favourably, indicating that the approximation for β is quite good.

A word of caution here. The Huygens-Fresnel principle does not take account of variations in the amplitude of each secondary wavelet which occur with changing θ. However, with small values of θ there is little need to worry.

Examination of equation (14.4) tells us that the distribution of light in the diffraction pattern clearly depends on the slit width b and the wavelength λ. For example, the second subsidiary maxima occur when $\beta = \pm\frac{5}{2}\pi$. But $\beta = \frac{\pi}{\lambda} b \sin\theta$, so $\sin\theta = \frac{\beta\lambda}{\pi b}$. It is clear from this equation that if b is reduced or λ is increased, the second order maxima must move further from the central maximum since θ will increase. This applies to the pattern in general, of course. If the slit is illuminated by white light each wavelength will give rise to its own diffraction pattern, and only at the centre will all the colours overlap to give white. The patterns will be wider for longer wavelengths, so the result will be a succession of colours leading off to red as θ increases.

Single slit Fraunhofer patterns can be demonstrated without resorting to specialised laboratory equipment. A slit can be formed by looking through a space between the prongs of a fork which may be rotated to reduce the projected width of the space. A sodium street lamp or a small incandescent bulb will serve as a source. Hold the fork close to the eye and look at the source with the eye focused at infinity. Widen and narrow the slit by rotating the fork a little and you will see the pattern contract and spread respectively.

14.1.2 Fraunhofer Diffraction at a Circular Aperture

Since most optical instruments, including the eye, have circular apertures we shall be especially interested in the diffraction pattern produced by circular stops. In a lens system which has been corrected for aberrations, the ultimate sharpness of the image is limited by diffraction which affects the reproduction of the object's intensity distribution in the image plane. Since diffraction causes an unwanted spreading of the light from an object, we must be aware when such spreading will degrade an image to the point of noticeably affecting its sharpness.

Plane wavefronts, which originate from a point source at infinity, of course, incident normally on a circular aperture produce a circular Fraunhofer diffraction pattern which is symmetrical about an axis through the centre of and perpendicular to the aperture, figure 14.10. A point source is imaged as a central circular disc of light surrounded by alternating dark and light annuli, as shown in figure 14.2. The central high intensity spot is called the *Airy disc* after George Biddell Airy (1801 - 1892), the British astronomer who first solved the analytical problem of describing the intensity distribution in this pattern in 1835. The radius of the Airy disc, r, is taken from the centre of the first dark ring to the centre of the central spot. Airy found the first intensity minimum subtended an angle θ, relative to the axis of symmetry, given by

$$\sin\theta \simeq \theta = 1.22\frac{\lambda}{d} \qquad (14.6),$$

for small θ, where d is the diameter of the aperture. From figure 14.10 the radius, r, of the Airy disc is

$$r = f'\theta = 1.22\frac{f'\lambda}{d} \qquad (14.7)$$

where f' is the focal length of the lens.

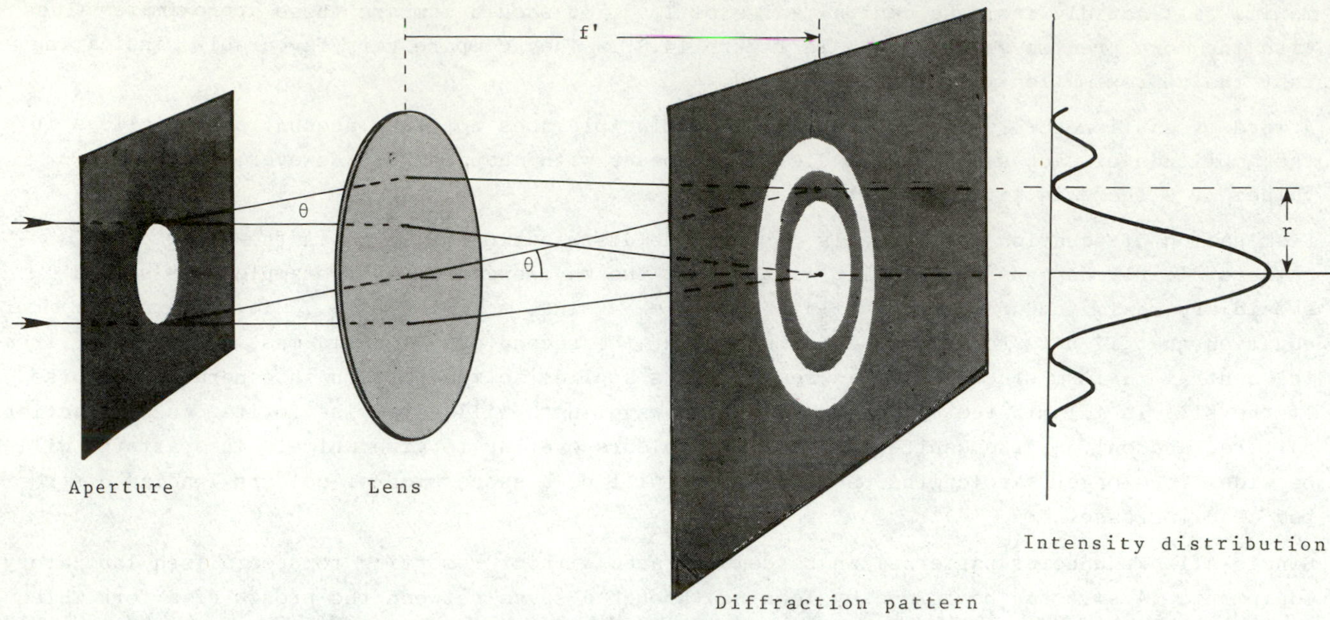

Fig. 14.10 Fraunhofer diffraction at a circular aperture.
$r = f'\tan\theta = f'\theta$ (if θ is small).

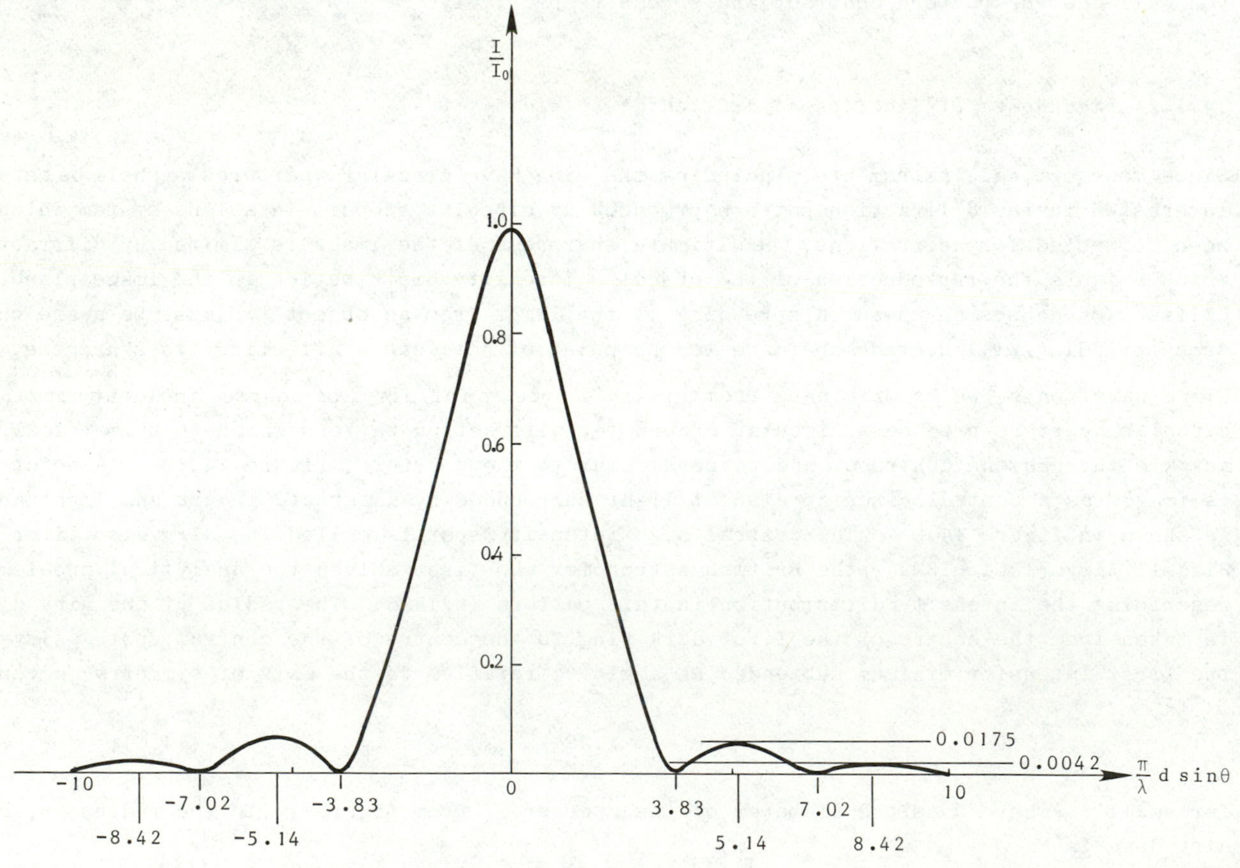

Fig. 14.11 The intensity distribution across the Airy pattern.

It is apparent from equation (14.7) that the size of the Airy disc is inversely proportional to the diameter of the aperture. Since a small Airy disc, rather than a large one, is a better representation of a point image, it is clearly desirable to have large aperture sizes in optical instruments where fine detail is to be resolved.

Figure 14.11 is a detailed plot of the intensity distribution across the centre of the Airy pattern. From this it is apparent that most of the light falls within the Airy disc. In fact, 84% of the light falls within the Airy disc and 91% within the area enclosed by the second dark ring.

14.1.3 Resolution in Imaging Systems

In any imaging system there will be a limit on the system's ability to distinguish detail in the image. In optical systems this can be stated quite simply by considering two point sources, emitting monochromatic light so there is only one Airy pattern for each source, and their images. Figure 14.12 shows two Airy discs overlapping.

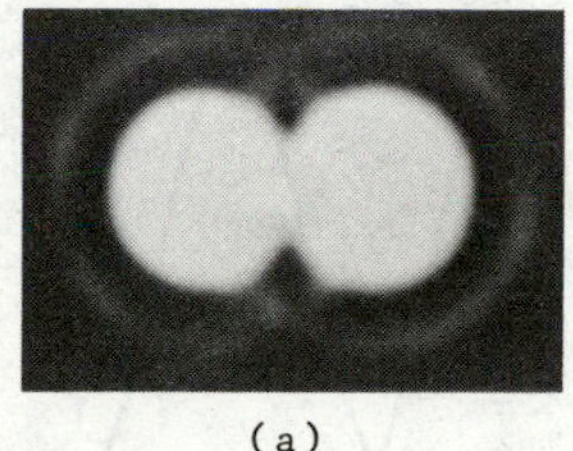

(a)

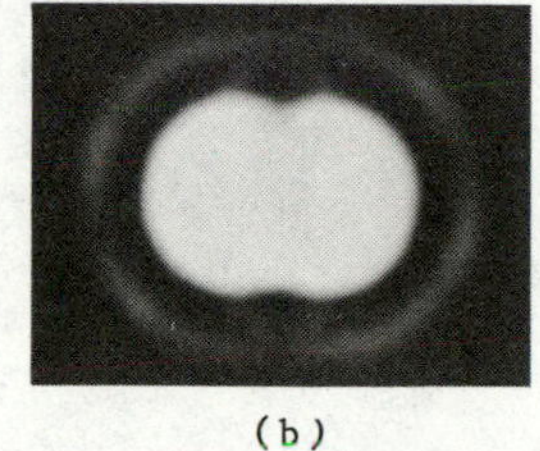

(b)

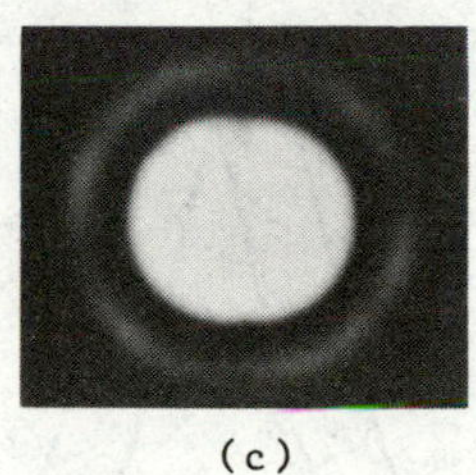

(c)

Fig. 14.12 Overlapping Airy discs from two point sources
(a) Easily resolved. (b) Just resolved. (c) Unresolved.

Figure 14.12(a) illustrates two easily distinguished images even though their Airy patterns are beginning to merge. The images are said to be resolved. As the object points move closer, so the Airy discs increasingly overlap until they are just resolved, figure 14.12(b). If the object points move closer still the two images mingle to such an extent as to make them unresolvable, figure 14.12(c). That is, the detector, which may be an eye or a photographic film, cannot pick out the separate images.

The size of the Airy disc is inevitably involved in resolution. Smaller Airy discs mean the objects and their images can approach each other more closely before overlapping becomes a problem. Lord Rayleigh (1842 - 1919) proposed a simple rule for the condition where two images of point sources can be regarded as 'just resolved'. The rule, which has become known as *Rayleigh's criterion*, states that two point images, say of two stars, will just be resolved if the centre of one Airy disc falls on the first minimum (dark ring) of the other. Figure 14.13 illustrates the geometry. This figure needs only a little further explanation. The upper part of the diagram shows two beams of parallel rays originating from two point objects at infinity. The intensity distributions across their Airy discs are indicated together with the overlapping Airy discs on the right. This is Rayleigh's criterion where the discs are just resolved. Parts (b) and (c) show the intensity distributions across the Airy discs when the angular separation of the objects is less and greater than in (a), respectively. In (b) the resultant intensity is the sum of the

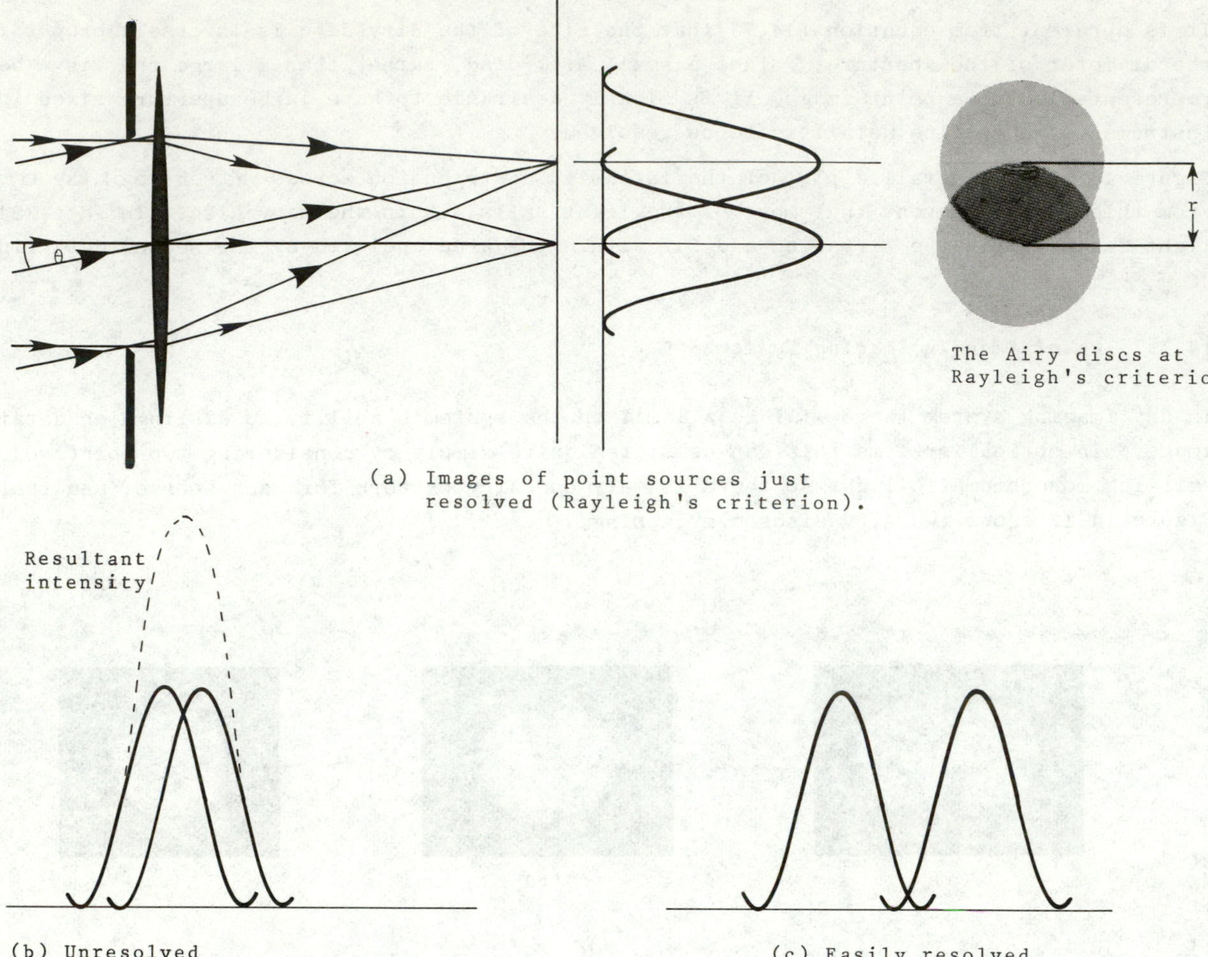

Fig. 14.13 The intensity distributions in the image plane of two point sources with an angular separation of θ. (a) θ is such that Rayleigh's criterion is satisfied. (b) θ smaller than in (a). (c) θ larger than in (a).

two separate intensities. It is possible to add these intensities because the two sources are incoherent. This is an application of equation (11.25) which implies that the resultant intensity is simply the addition of the individual intensities when the sources are incoherent. This was, of course, the principle used throughout the study of photometry. It is worthwhile emphasising that when the sources are coherent we must consider their relative phases. This is because the waves then interact in a sustained manner and produce observable interference effects, which is not the case with incoherent sources. With coherent sources then, we add the disturbances and square to obtain intensities, and this will produce a resultant intensity which is not simply the scalar sum of the individual intensities.

The Rayleigh criterion means that the two Airy disc centres are separated by the Airy disc radius r, and equation (14.6) therefore gives the angular separation of the object points when just resolved. At least, they are just resolved according to the Rayleigh criterion. The rule is arbitrary yet its simplicity makes it appealing. It is because of its simplicity that it has become well established, even though it is possible for a detector to sometimes do better than the Rayleigh criterion suggests. A more exact criterion will be mentioned in section 16.1.4 .

Examples

(1) Two stars have an angular separation of 44.73×10^{-7} radian. Find the minimum diameter of the telescope objective which can just resolve the stars in light of 550 nm.

Equation (14.6), rearranged, gives the minimum diameter as

$$d = 1.22\frac{\lambda}{\theta} = \frac{1.22 \times 550 \times 10^{-9}}{44.73 \times 10^{-7}} = 0.15\text{ m} \equiv 15\text{ cm}.$$

(2) Calculate the minimum angular subtense of two points which can just be resolved by an eye with a 6mm diameter pupil in light of 555 nm wavelength. Viewing the two stars in example (1) with the telescope there having a 150 cm focal length for the objective lens, find the focal length of the eyepiece.

The telescope must magnify the angular subtense of the stars to the eye's resolution limit which we calculate first from equation (14.6):

$$\theta = 1.22\frac{\lambda}{d} = 1.22 \times \frac{555 \times 10^{-9}}{6 \times 10^{-3}} = 1.1285 \times 10^{-4}\text{ radian.}$$

The angular magnification of the eyepiece must be

$$M = \frac{\beta}{\alpha} = \frac{-1.1285 \times 10^{-4}}{44.73 \times 10^{-7}} \simeq -25\,.$$

But $M = -\frac{f'_o}{f'_\varepsilon}$ where f'_o is the objective's focal length and f'_ε is the focal length of the eyepiece. Hence,

$$f'_\varepsilon = -\frac{f'_o}{M} = -\frac{150}{-25} = +6\text{ cm}.$$

14.1.3.1 The Resolving Power of a Microscope Objective

Resolving power is defined as $1/\theta$, where θ is given by $1.22\lambda/d$ in optical systems with a circular aperture stop. Clearly, resolving power is a measure of an instruments sensitivity. The smaller θ, the greater the resolving power, and the better the instrument for discriminating fine detail. However, although one talks about resolving power, it is common practice to quote the angular limit of resolution, θ.

In the case of microscopes the object is at a finite distance and it is useful to know the smallest linear size of detail which can be resolved by the microscope objective. With this in mind we refer to figure 14.14 where the separation of the object points is h.

B and Q are the object points, and $BQ = h$. B_0 and Q_0 are the positions of the zeroth or central maxima in the images of B and Q, respectively, whilst Q_1 is the position of the first zero in the Airy pattern in the image of Q. Since Q_1 occupies the same position as B_0, Rayleigh's criterion is satisfied and $\theta = 1.22\lambda/d$.

α is the half-angular diameter of the objective, seen from the point B. α was first defined by Ernst Abbe (1840 - 1905) whilst he was working in the Carl Zeiss microscope workshop.

From figure 14.14 it is evident that

$$\frac{h}{OB} \simeq \theta = 1.22\frac{\lambda}{d} .$$

If the object is not in air but in a medium of refractive index n, then the wavelength becomes λ/n, whence

$$h = \frac{1.22\,\lambda}{n}\cdot\frac{OB}{d} .$$

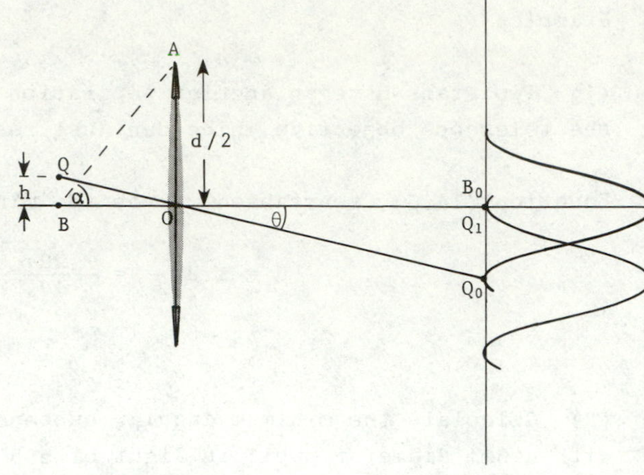

Fig. 14.14 Rayleigh's criterion in a microscope objective.

Now, $d/2 = AB\sin\alpha$, or $d = 2.AB\sin\alpha$,

so $$h = \frac{1.22\,\lambda}{n}\cdot\frac{OB}{2\,AB\sin\alpha} ,$$

or, equivalently,

$$h = \frac{\lambda}{2n\sin\alpha}\cdot\frac{1.22\,OB}{AB} .$$

The term 1.22 OB/AB is very nearly unity in most practical situations, so we can write the linear limit of resolution as

$$h = \frac{\lambda}{2n\sin\alpha} \qquad (14.8)$$

which agrees with a 'working rule' due to Abbe.

The product $n\sin\alpha$ was called by Abbe the *numerical aperture (N.A.)*. It ranges from about 0.07, for low power objectives, to about 1.4 to 1.6 for high power ones. The N.A. is normally printed on the barrel of the microscope. For a given microscope the N.A. will be fixed, of course, and we see that the linear limit of resolution can only be improved upon by using shorter wavelengths Ultraviolet light is now in use for this reason but the final image must be photographed, of course.

14.2 DIFFRACTION BY SMALL PARTICLES

Imagine an opaque sheet from which a small aperture is cut, figure 14.15 . Σ_1 and Σ_2 are said to be *complementary apertures*. Mounted concentrically and in the same plane the combination is obviously opaque. Suppose now that monochromatic plane wavefronts are incident normally on Σ_1 and the diffracted light is imaged on a screen. In some direction θ to the normal let the amplitude of the electric vector be E_1. Next, if we replace Σ_1 with Σ_2, let E_2 be the corresponding electric vector in the same direction. With both 'apertures' present the amplitude in the θ direction is evidently 0. Thus, we can write the vector addition

$$E_1 + E_2 = 0 , \quad \text{or} \quad E_1 = -E_2 .$$

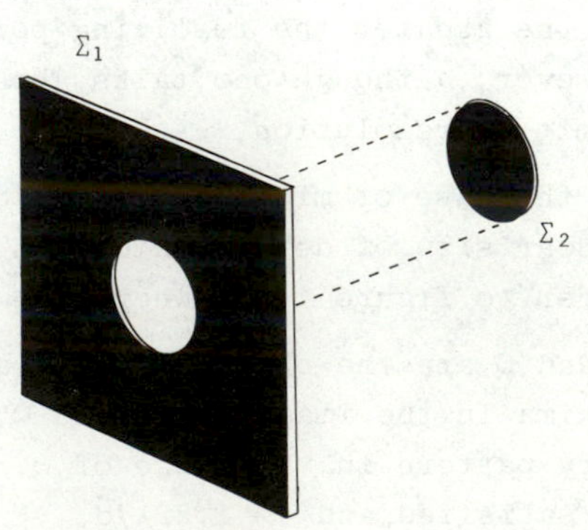

Fig. 14.15 Complementary apertures.

Because intensity ∝ amplitude squared, squaring both sides gives $I_1 = I_2$ which is known as *Babinet's Principle*. It means that the diffraction patterns for Σ_1 and Σ_2 will be identical.

Figure 14.16 is a Fraunhofer diffraction pattern which is produced either with a random distribution of small holes in an opaque screen or with the same distribution of small opaque dots on a transparent sheet. In both cases the holes and dots are similar in size and circular in shape.

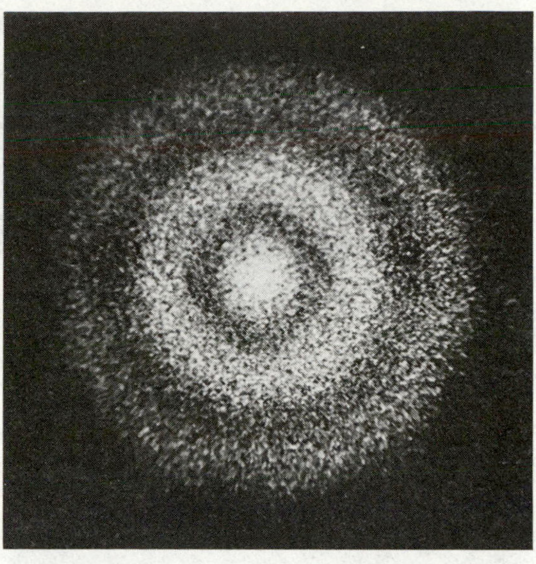

Fig. 14.16 Fraunhofer diffraction pattern formed by a random array of circular apertures.

An identical pattern, commonly called a halo, can be demonstrated by sprinkling lycopodium powder on a microscope slide and holding it close to the eye. Looking at a very small source results in a diffraction pattern focused on the retina. Estimation of the angular radius of the first dark ring allows an approximate calculation of the particle diameter. For example, in such an experiment the angular diameter of the first dark ring was roughly estimated to be 2.7^0 with sodium light illumination.
The diameter of the lycopodium particles is therefore approximately

$$d = \frac{1.22\,\lambda}{\sin\theta} = \frac{1.22 \times 589.3 \times 10^{-9}}{\sin \frac{2.7^0}{2}} \simeq 30 \times 10^{-6} \text{ m.}$$

Haloes produced with a small white source are coloured, of course. They are occasionally seen around the full moon when it is partially obscured by a faint haze of water droplets (cloud). Again, the size of the drops may be estimated using a mean value for λ of 550 nm. Similar haloes are seen when the corneal epithelium becomes waterlogged (oedema) in a condition called glaucoma (raised intraocular pressure). The swelling causes numerous minute raised 'droplets' which are, in fact, waterlogged epithelium cells on the outer surface of the eye.

14.3 FRAUNHOFER DIFFRACTION AT TWO SLITS

Figure 14.9 suggests that the diffraction pattern is centred on a line perpendicular to and passing through the centre of the slit. This is not the case, however, since all the rays parallel to the lens' principal or optical axis converge on the second focal point. Thus, displacing the slit parallel to the lens has no effect on the position of the diffraction pattern; it still centres on the optical axis of the lens. Suppose now we consider two identical parallel slits of width b and a centre-to-centre separation a. Each aperture by itself would generate the same diffraction pattern on the screen in the focal plane of the lens. Together there will be interference effects because the two slits are illuminated by the same primary wave which gives rise to coherent secondary wavelets. The interference pattern seen on the screen in a given direction θ is determined by the OPD between the secondary wavelets arising at symmetrical positions in the slits. The overall effect is a two-slit interference pattern modulated by a single slit diffraction pattern, figure 14.17 . This statement is illustrated in figure 14.21; the two slit interference pattern intensity distribution is shown enclosed by a single slit diffraction pattern intensity distribution envelope. Figure 14.21 will be discussed in more detail later.

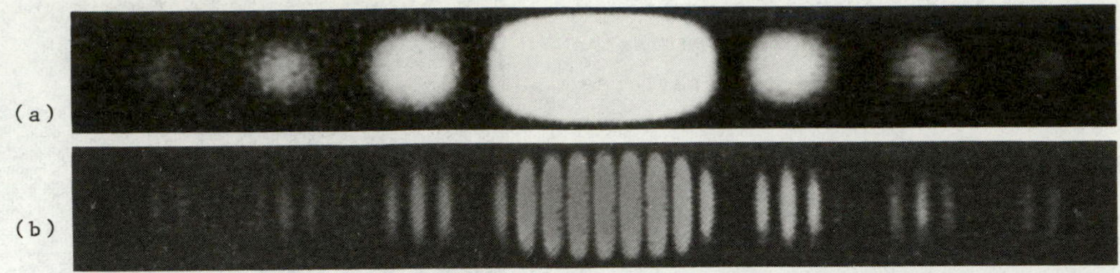

Fig. 14.17 (a) Single slit and (b) two slit Fraunhofer diffraction patterns. In (b) the slit separation is 5 × slit width.

Clearly, minima occur in the two-slit pattern in the same place as in the single slit pattern. The additional minima are due to the two-slit interference effect. We must now quantify these observations and find an expression for the intensity variation across the screen.

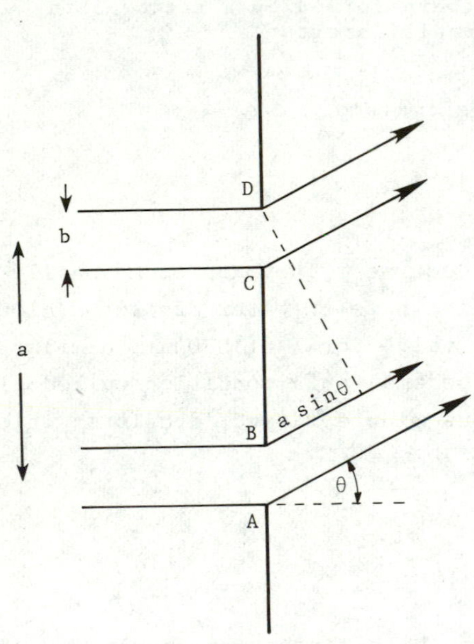

Fig. 14.18 Fraunhofer diffraction at two slits.

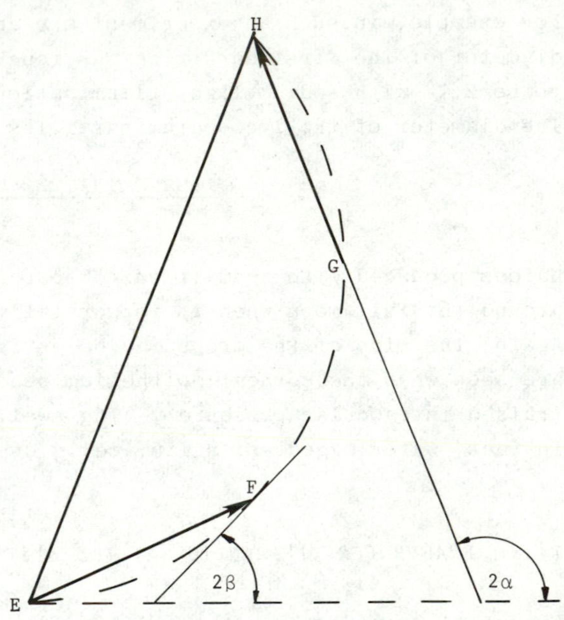

Fig. 14.19 Vector diagram for two slits.

Consider figure 14.18 which shows parallel light rays incident normally on two slits of equal width b and a centre-to-centre separation a. Initially, suppose we have a single slit of width AD. As in section 14.1.1 , the resultant amplitude in the θ direction would be EH in figure 14.19 . If we now occlude the section BC of the single slit we are left with the two slits AB and CD. The corresponding vector diagram now has the component resultants of the two slits and each resultant vector is EF and GH. The path difference between these resultant waves is a sinθ, shown in figure 14.18 . The resultant amplitude in the θ direction due to both slits is EH, the vector addition of EF and GH in figure 14.20 .

We know that the lengths EF and GH are equal, since the slits are identical and we assume uniform illumination. From equation (14.1), we have

$$EF = GH = E_0 \frac{\sin\beta}{\beta}, \quad \text{where } \beta = \frac{\pi}{\lambda} b \sin\theta .$$

Solving the triangle in figure 14.20 for EH, and noting that $2\alpha = \frac{2\pi}{\lambda} a \sin\theta$ (the phase difference between EF and GH), then using the cosine rule

$$EH^2 = E_\theta^2 = EF^2 + GH^2 - 2.EF.GH.\cos\widehat{EFH}$$

$$= \left(E_0 \frac{\sin\beta}{\beta}\right)^2 + \left(E_0 \frac{\sin\beta}{\beta}\right)^2 - 2E_0 \frac{\sin\beta}{\beta} \cdot E_0 \frac{\sin\beta}{\beta} (\cos 180^0 - 2\alpha)$$

$$= 2E_0^2 \left(\frac{\sin\beta}{\beta}\right)^2 (1 + \cos 2\alpha)$$

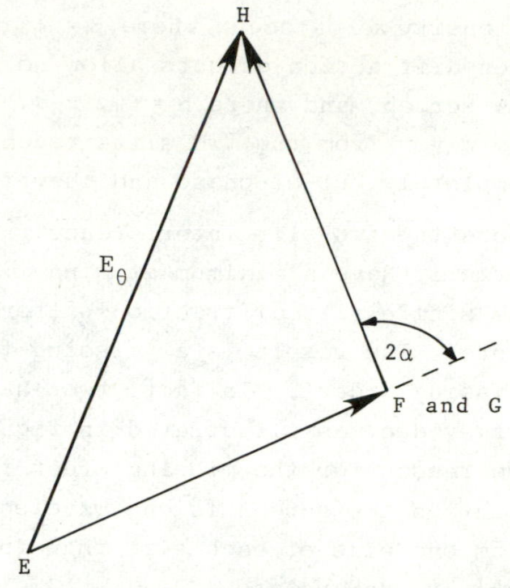

Fig. 14.20 Vector addition of the two separate slit amplitudes in the θ direction.

or $E_\theta^2 = 4E_0^2 \left(\frac{\sin\beta}{\beta}\right)^2 \cos^2\alpha$, since $1 + \cos 2\alpha = 2\cos^2\alpha$.

So, $\frac{E_\theta^2}{E_0^2} = 4\left(\frac{\sin\beta}{\beta}\right)^2 \cos^2\alpha$, whence

$$\frac{I_\theta}{I_0} = 4\left(\frac{\sin\beta}{\beta}\right)^2 \cos^2\alpha \qquad (14.9).$$

In the undeviated $(\theta = 0)$ direction, $\beta = \alpha = 0$ and $I_\theta = 4I_0$, where I_0 is the contribution from either slit alone in the $\theta = 0$ direction. The factor 4 occurs because the amplitude of the field is twice what it would be with one slit alone, so the intensity is $2^2 = 4$ times what it would be with one slit only.

If b becomes very small then $\frac{\sin\beta}{\beta} \simeq 1$, and the expression reduces to equation (13.10) which is the intensity distribution from two parallel line sources (Young's two-slit interference experiment). Alternatively, if a reduces until the two slits coalesce into one, $I_\theta = 4I_0\left(\frac{\sin\beta}{\beta}\right)^2$, which is equation (14.3) where the source strength has doubled due to a slit twice as wide.
Transposing equation (14.9) to

$$I_\theta = 4 I_0 \left(\frac{\sin\beta}{\beta}\right)^2 \cos^2\alpha \qquad (14.10)$$

we can imagine it as a $\cos^2\alpha$ two-slit interference distribution, as figure 13.4, modulated by a single slit Fraunhofer diffraction term $(\sin\beta/\beta)^2$, figure 14.21 .

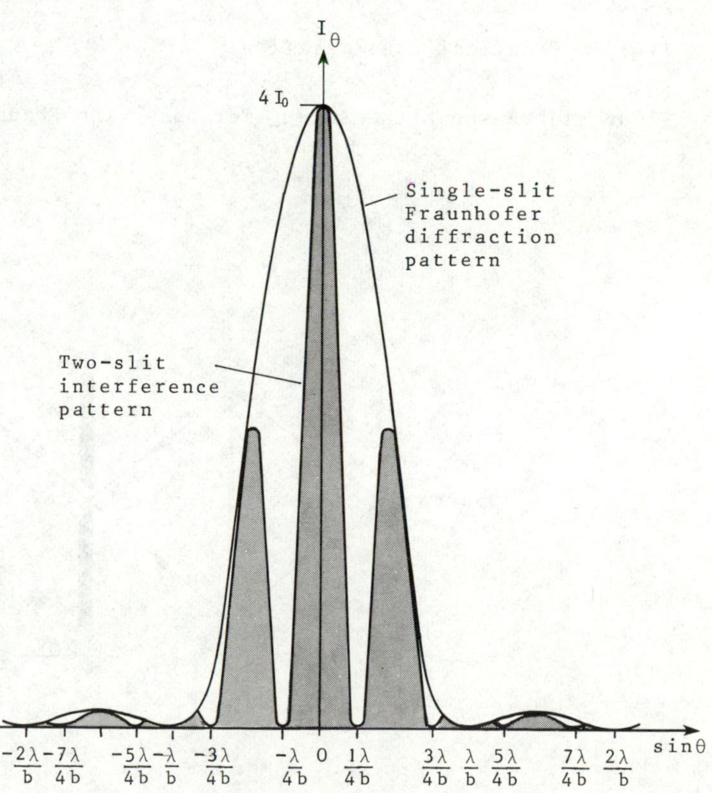

Fig. 14.21 A two-slit Fraunhofer pattern intensity distribution: a = 2b.

Intensity zeros occur where $\beta = \pm\pi, \pm2\pi, \pm3\pi, \ldots$, when diffraction effects allow no light to reach the screen, and where $\alpha = \pm\frac{\pi}{2}, \pm\frac{3}{2}\pi, \pm\frac{5}{2}\pi, \ldots$, when the waves from the two slits reach the screen completely out-of-phase and therefore cancel.

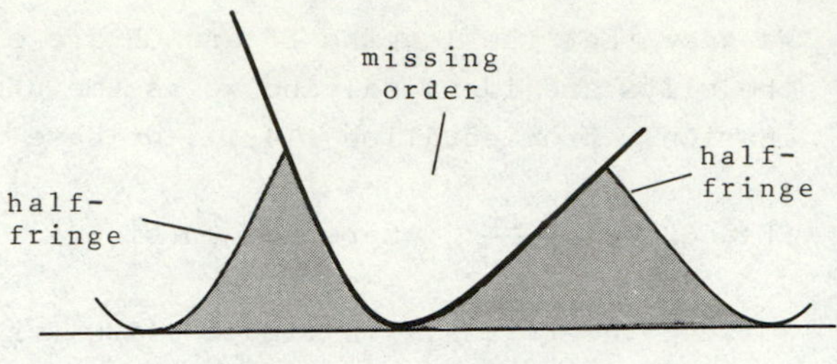

Fig. 14.22 Half-fringes at a single-slit diffraction envelope minimum.

Where the two-slit interference pattern would normally have a maximum at $\sin\theta \simeq \pm\frac{\lambda}{b}$ $\pm\frac{2\lambda}{b}, \ldots$, the single slit diffraction pattern has a zero there. The result is a 'missing fringe' or a 'missing order'. In fact, two 'half-fringes' are evident as illustrated in figure 14.22 . The reason for the missing order is that this point on the screen is one wavelength further from one edge of each slit than from the other edge. Thus, the contributions from the slits is zero at such points.

Figure 14.21 is a particular pattern occurring when $a = 2b$, or $\alpha = 2\beta$ equivalently. If $a = mb$ there will be 2m bright fringes within the central single slit diffraction peak, if one adds up half-fringes as well. Alternatively, since the half-fringes are often not discernible in many cases, there will be $(2\frac{a}{b} - 1)$ full bright fringes beneath the central diffraction peak. If you refer to figure 14.17(b) you can count 9 bright fringes occupying the same space occupied by the central fringe in (a). This tells us that in this particular case $\frac{a}{b} = 5$, or, the distance from slit-centre to slit-centre is five times the slit width.

14.3.1 *Practical Considerations*

Figure 14.23 shows the set-up for observing Fraunhofer diffraction at a double-slit.

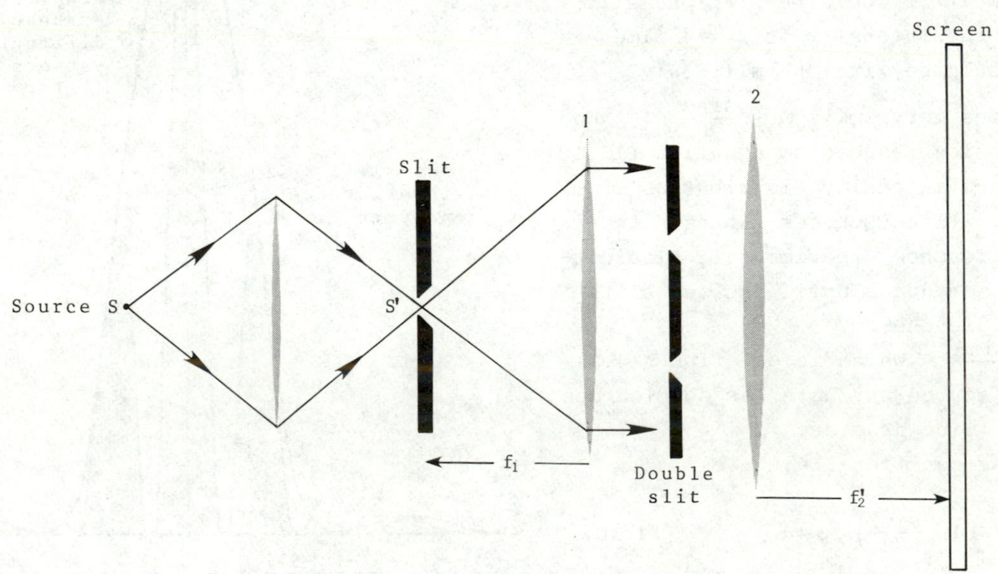

Fig. 14.23 Double-slit Fraunhofer pattern set-up: note that S' will be point-like for a small source S.

The waves arriving at the double-slit must be coherent and this implies a negligible width of the source slit, S'. If the width is not negligible different sets of waves will arrive at the double-slit, each producing its own set of fringes on the screen and displaced laterally with respect to each other. We can deduce a maximum slit width from figure 14.24 .

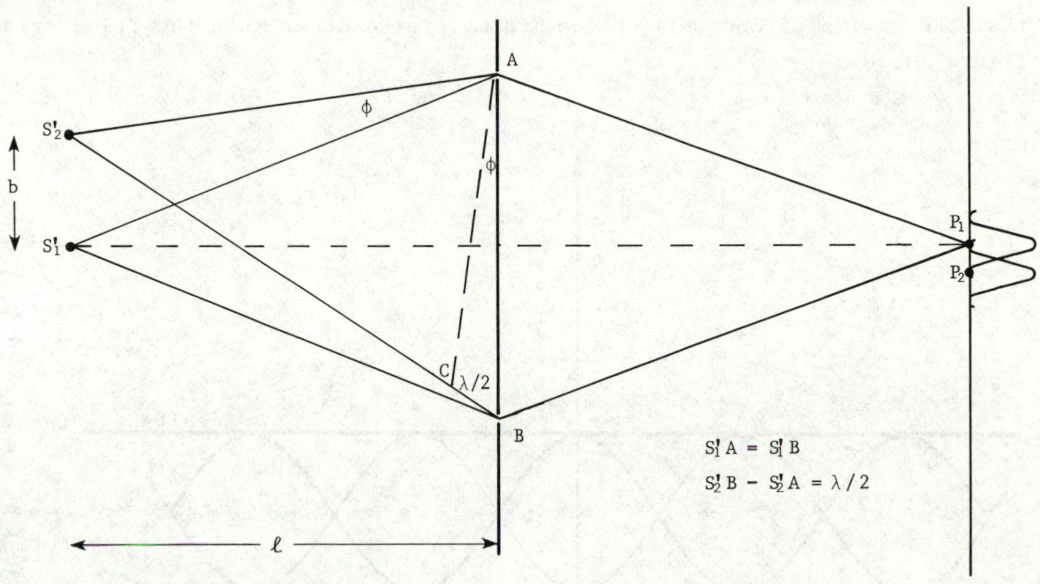

Fig. 14.24 Imagine AB rotates through ϕ into AC whilst AS'_1 rotates through ϕ into AS'_2 such that $BC = \lambda/2$.

Suppose two incoherent point sources S'_1 and S'_2 are a distance $S'_1S'_2 = b$ apart. They will give rise to two double-slit interference patterns on the screen, where P_1 and P_2 indicate their central maxima. If S'_1 and S'_2 are sufficiently close, then so will be P_1 and P_2 as to produce recognisable $\cos^2\alpha$ fringes, figure 14.25 .

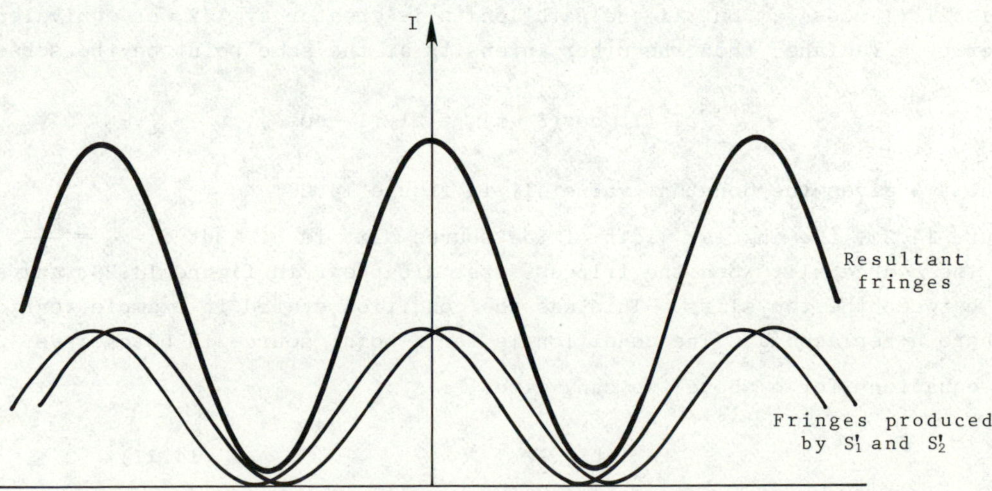

Fig. 14.25 $\cos^2\alpha$ fringes produced by incoherentpoint sources S'_1 and S'_2 : diffraction modulation effects ignored.

Although the resultant's minima are not quite zero the visibility is still appreciable. By the way, the definition of visibility is a purely physical quantity. It does not necessarily mean that the fringes are detectable by the eye even if the visibility is unity, for it only needs the maximum intensity to be below the viewing eye's sensitivity threshold for no light at all to be detected! Returning to figure 14.25 , when the individual fringe patterns on the screen are such that the maxima of one fall on the minima of the other then the fringe visibility will be zero, figure 14.26 .

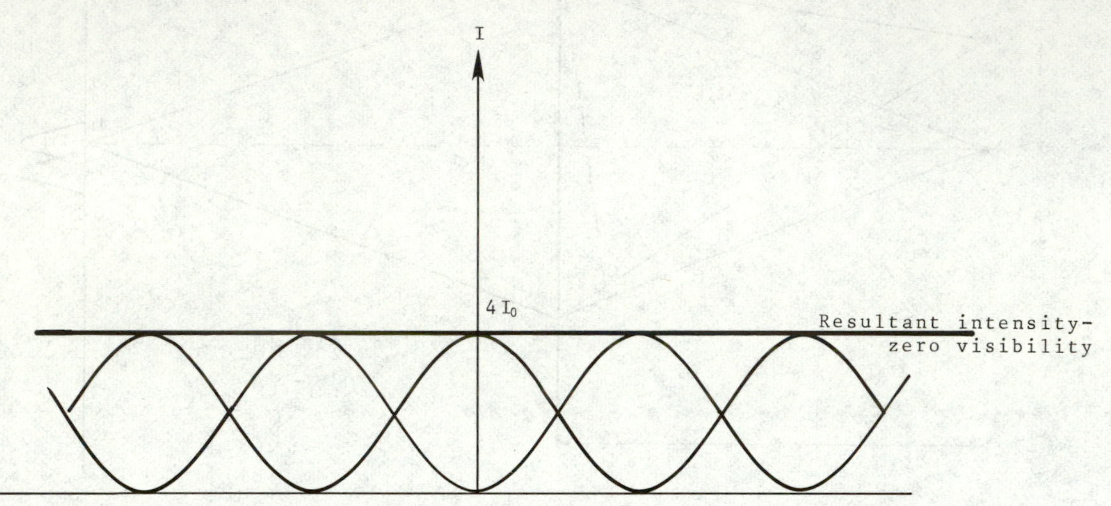

Fig. 14.26 Zero visibility of the two sets of fringes produced by the line sources S_1' and S_2' .

The condition of zero visibility of the resultant intensity pattern will first occur when the distance CB in figure 14.24 is $\lambda/2$; that is, the central maximum formed by S_2' is at P_2 and a minimum is formed at P_1 , and so on. Recall from section 13.3.3 that the intensity of either pattern is $2I_0(1+\cos\delta)$. But if one path length is greater by $\lambda/2$, or equivalently, the phase is greater by π radians, then the other intensity at the same point on the screen will be

$$2I_0(1+\cos(\delta+\pi)) = 2I_0(1-\cos\delta).$$

Adding the two gives the constant value $4I_0$ in figure 14.26 .

From figure 14.24, the angular width of the source slit is ϕ. But $\phi = \frac{b}{\ell} = \frac{\lambda/2}{a}$, where b is the width of the source slit when the fringes first disappear in figure 14.26, and a = AB is the distance between the two slits. This was the condition stated in example four of section 13.7.2 where h replaced a. The condition is for a point source in both cases.

From the equations for ϕ above, we can write

$$b = \frac{\ell\lambda}{2a} \qquad (14.11)$$

which is the condition when the fringes first disappear. They will become visible again as b increases, disappearing again when CB $= \frac{3}{2}\pi$, $\frac{5}{2}\pi$,.... However, they will never be as distinct as when $b \ll \frac{\ell\lambda}{2a}$.

In figure 14.23 $f_1' \simeq \ell$, so the condition for high visibility fringes is $b \ll \frac{f_1' \lambda}{2a}$. A good practical rule is that the width of the source slit should not exceed

$$b = \frac{f_1' \lambda}{4a} \qquad (14.12).$$

Example

In the arrangement of figure 14.23 the source slit is 0.05 mm wide and each lens has a focal length of 50 cm. For what distance between the two slits do the fringes first disappear if the source wavelength is 589.3 nm?

From equation (14.11)

$$a = \frac{\ell\lambda}{2b} = \frac{f_1'\lambda}{2b} = \frac{50 \times 10^{-2} \times 589.3 \times 10^{-9}}{2 \times 0.05 \times 10^{-3}} = 2.95 \times 10^{-3}\,\text{m}$$

$$\equiv 2.95\ \text{mm}.$$

Compare this result, where the source slit has an approximate point secondary source focused in in it, with problem (2) of section 13.7.2 where the source slit was uniformly illuminated. Here we have zero visibility and there we had maximum visibility, both with the same source slit width!

14.4 DIFFRACTION AT MANY SLITS

Once again, as we did with one and two slits, we wish to determine the intensity distribution on a screen. The procedure is similar to that used for two slits. The width of each slit is b and the centre-to-centre distance between the slits is again a. Suppose we commence with three slits, for simplicity. In a direction θ with the normal to the plane of the slits the amplitude contribution due to each slit will be $E_0 \frac{\sin\beta}{\beta}$, as given by equation (14.1). Recall once again that $\beta = \frac{\pi}{\lambda} b \sin\theta$. We are assuming here that the monochromatic light is incident normally on the three slits, so any path differences arise to the right of the slits. There will be a path difference of $a \sin\theta$ between the contributions from adjacent slits, exactly as in figure 14.18. Again, we have $\alpha = \frac{\pi}{\lambda} a \sin\theta$.

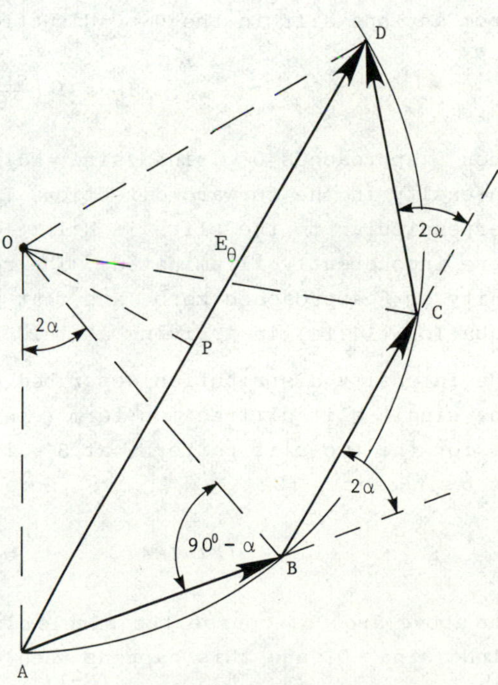

Fig. 14.27 Vector addition of amplitude contributions from 3 slits: $AB = BC = CD = E_0 \sin\beta/\beta$.

The resultant amplitude in the θ direction, E_θ, is the vector sum of each contribution. Figure 14.27 shows the three slit contributions added vectorially with a phase difference 2α between each. O is the centre of the regular polygon of which AB, BC, and CD are sides. Thus, $OA = OB = OC = OD$ and each side subtends an angle 2α at O. Hence, $\widehat{AOD} = 3 \times 2\alpha = 6\alpha$.

If OP is a perpendicular construction to AD then $\widehat{AOP} = \frac{1}{2}\widehat{AOD} = 3\alpha$, and $E_\theta = AD = 2\,AP = 2\,OA\sin 3\alpha$.

But $\dfrac{OA}{\sin\widehat{OBA}} = \dfrac{AB}{\sin 2\alpha}$, having used the sine rule, and this transposes to give

$$OA = AB.\frac{\sin\widehat{OBA}}{\sin 2\alpha} = AB.\frac{\sin(90^0-\alpha)}{2\sin\alpha\cos\alpha} = AB.\frac{\cos\alpha}{2\sin\alpha\cos\alpha}$$

$$= E_0\frac{\sin\beta}{\beta}.\frac{1}{2\sin\alpha}\quad,\quad \text{since } AB = E_0\frac{\sin\beta}{\beta}.$$

Substituting for OA in the expression for E_θ,

$$E_\theta = 2.OA\sin 3\alpha = 2.E_0\frac{\sin\beta}{\beta}.\frac{1}{2\sin\alpha}.\sin 3\alpha$$

$$\text{or}\qquad E_\theta = E_0\frac{\sin\beta}{\beta}.\frac{\sin 3\alpha}{\sin\alpha} \qquad (14.12).$$

Suppose now there were N slits instead of 3, then $\widehat{AOP}$ would be $N\alpha$ instead of 3α, and the equation for E_θ would be

$$E_\theta = E_0\frac{\sin\beta}{\beta}.\frac{\sin N\alpha}{\sin\alpha} \qquad (14.13).$$

Squaring both sides gives the intensity distribution in terms of I_0, where I_0 is the contribution from any one slit in the $\theta = 0$ direction: thus

$$I_\theta = I_0\left(\frac{\sin\beta}{\beta}\right)^2\left(\frac{\sin N\alpha}{\sin\alpha}\right)^2 \qquad (14.14).$$

When θ approaches 0, $\sin N\alpha/\sin\alpha \simeq N\alpha/\alpha$, and in the limit equals N when $\theta = 0$. Therefore the intensity in the forward direction, $I_{\theta=0}$, at a point P on the screen on an axis of symmetry perpendicular to the slits is $N^2 I_0$. Compare this with the intensity on the axis if all the slits were incoherently illuminated. The result would then be NI_0. Remember that $\sin\beta/\beta$ approaches unity as β approaches zero, and that $\beta = 0$ when $\alpha = 0$. These facts were taken into account in equation (14.14) in arriving at this result.

The intensity distribution described by equation (14.14) is an interference pattern modulated by the single slit diffraction term $(\sin\beta/\beta)^2$. Minima, of zero intensity, occur where $\sin\beta/\beta = 0$, as for the two-slit pattern, at $\beta = \pm\pi, \pm 2\pi, \pm 3\pi, \ldots$, or, equivalently, writing $\beta = \frac{\pi}{\lambda} b\sin\theta$, at $b\sin\theta = \pm\lambda, \pm 2\lambda, \pm 3\lambda, \ldots$. More generally, minima occur at

$$\textit{(MINIMA)}\qquad b\sin\theta = m\lambda, \qquad m = \pm 1, \pm 2, \pm 3, \ldots \qquad (14.15).$$

The above are, of course, the single slit diffraction minima. Zero minima also occur where $\sin N\alpha/\sin\alpha = 0$, and this happens when $N\alpha = \pm\pi, \pm 2\pi, \pm 3\pi, \ldots$, which after dividing through by N gives $\alpha = \pm\frac{\pi}{N}, \pm\frac{2\pi}{N}, \pm\frac{3\pi}{N}, \ldots, \pm\frac{(N-1)\pi}{N}, \pm\frac{(N+1)\pi}{N}, \ldots$ (14.16).

Putting $\frac{\pi}{\lambda} a\sin\theta$ for α gives the condition for minima

$$\textit{(MINIMA)}\qquad a\sin\theta = \pm\frac{\lambda}{N}, \pm\frac{2\lambda}{N}, \pm\frac{3\lambda}{N}, \ldots, \pm\frac{(N-1)\lambda}{N}, \pm\frac{(N+1)\lambda}{N}, \ldots \qquad (14.17).$$

Principal maxima can be shown to occur when $\sin N\alpha/\sin\alpha = N$; that is, when

$$\textit{(PRINCIPAL MAXIMA)} \qquad \alpha = 0,\ \pm\pi,\ \pm2\pi, \ldots \qquad (14.18),$$

or, equivalently, using $\alpha = \frac{\pi}{\lambda} a \sin\theta$, when

$$\textit{(PRINCIPAL MAXIMA)} \qquad a \sin\theta_m = m\lambda\ , \quad m = 0,\ \pm1,\ \pm2, \ldots \qquad (14.19).$$

This condition applies for all cases where $N \geq 2$. The subscript m on θ in equation (14.19) is called the *order*. Hence, θ_3 is the angular position of the third order principal maximum, θ being measured as usual from a central perpendicular to the array of slits.

Between the values of α for principal maxima, equation (14.18), subsidiary or secondary maxima occur because the term $(\sin N\alpha/\sin\alpha)^2$ has a rapidly varying numerator compared to its denominator. The secondary maxima occur approximately where $\sin N\alpha$ has its maximum value at

$$\textit{(SECONDARY MAXIMA)} \qquad \alpha = \pm\frac{3\pi}{2N},\ \pm\frac{5\pi}{2N},\ \pm\frac{7\pi}{2N}, \ldots \qquad (14.20).$$

Between successive principal maxima there will be N-1 minima and N-2 secondary maxima. Figure 14.28(a) shows the single slit diffraction pattern and figure 14.28(d) is the pattern produced by 4 slits of the same width. Notice there are (4 - 2) = 2 secondary maxima and (4 - 1) = 3 minima clearly visible between principal maxima under the central diffraction peak.

Inspection of the patterns in figures 14.28 (c) and (d) secondary maxima and minima in accord with the statement above.

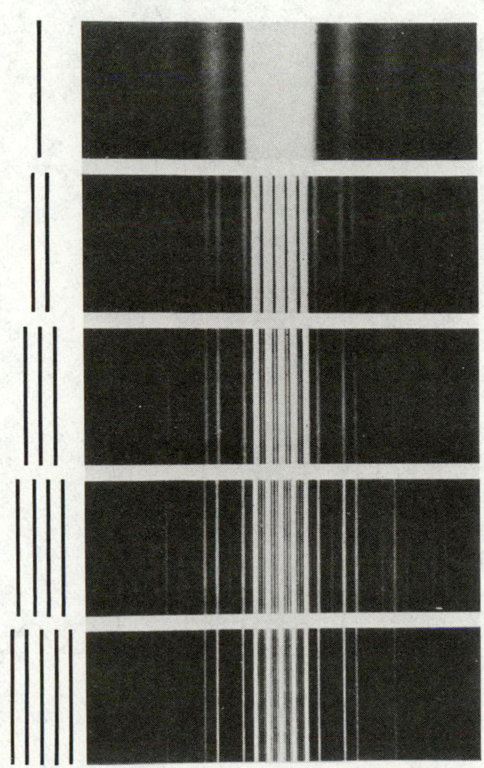

Fig. 14.28 .

(a) Fraunhofer diffraction pattern of a single slit.
(b), (c), (d), and (e) are interference patterns of 2, 3, 4, and 5 slits, respectively, modulated by the single slit diffraction term in equation (14.14). The slit systems are indicated at the left.

(Hecht/Zajac, OPTICS, © 1974, Addison-Wesley Publishing Company, Inc., Chapter 10, page 345, figure 10.20, "Diffraction Patterns for Slit Systems". Reprinted by permission.

14.4.1 *Width and Spacing of the Principal Maxima*

Considering each principal maximum to extend between its adjacent zeros, the width of each maximum is, from equation (14.16),

$$\Delta\alpha = \frac{(N+1)\pi}{N} - \frac{(N-1)\pi}{N} = \frac{2\pi}{n} \qquad (14.21).$$

Note that equation (14.18) states that a principal maximum occurs at $\alpha = \pi$, and this is midway between $\alpha = \frac{(N-1)\pi}{N}$ and $\alpha = \frac{(N+1)\pi}{N}$ at which minima occur in equation (14.16). Differentiating the equation $\alpha = \frac{\pi}{\lambda} a \sin\theta$ gives $\Delta\alpha = \frac{\pi}{\lambda} a \cos\theta . \Delta\theta$, and substituting for $\Delta\alpha$ in equation (14.21)

$$\frac{\pi}{\lambda} a \cos\theta . \Delta\theta = \frac{2\pi}{N} \quad \text{or} \quad \Delta\theta = \frac{2\lambda}{Na\cos\theta_m} \qquad (14.22),$$

for the mth order. If θ is small, and therefore $\cos\theta \simeq 1$, this becomes $\Delta\theta = \frac{2\lambda}{Na}$. Incidentally, notice that Na is the width of the grating of slits, which makes $\Delta\theta \propto 1/\text{width of grating}$.

$\Delta\theta$ is the *angular width of a principal maximum* measured at the slits. Notice how a and N affect the width of the principal maxima. As N increases the principal maxima become narrower although their angular spacing, from equation (14.19), remains λ/a for small θ. Figure 14.29 shows the case for $N = 4$.

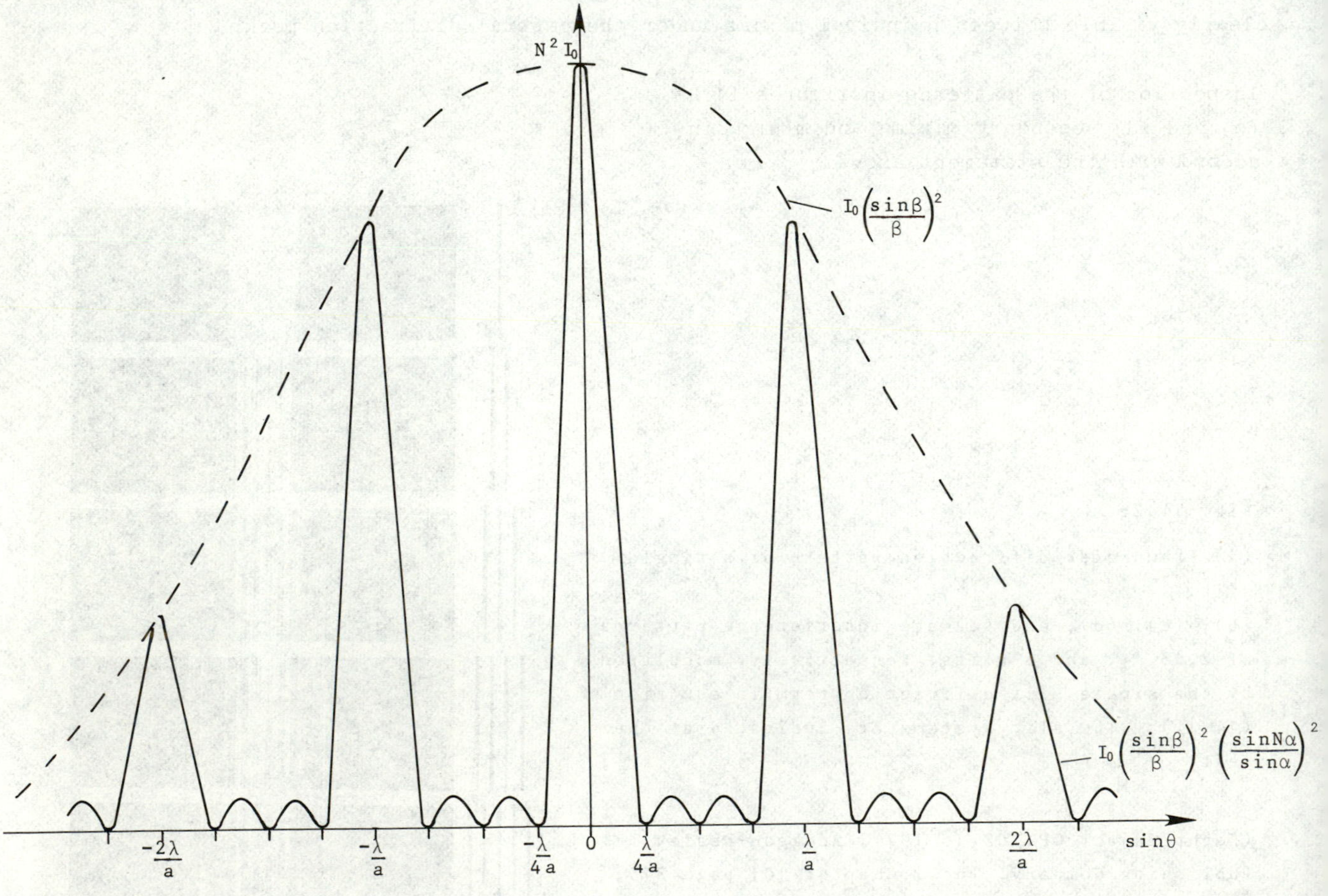

Fig. 14.29 The Fraunhofer intensity distribution for 4 slits ($N = 4$).

In this figure, suppose the single slit diffraction intensity distribution, $I_0\left(\frac{\sin\beta}{\beta}\right)^2$, has its first minimum at $\sin\theta = \frac{3\lambda}{a}$, then since the first diffraction minimum occurs at $\sin\theta = \lambda/b$, from equation (14.15), we would have $\sin\theta = \frac{3\lambda}{a} = \frac{\lambda}{b}$, or $a = 3b$. Note also in figure 14.29 there are 2 subsidiary maxima and 3 minima between each principal maximum.

14.5 THE RECTANGULAR APERTURE

In the foregoing treatment of slits, their widths were taken to be neglible compared to their lengths. The secondary wavelets were all imagined as arising from very narrow elementary strips parallel to the length of the slit. Knowing now that narrowing a slit widens the diffraction pattern and makes the intensity variations more noticeable, it seems reasonable to ignore the effects from the top and bottom of a very long slit. However, suppose the slit's length is not so large compared to its width. Well, then we will have observable diffraction patterns in two dimensions: perpendicular to both the length b' and the breadth b of the aperture, figure 14.30 .

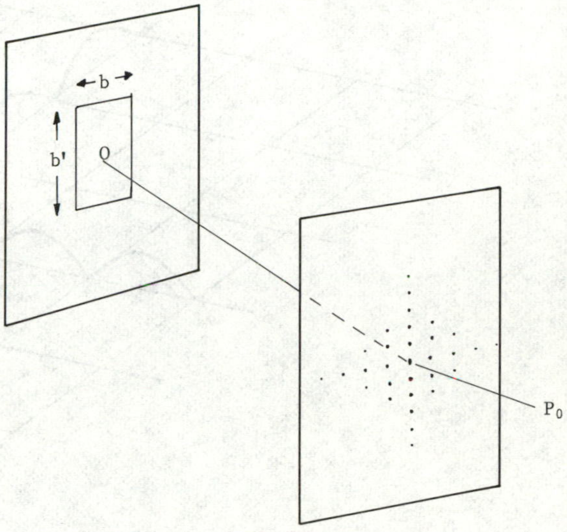

Fig. 14.30 Rectangular aperture of length b' and breadth b and its diffraction pattern.

Figure 14.31 is a photograph of a rectangular aperture Fraunhofer pattern. Notice that the rectangle was not quite perfect because the horizontal and vertical lines of maxima are not precisely mutually perpendicular.

In figure 14.30, OP_0 is perpendicular to the plane of the rectangular aperture and the screen, where O is the centre of the aperture. The direction of a point P on the screen can be designated by the angles θ and θ' which are measured from the normal OP_0 in planes through the normal parallel to the sides b and b'. It is possible to show that the intensity on the screen at P is given by

$$I \propto b^2 b'^2 \left(\frac{\sin\beta}{\beta}\right)^2 \left(\frac{\sin\beta'}{\beta'}\right)^2 \qquad (14.24)$$

where $\beta = \frac{\pi}{\lambda} b \sin\theta \qquad \beta' = \frac{\pi}{\lambda} b' \sin\theta'$. In terms of the intensity I_0 at the point P_0, we can rewrite the last proportionality thus

$$I(\theta,\theta') = I_0 \left(\frac{\sin\beta}{\beta}\right)^2 \left(\frac{\sin\beta'}{\beta'}\right)^2 \qquad (14,24a).$$

Figure 14.32 is a plot of the intensity distribution when $b = b'$.

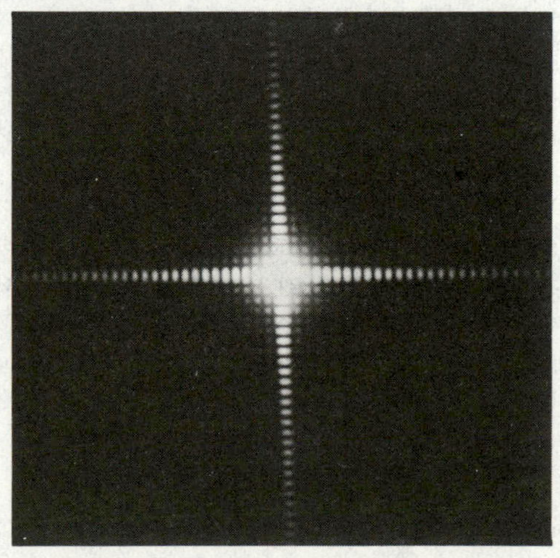

Fig. 14.31

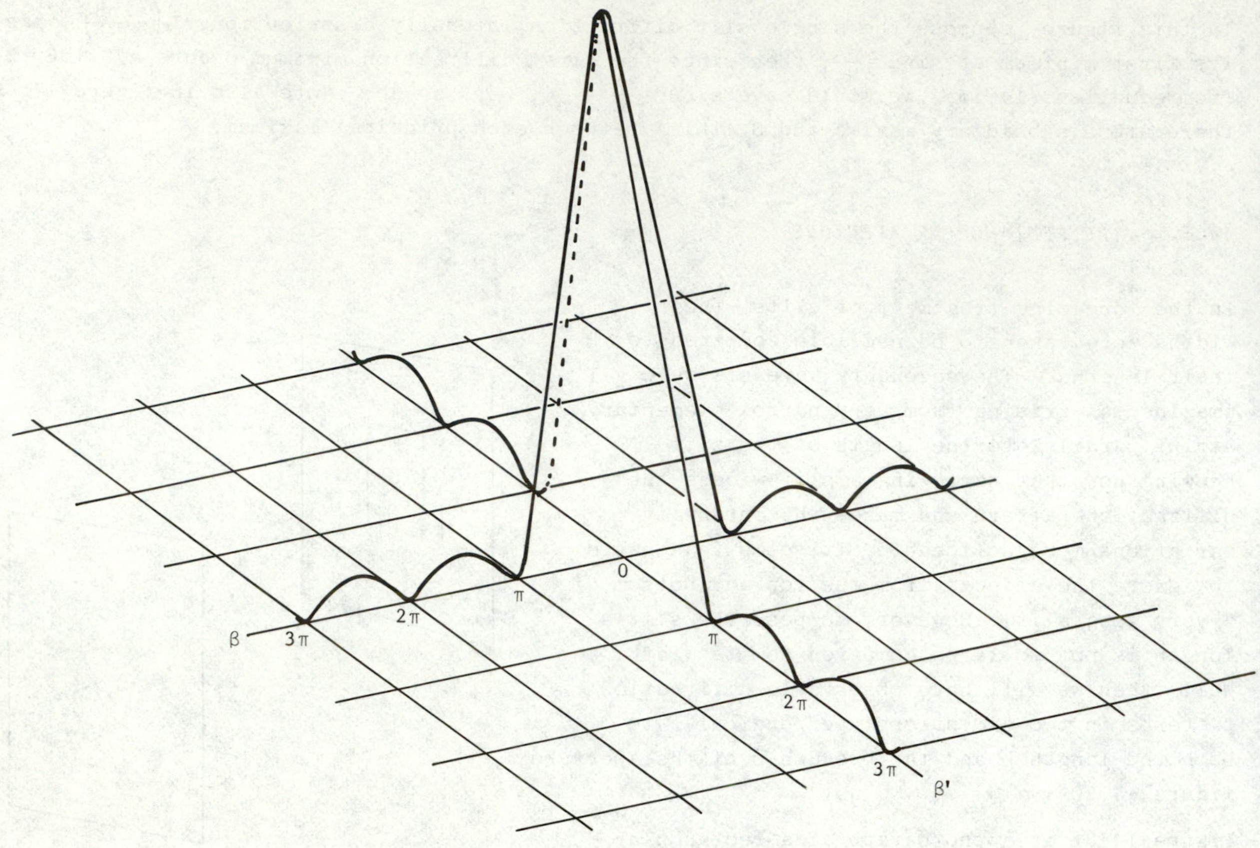

Fig. 14.32 The intensity distribution of a square aperture.

The similarity to the single slit intensity distribution is immediately apparent. Reducing the dimension of the aperture in one direction causes the pattern to spread in the direction at 90^0 to the dimension reduced.

14.6 THE DIFFRACTION GRATING

Any periodic array of linear diffracting elements, either apertures or obstacles, is called a *diffraction grating*. The multiple slit configuration we have already considered is one example of a diffraction grating. The earliest gratings, made by Fraunhofer, consisted of fine wire wound around two parallel spacers. A wavefront traversing such a system meets alternate transparent and opaque regions which results in the amplitude of the wavefront being modulated. If the grating is uniformly illuminated, then zero intensity is transmitted at an opaque strip and maximum intensity is transmitted through the spaces. Such gratings are not unnaturally said to be *amplitude transmission gratings*.

One form of modern grating consists of parallel grooves scratched or ruled into the surface of a clear, flat glass plate. The grooves act as sources of scattered light forming a periodic array of parallel line sources, figure 14.33 .

If the grating is entirely transparent so that no appreciable change in the amplitude of the incident wave occurs, but there is a periodic phase change across the grating, then such gratings are called *phase transmission gratings*, figure 14.34 . The resultant diffraction pattern depends on the relative phase difference between the advanced and retarded parts of the wavefront. Suffice it to say that such gratings do produce diffraction patterns when coherent plane waves are incident upon them.

Gratings can be designed to operate by reflection rather than transmission and, as might be expected, they are called *reflection gratings*, figure 14.35.

In the early eighteen-eighties, Henry Augustus Rowland (1848 - 1901) invented a machine for ruling gratings, and by 1885 he had developed it to the extent that he was able to rule 6000 grooves per centimetre. The earliest ruled gratings were reflection gratings made from an alloy of tin and copper. In the nineteen-thirties it became possible to make ruled gratings on aluminium evaporated onto an

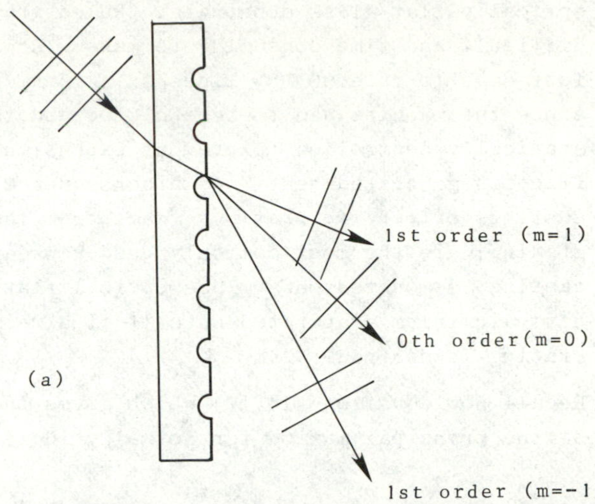

Fig. 14.33(a) Diffraction at a ruled amplitude transmission grating.

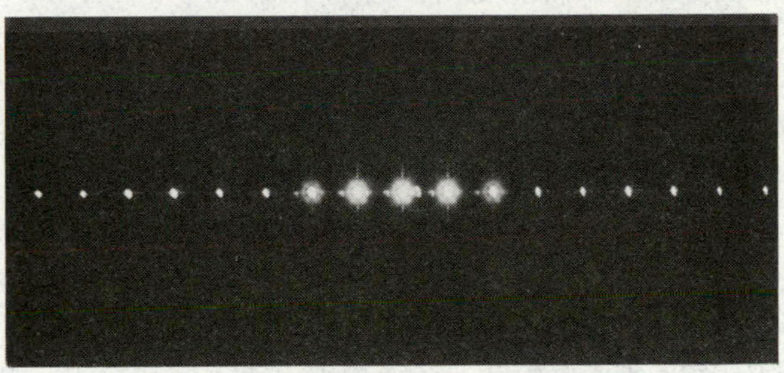

Fig. 14.33(b) A Fraunhofer pattern formed by a grating illuminated by a helium-neon laser.

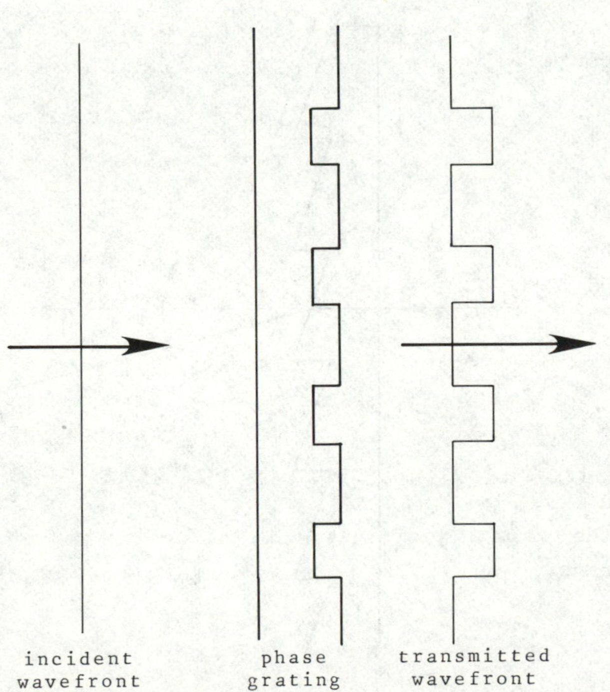

Fig. 14.34 Phase transmission grating.

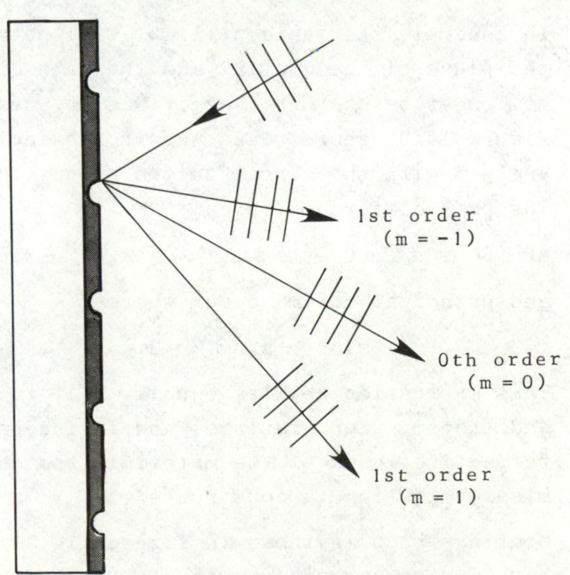

Fig. 14.35 Reflection grating.

optically flat glass substrate. Ruled gratings are difficult and time consuming to make. Rowland took four days to rule 90 000 lines on a 15 cm grating since the machine had to be enclosed and thermostatically controlled to prevent expansion from the friction-generated heat. As a consequence, plastics castings or *replica gratings*, made from the ruled grating, are the most commonly used nowadays. The casting is mounted on a glass optical flat which can have aluminium evaporated onto it if a reflection grating is desired.

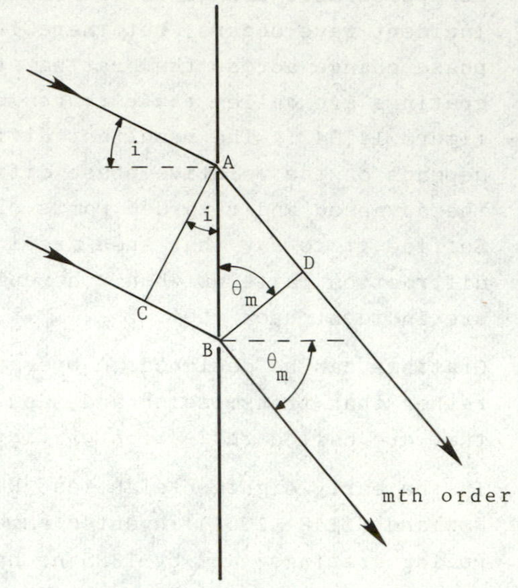

Fig. 14.36 Diffraction for an angle of incidence i.

Recall now equation (14.19) which gives the positions of the principal maxima for normally incident light:

$$a \sin\theta_m = m\lambda .$$

This is known as the *grating equation for normal incidence*, and a is the *grating interval* or *constant*. The values of m (0, ±1, ±2,...) specify the *order* of the principal maxima, m = 0 being the undiffracted zeroth order. In incident light with a broad continuous spectrum, such as daylight or light from a tungsten filament lamp, in all except the zeroth order the principal maxima for each λ spread out and thereby display a series of continuous spectra. Between each order, m = 1 and m = 2, say, there is an apparently dark region where the subsidiary maxima are all but invisible. Thus, there exists a series of continuous spectra on either side of the zeroth order. Reducing a causes the intervals between the orders to spread.

In general, the incident light will not be normal to the plane of the grating and the path difference will no longer be $a \sin\theta_m$, as it was in equation (14.19). Figure 14.36 shows plane wavefronts incident at an angle i with the normal to the plane of the grating. The path difference is now

$$AD - BC = a \sin\theta_m - a \sin i = a(\sin\theta_m - \sin i) ,$$

and principal maxima occur where

$$a(\sin\theta_m - \sin i) = m\lambda \qquad (14.25).$$

This expression applies equally well to reflection and transmission gratings, and is independent of the refractive index of the material from which the transmission grating is constructed.

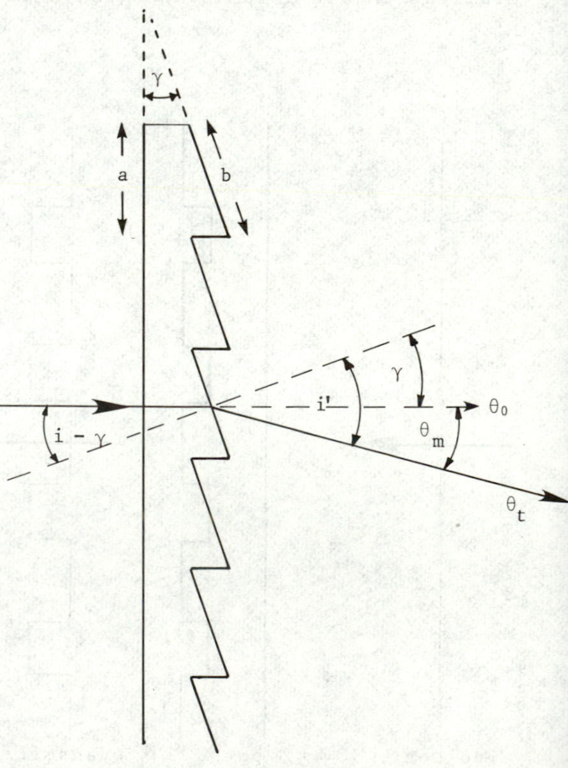

Fig. 14.37 Detail from a blazed grating: γ = blaze angle.

Gratings such as those in figures 14.33 and 14.35 have a singular disadvantage. Most of the incident light undergoes undeviated refraction or specular reflection and this concentrates a large part of the energy in the zeroth order where the λ components overlap. For spectroscopic purposes this is wasted, and it is for this reason that *blazed gratings* are preferred. Robert W. Wood (1868 - 1955) in 1910 produced ruled gratings with the controlled shape shown in figure 14.37 . Blazed gratings are able to concentrate more than 70% of the light into one non-zero order spectrum.

In figure 14.37, a normally incident wave will be transmitted in the θ_t direction. Without the tilting of the facets most of the energy would have gone in the θ_0 direction. From the diagram $\theta_m = i' - \gamma$, and if we want to channel most of the energy into the mth order the blaze angle γ must satisfy the grating equation

$$a \sin\theta_m = m\lambda \quad \text{or} \quad \sin\theta_m = \sin(i' - \gamma) = \frac{m\lambda}{a} .$$

14.7 GRATING SPECTROSCOPY

14.7.1 Dispersive Power or Angular Dispersion

A useful parameter for describing the angular spread for a small range of wavelengths is the *angular dispersion* or *dispersive power* of a grating. The dispersive power $\mathcal{D}$ is defined as

$$\mathcal{D} = \frac{\Delta\theta}{\Delta\lambda} \qquad (14.26).$$

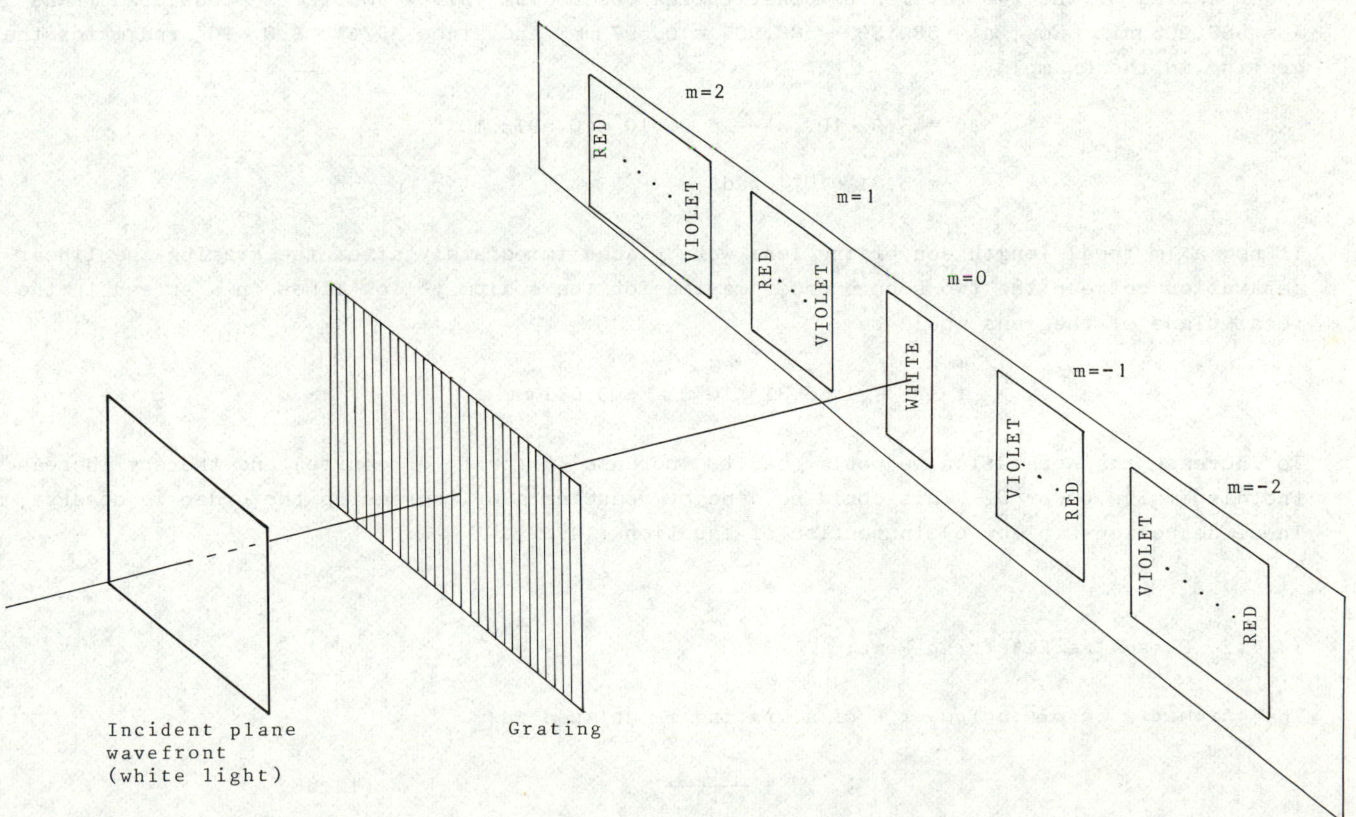

Fig. 14.38 Spectra produced by a diffraction grating.

From the grating equation for normal incidence, $a \sin \theta_m = m\lambda$, it is evident that principal maxima occur at different values of θ for each wavelength in a continuous spectrum. This gives rise to the broad bands of colours, or spectra, around each principal maximum at $m = \pm 1, \pm 2$, etcetera. Figure 14.38 illustrates the first and second order spectra.

Differentiating the grating equation produces $\frac{\Delta\theta}{\Delta\lambda} = \frac{m}{a \cos\theta_m}$, so the dispersive power or angular dispersion of a grating is

$$\mathcal{D} = \frac{\Delta\theta}{\Delta\lambda} = \frac{m}{a \cos \theta_m} \qquad (14.27).$$

Example

A grating has 4000 grooves or lines per centimetre. Calculate the dispersive power in the second order spectrum in the visible range.

The grating constant is $a = \frac{1}{4000}$ cm $\equiv 2.5 \times 10^{-6}$ m, since 4000 lines occupy each centimetre. The angular position of the second order is θ_2 given by the grating equation $a \sin \theta_2 = 2\lambda$,

$$\text{or} \qquad \sin\theta_2 = \frac{2\lambda}{a} = \frac{2 \times 550 \times 10^{-9}}{2.5 \times 10^{-6}} = 0.440$$

whence $\theta_2 = \arcsin 0.440 = 26.10^\circ$, having used 550 nm as a mean value in the visible spectrum.

$$\text{Now,} \quad \mathcal{D} = \frac{\Delta\theta}{\Delta\lambda} = \frac{m}{a \cos \theta_2} = \frac{2}{2.5 \times 10^{-6} \times \cos 26.10^\circ} = 8.9 \times 10^5 \text{ rad/m.}$$

To see the physical significance of this result consider the angular separation of the second order maxima of the two wavelenths constituting the sodium yellow doublet $\lambda_1 = 589.592$ nm and $\lambda_2 = 588.995$ nm. Now, $\Delta\lambda = 589.592 - 588.995 = 0.597$ nm, and since $\Delta\theta/\Delta\lambda = 8.9 \times 10^5$ rad/m for the grating in the example,

$$\Delta\theta = 8.9 \times 10^5 . \Delta\lambda = 8.9 \times 10^5 \times 0.597 \times 10^{-9}$$

$$= 5.31 \times 10^{-4} \text{ rad.}$$

If now a 1m focal length converging lens were placed immediately after the grating the linear separation between the two second order maxima for the sodium yellow lines on a screen in the focal plane of the lens would be

$$f'.\Delta\theta = 1 \times 5.31 \times 10^{-4} \text{ m} \equiv 0.531 \text{ mm.}$$

To increase the separation we would need to increase $\Delta\theta$ (or f', of course) and thereby increase the dispersive power $\mathcal{D}$. This could be done by reducing a or increasing the order we observe, m. These deductions follow by inspection of equation (14.27).

14.7.2 *Chromatic Resolving Power*

The *chromatic resolving power* R of a grating is defined as

$$R = \frac{\lambda}{(\Delta\lambda)_{min}} \qquad (14.28)$$

where λ is the mean wavelength and $(\Delta\lambda)_{min}$ is the least resolvable wavelength difference between two adjacent spectral lines (principal maxima for different λ of the same order). We can use Rayleigh's criterion, which you will recall, states that *two fringes are just resolvable when the maximum of one coincides with the first minimum of the other*. This is illustrated in figure 14.39 . This means the separation of the two peaks is equal to the half-width of either, very nearly. From equation (14.22), the half-width of a principal maximum or fringe is

$$(\Delta\theta)_{min} = \frac{\Delta\theta}{2} = \frac{\lambda}{Na\cos\theta_m} .$$

But, from equation (14.27)

$$(\Delta\theta)_{min} = \frac{m.(\Delta\lambda)_{min}}{a\cos\theta_m} .$$

Equating the right hand sides of these two equations (the left hand sides are equal!) gives

$$\frac{m(\Delta\lambda)_{min}}{a\cos\theta_m} = \frac{\lambda}{Na\cos\theta_m}$$

or $$R = \frac{\lambda}{(\Delta\lambda)_{min}} = mN \qquad (14.29)$$

for the resolving power of the mth order.

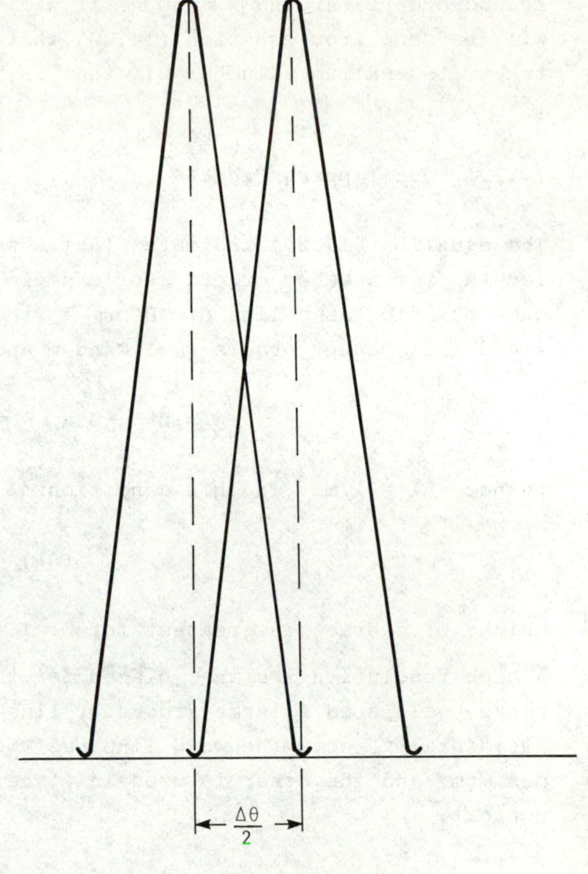

Fig. 14.39 Rayleigh's criterion for the resolution of fringes.

Example

Find the number of lines (grooves) required on a grating to just resolve the two sodium lines, $\lambda_1 = 589.592$ nm and $\lambda_2 = 588.995$ nm , in the second order spectrum of a grating.

Using equation (14.29), where λ is the mean of λ_1 and λ_2, i.e. $\lambda = (\lambda_1 + \lambda_2)/2$, and $(\Delta\lambda)_{min} = \lambda_1 - \lambda_2$ we have

$$N = \frac{\lambda}{m.(\Delta\lambda)_{min}} = \frac{(589.592 + 588.995)/2}{2\times(589.592 - 588.995)}$$

$$= 493.5$$

or N = 494 lines. Note that 987 lines would be required for resolution in the first order spectrum!

We can rewrite equation (14.29) by substituting for m from equation (14.25), whence the resolving power is

$$R = \frac{\lambda}{(\Delta\lambda)_{min}} = \frac{Na(\sin\theta_m - \sin i)}{\lambda} \qquad (14.30).$$

Thus we see that resolving power is a function of grating width, Na, the angle of incidence, and λ. A grating 15 cm wide, with 90 000 lines, will have a resolving power of 1.8×10^5 in the second order. Around $\lambda = 550$ nm it will resolve a wavelength difference of 0.0031 nm. It will be seen from equation (14.30) that the maximum resolving power occurs when $(\sin\theta_m - \sin i)$ takes its maximum value of 2; that is, where $i = -\theta_m = 90^0$.

14.7.3 *Overlapping Orders*

The equation (14.25) indicates that a wavelength from one order may fall on some other wavelength from another order. For example, a line of 350 nm in the second order will occupy the same position as a line of 700 nm in the first order. Suppose two lines formed by wavelengths λ and $\lambda+\Delta\lambda$ occupy orders (m+1) and m and overlap exactly, then

$$a(\sin\theta_m - \sin i) = (m+1)\lambda = m(\lambda+\Delta\lambda),$$

whence $\Delta\lambda = \lambda/m$. In this condition $\Delta\lambda$ is known as the *free spectral range*, and we can write

$$(\Delta\lambda)_{fsr} = \lambda/m \qquad (14.31)$$

which, of course, is greatest for $m = 1$.

A high resolution grating, blazed for the first order so as to have the highest free spectral range, will need a large groove or line density of perhaps 12 000 grooves per centimetre. Incidentally, note the word line has two uses: one is synonymous with fringe or principal maximum, and the other is used in place of groove. Its meaning is usually apparent from the context.

14.8 TWO-DIMENSIONAL GRATINGS

Rectangular arrays of diffracting elements such as circular or square apertures are effectively the same as a pair of gratings superpositioned at right angles. Under plane wave, normal incidence illumination each small element behaves as a coherent source and the diffracted waves emerging from each source bear a constant phase relationship with the others. In certain directions constructive interference will occur and a Fraunhofer diffraction pattern will be evident when the emerging light is focused on a screen, figure 14.40 .

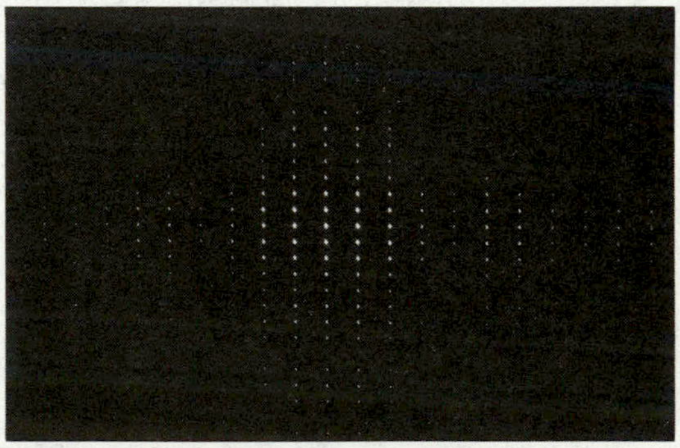

Fig. 14.40 Fraunhofer diffraction pattern formed by a rectangular mesh illuminated by a helium-neon laser.

The phenomenon may be observed by looking at a small source through rectangular woven net curtains, the material of an umbrella, or through a tea-strainer. Crossing two combs may achieve the same result. Use a sodium street light as a source. A single comb acts as a one-dimensional grating, of course.

14.9 MOIRĒ FRINGES OR PATTERNS

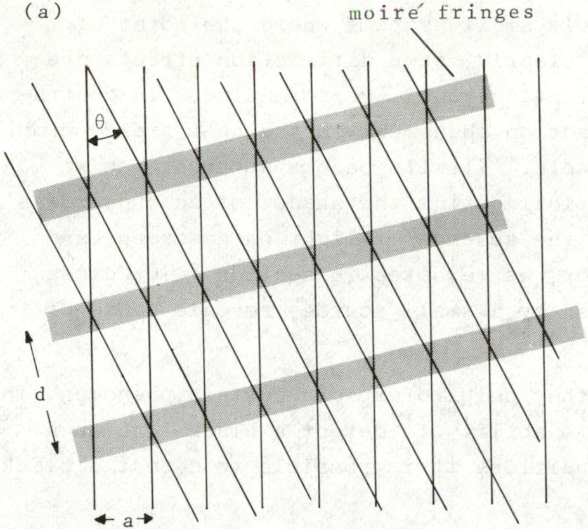

Fig. 14.41 Moiré fringes.

(b)

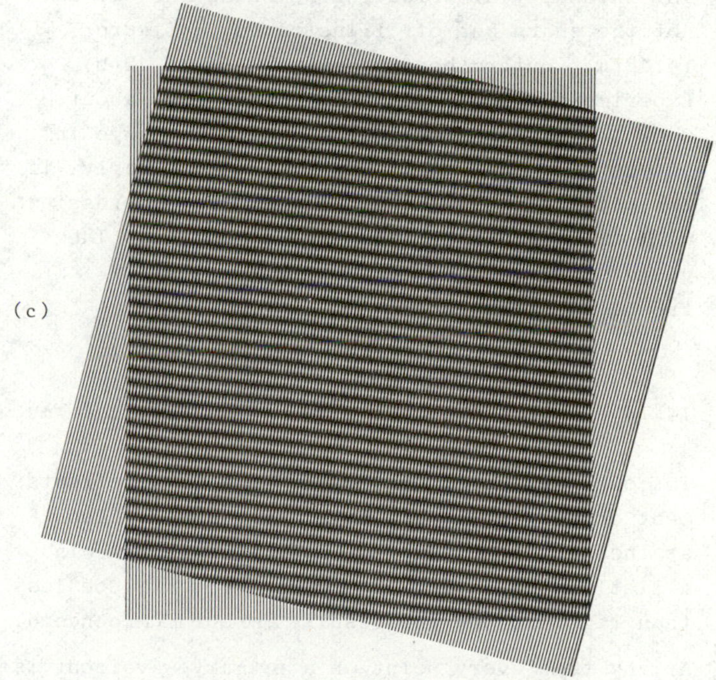

(c)

Fig. 14.41 (b) Two gratings with a=0.6mm crossed at approximately $2\frac{3}{4}^{0}$.
(c) The same gratings at 14^{0}.

Check the fringe widths by measurement and compare with the results from the relationship $d=a/\tan\theta$.

The word moiré is derived from the French and means watered. The word has been used for many years to describe the wavy, watery patterns obtainable in silk material. In physics it is used to describe the interference patterns seen when looking through the folds of a fine mesh nylon curtain, for example. If two similar line gratings are placed in contact with their grooves parallel and one is rotated through a small angle θ with respect to the other, fringes may be formed, figure 14.41 .

The principle is used for measuring small displacements in mechanical systems such as strain incurred under a shearing stress. Since θ is small, the periodicity of the fringes is $d \simeq a/\tan\theta$, where a is the grating constant for both gratings. With $\theta = 4'$, and $a = 0.001$ cm, fringes are formed with a spacing of just under 1 cm. A displacement of a fraction of a fringe width is discernible which allows the monitoring of movements of as little as 0.001 mm. Figures 14.41(b) and (c) show the effect of rotating two gratings from $\theta \simeq 2\frac{3}{4}^{0}$ to $\theta \simeq 14^{0}$.

14.10 FRESNEL DIFFRACTION

14.10.1 Shadows

Recall that Fraunhofer diffraction dealt with relatively small systems where the point of observation was very distant and in the image plane. Clearly, such diffraction effects are of great importance to us since they have visual implications in image formation. In contradistinction, Fresnel diffraction deals with the diffraction phenomena in the near field, which can mean right up to the diffracting aperture or obstacle. It will become apparent, as we proceed, that we are dealing with shadows. In geometrical optics the shadow of an object was considered to demarcate the distribution of light and the absence of light on a screen, and this boundary was imagined as a sharp cut-off. However, as we noted in section 14.0, close observation of the shadows of a narrow rod, illuminated by a small source, reveals bands of light within the geometrical shadow.

These effects are of particular interest to us since they help to explain certain phenomena in physiological optics. For example, the visual system's ability to detect a black line on a white background is quite remarkable. Under ideal conditions it is possible to detect a black line subtending only a few seconds of arc at the eye. The explanation of this almost incredible achievement depends on the distribution of light within the geometrical image, figure 14.42 .

The intensity distribution reaching the eye is not the sharp cut-off illustrated in figure 14.42(a), but rather the pattern in 14.42(b). Experimentally, if $\Delta I/I$ is greater than about 0.02 the visual system will detect a change in intensity across the field of view. Clearly, if the width of the line is reduced the 'depression' will fill in and ΔI will become smaller. The value of ΔI can be obtained by considering Fresnel diffraction.

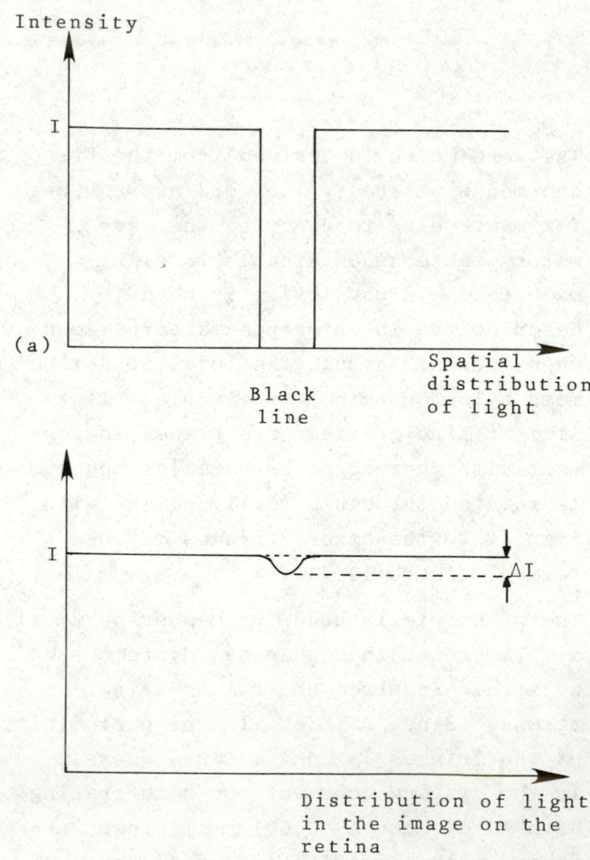

Fig. 14.42 (a) Distribution of light in the object plane - black line on a white background.
(b) Actual distribution of the light across the image.

14.10.2 The Free Propagation of a Spherical Wave

To consider the effects observed when wavefronts meet obstacles we must first take another look at the Huygens-Fresnel principle, this time in a little more detail. Our treatment will be less than rigorous but the results are our main concern.

At any time every point on a primary wavefront is regarded as a continuous emitter of spherical secondary wavelets radiating in all directions. Clearly, this secondary radiation cannot be equal in all directions for, if it were, we would see both a reverse wave travelling backwards as well as the forward progressing wave. To account for the lack of a backward travelling wave arising from the secondary wavelets, Kirchhoff formulated the obliquity factor

$$K(\theta) = \tfrac{1}{2}(1 + \cos\theta) \qquad (14.32).$$

The amplitude of the envelope of the secondary waves is modulated according to this factor, where θ is the angular direction measured from the normal to the primary wave, figure 14.43 . From equation (14.32) it is clear that when $\theta = 0$ $K(\theta) = K(0) = 1$, and when $\theta = \pi$ $K(\theta) = K(\pi) = 0$. That is, the obliquity factor is a maximum of unity in the forward direction and reduces to zero in the backwards direction. Figure 14.44 will now be used to illustrate the free propagation of a spherical wave from a point source S. The aim is to find the amplitude of the unobstructed primary wave reaching the point P. The obliquity factor will crop up in this investigation.

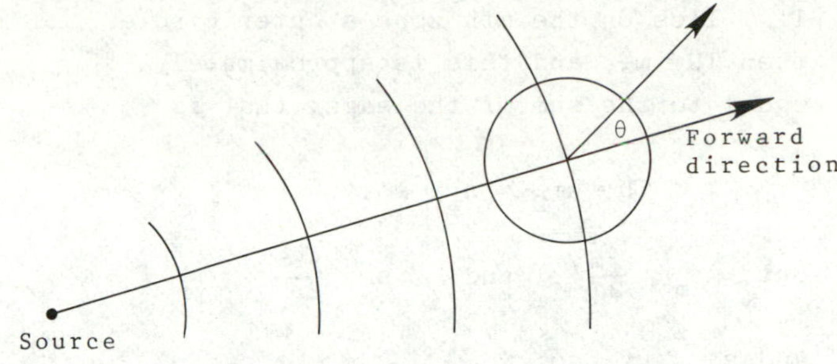

Fig. 14.43 Kirchhoff's obliquity factor.

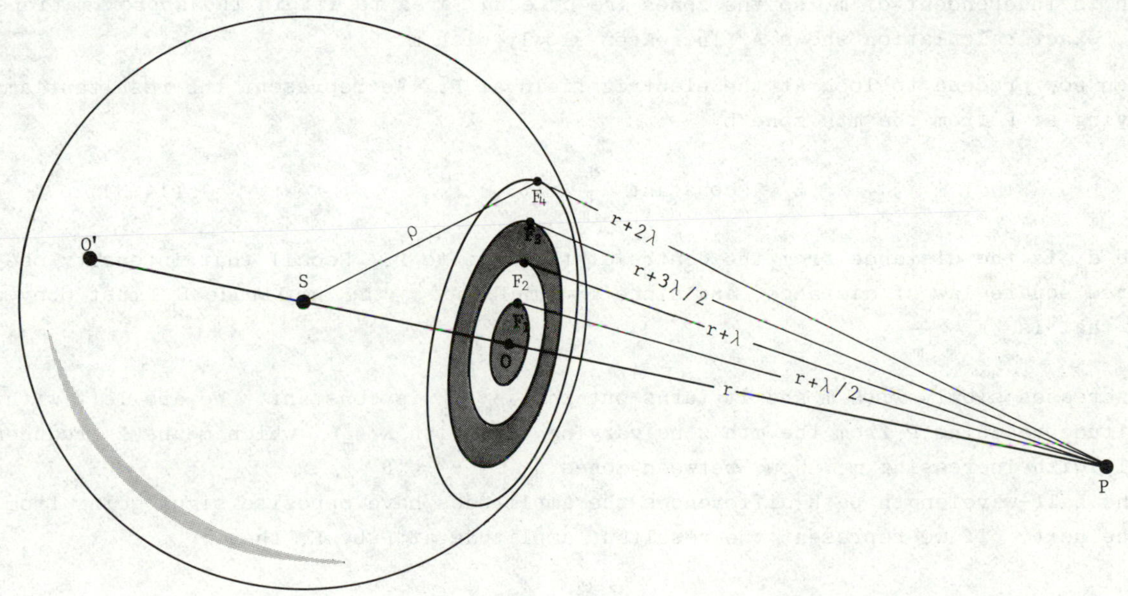

Fig. 14.44 Fresnel half-period zones on a primary wavefront.

Figure 14.44 represents the primary spherical wavefront, of radius ρ, divided into a number of annular regions the boundaries of which are distances $r+\frac{\lambda}{2}$, $r+\lambda$. $r+\frac{3}{2}\lambda$, and so on from the point P, where $OP = r$. These annular regions are the *Fresnel* or *half-period zones* of the primary wavefront. For a secondary point source in one zone there will be a secondary point source in an adjacent outer zone exactly $\lambda/2$ further from P. If we divide the primary wavefront into m zones, their areas are practically equal. We can show this is true by considering figure 14.45 where the sags of the spherical arcs TO and UO are small.

Let S_m and S' be the sags of the two arcs TO and UO, and y_m and y' their respective semi-chords.

If T lies on the mth zone's outer circle then $TU = m\frac{\lambda}{2}$, and this is approximately equal to the sum of the sags; that is

$$TU = m\frac{\lambda}{2} \simeq s_m + s'.$$

But, $s_m \simeq \frac{y_m^2}{2\rho}$ and $s' \simeq \frac{y'^2}{2r}$,

and $y_m \simeq y'$, so

$$m\frac{\lambda}{2} \simeq \frac{y_m^2}{2\rho} + \frac{y_m^2}{2r} = y_m^2 \cdot \frac{\rho+r}{2\rho r}$$

or $$y_m^2 = \frac{m\rho r\lambda}{\rho+r}$$

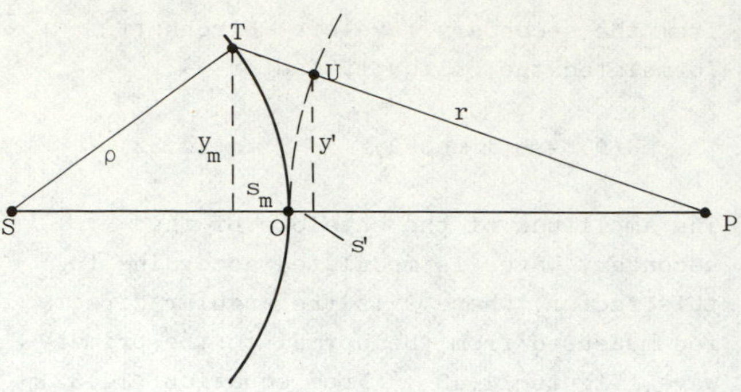

Fig. 14.45 See text for details.

The area of the mth zone is then

$$A_m = \pi(y_m^2 - y_{m-1}^2) = \pi\left(\frac{m\rho r\lambda}{\rho+r} - \frac{(m-1)\rho r\lambda}{\rho+r}\right) = \frac{\pi\rho r\lambda}{\rho+r} \qquad (14.33)$$

which is independent of m, so the zones are of equal area to within the approximations used. In fact, exact calculation shows A_m increases slowly with m.

We can now proceed to look at the electric field at P. We represent the resultant amplitude arriving at P from the mth zone by

$$E_m = \text{constant} \times \frac{A_m}{d_m} \times K(\theta_m) \qquad (14.34)$$

where d_m is the distance from the centre of the zone to P. Recall that intensity obeys an inverse square law of distance, and since $I \propto \text{amplitude}^2$, the amplitude E_m must obey an inverse law; that is $E_m \propto \frac{1}{d_m}$.

d_m increases slowly with m and it turns out that A_m/d_m is constant. We are left with the amplitude reaching P from the mth zone varying only with $K(\theta_m)$, which means E_m reduces only slowly with increasing m. Now, between zones $\theta_{m-1} \simeq \theta_m \simeq \theta_{m+1}$, so $E_{m-1} \simeq E_m \simeq E_{m+1}$, but because of the half-wavelength path differences the amplitudes have opposite signs going from one zone to the next. If we represent the resultant amplitude at P by E, then

$$E = E_1 - E_2 + E_3 - E_4 + \ldots\ldots(-1)^{m-1}E_m \qquad (14.35).$$

This series can be rewritten in two ways as below, assuming m to be an odd number,

$$E = \frac{E_1}{2} + \left(\frac{E_1}{2} - E_2 + \frac{E_3}{2}\right) + \left(\frac{E_3}{2} - E_4 + \frac{E_5}{2}\right) + \ldots\ldots + \left(\frac{E_{m-2}}{2} - E_{m-1} + \frac{E_m}{2}\right) + \frac{E_m}{2} \qquad (14.36)$$

or as

$$E = E_1 - \frac{E_2}{2} - \left(\frac{E_2}{2} - E_3 + \frac{E_4}{2}\right) - \left(\frac{E_4}{2} - E_5 + \frac{E_6}{2}\right) + \ldots - \left(\frac{E_{m-3}}{2} - E_{m-2} + \frac{E_{m-1}}{2}\right) - \frac{E_{m-1}}{2} + E_m \qquad (14.37).$$

The quantities in brackets in equations (14.36) and (14.37) are all positive, so from equation (14.36) we must have

$$E > \frac{E_1}{2} + \frac{E_m}{2} \qquad (14.38),$$

and from equation (14.37)

$$E < E_1 - \frac{E_2}{2} - \frac{E_{m-1}}{2} + E_m \qquad (14.39).$$

Because $E_1 \simeq E_2$ and $E_{m-1} \simeq E_m$, we can rewrite the inequality (14.39) as

$$E < \frac{E_1}{2} + \frac{E_m}{2} \qquad (14.40),$$

whence we must have

$$E = \frac{E_1}{2} + \frac{E_m}{2} \qquad (14.41).$$

When the last term of the series in equation (14.35) corresponds to m being an even integer, a similar treatment results in

$$E = \frac{E_1}{2} - \frac{E_m}{2} \qquad (14.42).$$

If m is large enough so that $K(\theta_m)$ approaches zero, then E_m approaches zero and equations (14.41) and (14.42) reduce to

$$E = \frac{E_1}{2} \qquad (14.43).$$

Thus, the amplitude at P, due to the whole unobstructed wave, is equal to half the amplitude of the contribution from the first (central) zone! This almost incredible result will soon be used to investigate a variety of intensity distributions for differing apertures and obstacles.

14.10.3 The Vibration Curve

Suppose we divide the first Fresnel zone, centred on O in figure 14.44, into N subzones where the outer edge of each is at successive distances

$$r + \frac{\lambda}{2N}, \quad r + \frac{\lambda}{N}, \quad r + \frac{3\lambda}{2N}, \ldots\ldots\ldots\ldots, \quad r + \frac{\lambda}{2},$$

from P. Each subzone contributes to the field at P giving this zone's resultant amplitude contribution E_1. The phase difference between each subzone component must be π/N since the zone central to outer edge path difference, $F_1P - OP$ equals half a wavelength and corresponds to π radians.

If N were 20, the vector addition of the amplitudes from the subzones would be as depicted in figure 14.46 . The curve deviates from a semicircle because the Kirchhoff obliquity factor shrinks successive vector components.

If we now subdivide each successive Fresnel zone and allow $N \to \infty$, then the polygon smooths out and the addition of each zone causes the curve to rotate through π radians.

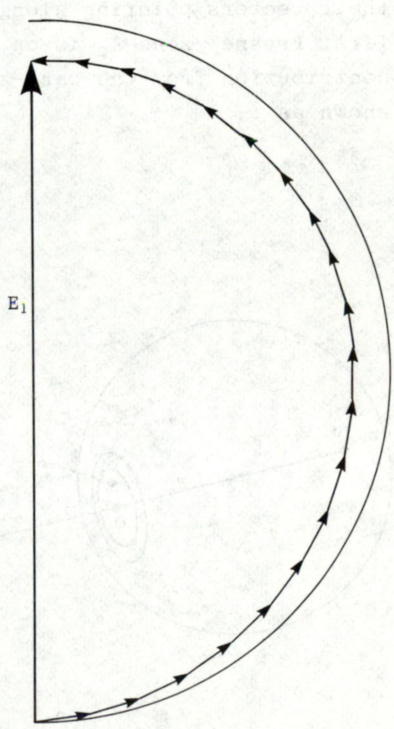

Fig. 14.46 Vector addition of 20 subzones of the central Fresnel zone. Note the angle (the phase difference) between adjacent subzone vectors is $\pi/20$.

Figure 14.47 shows the contributions from the Fresnel zones marked OF_1, F_1F_2, F_2F_3, ..., in figure 14.44 . O_cF_{c1} on the vibration curve corresponds to the contribution from OF_1, and so on. Each Fresnel zone adds a half-turn to the vibration curve which spirals in to O'_c.

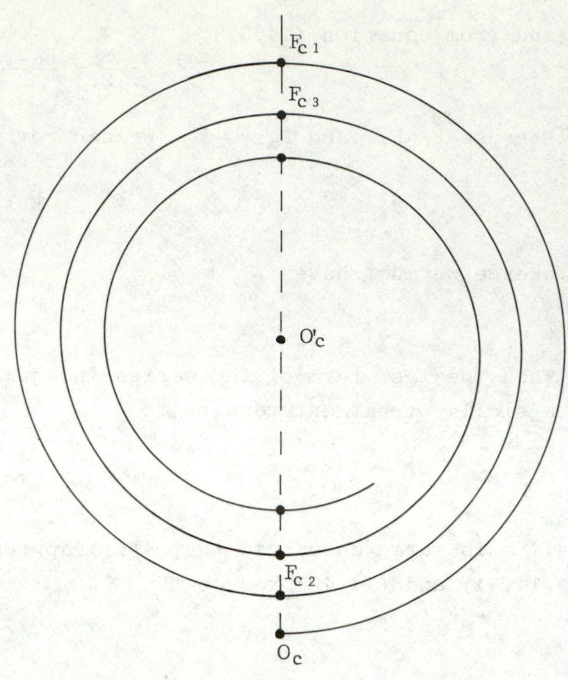

Fig. 14.47 The vibration curve.

When all the contributions from all the zones are added, the resultant is the vector $O_cO'_c$ which is approximately equal to $E_1/2$ since E_1 is the vector O_cF_{c1}. This is the result given by equation (14.43).

If we wish to know the relative phase difference between contributions at P from any two points on the wavefront, we simply find the angle between the tangents at corresponding points on the vibration curve. For example, take the points O and A on the wavefront in figure 14.48. The corresponding points on the vibration curve in figure 14.49 are O_c and A_c. Drawing tangents at these points, the angle required is clearly β. We draw tangents because the subzone amplitude contributions at O and A on the wavefront have their vectors pointing along the tangent on the vibration curve. Now, because A is within the first Fresnel zone A_c is on the first half-turn of the vibration curve. The vector O_cA_c is the contribution from the part-zone with radius approximately OA in figure 14.48. Its phase is shown as δ.

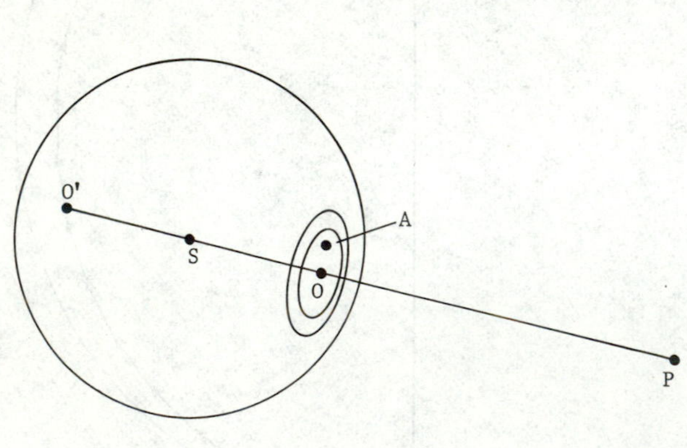

Fig. 14.48 Spherical wavefront with the point A in the first Fresnel zone.

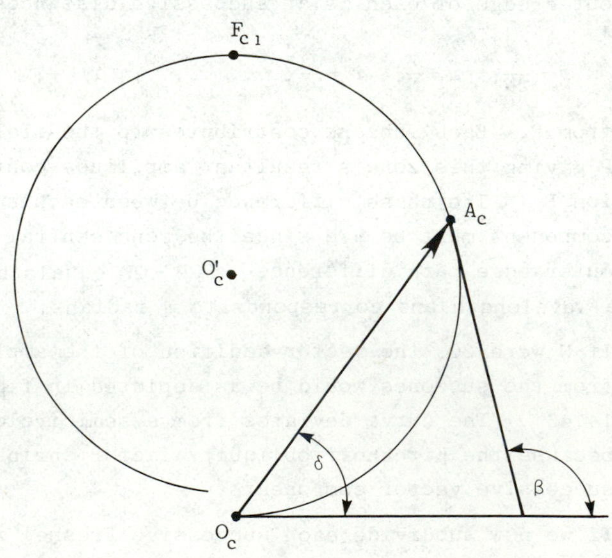

Fig. 14.49 Vibration curve corresponding to figure 14.48 .

14.10.4 *Circular Apertures*

We are now in a position to examine Fresnel diffraction at circular apertures. We hinted as much in the last section when considering the zone of radius OA on the wavefront in figure 14.48. So, let a circular aperture be centred at O in figure 14.50. At P a detector sees m zones on the wavefront from S just filling the aperture. If m is even, then the resultant amplitude at P is

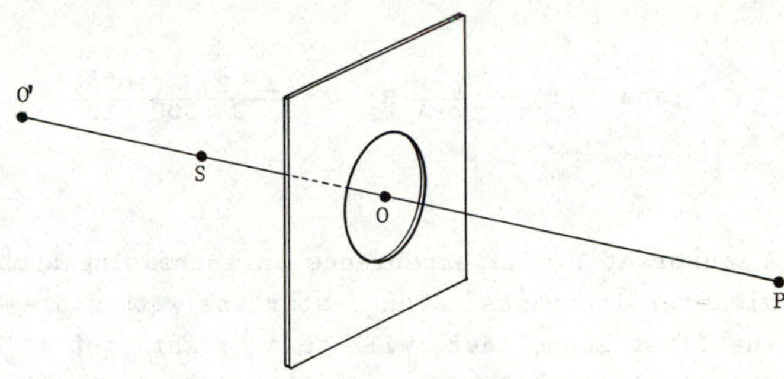

Fig. 14.50 A circular aperture: point source S.

$$E = E_1 - E_2 + E_3 - E_4 + \ldots . + E_{m-1} - E_m \, .$$

Because adjacent contributions are nearly equal we have

$$E = (E_1 - E_2) + (E_3 - E_4) + \ldots\ldots\ldots + (E_{m-1} - E_m) \simeq 0 \, .$$

If $E = 0$, then the intensity $I = 0$, of course. If instead of m being even it is odd, then

$$E = E_1 - (E_2 - E_3) - (E_4 - E_5) - \ldots . - (E_{m-1} - E_m) \simeq E_1 \, .$$

Recall that the amplitude at P from the unobstructed wave is $E_1/2$, so the intensity at P with the circular aperture in place is now 4 times the intensity there when the wave is unobstructed! This is because doubling the magnitude of the electric vector quadruples the intensity. Clearly, elsewhere along the direction SP the intensity must be different. Moving a sensor along the direction SP from O will successively detect intensities varying from zero to a maximum.

The amplitude at P can be obtained from the vibration curve. If A is on the circumference of the hole, then the amplitude of the field at P is O_cA_c on the vibration curve of figure 14.49. Obviously, to locate A_c on the vibration curve we must know how many Fresnel zones are enclosed in the aperture, since each zone corresponds to a half-turn around the spiral of the vibration curve.

Assuming monochromatic light, then the area of each zone is given by equation (14.33) as

$$A = \frac{\rho}{\rho + r} \cdot \pi r \lambda \, .$$

If the radius of the aperture is R, then its area is πR^2, and the number of zones enclosed within the aperture is

$$m = \frac{\pi R^2}{A} = \frac{(\rho + r)}{\rho r \lambda} R^2 \qquad (14.44).$$

Example

A circular aperture is illuminated by monochromatic light of wavelength $\lambda = 500$ nm from a point source 2 m away on the normal to the centre of the aperture. Calculate the number of zones enclosed in the aperture if it has (a) a 1 mm, and (b) a 1 cm radius, and the point P is 2 m from the aperture.

The number of zones is given by equation (14.44), so we have

$$\text{(a)} \quad \frac{\rho+r}{\rho r\lambda}\cdot R^2 = \frac{(2+2)\times(10^{-3})^2}{2\times 2\times 500\times 10^{-9}} = 2 \text{ zones}$$

$$\text{and} \quad \text{(b)} \quad \frac{\rho+r}{\rho r\lambda}\cdot R^2 = \frac{(2+2)\times(10^{-2})^2}{2\times 2\times 500\times 10^{-9}} = 200 \text{ zones.}$$

A sensor at P will experience an increasing number of amplitude components as the aperture's diameter increases. Thus, starting with a diameter just sufficient to fill the aperture with the first zone, there will be a bright spot at P. Increasing the diameter until two zones just fill it will give a resultant field amplitude of $E = E_1 - E_2 = 0$, and there will be a dark spot. Clearly, the sensor will experience successive light and dark spots as the diameter increases. Eventually, when the hole is large A_c approaches O_c' on the vibration curve, and no further change at P will take place.

Example

Plane waves of $\lambda = 550$ nm are incident normally on a circular aperture of radius $\sqrt{11}$ mm. Does a bright or dark spot appear at the point P on the axis 4 m from the hole?

The spot at P will be bright or dark according as the number of zones in the aperture is odd or even. Rearranging equation (14.44), the number of zones is

$$m = \frac{\rho+r}{\rho r\lambda}\cdot R^2 = \frac{\rho R^2}{\rho r\lambda} + \frac{rR^2}{\rho r\lambda} = \frac{R^2}{r\lambda}, \quad \text{since } \rho = \infty.$$

Hence, the number of zones is

$$m = \frac{R^2}{r\lambda} = \frac{(\sqrt{11}\times 10^{-3})^2}{4\times 550\times 10^{-9}} = \frac{11\times 10^{-6}}{2.2\times 10^{-6}} = 5,$$

and a bright spot will be present.

Example

In the last example, if the intensity of the incident light is I_0, calculate the intensity at P.

There are five Fresnel zones 'seen' by the point P, so the amplitude of the electric field at P is given by $E = E_1 - E_2 + E_3 - E_4 + E_5$. Since the amplitudes from successive zones reduce only slowly and are therefore nearly equal, we can write

$$E = (E_1 + E_3 + E_5) - (E_2 + E_4) \simeq 3E_1 - 2E_1 = E_1$$

If the wave were unobstructed the amplitude at P would be $E_1/2$, so the intensity I at P, relative to the intensity with the unobstructed wave would be

$$\frac{I}{I'} = \left(\frac{E}{E_1/2}\right)^2 = 4, \text{ since } E = E_1.$$

This rearranges to $I = 4I'$. But, because the incident waves are plane, the incident intensity I_0 is equal to the unobstructed wave intensity I' and we have $I = 4I_0$.

Example

A 4 mm diameter circular hole in an opaque screen is illuminated by plane waves of wavelength 500 nm. If the angle of incidence is zero, find the positions of the first two intensity maxima and the first intensity minimum along the central axis.

As in figure 14.50, maxima occur at P when an odd number of Fresnel zones is seen from P. Minima occur when an even number of zones is seen. The first two maxima will occur at distances OP when m equals 1 and 3, respectively, in the equation for the number of zones. Since the incident waves are plane we use the equation $m = \frac{R^2}{r\lambda}$ rearranged for r. Thus, for the maxima

$$r = \frac{R^2}{m\lambda} = \frac{(2 \times 10^{-3})^2}{1 \times 500 \times 10^{-9}} = 8\ \text{m} \quad \text{for the maximum when } m = 1,$$

and $$r = \frac{R^2}{m\lambda} = \frac{(2 \times 10^{-3})^2}{3 \times 500 \times 10^{-9}} = 2.67\ \text{m} \quad \text{for the maximum when } m = 3.$$

Clearly, the first minimum occurs when $m = 2$ and we need only divide the first answer by 2 to obtain the result: $r = 4$ m.

Incidentally, we ought to mention that when both ρ and r increase to such an extent that only a fraction of the first zone fills the aperture, then a Fraunhofer pattern occurs on the screen.

To complete our look at the diffraction at a circular aperture, let us investigate the pattern on the screen away from the central point P.

At P_1 on the screen in figure 14.51(a), the zones about the axis SP_1 appear displaced as in figure 14.51(b), say. Here, zone 1 and the partial zone 3 have in-phase components which add. The partial zone 2 subtracts and the effect at P_1 will be a bright region if the effects of zones 1 and 3 are greater than the effect of zone 2. We can expect, therefore, as we move further in the direction PP_1P_2, that a sensor will detect alternate light and dark regions. Because of the circular symmetry about P the pattern on the screen will consist of light and dark circular fringes.

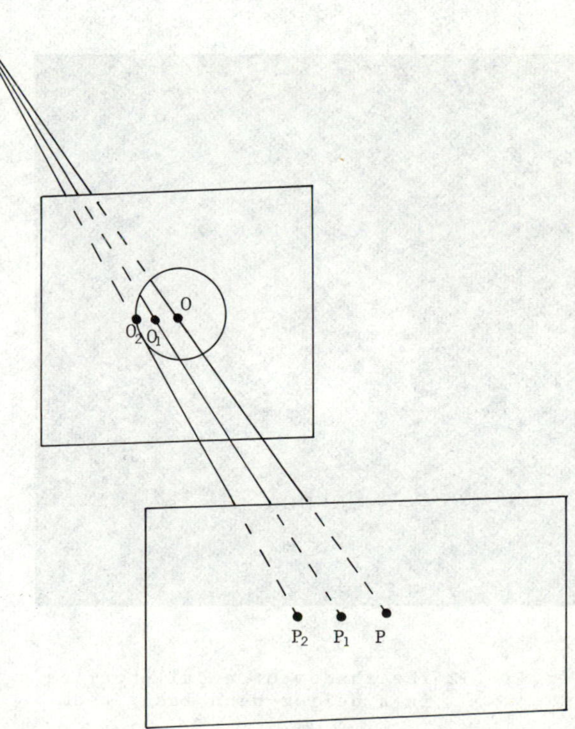

Fig. 14.51a The view of the circular aperture from the points P, P_1, and P_2.

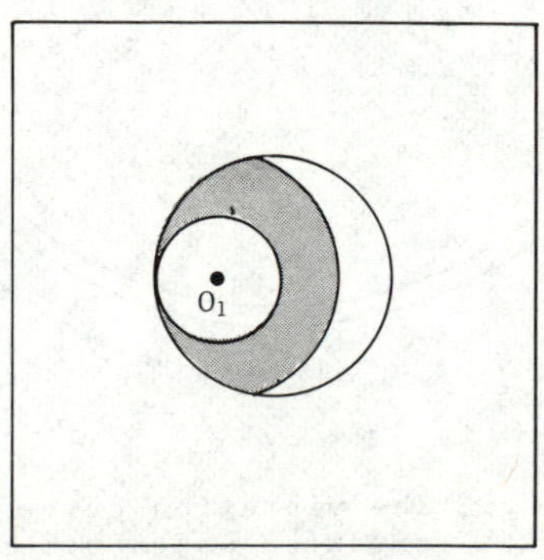

Fig. 14.51b The zones seen in the aperture from the point P_1.

These circular fringe patterns resemble the Fraunhofer pattern for a point source and a circular aperture. However, whereas there the central region was always a bright disc, the Airy disc, with Fresnel diffraction the centre of the pattern may be either bright or dark. We saw in the examples in this section how a bright or dark centre to the pattern depends on seeing an odd or an even number of Fresnel zones filling the aperture from the point of view, P, and that equation (14.44) determines the number of half-period zones.

14.10.5 *Circular Obstacles*

An unobstructed spherical wave produces an amplitude $E \simeq E_1/2$ at a point P, as in figure 14.44. If, now, a circular obstacle exactly covers the first Fresnel zone, the amplitude at P will be $-E_2/2$. That is, a bright spot will appear at the centre of the geometrical shadow. Figure 14.52 depicts this with the shadow of a ball-bearing. Suppose, now, a circular obstacle just obscures the first ℓ zones in the spherical wavefront so the amplitude at P is

$$E = E_{\ell+1} - E_{\ell+2} + \ldots\ldots\ldots\ldots + E_m.$$

As before, when m becomes large we have

$$E \simeq E_{\ell+1}/2 \qquad (14.45).$$

Now, $E_{\ell+1}$ differs only slightly from E_1 so there will be a bright spot everywhere on the axis except just behind the obstacle. This spot, known as *Poisson's spot*, will be most readily observable if the obscuring object is smooth and circular since, otherwise, irregularities will obscure parts of zones.

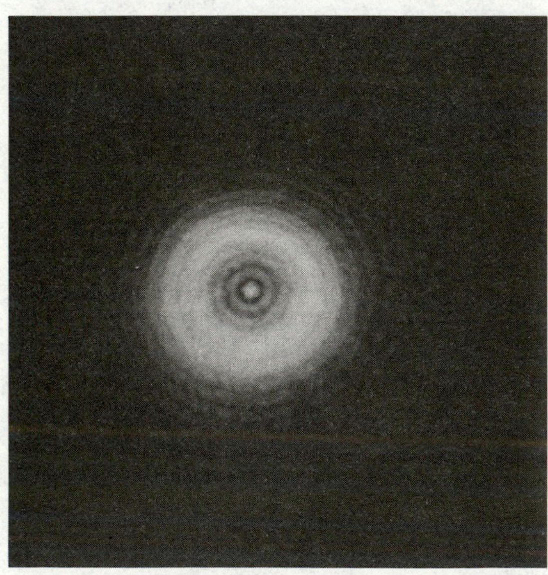

Fig. 14.52 The shadow of a ball-bearing in a helium-neon laser beam.

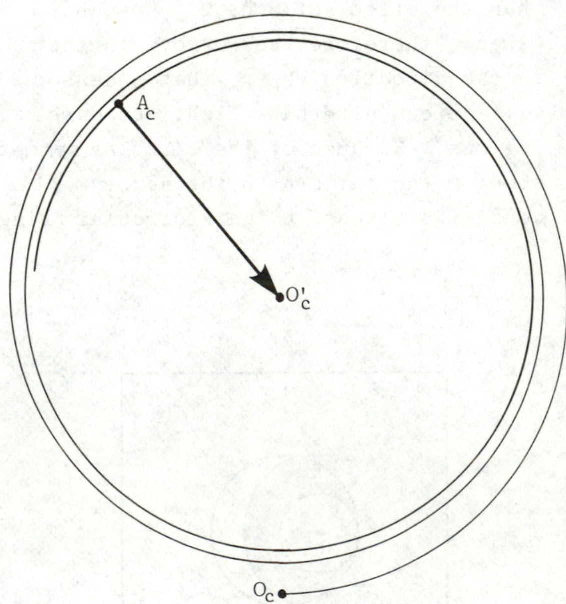

Fig. 14.53 The vibration curve for a circular obstruction - a disc.

The vibration curve can be used to determine qualitatively the change in intensity at a fixed point P. If A_c on the vibration curve corresponds to a point A on the circumference of the obstacle, then $A_cO'_c$ represents the amplitude at P, figure 14.53 . As the diameter of the obstacle increases A_c moves anticlockwise around the vibration curve and $A_cO'_c$ reduces in magnitude until Poisson's spot disappears when all the zones on the wavefront are covered.

Off the axis some zones will be partially exposed and there will result a series of bright and dark rings within the geometric shadow. In effect, the source S is imaged at P so that a small opaque disc acts as a crude lens.

14.10.6 The Fresnel Zone Plate

Recall that successive Fresnel half-period zones tend to cancel one another. If then we construct an obstacle which just lets through the light from alternate zones on the wavefront the intensity at P will be increased tremendously. For example, suppose the obstacle, called a *Fresnel zone plate*, allows through the five components E_1, E_3, E_5, E_7, E_9. Then the field at P will be $E \simeq 5E_1$ compared with $E \simeq E_1/2$ for the unobstructed wave. The amplitude is 10 times greater and the intensity is therefore 100 times as great as for the unobstructed wave. The zone plate would look like figure 14.54 with the white parts transparent.

The outer radius, y_m, of the mth Fresnel zone can be obtained from the relationship

$$m\frac{\lambda}{2} = \frac{y_m^2}{2\rho} + \frac{y_m^2}{2r}$$

developed in section 14.10.2 . Rearranging this relationship gives

$$\frac{1}{\rho} + \frac{1}{r} = \frac{m\lambda}{y_m^2} \qquad (14.46).$$

Equation (14.46) looks very like the thin lens equation and if the Cartesian sign convention were applied to ρ and r, measured from the wavefront, it would become

$$\frac{1}{r} - \frac{1}{\rho} = \frac{1}{y_m^2/(m\lambda)} \qquad (14.47),$$

looking even more like $\frac{1}{\ell'} - \frac{1}{\ell} = \frac{1}{f'}$.

This is not a coincidence, since S is imaged at P. Accordingly, $f'_1 = y_m^2/(m\lambda)$,

which gives $f'_1 = y_1^2/\lambda$ (14.48),

for $m = 1$. f'_1 is said to be the *primary focal length*. There will be other intensity maxima at focal lengths

Fig. 14.54 A Fresnel zone plate. The outer radii are given by $y_m = (mr\lambda)^{\frac{1}{2}}$ for plane incident wavefronts (see equation (14.48)). Thus $y_m \propto m^{\frac{1}{2}}$, m = 1,2,3,...

$f'_3 = \frac{f'_1}{3}$, $f'_5 = \frac{f'_1}{5}$, and so on for $m = 3, 5, \ldots$.

For plane wavefronts $\rho = \infty$ and equation (14.46) reduces to

$$y_m^2 = mr\lambda \qquad (14.49).$$

Zone plates are usually made by making a large scale drawing and reducing it photographically. Made of metal annuli supported by a spoked structure they function as lenses for ultraviolet or x-rays where ordinary glass is opaque.

14.10.7 The Free Propagation of a Cylindrical Wave

We now want to examine the diffraction effects when the apertures and the obstacles have straight edges. If we replace the point source with a uniformly illuminated slit, and keep the SOP notation from section 14.10.4 , cylindrical waves will be assumed to emerge from the source slit at S, figure 14.55 .

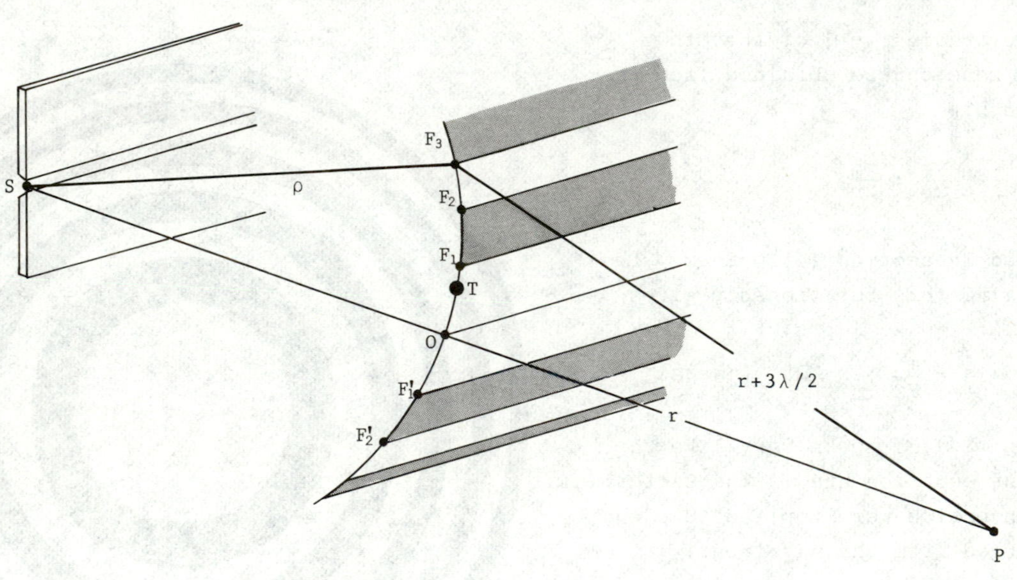

Fig. 14.55 A cylindrical wavefront from a coherently illuminated slit S.

The cylindrical wave is divided into half-period strips, relative to P, in the manner pursued with the spherical wave. The coherent illumination is assumed to be a parallel beam of monochromatic light incident normally on the source slit. The radius of the cylindrical wave is ρ and it is a distance r from P. F_1 , F_2 , F_3 , and so on, are distances $r+\frac{\lambda}{2}$, $r+\lambda$, $r+\frac{3\lambda}{2}$, ... from P. The widths of the strips are therefore approximately OF_1 , F_1F_2 , F_2F_3 , and so on. The contributions from these strips to the field at P are proportional to their areas, which decrease rapidly. Contrast this with the areas of the circular zones which were very nearly constant. Each strip zone is subdivided into N substrips with relative phase differences of π/N between adjacent substrips. Figure 14.56 shows the amplitude vector diagram for the cylindrical wavefront: the curve is smooth since N is taken to be very large. OF_1 on the wavefront is represented by the portion of the curve O_cF_{c1}. Similarly, F_1F_2 on the wavefront is represented by the infinity of subzones amplitude vectors adding to make up that portion of the curve $F_{c1}F_{c2}$. With the addition of amplitude vectors for successive subzones the curve spirals in to B, in the upper half.

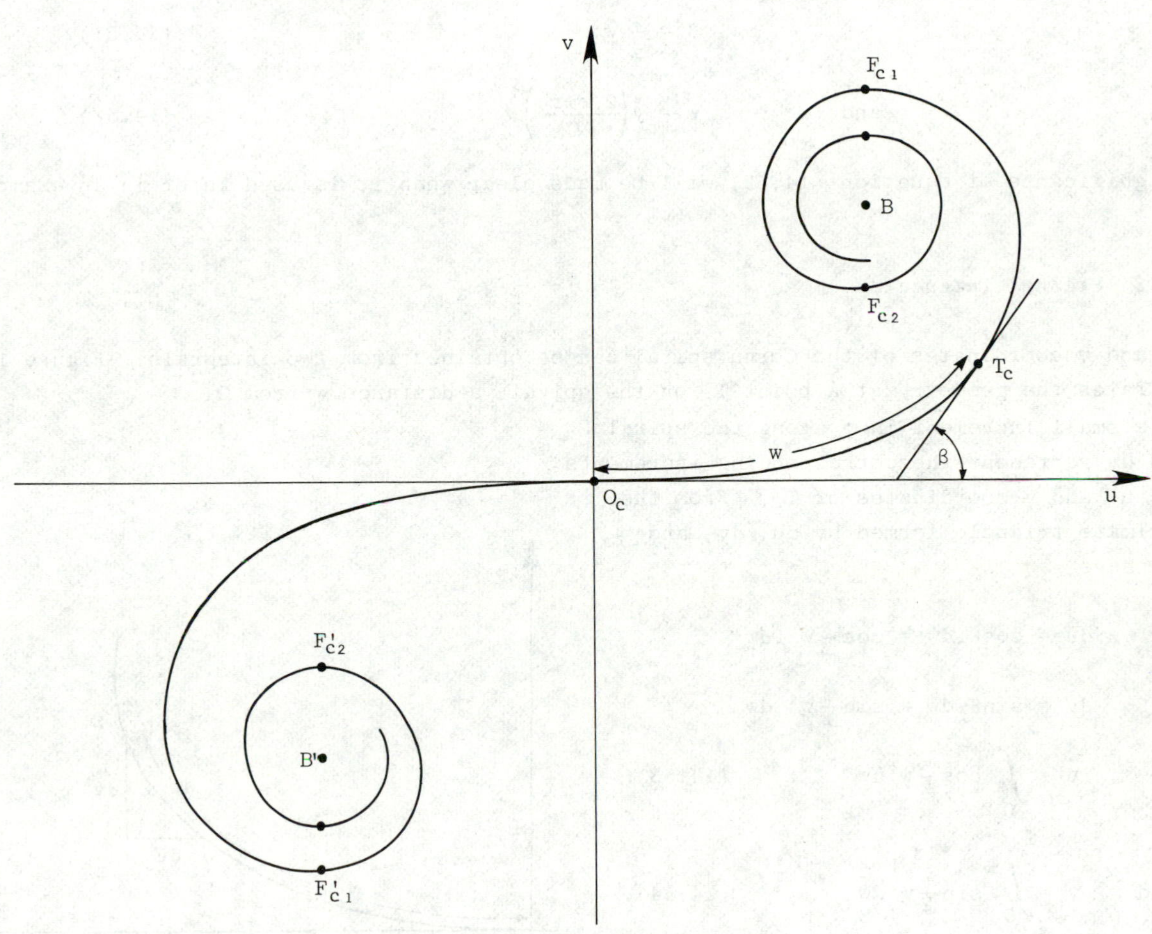

Fig. 14.56 Cornu spiral for the cylindrical wavefront in figure 14.55.

Adding the contributions from the zones OF_1', $F_1'F_2'$, and so on, produces the complete curve known as *Cornu's spiral* after its creator M.A. Cornu (1841 - 1902). The resultant amplitude of the field at P for the unobstructed upper half of the wave is O_cB. Similarly, the resultant due to the lower half of the wave is $B'O_c$, and the total field at P is therefore given by B'B.

The curve has the property that if a tangent at some point T_c on the spiral, figure 14.56, makes an angle β with the u-axis, then $\beta \propto w^2$, where w is the distance O_cT_c along the spiral.

Realise that β is the phase difference between the contributions to the field at P from elements at O and T on the wavefront of figure 14.55 . This phase difference corresponds to a path difference TU, as in figure 14.45, where $TU \simeq \frac{\rho+r}{2\rho r}\cdot y^2$, and y is the distance from T to the SOP line in figure 14.55 . Now, since phase difference = $\frac{2\pi}{\lambda}$ × path difference,

$$\beta = \frac{2\pi}{\lambda}\cdot TU = \frac{2\pi}{\lambda}\cdot\frac{\rho+r}{2\rho r}\cdot y^2 = \text{constant} \times w^2 \qquad (14.50),$$

from the geometry of the wavefront and the property of the spiral $\beta \propto w^2$. We can deduce the constant of proportionality in equation (14.50) by taking T to be the point F_1 in figure 14.55 when it is known that $w = \surd 2$ and $\beta = \pi$. The constant is therefore $\pi/2$, and we can write

$$\beta = \frac{\pi}{2}\, w^2 \qquad (14.51),$$

and

$$w = y\left(\frac{2(\rho+r)}{\rho r\lambda}\right)^{\frac{1}{2}} \qquad (14.52).$$

The significance of equation (14.52) will be made clear when it is used later in an example.

14.10.8 *Fresnel Integrals*

The u and v coordinates of the Cornu spiral can be obtained from two integrals. Figure 14.57 illustrates the geometry at a point T_c on the spiral, a distance w from O_c.
dw is a small increment in w along the spiral.
du and dv represent the corresponding increments in the u- and v-coordinates of T_c. From the approximate triangle formed by du, dv, and dw, we have

$$du = \cos\beta.dw = \cos\frac{\pi}{2}w^2.dw$$

and

$$dv = \sin\beta.dw = \sin\frac{\pi}{2}w^2.dw$$

whence

$$u = \int_0^w \cos\frac{\pi}{2}w^2 dw \qquad (14.53)$$

and

$$v = \int_0^w \sin\frac{\pi}{2}w^2 dw \qquad (14.54).$$

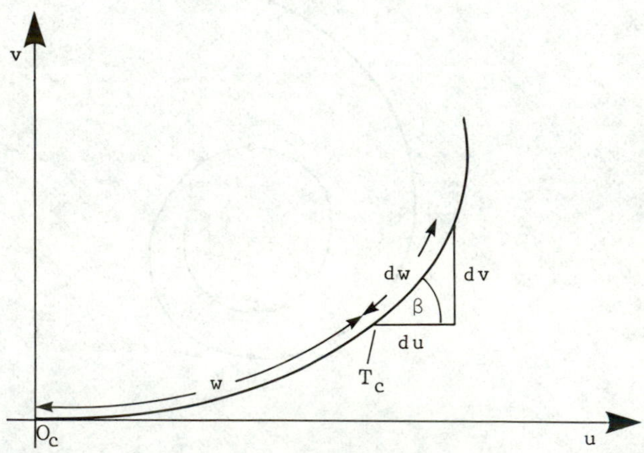

Fig. 14.57 Derivation of Fresnel's integrals.

Equations (14.53) and (14.54) are *Fresnel's integrals*. They can be integrated by expanding the cosine and sine functions as infinite series which result in the values listed in table 14.1 .

Figure 14.58 is a plot of Cornu's spiral showing some of the quantitative features. Note that w in equation (14.52) is dimensionless. This allows its use in calculations with any set of numbers with consistent units.

Table 14.1 Fresnel integrals

w	u	v	w	u	v	w	u	v
0.00	0.0000	0.0000	3.00	0.6058	0.4963	5.50	0.4784	0.5537
0.10	0.1000	0.0005	3.10	0.5616	0.5818	5.55	0.4456	0.5181
0.20	0.1999	0.0042	3.20	0.4664	0.5933	5.60	0.4517	0.4700
0.30	0.2994	0.0141	3.30	0.4058	0.5192	5.65	0.4926	0.4441
0.40	0.3975	0.0334	3.40	0.4385	0.4296	5.70	0.5385	0.4595
0.50	0.4923	0.0647	3.50	0.5326	0.4152	5.75	0.5551	0.5049
0.60	0.5811	0.1105	3.60	0.5880	0.4923	5.80	0.5298	0.5461
0.70	0.6597	0.1721	3.70	0.5420	0.5750	5.85	0.4819	0.5513
0.80	0.7230	0.2493	3.80	0.4481	0.5656	5.90	0.4486	0.5163
0.90	0.7648	0.3398	3.90	0.4223	0.4752	5.95	0.4566	0.4688
1.00	0.7799	0.4383	4.00	0.4984	0.4204	6.00	0.4995	0.4470
1.10	0.7638	0.5365	4.10	0.5738	0.4758	6.05	0.5424	0.4689
1.20	0.7154	0.6234	4.20	0.5418	0.5633	6.10	0.5495	0.5165
1.30	0.6386	0.6863	4.30	0.4494	0.5540	6.15	0.5146	0.5496
1.40	0.5431	0.7135	4.40	0.4383	0.4622	6.20	0.4676	0.5398
1.50	0.4453	0.6975	4.50	0.5261	0.4342	6.25	0.4493	0.4954
1.60	0.3655	0.6389	4.60	0.5673	0.5162	6.30	0.4760	0.4555
1.70	0.3238	0.5492	4.70	0.4914	0.5672	6.35	0.5240	0.4560
1.80	0.3336	0.4508	4.80	0.4338	0.4968	6.40	0.5496	0.4965
1.90	0.3944	0.3734	4.90	0.5002	0.4350	6.45	0.5292	0.5398
2.00	0.4882	0.3434	5.00	0.5637	0.4992	6.50	0.4816	0.5454
2.10	0.5815	0.3743	5.05	0.5450	0.5442	6.55	0.4520	0.5078
2.20	0.6363	0.4557	5.10	0.4998	0.5624	6.60	0.4690	0.4631
2.30	0.6266	0.5531	5.15	0.4553	0.5427	6.65	0.5161	0.4549
2.40	0.5550	0.6197	5.20	0.4389	0.4969	6.70	0.5467	0.4915
2.50	0.4574	0.6192	5.25	0.4610	0.4536	6.75	0.5302	0.5362
2.60	0.3890	0.5500	5.30	0.5078	0.4405	6.80	0.4831	0.5436
2.70	0.3925	0.4529	5.35	0.5490	0.4662	6.85	0.4539	0.5060
2.80	0.4675	0.3915	5.40	0.5573	0.5140	6.90	0.4732	0.4624
2.90	0.5624	0.4101	5.45	0.5269	0.5519	6.95	0.5207	0.4591

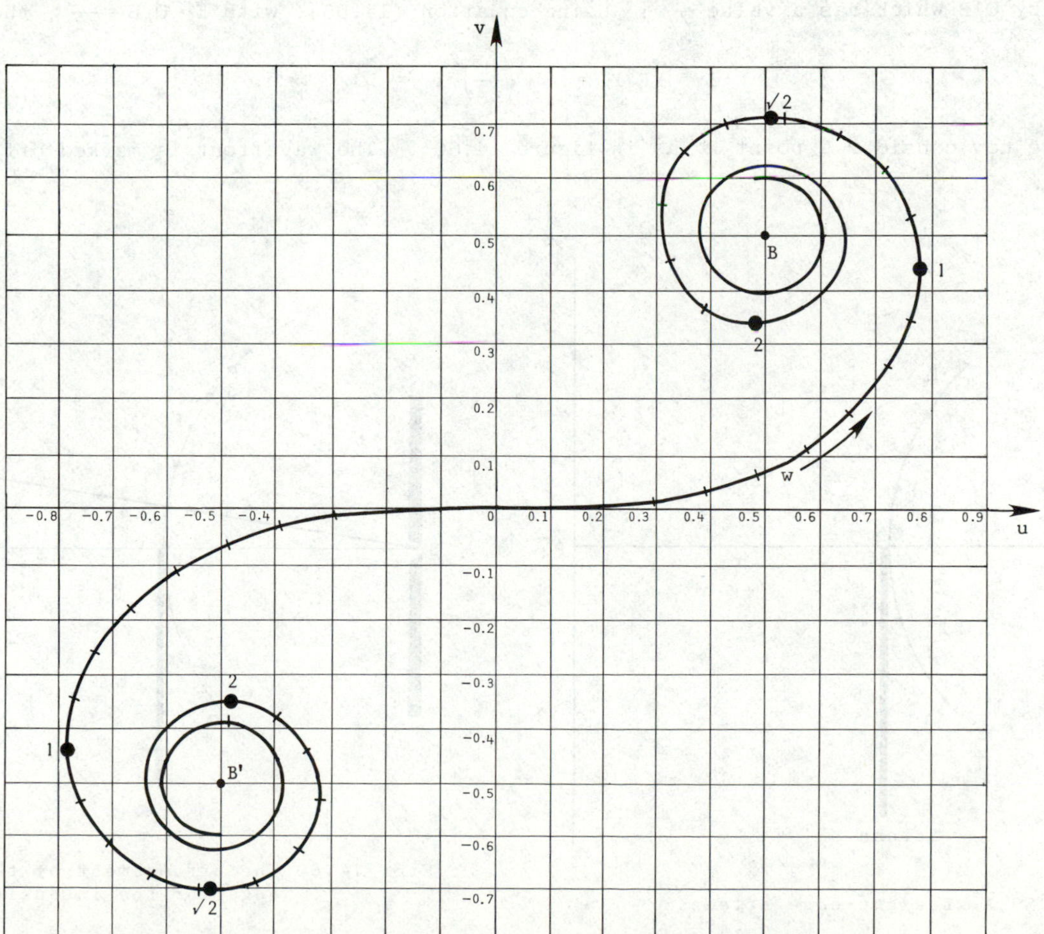

Fig. 14.58 A u,v, and w plot of Cornu's spiral.

w is marked on the spiral in equal increments and some useful points are marked in particular. For example, $w = 1$, $\sqrt{2}$, and 2 correspond to the limits of one half, one, and two half-period zones. The coordinates of the end-points are $B(\frac{1}{2},\frac{1}{2})$ and $B'(-\frac{1}{2},-\frac{1}{2})$.

The amplitude resulting from any portion of the cylindrical wavefront may be found from the curve and squaring this gives the intensity. However, note that the intensity so calculated is relative to the value 2 for the unobstructed wave. It is clear from figure 14.58 that the amplitude E_0 for the unobstructed wave is given by $B'B = \sqrt{2}$. So, the intensity I resulting from a part of the wave with amplitude E can be expressed as a fraction of the incident intensity I_0 due to the unobstructed wave: thus

$$\frac{I}{I_0} = \frac{E^2}{E_0^2} = \frac{E^2}{(\sqrt{2})^2} \quad \text{and this rearranges to give} \quad \frac{I}{I_0} = \tfrac{1}{2}E^2 \qquad (14.55)$$

14.10.9 Fresnel Diffraction by a Straight-edge

As we did with a spherical wavefront, we are now going to investigate the effects of using a variety of obstacles in the path of a cylindrical wavefront. Figure 14.59 represents the partial obstruction of a cylindrical wave by a screen with a straight edge parallel to the source slit S. The half-period strips corresponding to the point P are marked off on the wavefront. Note that only the upper half of the wavefront is effective. The intensity at P is due entirely to the half-period zones above O. On the Cornu spiral the amplitude at P is given by O_cB which has a value $\frac{1}{\sqrt{2}}$. Using equation (14.55), with $E = O_cB = \frac{1}{\sqrt{2}}$, the intensity at P is

$$I = \tfrac{1}{2}E^2 I_0 = \tfrac{1}{2} \times \left(\frac{1}{\sqrt{2}}\right)^2 I_0 = \tfrac{1}{4}I_0$$

Suppose we now consider a point P_1, as in figure 14.60 . The wavefront is marked off in half-

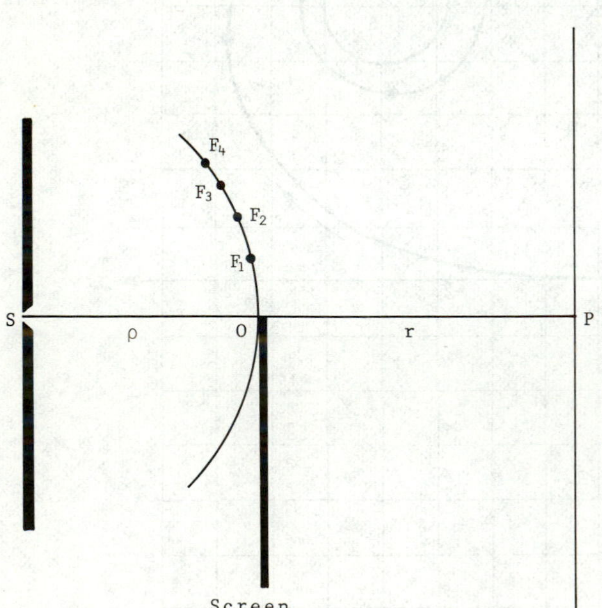

Fig. 14.59 A straight-edge screen.

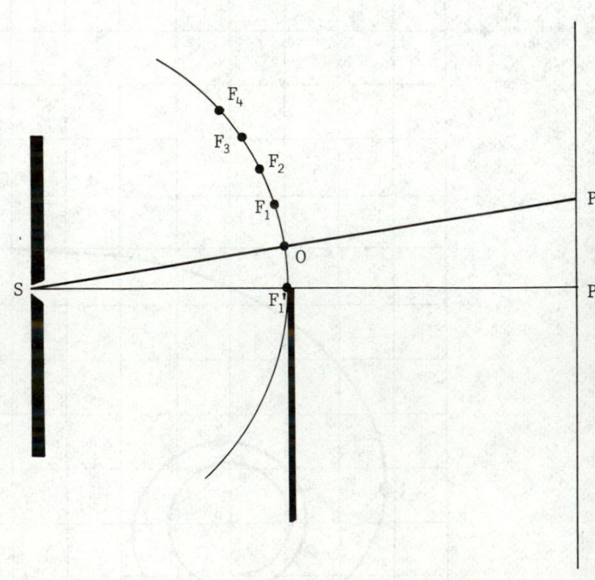

Fig. 14.60 The arrangement of the straight-edge for the intensity at P_1 .

period strip zones relative to P_1. The centre of the strips, O, lies on the line SP_1 and P_1 has been chosen so that one complete half-period strip is exposed below O. The resultant amplitude at P_1 is represented by the line joining F'_{c1} and B in figure 14.56. Clearly, the line $F'_{c1}B$ is more than twice as long as O_cB, so the intensity at P_1 is more than 4 times greater than at P. Considering points above P_1, increasing amounts of the wavefront will be exposed and the tail of the amplitude vector will rotate anticlockwise about the lower half of the spiral whilst its head remains at B. The length of the amplitude vector will oscillate between local maximum and minimum values approaching the constant value B'B for a point where the whole wavefront is exposed.

Going down from P, in figure 14.60, the tail of the amplitude vector moves along the spiral to the right from O_c and the amplitude reduces steadily as the tail spirals in towards B when the whole wavefront is occluded and the amplitude at the point considered is BB = 0. Figure 14.61 is a plot of I/I_0 against w, w being taken as positive from O_c to B and negative from O_c to B'. Notice how this plot indicates a steady dropping off of the intensity in what is the geometrical shadow, and the intensity in the illuminated part of the field oscillates before settling down to a steady intensity. Figure 14.62 is a photograph of a razor blade illuminated by a helium-neon laser so as to demonstrate the shadow of a straight edge. Given the highly coherent source, you will agree, the effect is quite startling.

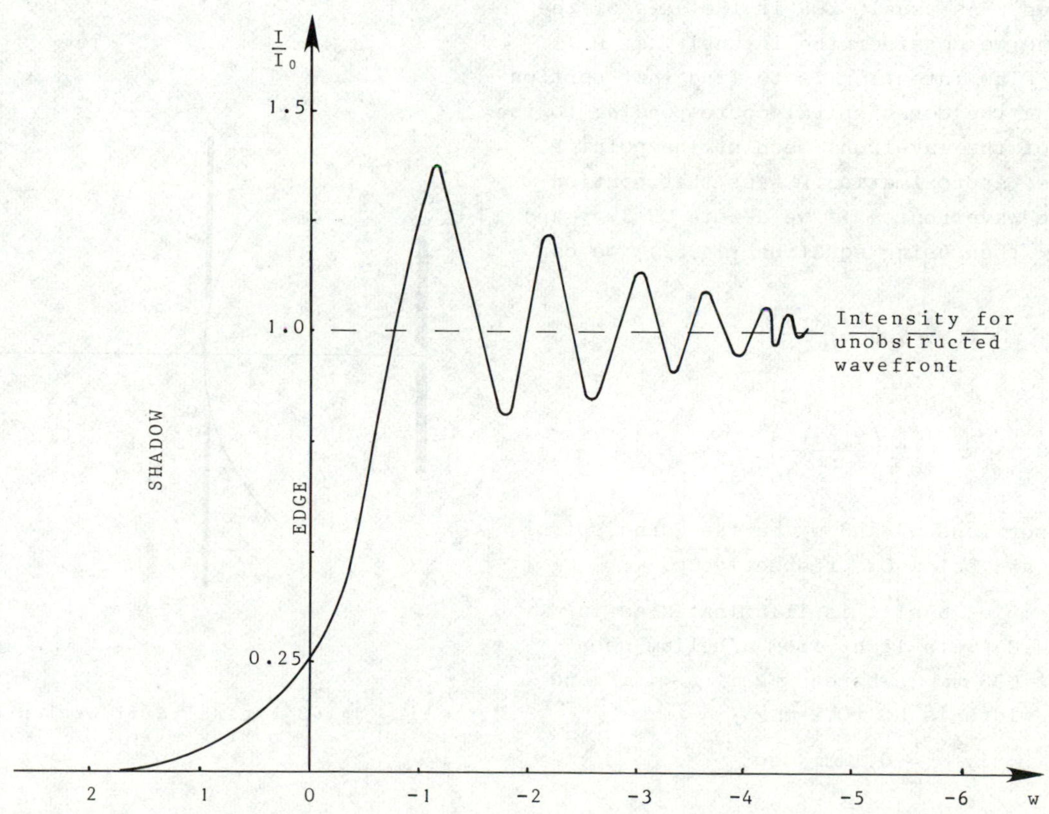

Fig. 14.61 Intensity distribution for a straight-edge.

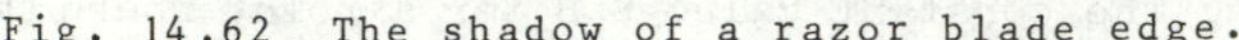

Fig. 14.62 The shadow of a razor blade edge.

14.10.10 Fresnel Diffraction by a Slit

The method used to examine the intensity pattern on the screen in the last section can be adapted very straightforwardly for a slit. Figure 14.63 depicts a slit GH and two points P and P_1 on the screen. As usual, SOP is the axis of the system and we consider the intensity at P firstly. The intention is to find that portion of w along the Cornu spiral corresponding to the portion of the wavefront seen by the point P. To a close approximation P sees that portion GH of the wavefront. If we denote OH by y_1 and OG by y_2, then using equation (14.52), we can write

$$w_1 = y_1\left(\frac{2(\rho+r)}{\rho r\lambda}\right)^{\frac{1}{2}}$$

and
$$w_2 = y_2\left(\frac{2(\rho+r)}{\rho r\lambda}\right)^{\frac{1}{2}}$$

for the portions w_1 and w_2 on the Cornu spiral above O_c and below O_c, respectively.

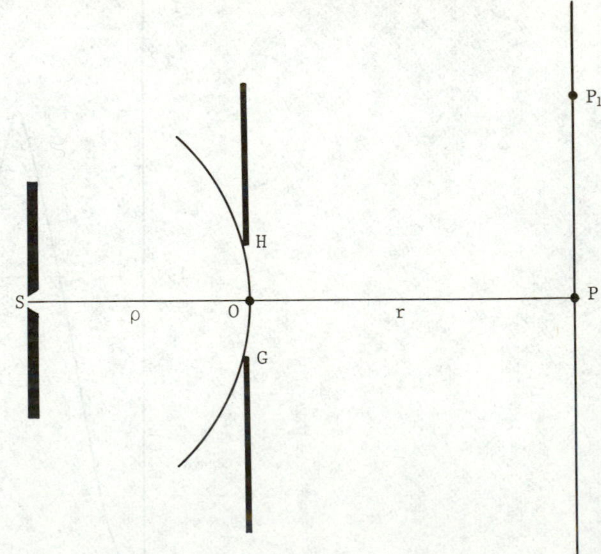

Fig. 14.63 Fresnel diffraction at a narrow slit.

As an example, a slit is illuminated as in figure 14.63 with light from a helium-neon laser ($\lambda = 633$ nm), where $\rho = 2$ m, $r = 4$ m, and the slit width is HG = 0.2 mm.

Now, $|y_1| = |y_2| = 0.1$ mm, so

$$|w_1| = |w_2| = y_1\left(\frac{2(\rho+r)}{\rho r\lambda}\right)^{\frac{1}{2}} = 0.1\times10^{-3}\left(\frac{2(2+4)}{2\times4\times633\times10^{-9}}\right)^{\frac{1}{2}} = 0.154\ .$$

Remember w is dimensionless and therefore the results for w_1 and w_2 require no units. w_1 and w_2 are illustrated in figure 14.64 .

The amplitude at P is given by the chord joining the end-points H_c and G_c of w_1 and w_2, respectively. This chord length is the amplitude $E \simeq \Delta w = |w_1| + |w_2| = 0.308$

Now, the intensity at P is

$$I = I_0 \times \frac{E^2}{2} = I_0 \times \frac{(0.308)^2}{2} = 0.047 I_0 ,$$

having used equation (14.55) . To find the intensity at P_1 we must redraw the wavefront showing the strip division from P_1's point of view, figure 14.65 . Note that $OP_1 \simeq r$ if $\widehat{PSP_1}$ is small. The exposed wavefront in the slit aperture is still equal to the slit width, so the length on the spiral is still $\Delta w = 0.308$. However, since the exposed wavefront lies below O, the length Δw lies somewhere on the lower half of the spiral, figure 14.66 . Clearly, $E = G_cH_c$ is no longer necessarily approximated by Δw and we must calculate the position of H_c or G_c. Suppose we decide to find the position of H_c, then we must find $w_1 = O_cH_c$. Equation (14.52) enables us to do this if we can find the value of y_1 indicated in figure 14.67 .

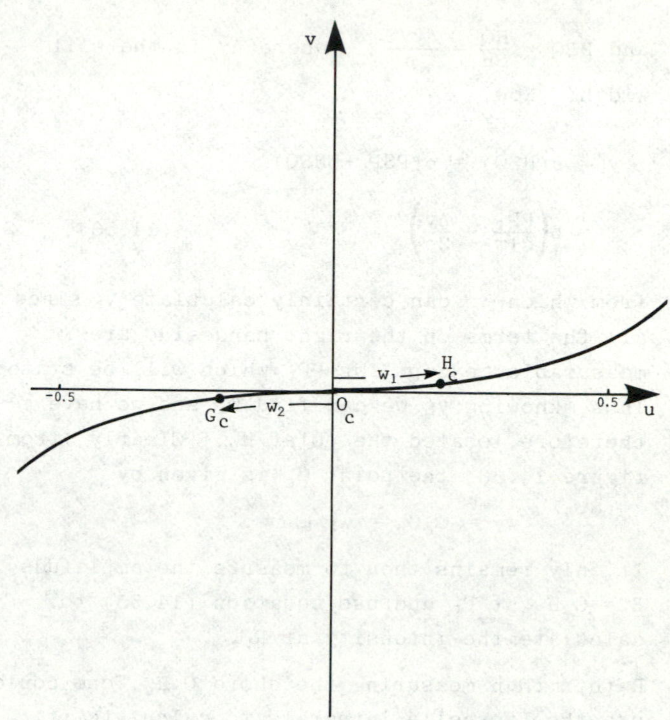

Fig. 14.64 Cornu spiral used with a slit aperture.

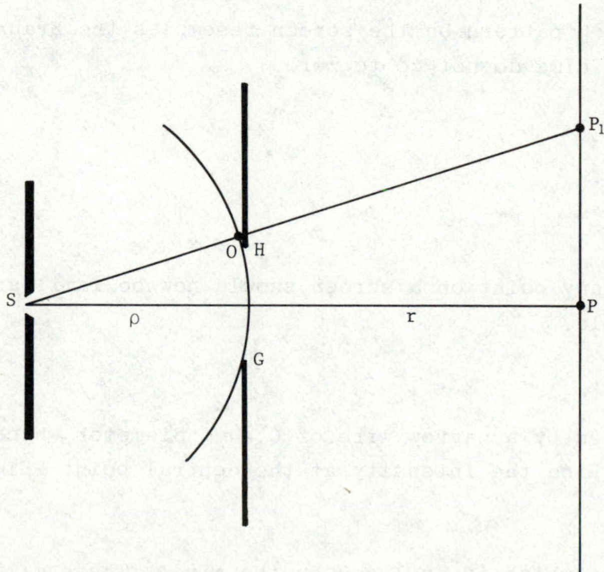

Fig. 14.65 Wavefront division for the

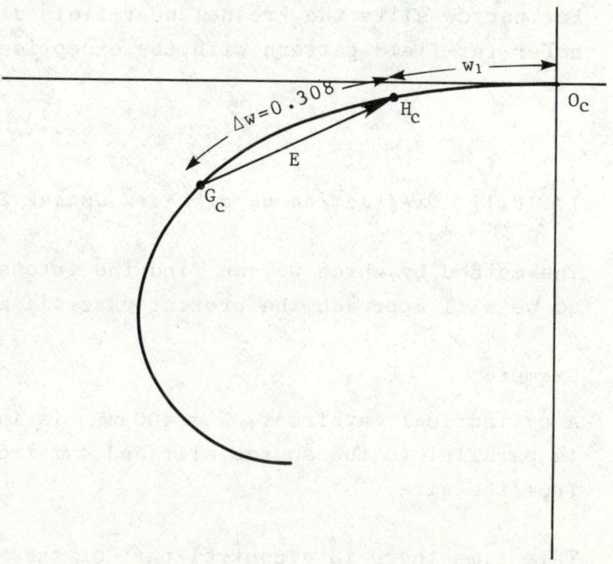

Fig. 14.66 Δw on the spiral corresponding to the wavefront exposed in figure 14.65 .

In figure 14.67, $SQ \simeq \rho$, $\widehat{PSP_1} \simeq \frac{PP_1}{\rho+r}$,

and $\widehat{HSQ} \simeq \frac{HQ}{SH} = \frac{\Delta y/2}{\rho}$, where Δy is the slit width. Now,

$$y_1 \simeq \rho(\widehat{HSO}) = \rho(\widehat{PSP_1} - \widehat{HSQ})$$

$$= \rho\left(\frac{PP_1}{\rho+r} - \frac{\Delta y}{2\rho}\right) \qquad (14.56),$$

from which we can certainly calculate y_1 since all the terms on the right hand side are measurable, except for PP_1 which will be stated. Thus, knowing y_1 we can find w_1 and we have therefore located the point H_c. Clearly, from figure 14.66, the point G_c is given by

$$w_2 = O_cG_c = w_1 + \Delta w .$$

It only remains then to measure the amplitude $E = G_cH_c$ at P_1 and use equation (14.55) to calculate the intensity at P_1.

Rather than measuring the chord G_cH_c, one could use the Fresnel's integrals to calculate it. For example, having obtained w_1 and w_2 as described, it is a simple matter to find the corresponding u- and v-coordinates (u_1, v_1) and (u_1, v_2). Then by Pythagoras' theorem,

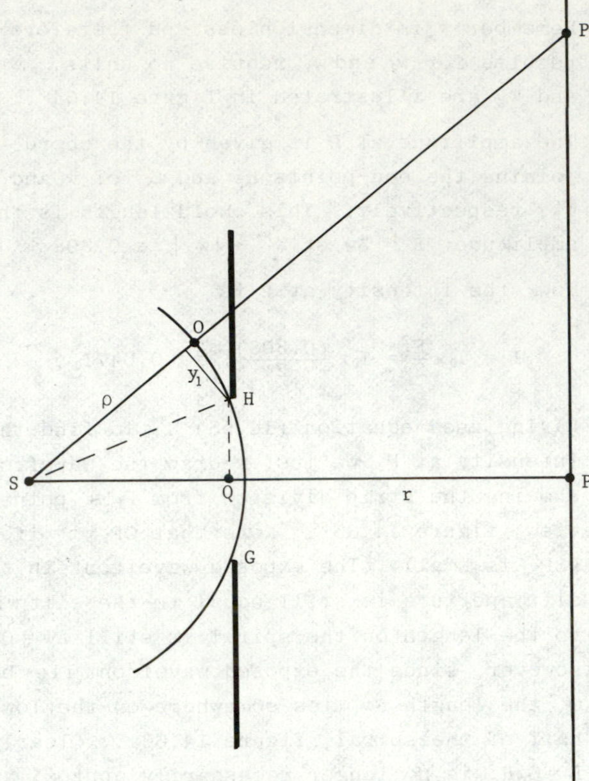

Fig. 14.67 The geometry of the wavefront and the slit for the intensity at P_1. (Angles exaggerated).

$$E^2 = (u_2 - u_1)^2 + (v_2 - v_1)^2$$

and this affords a more accurate result for E.

For narrow slits the Fresnel near-field diffraction pattern on the screen resembles the Fraunhofer far-field pattern with the exception that minima do not go to zero.

14.10.11 *Diffraction by a Narrow Opaque Strip*

The method by which we can find the intensity at any point on a screen should now be familiar, so we will approach the present case via an example.

Example

A cylindrical wavefront, $\lambda = 400$ nm, is interrupted by a narrow wire of 0.4 mm diameter which is parallel to the source slit and 4 m from it. Find the intensity at the central point P 1m from the wire.

This time there is a central part of the wavefront which is occluded by the wire. There will be two contributing amplitude vectors shown in figure 14.68 as $B'G_c$ and H_cB. These are

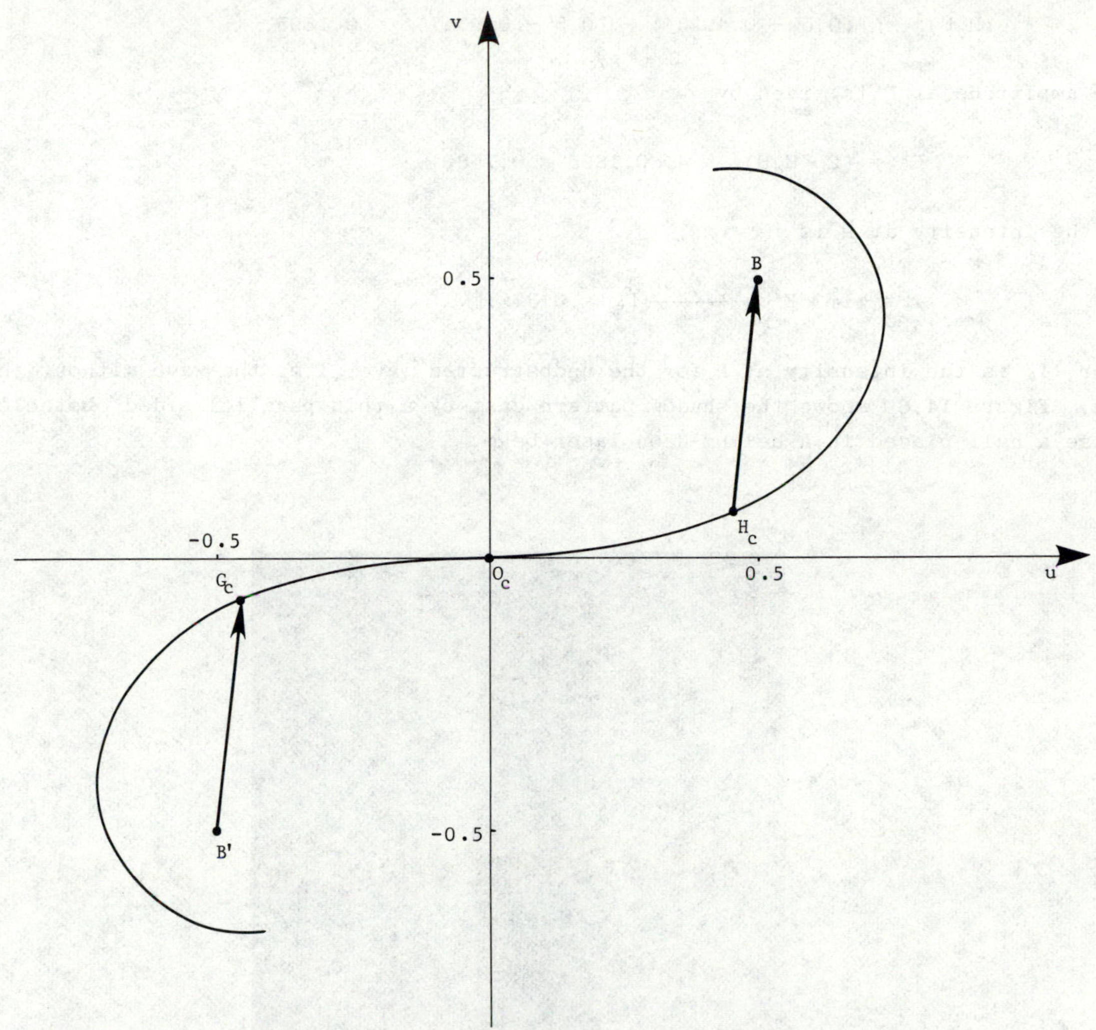

Fig. 14.68 The Cornu spiral used in the case of a straight narrow wire obstacle. Not drawn to scale.

associated with the parts of the wavefront below and above the wire, respectively. The points G_c and H_c correspond to the points G and H on the wavefront such that GH = 0.4 mm, the thickness of the wire. By symmetry, $B'G_c$ is equal and parallel to H_cB so we need only find w_1 in order to locate H_c. The amplitude vector at P is then

$$E = B'G_c + H_cB = 2.H_cB\ .$$

Proceeding as for the slit,

$$w_1 = y_1\left(\frac{2(\rho+r)}{\rho r\lambda}\right)^{\frac{1}{2}} = 0.2\times10^{-3}\left(\frac{2(4+1)}{4\times1\times400\times10^{-9}}\right)^{\frac{1}{2}} = 0.5$$

The u- and v-coordinates for H_c, obtained from the table of Fresnel integrals, are 0.4923 and 0.0647, respectively. The u- and v-coordinates for B are each 0.5, so

$$(H_cB)^2 = (0.5 - 0.4923)^2 + (0.5 - 0.0647)^2 = 0.1895,$$

and the amplitude at P is given by

$$E^2 = (2 \times H_cB)^2 = 4 \times 0.1895 = 0.7580 \ .$$

Hence, the intensity at P is

$$I = \tfrac{1}{2}I_0 \times E^2 = \frac{0.7580}{2} I_0 = 0.379I_0 \, .$$

Remember, I_0 is the intensity at P for the unobstructed wave, i.e. the wave without the wire present. Figure 14.69 shows the shadow pattern cast by a thin parallel sided obstacle - in this case a nail placed in a helium-neon laser beam.

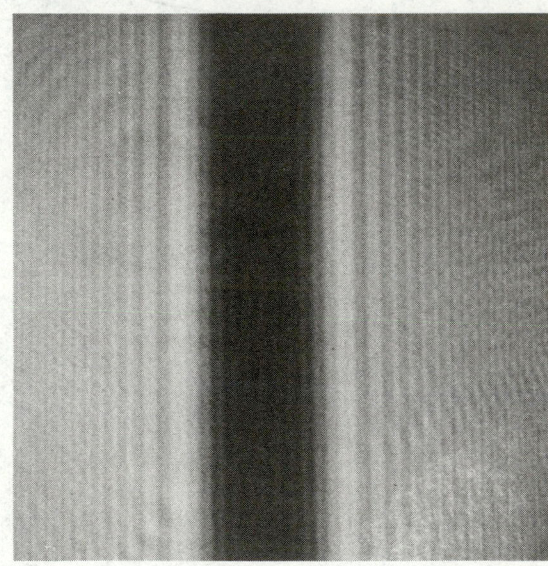

Fig. 14.69 The shadow cast by a nail placed in a laser beam.

1 Plane waves of 550 nm wavelength are incident normally on a narrow slit of width 0.25 mm. Calculate the distance between the first minima on either side of the central maximum when the Fraunhofer diffraction pattern is imaged by a lens of focal length 60 cm. The lens should be 'large'. Why is this?

Ans. 2.64 mm. See W.P.O. page 125, question 1.

2 Plane waves ($\lambda = 550$ nm) fall normally on a slit 0.25 mm wide. The separation of the fourth order minima of the Fraunhofer diffraction pattern in the focal plane of the lens is 1.25 mm. Calculate the focal length of the lens.

Ans. 7.10 cm. See W.P.O. page 126, question 2.

3 Light from a distant point source enters a converging lens of focal length 22.5 cm. How large must the lens be if the Airy disc is to be 10^{-6} m in diameter? $\lambda = 450$ nm.

Ans. 24.7 cm. See W.P.O. page 131, question 11.

4 A telescope objective is 12 cm in diameter and has a focal length of 150 cm. Light of mean wavelength 550 nm from a star is imaged by the objective. Calculate the size of the Airy disc.

Ans. 0.01678 mm. See W.P.O. page 131, question 10.

5 Find the radius of the first bright ring in the Airy pattern of the 91 cm diameter Lick Observatory refracting telescope, the focal length of which is 17 m. Take the mean wavelength of white light as 550 nm. Note from figure 14.11 that $\frac{\pi}{\lambda} d \sin\theta = 5.14$.

Ans. 0.017 mm.

6 Assuming Rayleigh's criterion can be applied to the eye, how far apart must two small lights be to be seen as two at a distance of 1 000 m? Take the pupil diameter as 2.5 mm, the wavelength to be 555 nm, and the eye's refractive index 1.333 .

Ans. 27.1 cm. See W.P.O. page 133, question 13.

7 A microscope is just able to resolve the rulings on a grating with 5 000 lines per cm. Using a mean white light wavelength of 550 nm, calculate the objective's numerical aperture.

Ans. 0.1375 .

8 What would be the angle of the cone of light which would just fill the objective in the previous question?

Ans. 15.8°.

9 Show that $2\frac{a}{b} - 1$ full principal maxima occur under the central diffraction maximum in a two-slit Fraunhofer pattern.

10 For a grating with N slits, show that there will be (N - 1) minima and (N - 2) secondary maxima between successive principal maxima.

11 This is a simple one! Use the grating equation $a \sin\theta_m = m\lambda$ to show that the limit of

the number of principal maxima in the far-field diffraction pattern is $m \leq \frac{a}{\lambda}$. Use the fact that $\sin\theta$ cannot exceed unity.

12 White light, consisting of wavelengths from 400 nm to 750 nm, is incident normally on a grating containing 5 000 lines per cm. A 1 m focal length converging lens focuses the Fraunhofer pattern in its back focal plane. What is the extent of (i) the first order and (ii) the second order spectra on the screen?

Ans. (i) 20.1 cm, (ii) 69.7 cm.

13 Prove that the maximum value of a grating's resolving power is aN/λ. Hint: use the result from question 11.

14 A beam of polychromatic light falls normally on a transmission grating with a grating constant of 2.12×10^{-6} m. If the second order spectrum is 12.5 mm in breadth, and the extreme wavelengths are 450 nm and 650 nm, find the focal length of the converging lens used to focus the Fraunhofer pattern on the screen.

Ans. 6.625 cm. See W.P.O. page 127, question 4.

15 What will be the angular separation of the two sodium lines $\lambda = 589.0$ nm and $\lambda' = 589.6$ nm in the first order spectrum produced by a diffraction grating of 500 lines per mm, the light being incident normally on the grating?

Ans. 0.02^0. See W.P.O. page 127, question 5.

16 Parallel light from a hydrogen tube is incident normally on a diffraction grating and the angles for the C and F lines in the second order spectrum are 41.017^0 and 29.083^0, respectively. If the wavelength of the C line is $\lambda_C = 656.3$ nm, find (i) the number of lines per cm on the grating and (ii) the wavelength for the F line.

Ans. (i) 5 000. (ii) 486.1 nm. See W.P.O. page 128, question 6.

17 Parallel monochromatic light falls normally on a diffraction grating and is focused on a screen by a converging lens which follows the grating. How is the Fraunhofer pattern affected by (a) the line spacing, a, on the grating, and (b) the number of lines on the grating?

Ans. See W.P.O. page 128, question 7.

18 21 cm microwaves fall on a circular diaphragm from a source at infinity. If the aperture is slowly increased from its closed position, at what radius will the first maximum intensity value occur on the axis 3 m from the diaphragm? Hint: use equation (14.44).

Ans. 79 cm.

19 Compute the intensity at a point P_1 in the geometrical shadow of a straight edge, such that one whole half-period zone above O is occluded. Hint: the tail of the amplitude vector will be on the upper curve at the point marked $w = \sqrt{2}$ in figure 14.58 .

Ans. $0.024\ I_o$.

15 ATOMS, SPECTRA, AND SOURCES

15.0 INTRODUCTION

This chapter will deal briefly and elementarily with sources of light and the appearance they present when they are viewed through a spectroscope. In order to understand something about the emission of light by matter we must look at the structure of atoms and molecules. The emission and absorption of light will form a natural part of our discussion and will, in turn, enable us to understand the manner in which light sources work.

15.1 THE ATOMIC STRUCTURE OF MATTER

Everyday experience tells us that matter exists in gaseous, liquid, and solid states. Just why a particular substance exists in any one of these states at a particular time is not a subject we shall pursue here. What interests us is the fact that matter is composed of very small particles called *atoms* and these atoms, and combinations of atoms, are responsible for electromagnetic radiation.

15.1.1 The Size of Atoms

Atoms are small, but how small? How should we picture atoms? These two questions are related since to measure the size of an atom we must have some idea of its form. Modern theory suggests we should not picture the atom as a body with a sharp boundary like a billiard ball, say. Rather we should imagine it like a cloud with an indistinct boundary. Given this model of the atom the problem then is to measure its size.

Clearly, this cannot be done exactly but we can determine an effective size. If it is assumed that atoms occupy a spherical volume methods can be designed to measure the closest approach between the centres of atomic spheres, and this distance defines the *atomic diameter*. This diameter is found to lie between 0.07 and 0.6 nm*.

X-ray diffraction is normally used to determine atomic diameters in the same way that the grating constant of a diffraction grating can be obtained from the diffraction pattern for visible electromagnetic radiation. However, a fairly simple experiment allows one to obtain an 'order of magnitude' judgement of atomic diameters. Suppose we take a known volume of oil and allow it to spread out to its maximum extent on the surface of a pool of water. If we then measure its area we can calculate its thickness since we know its volume. The experiment results in an oil layer about 1 nm thick when its reaches its maximum extent. This thickness is due to a single layer of oil molecules (combinations of atoms forming the oil) so the constituent atoms must have a diameter less than 1 nm.

* The Ångström unit is convenient for stating atomic diameters: 1 Ångström $\equiv$ 0.1 nm $\equiv 10^{-10}$ m. Interatomic distances in solids and liquids lie between 1 and 10 Ångström.

15.1.2 The Structure of Atoms

So far, you may have the impression that atoms are structureless cloud formations. Despite their small size this could not be further from the truth; they have a complex structure. Atoms are electrically neutral overall although they possess a positively charged nucleus and a negatively charged electron cloud. J.J. Thomson (1856 - 1940) discovered the electron in 1897, and in a famous series of experiments on the conduction of electrons in a vacuum he was able to show that electrons are emitted from a piece of metal heated in a vacuum. He was able to measure the ratio of the electron's charge to its mass. R.A. Millikan (1868 - 1953) measured electric charges on oil drops and found the charges were integer multiples of a basic charge which he called the electronic charge. This negative charge had the same magnitude as the positive charge on a hydrogen atom which had lost its sole electron. The value of the electronic charge, taken with Thomson's value of the electron's charge/mass ratio, gave the mass of the electron as 1/1837th of the total mass of the hydrogen atom. Thus, Thomson and Millikan identified the negatively charged subatomic particle, the electron, which has a much smaller mass than the hydrogen atom although it possesses the same, though opposite, charge.

An experiment conceived by Ernest Rutherford (1871 - 1937), and carried out with the assistance of his students H.Geiger and E.Marsden, led to the conceptual relationship between the positively and negatively charged parts of the atom. Rutherford was using a beam of alpha particles, the positively charged particles which turned out to be helium atoms stripped of their two electrons, to probe the atom. Alpha particles emitted from radium emerged at a high velocity from a hole in a lead container. This beam of positively charged particles was shot at a very thin gold foil. Rutherford expected the beam to traverse the thin gold foil and thereby be dispersed slightly. To his surprise a variety of deflections were detected on the fluorescent screen surrounding the gold foil. Some particles were even deflected backwards, figure 15.1 .

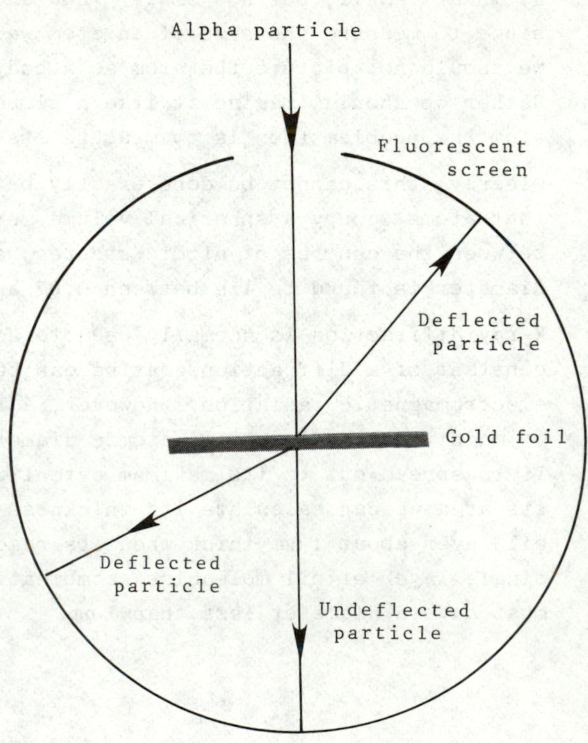

Fig. 15.1 Rutherford's alpha particle scattering experiment.

Rutherford realised that the positively charged part of the gold atoms deflecting the alpha particles must be located in a relatively small volume of the atom. To account for the range of deflections observed Rutherford calculated that the positively charged part of the atom formed a nucleus with a diameter 1/100 000th the atomic diameter. That is, if the atomic diameter is 10^{-10} m then the diameter of the nucleus is 10^{-15} m.

To obtain some idea of the significance of this relationship imagine we blow up the atom so that it is the size of a soccer ball, then the nucleus would be a ball about 0.002 mm in diameter! It has been said of matter that it is almost entirely empty. If it were possible to compress the subatomic particles of 5 000 battleships together so as to remove all the space in the atoms, the result would be the size of a cricket ball!

15.1.3 The Nucleus: Protons and Neutrons

The nucleus can be regarded as consisting of positively charged particles called *protons* and electrically neutral particles of almost equal mass called *neutrons*. The nucleus of the hydrogen atom, the lightest atom, consists of a single proton. Helium has two protons and two neutrons in its nucleus. The number of neutrons is generally about equal to the number of protons but the neutron/proton ratio increases slowly as the number of protons increases. The nucleus interests us only in that gamma rays, those very high frequency radiations at the short wavelength end of the electromagnetic spectrum, arise from energy level changes there.

Since atoms are electrically neutral, an atom must contain equal numbers of electrons and protons. This number is known as the *atomic number*.

15.1.4 Elements, Compounds, and Ions

All atoms which have the same atomic number have, by definition, the same number of electrons. The chemical behaviour of atoms depends on their electrons, so atoms with the same atomic number have the same chemical behaviour. This leads to the concept of an *element* as a substance in which all the atoms have the same atomic number. Elements are the basic substances from which all matter is made, and to date there are 105 of which only 92 occur naturally. The rest have been made in the laboratory.

Two or more elements can combine to form another substance known as a *compound*. There are about four million known compounds and the number of possible compounds is very much higher. For reasons we shall not discuss, an atom may gain or lose electrons which results in a negatively or positively charged particle called an *ion*. The word is derived from the Greek meaning wanderer, a name which describes the fact that an ion will move in or against the direction of an applied electric field according as the charge is negative or positive.

5.2 SPECTRA

15.2.1 Introductory note

Along with the chemical properties, the spectroscopic properties of atoms are primarily determined by the numbers and the arrangements of the electrons with respect to their nuclei. In the case of molecules there are the additional aspects of vibrational and rotational modes of the molecules which give rise to different spectra from those produced by atoms alone. We shall now look at the different types of spectra produced by atoms and molecules.

15.2.2 Dispersion of Light

Recall that white light is dispersed when it is passed through a prism. It is possible to show this dispersion in the laboratory with the apparatus shown in figure 15.2 . Light from a source illuminates the slit which acts as a secondary source. Lens 1 collimates the beam which then traverses the prism, thereupon being dispersed into its component frequencies. Rays of like frequency are still parallel on emerging from the prism and are focused on the screen

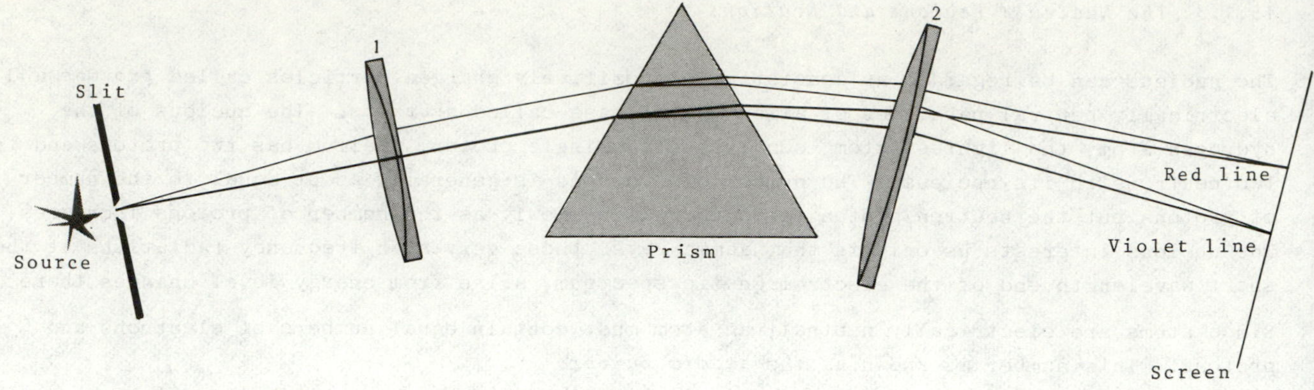

Fig. 15.2 The prism spectrograph.

as line images of the slit. Different colours form different line images in different places on the screen with red and violet lines at the extremes. The result on the screen, that is all the line images, is called a spectrum.

The camera may be replaced by an astronomical telescope when the eye can be used as the detector. Provided with a Ramsden eyepiece fitted with cross-wires the instrument becomes a spectrometer and can be used to measure the deviations of the frequency components.

A diffraction grating may be used as an alternative method of displaying spectra. The details have been covered in section 14.7.1, and first and second order spectra were shown in figure 14.38 . Light dispersed in such ways is also dispersed in terms of its energy. As noted in section 12.1.5, electromagnetic radiation is emitted in 'packets' or quanta of energy. These quanta are called *photons* from the Greek photos = light. The term came into use in 1926 following Arthur Compton's demonstration of the corpuscular nature of X-rays in 1923. The energy of the photon depends upon the frequency of the light and equation (12.28), $E = h\nu$, indicates that violet light with its higher frequency will have a greater energy per photon than will red light. In terms of wavelengths, we can combine the equations $E = h\nu$ and $c = \nu\lambda$ to give

$$E = \frac{hc}{\lambda} \qquad (15.1)$$

which demonstrates that the photon energy is less for larger wavelengths.

Using a spectrometer or a simple direct vision spectroscope (a slit, a convex lens, and alternate crown and flint prisms) some immediate differences are noticed in the spectra produced by different sources. Looking in turn at a tungsten filament lamp, a sodium street lamp, and finally a mercury street lamp (the blueish-white ones), one sees completely different spectra. In the case of the tungsten lamp there is an unbroken progression of colours from red to violet. The line images of the slit merge to produce what is known as a *continuous spectrum*. Figure 15.3 illustrates a continuous spectrum taken with a camera attached to a prism spectrograph. When displayed in colour the slit images for each wavelength are not distinguishable, one colour seems to change gradually into the next with no definite demarcations visible.

In the case of sodium and mercury lamps the line images are discrete, and such spectra are called *atomic spectra*. Atomic spectra and continuous spectra have different origins which we shall now discuss.

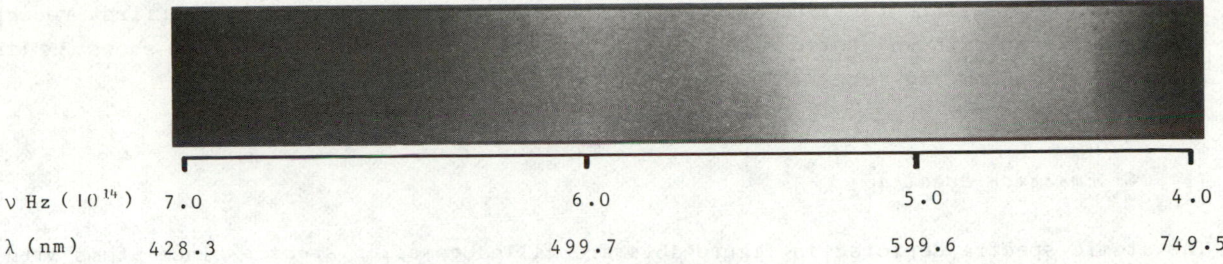

Fig. 15.3 The continuous spectrum from an incandescent lamp.

15.2.3 Atomic Spectra

We suggested looking at sodium and mercury street lamps with a spectroscope. If you do this you will see discrete coloured lines as opposed to the continuum of lines merging to form the continuous spectrum of figure 15.3 .

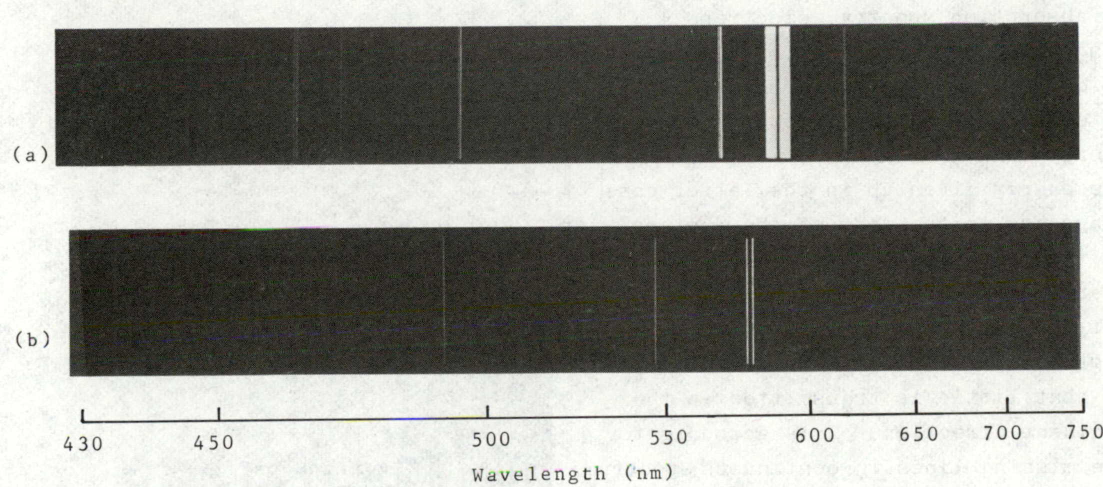

Fig. 15.4 Atomic line spectra for (a) sodium and (b) mercury.

In the case of the sodium light a bright yellow line is the most noticeable feature. This is, in fact, two lines close together with wavelengths 588.995 nm and 589.592 nm, but unless your spectroscope can resolve them they will appear as one. Other lines are less intense and may not be visible. Looking at a mercury lamp, you will see a variety of coloured lines, and if you look at a common fluorescent tube you will see these lines against the background of a continuous spectrum. These bright lines are the atomic spectra of sodium and mercury atoms, respectively. They result from atoms being heated, after which they give up the energy they gained in the form of light. When an atom gains energy, in the cases considered here, it raises electrons to higher energy levels and they are able to give it out again on returning to a lower energy level. Since the images of the slit in the spectrograph appear as well-defined lines, they are called *line spectra*. Each atom has its own characteristic line spectrum. In fact, the existence of the element helium, named from the Greek word helios = sun, was first recognised by detecting an unknown atomic spectrum in the solar spectrum. Figure 15.4 shows the line spectra for sodium and mercury lamps.

15.2.4 *Emission Spectra*

The atomic spectra depicted in figure 15.4 are called *emission spectra* since atoms with their electrons raised to higher energy levels emit photons as they return to lower levels. If light consisting of a continuous spectrum is shone through an atomic vapour, under suitable conditions it is possible to observe dark lines in the spectrum. Some light has been absorbed in passing through the vapour. Electrons in the atoms of the vapour have absorbed specific amounts of energy, removing in the process certain frequencies associated with the energy so absorbed. These dark lines on the background of a continuous spectrum are referred to as an *absorption spectrum*.

15.2.5 *Absorption Spectra*

In absorption spectra, electrons absorb the energy of photons and, in so doing, move further from the nucleus. In emission spectra electrons lose energy and move nearer to the nucleus. The energy given up in the latter case reappears in the form of the photons emitted. The electrons absorb the energy associated with single photons but, in giving up the energy the photons are emitted in random directions so that little is transmitted in the incident direction. This accounts for the missing lines in continuous spectrum light passing through an atomic vapour.

The solar spectrum shows dark absorption lines known as *Fraunhofer lines*, after their discoverer. They are caused by the light from the sun's surface passing through the atomic vapours of the sun's atmosphere, figure 15.5, where specific photons are absorbed. Solar scientists use absorption and emission spectra to learn something about the composition of the sun's atmosphere.

Absorption spectra can be demonstrated in the laboratory with the apparatus shown in figure 15.6(a).

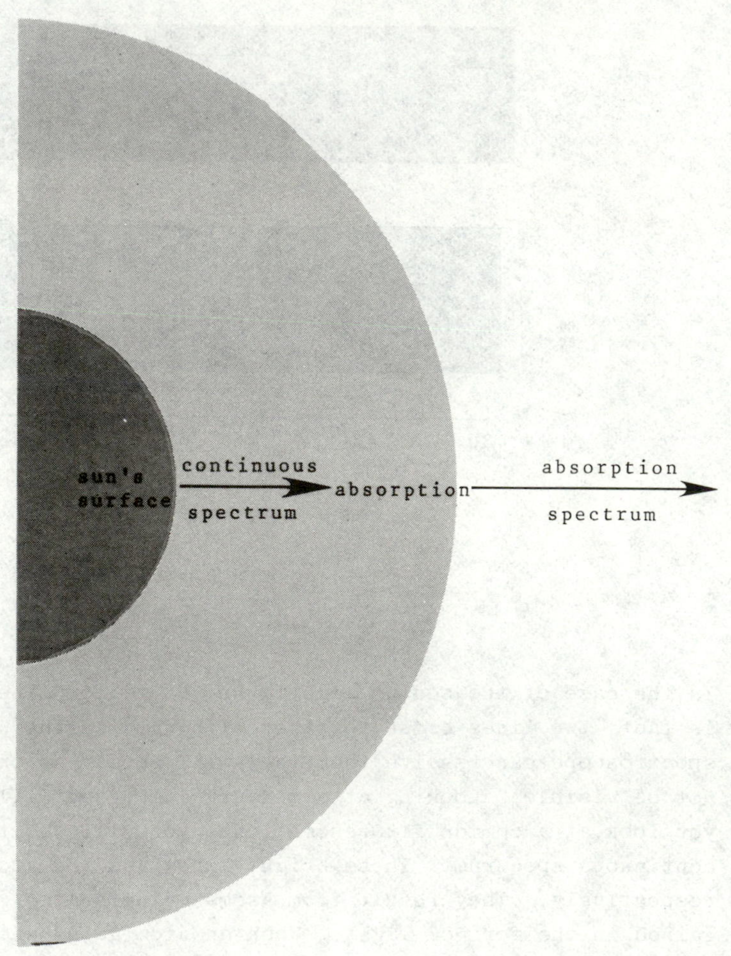

Fig. 15.5 Absorption of light by the sun's atmosphere.

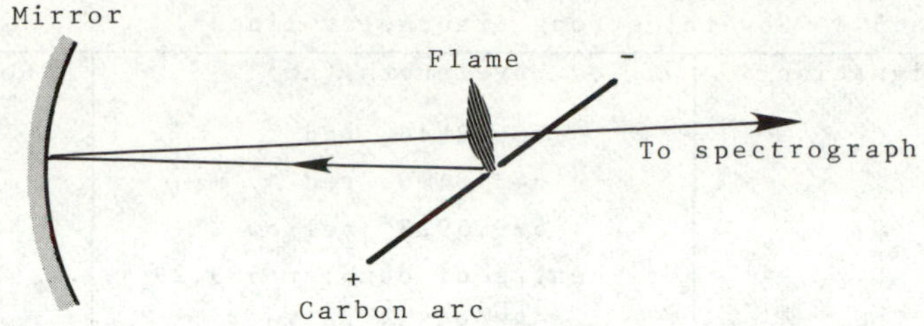

Fig. 15.6(a) Arrangement for viewing the absorption spectrum of the sodium D line.

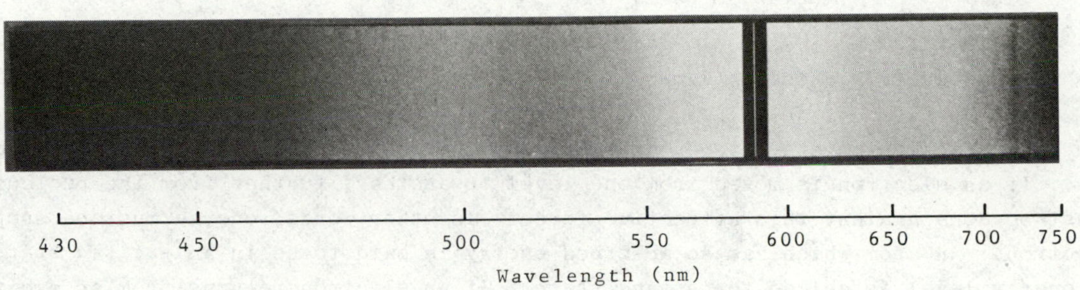

Fig. 15.6(b) Absorption spectrum for sodium.

A carbon arc with a sodium chloride core is run on a fairly large current. A bright yellow flame of hot sodium vapour gives rise to a sodium emission spectrum. If a concave mirror is positioned so as to image the bright positive pole of the anode on the slit of a spectrograph, and the light from the pole is made to traverse the flame on its way to the slit, then absorption of the frequencies corresponding to the sodium D lines will occur in the flame. Figure 15.6(b) shows the dark lines of the sodium yellow doublet. The sodium atoms in the flame are able to absorb the sodium light from the pole because the latter is at a higher temperature.

Transparent chambers can serve as absorption cells for the production of absorption spectra. Light from a source emitting a continuous spectrum is passed through the chamber, into which a gas or vapour is introduced for analysis. The change in the transmitted spectrum is the absorption spectrum of the vapour. The cells may be enclosed in an oven to accommodate substances which only vapourise at high temperatures.

Table 15.1 gives some strong Fraunhofer lines. The letter designation arose in the early days of the investigation of these absorption spectra when Joseph von Fraunhofer (1787 - 1826) from about 1814 mapped more than 500 of them. The brightest were designated by letters, a system still in use. About 25 000 Fraunhofer lines are known to exist in the solar spectrum between wavelengths 295 and 1000 nanometres.

Table 15.1 Several strong Fraunhofer lines

Designation	Wavelength (nm)		Source
C	656.2816	red	H
C'	643.8470	red	Cd
D_1	589.5923	yellow	Na
D	centre of doublet 589.29		Na
D_2	588.9953	yellow	Na
D_3 or d	587.5618	yellow	He
c	495.7609	green	Fe
F	486.1327	blue	H
F'	479.9912	blue	Cd
f	434.0465	violet	H
g	422.6728	violet	Ca
K	393.3666	violet	Ca

15.2.6 Energy Levels - Electron Jumps

A force of attraction exists between a negatively charged electron and a positively charged nucleus. If an electron is moved from one level to another, further from the nucleus, then work must be done against this attractive force. In other words, energy must be supplied to the electron. An atom which has so absorbed energy is said to be in an *excited state*. The lowest energy level is called the *ground state*. If an electron returns from an excited state to its former level, assisted by the attractive force of the nucleus, it will release the energy previously gained. The energy released appears as a photon. Both these changes in energy level of the electron are spoken of as 'jumps'. They are at once 'energy jumps' and 'position jumps'. Atomic spectra are direct evidence of energy jumps.

We shall now consider the atomic spectrum of hydrogen which is the simplest element, possessing, as it does, only one electron and a nucleus of one proton. The spectrum is shown in figure 15.7 .

Fig. 15.7 Three lines from the hydrogen spectrum in the visible region.

The spectrum consists of a series of sharp lines which suggest a limited number of energy jumps, and these must be of a specific magnitude. Each of the lines is caused by an electron jumping from an initial, higher energy level to a final, lower energy level, emitting as it does a photon with energy given by

$$E = E_i - E_f = h\nu \qquad (12.28).$$

This was the equation we met in section 12.1.5 .

Table 15.2 shows four energy values from the hydrogen spectrum.

Table 15.2 Hydrogen spectrum energy values

Line	red	light blue	1st violet	2nd violet
ν(Hz)	4.568×10^{14}	6.167×10^{14}	6.907×10^{14}	7.309×10^{14}
E(joule)	3.027×10^{-19}	4.086×10^{-19}	4.577×10^{-19}	4.843×10^{-19}

These values are calculated using $h = 6.626 \times 10^{-34}$ joule seconds and the appropriate ν value in equation (12.28).

15.2.7 *Molecular Spectra*

The spectra emitted by molecules are generally much more complicated than atomic spectra because there is the additional energy contribution from the relative motion of the atomic nuclei. Molecules are made up from two or more atoms. They therefore consist of nuclei surrounded by a cloud of electrons which are attracted by the nuclei and thereby bind the atoms together. Atmospheric oxygen, for example, has molecules consisting of two oxygen atoms.

The energy of a molecule arises from the kinetic and potential energy of the electrons and from the potential energy originating in the mutual repulsions of the positively charged nuclei. These energies are changed under electron or molecular impact, and by absorption of X-rays, ultraviolet, and visible light. X-rays affect the inner atomic electrons, whilst ultraviolet and visible light affect the outer electrons.

Molecules with several nuclei can exhibit vibrational movements of the nuclei with respect to each other which give rise to infrared emission, or the absorption of infrared may start the vibration. In the gaseous state molecules can also possess rotational energy. In order of magnitude a molecule will absorb rotational energy the least, followed by vibrational energy, outer electron energy, and the greatest being that absorbed by the inner electrons. Just as it is for the energies in atomic spectra, these molecular energies are quantized. That is, they occur in certain allowed discrete amounts. An incandescent liquid or solid, such as a lamp filament, emits a broad continuous range of frequencies because the many individual energy levels merge together into overlapping levels; the emitted lines then overlap and merge into what is then seen as a continuous spectrum.

15.3 ASPECTS OF THE QUANTUM NATURE OF LIGHT

15.3.1 *Blackbody Radiation - Planck's Quantum Hypothesis*

At the turn of the nineteenth century Max Planck (1858 - 1947) was studying the process of emission of thermal radiation by hot bodies. This radiation is electromagnetic radiation and figure 15.8 shows its spectral distribution; that is, intensity I plotted against wavelength λ for several temperatures. The phenomenon is known as *blackbody radiation*. If an object is in thermal equilibrium with its environment it must emit as much radiant energy as it absorbs. A good absorber is therefore a good emitter. A perfect absorber, one which absorbs all radiant energy incident upon it, is said to be a *black body*.

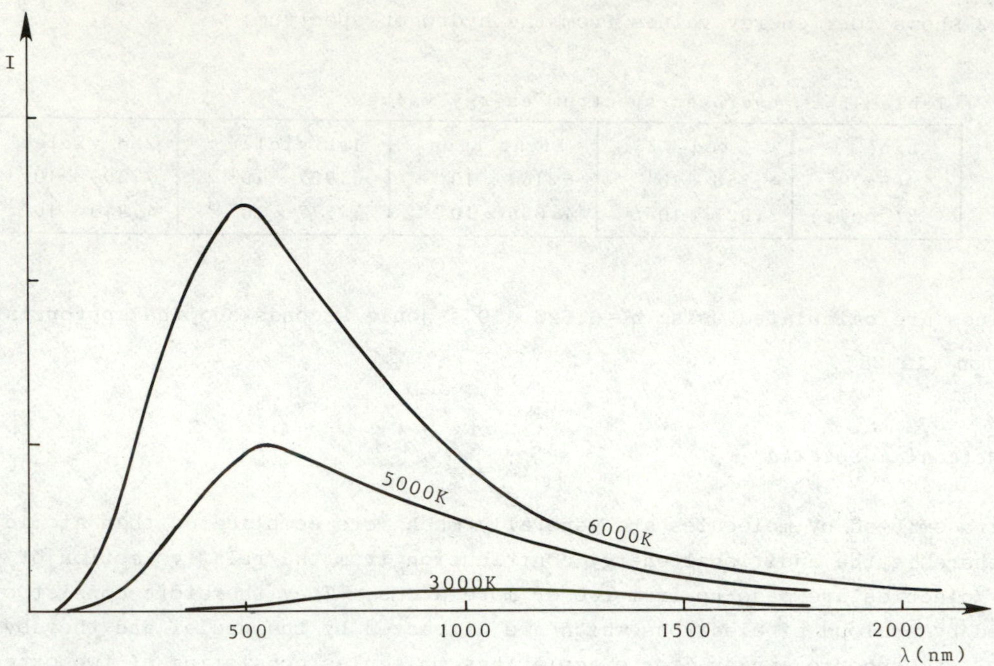

Fig. 15.8 Blackbody radiation curves.

One approximates to a blackbody with an insulated oven with a small hole in one wall. Radiant energy entering the hole has little chance of escaping, so the oven is a near perfect absorber. Alternatively, heating the oven allows it to emit radiation through the hole. As the temperature increases the radiation will initially be in the infrared, eventually glowing dull red, becoming bright red, then yellow, and finally white to blue-white.

The classical theory for the process, depicted in figure 15.8, was formulated by Rayleigh and Jeans, and completely failed to reproduce experimental results except in the long wavelength limit. Planck took a very practical approach. He first matched the curves with a mathematical expression. His expression does not concern us here, but the significant hypothesis to emerge from the expression was the quantum of energy. The hypothesis was that energy is emitted and absorbed in quanta of $h\nu$, and that emission and absorption are not therefore continuous processes. In the long wavelength limit the frequency decreases so that the energy changes appear to be nearly continuous.

15.3.2 The Photoelectric Effect

When a beam of short wavelength visible or ultraviolet light falls on a metal plate, electrons are emitted from the plate. This is known as the *photoelectric effect*. The apparatus used for studying this effect is shown in figure 15.9 .

In 1905, Albert Einstein (1879 - 1955) applied Planck's quantum theory to the photoelectric effect, although this was no sudden flash of insight. Einstein confided in a friend that it was the result of five years of thinking about Planck's hypothesis. He made the following assumptions:

(1) Electromagnetic radiation of frequency ν consists of quanta of energy, later called photons, with energy $E = h\nu$.

(2) Photons travel at the speed of light in vacuum.

(3) In the photoelectric effect one photon is completely absorbed by one electron which gains the quantum of energy and may be emitted from the metal.

The electrons are bound in the metal, but if the photon energy absorbed by an electron is greater than the binding energy it will be released from the metal. The amount by which the photon's energy exceeds the binding energy of the electron appears as kinetic energy in the released electron. If W, called the *work function*, is the minimum energy required to release an electron from the surface of the metal, then the electron's maximum kinetic energy is

$$\tfrac{1}{2}mv^2_{max} = h\nu - W \qquad (15.2)$$

where m and v_{max} are the released electron's mass and maximum velocity. Clearly, the maximum kinetic energy is directly proportional to the frequency. Also, from equation (15.2) there must be a threshold frequency ν_0 such that

$$h\nu_0 = W \qquad (15.3)$$

when the photon energy is just sufficient to overcome the binding energy of the least bound electrons. Below this frequency no electrons will be emitted no matter how intense the

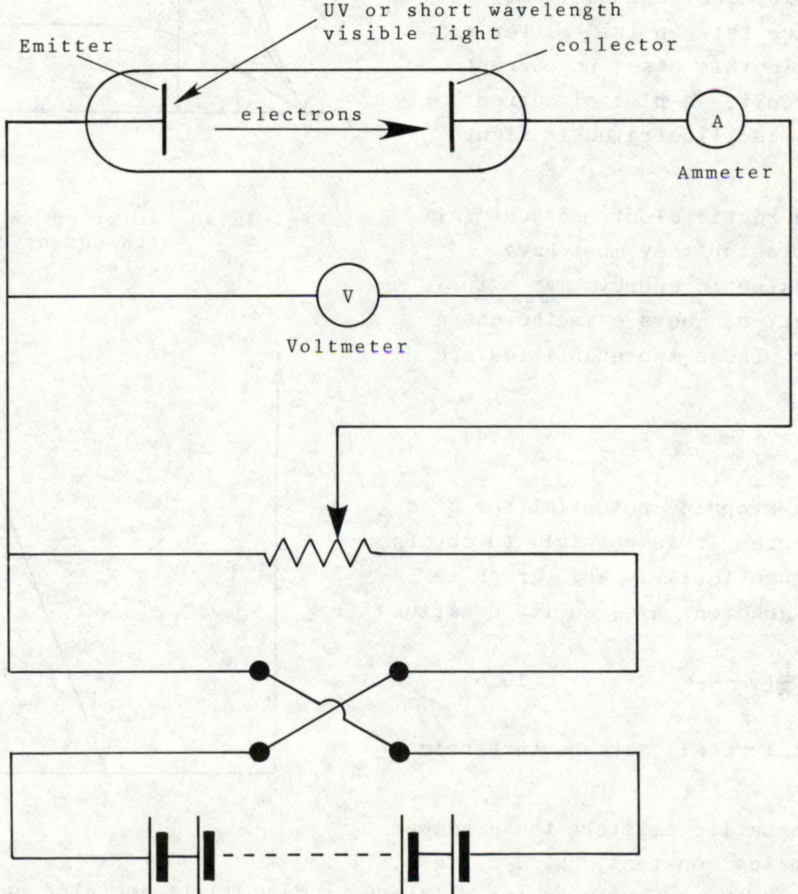

Fig. 15.9 The apparatus used to study the photoelectric effect.

irradiation; that is, no matter how many photons per square metre per second are incident on the metal. However, for photons where $\nu > \nu_0$ there is always a chance that an electron will be emitted even in an incident beam of very low intensity.

Equation (15.2) gives the maximum kinetic energy of the photoelectrons (the emitted electrons). Photoelectrons with lower kinetic energies may have suffered losses by collision on the way out of the metal, or they may have had a binding energy slightly greater than W.

Using the apparatus shown in figure 15.9 equation (15.2) can be verified.

Firstly, the collecting plate potential, V, is made positive relative to the emitting plate. It is then possible to draw a steady current of electrons which are attracted to the positively charged collector. For radiation of a fixed frequency incident upon the emitter, it is then possible to study the effect of varying the intensity of the incident beam. Secondly, making the collector potential V negative with respect to the emitter, the electrons are now repelled by the collector, and for some potential V_s, *the stopping potential*, even the most energetic electrons will fail to traverse the space between the emitter and the collector. In this case, no current flows in the circuit. A plot of current versus potential is illustrated in figure 15.10 .

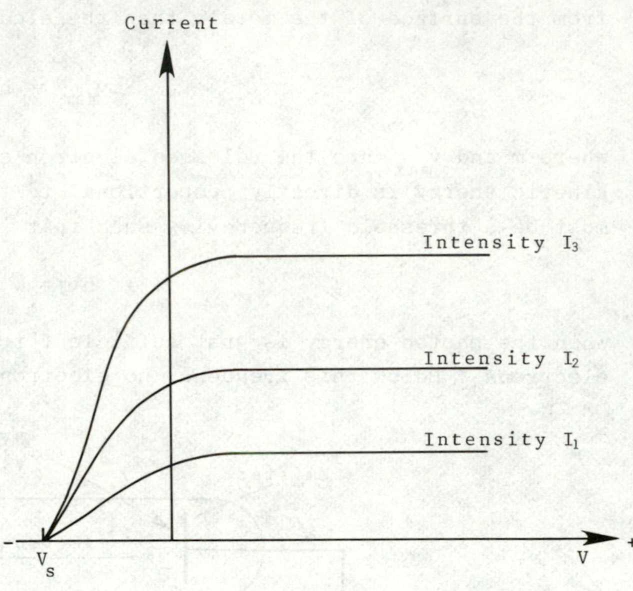

Fig. 15.10 Variation in photoelectric current with potential on the collector plate.

When the most energetic electrons just fail to reach the collector they must have converted their kinetic energy $\frac{1}{2}mv^2_{max}$ to potential energy $V_s e$, where e is the charge on the electron. These two quantities are equal, so

$$V_s e = \tfrac{1}{2}mv^2_{max} \qquad (15.4).$$

By measuring the stopping potential for a range of frequencies it is possible to obtain the graph in figure 15.11 . This graph is found to have a gradient h/e, so its equation is

$$V_s = \frac{h}{e}(\nu - \nu_0) \qquad (15.5)$$

where ν_0, you will recall, is the threshold frequency.

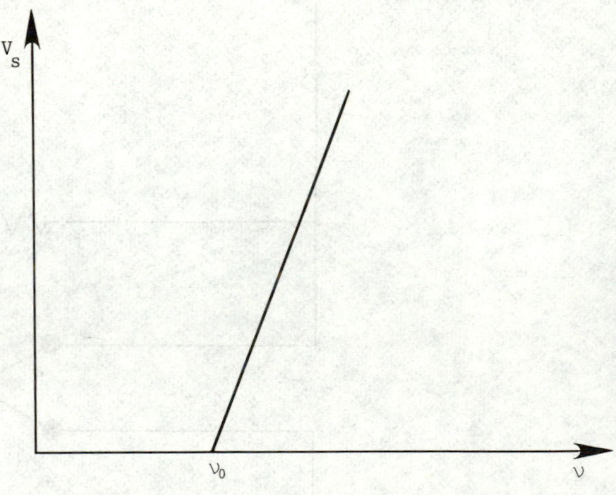

Fig. 15.11 Stopping potential versus frequency.

With different metallic emitters the gradient of the graph remains constant, but ν_0 differs. It should be noted the emission of one electron results from the absorption of one photon and the experiment verifies Einstein's equation (15.2): electromagnetic energy is quantized.

15.3.3 The Wave-Particle Duality of Electromagnetic Radiation

According to the classical theory of the propagation of electromagnetic radiation, the energy carried by a wave motion is spread out uniformly over the surface of the wave. Accordingly, when radiation falls on a metal plate the energy should be transferred uniformly to electrons in the surface, and the energy transfer per unit time should be proportional to the intensity of the wave. Given a low intensity an electron should, on the strength of this theory, acquire sufficient energy to escape the metal providing enough time is allowed. This does not occur, however, if the frequency of the incident light is less than ν_0, and the classical theory fails. The material of the chapters on polarisation, interference, and diffraction none-the-less indicate that electromagnetic radiation has very definite wave properties, yet the photoelectric effect implies particle properties. We do not have here an either/or situation. In one experiment we use the wave properties, whilst in another we use the particle properties. Both are equally justified. Paul Dirac, a British physicist, made a famous statement on this question:

> " Each photon interferes only with itself: interference between different photons never occurs."

Wave properties of light are not attributable to the beam as a whole, but each photon possesses a wave nature. In Young's two-slit experiment each photon simultaneously interacts with both slits. Closing either slit causes the interference pattern to disappear. In 1909 Geoffrey Taylor, a student at Cambridge University, demonstrated the wave nature of individual photons. Using a lightproof box, a gas flame illuminating an entrance slit, and a number of attenuating smoked glass filters, he photographed the diffraction pattern in the shadow of a needle. Using exposure times of up to three months so that exceedingly low flux densities could be used, he was able to obtain photographs of the diffraction pattern when only one photon at a time was present in the box!

15.3.4 Absorption, Emission, and Scattering

Suppose a photon of frequency ν collides with and is absorbed by an atom, and energy is transferred to a bound electron resulting in an excited atom. In dense gases, liquids, and solids, where the molecules lie in close proximity to each other, absorption occurs over a range of frequencies and the energy is generally dissipated through intermolecular collisions. By way of contrast, the excited atoms of a low pressure gas can reradiate a photon of the frequency ν in a random direction. The process was first observed by R.W. Wood in 1904, and is known as *resonance radiation*. Wood used an evacuated glass bulb containing a piece of pure sodium. When heated the sodium vapourises and subsequent irradiation with light from a sodium arc lamp causes the vapour to glow with the characteristic yellow sodium light. If the beam of incident light is restricted to a portion of the sodium vapour within the bulb, only that portion will glow.

Scattering can also occur at frequencies other than those corresponding to the atom's quantum energy levels. In such cases a photon will be reradiated without appreciable delay and with the same energy as the incident photon. The process is called *coherent scattering* because the incident and scattered photons have a constant phase relationship. This is the process referred to as Rayleigh scattering in section 12.5.1 .

Suppose an atom is excited into some higher energy level by absorption of a photon. It is possible for it not to return to its initial state after the emission of a photon. The electron drops from its excited level to some interim level, but still above the ground state. The emitted photon has a lower energy than the incident, absorbed photon. From the equation $E = h\nu$, the reradiated photon must therefore possess a lower frequency. This process is called a *Stokes' transition*, after George Stokes who studied the phenomenon. If the transition takes place rapidly, in about 10^{-8} s, it is called *fluorescence*. When there is a delay, which may be as long as several hours before the photons are emitted, it is known as *phosphorescence*. Using ultraviolet and shortwave visible radiation for the exciting photons many commonplace substances will re-emit longer wavelengths in the visible region and appear to glow. Organic dyes and detergents are examples. Sodium fluorescein is an organic dye used in hard contact lens practice where, dissolved in the tears, it fluoresces intensely green under ultraviolet and short-wave visible illumination. One use is then to check the cornea-tear-lens relationship.

The difference between fluorescence and phosphorescence can be illustrated by the use of energy level diagrams, figure 15.12 .

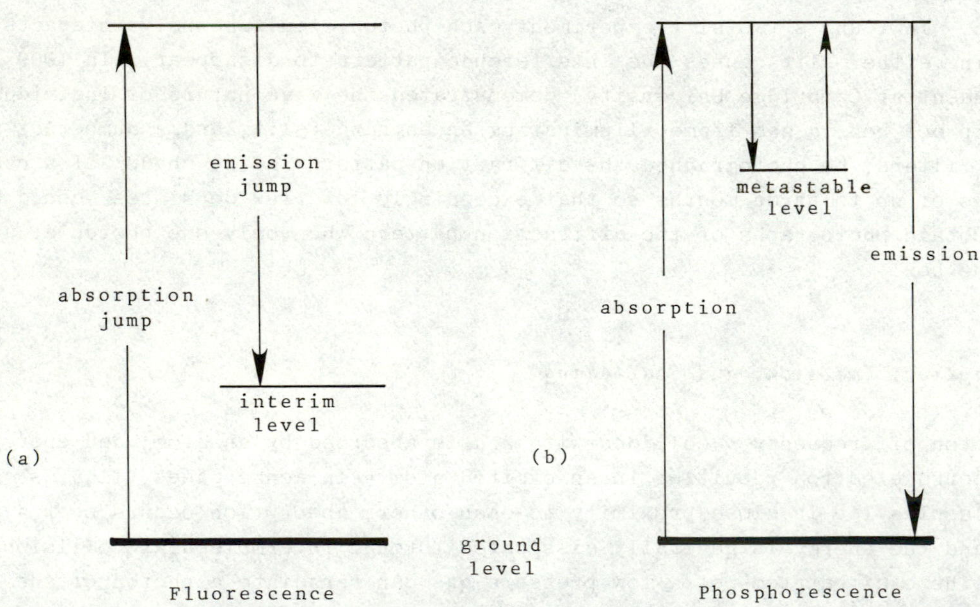

Fig. 15.12 Energy level diagrams illustrating fluorescence and phosphorescence.

In figure 15.12(a) there is a speedy re-emission of a photon as the electron jumps from the excited level down to an interim level. In 15.12(b), however, the electron falls from the excited level to a metastable level and remains there until it is further excited and raised back to the higher level, whence it can fall to the ground level and emit a photon. This excitation may come about from thermal agitation of neighbouring atoms or molecules, when it is known as *thermoluminescence*, or through infrared radiation absorption when it is called *optically stimulated luminescence*. The time spent in the metastable state determines the time that phosphorescence persists.

Phosphors, solid materials which exhibit phosphorescence, are used in television tubes where beams of electrons bombard them and excite them to emit light. This is the basis of the image in television and oscilloscope cathode ray tubes. Used in fluorescent tubes for general lighting, ultraviolet and visible light stimulate the process.

15.3.4.1 The Spontaneous Raman Effect

Light scattered from a substance mainly consists of the same frequency as the incident light. However, it is possible to observe weak additional components with higher and lower frequencies. These components are called side bands, a term borrowed from electronics. The difference between the incident frequency ν_i and the sideband frequencies is characteristic of the scattering substance. In 1928, C.V. Raman (1888 - 1970) observed what is now known as the *spontaneous Raman effect*. Strong sources are needed and it was not until the advent of the laser in the nineteen sixties that the effect was put to practical use in spectroscopy.

A molecule able to rotate can absorb radiant energy in the far infrared and the microwave regions. This energy is converted into rotational energy. A molecule which can absorb the energy from photons with a wavelength from about 10^{-5} m down into the visible spectrum to about 700 nm will convert this energy into vibrational motion. UV and visible light energy, of the correct quantization, is used to raise electrons to an excited level. There is a clear relationship between the absorbed energy and the molecule's subsequent behaviour which can be used to gain insights into molecular structure. Suppose, for example, a molecule absorbs a photon of energy $h\nu_i$ thereby raising an electron to some higher level. If the electron drops back almost immediately, but only to some intermediate level, then a lower energy and therefore lower frequency photon will be emitted. Since the emission of this photon is stimulated by the incident photon, let us call the emitted photon's energy $h\nu_s$. The difference $h\nu_i - h\nu_s$ goes to excite the molecule to greater vibrational motion, or electronic or rotational energy. If the molecule was initially in a state above ground level, it is possible for more energy to be emitted with the photon than was incident. In such cases the nett loss of energy can mean a reduction in vibrational motion. In both cases the emission of stimulated side band frequencies can be used to investigate molecular structure.

15.4 SOURCES OF LIGHT

15.4.1 Classification of Sources

Sources may be conveniently divided into two types dependent on the manner in which the light is generated:

(1) *thermal sources*, where the radiation results from a high temperature,

and (2) *discharge sources*, where an electrical discharge is passed through a gas.

Examples of the first class are the sun with its surface temperature of about 6000°C, the tungsten-filament lamp, arc lamps, and the flame. Discharge lamps usually comprise a glass chamber, containing a gas or vapour at low pressure, through which an electric discharge (electrons) is passed. The discharge transfers energy to the atoms of the gas which raises atomic electrons to excited states from which they fall back and emit photons.

15.4.2 High Temperature Sources

Incandescence is the name given to the emission of light by a body at a high temperature, and *incandescent lamp* is the general term for the tungsten-filament lamp in everyday use. It comprises a wire, electrically conducting filament through which a current is passed which causes it to glow at white heat. To prevent oxydation the filament is enclosed in a bulb, filled with a mixture of nitrogen and argon gases, and sealed. The early tungsten bulbs were either evacuated or contained solely nitrogen. They suffered from migration of the tungsten to the glass with the consequent blackening of the bulb, loss of light, and thinning of the filament until it broke. The addition of the inert gas argon prevented tungsten migration and allowed the filament to be run at a higher temperature which increased the contribution of shorter wavelengths in the visible range to the lamp's emission spectrum. The tungsten lamp emits a continuous spectrum which is somewhat yellower than daylight.

The most recent development is the introduction of the tungsten-halogen-filament lamp in which the presence of a small amount of the halogen iodine further reduces the migration of tungsten, and allows the filament to be run at a higher temperature still. Because iodine reacts chemically with glass, the filament must be enclosed in a quartz bulb. Though these bulbs have a whiter light and longer life they have not yet been used for domestic lighting, although they are finding increasing application in automobile headlights.

The incandescent lamp was not the first electric lamp; lights using an electric arc between carbon electrodes were in use before the invention of the first incandescent lamp in 1870. An electric arc is a continuous, high density electric current between two separated conductors in a gas or vapour, with a relatively low potential difference across the conductors. Light comes from the heated ends of the conductors, usually carbon rods, as well as from the arc itself. They are used in applications requiring a very bright source, such as in floodlights and large film projectors. The continuous spectrum produced is nearer to the solar spectrum than the tungsten lamp's.

Prior to the invention of electric lamps the flame was the only artificial source of light. The principle behind all flame sources, whether they be burning brands, oils, waxes, fats, or gases, is the excitation of molecules which occurs in the high temperatures evolved in the chemical processes involved in burning.

The development of the gas light in the 19th century was a major development in artificial lighting. The first street installation, using coal-gas, was in Pall Mall in London in 1820. Initially, light was derived from the yellow gas flame alone. During the 1820s air was introduced into the gas which reduced the flame's luminosity but increased its temperature. The higher temperature was used to heat a non-conbustible material to incandescence. The 1880s saw the introduction of the *Welsbach gas mantle* as the non-combustible incandescent material. It consisted of a woven cotton net, of spherical, cylindrical, or planar construction, impregnated with the salts of thorium or cesium. Despite their fragile nature they were soon in use everywhere where there was a distributive network of natural or coal-gas. The incandescent light from the thorium and cerium salts had a preponderance of green which was not very kind to the complexion.

15.4.3 Discharge Sources

The electric discharge lamp is a lighting device consisting of a glass or quartz container within which a gas or vapour is ionised by an electric current passed between two electrodes. Gaseous ions are attracted to the charged electrodes and collisions with other ions cause excitation and further ionisations, light being produced when excited atoms and ions return to the ground state. Neon gas in the container, first used about 1910, gives a red light, and mercury vapour a blueish light. Combinations of gases and coloured glass containers allow a variety of coloured lights. For example, helium in an amber glass glows gold, mercury in a yellow glass glows green, and combinations of gases produce white light. The sodium vapour lamp, developed in Europe about 1931, produces a yellow light.

Three particular lamps are in everyday use: the sodium lamp, the mercury lamp, and the domestic fluorescent tube. The sodium lamp, used for street lighting, emits the characteristic sodium yellow 589.3 nm wavelength. The lamp contains some neon gas which is bombarded by electrons emitted from the negative electrode. The neon atoms gain energy from the accelerated electron current and in turn bombard the metallic sodium contained in the bulb. The sodium soon vapourises in the low pressure bulb and is thereafter excited to emit its atomic spectrum by the bombarding electrons passing from cathode to anode. The presence of neon accounts for the red glow seen before the sodium has vapourised.

The mercury lamp does not need a starter gas since it is a liquid at temperatures above -38.37^{0}C, and in an evacuated bulb some mercury vapour will exist at all times. Switching on this lamp soon causes further vapourisation and mercury atoms are excited to produce their atomic spectrum. The spectrum consists of emissions in the ultraviolet and the green and blue parts of the visible spectrum. Were that the whole story the mercury lamp would not have been viable. The ultraviolet emission is converted to visible light by coating the inside of the bulb with fluorescent or phosphorescent materials which are excited to emit longer wavelengths. In their earlier form the overall effect was a white light with a predominance of blue.

The fluorescent lamp used indoors is a mercury discharge lamp with a fluorescent coating of metallic salts such as calcium tungstate, zinc sulphide, and zinc silicate. They are now available in 'daylight' and 'warm white' versions due to subtle application of the fluorescent coating. A 40 watt fluorescent tube gives out as much light as a 150 watt tungsten-filament bulb, and being larger it casts more diffuse or softer shadows.

15.4.4 Laboratory Sources

Lamps used in the laboratory are most often required to emit specific wavelengths. These sources are the hollow cathode, the Geissler tube, the carbon arc, sparks, and flames.

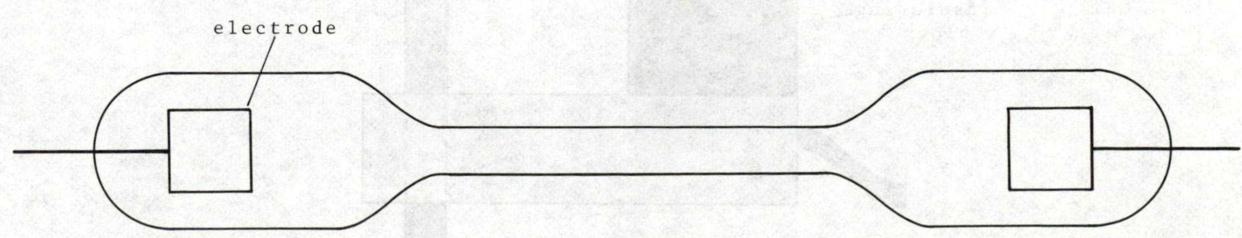

Fig. 15.13 The Geissler tube.

The *hollow cathode* discharge lamp consists of two electrodes mounted inside a cylindrical glass bulb. The negative electrode (cathode) is usually a hollow cylinder constructed of or coated with the substance the spectrum of which is to be investigated. The bulb usually contains helium, neon, or argon at a pressure of about 1 torr (760 torr ≡ 1 atmosphere). A direct current of 1 to 100 amperes at several hundred volts is passed through the bulb.

The *Geissler discharge tube* is illustrated at about actual size in figure 15.13 . It consists of a glass tube with electrodes at each end, and filled with a gas at 0.1 to 10 torr pressure. Often, a portion of the tube is constricted to provide a greater current density with a consequently greater brightness in that region. They operate on alternating current of 10 - 50 milliamperes, at voltages stepped up to 1500 to 10 000 volts. In the long term, the inert gases helium, neon, and argon are the most satisfactory fillers, but hydrogen, nitrogen, and carbon dioxide filled tubes are available. The latter three gases gradually disappear in combining chemically with the electrodes or the walls.

Electric arcs are convenient sources for spectroscopic analysis of solid materials. Graphite and copper are typical electrode materials. A small amount of the material to be analysed is powdered, mixed with powdered electrode material, and formed into the shape convenient for one electrode. An alternative method is to core one electrode and fill it with the powdered sample. Arcs are run on currents of 1 to 100 amperes, and at voltages from 10 to 100 volts. They may be run in an atmosphere of air or some other gas, and temperatures in the arc range from 3500^{0}C to 5000^{0}C. Figure 15.14 illustrates the electric arc.

Spark sources are used when higher excitation is required than is available with an arc. The simplest spark source is formed when a discharge of a strong current occurs across a gap between electrodes at a potential difference of about 5000 volts. Materials in the spark are often ionised and the temperature may reach several tens of thousands of degrees centigrade.
Flames were the first man-made sources to which small amounts of substances were added to be vapourised and excited by collisions with the constituents of the flame.

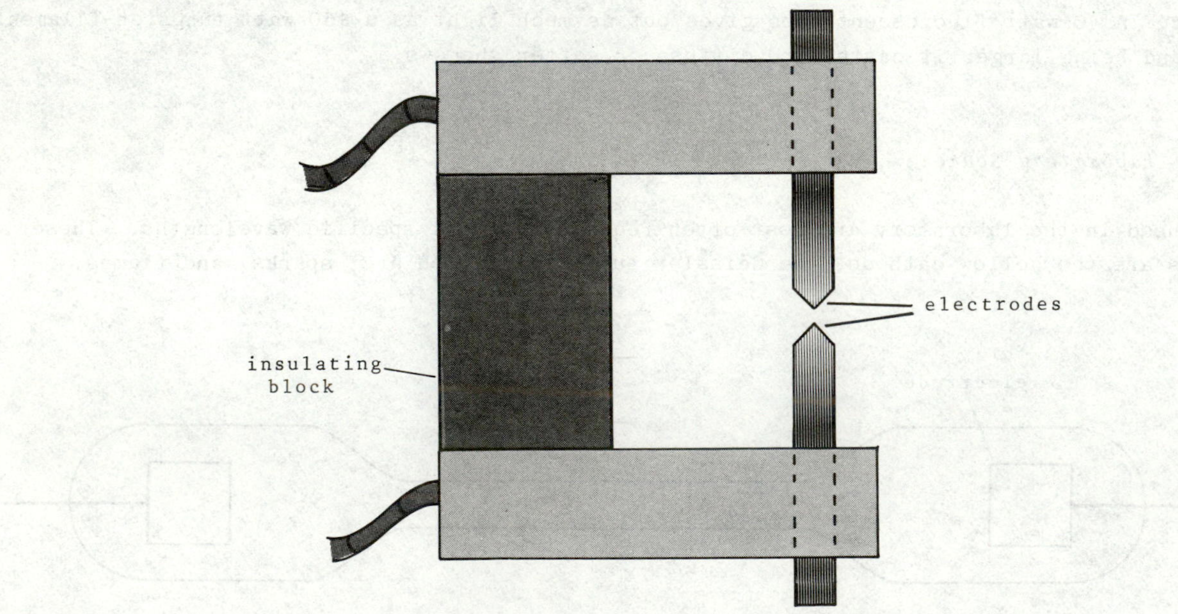

Fig. 15.14 Electric arc.

15.5 *Multiple-choice Self-assessment Questions*

Choose ONE response only to questions 1 to 4. The answers are given at the end of the section.

15.5.1 The term *atomic diameter* means:

A) the diameter of the atomic nucleus
B) the distance between atoms in a molecule
C) the diameter of an imaginary sphere into which the atom would just fit
D) the distance between atoms when packed together in solids.

15.5.2 The charge on an electron is numerically

A) greater than that on a proton
B) less than that on a proton
C) equal to that on a proton.

15.5.3 The Rutherford scattering experiment showed that

A) gold is penetrable by alpha particles
B) nearly all the atomic mass is concentrated in a small nucleus
C) gold atoms are largely empty space
D) helium atoms are less massive than gold atoms.

15.5.4 The diameter of the nucleus of the atom is which fractional atomic diameter?

A) 1/100 000
B) 1/10 000
C) 1/1 000 .

15.5.5 Figure 15.15 represents an energy level diagram of an atom. Which arrow represents an emission process?

A) The arrow marked 2.

B) The arrow marked 1.

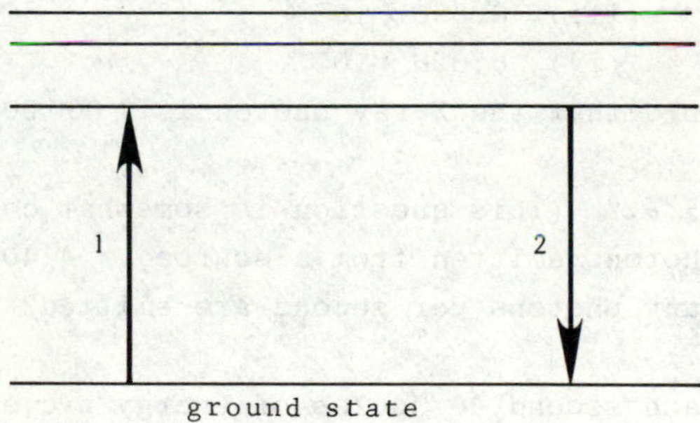

Fig. 15.15

15.5.6 Indicate the type of spectrum (1 - 6) you would see if you looked at the following light sources with a spectroscope.

A) A sodium lamp.
B) White light through an atomic vapour.
C) Diffuse daylight.
D) A car headlamp.

1) Line spectrum. 2) Emission spectrum. 3) Continuous spectrum.
4) Solar spectrum. 5) Atomic spectrum. 6) Absorption spectrum.

ANSWERS

15.5.1 C 15.5.2 C 15.5.3 B

15.5.4 A

15.5.5 A. An arrow going from a higher energy level to a lower one represents an electron losing energy which is emitted as a photon

15.5.6 A) 1,2, and 5.

B) 6. It might also be described as a line, atomic, absorption spectrum.

C) 4 and 6. Remember, the sun's atmosphere absorbs many line spectra.

D) 2 and 3.

15.6 WORKED PROBLEMS

15.6.1 Calculate the quantum energies for the following radiations. (Planck's constant $h = 6.626 \times 10^{-34}$ joule seconds; speed of light in vacuum $c = 3 \times 10^{8}$ metres per second).

(i) Infrared, $\lambda = 0.3$ mm.

(ii) Visible, $\lambda = 600$ nm.

(iii) Ultraviolet, $\lambda = 300$ nm.

(iv) X-rays, $\lambda = 0.3$ nm.

In each case it is simply a matter of using equation (15.1):

(i) $E = \frac{hc}{\lambda} = \frac{6.626 \times 10^{-34} \times 3 \times 10^{8}}{0.3 \times 10^{-3}} = 6.626 \times 10^{-22}$ Joule.

The answers for the remainder are

(ii) 3.313×10^{-19}

(iii) 6.626×10^{-19}

(iv) 6.626×10^{-16}.

Note that the X-ray photon is 1 000 000 times more energetic than the infrared photon.

15.6.2 (This question is somewhat contrived but it will give you a feel for the number of photons emitted from a source). A 40W light bulb emits radiation of wavelength 500 nm. How many photons per second are emitted? Note: 1 watt $\equiv$ 1 joule/second.

Each second 40 joules of energy are expended. Each photon possesses

$$E = \frac{hc}{\lambda} = \frac{6.626 \times 10^{-34} \times 3 \times 10^{8}}{500 \times 10^{-9}} = 3.9756 \times 10^{-19} \text{ J.}$$

Thus, the number of photons per second is

$$\frac{40}{3.9756 \times 10^{-19}} \simeq 10^{20},$$

or, one hundred million million million!

15.6.3 A tungsten target is bombarded with electrons each accelerated to possess a kinetic energy of 1.601×10^{-14} J. All of the kinetic energy from one accelerated electron is transferred to the inner electron of a tungsten atom which, as a result, jumps to a higher level. In falling back it emits an X-ray photon. What is its wavelength?

From equation (15.1)

$$\lambda = \frac{hc}{E} = \frac{6.626 \times 10^{-34} \times 3 \times 10^{8}}{1.601 \times 10^{-14}} = 12.42 \times 10^{-12} \text{ m.}$$

EXERCISES

1 Explain briefly what is meant by atom, proton, nucleus, electron, and ion.

2 Describe how the visible spectrum may be displayed in the laboratory using (a) a prism spectrograph and (b) a grating.

3 Compare the energies of photons with frequencies (a) 3×10^{18} Hz, (b) 3×10^{12} Hz, and (c) 3×10^{16} Hz. From which regions of the electromagnetic spectrum do these radiations come?

4 Distinguish between continuous, atomic, and absorption spectra.

5 What are energy jumps in an atom, and how are they related to the emission of photons?

6 Describe the photoelectric effect and how it is possible to conclude from it that electromagnetic radiation is quantized.

7 Briefly explain what is meant by fluorescence and phosphorescence.

8 How do (a) incandescent lamps and (b) fluorescent tubes work?

9 Write a short essay on the atom and the manner in which radiation is emitted from it. What are the differences between atomic and molecular spectra?

10 Explain the terms quantum and photon. How does the quantum theory of radiation differ from the wave theory?

11 Light from an incandescent lamp is passed through mercury vapour and is viewed through a prism spectrometer. With the aid of figure 15.4(b), describe the appearance of the spectrum.

12 Explain why the sun's spectrum is not a continuous spectrum.

13 Write a brief account of the development of artificial lighting.

14 By considering the energy of ultraviolet and infrared photons, explain why exposure for the same time to short wave ultraviolet is more harmful than exposure to short wave infrared.

15 Ultraviolet radiation is sometimes described as ionising radiation. Why is this?

16 A photoelectric cell, whose work function is $W = 5.0 \times 10^{-15}$ J, is irradiated by light of wavelength 400 nm. No light is detected. Why is this?

17 Explain why a photoelectric current does not flow when the irradiating light does not possess photon energies in excess of $W = h\nu_0$ no matter how intense the irradiation might be.

18 What is meant by an atom in (a) its ground state and (b) an excited state?

19 Rank the following electromagnetic waves in increasing order of their photon energies. (i) visible light, (ii) infrared, (iii) X-rays, (iv) radio waves, (v) gamma radiation.

16 IMAGERY AND LASERS

16.0 INTRODUCTION

The topics introduced in this chapter are very much in the forefront of contemporary optics. Enhancement of images, the invention of the laser, and its subsequent application to holography, all date from the early sixties. Diffraction looms large in the theory of image formation, whether it be electromagnetic radiation, accelerated electrons, or ultrasound which carries the information. The subject matter will seem far removed from the subject of images dealt with in geometrical optics, but it is so important that an introduction must be made in a text of this sort. We cannot do more than give a taste of the subject matter since to do it real justice would require a volume devoted entirely to the topic. However, it is hoped that some students will be encouraged to probe more deeply into the subject via the specialist books which are available.

16.1 IMAGERY

In the chapters on geometrical and physical optics little has been said of the intensity distribution across an illuminated or self-luminous object. In any imaging system such as the eye-brain, a camera, television, or whatever, the quality of the simage depends on how well these object intensity variations are reproduced in the image. If we have some quantitative knowledge of how the intensity across an object is reproduced in the image, then we can hope to optimise the system to produce the best possible image. With this in mind an imaging system may be conveniently divided into four parts:

(1) the object and its illumination,
(2) the imaging components,
(3) detection of the image,
and (4) any subsequent processing.

In a photographic system the object, the external light, and the lenses and stop form the parts (1) and (2) of the system. The detection of the image is a function of the film and the action of light upon it. Finally, the processing involves the development of the negative and the printing of the positive. We are all aware that, perfect though the first three stages may be, the processing may ruin the final image. In the case of vision, stage (3) involves the cones (and rods) of the retina, and processing is all about the encoding and the modification of the optical information in the retina and brain. Of course, with vision we are not able to specify fully the events in stage (4) but this does not prevent some useful applications being foreseeable.

16.1.1 Spatial Frequencies

You need only look about you to realise that intensity variations in almost any object are many and irregular. That is, irregular in the sense of not generally presenting a repeated pattern. To simplify things we shall start with a particular type of periodically varying intensity in the form of a cosinusoidal grating, figure 16.1 .

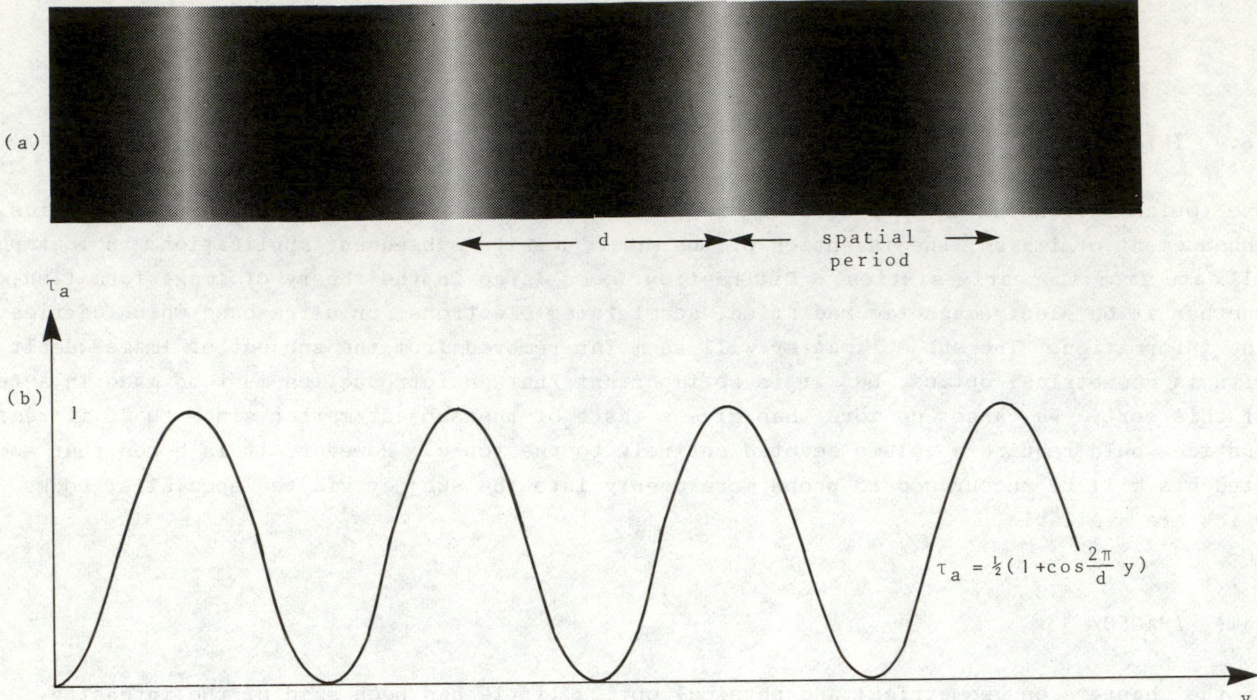

Fig. 16.1 (a) Cosinusoidal grating. (b) Amplitude transmittance of the grating.

The reason for choosing this particular form of grating will only become apparent as we proceed. Illuminating this grating uniformly produces an amplitude or field output which varies cosinusoidally as in figure 16.1(b). Contrast this with the multiple slit grating which has an output intensity which varies as shown in figure 16.4 . Recall that the multiple slit grating produced a series of spots in the focal plane of a converging lens when illuminated by plane monochromatic waves. In contrast, the cosinusoidal grating forms three spots only in the lens' focal plane as shown in figure 16.2 .

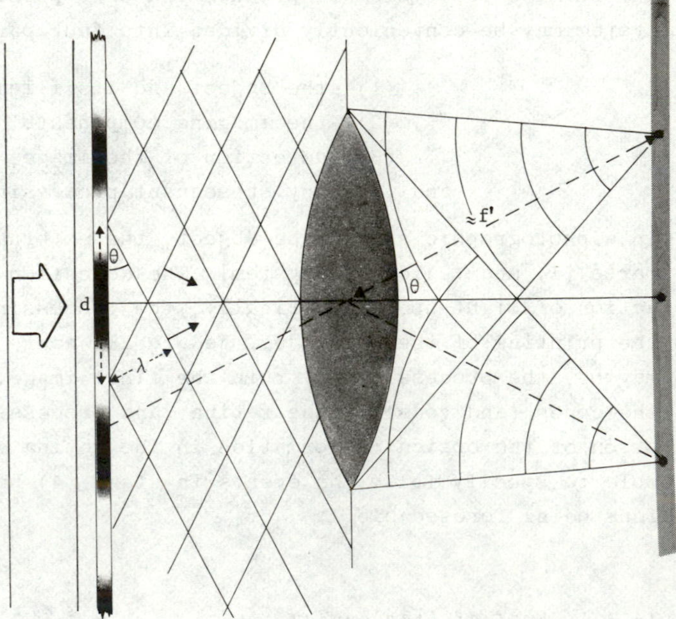

Fig. 16.2 The three-spot diffraction pattern formed by a cosinusoidal grating.

We now wish to analyse this phenomenon. Figure 16.1(b) is a plot of the field amplitude on the output side of a cosinusoidal grating which is uniformly illuminated with plane monochromatic waves. The distance over which the field repeats its value is called the *spatial period*, the *spatial wavelength*, or simply the *repeat distance* which we give the symbol d. The fact that three spots are produced in the second focal plane of the lens in figure 16.2 suggests that three sets of plane waves propagate from the grating; an undeviated central wave, and two diffracted waves, one upwards and one downwards. We can relate this to figure 16.1(b) as follows. If τ_a represents the amplitude transmittance of the grating, where

$$\tau_a = \frac{\text{emerging field amplitude}}{\text{incident field amplitude}},$$

and we arbitrarily set the incident field amplitude to unity, then by inspection of figure 16.1(b) we have

$$\tau_a = \tfrac{1}{2}\left(1 + \cos\frac{2\pi}{d}y\right) \qquad (16.1).$$

Now, in general $\cos A = \tfrac{1}{2}\cos A + \tfrac{1}{2}\cos(-A)$, so we can write equation (16.1) as

$$\tau_a = \tfrac{1}{2} + \tfrac{1}{4}\cos\left(\frac{2\pi}{d}y\right) + \tfrac{1}{4}\cos\left(-\frac{2\pi}{d}y\right).$$

This tells us that the output field from the cosinusoidal grating is the sum of three fields. There is the constant component with a value of ½ and the two cosine varying fields $\tfrac{1}{4}\cos\left(\frac{2\pi}{d}y\right)$ and $\tfrac{1}{4}\cos\left(-\frac{2\pi}{d}y\right)$. The sum of these components is shown graphically in figure 16.3 .

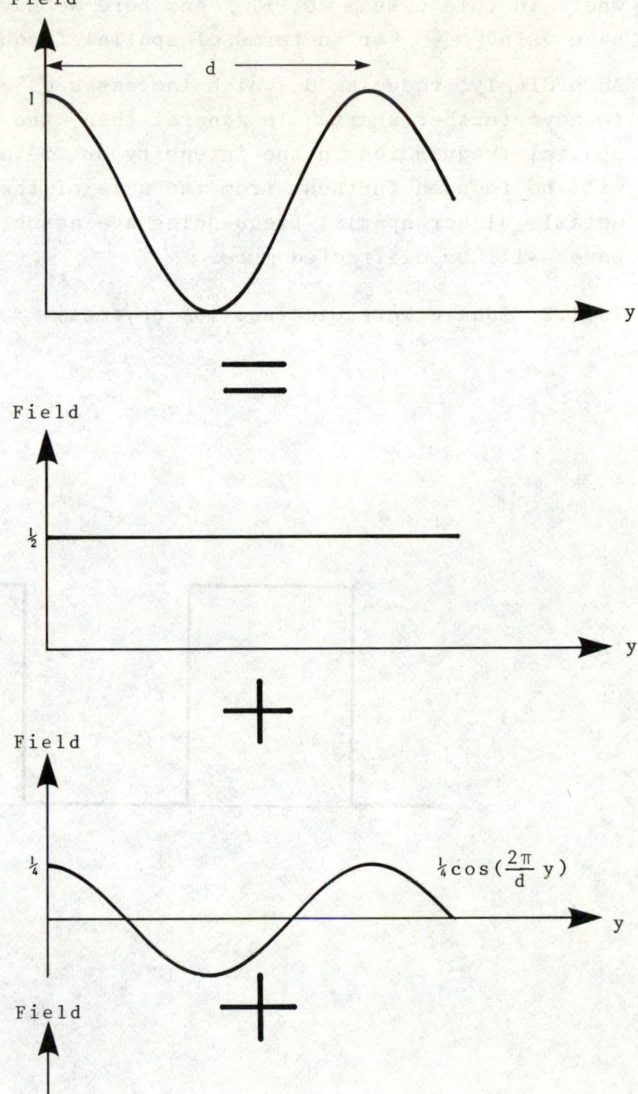

Fig. 16.3 Components of the field emerging from the cosinusoidal grating.

These three component fields have repeat distances of ∞, d, and d, respectively. The reciprocal of the repeat distance is known as the *spatial frequency*, q; so

$$q = \frac{1}{d} \qquad (16.2)$$

for the two diffracted waves. For the constant value component the spatial frequency is $\frac{1}{\infty} = 0$. The waves associated with these spatial frequencies are separated out on the screen in figure 16.2 . That is, each spatial frequency is represented on the screen in the focal plane of the lens.

From our earlier treatment of diffraction we have

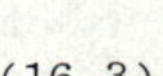

$$d \sin\theta_m = m\lambda \qquad (16.3)$$

where in this case $m = 0, \pm 1$, and here we have used d instead of a. Putting $m = 1$ we clearly have $\sin\theta_1 = \frac{\lambda}{d}$, or in terms of spatial frequency, $\sin\theta_1 = q\lambda$.

Accordingly, reducing d, which increases q, causes θ to increase and the spots on the screen to move further apart. In general then, the diffracted waves associated with the highest spatial frequencies in the intensity or field distribution across the object (the grating) will be focused furthest from the axis of the system. Since small repeat distances mean finer detail, higher spatial frequencies are associated with the fine detail in an object and these waves will be diffracted more.

16.1.2 Square Wave Gratings (or objects)

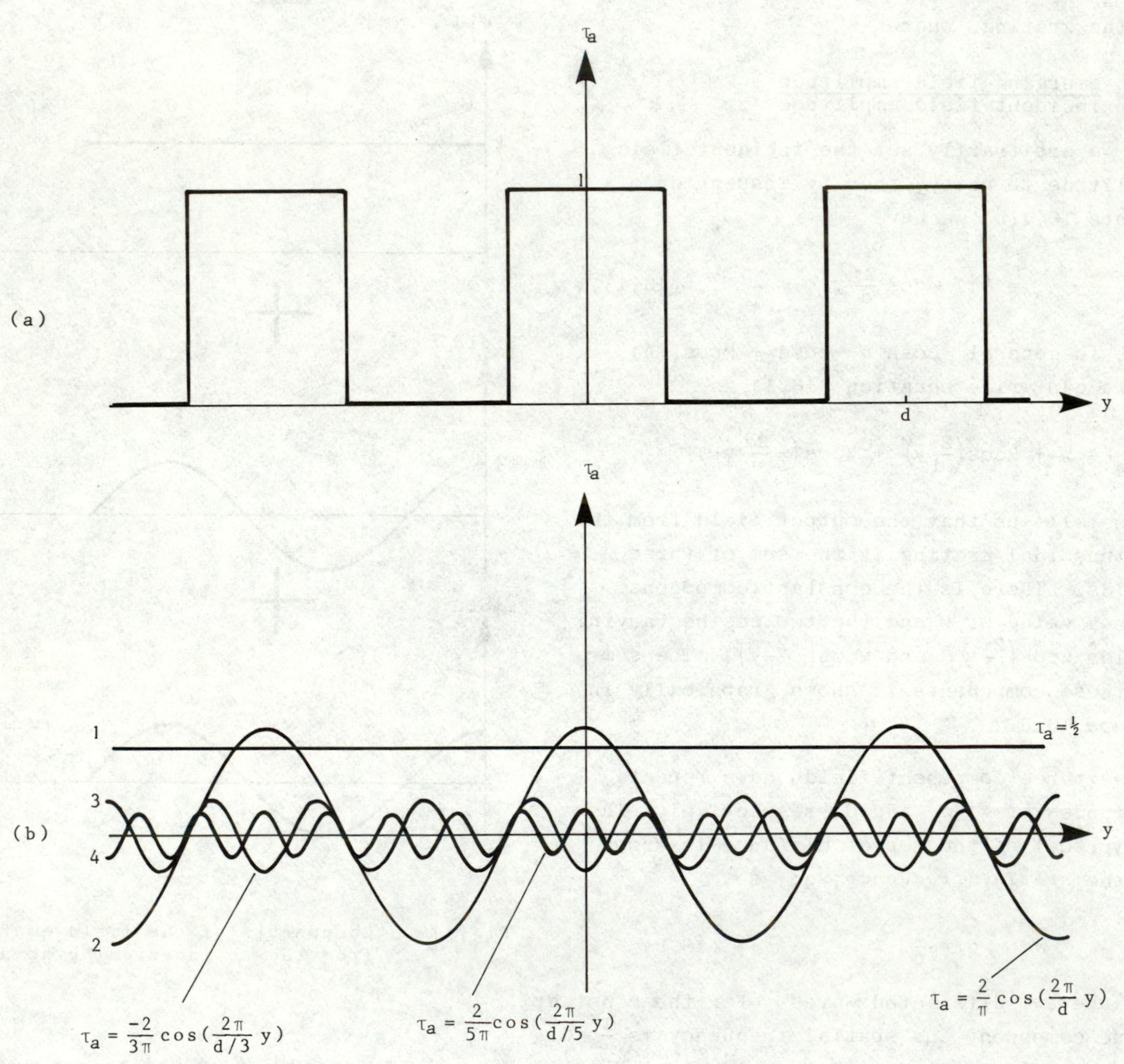

Fig. 16.4(a) A square wave amplitude transmittance: spatial period d.
(b) The first four sinusoidal components which add to give the square wave in (a). Note that the function $\tau_a = \frac{1}{2}$ may be regarded as a sinusoidal function with a zero frequency or an infinite wavelength.

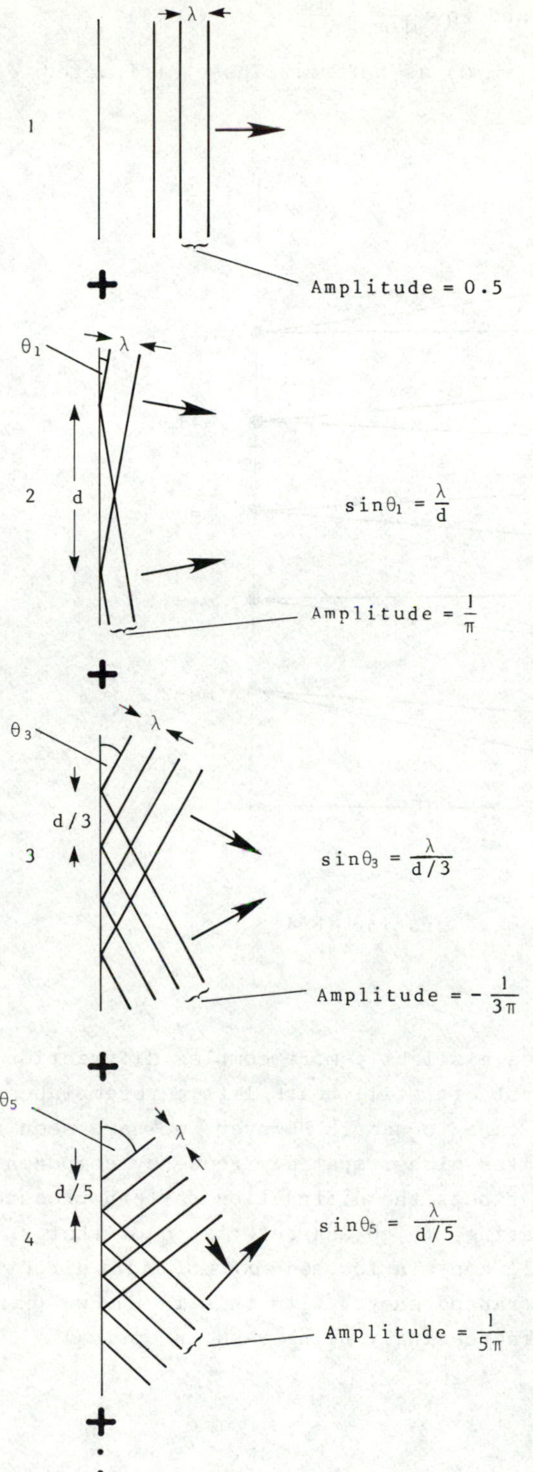

A cosinusoidal grating is certainly a very particular form of object. We have introduced it first because it has a special significance; any periodic intensity pattern can be expressed as a sum of an infinite number of cosine and sine functions. This means we can analyse any periodic intensity distribution into its sinusoidal components, and a knowledge of how an imaging system handles different spatial frequencies will allow us to judge how well an object will be imaged.

Before we define the performance of imaging systems let us look at another particular object - a square wave grating. Figure 16.4 shows the amplitude transmittance of the square wave grating and its sinusoidal components; the latter are all cosines since the ordinate passes symmetrically through one 'square peak'. If you wish to pursue this point further you may care to tackle Appendix 1 which looks at the analysis of periodic functions into their sinusoidal components. This figure implies that the square wave field amplitude distribution is composed of the sum of an infinite number of cosine varying components shown in the lower half of the figure. Notice there is the zero frequency, undiffracted component shown as $\tau_a = ½$, and other components with amplitudes and spatial frequencies

$$\left(\frac{2}{\pi}, \frac{1}{d}\right), \quad \left(-\frac{2}{3\pi}, \frac{3}{d}\right), \quad \left(\frac{2}{5\pi}, \frac{5}{d}\right),$$

and so on. d is the repeat distance of the square wave. If we were to replace the cosinusoidal grating of figure 16.2 with the square wave grating, there would no longer be only three sets of plane waves leaving the object (the grating). Figure 16.5 illustrates the waves which would leave the grating and which would each be focused to a spot of light by the lens. Only the first four spatial frequencies are shown, of course.

Again, from our earlier treatment of diffraction, we can write

$$d \sin \theta_m = m\lambda$$

Fig. 16.5
The waves emerging from the square wave grating of figure 16.4. Note that the grating is illuminated by monochromatic light with plane wavefronts. Also, each cosine component in figure 16.4 is represented by two sets of waves propagating in the θ and -θ directions as they were in figure 16.3: their amplitudes are therefore half of their resultant's. The numbers 1,2,3,4 correspond to the numbers against the functions in the previous figure.

or $\quad \sin\theta_m = \frac{m}{d}\lambda = q\lambda, \quad$ where $m = 0, \pm1, \pm2, \ldots\ldots,$ and $q = \frac{1}{d/m} = \frac{m}{d}$ (16.4).

Thus, θ_m increases with increasing spatial frequency q (=m/d) as before. These diffracted component plane waves are shown in figure 16.6 .

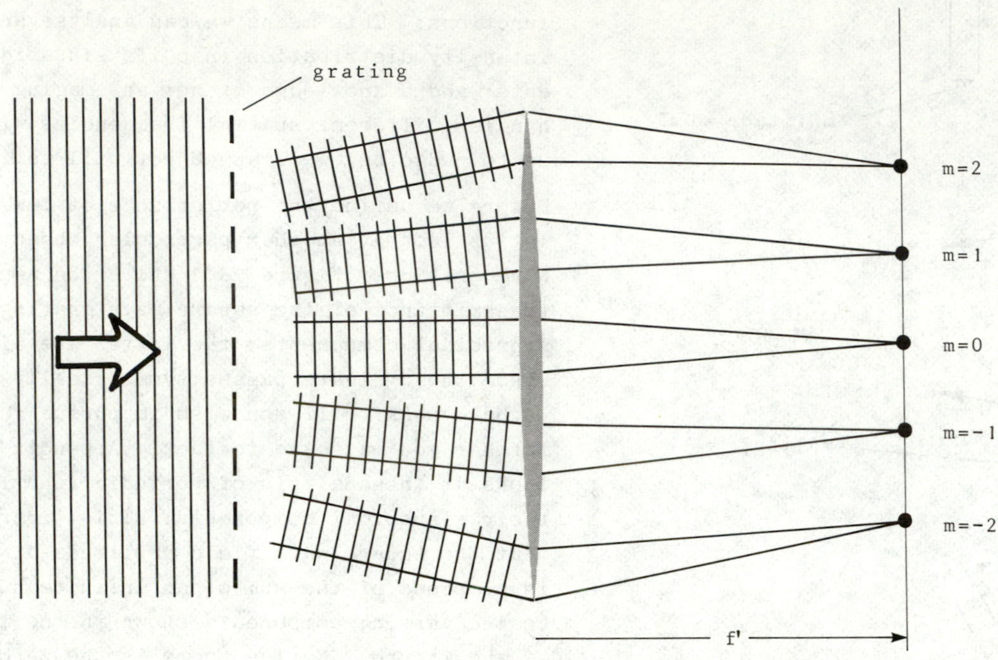

Fig. 16.6 Diffraction pattern for a square wave grating (only the orders m=-2 to m=2 shown).

Although the square wave intensity distribution produces a slightly more complex diffraction pattern than the cosinusoidal grating we considered first, it still bears little resemblance to almost any common object intensity distribution you care to name. However, we have seen that increasing complexity of detail in the object creates higher spatial frequency components in the light diffracted by the object. Also, we might expect the diffraction pattern produced by a photographic transparency, acting as an object grating, to be characteristic of that object. The focal plane of the lens in figure 16.2 will contain focused spots of light directly related to spatial frequency components in the object transparency. With this in mind we shall look at some interesting aspects of imaging which are increasingly being used in optical information processing.

16.1.3 Abbe's Theory of Image Formation

We are now going to carry figure 16.6 a stage further. Suppose the light is allowed to continue to the right, as shown in figure 16.7 with a little elaboration. Plane waves leaving the grating are diffracted in directions θ_m dependent on $\sin\theta_m = \frac{m\lambda}{d} = q\lambda$.

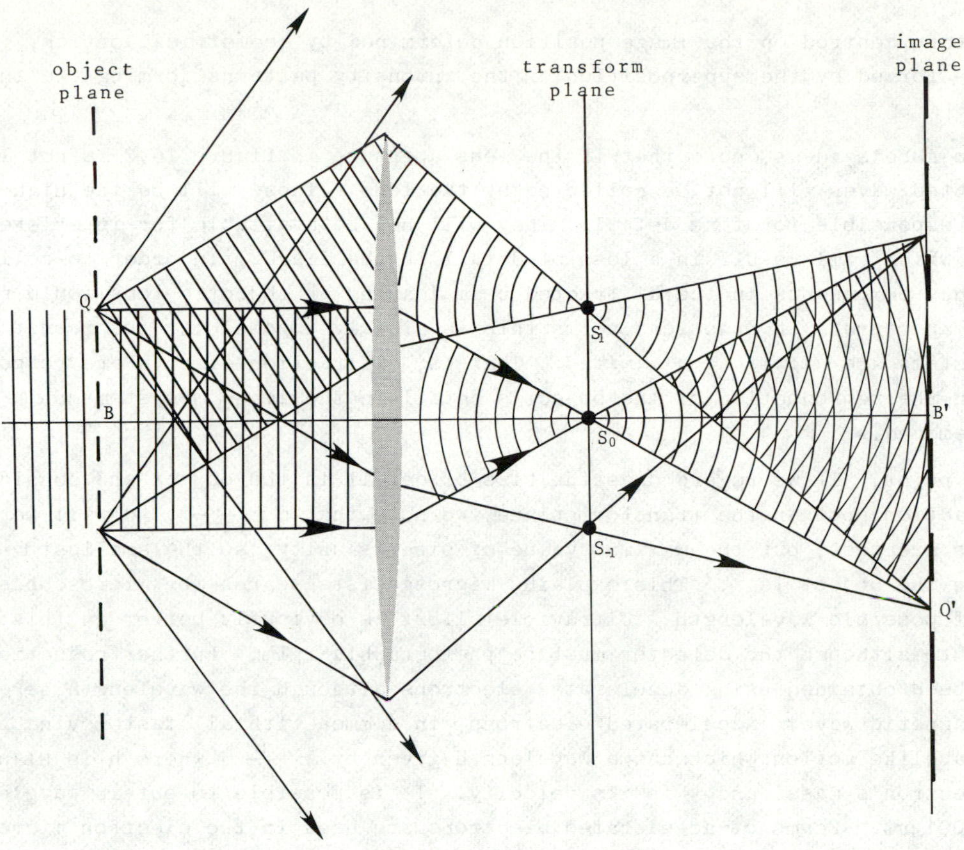

Fig. 16.7 Image formation.

The lens is called a *transform lens* since it transforms the intensity distribution immediately to the right of the object grating into a spatial frequency distribution in its back focal plane. The focal plane, with the focused spots S_0, S_1, etcetera, is called the *transform plane*. Incidentally, S_0, S_1, ..., are images of the source of the plane waves incident upon the grating. The waves continue to the right after focusing in the transform plane and arrive at the image plane. Here, the superposition of all the waves which have passed through the lens gives rise to the intensity distribution which is the image. Thus B is imaged at B', and Q at Q'. The transform lens forms two distinctly interesting patterns; one is the diffraction pattern in the transform plane, and the other is the image in the image plane.

Ernst Abbe (1840 - 1905) first propounded his ideas on image formation in 1873. He regarded the image formation as a double diffraction process. The incoming wave is diffracted firstly at the object and again at the lens. If the lens were taken away a diffraction pattern of the object would appear in the image plane instead of the image. The job of the lens system then is to recombine the diffracted waves leaving the object into a spatially distributed field amplitude or intensity pattern as nearly as possible similar to that in the object plane.

Lord Rayleigh proposed an alternative view which is equivalent. He envisaged each point on the object as a coherent emitter of secondary wavelets which were diffracted by the lens into an Airy pattern centred on the image position determined by geometrical optics. The image is therefore formed by the superposition of the intensity patterns forming the innumerable Airy discs.

Returning to Abbe's ideas, note that if the lens aperture in figure 16.7 is not large enough some diffracted waves will not be collected by the lens. These will be the high frequency components responsible for fine detail. They will not be available for interference in the image plane which will result in a loss of detail in the image. In order to collect all the high frequency components in the diffracted beam leaving an object a lens would need to have an infinite aperture. In practical terms this is clearly impossible, the result being that all lens systems are *diffraction limited*. That is, diffraction effects are responsible for the limit on the reproduction of the object's detail in the image in systems where aberrations have been removed.

One further point. If we regard d as the size of detail in the object and consider the first order diffraction spot in the transform plane, we note that $d = \frac{\lambda}{\sin\theta}$. d will be smallest when $\sin\theta$ is greatest, but the maximum value of $\sin\theta$ is unity, so the smallest resolvable size of detail in the object is λ. This explains microscopists search for microscopes employing the shortest possible wavelength. Ultraviolet light is obviously better in this respect than visible light, although the detector must be photographic film. Further reduction in wavelength has been obtained using accelerated electrons although the wavelength here is not of an electromagnetic wave. Accelerated electrons, in common with all fast moving particles, exhibit a wavelike motion which has a wavelength given by $\lambda = \frac{h}{mv}$, where h is Planck's constant, m is the electron's mass, and v is its velocity. It is possible to obtain wavelengths as small as 0.001 nm . Beams of accelerated electrons are used in the electron microscope where they are focused by magnets.

16.1.4 The Spatial Modulation Transfer Function

When evaluating the quality of an optical system we have previously determined its limit of resolution. This is equivalent to stating its cut-off spatial frequency; that is, the spatial frequency corresponding to the smallest detail in the object which can just be resolved.

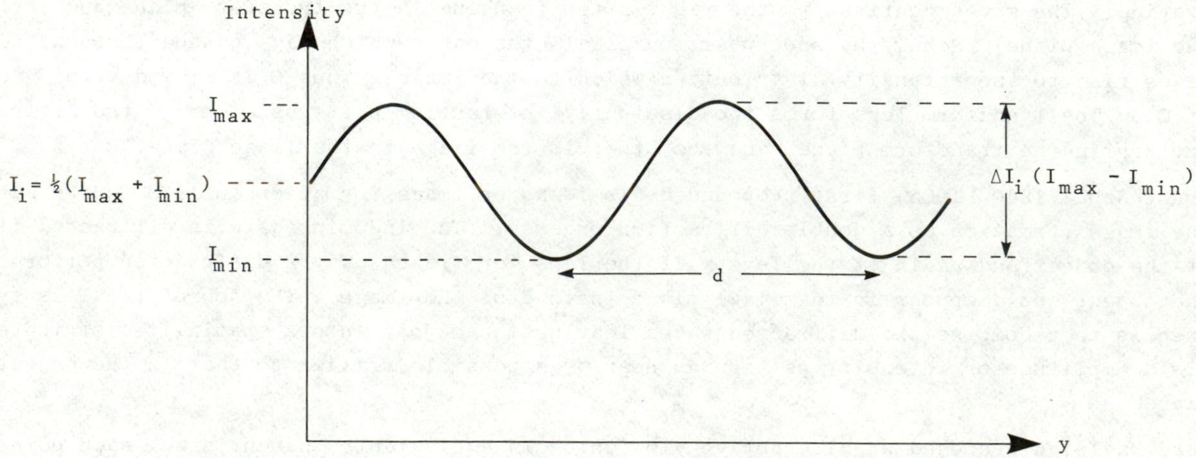

Fig. 16.8 Intensity variation across a sinusoidal grating(object). ($i \equiv$ input).

It was presumed that an optical system was better if the resolution spatial frequency was greater. This is somewhat unsatisfactory since it does not tell us how a system handles lower spatial frequencies. The fact that complex intensity patterns can be synthesised from sinusoidally varying components suggests that we should measure a system's response to intensity patterns with different spatial frequencies. Consider figure 16.8 which is the intensity distribution across a sinusoidal grating (object); that is, the intensity leaving the grating when it is uniformly illuminated. This forms the object for, or the input to, an imaging system. In order to quantify an imaging system's performance with different spatial frequencies, we now define a parameter called the *modulation* or *contrast* in the sinusoidally varying object intensity as

$$\frac{\Delta I_i}{I_i} = \left(\frac{I_{max} - I_{min}}{\frac{1}{2}(I_{max} + I_{min})}\right)_i \qquad (16.5).$$

Clearly, if $(I_{max} - I_{min})_i = 0$ we have an object with an intensity everywhere equal to I_i in the meridian we are considering. Such an object has no detail or information: the contrast is zero. Contrast will be a maximum (=1) when when $I_{min} = 0$; such an intensity distribution would be as shown in figure 16.9 .

So, as defined, the contrast can take values from zero to one. Now suppose we define the output contrast in the image plane in a similar manner; that is

$$\frac{\Delta I_o}{I_o} = \left(\frac{I_{max} - I_{min}}{\frac{1}{2}(I_{max} + I_{min})}\right)_o \qquad (16.6).$$

The ratio of the output to the input contrast,

$$\frac{\Delta I_o / I_o}{\Delta I_i / I_i} \qquad (16.7),$$

is known as the *spatial Modulation Transfer Function*, or spatial MTF, when plotted for spatial frequencies from zero to the system's cut-off frequency. The cut-off frequency is that spatial frequency in the object for which the contrast in the image is zero. That is, $\Delta I_o = 0$. (Remember the subscripts o and i refer to output and input, respectively, and output is in the image plane and input in the object plane).

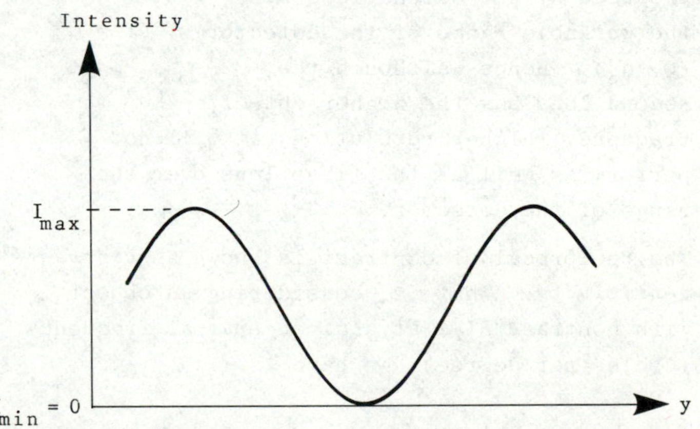

Fig. 16.9 A sinusoidal intensity distribution with maximum contrast (unity).

Ideal optical systems transform a sinusoidal input into an undistorted sinusoidal output. However, in practice input and output intensity distributions will not be identical. Diffraction by the optical components and aberrations reduce the contrast in the image. Off axis aberrations and poor centring of components cause a shift in the position of the output sinusoid which is equivalent to the introduction of a phase shift. As a consequence, considering an object's intensity distribution to consist of sinusoidal components and a zero frequency component, the way in which the system transforms the sinusoids into the corresponding sinusoids of the image determines the image quality.

Figure 16.10 shows the MTF for two hypothetical lenses with cut-off frequencies q_{c1} and q_{c2}.

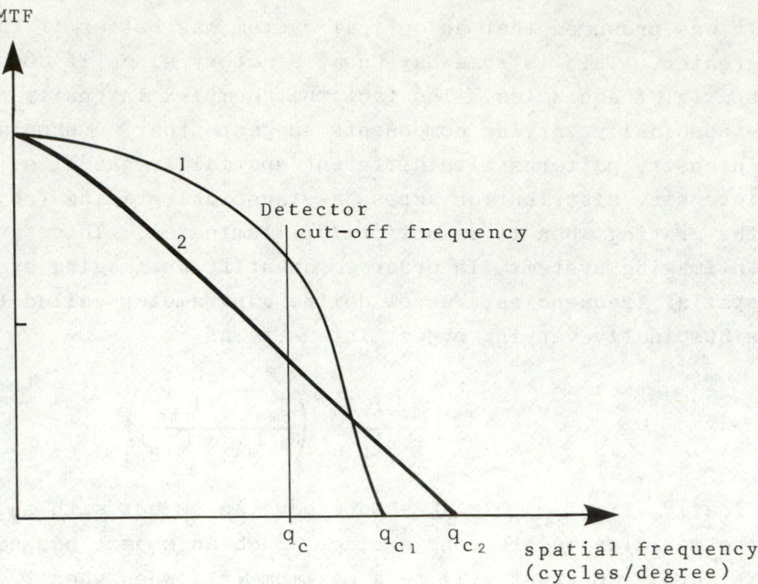

Fig. 16.10 MTF versus spatial frequency for two lenses.

The subscript c is for cut-off, of course. The detector cut-off frequency q_c is shown to be less than either q_{c1} or q_{c2}, so in neither lens is it essential to pass spatial frequencies as high as these latter two. The spatial frequencies are shown in cycles per degree on the abscissa. That is, the number of repeat distances on the object subtending 1^0 of arc at the first nodal point of the lens. The same number of cycles per degree would appear in the image of an ideal lens measured at the second nodal point. In the example here, the contrast in the image produced by the first lens is considerably greater than that produced by the second lens over the workable range of the detector ($0 - q_c$). Hence, although the second lens has the higher cut-off frequency (higher resolution) it will not perform as well as the first lens over the range of the detector.

The reciprocal of contrast is known as *contrast sensitivity*. That is, considering an object with contrast $\Delta I/I$ at a given spatial frequency (cycles per degree), we have

$$\text{contrast sensitivity} = \frac{1}{\text{contrast}} = \frac{1}{\Delta I/I} = \frac{I}{\Delta I} \tag{16.8}$$

where the object contrast is the minimum detectable by the detector.

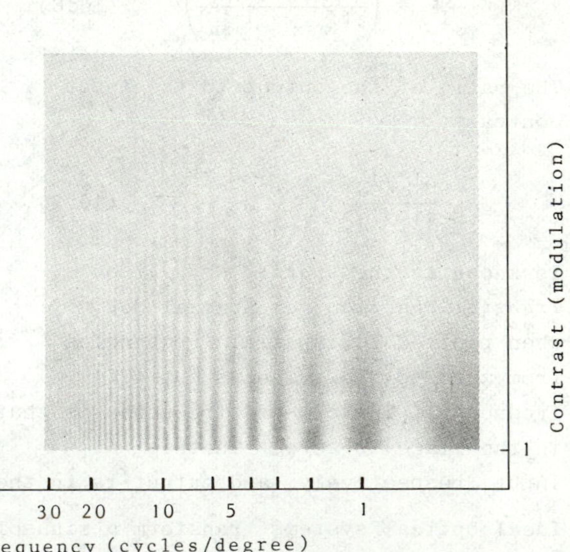

Fig. 16.11 Pattern for measuring contrast sensitivity.

This can be measured for the eye using the object in figure 16.11 . This figure consists of a pattern of sinusoidally varying stripes with spatial frequencies changing from one cycle to about 30 cycles per degree when held at ¾ m distance. The contrast, or modulation, gradually increases from top to bottom. If you place the figure at ¾ m distance and mark some 10 points, at different frequencies, for which you cannot detect any contrast, you will be able to draw a curve something like figure 16.12 . Perhaps you should cover the diagram with a sheet of glass and mark the points on the glass. The slight reduction in the air-equivalent distance due to the glass thickness will not matter for our illustration purposes.

Since we cannot measure $\Delta I_o/I_o$ for the eye we conventionally set it to 1 whence the

$$\text{MTF} = \frac{\Delta I_o/I_o}{\Delta I_i/I_i} = \frac{1}{\Delta I_i/I_i} \qquad (16.9)$$

which is, of course, the contrast sensitivity at the frequencies for which it is measured. So, the curve in figure 16.12 is the MTF for the eye-brain system. The shape of the curve shown is the MTF for a 'normal' eye-brain system. Current research is showing changes in the MTF occur early in eye diseases such as glaucoma, diabetic retinopathy, and multiple sclerosis affecting the optic nerve. Here then is the indication that optometrists and opticians will be measuring patient's MTFs in the not-too-distant future.

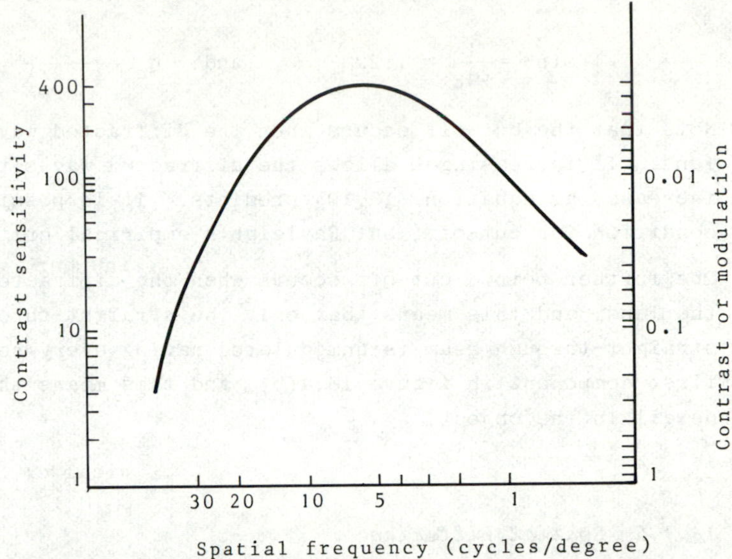

Fig. 16.12 The human contrast sensitivity curve.

The MTF can be used to specify the performance of systems as diverse as magnetic tape, film, lenses, telescopes, and the eye-brain. Further, if the MTFs of individual components in a system are known, then the system's MTF is the product of the component MTFs.

Incidentally, we can relate the cut-off frequency for a lens to the Rayleigh criterion $\sin\theta = 1.22\frac{\lambda}{D}$, where here D is the diameter of the lens. Consider figure 16.13 . If d_c is the repeat distance on a cosinusoidal grating for which the lens just cuts off the image modulation, then the cut-off frequency is found as follows:

$$\sin\theta = \tan\theta = \frac{d_c}{\ell} = \frac{1/q_c}{\ell} = \frac{1}{\ell q_c}, \text{ for small } \theta,$$

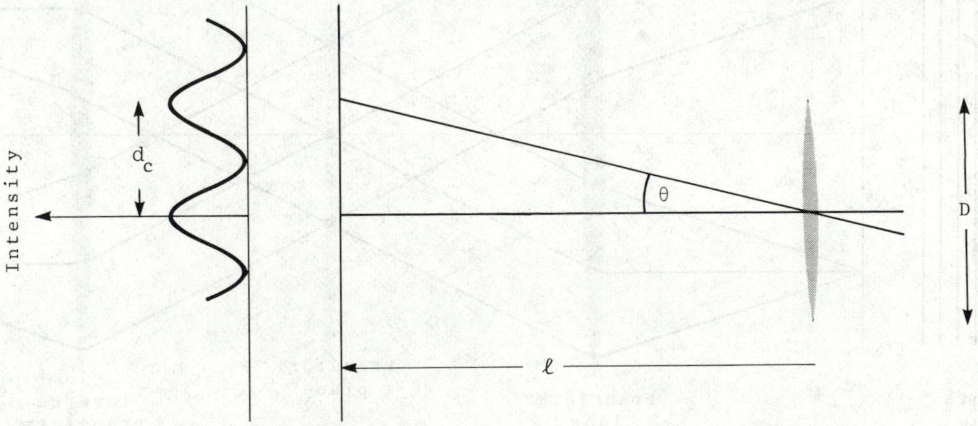

Intensity

d_c

θ

D

ℓ

object cosinusoidal grating

Fig. 16.13 Cut-off frequency vis-à-vis Rayleigh's criterion.

and where ℓ is the object distance. But $\sin\theta$ here is the same as $\sin\theta$ in Rayleigh's criterion, so

$$\sin\theta = \frac{1}{\ell q_c} = 1.22\frac{\lambda}{D}, \qquad \text{and} \qquad q_c = \frac{D}{1.22\ell\lambda} \qquad (16.10).$$

Note that the cut-off occurs when the diffracted waves in figure 16.7 do not pass through the lens. If increasing D allows the diffracted waves through the lens then clearly q_c will increase, as equation (16.10) predicts. It is possible to show that $q_c = \frac{D}{\ell\lambda}$ is a more exact condition for cut-off, but Rayleigh's empirical criterion is very close!

One further point, cut-off occurs when the diffracted waves are prevented from passing through the lens, and this means that only the straight-through beam reaches the image plane. The straight-through beam is unmodulated having everywhere the same intensity, as shown by the first component in figure 16.4(b), and this means that it carries no information about the detail in the object.

16.1.5 Spatial Filtering

The fact that the optical information in a transmittance grating, which may be a photographic transparency, is contained in the spatial frequencies which focus as spots in the transform plane has led to a simple but elegant method of improving images. Consider the system in figure 16.14 which is occasionally referred to as a *coherent optical computer*. The lenses usually have focal lengths of 30 or 40 cm and are able to resolve 150 line-spaces per mm in a bar grating (a transparency with equally wide opaque and clear strips; one opaque and one

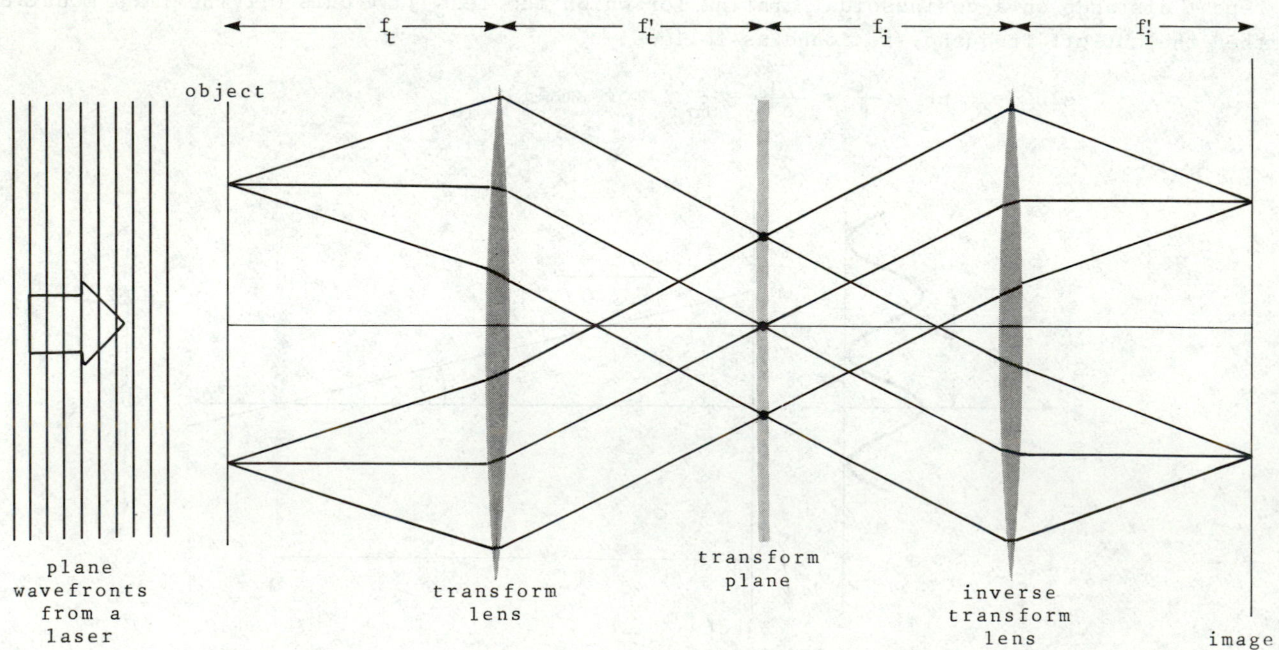

Fig. 16.14 Object, transform, and image planes in the coherent optical computer. (The subscripts t and i refer to the transform and inverse transform lenses)

clear strip form a line pair, the width of the latter being the repeat distance). The frequency spectrum spots, as they are called, in the transform plane can be intercepted by placing an obstruction of the correct size and shape in that plane. Such an obstruction is called a *mask*. Removing the diffracted waves of specific spatial frequencies is called *spatial filtering*. For example, space probe photographs are received as horizontal strips which are put together to form a whole. In so doing the boundaries of the strips are visible as artefacts on the planet or satellite. Placing a photographic transparency of the planet in the object plane of the coherent optical computer produces a diffraction pattern in the transform plane which consists of two distinct parts, figure 16.15 .

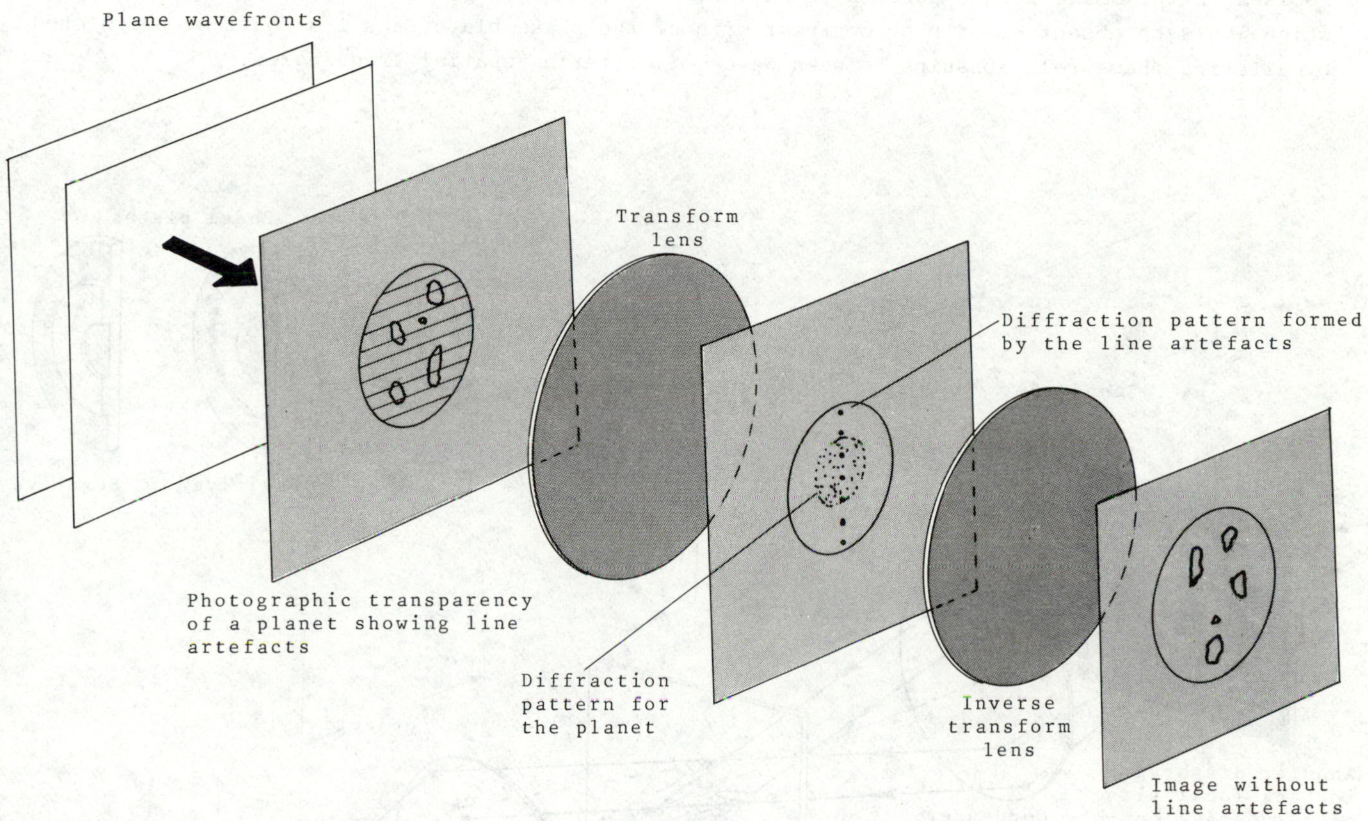

Fig. 16.15 Spatial filtering.

The spatial frequencies of the light carrying information about the planet's detail are focused in innumerable spots in a circular pattern, whilst the line artefacts generate a series of diffraction spots perpendicular to the lines. One has only to place a sheet of optical glass in the transform plane, place black masking spots where the spatial frequencies for the lines focus, and this removes or filters this information. If the second lens is a camera lens, and the image plane is the film, the new transparency will be devoid of the line artefacts. Again, this is illustrated in figure 16.15 .

16.1.6 Phase Contrast

The intensity distribution in the image depends on the amplitude of the waves in the image plane and on their phase relationships. It is possible to enhance the detail in transparent objects, which do not change the amplitude of the transmitted waves, by modifying the phase relationships with the aid of *phase filters* in the back focal plane of a microscope objective lens, figure 16.16 . The sort of object we have in mind here would be a microscope slide with transparent cells on it.

Plane waves arriving at the object slide are diffracted by the object. The diffracted wave components and the undiffracted or direct wave component have their phase relationships changed on passing through different thicknesses in the phase plate. The waves are then focused in the image plane where they interfere to form the phase contrast image. Detail, which would be absent or of poor contrast without the phase plate, has its contrast heightened by altering phase relationships between waves of differing spatial frequencies.

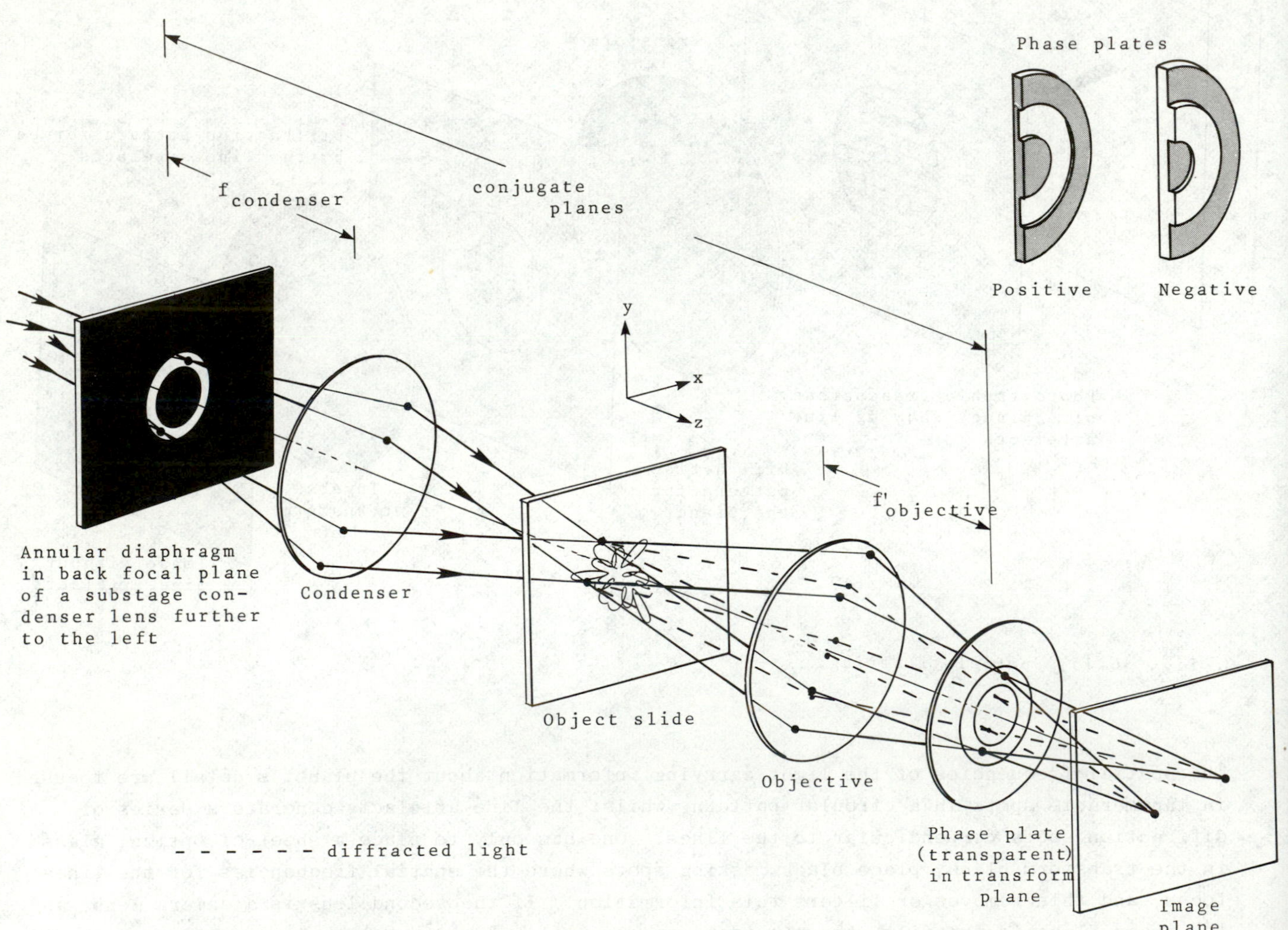

Fig. 16.16 A phase-contrast system.

Now assume incident plane waves of monochromatic light travelling in the z-direction in figure 16.16 . The incident wave may be represented by the time-dependent function

$$\psi_i(t) = E \sin \omega t \qquad (16.11).$$

A transparent object with variations in optical path lengths across its extent will modify this wave on emerging by a phase term which will vary from point to point in the x-y-plane. Call the phase term $\varepsilon(x,y)$ or, implying the x-y dependence, simply ε. Then the wave emerging from the object slide is

$$\psi(x,y,t) = E \sin(\omega t + \varepsilon) \qquad (16.12),$$

which is x, y, and t dependent. Expanding equation (16.12) gives

$$\psi = E(\sin \omega t \cos \varepsilon + \cos \omega t \sin \varepsilon) \qquad (16.13).$$

For small ε we can write $\sin \varepsilon \simeq \varepsilon$ and $\cos \varepsilon \simeq 1$, whereupon equation (16.13) becomes

$$\psi = E \sin \omega t + E\varepsilon \cos \omega t \qquad (16.14).$$

The two terms on the right hand side represent a strong ($E \gg E\varepsilon$!) unchanged, direct wave $E \sin \omega t$, and a weak diffracted wave $E\varepsilon \cos \omega t$. The latter is object dependent since ε occurs in its amplitude term. Note that $\varepsilon(x,y)$ can be Fourier analysed (see appendix 1) into spatial frequencies associated with the object. This is done by the objective lens in figure 16.16 which separates the direct and diffracted waves in its back focal plane, the transform plane. Before reaching the phase plate these two waves have the relationship shown in figure 16.17 . The two waves follow different paths through the system but are brought

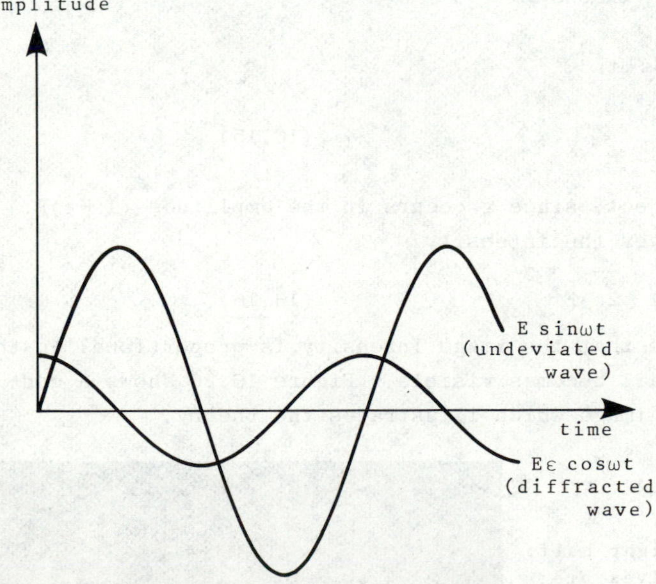

Fig. 16.17 The diffracted and direct waves leaving the object slide; phase difference 90°

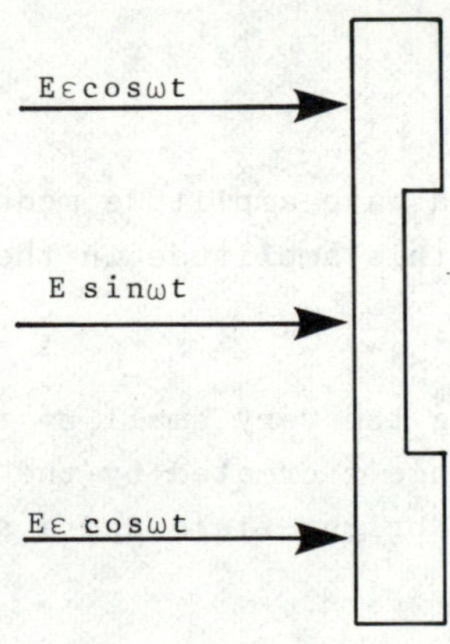

Fig. 16.18 Positive phase plate.

together in the image plane. Without the phase plate they would recombine to form the wave ψ which differs only from ψ_i by a phase change due to the object. Since the eye cannot detect a phase change, and the intensity in the image plane remains uniform, the transparent object remains invisible. With the phase plate in the system their phase and amplitude relationship is changed and this provides a change in the intensity in the image plane which renders the object visible. For example, the phase plate in figure 16.18 retards the diffracted wave $E\varepsilon \cos \omega t$ relative to the direct wave $E \sin \omega t$, or if you like, advances the direct wave relative to the diffracted wave. If the optical path difference of the two waves through the phase plate is made equal to $\lambda/4$, that is $(n_g - 1).\Delta t = \lambda/4$ where Δt is the thickness difference in the two parts of the plate and n_g is the plate's refractive index, then the phase relationship will be changed to that shown in figure 16.19 . When these waves are recombined in the image plane the intensity pattern will be different to that evident without the phase plate. To see this, simply compare figures 16.17 and 16.19 which represent the waves in the image plane without and with the phase plate, respectively.

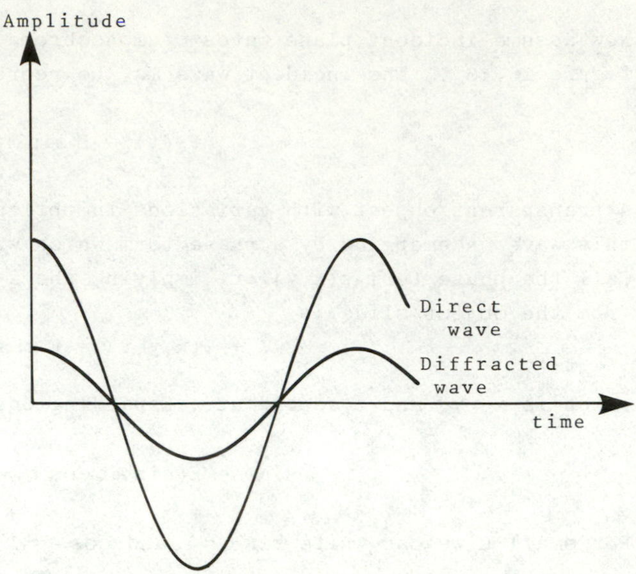

Fig. 16.19 Diffracted wave's phase advanced 90^0 relative to the direct wave: maximum contrast results from a 90^0 or $\pi/2$ phase change.

Advancing the direct beam given in equation (16.14) by $\pi/2$ results in

$$\begin{aligned} \psi &= E \sin(\omega t + \pi/2) + E\varepsilon \cos \omega t \\ &= E \cos \omega t + E\varepsilon \cos \omega t \\ &= (1 + \varepsilon) E \cos \omega t \qquad (16.15) \end{aligned}$$

which is a wave amplitude modulated by the object, since ε occurs in the amplitude $(1 + \varepsilon)E$. Squaring this amplitude in the image plane gives the intensity

$$I \propto ((1 + \varepsilon)E)^2 \simeq (1 + 2\varepsilon)E^2 \qquad (16.16),$$

neglecting the very small ε^2 term, which means that the image intensity is proportional to the phase change ε created by the object, and detail becomes visible. Figure 16.20 shows a conventional bright-field and a a phase contrast image which illustrates the theory.

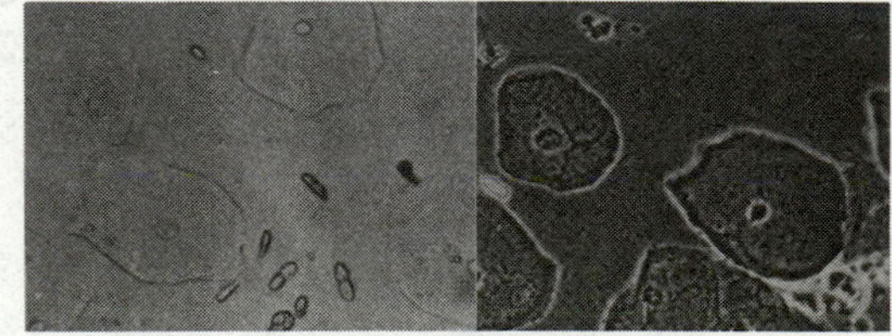

Fig. 16.20 Left half: bright field image. Right half: phase contrast image. The two halves are in fact the left and right half shots of the same microscope slide of epithelial cells.

Advancing the direct wave as discussed above results in what is called *positive phase contrast*. Since $I \propto (1+2\varepsilon)$ and ε will be positive for thinner and lower refractive index portions of the object, these portions will appear brighter than their surrounds. On the other hand, replacing the phase plate with one where the centre is raised rather than depressed leads to the result

$$I \propto (1-2\varepsilon)E^2 \qquad (16.17).$$

Here, using equation (16.17), thinner and lower refractive index portions of the object have a lower intensity than their surrounds since ε is still positive. The contrast in the image can be improved by reducing the intensity of the much stronger direct beam. This has the effect of reducing I but leaving ΔI the same in the contrast relationship $\Delta I/I$.

16.1.7 *Dark Field or Dark Ground Methods*

Because the detection of lens defects such as scratches, waves, and striae is vital in optics, a simple method of observing them will be introduced here. Figure 16.21 shows the method for transmitted and reflected light.

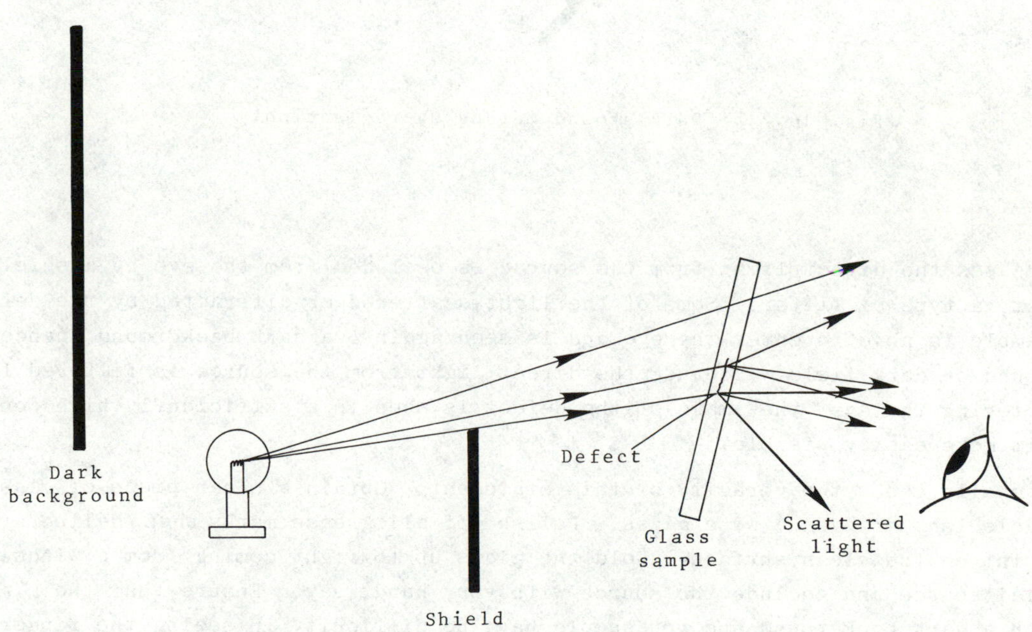

Fig. 16.21a Dark ground method by transmission.

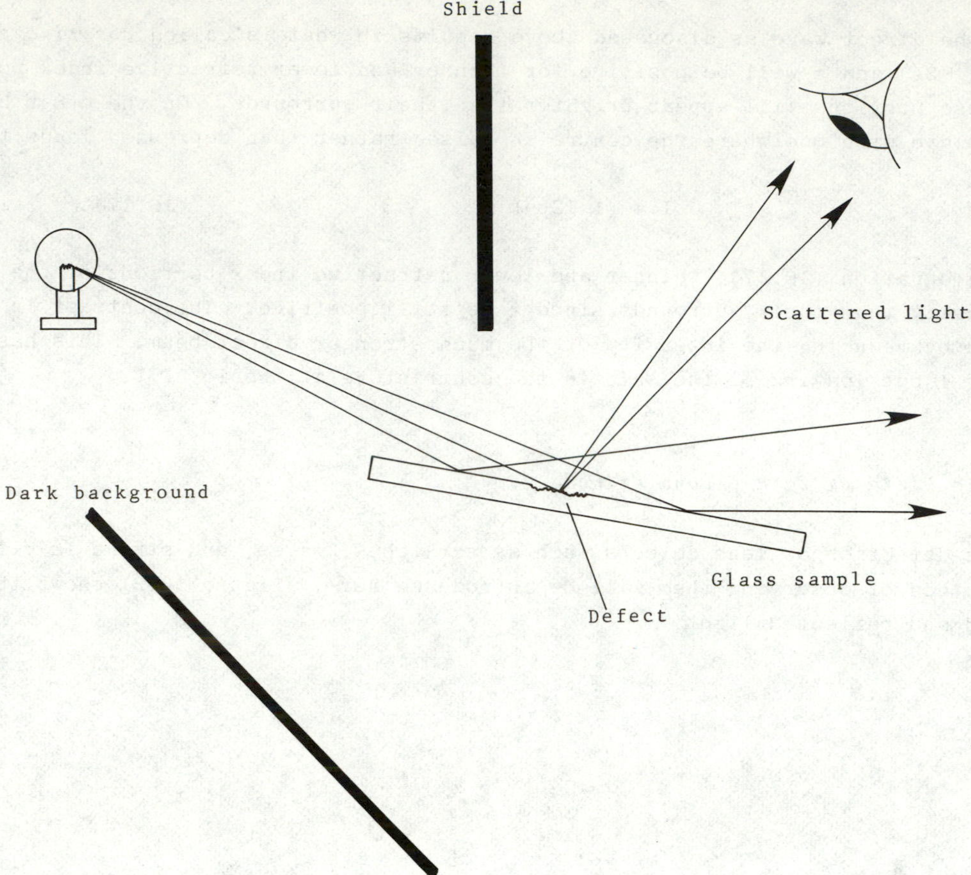

Fig. 16.21b Dark ground method by reflection.

In both cases the direct light from the source is occluded from the eye by a shield which is, in effect, a type of filter. Some of the light scattered or diffracted by the defect in the glass sample is able to enter the eye and is seen against a dark background; hence the name dark ground or dark field. Because the direct light from the source is filtered from the light entering the eye, the image of the defect is seen in a sufficiently high contrast to render it more easily visible.

You can easily check the veracity of this statement. Obtain a clean piece of glass such as a spectacle lens or even a wine glass. Polish off all grease marks then deliberately place a fingerprint on the clean surface. Hold the glass up to light coming from a window or from an artificial source and occlude the source with your hand, say. Ensure that the glass is in line with a dark background and you should have no difficulty in seeing the fingerprint. If you then allow the light from the source to enter the eye directly you will experience some considerable difficulty in detecting the fingerprint. Try the effect with both transmitted and reflected light. Often, reflected light is the better to use since it is easier to block out the light from the source in this method. This is because the source is usually away to one side of or maybe above the observer and is already out of the line of direct vision.

Figure 16.22 illustrates the difference in the contrast when the direct light is or is not

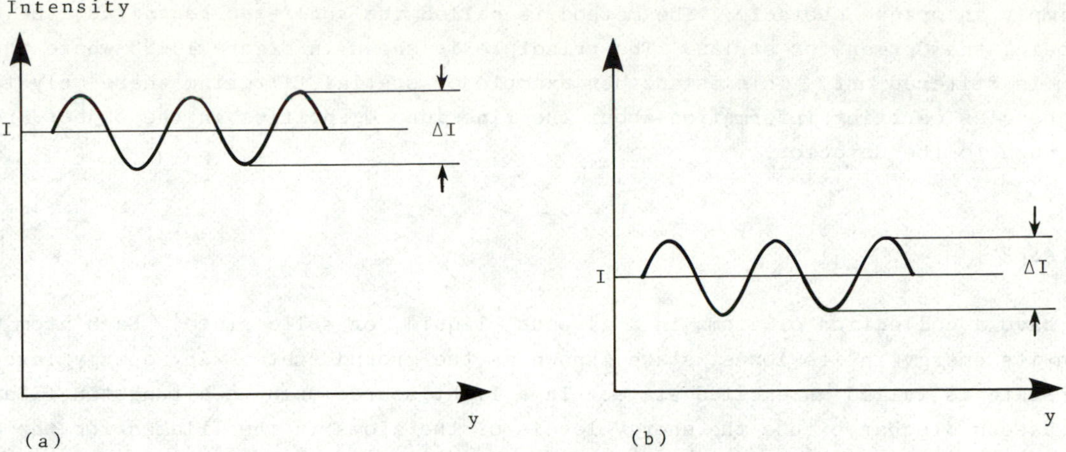

Fig. 16.22 Increased contrast by reducing the mean intensity I.

present in the beam reaching the eye. The contrast in both cases is $\Delta I/I$. Suppose ΔI, the depth of modulation, is identical in the two cases, then clearly $\Delta I/I$ is greater in (b) than in (a) since I is smaller in the former. To detect a change in intensity across an object $\Delta I/I$ must exceed about 0.02 for the eye. Thus, in a case where I is too large $\Delta I/I$ may be less than 0.02 and the defect in the lens will go undetected.

Several systems involving lenses or mirrors have been developed to bring the scattered and direct beams to a focus. In the case of a converging lens the two beams will focus in two distinct places in the back focal plane of the lens if parallel incident light is used. It is then a simple matter to filter out the direct beam with what is known as a *schlieren filter*,

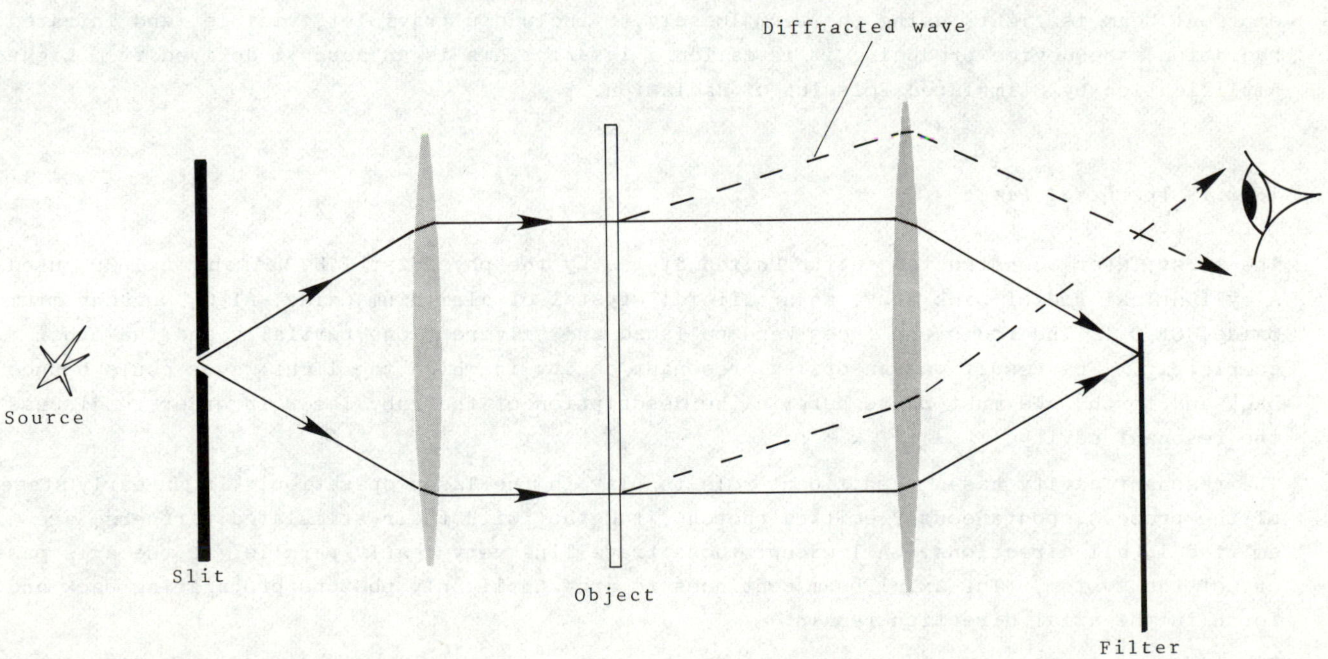

Fig. 16.23 A schlieren filtering system.

which is simply an opaque obstacle. The method is called the *schlieren technique*, the word schlieren being the German for striae. The principle is shown in figure 16.23 where the direct beam is filtered out. This is another example of spatial filtering where only the higher frequencies carrying information about the fine inhomogeneities in the object are allowed through to the detector.

16.2 THE LASER

Suppose we have a collection of atoms in a gaseous, liquid, or solid state. Each atom tends to maintain its energy in its lowest state, known as the ground state. Any energy level above the ground state is called an excited state. In a light source such as a tungsten filament lamp or a gaseous discharge tube the energy levels of the atoms in the filament or the gas are raised by 'pumping' energy into the system. Each atom drops spontaneously back to the ground state emitting, as it does, a randomly directed photon. The radiating atoms in the source bear no phase relationship which is constant and enduring with the result that the light emitted is incoherent.

In 1917, Einstein pointed out that an excited atom can revert to a lower state by emitting the photon spontaneously, or it can be triggered by the presence of another photon of the same frequency. This latter method is known as *stimulated emission*. What makes the phenomenon vitally important is the fact that the emitted photon is in phase with, has the polarisation of, and propagates in the same direction as the stimulating photon. If the majority of the atoms in a gas, a liquid, or a solid could be maintained in an excited state, stimulated emission could then trigger a flood of in-phase photons. A state in which the majority of atoms are excited is called a *population inversion*. Provided the population inversion can be maintained the wave would continue to build like an avalanche. Energy would have to be pumped in to sustain the inversion and the coherent beam could be extracted after traversing the medium. When the emergent beam is light, using the term loosely to include ultraviolet, visible, and infrared radiation, the device producing it is called a *laser*. This is an acronym derived from Light Amplification by Stimulated Emission of Radiation.

16.2.1 The First Laser

The first laser, constructed in the United States by the physicist T.H. Maiman in 1960, used a cylindrical rod of pink ruby, an artificial crystal of aluminium oxide, Al_2O_3, and chromium oxide, Cr_2O_3. The rod's end faces were polished and silvered, one partially and the other completely. The result was an optical resonant cavity in which the light waves could bounce back and forth. We must pause here in the description of the ruby laser in order to discuss the resonant cavity.

The resonant cavity has a significant role to play in the laser operation. In the early stages of the process spontaneously emitted photons, together with their stimulated partners, are emitted in all directions. All except those travelling very nearly parallel to the axis pass out of the system. The axial beam continues to grow until only photons propagating back and forth in the axial direction remain.

The wave propagating back and forth within the cavity takes on a standing wave profile which is determined by the distance d between the mirrors. That is, the cavity resonates when an integer number m of half-wavelengths spans the distance d. Thus

$$m = \frac{d}{\lambda/2} \qquad (16.18).$$

Substituting $\lambda = v/\nu$ in equation (16.18), then

$$\nu_m = \frac{mv}{2d} \qquad (16.19),$$

where ν_m is called a *frequency mode*. Consecutive frequency modes are separated by

$$\Delta\nu = \nu_{m+1} - \nu_m = \frac{(m+1)v}{2d} - \frac{mv}{2d} = \frac{v}{2d} \qquad (16.20).$$

For a gas laser ½m long, $\Delta\nu = \frac{v}{2d} = \frac{3 \times 10^8}{2 \times \frac{1}{2}} = 3 \times 10^8$ Hz.

The cavity can be designed to produce one or more frequency modes and the beam is therefore restricted to a region close to those frequencies. Where one frequency only is selected this gives rise to the extreme monochromaticity of the laser.

We can now continue the description of the first laser. Surrounding the ruby cylinder was a gaseous flashtube which provided the optical pumping - the energy to create the population inversion. Firing the flashtube produces a short but intense burst of light lasting several milliseconds. Many of the chromium Cr^{3+} ions are excited and absorb frequencies in the blue and green regions of the visible spectrum. They rapidly give up energy to the crystal lattice and make non-radiative jumps down to a relatively long-lived metastable state above the ground level. They remain in this interim state for several milliseconds, thereafter dropping to the ground state and radiating the characteristic red fluorescence of ruby. The emission is centred at 694.3 nm, is randomly directed, and is incoherent. Increasing the pumping rate creates the population inversion and the avalanche of in-phase photons begins. The wave grows as it sweeps backwards and forwards across the length of the cylinder. Since one end is only partially silvered, an intense pulse of red laser light lasting some 0.5 milliseconds emerges from that end.

16.2.2 The Helium-neon Laser

Maiman announced the first laser to the world in July 1960. The first helium-neon gas discharge laser was constructed by Donald R. Herriott, Ali Javan, and William R. Bennett, and put in operation by late 1960. It produced a continuous wave, rather than pulses, at 1152.3 nm. The modern helium-neon laser, figure 16.24, provides a continuous wave in the visible region, centred at 632.8 nm. It is the type most likely found in teaching laboratories, typically producing less than a watt of power and being robust and reliable.

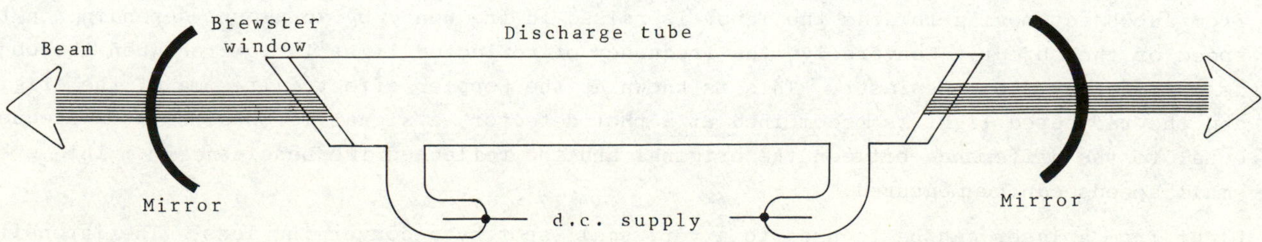

Fig. 16.24 A helium-neon laser (schematic).

The mirrors have a multilayered, 99% plus, reflecting film. Inclining the end windows, Brewster windows, at the polarisation angle makes the emergent beam linearly polarised and allows 100% transmission of light polarised in the plane of incidence (the plane of the paper). The polarisation state rapidly becomes dominant because the component vector perpendicular to the plane of incidence is partially reflected off axis at each window. Since the stimulated emission is by a preponderance of photons polarised in the plane of incidence, these predominate to the ultimate exclusion of those polarised in the mutually orthogonal plane. The emergent beam is therefore linearly polarised. Rotating a piece of polaroid in the beam, in a plane perpendicular to the axis, shows the intensity to vary according to Malus' law.

16.2.3 Laser Applications

Light produced by lasers is generally far more monochromatic, directional, powerful, and coherent than light from any other type of source. A continuous visible beam from a helium-neon gas laser, say, provides a near perfect straight line for all kinds of alignment work. The beam diverges by less than one part in a thousand, or about 1 minute of arc, and is typically a few millimetres in diameter. Lasers have been used for alignment in drilling tunnels and laying pipes. They are used to align jigs employed in the building of large aircraft, permitting an accuracy of ±0.25 mm over 60 m.

A pulsed laser can be used in a similar fashion to radar*. Called *LIDAR*, the narrowness of the beam permits sharp definition of targets. The distance of the object is determined by the time taken for the light to reach and return from it. The first astronauts to land on the moon placed a multiprism reflector on the surface which was used with lidar echoes to measure the Earth-Moon distance to within 30 cm. Simultaneous measurements of the distance and direction from two observatories on Earth allow an accurate calculation of the distance between the observatories. Series of such measurements from observatories on different continents have enabled Earth scientists to measure continental drift!

Vertically directed lidar from an aircraft serves to detect contour details such as steps in outdoor public places or the shapes of buildings. Reflections from clouds can measure the height of cloud cover over airports.

The high spatial and temporal coherence of the laser beam makes it suitable for interferometric applications over large distances. If the beam is divided into two parts that travel different paths, when the beams come together they will exhibit interference fringes. Changing the path length difference by one half-wavelength shifts a bright fringe into the space previously occupied by a dark fringe. Thus, small displacements in path length can be measured. Laser interferometers are used to monitor changes in the Earth's crust across geological faults. They are used in manufacturing to monitor the products of automated machine tools and to test optical components.

Lasers can be so monochromatic that a small shift in frequency can be detected. Light reflected from an object moving towards the laser is raised in frequency by an amount depending on the speed of the object. Conversely, the frequency of reflected light is lowered when the object is moving away from the laser. This is known as the Doppler effect. If some of the original and the reflected light is recombined at a photodetector, a signal is obtained at a frequency equal to the difference between the original and the reflected frequencies. From this even small speeds can be measured.

Light from a laser can be focused to a very small spot by a converging lens. The intensity in

**RADAR - RAdio Detecting And Ranging*

the spot will be relatively very great so that even moderately powered lasers can vapourise any substance. For example, a carbon dioxide laser beam of a few kilowatts continuous wave can burn a hole through a six millimetre thick stainless steel plate in about ten seconds. At the other extreme, a finely focused laser can vapourise parts of a single cell, thus permitting microsurgery on chromosomes.

A pulsed laser can be focused on and vapourise the ink on some paper whilst leaving the paper unscathed. Here then is the basis of a laser eraser. Such localised heating effects have been exploited in surgery on the retina of the eye. It is used to produce sterile burns around a hole in the retina, thereby sealing it and preventing a detachment. Small balloon-like defects (aneurisms) in arteries can be 'knocked out' in retinal vessels, thus preventing possible haemorrhaging.

Laser beams can also be used for communications. Because the frequency of light is so high the intensity can be rapidly varied to encode very complex signals. One laser beam could carry as much information as all the existing radio channels. However, the laser beam would have to be enclosed in pipes to prevent scattering by rain, fog, or snow. Even atmospheric turbulence would be a nuisance. In space, free from atmospheric scattering, the laser would make an ideal low power communication beam.

16.2.4 Speckle Patterns

The spatial coherence of laser light is responsible for the striking pattern observable on reflection from a diffusing surface. A piece of white blotting paper and a helium-neon laser beam, expanded through a reversed Galilean telescope system, illustrates the phenomenon well. The spatially coherent light scattered by the diffusing surface fills the surrounding space with a stationary interference pattern. The pattern takes on a granular or speckled appearance, the granules appearing larger in size as the distance from the surface increases. For the pattern to be sustained as it is, there must be a constant relative phase relationship between the interfering waves at any point in space where the waves from different scattering regions superpose.

A real system of interference fringes (speckles) is formed in front of the diffusing surface and these can be imaged on a screen, as would be expected. After forming the real images the rays diverge and can subsequently be focused by the eye. Rays initially diverging from the surface can also be focused by the eye, of course, and appear to come from a virtual image behind the surface. Since the negative vergence arriving at the eye will have a greater magnitude for the real image, uncorrected myopes focus on this one. Conversely, uncorrected hypermetropes focus on the virtual image. If the subject moves his head to the left, the pattern will move in the same direction for a hypermetrope and in the opposite direction for a myope. If the subject is very close to the surface the pattern will follow the head movement no matter which is focused upon.

16.3 HOLOGRAPHY

Holography is the means of creating an image without the use of a lens. In normal photographic images a record of the intensity distribution across the object is made in the image plane of the lens. The image is a two-dimensional record of a three-dimensional object and, as such, must lack something when viewed by an observer. Depth in the object is recorded in the geometrical perpective, light and shade variations, overlapping, and relative size clues in the image. Although a sense of depth can be obtained by taking two slightly dissimilar photographs from two camera positions horizontally about 6 - 8 centimetres apart and presenting these stereo-photographs separately to each eye, there is still something lacking in the perceived image. Photographs or stereo-photographic pairs do not require a change of focus when looking from the image of a far distant object to the image of a near object. Neither can they reproduce the parallax effects that existed in the real object. That is to say, if the observer moves laterally with respect to the photograph, he does not see objects in the foreground moving in the opposite direction. Thus, photographs or stereo-photographs do not reproduce the depth relationships which exist in the real object.

Holography was invented in 1948 by the Hungarian born scientist Denis Gabor, and it gained him the 1971 Nobel prize for physics. He coined the word from the Greek *holos (= whole)* and *gram (= message)* because the image forming system contained all the optical information in the original object. Whereas the conventional photograph contains the intensity distribution in the image plane, a hologram when correctly illuminated reproduces the optical field with its amplitude and phase relationships which were present in the light leaving the original object. The development of the laser in the 1960s caused a sudden increase of interest in holography since, as we shall see, a highly coherent source of illumination with a high intensity is required for the production of good holograms. The continuous wave and pulsed types of lasers are of especial interest. The former emits a continuous beam of very nearly monochromatic light whilst the pulsed laser emits an extremely intense, short flash of light which lasts about 10^{-8} seconds. Two physicists in the United States, E.N. Leith and J. Upatnieks at Michigan University, applied the continuous wave laser to holography. Their success paved the way for further interest and research.

16.3.1 Continuous-wave Holography

In a darkened room a beam of continuous wave laser light is directed from a source onto an object. The beam is reflected, scattered, and diffracted by the physical features of the object and some of this light subsequently arrives at the photographic plate, figure 16.25 . Part of the beam from the laser is earlier split off to be reflected at a mirror onto the photographic plate. The two coherent beams interfere creating a complex pattern of interference fringes which are recorded on the plate. The plate, when developed, is called a *hologram*.

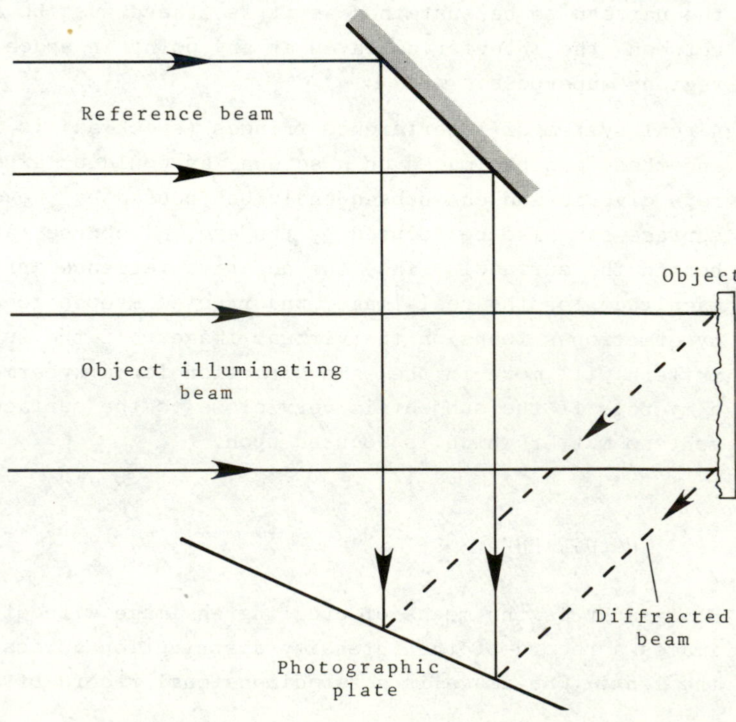

Fig. 16.25 Producing a hologram.

A hologram bears no resemblance to the object but it does contain a record of the optical field amplitude and phase data which can be used to reproduce an image indistinguishable from the object.

If the hologram is illuminated by the reconstructing or playback laser beam, as in figure 16.26, most of the light passes straight through it as a central beam which is not used.

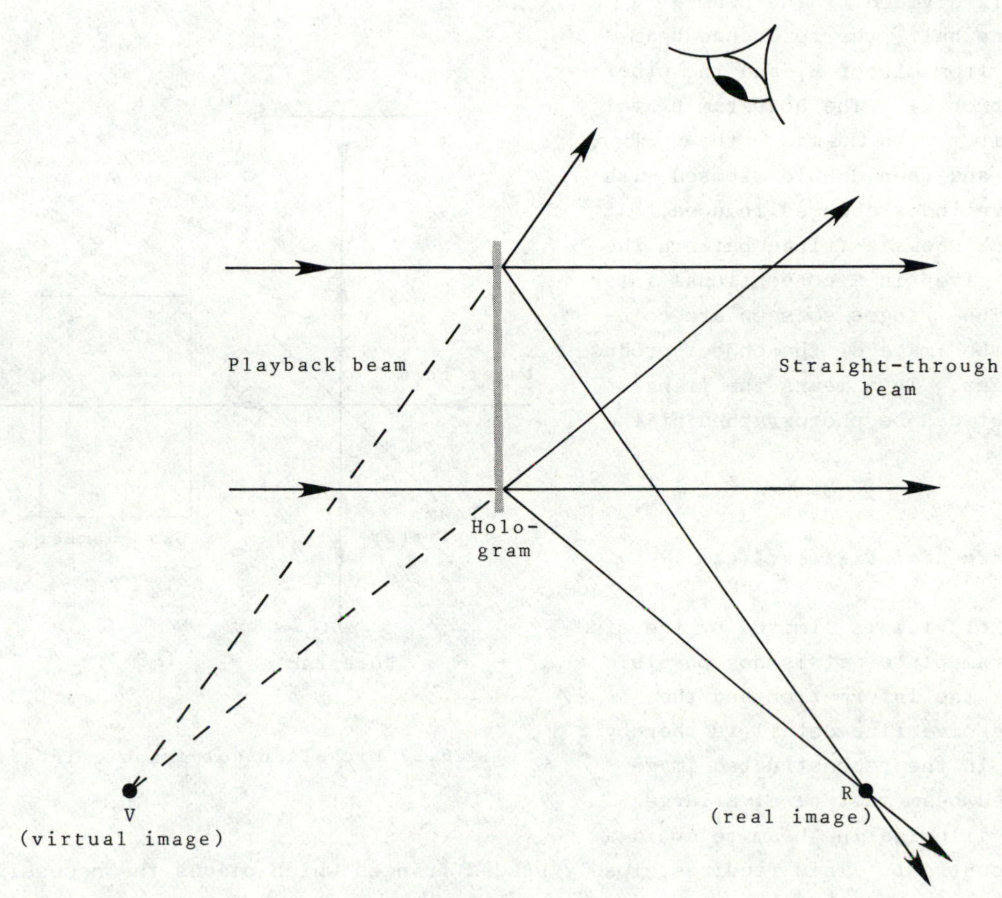

Fig. 16.26 Reproducing the image.

The fine detailed interference pattern on the plate acts as a diffraction grating diffracting the remaining light at a wide angle from the straight-through beam. This diffracted light recreates the optical field at V which existed due to the light originally reflected from the object. At R a pseudoscopic field, reversed back to front, is produced. The virtual image V has all the properties an observer would expect of a real object; parallax and focusing changes are exhibited between parts of the image which are at different distances from the eye.

16.3.2 Pulsed-laser Holography

A moving object can be made to appear at rest when a hologram is produced with an extremely rapid and high intensity pulsed laser. Providing the object does not move more than a distance of about $\lambda/10$ during the 10^{-8} or 10^{-7} second duration of the flash, a useable hologram can be made.

Pulsed laser holography has been used in wind-tunnel experiments. It records interferometrically the changes in refractive index in the air flow created by pressure changes in the gas as air flows around the object under inspection. In effect, the system is a Michelson holograph system, figure 16.27 .

Laser light is divided by the beam-splitter. One half, the reference beam, is reflected from mirror M_1 and the other half from mirror M_2. The hologram plate is exposed first with the gas in the chamber undisturbed, and then double exposed with the refractive index changes induced. A later playback shows a fringe pattern the same as that given in a conventional interferometer. The fringes so seen are coincident with the image of the object producing the fringes. This means the fringes and the object can be photographed at the same time.

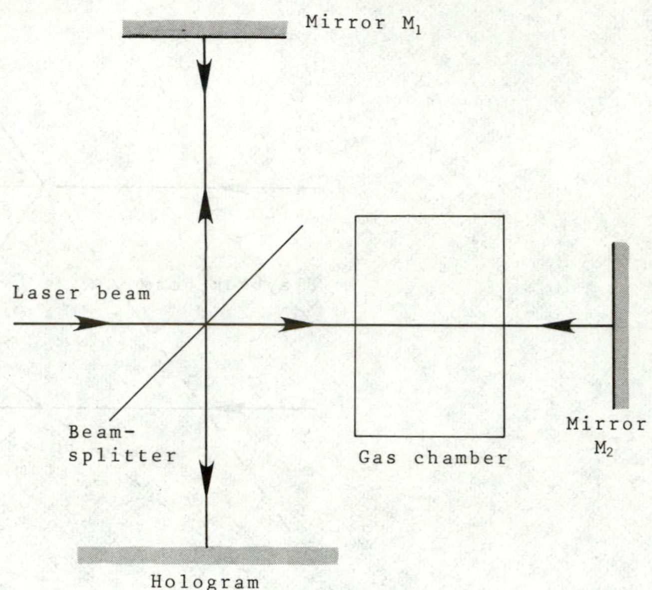

Fig. 16.27 Michelson holographic interferometer.

16.3.3 *Experimental Difficulties*

As the field of view is limited by the size of the hologram plate, it is not possible to record all the information and the ability to resolve fine detail is thereby restricted. In the reconstructed image the diffracted beam must be at a large angle to the illuminating beam to avoid a lowering of contrast. This requires closely spaced fringes which places the necessity of a high resolution on the photographic plate. No matter how good the plate is there will inevitably be some loss of detail.

The speckle pattern produced when laser light is reflected from a diffusing surface causes a further loss of detail. The interference creating the speckle pattern is recorded in the hologram and reproduced in the play-back where it acts as 'noise' in the system.

Slight alterations in the path lengths of the reference and diffracted beams can cause distortion of the fringes in the hologram. Such fluctuations can occur from vibrations or even from index changes caused by draughts.

16.3.4 *The Mathematics of Holography*

Suppose we represent the reference beam which reaches the photographic plate by

$$\psi_R = E_R \cos(\omega t + \varepsilon_R(x,y)) \qquad (16.21),$$

where E_R is the amplitude of the reference wave, and $\varepsilon_R(x,y)$ is a phase term at all points (x,y) on the photographic plate.

Similarly, let the wave arriving at the plate from the object be

$$\psi_0 = E_0 \cos(\omega t + \varepsilon_0(x,y)) \qquad (16.22),$$

where the amplitude E_0 and the phase term ε_0 are functions of x and y determined by the physical characteristics of the object.

There are two significant differences between the reference and object waves at the photographic plate. Firstly, the amplitude of the reference wave, E_R, is constant, whereas the amplitude of the object wave, E_0, varies from point to point across the plate because of the diffraction by the object; it is therefore a function of x and y. Secondly, the term ε_R will be zero if the reference beam is normal to the plate, or it will vary in some simple way with x and/or y. For example, suppose a wavefront could be brought into coincidence with the plate by rotating the beam through an angle θ about the y-axis. Then we would have for any point on the plate $\varepsilon_R = \frac{2\pi}{\lambda} x \sin\theta$, figure 16.28 .

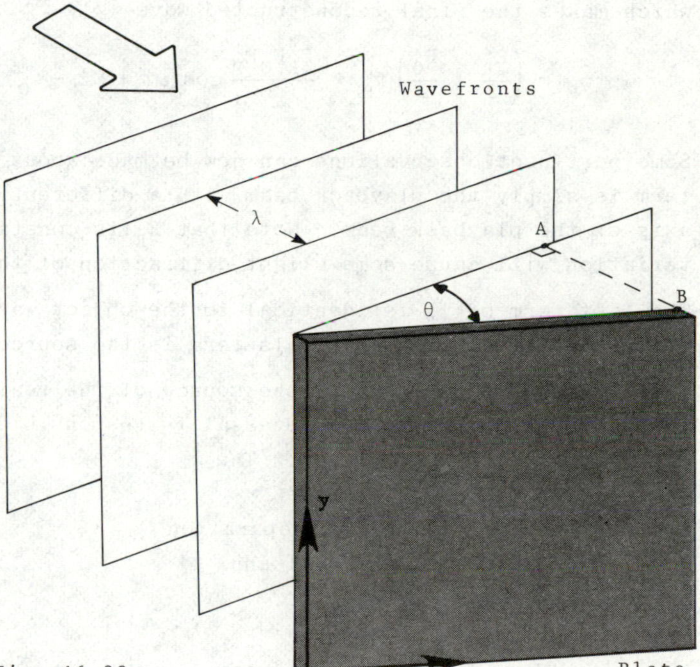

Suppose the point B on the plate has x-coordinate x_1, then $AB = x_1 \sin\theta$ where AB is the distance the wavefront at A must travel to reach the plate. If at $x = 0$ the wavefront is at the plate, then the path difference is AB for the two points on the wavefront, and the phase difference is $\varepsilon_R = \frac{2\pi}{\lambda} \cdot AB = \frac{2\pi}{\lambda} x_1 \sin\theta$.

In contrast, the phase term ε_0 will be a very complicated function of x and y.

Fig. 16.28

When these two waves superpose at the photographic plate the intensity will be given by

$$I(x,y) = \langle(\psi_R + \psi_0)^2\rangle \qquad (16.23),$$

where this expression indicates the time average of the sum of the waves squared. Putting the expressions for ψ_R and ψ_0 into equation (16.23), and using the identity

$$\cos(A+B) + \cos(A-B) = 2 \cos A \cos B$$

gives

$$I(x,y) = \frac{E_R^2}{2} + \frac{E_0^2}{2} + E_R E_0 \cos(\varepsilon_R - \varepsilon_0) \qquad (16.24).$$

Note that the cosine term, which can vary between -1 and +1, contains ε_0. So the object wave's phase determines the maxima and minima of intensity on the photographic plate!

Suppose now the plate is developed so that its amplitude transmittance is proportional to the incident intensity and we use a playback wave ψ_P with a phase identical to the reference beam's, then the final reconstructed field is

$$\frac{\psi_F}{\psi_P} \propto I \qquad (16.25)$$

or, using I from (16.24),

$$\psi_F \propto I\psi_P = \left(\frac{E_R^{\,2}}{2} + \frac{E_0^{\,2}}{2}\right)\psi_P + E_R E_0 \psi_P \cos(\varepsilon_R - \varepsilon_0) \qquad (16.26).$$

Let $\psi_P = E_P\cos(\omega t + \varepsilon_R)$, whence the right hand term of equation (16.26) becomes $E_R E_0 E_P \cos(\omega t + \varepsilon_R)\cos(\varepsilon_R - \varepsilon_0)$. Using the identity $\cos(A+B) + \cos(A-B) = 2\cos A\cos B$ again, this becomes

$$\frac{E_R E_0 E_P}{2}\cos(\omega t + 2\varepsilon_R - \varepsilon_0) + \frac{E_R E_0 E_P}{2}\cos(\omega t + \varepsilon_0)$$

which makes the final reconstructed wave

$$\psi_F \propto \left(\frac{E_R^2}{2} + \frac{E_0^2}{2}\right)\psi_P + \frac{E_R E_0 E_P}{2}\cos(\omega t + 2\varepsilon_R - \varepsilon_0) + \frac{E_R E_0 E_P}{2}\cos(\omega t + \varepsilon_0) \qquad (16.27).$$

Some pertinent observations can now be made about the final reconstructed field ψ_F. The first term is simply the playback beam with a different amplitude. As such it will lie along the axis of the playback beam. Note that E_R is constant, whereas E_0 is x,y dependent and this variation will cause some slight diffraction of the straight-through beam.

The last term of ψ_F is identical to the object wave, equation (16.22), except for the constant of multiplication $E_R E_P/2$. This term is the source of the virtual image in figure 16.26 .

Finally, the middle term is the source of the real image. Note that the phase contains $-\varepsilon_0$ instead of ε_0. This phase reversal is responsible for the pseudoscopic, i.e. back-to-front, appearance of the real image. The term $2\varepsilon_R$ in the phase is here responsible for the angular separation of the waves forming the real and virtual images, figure 16.29 .

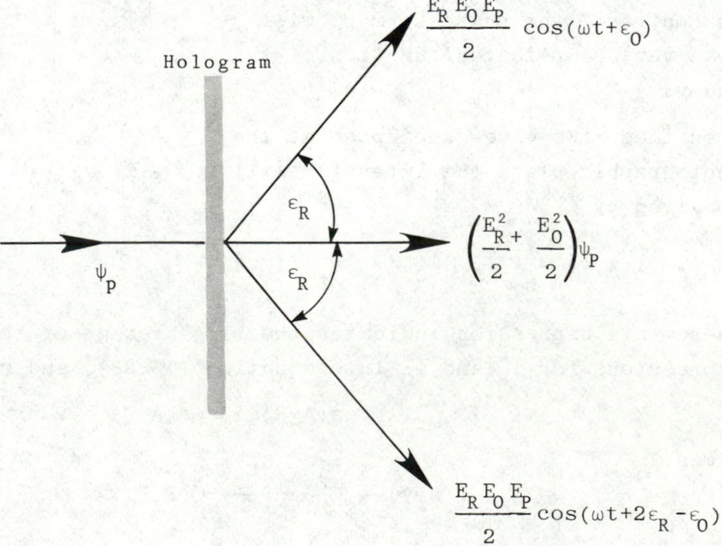

Fig. 16.29 Angular relationships between the directions of the playback beam, the real image wave, and the virtual image wave.

EXERCISES

1 What is the amplitude transmittance of (a) a completely transparent substance, (b) a completely opaque substance, and (c) a substance which reduces the amplitude of the incident light by ⅓ on emerging from it?

Ans. (a) 1. (b) 0. (c) ⅓.

2 Plane waves of amplitude 8 arbitrary units are incident normally on a cosinusoidal grating. What are the amplitudes of the straight-through beam and the two diffracted beams?

Ans. 4, 2, and 2.

3 If the square wave grating in figure 16.4 has a repeat distance of 10^{-4} m, what are the spatial frequencies of the first three diffracted waves?

Ans. (i) $\frac{1}{d} = \frac{1}{10^{-4}} = 1\,000$. (ii) 30 000. (iii) 50 000. (Units of cycles per metre).

4 Briefly state Abbe's and Rayleigh's theories of image formation.

5 Write an account of contrast sensitivity referring especially to its measurement in the visual system.

6 Design a rudimentary schlieren filtering system with a small light source, a concave mirror, and a circular opaque disc for a filter.

7 Write an account of the laser and its applications.

8 What is holography?

APPENDIX 1 - Fourier Analysis

Jean Baptiste Joseph, Baron de Fourier (1768 - 1830), French physicist and Egyptologist, devised a mathematical technique for representing any periodic wave as the sum of an infinite number of sine and cosine functions. His theorem states that

a function f(x), with spatial period or wavelength λ, can be synthesised by the superposition of harmonic functions which have wavelengths λ, λ/2, λ/3, and so on.

Thus,

$$f(x) = \frac{A_0}{2} + \sum_{m=1}^{\infty} A_m \cos\frac{2\pi}{\lambda}mx + \sum_{m=1}^{\infty} B_m \sin\frac{2\pi}{\lambda}mx \qquad (1).$$

The first term is written as $A_0/2$ since it leads to a simplification later in the treatment. Given f(x), it will be possible to determine the coefficients A_0, A_m, and B_m. The process of determining the coefficients is known as *Fourier Analysis*.

We will derive firstly some equations which will allow the calculation of the coefficients. We begin by integrating equation (1) over any wavelength interval, say 0 to λ.

Since $\int_0^\lambda \sin\frac{2\pi}{\lambda}mx\,dx = \int_0^\lambda \cos\frac{2\pi}{\lambda}mx\,dx = 0$, there is only one non-zero term which gives

$$\int_0^\lambda f(x)\,dx = \int_0^\lambda \frac{A_0}{2}\,dx = A_0\frac{\lambda}{2}, \text{ whence } A_0 = \frac{2}{\lambda}\int_0^\lambda f(x)\,dx \qquad (2).$$

To find A_m and B_m we use the facts that

$$\int_0^\lambda \sin\frac{2\pi}{\lambda}ax\cos\frac{2\pi}{\lambda}bx\,dx = 0 \qquad (3),$$

$$\int_0^\lambda \cos\frac{2\pi}{\lambda}ax\cos\frac{2\pi}{\lambda}bx\,dx = \frac{\lambda}{2}\delta \qquad (4),$$

and

$$\int_0^\lambda \sin\frac{2\pi}{\lambda}ax\sin\frac{2\pi}{\lambda}bx\,dx = \frac{\lambda}{2}\delta \qquad (5),$$

where a and b are non-zero positive integers, and $\delta = 1$ when $a = b$ or $\delta = 0$ when $a \neq b$.

If we multiply both sides of equation (1) by $\cos\frac{2\pi}{\lambda}px$, where p is a positive integer, and then integrate over 0 to λ, only one term is non-zero and that occurs when $p = m$, whence

$$\int_0^\lambda f(x)\cos\frac{2\pi}{\lambda}mx\,dx = \int_0^\lambda A_m\cos^2\frac{2\pi}{\lambda}mx\,dx = \frac{\lambda}{2}A_m,$$

having used equation (4) to integrate the $\cos^2$ term with $a = b = m$.

Therefore,

$$A_m = \frac{2}{\lambda}\int_0^\lambda f(x)\cos\frac{2\pi}{\lambda}mx\,dx \qquad (6).$$

Similarly, multiplying both sides of equation (1) by $\sin\frac{2\pi}{\lambda}px$ and integrating again gives

$$B_m = \int_0^\lambda f(x) \sin \frac{2\pi}{\lambda} mx \, dx \qquad (7).$$

To recap then, we can represent any infinite periodic function by the Fourier series in equation (1) where the coefficients A_0, A_m and B_m are calculated from equations (2), (6), and (7), respectively.

In section 16.1.2 , we met the square wave grating transmittance where the repeat distance was d instead of λ, and the variable was y instead of x. Accordingly, the Fourier series is

$$f(y) = \frac{A_0}{2} + \sum_{m=1}^{\infty} A_m \cos \frac{2\pi}{d} my + \sum_{m=1}^{\infty} B_m \sin \frac{2\pi}{d} my ,$$

assuming the square wave grating has infinite extent, something we overlooked in section 16.1.2 . In calculating the coefficients one should note that where f(y) is an even function, that is $f(y) = f(-y)$, then the B_m coefficients will be zero since the sine function is odd. Also, when f(y) is an odd function, that is $f(y) = -f(-y)$, the coefficients A_m will be zero since the cosine function is even and will not contribute to the series. This observation saves the effort of unnecessary computation.

In figure (1) the square wave is shown as an even function, as it was in figure 16.4 .

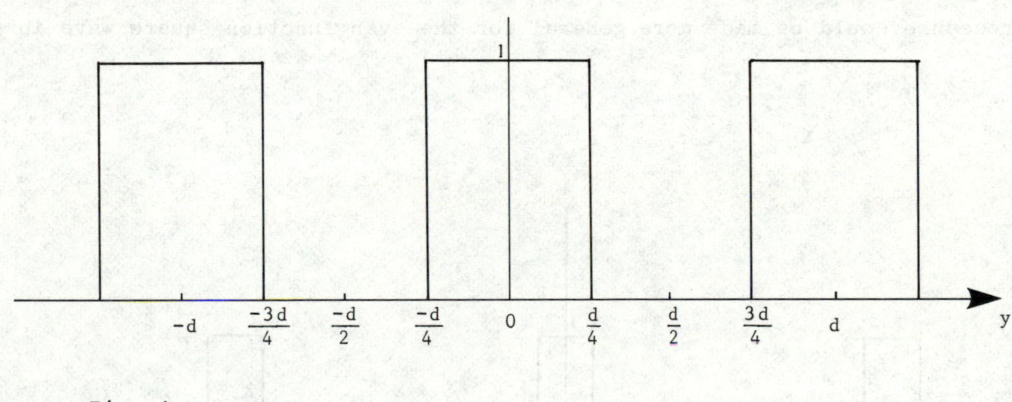

Fig. 1

Since it is an even function all the B_m terms will be zero. Now, since $f(y) = 1$ for $0 \le y \le \frac{d}{4}$ and for $\frac{3}{4}d \le y \le d$, and $f(y) = 0$ for $\frac{d}{4} < y < \frac{3}{4}d$, then

$$A_0 = \frac{2}{\lambda}\int_0^\lambda f(y)\,dy = \frac{2}{d}\left\{\int_0^{\frac{d}{4}} 1.dy + \int_{\frac{d}{4}}^{\frac{3}{4}d} 0.dy + \int_{\frac{3}{4}d}^{d} 1.dy\right\}$$

$$= \frac{2}{d}\left\{\frac{d}{4} + 0 + \frac{d}{4}\right\} = 1 .$$

That is, $A_0 = 1$.

Remembering that all the B_m terms are zero in an even function, we now calculate the A_m terms.

$$A_m = \frac{2}{\lambda}\int_0^{\lambda} f(y)\cos\frac{2\pi}{\lambda}my\,dy \quad = \frac{2}{d}\left\{\int_0^{\frac{d}{4}} 1.\cos\frac{2\pi}{d}my\,dy \; + \int_{\frac{3}{4}d}^{d} 1.\cos\frac{2\pi}{d}my\,dy\right\}$$

$$= \frac{2}{d}\left\{2\int_0^{\frac{d}{4}} 1.\cos\frac{2\pi}{d}my\,dy\right\},$$ since the area under the cosine from 0 to d/4 equals the area from $\frac{3}{4}d$ to d,

$$= \frac{4}{d}\cdot\frac{d}{2\pi m}\sin\frac{2\pi}{d}my\Big|_0^{\frac{d}{4}}$$

that is, $A_m = \frac{2}{m\pi}\sin\left(m\frac{\pi}{2}\right)$.

So, the series for the square wave is

$$f(y) = \frac{A_0}{2} + \sum_{m=1}^{\infty} A_m\cos\frac{2\pi}{d}my\,,$$ which for m = 1, 2, 3,....., gives

$$f(y) = \frac{1}{2} + \frac{2}{\pi}\cos\frac{2\pi}{d}y - \frac{2}{3\pi}\cos\frac{2\pi}{d}\cdot 3y + \frac{2}{5\pi}\cos\frac{2\pi}{d}\cdot 5y - \ldots \qquad (8).$$

The above procedure could be made more general for the even function square wave in figure 2.

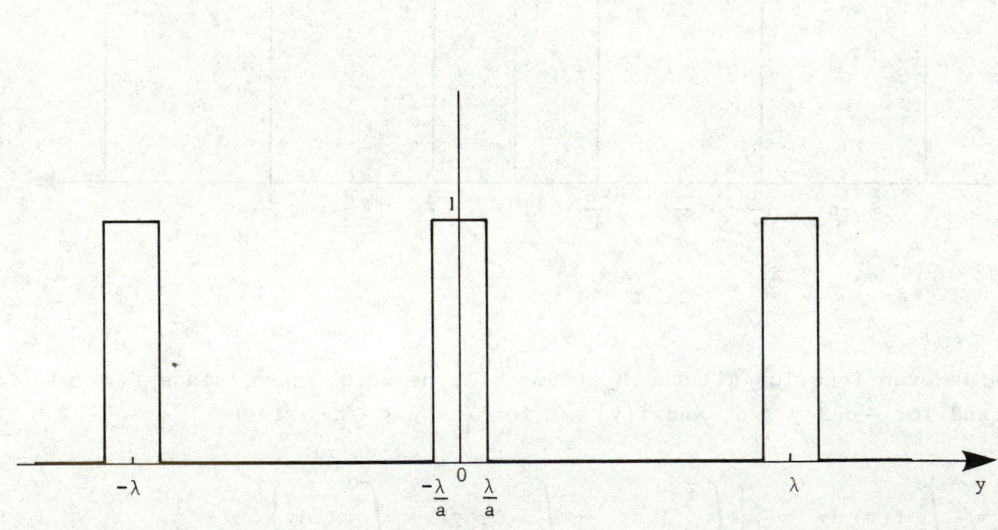

Fig. 2

We can make the square peak as narrow as we wish by varying our choice of a. In the previous case we chose a = 4, so the width of the peak was d/2 (λ was given the symbol d). Repeating the integration for a peak width 2λ/a gives the coefficients

$$A_0 = 4/a \qquad (9)$$

and $A_m = \frac{4}{a}\left(\frac{\sin 2\pi m/a}{2\pi m/a}\right)$ ....(10). The Fourier series for this funcion is now

$$f(y) = \frac{2}{a} + \sum_{m=1}^{\infty} \frac{4}{a}\left(\frac{\sin 2\pi m/a}{2\pi m/a}\right) \cos \frac{2\pi}{\lambda} my .$$

This applies to any wave of the form shown in figure 2. If y were replaced by time t, then λ would simply be replaced by the temporal period τ.

As the width of the square peak is reduced the number of terms in the series needed to present a general resemblance of the square wave increases. This is simply due to the increase in the relative amplitude of higher terms. Comparing A_m with A_1 gives

$$\frac{A_m}{A_1} = \frac{\sin 2\pi m/a}{m \sin 2\pi/a} \qquad (11).$$

For a wide peak, say $a = 4$, then for the 99th term $A_{99} \simeq 0.01\, A_1$. However, if $a = 400$ say, then $A_{99} \simeq 0.64\, A_1$. Evidently, the narrower the peak the greater the number of terms which will be needed to present a reasonable resemblance to the square wave since the higher terms are not negligible.

FOURIER INTEGRALS

If, in figure 2, the width of the square peak is kept constant whilst λ is made to increase, as $\lambda \to \infty$ then there results a single square pulse standing alone, and the function is no longer periodic. Since single 'pulses' may represent photons the need for a Fourier representation is clear.

Recall that we wrote $k = 2\pi/\lambda$ in chapter 10. Since $1/\lambda$ is here called the spatial frequency, k is called the angular spatial frequency. It allows a slightly more compact written notation which, in any case, is in general use. In general, mk will be the angular spatial frequency of the components. This can be appreciated by noting that

$$mk = \frac{2\pi}{\lambda}, \quad \text{for } m = 1,$$

$$mk = \frac{2\pi}{\lambda/2} = 2\cdot\frac{2\pi}{\lambda}, \quad \text{for } m = 2,$$

$$mk = \frac{2\pi}{\lambda/3} = 3\cdot\frac{2\pi}{\lambda}, \quad \text{for } m = 3,$$

and so on.

Setting $a = 8$, say, and $\lambda = 2$ cm gives a peak width of $2\lambda/a = ½$ cm. Figure 3(a) shows the result-plot of A_m against mk. Such a plot is called a *frequency spectrum*; that is, it shows the spatial frequencies and the amplitudes associated with the cosine functions in the Fourier series. A_m can be thought of as a function of mk which is zero except at values of $m = 0, 1, 2, ..$ Now, if we make $a = 16$ and $\lambda = 4$ cm, the peak widths are still ½ cm whilst their separation is increased. Note how the density of the frequency spectrum terms has increased in figure 3(b). Indeed, the shape of the envelope formed by the A(mk) amplitudes is quite evident. It is no coincidence that if we were to square these amplitudes we should have an envelope resembling a single slit diffraction intensity distribution. The envelopes in figures 3(a) and (b) are identical except for a scaling factor evident on the ordinates.

It should now be apparent that as λ increases and the function approaches a single square

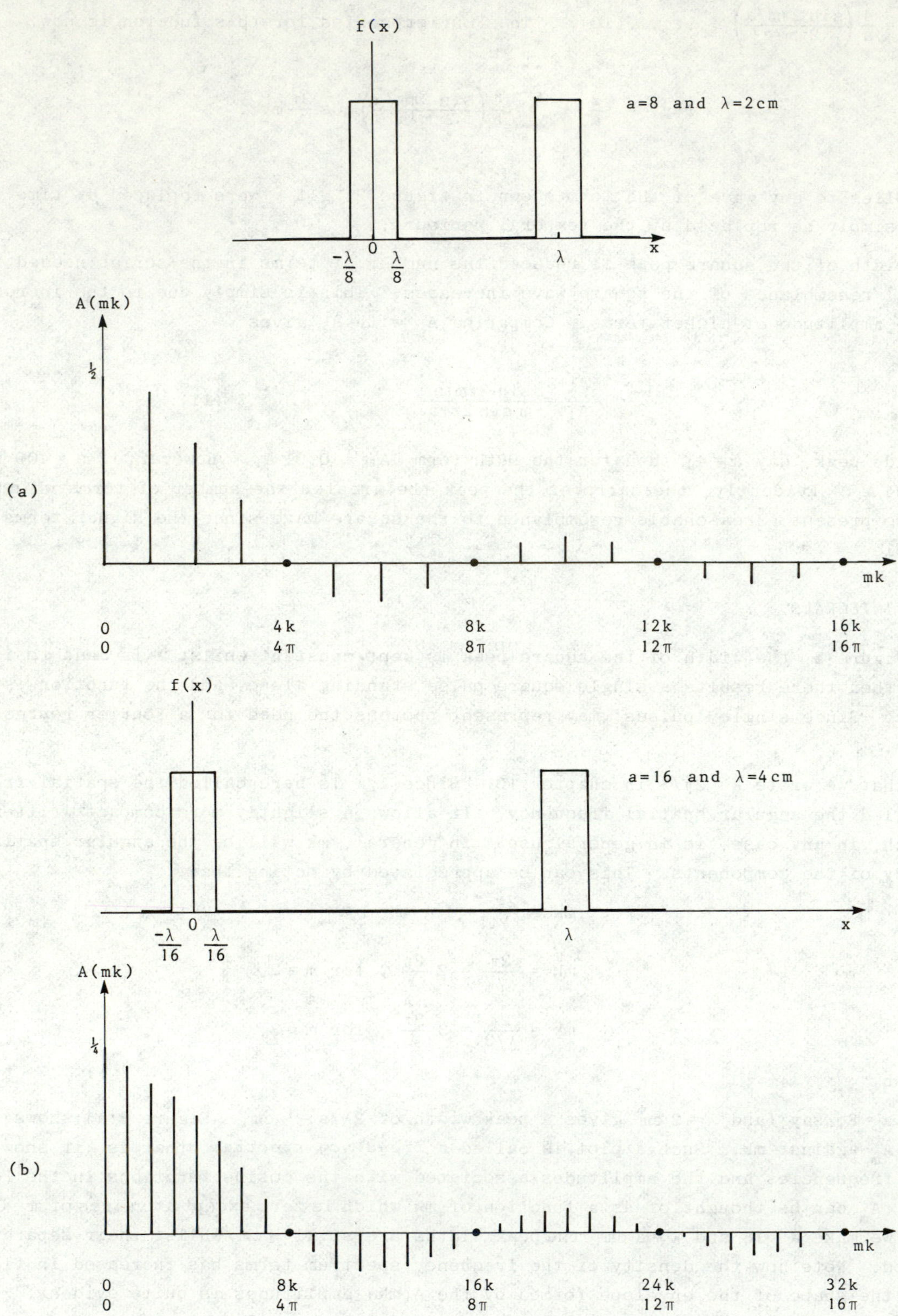

Fig. 3 Frequency spectrum for a square wave.

pulse the density of the A(mk) terms increases. In the limit, as λ becomes very large, k must become very small, becoming eventually a continuous function. The function A(k) is the shape of the envelope in figure 3. At this stage the Fourier series gives way to the Fourier integral which resembles the series and is stated below.

$$f(x) = \frac{1}{\pi}\left[\int_0^\infty A(k)\cos kx\,dk + \int_0^\infty B(k)\sin kx\,dk\right] \qquad (12).$$

where $A(k) = \int_{-\infty}^{\infty} f(x)\cos kx\,dx$ and $B(k) = \int_{-\infty}^{\infty} f(x)\sin kx\,dx$ (13).

A(k) and B(k) are known as the *Fourier cosine and sine transforms*, respectively. They are the amplitudes of the cosine and sine contributions in the synthesis of the pulse given in equation (12).

In conclusion, the Fourier series allows us to express a periodic function of infinite extent as the sum of an infinite series of cosine and sine terms. The Fourier integral does for the pulse what the Fourier series does for the periodic function of infinite extent.

APPENDIX 2 - Useful Trigonometrical Identities

$\sin(A+B) = \sin A\cos B + \cos A\sin B$ (1). $\sin(A-B) = \sin A\cos B - \cos A\sin B$ (2).

$\cos(A+B) = \cos A\cos B - \sin A\sin B$ (3). $\cos(A-B) = \cos A\cos B + \sin A\sin B$ (4).

$\sin A + \sin B = 2\sin\frac{A+B}{2}\cos\frac{A-B}{2}$ (5). $\sin A - \sin B = 2\cos\frac{A+B}{2}\sin\frac{A-B}{2}$ (6).

$\cos A + \cos B = 2\cos\frac{A+B}{2}\cos\frac{A-B}{2}$ (7). $\cos A - \cos B = 2\sin\frac{A+B}{2}\sin\frac{A-B}{2}$ (8).

Writing $B = A$ in equation (1) gives $\sin 2A = 2\sin A\cos A$ (9).

Similarly, $\cos 2A = \cos^2 A - \sin^2 A$ (10), and $\cos^2 A + \sin^2 A = 1$ (11), having put $B = A$ in equations (3) and (4) to deduce equations (10) and (11), respectively.

APPENDIX 3

Apparent displacement

Let the object B be placed some distance behind a parallel sided layer of glass, of thickness d.

The ray BM'MN, which is normal to the surfaces, is transmitted through the refracting medium without deviation. The ray BP'PR, which is close to the normal, is refracted along the direction PP' in the medium and emerges in air along PR in a direction parallel to BP' (see section 2.1.6). The observer in front of the glass thus sees the object at B', which is the intersection of NB and RP produced.

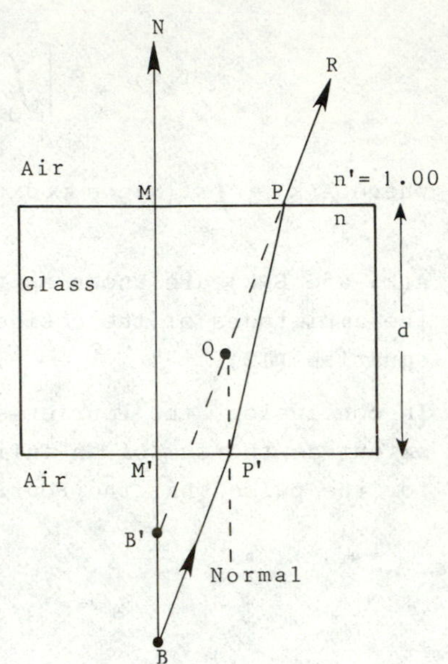

Now, it is clear that BB'QP' is a parallelogram. Hence, BB' = P'Q. But BB' is the apparent displacement of the object. Hence P'Q is equal to the apparent displacement. But, from figure 2.11, the apparent position of an object at P' is Q. Hence, the apparent displacement BB' of the object is independent of the position of O behind the refracting medium, and is given by

$$BB' = P'Q = \text{real depth} - \text{apparent depth}$$
$$= d - \frac{d}{n} = d(1 - \frac{1}{n}).$$

Deviation by a Reflecting Prism (Achromatic Prism)

We can show that it is possible for a prism to cause internal reflection without any accompanying dispersion. ABC represents a principal section of a prism, of apical angle a. The ray is incident at the first face such that the angle of incidence inside the prism at the second face is greater than the critical angle i_c, defined by $\sin i_c = \frac{1}{n_g}$.

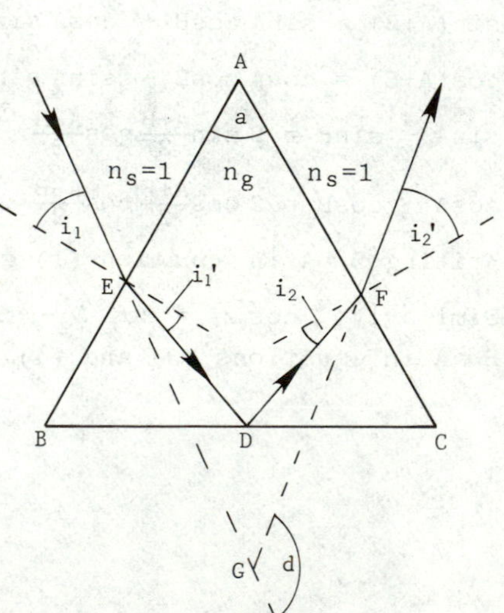

The angle of deviation between the incident and emergent rays is

$$d = 180^0 - E\hat{G}F \qquad \text{(i)}.$$

Now, in the polygon AEGF we have

$$a + A\hat{F}G + E\hat{G}F + A\hat{E}G = 360^0 \qquad \text{(ii)}.$$

Also, at E, $A\hat{E}G = 90^0 + i_1$, and at F, $A\hat{F}G = 90^0 + i_2'$.

Substituting in (ii) we get

$$a + (90^0 + i_2') + E\hat{G}F + (90^0 + i_1) = 360^0.$$

That is, $E\hat{G}F = 180^0 - a - (i_1 + i_2')$.

Substituting for $E\hat{G}F$ in (i) we get $d = 180^0 - (180^0 - a - (i_1 + i_2'))$

or $$d = a + i_1 + i_2' \qquad \text{(iii)}.$$

Now, at D the angles of incidence and reflection are equal. $\therefore E\hat{D}B = F\hat{D}C$. Thus, if the prism is isosceles, $E\hat{B}D = F\hat{C}D$ and the triangles EBD and FCD are similar.

Hence, $B\hat{E}D = D\hat{F}C$ and thus $i_1' = i_2$. From Snell's law this is equivalent to saying $i_1 = i_2'$. Equation (iii) becomes $d = a + 2i_1$, which is independent of wavelength (colour) and the refractive index. Reflection occurs without colour preference and the prism is said to be achromatic.

The Fundamental Paraxial Equation

The derivation given in section 2.2.3 was for the specific case of a converging surface forming a real image of a real object. This is only one of several possible cases for a single refracting surface and the illustrations below show some of the other possible refractions involving converging and diverging surfaces, including real and virtual objects. In all cases $n' > n$, and it is left to the reader to derive the expressions shown, in a similar fashion to the derivation in section 2.2.3 . It is seen that application of the sign convention yields equation (2.25) for all cases of a single spherical refracting surface.

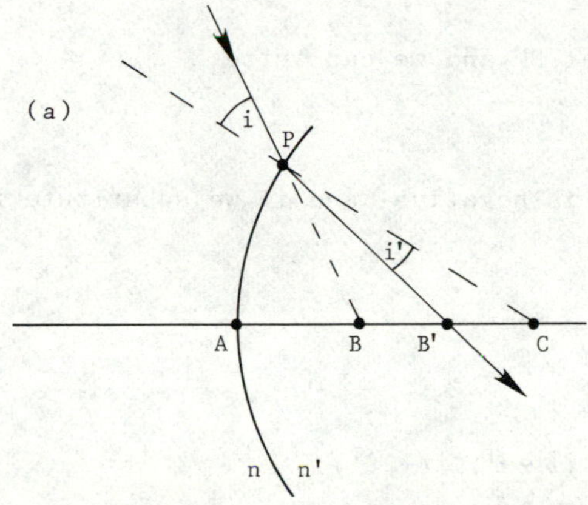

(a) Converging surface: virtual object B, real image B'.

$$\frac{n'-n}{AC} = \frac{n'}{AB'} - \frac{n}{AB}$$

$$\frac{n'-n}{r} = \frac{n'}{\ell'} - \frac{n}{\ell} \qquad (2.25)$$

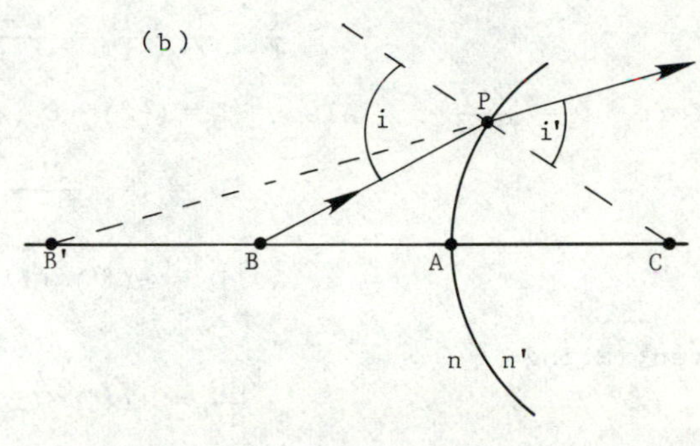

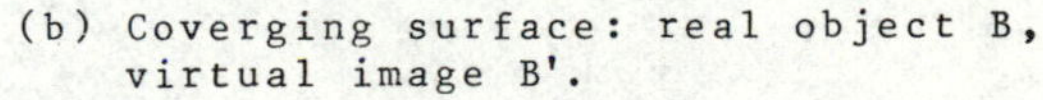

(b) Coverging surface: real object B, virtual image B'.

$$\frac{n}{BA} - \frac{n'}{B'A} = \frac{n'-n}{AC}$$

$$\frac{n}{-\ell} - \frac{n'}{-\ell'} = \frac{n'-n}{r}$$

$$\frac{n'}{\ell'} - \frac{n}{\ell} = \frac{n'-n}{r} \qquad (2.25)$$

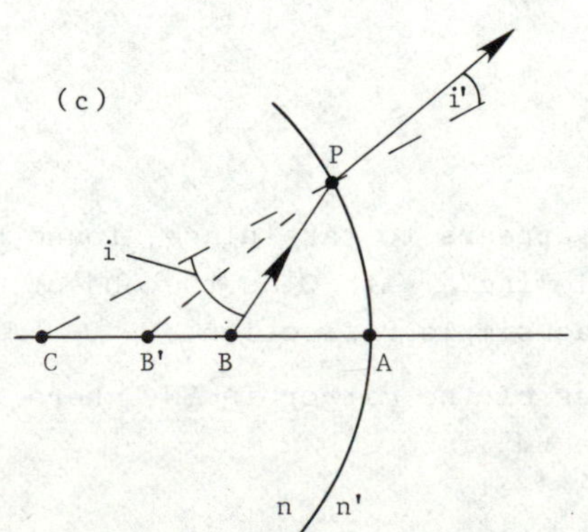

(c) Diverging surface: real object B, virtual image B'.

$$\frac{n'}{B'A} - \frac{n}{BA} = \frac{n'-n}{CA}$$

$$\frac{n'}{-\ell'} - \frac{n}{-\ell} = \frac{n'-n}{-r} \qquad (2.25).$$

$$\frac{n'}{\ell'} - \frac{n}{\ell} = \frac{n'-n}{r}$$

Minimum separation of an Object and its Real Image

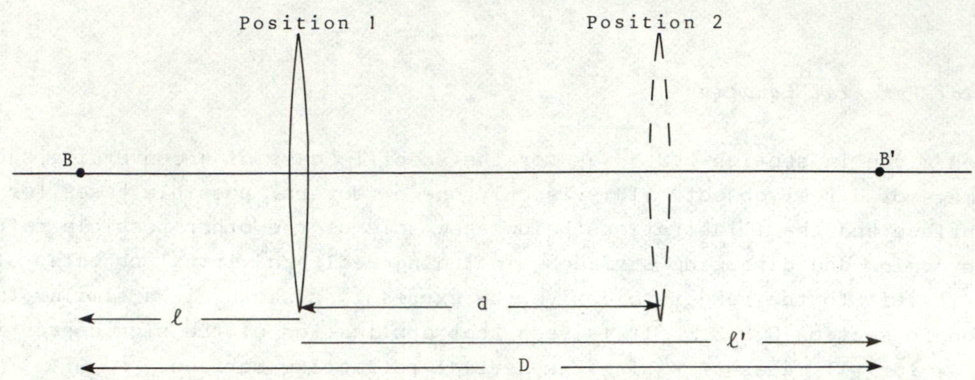

With the lens in position (1) an object B is imaged at B' and we can write

$$\frac{1}{\ell} = \frac{1}{\ell'} - \frac{1}{f'} = \frac{f'-\ell'}{\ell' f'} .$$

Therefore $\ell = \frac{\ell' f'}{f'-\ell'}$. Also, $D = (-\ell) + \ell'$, since ℓ is negative, and if we substitute for ℓ

$$D = \ell' - \frac{\ell' f'}{f'-\ell'} = \frac{-\ell'^2}{f'-\ell'}$$

$$\therefore \quad D = -(\ell')^2 (f'-\ell')^{-1} .$$

Differentiating,

$$\frac{dD}{d\ell'} = +(\ell')^2 (f'-\ell')^{-2} + (f'-\ell')^{-1}(-2\ell')$$

$$= \frac{\ell'^2 - 2\ell' f' + 2\ell'^2}{(f'-\ell')^2} = \frac{\ell'^2 - 2\ell' f'}{(f'-\ell')^2} .$$

For maximum or minimum, $\frac{dD}{d\ell'} = 0$. Therefore, $\ell'^2 - 2\ell' f' = 0$, or $\ell'(\ell' - 2f') = 0$. Since $\ell' = 0$ is impossible, we have $\ell' = 2f'$. Hence, using the paraxial equation and substituting for ℓ',

$$\frac{1}{2f'} - \frac{1}{\ell} = \frac{1}{f'} \qquad \text{which gives} \quad \ell = -2f' .$$

Since $D = -\ell + \ell'$, minimum $D = -(-2f') + 2f' = 4f'$!

The Bright Image in a Thick Mirror

Let MM' represent the plane at which reflection appears to take place, found by producing OA and DC until they intersect, see the figure following. With O as the object, I_2 will be as far behind MM' as O is in front, treating this as simple reflection.

Now, from equation (2.10), the apparent thickness of the mirror is MN, where $MN = \frac{t}{n_g}$. Hence, the distance of the object in front of the plane MM' is

$$ON + MN = d + \frac{t}{n_g} .$$

Thus, the image distance behind the front surface of the mirror is

$$d' = d + \frac{t}{n_g} + \frac{t}{n_g}\ .$$

That is,

$$d' = d + \frac{2t}{n_g} \qquad (5.2)$$

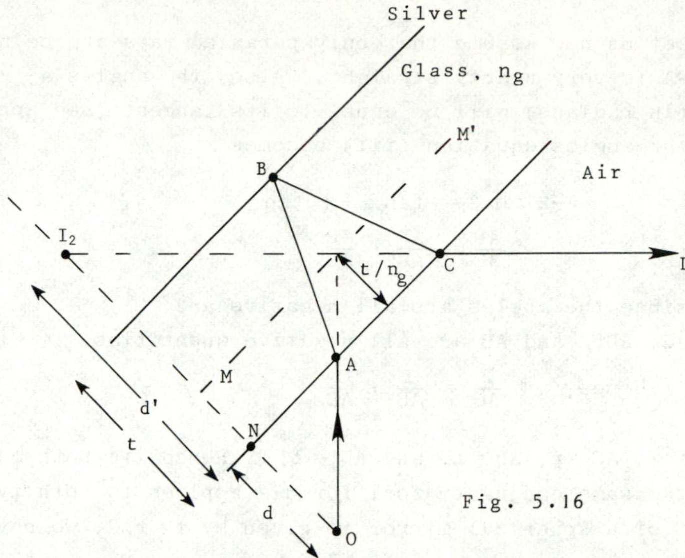

Fig. 5.16

Conjugate Foci and Magnification Formulae for a Spherical Mirror

We shall consider the reflection of a ray of light from an axial object point by a concave and and a convex spherical mirror. The derivations of the formulae are set out side by side for the two types of mirror for ease of comparison.

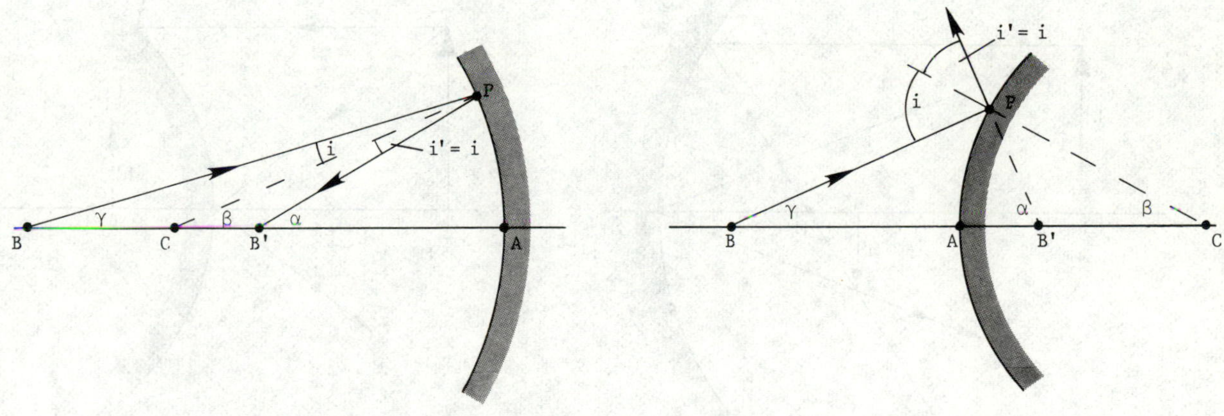

Concave mirror

Convex mirror

C is the centre of curvature of the mirror surface, and CP is the normal at the point P. The angle of reflection i' = the angle of incidence i.

Concave mirror:

In triangle BB'P, $\alpha = 2i + \gamma$ (i).

In triangle BCP, $\beta = i + \gamma$

that is, $i = \beta - \gamma$ (ii).

From (i) and (ii),

$$\alpha = 2(\beta - \gamma) + \gamma$$

giving $2\beta = \alpha + \gamma$ (iii)

Convex mirror:

In triangle BB'P, $2i = \alpha + \gamma$ (i).

In triangle BCP, $i = \beta + \gamma$ (ii).

From (i) and (ii),

$$2(\beta + \gamma) = \alpha + \gamma$$

giving $2\beta = \alpha - \gamma$ (iii).

Let us now assume that only paraxial rays are being considered so that P is close to A, and PA is very nearly straight. Also, the angles α, β, γ will be small and, hence, each angle (in radians) will be equal to its tangent (see appendix 4). Applying the sign convention to the angles equation (iii) becomes

$$-2\tan\beta = -\tan\alpha + (-\tan\gamma)$$
$$2\frac{AP}{AC} = \frac{AP}{AB'} + \frac{AP}{AB}$$

since the angles are all negative and AC, AB', and AB are all negative quantities.

$$\text{Hence,}\quad \frac{2}{AC} = \frac{1}{AB'} + \frac{1}{AB}$$

$$2\tan\beta = \tan\alpha - (-\tan\gamma)$$
$$2\frac{AP}{AC} = \frac{AP}{AB'} + \frac{AP}{AB}$$

since γ and AB are negative.

$$\text{Hence,}\quad \frac{2}{AC} = \frac{1}{AB'} + \frac{1}{AB}.$$

Now, AC = r, AB = ℓ, and AB' = ℓ'. Hence, in both cases, $\frac{2}{r} = \frac{1}{\ell'} + \frac{1}{\ell}$. Thus, it is seen that the same conjugate foci formula applies to both types of mirror. Also, since the focal length f of a spherical mirror is given by f = r/2, we have

$$\frac{1}{\ell'} + \frac{1}{\ell} = \frac{2}{r} = \frac{1}{f} \qquad (5.4)$$

Let us now consider the magnification effect. The following diagrams show an object BT of size h, and the corresponding image B'T' of size h'.

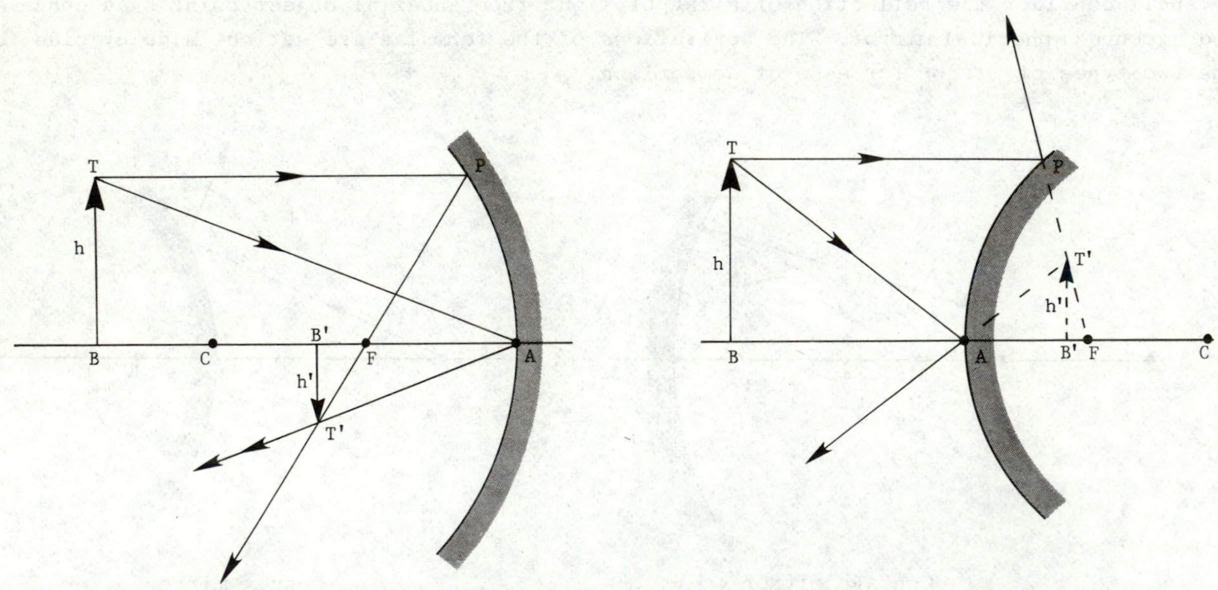

In each diagram the triangles ABT and AB'T' are similar. Hence, ratios of corresponding pairs of sides are equal:

$$\frac{T'B'}{BT} = \frac{B'A}{BA}$$

Now, BT = h, T'B' = -h', BA = -ℓ, and B'A = -ℓ'. Therefore

$$\frac{-h'}{h} = \frac{-\ell'}{-\ell}$$

$$\frac{B'T'}{BT} = \frac{AB'}{BA}$$

Now, BT = h, B'T' = h, BA = -ℓ, and AB' = ℓ. Therefore

$$\frac{h'}{h} = \frac{\ell'}{-\ell}$$

which gives, for both cases $\frac{h'}{h} = -\frac{\ell'}{\ell}$. Thus, the same magnification formula applies to both types of mirror. Now, linear magnification is defined by the ratio $m = \frac{\text{size of image}}{\text{size of object}}$. So, we may write

$$m = \frac{h'}{h} = -\frac{\ell'}{\ell} \qquad (5.5).$$

Hyperfocal Distance

With reference to figure 8.10, repeated here, we require to show that if $\ell_1 = \frac{\ell}{2}$, where ℓ is the hyperfocal distance and the lens is focused for this distance, all objects lying between infinity and $\frac{\ell}{2}$ are tolerably in focus on the image plane. See the text in section 8.1.1.5 for details.

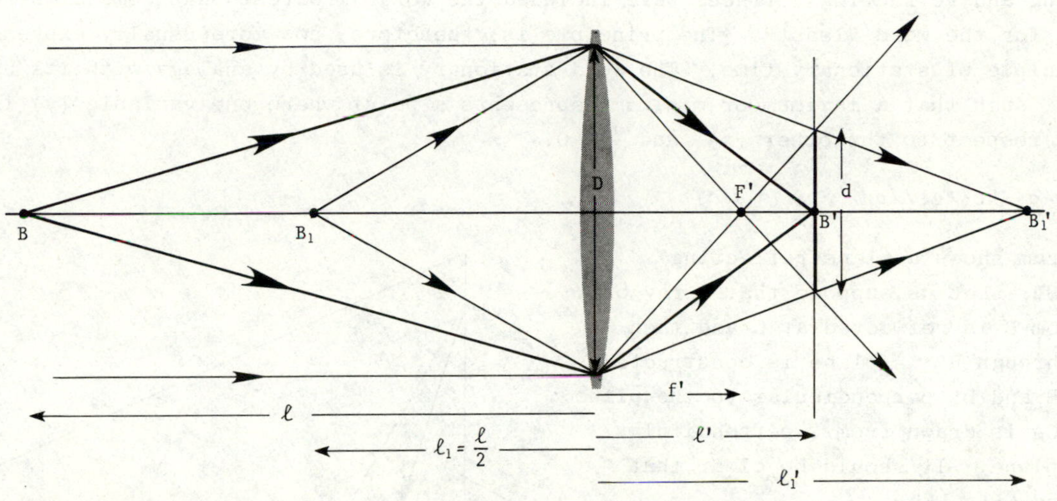

Fig. 8.10

Now, in figure 8.10, from similar triangles:

$$\frac{d}{D} = \frac{\ell' - f'}{f'} \qquad \text{and} \qquad \frac{d}{D} = \frac{\ell_1' - \ell'}{\ell_1'}.$$

$$\text{Therefore} \quad \frac{\ell'}{f'} - 1 = 1 - \frac{\ell'}{\ell_1'} \qquad \text{(i)}.$$

Eliminating ℓ' and ℓ_1',

$$\frac{1}{\ell'} = \frac{1}{f'} + \frac{1}{\ell} = \frac{\ell + f'}{\ell f'}.$$

Therefore, $\ell' = \frac{\ell f'}{\ell + f'}$. Similarly, $\ell_1' = \frac{\ell_1 f'}{\ell_1 + f'}$.

Substituting in (i),

$$\frac{\ell f'}{(\ell + f')f'} - 1 = 1 - \frac{\ell f'}{(\ell + f')} \cdot \frac{(\ell_1 + f')}{\ell_1 f'}$$

$$\text{or} \qquad \frac{\ell}{\ell + f'} + \frac{\ell(\ell_1 + f')}{\ell_1(\ell + f')} = 2.$$

This can be rearranged with a common denominator as follows:

$$\frac{\ell\ell_1 + \ell(\ell_1 + f')}{\ell_1(\ell + f')} = 2 .$$

Multiplying out gives

$$\ell\ell_1 + \ell\ell_1 + \ell f' = 2\ell\ell_1 + 2\ell_1 f'$$

which reduces to

$$\ell_1 = \frac{\ell}{2} .$$

Fermat's Principle

It is possible to derive the laws of reflection (section 5.1.1) and refraction (section 2.1.1) from a general principle first stated by the French mathematician Pierre Fermat in 1658. In its original form the principle stated that the path of a ray in travelling between two points during reflection or refraction was the path of least time. Later workers showed that when curved reflecting and refracting surfaces were included the word "greatest" has sometimes to be substituted for the word "least". The principle is, therefore, now more usually expressed as the principle of stationary time. The word stationary is used by analogy with its use in calculus, such that a maximum or minimum represents a point where one variable (y) is stationary with respect to the other (x), and $\frac{dy}{dx} = 0$.

The Laws of Reflection

The diagram shows a plane reflecting surface MM. Let us suppose that a ray of light from B is reflected at C and then passes through B'. A plane is constructed through B and B' perpendicular to the plane MM, and CA is drawn from C perpendicular to this plane. It should be clear that BC > BA and CB' > AB'.

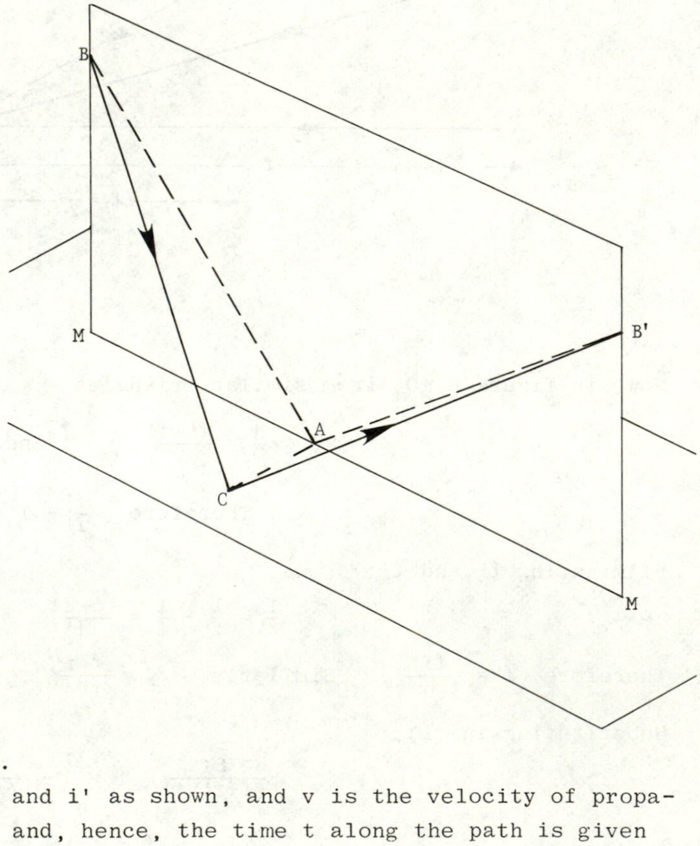

Hence, the time taken for the light to travel along BCB' is greater than the time along BAB', and this is contrary to Fermat's principle. Therefore, it follows that C and A must coincide and the rays BA and AB', and the normal at A all lie in the plane BAB', which was to be proved.

We can now determine the position of A which corresponds with a minimum time for the light to travel from B to B'. In the next diagram the lettered points are in the same positions as in the first diagram. For the moment, let us assume that A may lie anywhere along the reflecting surface MM. The angles of incidence and reflection are i and i' as shown, and v is the velocity of propagation. The length of the ray path is s + s' and, hence, the time t along the path is given by

$$t = \frac{s + s'}{v} \qquad \text{(i)}.$$

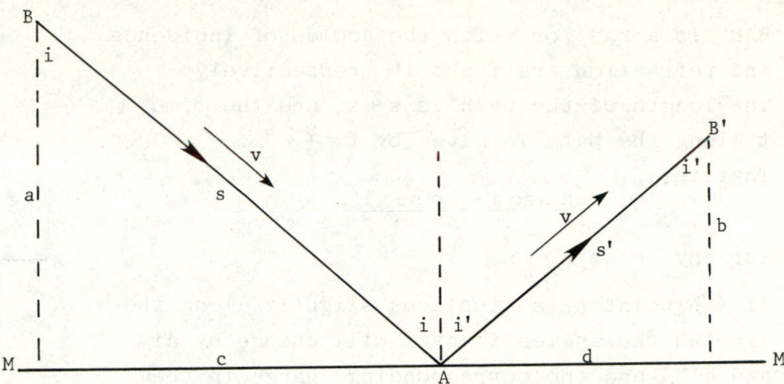

It can be seen that $\frac{a}{s} = \cos i$.
$\therefore$ $s = a \sec i$. Also, it is evident that $\frac{b}{s'} = \cos i'$, whence $s' = b \sec i'$.
Thus, substituting in (i) gives

$$t = \frac{1}{v}(a \sec i + b \sec i')$$

for any point A. Now, if the point A is displaced slightly along the line MM, the angles i and i' will change by di and di' respectively, and the corresponding change in the time dt is given by

$$dt = \frac{1}{v}(a \sec i . \tan i . di + b \sec i' . \tan i' . di') .$$

For a minimum time, $dt = 0$ and so it follows that

$$a \sec i . \tan i . di = -b \sec i' . \tan i' . di' \qquad \text{(ii)}.$$

But, for any position of A, $c + d = \text{constant}$. That is, $a \tan i + b \tan i' = \text{constant}$.
Differentiating, we get

$$a \sec^2 i . di + b \sec^2 i' . di' = 0,$$

$$\text{or} \qquad a \sec^2 i . di = -b \sec^2 i' . di' \qquad \text{(iii)}.$$

Dividing (iii) by (ii) gives

$$\frac{a \sec^2 i . di}{a \sec i . \tan i . di} = \frac{-b \sec^2 i' . di'}{-b \sec i' . \tan i' . di'}$$

which reduces to

$$\frac{\sec i}{\tan i} = \frac{\sec i'}{\tan i'}$$

But writing $\sec = \frac{1}{\cos}$ and $\tan = \frac{\sin}{\cos}$ we get

$$\frac{1}{\sin i} = \frac{1}{\sin i'} \qquad \text{or } i = i'.$$

That is, the ray path BAB', for which the angles of incidence and reflection are equal, is the one traversed in the least time.

The Laws of Refraction

The Laws of Refraction may now be considered. The coplanar nature of the incident ray, the refracted ray, and the normal is proved in a similar fashion to the case of reflection and is left to the reader. However, we can now derive Snell's law with the aid of the following diagram. MM is the boundary between two transparent media of refractive indices n and n', with $n' > n$, and the corresponding velocities of propagation are v and v'.

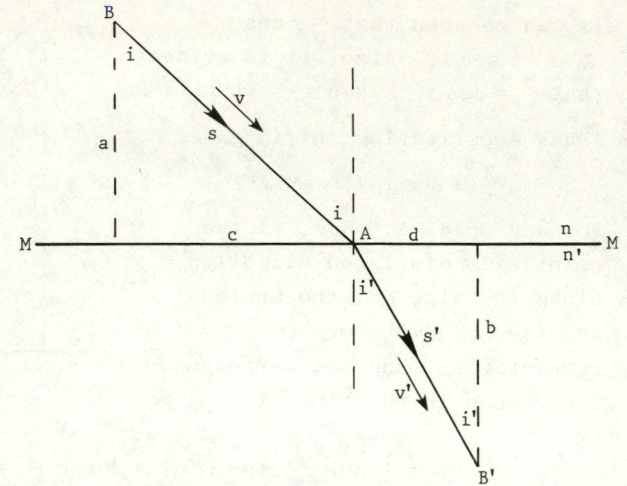

BAB' is a ray for which the angles of incidence and refraction are i and i', respectively. The length of the path is s + s', and the time t along the path is given by $t = \frac{s}{v} + \frac{s'}{v'}$.
That is,

$$t = \frac{a \sec i}{v} + \frac{b \sec i'}{v'}$$

for any point A.

If the point A is displaced slightly along the line MM the angles i and i' will change by di and di', and the corresponding change in the time dt is given by

$$dt = \frac{a}{v}.\sec i.\tan i.di + \frac{b}{v'}.\sec i'.\tan i'.di'.$$

For a minimum time dt = 0, and it follows that

$$\frac{a}{v}.\sec i.\tan i.di = -\frac{b}{v'}.\sec i'.\tan i'.di' \qquad \text{(i)}.$$

But, for any position of A, c + d = constant. That is, a tan i + b tan i' = constant.
Differentiating, we get

$$a \sec^2 i.di = -b \sec^2 i'.di' \qquad \text{(ii)}.$$

Dividing (i) by (ii) gives

$$\frac{a \sec i.\tan i.di}{va \sec^2 i.di} = \frac{-b \sec i'.\tan i'.di'}{-v'b \sec^2 i'.di'}$$

which gives

$$\frac{\tan i}{v.\sec i} = \frac{\tan i'}{v' \sec i'}$$

But, writing tan = sin/cos and sec = 1/cos we get $\frac{\sin i}{v} = \frac{\sin i'}{v'}$. Now, since $v = \frac{c}{n}$ and $v' = \frac{c}{n'}$, where c is the velocity of light in vacuum, this reduces to

$$\sin i \times \frac{n}{c} = \sin i' \times \frac{n'}{c}$$

$$\text{or} \qquad n \sin i = n' \sin i'$$

which is the familiar form of Snell's law.

APPENDIX 4

Similar Triangles

Similar triangles are often met with in ray diagrams. It is useful to have available for reference a rule concerning such triangles to assist in certain calculations.

A pair of triangles is said to be similar if the three angles in one of the triangles are equal to the corresponding angles in the other triangle. In other words, one triangle may be thought of as a photographic enlargement of the other.

The following diagrams show various pairs of similar triangles. In each case, the side BC is parallel to the side DE.

In each pair:

angles marked × are equal

angles marked o are equal

angles marked) are equal or belong to both triangles.

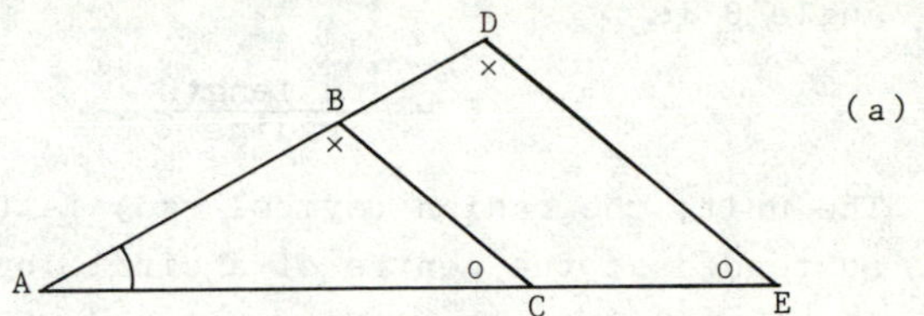

Rule: in a pair of similar triangles, ratios of corresponding pairs of sides are equal.

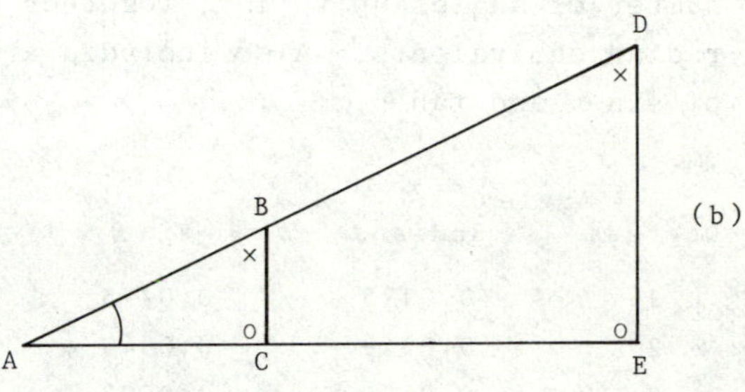

In (a) $\frac{BC}{DE} = \frac{AC}{AE}$

In (b) $\frac{BC}{DE} = \frac{AC}{AE}$

In (c) $\frac{BC}{DE} = \frac{AC}{AD} = \frac{AF}{AG}$

In (d) $\frac{BC}{DE} = \frac{AC}{AE} = \frac{AF}{AG}$.

Note that in (c) and (d) the distances AF and AG are not actually sides of the triangles.

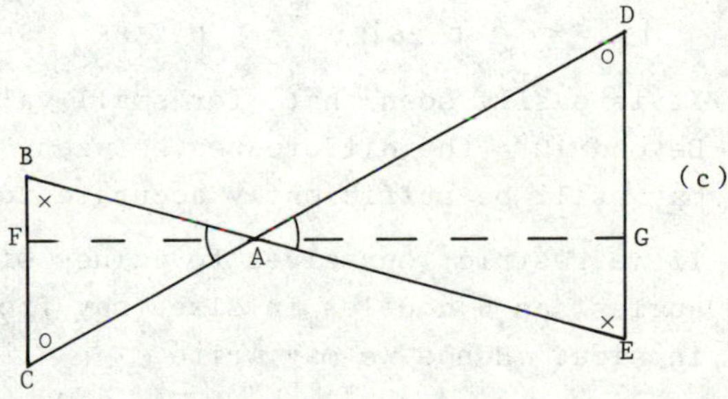

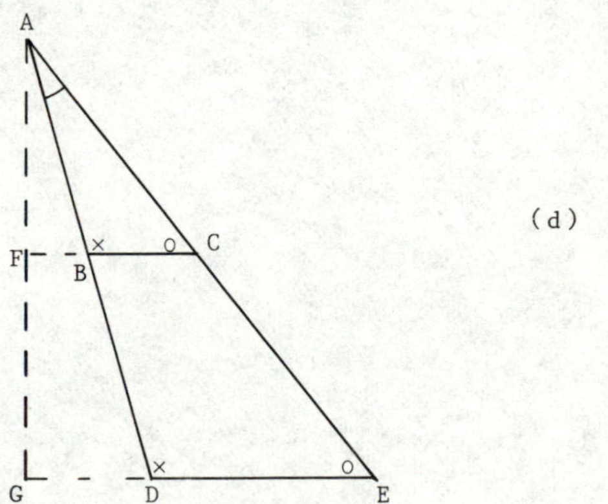

Small Angles

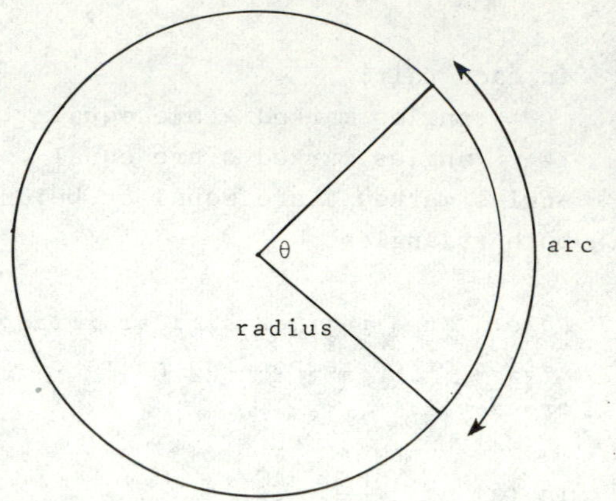

In circular measure, the defining equation for an angle θ is

$$\theta = \frac{\text{arc length}}{\text{radius}} \ .$$

The unit, the radian (symbol rad) is the angle subtended at the centre of a circle by an arc of length equal to the radius. 1 rad = 57.295 78° and 1° = 0.017 45 rad. In the table below are given a number of angles up to 15°, together with their radian equivalents. Also included are the values of sin θ and tan θ.

Angle θ			
Degrees	Radians	sin θ	tan θ
1	0.0175	0.0175	0.0175
2	0.0349	0.0349	0.0349
3	0.0524	0.0523	0.0524
4	0.0698	0.0698	0.0699
5	0.0873	0.0872	0.0875
8	0.1396	0.1392	0.1405
10	0.1745	0.1736	0.1763
15	0.2618	0.2588	0.2679

It is easily seen that, for small values of an angle, say less than 10°, θ rad = sin θ = tan θ. Beyond 10°, the differences between θ (rad), sin θ, and tan θ become larger, but the equality may still be sufficiently accurate for some purposes.

If we restrict ourselves to values of θ in degrees, then it may also be seen that for small angles, as θ doubles in size (say from 2° to 4°), the values of sin θ and tan θ also double in size. Hence we may write

$$\theta \text{ (degrees)} \propto \sin\theta \propto \tan\theta \ .$$

Solid Angles

We have previously stated the definition of the radian, namely:

1 radian (rad) is the angle subtended at the centre of a circle by a length of arc equal to the radius of the circle.

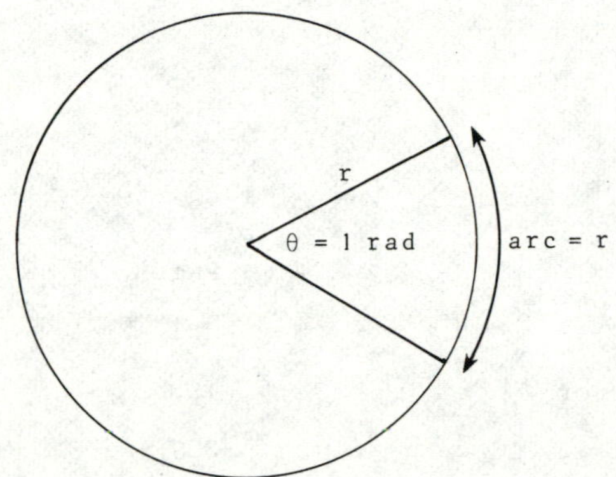

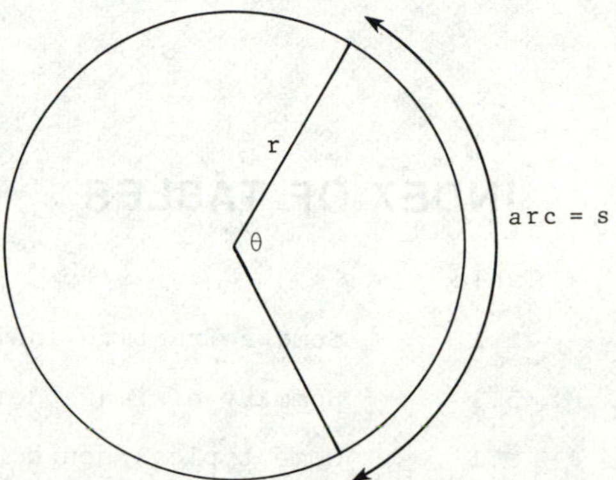

In general, the angle θ (rad) subtended by an arc of length s is given by $\theta = \frac{\text{arc length}}{\text{radius}} = \frac{s}{r}$.
The complete angle, in radians, surrounding a point is given by the number of arcs, each of length r, which fit into the whole circumference of the circle:

$$\text{that is,} \quad \frac{\text{whole circumference}}{\text{radius}} = \frac{2\pi r}{r} = 2\pi .$$

Therefore, full circle (360^0) = 2π radians.

Now, extending to a further dimension, the unit solid angle, called the steradian (sr), is the angle subtended at the centre of a sphere of unit radius by a portion of its surface of unit area. The defining equation is

$$\omega = \frac{\text{area}}{\text{radius}^2} .$$

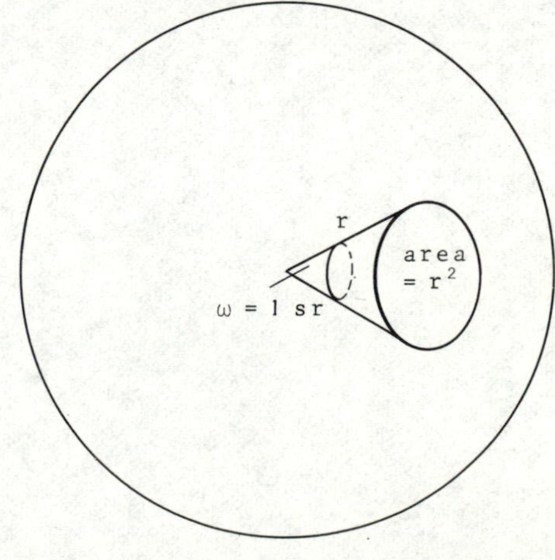

The total number of steradians surrounding a point is given by the number of areas, each of magnitude r^2, which can be fitted on to the full surface of a sphere: that is

$$\frac{\text{whole area of surface}}{r^2} = \frac{4\pi r^2}{r^2} = 4\pi \text{ sr}.$$

Therefore the total space around a point is 4π sr.

INDEX OF TABLES

INDEX